S0-BAF-062

Recommended Dietary Allowances (RDA) and Adequate Intakes (AI) for Vitamins

Age (yr)	Thiamin (mg/day)	Riboflavin (mg/day)	Niacin (mg/day)[a]	Biotin AI (µg/day)	Pantothenic acid AI (mg/day)	Vitamin B₆ (mg/day)	Folate (µg/day)[b]	Vitamin B₁₂ (µg/day)	Choline AI (mg/day)	Vitamin C (mg/day)	Vitamin A (µg/day)[c]	Vitamin D (IU/day)[d]	Vitamin E (mg/day)[e]	Vitamin K AI (µg/day)
Infants														
0–0.5	0.2	0.3	2	5	1.7	0.1	65	0.4	125	40	400	400 (10 µg)	4	2.0
0.5–1	0.3	0.4	4	6	1.8	0.3	80	0.5	150	50	500	400 (10 µg)	5	2.5
Children														
1–3	0.5	0.5	6	8	2	0.5	150	0.9	200	15	300	600 (15 µg)	6	30
4–8	0.6	0.6	8	12	3	0.6	200	1.2	250	25	400	600 (15 µg)	7	55
Males														
9–13	0.9	0.9	12	20	4	1.0	300	1.8	375	45	600	600 (15 µg)	11	60
14–18	1.2	1.3	16	25	5	1.3	400	2.4	550	75	900	600 (15 µg)	15	75
19–30	1.2	1.3	16	30	5	1.3	400	2.4	550	90	900	600 (15 µg)	15	120
31–50	1.2	1.3	16	30	5	1.3	400	2.4	550	90	900	600 (15 µg)	15	120
51–70	1.2	1.3	16	30	5	1.7	400	2.4	550	90	900	600 (15 µg)	15	120
>70	1.2	1.3	16	30	5	1.7	400	2.4	550	90	900	800 (20 µg)	15	120
Females														
9–13	0.9	0.9	12	20	4	1.0	300	1.8	375	45	600	600 (15 µg)	11	60
14–18	1.0	1.0	14	25	5	1.2	400	2.4	400	65	700	600 (15 µg)	15	75
19–30	1.1	1.1	14	30	5	1.3	400	2.4	425	75	700	600 (15 µg)	15	90
31–50	1.1	1.1	14	30	5	1.3	400	2.4	425	75	700	600 (15 µg)	15	90
51–70	1.1	1.1	14	30	5	1.5	400	2.4	425	75	700	600 (15 µg)	15	90
>70	1.1	1.1	14	30	5	1.5	400	2.4	425	75	700	800 (20 µg)	15	90
Pregnancy														
≤18	1.4	1.4	18	30	6	1.9	600	2.6	450	80	750	600 (15 µg)	15	75
19–30	1.4	1.4	18	30	6	1.9	600	2.6	450	85	770	600 (15 µg)	15	90
31–50	1.4	1.4	18	30	6	1.9	600	2.6	450	85	770	600 (15 µg)	15	90
Lactation														
≤18	1.4	1.6	17	35	7	2.0	500	2.8	550	115	1200	600 (15 µg)	19	75
19–30	1.4	1.6	17	35	7	2.0	500	2.8	550	120	1300	600 (15 µg)	19	90
31–50	1.4	1.6	17	35	7	2.0	500	2.8	550	120	1300	600 (15 µg)	19	90

NOTE: For all nutrients, values for infants are AI. The glossary on the inside back cover defines units of nutrient measure.

[a]Niacin recommendations are expressed as niacin equivalents (NE), except for recommendations for infants younger than 6 months, which are expressed as preformed niacin.

[b]Folate recommendations are expressed as dietary folate equivalents (DFE).

[c]Vitamin A recommendations are expressed as retinol activity equivalents (RAE).

[d]Vitamin D recommendations are expressed as cholecalciferol and assume an absence of adequate exposure to sunlight.

[e]Vitamin E recommendations are expressed as α-tocopherol.

Recommended Dietary Allowances (RDA) and Adequate Intakes (AI) for Minerals

Age (yr)	Sodium AI (mg/day)	Chloride AI (mg/day)	Potassium AI (mg/day)	Calcium (mg/day)	Phosphorus (mg/day)	Magnesium (mg/day)	Iron (mg/day)	Zinc (mg/day)	Iodine (µg/day)	Selenium (µg/day)	Copper (µg/day)	Manganese AI (mg/day)	Fluoride AI (mg/day)	Chromium AI (µg/day)	Molybdenum (µg/day)
Infants															
0–0.5	120	180	400	200	100	30	0.27	2	110	15	200	0.003	0.01	0.2	2
0.5–1	370	570	700	260	275	75	11	3	130	20	220	0.6	0.5	5.5	3
Children															
1–3	1000	1500	3000	700	460	80	7	3	90	20	340	1.2	0.7	11	17
4–8	1200	1900	3800	1000	500	130	10	5	90	30	440	1.5	1.0	15	22
Males															
9–13	1500	2300	4500	1300	1250	240	8	8	120	40	700	1.9	2	25	34
14–18	1500	2300	4700	1300	1250	410	11	11	150	55	890	2.2	3	35	43
19–30	1500	2300	4700	1000	700	400	8	11	150	55	900	2.3	4	35	45
31–50	1500	2300	4700	1000	700	420	8	11	150	55	900	2.3	4	35	45
51–70	1300	2000	4700	1000	700	420	8	11	150	55	900	2.3	4	30	45
>70	1200	1800	4700	1200	700	420	8	11	150	55	900	2.3	4	30	45
Females															
9–13	1500	2300	4500	1300	1250	240	8	8	120	40	700	1.6	2	21	34
14–18	1500	2300	4700	1300	1250	360	15	9	150	55	890	1.6	3	24	43
19–30	1500	2300	4700	1000	700	310	18	8	150	55	900	1.8	3	25	45
31–50	1500	2300	4700	1000	700	320	18	8	150	55	900	1.8	3	25	45
51–70	1300	2000	4700	1200	700	320	8	8	150	55	900	1.8	3	20	45
>70	1200	1800	4700	1200	700	320	8	8	150	55	900	1.8	3	20	45
Pregnancy															
≤18	1500	2300	4700	1300	1250	400	27	12	220	60	1000	2.0	3	29	50
19–30	1500	2300	4700	1000	700	350	27	11	220	60	1000	2.0	3	30	50
31–50	1500	2300	4700	1000	700	360	27	11	220	60	1000	2.0	3	30	50
Lactation															
≤18	1500	2300	5100	1300	1250	360	10	13	290	70	1300	2.6	3	44	50
19–30	1500	2300	5100	1000	700	310	9	12	290	70	1300	2.6	3	45	50
31–50	1500	2300	5100	1000	700	320	9	12	290	70	1300	2.6	3	45	50

NOTE: For all nutrients, values for infants are AI. The glossary on the inside back cover defines units of nutrient measure.

Tolerable Upper Intake Levels (UL) for Vitamins

Age (yr)	Niacin (mg/day)[a]	Vitamin B6 (mg/day)	Folate (µg/day)[a]	Choline (mg/day)	Vitamin C (mg/day)	Vitamin A (IU/day)[b]	Vitamin D (IU/day)	Vitamin E (mg/day)[c]
Infants
0–0.5	—	—	—	—	—	600	1000 (25 µg)	—
0.5–1	—	—	—	—	—	600	1500 (38 µg)	—
Children
1–3	10	30	300	1000	400	600	2500 (63 µg)	200
4–8	15	40	400	1000	650	900	3000 (75 µg)	300
9–13	20	60	600	2000	1200	1700	4000 (100 µg)	600
Adolescents
14–18	30	80	800	3000	1800	2800	4000 (100 µg)	800
Adults
19–70	35	100	1000	3500	2000	3000	4000 (100 µg)	1000
>70	35	100	1000	3500	2000	3000	4000 (100 µg)	1000
Pregnancy
≤18	30	80	800	3000	1800	2800	4000 (100 µg)	800
19–50	35	100	1000	3500	2000	3000	4000 (100 µg)	1000
Lactation
≤18	30	80	800	3000	1800	2800	4000 (100 µg)	800
19–50	35	100	1000	3500	2000	3000	4000 (100 µg)	1000

[a]The UL for niacin and folate apply to synthetic forms obtained from supplements, fortified foods, or a combination of the two.

[b]The UL for vitamin A applies to the preformed vitamin only.

[c]The UL for vitamin E applies to any form of supplemental α-tocopherol, fortified foods, or a combination of the two.

Tolerable Upper Intake Levels (UL) for Minerals

Age (yr)	Sodium (mg/day)	Chloride (mg/day)	Calcium (mg/day)	Phosphorus (mg/day)	Magnesium (mg/day)[d]	Iron (mg/day)	Zinc (mg/day)	Iodine (µg/day)	Selenium (µg/day)	Copper (µg/day)	Manganese (mg/day)	Fluoride (mg/day)	Molybdenum (µg/day)	Boron (mg/day)	Nickel (mg/day)	Vanadium (mg/day)
Infants
0–0.5	—	—	1000	—	—	40	4	—	45	—	—	0.7	—	—	—	—
0.5–1	—	—	1500	—	—	40	5	—	60	—	—	0.9	—	—	—	—
Children
1–3	1500	2300	2500	3000	65	40	7	200	90	1000	2	1.3	300	3	0.2	—
4–8	1900	2900	2500	3000	110	40	12	300	150	3000	3	2.2	600	6	0.3	—
9–13	2200	3400	3000	4000	350	40	23	600	280	5000	6	10	1100	11	0.6	—
Adolescents
14–18	2300	3600	3000	4000	350	45	34	900	400	8000	9	10	1700	17	1.0	—
Adults
19–50	2300	3600	2500	4000	350	45	40	1100	400	10,000	11	10	2000	20	1.0	1.8
51–70	2300	3600	2000	4000	350	45	40	1100	400	10,000	11	10	2000	20	1.0	1.8
>70	2300	3600	2000	3000	350	45	40	1100	400	10,000	11	10	2000	20	1.0	1.8
Pregnancy
≤18	2300	3600	3000	3500	350	45	34	900	400	8000	9	10	1700	17	1.0	—
19–50	2300	3600	2500	3500	350	45	40	1100	400	10,000	11	10	2000	20	1.0	—
Lactation
≤18	2300	3600	3000	4000	350	45	34	900	400	8000	9	10	1700	17	1.0	—
19–50	2300	3600	2500	4000	350	45	40	1100	400	10,000	11	10	2000	20	1.0	—

[d]The UL for magnesium applies to synthetic forms obtained from supplements or drugs only.

NOTE: An upper Limit was not established for vitamins and minerals not listed and for those age groups listed with a dash (—) because of a lack of data, not because these nutrients are safe to consume at any level of intake. All nutrients can have adverse effects when intakes are excessive.

SOURCE: Adapted with permission from the *Dietary Reference Intakes series,* National Academies Press. Copyright 1997, 1998, 2000, 2001, 2002, 2005, 2011 by the National Academies of Sciences.

Nutritional Sciences

Kent State University

Fourth Edition

Michelle "Shelley" McGuire| Kathy A. Beerman | Judith E. Brown

CENGAGE
Learning·

Australia • Brazil • Japan • Korea • Mexico • Singapore • Spain • United Kingdom • United States

Nutritional Sciences: Kent State University, Fourth Edition

Nutrition Now, 8th Edition Source Edition
Judith E. Brown

© 2017, 2014, 2011, 2008 Cengage Learning. All rights reserved.

Nutritional Sciences: From Fundamentals to Food, 3rd Edition
Michelle 'Shelley' McGuire | Kathy A. Beerman

© 2013, 2010, 2007 Cengage Learning. All rights reserved.

For product information and technology assistance, contact us at
Cengage Learning Customer & Sales Support, 1-800-354-9706

For permission to use material from this text or product,
submit all requests online at **cengage.com/permissions**
Further permissions questions can be emailed to
permissionrequest@cengage.com

This book contains select works from existing Cengage Learning resources and was produced by Cengage Learning Custom Solutions for collegiate use. As such, those adopting and/or contributing to this work are responsible for editorial content accuracy, continuity and completeness.

Compilation © 2016 Cengage Learning

ISBN: 9781337037976

Cengage Learning
20 Channel Center Street
Boston, MA 02210
USA

Cengage Learning is a leading provider of customized learning solutions with office locations around the globe, including Singapore, the United Kingdom, Australia, Mexico, Brazil, and Japan. Locate your local office at: **www.international.cengage.com/region.**

Cengage Learning products are represented in Canada by Nelson Education, Ltd.

For your lifelong learning solutions, visit **www.cengage.com/custom.**

Visit our corporate website at **www.cengage.com.**

Contents in Brief

Chemical, Biological, and Physiological Aspects of Nutrition

Your body is made of carbon, hydrogen, oxygen, nitrogen, and a few other assorted elements. When joined together, elements can form large, life-sustaining molecules, such as proteins, carbohydrates, lipids, and nucleic acids. The building blocks for these biological substances come from the foods you eat. Without proper nourishment, cells—the basic units of all living organisms—die. To satisfy its nutritional needs, your body extracts and utilizes nutrients from the multitude of complex foods you eat. The first step in this process takes place in the gastrointestinal tract, where food is physically and chemically broken down into its most basic nutrient components. Once absorbed, these nutrients circulate through the extensive network of arteries and veins that make up your vascular system. Nutrients taken up by cells then undergo a series of chemical transformations that often produce metabolic waste products. The kidneys, lungs, and skin assist in eliminating these potentially harmful substances from the body. To ensure that these activities take place under optimal conditions, your endocrine and nervous systems maintain a nonstop communication network. In this chapter, you will learn about the biochemical and physiological events that take place every time you eat, and you will gain an appreciation for the intricate and varied tasks required to nourish your body.

Living with Crohn's Disease

Paige is a typical high school student who enjoys school, good friends, and an active social life. An accomplished athlete, Paige plays volleyball and is on the high school swim team. These sports are physically demanding, so it was common for Paige to feel tired at the end of the day. However, the fatigue she experienced during her junior year of high school was unlike anything she had experienced before. Having little energy, she found it increasingly more difficult to make it through the day without feeling exhausted. Next came weight loss, stomachaches, and diarrhea. Everyone was quick with advice—you are probably anemic, you are having another growth spurt, your body is telling you that you need to get more sleep. Even her doctors were quick to dismiss Paige's symptoms as nothing out of the ordinary. But, what was going on inside Paige's body was anything but ordinary—especially for an active teenage girl.

Nobody took Paige's health concerns seriously, but she was convinced something was terribly wrong. The weight loss continued, her appetite diminished further, and the stomachaches only got worse. Her doctor first suspected a parasite, but testing eliminated that as a possible cause of Paige's health problems. More doctors, more specialists, and one by one other possible causes were ruled out. If it was not a food intolerance, celiac disease, food allergy, or irritable bowel syndrome, what could it be? After a colonoscopy, the answer became alarmingly clear—Paige had Crohn's disease.

Like many people, Paige and her family had never heard of Crohn's disease. They quickly learned that Crohn's disease causes inflammation of the small intestine and/or colon, and is often recurrent. The next piece of news, that Paige would require surgery, was equally frightening. Not only was her colon swollen and inflamed, but an abscess had developed, and it needed to be removed. To prepare for surgery, Paige endured 10 days on a clear liquid diet. Once her condition was stabilized, Paige underwent a surgical procedure during which 30 centimeters (about 12 inches) of her colon were removed. Recovery came quickly, and within a few days Paige was eating solid foods again. She was released from the hospital and was back in school the following week. Paige is hopeful that she will not experience a recurrence of Crohn's disease. Although no further treatment is needed at this time, Paige takes dietary supplements to make sure she is getting all the nutrients her body needs. She has resumed her active lifestyle, and the only thing that reminds her of this long, painful ordeal is a 3-inch scar. Even that is starting to fade away, along with the memory of how sick she had been.

muzsy/Shutterstock.com

Critical Thinking: Paige's Story

Having a chronic disease can impact almost every aspect of a person's life. If you are familiar with Crohn's disease, you are already aware of the many challenges a person may face. This is particularly true for Paige, who was 16 years old when she was diagnosed. After reading Paige's story, how would you respond to this diagnosis? How would it impact your daily routines and lifestyle?

How Does Chemistry Apply to the Study of Nutrition?

Chemistry is fundamental to the study of nutrition. The organization of atoms into simple molecules, simple molecules into complex molecules, complex molecules into cells, cells into tissues, tissues into organs, and organs into organ systems is indeed remarkable (Figure 1.1). This entire circuit is made of and fueled by the nutrients contained in food. To appreciate these life-sustaining functions, it is important that you first have a basic understanding of chemistry—the science that deals with matter.

FIGURE 1.1 Levels of Organization in the Body

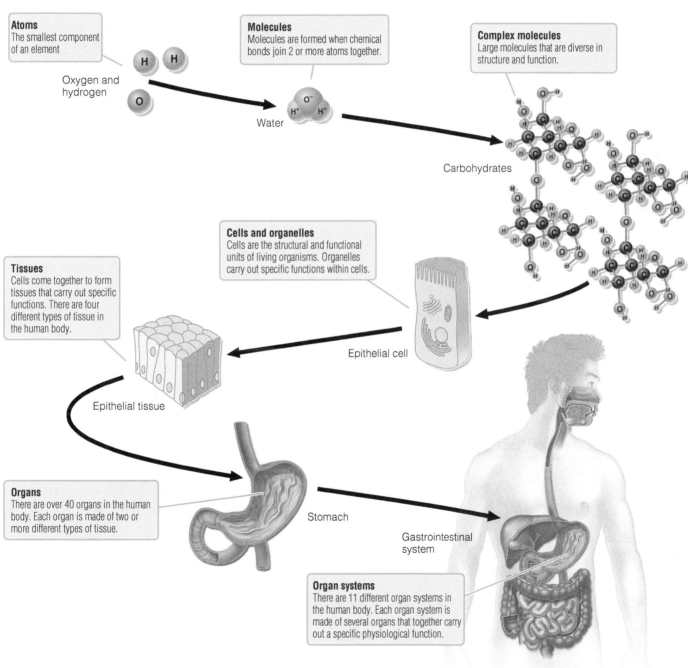

Atoms
The smallest component of an element

Oxygen and hydrogen

Molecules
Molecules are formed when chemical bonds join 2 or more atoms together.

Water

Complex molecules
Large molecules that are diverse in structure and function.

Carbohydrates

Cells and organelles
Cells are the structural and functional units of living organisms. Organelles carry out specific functions within cells.

Epithelial cell

Tissues
Cells come together to form tissues that carry out specific functions. There are four different types of tissue in the human body.

Epithelial tissue

Organs
There are over 40 organs in the human body. Each organ is made of two or more different types of tissue.

Stomach

Gastrointestinal system

Organ systems
There are 11 different organ systems in the human body. Each organ system is made of several organs that together carry out a specific physiological function.

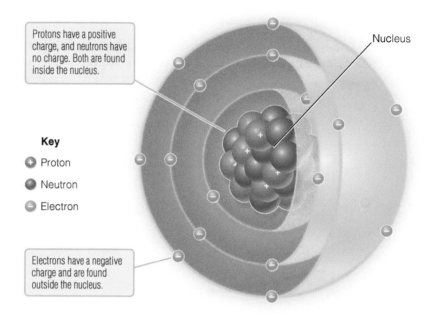

FIGURE 1.2 A Model of an Atom Atoms consist of subatomic particles—protons, neutrons, and electrons.

Protons have a positive charge, and neutrons have no charge. Both are found inside the nucleus.

Nucleus

Key

⊕ Proton

● Neutron

⊖ Electron

Electrons have a negative charge and are found outside the nucleus.

ATOMS ARE FUNDAMENTAL UNITS OF MATTER THAT MAKE UP THE WORLD AROUND US

It is difficult to imagine the existence of something that we cannot see, taste, touch, or hear. Yet particles called **atoms** are the fundamental units that make up the world around us. The atom itself, illustrated in Figure 1.2, consists of still smaller units—neutrons, protons, and electrons. Uncharged **neutrons** and positively charged **protons** are both housed in the center (*nucleus*) of the atom, whereas negatively charged **electrons** orbit the nucleus.

Anions and Cations Are Charged Atoms Most atoms have equal numbers of protons and electrons and, therefore, are neutral. However, it is possible for atoms to gain or lose electrons, as shown in Figure 1.3. When this occurs, the number of protons and electrons in the atom is no longer equal, resulting in an atom with a positive or negative charge. Atoms that have an unequal number of protons and electrons are called **ions.*** Ions with an overall positive charge are called **cations,** and ions with an overall negative charge are called **anions.** For example, a hydrogen atom typically has one proton and one electron. The loss of an electron results in a hydrogen ion with an overall positive charge—in other words, a cation.

When atoms such as chlorine, iodine, and fluorine gain an electron, the resulting anions undergo a name change; the suffix -ine becomes -ide. Therefore, when the atom fluor<u>ine</u> (F) gains an electron, it becomes the anion fluor<u>ide</u> (F^-). Important ions found in the human body include sodium (Na^+), potassium (K^+), calcium (Ca^{2+}), chloride (Cl^-), iodide (I^-), and fluoride (F^-). Ions serve many vital functions in the body. For example, calcium (Ca^{2+}) and magnesium (Mg^{2+}) are required for muscle contraction and nerve function.

Oxidation and Reduction The transfer of electrons between atoms and molecules is an important chemical event. The loss of one or more electrons is called **oxidation.** Atoms that lose one or more electrons are said

*Note that molecules such as hydroxide (OH^-) can also be ions.

atom The smallest portion into which an element can be divided into and still retain its properties.

neutron A subatomic particle, in the nucleus of an atom, with no electrical charge.

proton A subatomic particle, in the nucleus of an atom, that carries a positive charge.

electron A subatomic particle that orbits around the nucleus of an atom, and carries a negative charge.

ion An atom that has acquired an electrical charge by gaining or losing one or more electrons.

cation (CAT – i – on) An ion with a net positive charge.

anion (AN – i – on) An ion with a net negative charge.

oxidation (ox – i – DA – tion) The loss of one or more electrons.

FIGURE 1.3 Formation of Cations and Anions Atoms with unequal numbers of protons and electrons are ions. An ion can have a positive charge (cation) or a negative charge (anion).

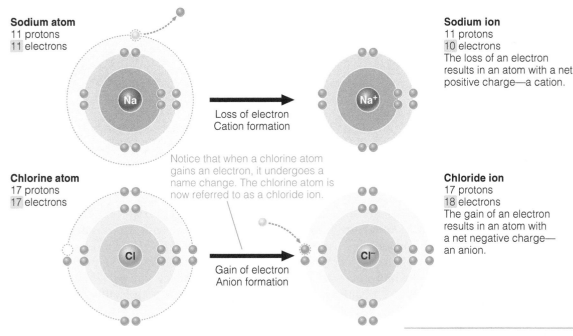

Sodium atom
11 protons
11 electrons

Loss of electron
Cation formation

Sodium ion
11 protons
10 electrons
The loss of an electron results in an atom with a net positive charge—a cation.

Notice that when a chlorine atom gains an electron, it undergoes a name change. The chlorine atom is now referred to as a chloride ion.

Chlorine atom
17 protons
17 electrons

Gain of electron
Anion formation

Chloride ion
17 protons
18 electrons
The gain of an electron results in an atom with a net negative charge—an anion.

to be oxidized and become more positively charged. For example, when iron (Fe) loses two electrons, the overall charge changes to Fe^{2+}. When Fe is fully oxidized, it has lost three electrons, becoming Fe^{3+}. Therefore, the oxidation of iron increases the net positive charge (Fe to Fe^{3+}), reflecting the loss of three negatively charged electrons. In fact, when iron-containing proteins in red meat are exposed to air for a prolonged period of time, they become oxidized, causing the meat to turn brown.

Conversely, the gain of one or more electrons is called **reduction,** and atoms become more negative during this process. **Reduction–oxidation** (or **redox**) **reactions** often take place simultaneously and are said to be coupled reactions (Figure 1.4). That is, the loss of an electron by one atom (oxidation) results in the gain of an electron by another (reduction). There are many important redox reactions in the body that involve the transfer of electrons between molecules.

CHEMICAL BONDS ENABLE ATOMS TO FORM MILLIONS OF DIFFERENT MOLECULES

An **element** is defined as a pure substance made up of only one type of atom. There are approximately 92 naturally occurring elements, 20 of which are essential for human health. In fact, just 6 elements—carbon, oxygen, hydrogen, nitrogen, calcium, and phosphorus—account for 99% of your total body weight. These important elements (Figure 1.5) provide the raw materials needed to form large, complex molecules found in living systems, such as proteins, carbohydrates, lipids, and nucleic acids.

Tony Cenicola/The New York Times/Redux

When red meat is exposed to the air too long, iron-containing proteins become oxidized, causing the surface of the meat to turn a grayish-brown color.

FIGURE 1.4 Electron Transfer in Reduction–Oxidation (Redox) Reactions Redox reactions are coupled reactions in which one or more electrons are transferred between molecules.

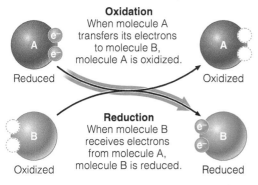

Oxidation
When molecule A transfers its electrons to molecule B, molecule A is oxidized.

Reduced → Oxidized

Reduction
When molecule B receives electrons from molecule A, molecule B is reduced.

Oxidized → Reduced

reduction The gain of one or more electrons.

reduction–oxidation (redox) reactions Chemical reactions that take place simultaneously whereby one molecule gives up one or more electrons (is oxidized) while the other molecule receives one or more electrons (is reduced).

element A pure substance made of only one type of atom.

FIGURE 1.5 The Most Abundant Elements in the Human Body

Just 6 elements account for 99% of body weight in humans.

Element	Atomic symbol	% of human body by weight	Distribution in the body
Oxygen	O	65	Found in water
Carbon	C	18	Found in all organic molecules
Hydrogen	H	10	Found in most molecules, including water
Nitrogen	N	3	Component of proteins
Calcium	Ca	2	Component of bones, teeth, and body fluids
Phosphorus	P	1	Found in cell membranes and bone matrix

FIGURE 1.6 Understanding Molecular Formulas Molecular formulas are used to indicate the number and types of atoms in a molecule.

Some molecules such as hydrogen are made of only 1 type of atom.

The subscript indicates there are 2 hydrogen atoms in this molecule.

H₂

Some molecules are made of 2 or more types of atoms.

Glucose is made of 6 carbon, 12 hydrogen, and 6 oxygen atoms.

C₆H₁₂O₆

A number in front of the molecule indicates the number of molecules.

There are 3 water molecules. Each molecule is made of 2 hydrogen atoms and 1 oxygen atom.

3H₂O

---

molecule A substance held together by chemical bonds.

chemical bonds The attractive force between atoms formed by the transfer, sharing, or interaction of electrons.

molecular formula (mo – LEC – u – lar) Indicates the number and types of atoms in a molecule.

compound A molecule made of two or more different types of atoms.

condensation A chemical reaction that results in the formation of water.

hydrolysis (hy – DRO – ly – sis) A chemical reaction whereby compounds react with water and are split apart.

Molecules are formed when chemical bonds join two or more atoms together. **Chemical bonds** are the attractive force between atoms that are formed by the transfer or sharing of electrons. These bonds enable a relatively small number of atoms to create millions of different molecules. Without chemical bonds, the molecular world would fall apart. In fact, you can think of chemical bonds as the "glue" that holds atoms together in a molecule. For example, the ions sodium (Na^+) and chloride (Cl^-) are always found together in food because they readily join together via chemical bonds to form a salt—sodium chloride (NaCl), which you know better as "table salt."

Understanding a Molecular Formula A **molecular formula,** such as H_2O, describes the number and type of atoms present in a molecule. For example, glucose, an important source of energy in your body, has a molecular formula of $C_6H_{12}O_6$. These numbers and letters tell you that one molecule of glucose consists of 6 carbon, 12 hydrogen, and 6 oxygen atoms. When more than one molecule of a substance is present, a number is placed before the molecular formula. For example, three molecules of water is written as $3H_2O$. Although some molecules, such as oxygen (O_2), consist of only one type of atom, most are made of several types of atoms. Molecules composed of two or more different types of atoms, such as water (H_2O) and glucose ($C_6H_{12}O_6$), are called **compounds** (Figure 1.6).

COMPLEX MOLECULES ARE VITAL TO CELL FUNCTION

So far, we have talked about molecules that are small and simple. However, molecules can also be very large, consisting of thousands of atoms bonded together. Carbohydrates, lipids, proteins, and nucleic acids (DNA and RNA) are large molecules and are vital to the functions of cells. The raw materials used to make these complex molecules come from the nutrients in foods that we eat.

Condensation and Hydrolysis: Make-and-Break Reactions Complex molecules such as proteins and lipids are often assembled and disassembled within cells. This is accomplished by chemical reactions that either form or break chemical bonds. One type of chemical reaction that joins molecules together is **condensation.** For example, a condensation reaction takes place when a chemical bond forms between two glucose molecules. This reaction results in the formation and release of a water molecule. The opposite type of reaction, called **hydrolysis,** can split molecules apart. During hydrolysis, a molecule of water is used to break chemical bonds. You can think of condensation and hydrolysis as opposite "make-and-break" reactions. Condensation and hydrolysis play important roles in digestion and metabolism, and they are illustrated in Figure 1.7.

Condensation

Condensation reactions result in the formation of a chemical bond that joins molecules together. When a condensation reaction occurs, a molecule of water is released.

Hydrolysis

Hydrolysis reactions break chemical bonds by the addition of a molecule of water.

ACID–BASE CHEMISTRY IS IMPORTANT TO THE STUDY OF NUTRITION

Grapefruits and lemons taste sour, baking soda tastes bitter, and pure water has no taste at all. These taste differences are attributed, in part, to the level of acidity, ranging from acidic (such as citrus) to neutral (such as water) to alkaline (such as baking soda). Acid–base chemistry is important to the study of nutrition. In fact, diseases that disrupt the acid–base balance of the body can cause serious health problems. To better understand acid–base chemistry, it is best to begin with an understanding of water, the medium in which chemical reactions take place.

Ionization of Water Molecules: The Basis of the pH Scale Although the chemical bonds holding together the hydrogen and oxygen atoms in a water molecule are quite strong, water molecules can also separate, or dissociate, into their charged (ionic) components. Water molecules dissociate to form hydrogen (H^+) and hydroxide (OH^-) ions, which can reform back into water molecules (H_2O).[†] If you analyzed a sample of pure water, you would find that the numbers of hydrogen and hydroxide ions are extremely small. That is, in a sample of pure water, only a small fraction of the water molecules are in an ionized state. You would also find that the concentration of hydrogen ions equals that of hydroxide ions. Thus, water is neither acidic nor basic, but neutral.

The ionization of water molecules is the basis for the **pH scale,** which ranges from 0 to 14. Because pure water has equal amounts of hydrogen (H^+) and hydroxide (OH^-) ions, it is neutral and has a pH of 7. However, when an acid is added to water, hydrogen ions (H^+) are released. This increases the concentration of hydrogen ions (H^+) in the solution, causing it to become more **acidic.** Fluids that contain a higher concentration of hydrogen ions (H^+) than hydroxide ions (OH^-) have a pH less than 7. As a solution becomes more acidic, its pH decreases even further.

When a base is added to water, hydroxide ions (OH^-) are released. Fluids that contain more hydroxide ions (OH^-) than hydrogen ions (H^+) have a pH greater than 7, making them **basic** or **alkaline.** The pH value of a solution increases as it becomes more basic.

Each consecutive number on the pH scale represents a 10-fold increase or decrease in the concentration of hydrogen ions (H^+). Thus, a fluid with a pH of 3 is 10 times more acidic than a fluid with a pH of 4. The various fluids in your body all have specific pH ranges. For example, the pH of urine

pH scale A scale, ranging from 0 to 14, that signifies the acidity or alkalinity of a solution.

acidic Having a pH less than 7.

basic (also called alkaline) Having a pH greater than 7.

[†]Technically, this reaction involves two water molecules and forms a hydronium ion (H_3O^+), and the ionization reaction is written as $2H_2O \leftrightarrow H_3O^+ + OH^-$; however, for simplicity we will describe water ionization as forming H^+ and OH^-.

typically ranges between 5.5 and 7.5, whereas the pH of blood is usually between 7.3 and 7.5. It is vitally important for the various fluids in your body to maintain their proper pH. Normal physiological function depends on it. To accomplish this, your body has built-in buffering systems designed to prevent changes in pH. A **buffer** is a substance that releases or binds hydrogen ions that enables fluids to resist changes in pH. Buffers can react with both acids and bases to maintain a constant pH.

How Do Biological Molecules Form Cells, Tissues, Organs, and Organ Systems?

Biological molecules such as carbohydrates, proteins, lipids, water, and nucleic acids are basic to life. However, these molecules by themselves cannot function in useful ways. Rather, some of these building blocks are used to form structural and functional units called cells. Cells make up tissues, which in turn organize to form organs. Last, organs work together as part of an organ system. The human body has 11 organ systems, all of which are pertinent to the study of nutrition.

SUBSTANCES CROSS CELL MEMBRANES BY PASSIVE AND ACTIVE TRANSPORT

Cells are like microscopic cities, full of activity. These activities are carried out by structures called **organelles,** which are distributed in a gel-like matrix called the **cytoplasm** (also called the cytosol). Cells are surrounded by a protective cell membrane that provides a boundary between the **extracellular** (outside the cell) and **intracellular** (within the cell) environments. Thus, it is the cell membrane that regulates the movement of substances into and out of cells.

Transport across Cell Membranes Cell membranes are selectively permeable, meaning that they allow some substances to cross them more readily than others. The movement of nutrients and other substances across a cell membrane occurs through a variety of processes, which are referred to collectively as *transport mechanisms*. Broadly speaking, a process that does not require energy (ATP) is called a **passive transport mechanism,** whereas one that does require energy (ATP) is called an **active transport mechanism.** Both passive and active transport mechanisms are utilized for the passage of nutrients into and out of all the cells in your body, including those that line the small intestine.

Passive Transport Mechanisms The three main types of passive transport mechanisms are simple diffusion, facilitated diffusion, and osmosis (Figure 1.8). **Simple diffusion** enables substances to cross cell membranes from a region of higher concentration to a region of lower concentration without using energy (ATP) in the process. When this occurs, the substance is said to move passively "down its concentration gradient," similar to a floating raft moving downstream. Once the concentration of the substance is equal on both sides of the cell membrane, a state of *equilibrium* has been reached. **Facilitated diffusion** also involves the passive movement of a substance down its concentration gradient (high concentration to low concentration) but differs from simple diffusion in that it requires the assistance of a membrane-bound transport protein that "escorts" materials across cell membranes.

Another type of passive transport is called **osmosis,** which is defined as the movement of water molecules across a selectively permeable membrane, such as those that surround cells. However, because cells rupture if they contain too

buffer A substance that releases or binds hydrogen ions in order to resist changes in pH.

organelles (or – gan – ELLES) Cellular structures that have a particular function.

cytoplasm (also called cytosol) The gel-like matrix inside cells.

extracellular (ex – tra – CEL – lu – lar) Situated outside of a cell.

intracellular (in – tra – CEL – lu – lar) Situated within a cell.

passive transport mechanism Transport mechanism that enables substances to cross cell membranes without expenditure of energy (ATP).

active transport mechanism Transport mechanism that enables substances to cross cell membranes, requiring the expenditure of energy (ATP).

simple diffusion A passive transport mechanism whereby substances cross cell membranes from a region of higher concentration to a region of lower concentration without using energy (ATP).

facilitated diffusion A passive transport mechanism whereby substances cross cell membranes from a region of higher concentration to a region of lower concentration with the assistance of a transport protein.

osmosis Movement of water molecules from a region of lower solute concentration to that of a higher solute concentration, until equilibrium is reached.

FIGURE 1.8 Passive Transport Passive transport mechanisms—such as simple diffusion, facilitated diffusion, and osmosis—do not require energy (ATP).

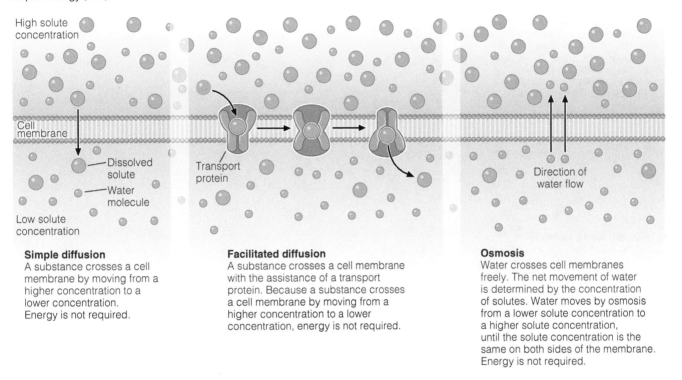

High solute concentration

Cell membrane

Dissolved solute

Water molecule

Low solute concentration

Transport protein

Direction of water flow

Simple diffusion
A substance crosses a cell membrane by moving from a higher concentration to a lower concentration. Energy is not required.

Facilitated diffusion
A substance crosses a cell membrane with the assistance of a transport protein. Because a substance crosses a cell membrane by moving from a higher concentration to a lower concentration, energy is not required.

Osmosis
Water crosses cell membranes freely. The net movement of water is determined by the concentration of solutes. Water moves by osmosis from a lower solute concentration to a higher solute concentration, until the solute concentration is the same on both sides of the membrane. Energy is not required.

much water and collapse if they contain too little, osmosis is carefully regulated. Fluids in your body contain different types of dissolved substances, such as proteins and electrolytes. Although the terms *ions* and *electrolytes* are often used interchangeably, **electrolytes** are substances such as salt that dissolve or dissociate into ions when put in water. Substances that are dissolved in fluids are called **solutes,** which, when uniformly dispersed, form a **solution.** The concentration of these dissolved substances inside and outside of the cell determines the direction and the amount of water that crosses cell membranes. A solution with a low solute concentration contains relatively more water (and fewer solutes) than a solution with a high solute concentration. Therefore, water moves from a region of lower solute concentration to that of a higher solute concentration, until equilibrium is reached.

Active Transport Mechanisms Similar to swimming upstream, some substances must cross cell membranes against the prevailing concentration gradient, moving from a region of lower concentration to that of a higher concentration. In cells, this uphill journey is accomplished by active transport mechanisms, which include carrier-mediated active transport and vesicular active transport (Figure 1.9). **Carrier-mediated active transport** requires both energy (ATP) and the assistance of a transport protein. Energy is used to "pump" molecules across cell membranes against their concentration gradients.

Another type of active transport used to move molecules into and out of cells is **vesicular active transport.** There are two types of vesicular active transport mechanisms: endocytosis and exocytosis. **Endocytosis** moves substances from the extracellular to the intracellular environment. In endocytosis, a portion of the cell membrane surrounds an extracellular particle, enclosing it in a saclike structure called a vesicle. The contents of the vesicle are then released to the inside of the cell. The reverse process, **exocytosis,** enables substances to leave cells by packaging them in vesicles, which are then released into the surrounding extracellular fluid.

electrolytes Substances such as salt that dissolve or dissociate into ions when put in water.

solute (SOL – ute) A substance that dissolves in a solvent.

solution A mixture of two or more substances that are uniformly dispersed.

carrier-mediated active transport An energy-requiring mechanism whereby a substance moves from a region of lower concentration to a region of higher concentration, requiring the assistance of a carrier protein.

vesicular active transport (ve – SIC – u – lar) An energy-requiring mechanism whereby large molecules move into or out of cells by an enclosed vesicle.

endocytosis (en – do – cy – TO – sis) A form of vesicular active transport whereby the cell membrane surrounds extracellular substances and releases them to the cytoplasm.

exocytosis (ex – o – cy – TO – sis) A form of vesicular active transport whereby intracellular cell products are enclosed in a vesicle and the contents of the vesicle are released to the outside of the cell.

FIGURE 1.9 Active Transport Active transport mechanisms—such as carrier-mediated active transport and vesicular active transport—require energy (ATP).

A. Carrier-mediated active transport
A solute crosses a cell membrane with the assistance of a transport protein. Energy (ATP) is required, because the substance moves against its concentration gradient, moving from a lower concentration to a higher concentration.

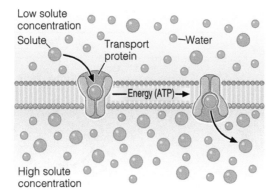

B. Vesicular active transport
Part of the cell membrane surrounds the substance, forming a vesicle. The vesicle then moves across the cell membrane and the substances are released inside (endocytosis) or outside (exocytosis) the cell.

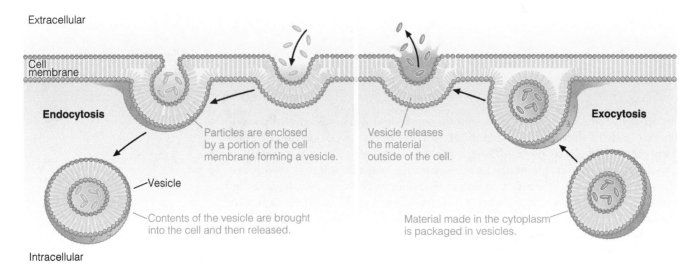

CELL ORGANELLES CARRY OUT SPECIALIZED FUNCTIONS CRITICAL FOR LIFE

Within the cell, small membrane-bound structures called organelles carry out specialized functions that are critical for life. Each type of organelle is responsible for a specific function. Some organelles produce substances needed for cellular activities, whereas others function as waste-disposal systems that assist with degrading and recycling worn-out cellular components. For example, organelles called **mitochondria** serve as power stations, converting the chemical energy in energy-yielding nutrients (glucose, fatty acids, and amino acids) into a form of energy (ATP) that is used by cells. Another organelle, called the **nucleus,** houses the genetic material DNA, which provides the "blueprint" for protein synthesis. Figure 1.10 provides an overview of cellular organelles and their functions.

mitochondria (mi – to – CHON – dri – a) Cellular organelles involved in generating energy (ATP).

nucleus A membrane-enclosed organelle that contains the genetic material DNA.

FIGURE 1.10 A Typical Cell

Cell membrane or plasma membrane
Cells are surrounded by a membrane that provides a protective boundary between the extracellular and intracellular environments.

Smooth endoplasmic reticulum
Region of the endoplasmic reticulum involved in lipid synthesis. Smooth endoplasmic reticula do not have ribosomes and are not involved in protein synthesis.

Rough endoplasmic reticulum
A series of membrane sacs that contain ribosomes that build and process proteins.

Golgi apparatus
The Golgi apparatus is a series of membrane sacs that process and package proteins.

Lysosome
Contains digestive enzymes that break down proteins, lipids, and nucleic acids. It also removes and recycles waste products.

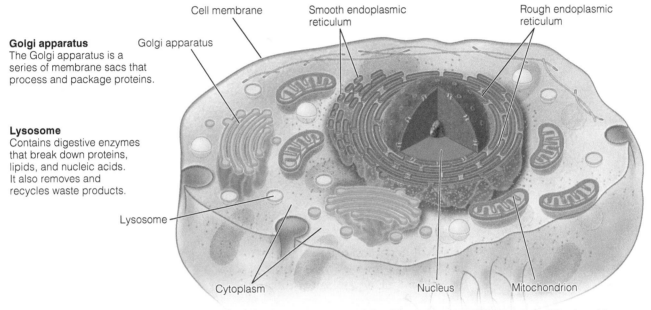

Cell membrane

Golgi apparatus

Smooth endoplasmic reticulum

Rough endoplasmic reticulum

Lysosome

Cytoplasm

Nucleus

Mitochondrion

Cytoplasm or cytosol
The cytoplasm is the gel-like substance inside cells. Cytoplasm contains cell organelles, proteins, electrolytes, and other molecules.

Nucleus
The nucleus contains the DNA in the cell. Molecules of DNA provide coded instructions used for protein synthesis.

Mitochondrion
Organelle that produces most of the energy (ATP) used by cells.

GROUPS OF CELLS MAKE UP TISSUES, TISSUES MAKE UP ORGANS, AND ORGANS MAKE UP ORGAN SYSTEMS

So far, you have learned that atoms make up molecules, molecules make up cells, and cells carry out the basic functions of life. The next level of complexity is the tissue, consisting of cells that carry out specialized functions. In the human body, four types of tissue make up more than 40 organs, which in turn make up 11 unique organ systems—all of which are relevant to the study of nutrition.

Four Different Types of Tissue **Tissue** is formed when a large number of cells with similar structure and function group together. The human body contains four different types of tissue: (1) epithelial, (2) connective, (3) muscle, and (4) neural, shown in Figure 1.11. **Epithelial tissue** (or epithelium) provides a protective layer on body surfaces (skin) as well as lining internal organs, ducts, and cavities. **Connective tissue** is the "glue" that holds the body together. Tendons, cartilage, and some parts of bones are examples of connective tissue. Blood is also a type of connective tissue, consisting of cells and platelets in a liquid called **plasma.** The body contains three major types of **muscle tissue**—skeletal muscle, smooth muscle, and cardiac muscle. Large muscles in the body consist of skeletal muscle tissue, which is needed for voluntary movement. Smooth muscle tissue is found in the lining of organs such as the esophagus, stomach, and small intestine, and is even found in the interior lining of

tissue (TIS – sue) An aggregation of specialized cells that are similar in form and function.

epithelial tissue (ep – i – THE – li – al) Tissue that forms a protective layer on bodily surfaces and lines internal organs, ducts, and cavities.

connective tissue (con – NEC – tive) Tissue that supports, connects, and anchors body structures.

plasma (PLAS – ma) The fluid component of blood.

muscle tissue Tissue that specializes in movement.

FIGURE 1.11 Four Basic Types of Tissue Epithelial, connective, neural, and muscle tissue make up all the organs in the human body.

Epithelial tissue

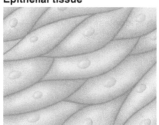

Epithelial tissue covers and lines body surfaces, organs, and cavities.

Connective tissue

Connective tissue provides structure to the body by binding and anchoring body parts.

Muscle tissue

Muscle tissue contracts and shortens when stimulated, playing an important role in movement.

Neural tissue

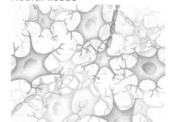

Neural tissue plays a role in communication by receiving and responding to stimuli.

blood vessels. Because smooth muscles are not under our conscious control, they are referred to as involuntary muscles. Cardiac muscle is found exclusively in the heart. The fourth type of tissue is **neural tissue,** which makes up the brain, spinal cord, and nerves. Neural tissue plays an important communicative role in the body.

Organs Form Organ Systems **Organs** consist of two or more different types of tissue, functioning together to perform a variety of related tasks. An **organ system** is formed when several organs work together, each organ carrying out important physiological functions. For example, the digestive system is composed of several organs that work collectively to physically and chemically break down food. The major organ systems and their basic functions are summarized in Table 1.1.

neural tissue Tissue that specializes in communication via nerves.

organ A group of tissues that combine to carry out coordinated functions.

organ system Organs that work collectively to carry out related functions.

TABLE 1.1 Organ Systems and Related Major Functions

System	Major Organs and Structures	Major Function
Integumentary	Skin, hair, nails, and sweat glands	Protects against pathogens and helps regulate body temperature.
Skeletal	Bones, cartilage, and joints	Provides support and structure to the body. The bone marrow of some bones produces blood cells. Also provides a storage site for certain minerals.
Muscular	Smooth, cardiac, and skeletal muscle	Assists in voluntary and involuntary body movements.
Nervous	Brain, spinal cord, nerves, and sensory receptors	Interprets and responds to information. Controls the basic senses, movement, and intellectual functions.
Endocrine	Endocrine glands	Produces and releases hormones that control physiological functions such as reproduction, hunger, satiety, blood glucose regulation, metabolism, and stress response.
Respiratory	Lungs, nose, mouth, throat, and trachea	Governs gas exchange between the blood and air. Also assists in regulating blood acid–base (pH) balance.
Circulatory	Heart, blood vessels, blood, lymph vessels, lymph nodes, and lymph organs	Transports nutrients, waste products, gases, and hormones. Also plays a role in regulating body temperature. Helps remove foreign substances and plays a role in immunity.
Digestive	Mouth, esophagus, stomach, small intestine, large intestine, liver, gallbladder, pancreas, and salivary glands	Governs the physical and chemical breakdown of food into a form that can be absorbed into the circulatory system. Eliminates solid wastes.
Reproductive	Gonads and genitals	Carries out reproductive functions and is associated with sexual characteristics, sexual function, and sexual behaviors.
Urinary	Kidneys, bladder, and ureters	Removes metabolic waste products from the blood; governs nutrient reabsorption, acid–base balance, and regulates water balance.
Immune	White blood cells, lymph vessels, bone marrow, and lymphatic tissue	Provides a defense against foreign bodies, such as bacteria and viruses, and unregulated cell growth.

Organ Systems Work Together to Maintain Homeostasis The ability of organ systems to work together to carry out common functions requires constant communication. In other words, the right hand must know what the left hand is doing. The body has two well-developed communication systems that coordinate physiologic processes—the nervous and endocrine systems. The nervous system receives and transmits information via electrical impulses and **neurotransmitters** between nerve cells, whereas the endocrine system communicates via chemical messengers, called **hormones,** in the blood. Hormones are released from glands or cells in response to various stimuli, and exert their effects by binding to receptors on specific tissues. When this occurs, tissues initiate an appropriate response to the initial stimulus. Together, the nervous and endocrine systems continuously monitor our internal environment, responding to change and restoring balance. These mechanisms allow complex organisms, such as humans, to adapt in an ever-changing environment, a process known as **homeostasis.**

Homeostasis, a state of equilibrium or balance, is an important concept in physiology and nutrition. **Negative feedback systems** play vital roles in homeostasis by opposing changes in the internal environment and by initiating corrective responses that restore balance. An example of a negative feedback system is the regulation of blood glucose. In response to a carbohydrate-rich meal, blood glucose levels rise (a change in the internal environment). The pancreas detects this change and initiates a response—the release of the hormone insulin. Insulin then binds to specific receptors on cell membranes, which facilitates the uptake of glucose. As a result, blood glucose levels are restored. This is why diseases that disrupt homeostatic responses can have serious health consequences.

How Does the Digestive System Break Down Food into Absorbable Components?

To nourish your body, your digestive system methodically disassembles the complex molecules in food into simpler basic components. This arduous task requires many organ systems, but primarily it is the digestive system that gets the job done. Your digestive system is made up of the digestive tract and accessory organs. The digestive tract, more commonly known as the **gastrointestinal (GI) tract** or alimentary tract, can be thought of as a hollow tube that runs from the mouth to the anus (Figure 1.12). Organs that make up the GI tract include the mouth, esophagus, stomach, small intestine, and large intestine.

The accessory organs, which participate in digestion but are not part of the GI tract, include the salivary glands, pancreas, and biliary system (liver and gallbladder). The accessory organs release secretions needed for the process of digestion into ducts, which empty into the **lumen,** the inner cavity that spans the entire length of the GI tract. Together, the GI tract and accessory organs carry out three important functions: (1) **digestion,** the physical and chemical breakdown of food; (2) **absorption,** the transfer of nutrients from the GI tract into the blood or lymphatic circulatory systems; and (3) **egestion,** the process whereby solid waste (feces) is expelled from the body.

THE GI TRACT HAS FOUR TISSUE LAYERS THAT CONTRIBUTE TO THE PROCESS OF DIGESTION

As shown in Figure 1.13, the digestive tract contains four major tissue layers—the mucosa, submucosa, muscularis, and serosa. Each tissue layer contributes to the overall function of the GI tract by providing secretions, movement, communication, and protection.

neurotransmitters Chemical messengers released from nerve cells that transmit information.

hormones Substances released from glands or cells in response to various stimuli that exert their effect by binding to receptors on specific tissues.

homeostasis (ho – me – o – STA – sis) A state of balance or equilibrium.

negative feedback systems Corrective responses that oppose change and restore homeostasis.

gastrointestinal (GI) tract A tubular passage that runs from the mouth to the anus that includes several organs that participate in the process of digestion; also called the digestive tract.

lumen (LU – men) The cavity inside a tubular structure in the body.

digestion The physical and chemical breakdown of food by the digestive system into a form that allows nutrients to be absorbed.

absorption The passage of nutrients through the lining of the GI tract into the blood or lymphatic circulation.

egestion The process whereby solid waste (feces) is expelled from the body.

FIGURE 1.12 Organs of the Digestive System The digestive system consists of the gastrointestinal tract and the accessory organs.

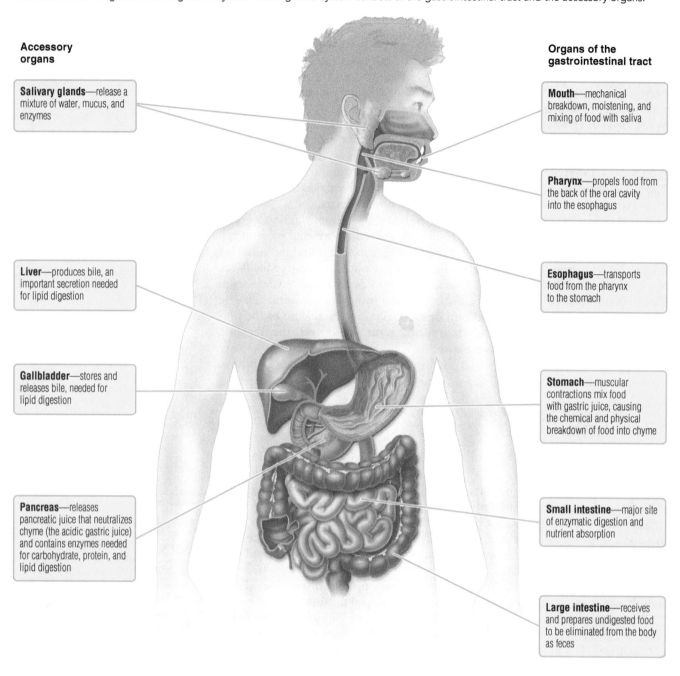

Accessory organs

Salivary glands—release a mixture of water, mucus, and enzymes

Liver—produces bile, an important secretion needed for lipid digestion

Gallbladder—stores and releases bile, needed for lipid digestion

Pancreas—releases pancreatic juice that neutralizes chyme (the acidic gastric juice) and contains enzymes needed for carbohydrate, protein, and lipid digestion

Organs of the gastrointestinal tract

Mouth—mechanical breakdown, moistening, and mixing of food with saliva

Pharynx—propels food from the back of the oral cavity into the esophagus

Esophagus—transports food from the pharynx to the stomach

Stomach—muscular contractions mix food with gastric juice, causing the chemical and physical breakdown of food into chyme

Small intestine—major site of enzymatic digestion and nutrient absorption

Large intestine—receives and prepares undigested food to be eliminated from the body as feces

The Mucosa: Source of GI Secretions The innermost lining of the digestive tract, called the **mucosa,** consists mainly of epithelial tissue, and carries out a variety of digestive functions. The mucosa, often called the mucosal epithelial lining, produces a variety of secretions such as enzymes and hormones, referred to collectively as **GI secretions,** that facilitate the chemical breakdown of food in the GI tract. Because the cells that make up the mucosal lining are continuously exposed to harsh digestive secretions within the GI tract, their lifespan is a mere two to five days. Once the mucosal epithelial cells wear out, they slough off and are replaced with new cells. Worn-out cells

mucosa (mu – CO – sa) The lining of the gastrointestinal tract that is made up of epithelial cells; also called mucosal lining.

GI secretions Substances released by organs that make up the digestive system that facilitate the process of digestion; also called digestive juices.

FIGURE 1.13 The Layers of the Gastrointestinal Tract The gastrointestinal tract consists of four tissue layers: the mucosa, submucosa, muscularis, and serosa. Each layer carries out specific functions—communication, secretion, movement, and protection—all of which facilitate digestion and nutrient absorption.

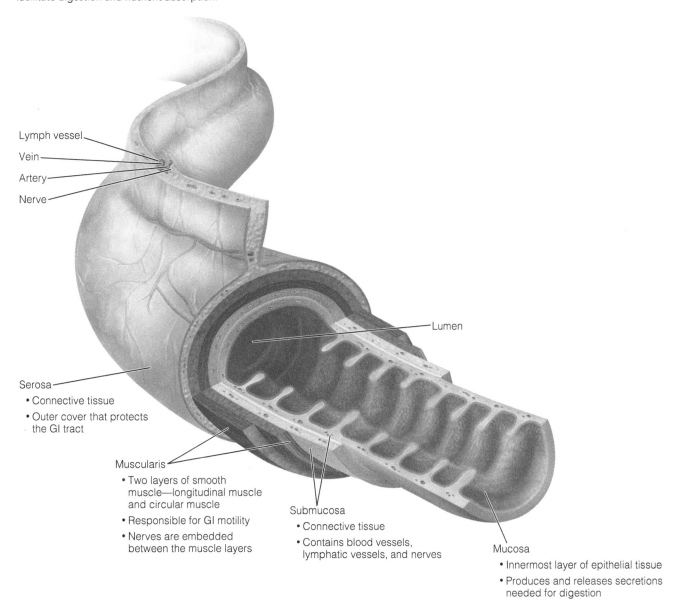

Lymph vessel

Vein

Artery

Nerve

Lumen

Serosa
- Connective tissue
- Outer cover that protects the GI tract

Muscularis
- Two layers of smooth muscle—longitudinal muscle and circular muscle
- Responsible for GI motility
- Nerves are embedded between the muscle layers

Submucosa
- Connective tissue
- Contains blood vessels, lymphatic vessels, and nerves

Mucosa
- Innermost layer of epithelial tissue
- Produces and releases secretions needed for digestion

from the mucosal lining are eliminated from the body in the feces. To support this rapid rate of cell turnover, the mucosa has high nutrient requirements. Nutrient deficiencies can profoundly affect the ability to maintain the mucosal lining, impairing digestion and nutrient absorption.

Blood Vessels and Nerves Embedded in the Submucosa A layer of connective tissue called the **submucosa** surrounds the mucosal layer. The submucosa contains a rich supply of blood vessels, delivering nutrients and oxygen to the inner mucosal layer and the next outward layer, called the muscularis. These blood vessels also circulate most of the nutrients absorbed from the small intestine away from the GI tract. In addition to blood vessels, the submucosa contains lymphatic vessels that are filled with a fluid called **lymph.** Lymph aids in the circulation of water-insoluble substances such as dietary fat away

submucosa (SUB – mu – co – sa) A layer of tissue that lies between the mucosa and muscularis tissue layers.

lymph A fluid found in lymphatic vessels.

from the GI tract. The submucosa also contains a network of nerves, which regulate the release of GI secretions from cells making up the mucosal lining.

The Muscularis Enables Food to Mix and Move through the GI Tract Moving outward from the submucosa, the next layer in the GI tract is the muscularis. The **muscularis** typically consists of two layers of smooth muscle: an outer *longitudinal* layer and an inner *circular* layer. Located between these two muscle layers is another network of nerves that control the contraction and relaxation of the muscularis. The movement of the muscularis promotes mixing of the food mass with digestive secretions and keeps food moving through the entire length of the GI tract.

The GI Tract Is Enclosed by the Serosa The outer layer of connective tissue that encloses the GI tract is the **serosa.** The serosa secretes a fluid that lubricates the digestive organs, preventing them from adhering to one another. In addition, much of the GI tract is anchored within the abdominal cavity by a membrane (called mesentery) that is continuous with the serosa.

How Do Gastrointestinal Motility and Secretions Facilitate Digestion?

The amount of time between the consumption of food and its elimination as solid waste is called **transit time.** It takes approximately 24 to 72 hours for food to pass from mouth to anus. Many factors affect transit time, such as composition of diet, illness, certain medications, physical activity, and emotions. Because each organ in the digestive system makes a unique contribution to the overall process of digestion, food must remain within each region long enough for all the digestive events to be complete. This is accomplished in part by circular bands of smooth muscle called **sphincters** that act like one-way valves, regulating the flow of the luminal contents from one organ to the next (Figure 1.14). The GI tract has several sphincters, which are often named according to their anatomical locations. For example, the gastroesophageal sphincter is located between the stomach and the esophagus. Note that the term *gastric* pertains or relates to the stomach.

muscularis (mus – cu – LAR – is) The layer of tissue in the gastrointestinal tract that consists of at least two layers of smooth muscle.

serosa (se – RO – sa) Connective tissue that encloses the gastrointestinal tract.

transit time Amount of time between the consumption of food and its elimination as solid waste.

sphincter (SFINK-ter) A muscular band that narrows an opening between organs in the GI tract.

FIGURE 1.14 Sphincters Regulate the Flow of Food Sphincters are circular bands of muscle located between organs that regulate the flow of material through the gastrointestinal tract.

The gastroesophageal sphincter, located between the esophagus and the stomach, relaxes briefly to allow food to enter the stomach.

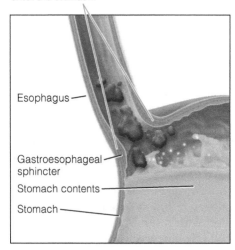

Esophagus

Gastroesophageal sphincter

Stomach contents

Stomach

After the food passes into the stomach, the gastroesophageal sphincter closes to prevent the stomach contents from re-entering the esophagus.

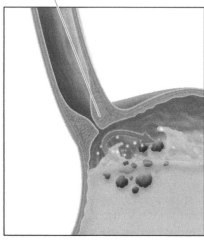

GASTROINTESTINAL (GI) MOTILITY MIXES AND PROPELS FOOD IN THE GI TRACT

The term **GI motility** refers to the mixing and propulsion of material by muscular contractions in the GI tract. These vigorous movements result from the contraction and relaxation of muscles that make up the muscularis. There are two types of movement in the GI tract: segmentation and peristalsis. **Segmentation** both mixes and slowly propels food, whereas **peristalsis** involves more vigorous propulsive movements. These movements serve different functions, as shown in Figure 1.15. For example, when segmentation occurs in the small intestine, circular muscles move the food mass back and forth in a bidirectional fashion. These mixing and propulsive movements increase the contact between food particles and digestive secretions, giving the intestine an appearance of a chain of sausages. In contrast, peristalsis involves rhythmic, wavelike muscle contractions that propel food along the entire length of the GI tract. The contraction of circular muscles behind the food mass causes the longitudinal muscles to shorten. When the longitudinal muscles subsequently lengthen, the food is

GI motility Mixing and propulsive movements of the gastrointestinal tract caused by contraction and relaxation of the muscularis.

segmentation A muscular movement in the gastrointestinal tract that moves the contents back and forth within a small region.

peristalsis (per – i – STAL – sis) Waves of muscular contractions that move materials in the GI tract in a forward direction.

FIGURE 1.15 Segmentation and Peristalsis GI motility involves both mixing and propulsive movements. Segmentation (A) both mixes and slowly propels food, whereas peristalsis (B) involves more vigorous propulsive movements.

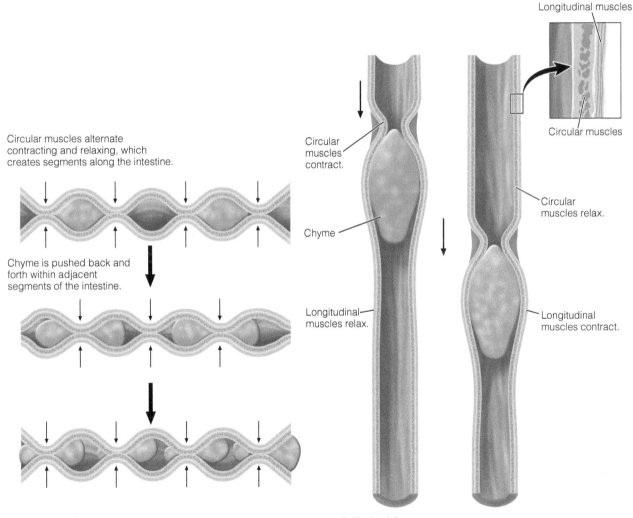

Circular muscles alternate contracting and relaxing, which creates segments along the intestine.

Chyme is pushed back and forth within adjacent segments of the intestine.

Longitudinal muscles

Circular muscles

Circular muscles contract.

Chyme

Longitudinal muscles relax.

Circular muscles relax.

Longitudinal muscles contract.

A. Segmentation
Segmentation both mixes and slowly moves food through the GI tract. The circular muscles contract and relax, creating a "chopping" motion.

B. Peristalsis
Peristalsis consists of a series of wavelike rhythmic contractions and relaxation involving both the circular and longitudinal muscles. This action propels food forward through the GI tract.

propelled forward. Thus, peristalsis is similar to the motion generated when a crowd at a sporting event does "the wave."

GASTROINTESTINAL (GI) SECRETIONS AID DIGESTION AND PROTECT THE GI TRACT

The organs that make up the digestive system release a variety of secretions—water, acid, electrolytes, mucus, salts, enzymes, bile, bicarbonate, and other substances—that are important for digestion and the protection of the GI tract. For example, **mucus** forms a protective coating that lubricates the mucosal lining. **Digestive enzymes** are biological catalysts that facilitate chemical reactions that break down complex food particles. Specifically, digestive enzymes catalyze hydrolytic chemical reactions (hydrolysis) that break chemical bonds by the addition of water. As a result, complex molecules such as starch and protein are broken down into smaller, simpler molecules.

Organs that release digestive secretions include the salivary glands, stomach, pancreas, gallbladder, small intestine, and large intestine. In fact, approximately 7 liters of secretions, most of which is water, are released daily into the lumen of your GI tract. Fortunately, the body has an elaborate recycling system that enables much of this water to be reclaimed. The major GI secretions and their related functions are summarized in Table 1.2.

NEURAL AND HORMONAL SIGNALS REGULATE GASTROINTESTINAL MOTILITY AND SECRETIONS

Like the conductor of an orchestra, neural and hormonal signals precisely coordinate GI motility and the release of GI secretions. These involuntary regulatory activities ensure that complex food particles are physically and chemically broken down and that the food mass moves along the GI tract at the appropriate rate. The GI tract has three regulatory control mechanisms—two of which provide neural control and the other hormonal.

mucus (MU – cus) A substance that coats and protects mucous membranes.

digestive enzymes Biological catalysts that facilitate chemical reactions that break chemical bonds by the addition of water (hydrolysis), resulting in the breakdown of large molecules into smaller components.

TABLE 1.2 Summary of Major Secretions Produced and Released by the Gastrointestinal Tract and Accessory Organs and Their Related Functions

Secretion	Source	Function
Mucus	Mucosal cells of the GI tract	• Protects and lubricates the GI tract
Saliva (contains water, mucus, and enzymes)	Salivary glands	• Moistens foods • Helps form the bolus • Facilitates taste • Aids swallowing • Chemically breaks down food via enzymes
Enzymes	Salivary glands, stomach, small intestine, and pancreas	• Chemically break down foods
GI hormones (e.g., gastrin, secretin, cholecystokinin)	Endocrine cells in the gastric pits, and endocrine cells in the mucosal lining of the small intestine	• Facilitate communication, and regulate GI motility and release of GI secretions
Bile	Made in the liver; stored and released from the gallbladder	• Enables lipid globules to disperse in water
Pancreatic juice (contains bicarbonate and enzymes)	Pancreas	• Neutralizes chyme • Provides enzymes needed for the chemical breakdown of carbohydrates, proteins, and lipids
Gastric juice (contains hydrochloric acid, enzymes, water, intrinsic factor)	Exocrine cells of the gastric pits (mucus-secreting cells, parietal cells, and chief cells)	• Provides enzymes needed for the chemical breakdown of some nutrients • Hydrochloric acid is needed for forming chyme and activating some enzymes • Intrinsic factor is needed for absorption of vitamin B_{12}

Neural Regulation of Digestive Functions The GI tract has its own "local" nervous system that consists of neural networks that are embedded in the submucosa and muscularis layers. Collectively, these networks of nerves are referred to as the **enteric nervous system.** Although the term *enteric* typically pertains to the intestine, the enteric nervous system actually spans the entire length of the GI tract. The enteric nervous system receives information from sensory receptors located within the GI tract. **Sensory receptors** monitor conditions and changes related to digestive activities. For example, **chemoreceptors** detect changes in the chemical composition of the luminal contents, whereas **mechanoreceptors** detect stretching or distension in the walls of the GI tract. As you might expect, the presence of food in the GI tract can stimulate both chemo- and mechanoreceptors. Information from sensory receptors is relayed to the enteric nervous system, which responds by communicating with the muscles and hormone-producing cells of the GI tract. In response, muscles and glands carry out the appropriate response to help digest food, such as increasing peristalsis or releasing secretions.

While the enteric nervous system controls digestive functions on the local level, the GI tract also communicates with the **central nervous system,** which consists of the brain and spinal cord. The neural network connecting the central and enteric nervous systems keeps the GI tract and the brain in close communication. Because the central nervous system can also initiate neural communication with the GI tract, sensory and emotional stimuli can affect GI function. For example, sensory stimuli such as the sight, smell, and thought of food stimulate GI motility and the release of GI secretions. Similarly, emotional factors such as sadness, anger, and anxiety can disrupt digestive functions, causing GI distress such as an upset stomach.

Hormonal Regulation of Digestive Functions Recall that the mucosal lining of the GI tract contains hormone-producing endocrine cells. **GI hormones** play an important communicative role in the process of digestion by acting as chemical messengers. Released into the blood in response to chemical and physical changes in the GI tract, these hormones alert other organs to the impending arrival of food. Like neural signals, hormones also influence transit time and the release of secretions that aid digestion. In addition, some GI hormones communicate with appetite centers in the brain—influencing the desire to eat. For example, release of the hormone ghrelin increases in response to fasting, stimulating hunger and food intake. The major roles of various GI hormones are summarized in Table 1.3 and discussed in more detail in subsequent chapters.

enteric nervous system (en – TER – ic) Neurons located within the submucosa and muscularis layers of the digestive tract.

sensory receptors Receptors that monitor conditions and changes in the GI tract.

chemoreceptor (CHE – mo – re – cep – tor) A type of sensory receptor that responds to a chemical stimulus.

mechanoreceptor (mech – A – no – re – cep – tor) A type of sensory receptor that responds to pressure, stretching, or mechanical stimulus.

central nervous system The part of the nervous system consisting of the brain and spinal cord.

GI hormones Hormones secreted by the mucosal lining of the GI tract that regulate GI motility and secretion.

TABLE 1.3 The Major GI Hormones and Their Related Functions

Hormone	Site of Production	Stimuli for Release	Major Activities
Gastrin	Stomach	• Food in the stomach • Stretching of the stomach walls • Alcohol • Caffeine • Cephalic stimuli (smell, taste)	• Stimulates gastric motility • Stimulates secretion of gastric juice • Increases gastric emptying
Secretin	Small intestine	• Arrival of chyme into the duodenum	• Inhibits gastric motility • Inhibits secretion of gastric juice • Stimulates release of pancreatic juice containing sodium bicarbonate and enzymes
Cholecystokinin (CCK)	Small intestine	• Arrival of partially digested protein and fat into the duodenum	• Stimulates gallbladder to contract and release bile • Stimulates release of pancreatic enzymes
Ghrelin	Stomach (and other tissues)	• Not well understood, but release is greater during fasting	• Stimulates hunger

How Does the Gastrointestinal Tract Coordinate Functions to Optimize Digestion and Nutrient Absorption?

The process of digestion begins with the intake, or ingestion, of food. Over the next 24 to 72 hours, the food is physically and chemically transformed. To optimize digestion and nutrient absorption, the intensity of GI motility and the timing of the release of digestive secretions must be synchronized with the arrival of food. For this reason, digestion is often divided into three phases: the cephalic, gastric, and intestinal phases.

The **cephalic phase** begins even before food enters your mouth. During this phase, the thought, smell, and sight of food stimulate the central nervous system, which in turn stimulates GI motility and the release of digestive secretions. This response serves as a "wake-up" call to your GI tract, preparing it to receive and digest food. The **gastric phase** of digestion begins with the arrival of food in the stomach. During this phase, muscular contractions (gastric motility) become more forceful and the release of gastric secretions increases, which prepare the stomach for its role in the digestive process. By the time food reaches the small intestine, it has undergone considerable physical and chemical change, and no longer resembles the food that you consumed. Yet the process of digestion is not complete. In preparation for the next phase of digestion, the **intestinal phase,** hormonal signals from the small intestine slow the churning action of the stomach (motility), decreasing the rate at which material passes out of the stomach and into the small intestine. As food enters the small intestine, hormonal responses alert the accessory organs (pancreas and gallbladder) of the digestive tasks that lie ahead, signifying that the intestinal phase of digestion is under way.

DIGESTION BEGINS IN THE MOUTH WITH CHEWING AND MIXING FOOD

Whereas the cephalic phase prepares the GI tract to receive and digest food, digestion actually begins when food enters the mouth (also called the oral cavity). The forceful grinding action of your teeth breaks food into manageable pieces. This process, called **mastication,** results in the physical (mechanical) breakdown of food. The presence of food in the mouth stimulates the salivary glands to release **saliva**—as much as a quart per day. Saliva consists of water, mucus, digestive enzymes, and antibacterial agents. As food is broken apart, it mixes with saliva, becoming moist and easier to swallow.

Saliva is also a necessary factor in taste sensation, because some food components must first be dissolved before they can be detected by taste buds. Taste buds, located on specific regions of the tongue, were initially thought to discriminate only four basic tastes: salty, sour, sweet, and bitter. More recently, a fifth taste called **umami** has been identified, and is often described as a meat-like taste sensation. When food is consumed, gustatory (taste) cells and olfactory (smell) cells are stimulated, sending neural signals to the brain. Together, these signals enable the brain to distinguish the thousands of different flavors we enjoy in our foods. Olfactory cells in particular have a profound effect on our ability to taste food, accounting for approximately 80% of taste. Blocked nasal passages make it difficult for the aroma molecules emitted by food to stimulate olfactory cells, which is

cephalic phase (ce – PHAL – ic) The response of the central nervous system to sensory stimuli, such as smell, sight, and taste, that occurs before food enters the GI tract; characterized by increased GI motility and release of GI secretions.

gastric phase The phase of digestion stimulated by the arrival of food into the stomach; characterized by increased GI motility and release of GI secretions.

intestinal phase The phase of digestion in which chyme enters the small intestine; characterized by both a decrease in gastric motility and secretion of gastric juice.

mastication Chewing and grinding of food by the teeth to prepare for swallowing.

saliva A secretion released into the mouth by the salivary glands; moistens food and starts the process of digestion.

umami (u – MAM – i) A taste, in addition to the four basic taste components, that imparts a savory or meat-like flavor.

why it is difficult to taste food when your nasal passages are congested.

Swallowing Moves Food from the Mouth to the Esophagus The tongue, made primarily of muscle, assists in chewing and swallowing. As food mixes with saliva, the tongue manipulates the food mass, pushing it up against the hard, bony palate making up the roof of the mouth. Infants born with an anatomical birth defect called a cleft palate have difficulty swallowing because the bones that form the hard palate are not fused properly. This results in an opening between the oral and nasal cavities. Fortunately, this is a treatable condition, and corrective surgery is typically performed within a few months of birth. Until then, special care must be given to ensure that the infant is not experiencing feeding problems and is adequately nourished.

Figure 1.16 shows that swallowing takes place in two phases. As you prepare to swallow, your tongue directs the soft, moist mass of food, now referred to as a **bolus,** toward the back of your mouth, an area known as the **pharynx.** The pharynx is the shared space between the oral and nasal cavities. This phase of swallowing is under voluntary control, but once

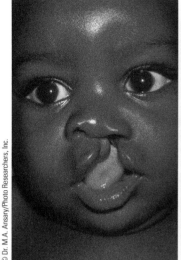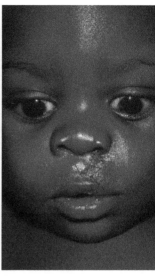

© Dr. M. A. Ansary/Photo Researchers, Inc.

A cleft palate occurs when the bones that make up the hard palate do not completely fuse together. This birth defect can be corrected surgically, often leaving minimal visible signs.

bolus (BO – lus) A soft, moist mass of chewed food.

pharynx (PHAR – nyx) Region toward the back of the mouth that is the shared space between the oral and nasal cavities.

FIGURE 1.16 Voluntary and Involuntary Phases of Swallowing The first phase of swallowing is under our control (voluntary), whereas the second phase is not (involuntary).

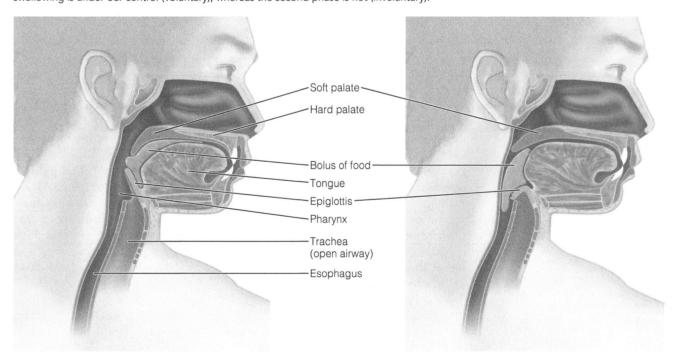

Soft palate
Hard palate
Bolus of food
Tongue
Epiglottis
Pharynx
Trachea (open airway)
Esophagus

A. Voluntary phase
The tongue pushes the bolus of food against the hard palate. Next, the tongue pushes the bolus against the soft palate, triggering the swallowing response.

B. Involuntary phase
The soft palate rises, in turn preventing the bolus from entering the nasal cavity. The epiglottis covers the trachea, blocking the opening to the lungs. The bolus enters the esophagus and is propelled toward the stomach by peristalsis.

the bolus reaches the pharynx the involuntary phase of swallowing begins. At this point, the bolus is ready to enter the **esophagus,** a narrow muscular tube that leads to the stomach.

THE ESOPHAGUS DELIVERS FOOD TO THE STOMACH

During the involuntary phase of swallowing, the upper, back portion of your mouth (called the soft palate) rises, blocking the entrance to the nasal cavity. This helps guide the bolus into the correct passageway—the esophagus. Swallowing is also facilitated by the movement of the soft palate that pulls the larynx (vocal cords) upward, causing the **epiglottis** (a cartilage flap) to cover the trachea—the airway leading to the lungs. Once the bolus moves past this dangerous intersection, the muscles relax and prepare for the next swallow. Disorders that affect skeletal muscles or nerves, such as Parkinson's disease and strokes, can affect our ability to swallow. Impaired swallowing, or what is called **dysphagia,** can make it difficult for someone to obtain adequate nourishment.

The esophagus is lubricated and protected by a thin layer of mucus, which facilitates the passage of food. Peristalsis propels the food toward the stomach, where it encounters the first of several sphincters in the GI tract. The **gastroesophageal sphincter** (also called the lower esophageal sphincter or the cardiac sphincter) forms a juncture between the esophagus and the stomach. To prevent the contents of the stomach from re-entering the esophagus, the gastroesophageal sphincter remains closed. However, as the bolus travels toward the stomach, stretch receptors trigger a neural response that signals the gastroesophageal sphincter to relax long enough for the bolus to enter the stomach. This entire trip, from the pharynx to the stomach, takes less than 10 seconds.

FUNCTIONS OF THE STOMACH INCLUDE STORAGE, RELEASE OF GASTRIC SECRETIONS, AND MIXING

The stomach is a large, muscular, J-shaped sac composed of three regions: the fundus, the body, and the antrum (Figure 1.17). The fundus is the top portion of the stomach that extends upward, above the lower portion of the esophagus. The middle portion of the stomach is called the body, and the lower portion of the stomach is called the pyloric region or antrum. The **pyloric sphincter,** located at the base of the antrum, regulates the movement of food from the stomach into the duodenum (the first portion of the small intestine).

The stomach is uniquely equipped to carry out three important functions: (1) temporary storage of food, (2) production of gastric secretions needed for digestion, and (3) mixing of food with gastric secretions. By the time food leaves the stomach, the bolus has been transformed into a semi-liquid paste called **chyme.**

Stretching of the Stomach Walls Allows Temporary Food Storage Your stomach has an amazing capacity to accommodate large amounts of food. When empty, its volume is quite small—approximately one quarter of a cup. When we eat, the walls of the stomach expand to increase its capacity to 1 to 2 quarts. The ability to expand to this extent is due to the special interior lining of the stomach, which is folded into convoluted pleats called **rugae.** Like an accordion, the rugae unfold and flatten, allowing the stomach walls to stretch as it fills with food. This stretching triggers a neural response (via mechanoreceptors), signaling the brain that the stomach is becoming full. Shortly thereafter, hunger diminishes, causing a person to stop eating.

esophagus (e – SOPH – a – gus) The passageway that begins at the pharynx and ends at the stomach.

epiglottis (ep – i – GLOT – tis) A cartilage flap that covers the trachea while swallowing.

dysphagia (dys – PHA – gia) Difficulty swallowing.

gastroesophageal sphincter (gas – tro – e – soph – a – GEAL) A circular muscle that regulates the flow of food between the esophagus and the stomach; also called lower esophageal sphincter or cardiac sphincter.

pyloric sphincter (py – LOR – ic) A circular muscle that regulates the flow of food between the stomach and the duodenum.

chyme The thick fluid resulting from the mixing of food with gastric secretions in the stomach.

rugae (RU – gae) Folds that line the inner stomach wall.

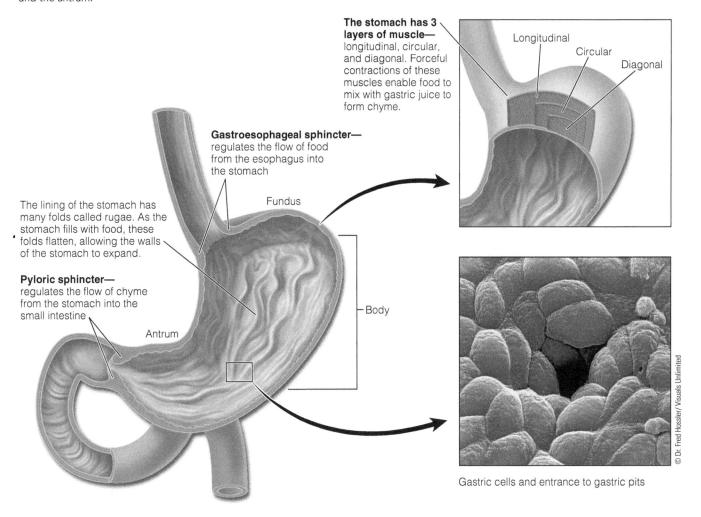

FIGURE 1.17 Anatomy of the Stomach and Its Role in Digestion The stomach is divided into three regions: the fundus, the body, and the antrum.

The stomach has 3 layers of muscle— longitudinal, circular, and diagonal. Forceful contractions of these muscles enable food to mix with gastric juice to form chyme.

Longitudinal

Circular

Diagonal

Gastroesophageal sphincter— regulates the flow of food from the esophagus into the stomach

Fundus

The lining of the stomach has many folds called rugae. As the stomach fills with food, these folds flatten, allowing the walls of the stomach to expand.

Pyloric sphincter— regulates the flow of chyme from the stomach into the small intestine

Antrum

Body

Gastric cells and entrance to gastric pits

© Dr. Fred Hossler/ Visuals Unlimited

Gastric Secretions Help Liquefy Solid Food At first glance, the stomach lining appears to be covered with numerous small holes. When the holes are magnified, you can see that they penetrate deep into the mucosal layer and form structures called **gastric pits** (Figure 1.18). The stomach contains several million gastric pits, that are formed by cells that produce and release a variety of gastric secretions. Some of these cells are **endocrine** cells that release their secretions (hormones) into the blood. Others, called **exocrine** cells, release their secretions into ducts that empty directly into the cavity of the gastric pit. These secretions, collectively called **gastric juice,** consist mainly of water, hydrochloric acid, digestive enzymes, mucus, and intrinsic factor—a substance needed for vitamin B_{12} absorption. Your stomach produces more than 2 liters (roughly 2 quarts) of gastric juice daily.

The presence of food in the stomach causes the endocrine cells in the gastric pits to release the hormone **gastrin.** Even the thought, smell, taste, and anticipation of food can trigger its release. Gastrin stimulates exocrine cells to release hydrochloric acid (HCl) and intrinsic factor from **parietal cells** and digestive enzymes from neighboring **chief cells.** HCl, a major component of gastric juice, dissolves food particles, destroys bacteria that may be present in food, and provides an optimal acidic environment (pH 2) for digestive enzymes produced in the stomach to function.

gastric pits Invaginations of the mucosal lining of the stomach that contain specialized endocrine and exocrine cells.

endocrine cells Those that produce and release hormones into the blood.

exocrine cells Those that produce and release their secretions into ducts.

gastric juice Digestive secretions produced by exocrine cells that make up gastric pits.

gastrin (GAS – trin) A hormone secreted by endocrine cells that stimulates the production and release of gastric juice.

parietal cells (pa – RI – e – tal) Exocrine cells within the gastric mucosa that secrete hydrochloric acid and intrinsic factor.

chief cells Exocrine cells in the gastric mucosa that produce digestive enzymes.

FIGURE 1.18 **Gastric Pits** The mucosal lining of the stomach is made up of exocrine and endocrine cells. The exocrine cells (mucus-secreting cells, chief cells, and parietal cells) secrete gastric juices, whereas the endocrine cells secrete hormones.

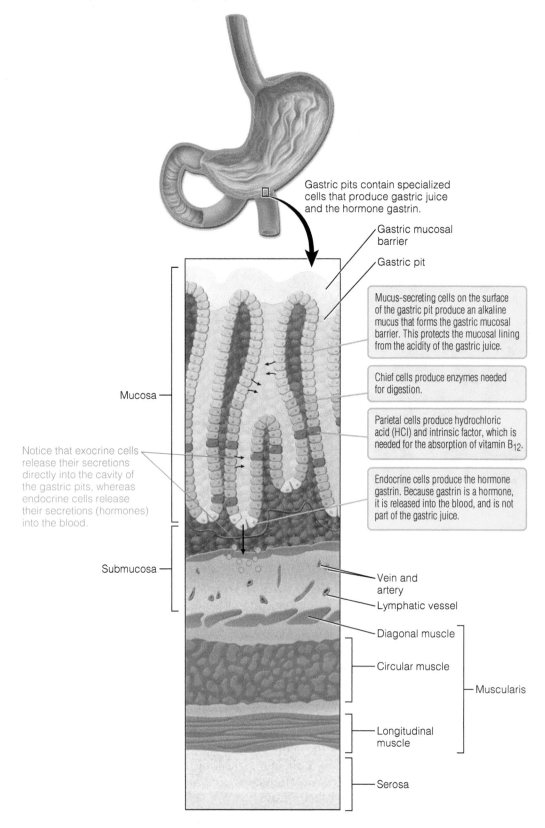

Gastric pits contain specialized cells that produce gastric juice and the hormone gastrin.

Gastric mucosal barrier

Gastric pit

Mucus-secreting cells on the surface of the gastric pit produce an alkaline mucus that forms the gastric mucosal barrier. This protects the mucosal lining from the acidity of the gastric juice.

Chief cells produce enzymes needed for digestion.

Parietal cells produce hydrochloric acid (HCl) and intrinsic factor, which is needed for the absorption of vitamin B_{12}.

Endocrine cells produce the hormone gastrin. Because gastrin is a hormone, it is released into the blood, and is not part of the gastric juice.

Mucosa

Notice that exocrine cells release their secretions directly into the cavity of the gastric pits, whereas endocrine cells release their secretions (hormones) into the blood.

Submucosa

Vein and artery

Lymphatic vessel

Diagonal muscle

Circular muscle

Muscularis

Longitudinal muscle

Serosa

Located near the entrance of the gastric pits are numerous secretory cells that release a thin, watery mucus. The mucus forms a protective layer called the **gastric mucosal barrier,** which prevents the acidic gastric juice from damaging the delicate lining of the stomach. Without this layer, the mucosal lining could not withstand its harsh environment, resulting in inflammation and the formation of sores or ulcers. Similarly, a condition called gastroesophageal reflux disease results when the unprotected lining of the esophagus is repeatedly exposed to gastric juice.[1]

Gastric Ulcers and Gastroesophageal Reflux Disease Although the mucosal barrier is usually successful in protecting the stomach and esophagus from the harsh gastric juices, this protection sometimes fails, resulting in two common GI disorders: gastric ulcers and gastroesophageal reflux disease (GERD). As you can see from Figure 1.19, an ulcer looks similar to an open canker sore in the mouth and, if left untreated, can erode through the various tissue layers. This can result in severe complications such as bleeding and infection in the abdominal cavity. Ulcers can occur when gastric juice erodes areas of the mucosal lining in the esophagus (esophageal ulcer), stomach (**gastric ulcer**), or duodenum (duodenal ulcer). Collectively, GI ulcers (stomach, duodenum, or esophagus) are referred to as **peptic ulcers.**

The primary cause of GERD is the relaxation of the gastroesophageal sphincter, which enables the stomach contents (chyme) to move from the stomach back into the esophagus, or what is referred to as reflux (Figure 1.20). Repeated exposure to the acidic gastric juice can cause the delicate lining of the esophagus to become inflamed. You can read more about these two common GI disorders (peptic ulcers and GERD) in the Focus on Clinical Applications feature.

gastric mucosal barrier A thick layer of mucus that protects mucosal lining of the stomach from the acidic gastric juice.

gastric ulcer A sore in the lining of the stomach.

peptic ulcer An irritation or erosion of the mucosal lining in the stomach, duodenum, or esophagus.

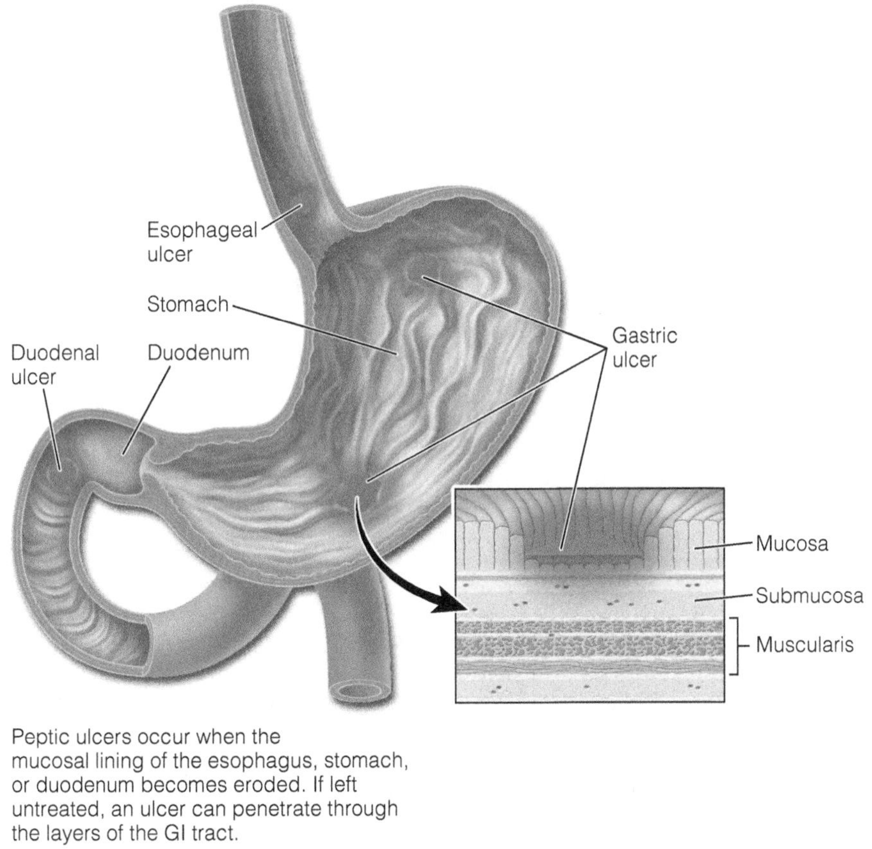

FIGURE 1.19 Peptic Ulcers
Peptic ulcers are erosions that occur in the mucosal lining of the esophagus, stomach, or duodenum.

Peptic ulcers occur when the mucosal lining of the esophagus, stomach, or duodenum becomes eroded. If left untreated, an ulcer can penetrate through the layers of the GI tract.

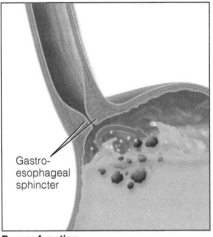

Proper function.

To prevent food from flowing back into the esophagus, the gastroesophageal sphincter normally remains contracted.

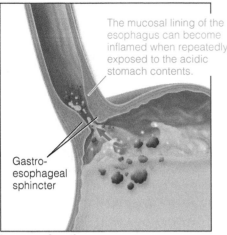

The mucosal lining of the esophagus can become inflamed when repeatedly exposed to the acidic stomach contents.

Improper function.

If the gastroesophageal sphincter weakens, the stomach contents flow back into the esophagus. The reflux of stomach contents into the esophagus is called gastro-esophageal reflux disease.

Gastric Mixing Helps Form Chyme In addition to storing food and producing secretions, the stomach also mixes the food with the gastric juices to form chyme. Unlike the rest of the GI tract, your stomach has a third (diagonal) layer of smooth muscle that generates a forceful churning action, much like the kneading of bread. The strength of these contractions increases under the influence of the hormone gastrin. Within three to five hours after ingestion, the partially digested food is thoroughly mixed with the gastric juices. The resulting chyme, now the consistency of a soupy paste, is pushed toward the pyloric sphincter, and slowly released into the small intestine. With each peristaltic wave, a few milliliters (less than a teaspoon) of chyme squeezes through the pyloric sphincter as it briefly opens; the remaining chyme tumbles back and forth, allowing for even more mixing.

Regulation of Gastric Emptying The rate of **gastric emptying,** or the time it takes for food to leave the stomach and enter the small intestine, is influenced by several factors, including the volume, consistency, and composition of chyme. For example, large volumes of chyme increase the force and frequency of peristaltic contractions, which in turn increase the rate of gastric emptying. Therefore, large meals leave your stomach at a faster rate than small meals. The consistency of food (liquid versus solid) also affects the rate of gastric emptying. Because the opening of the pyloric sphincter is small, only fluids and small particles (<2 mm in diameter) can pass through. Solid foods take more time to be liquefied than fluids and, therefore, remain in the stomach longer. The nutrient composition of your meal also impacts the rate of gastric emptying. In general, foods high in fat slow gastric emptying.[2]

The small intestine itself also influences the rate of gastric emptying. Although the duodenum receives approximately 10 quarts (roughly 40 cups) of ingested food, drinks, saliva, gastric juice, and so forth over the course of one day, only small amounts of chyme can be processed at a time. To prevent being overwhelmed by too much chyme, the small intestine releases a hormone called **cholecystokinin (CCK).** Although CCK has several functions, one of its most important functions is to slow gastric emptying. This delay enables the small intestine to properly digest the chyme that it receives. When the small intestine is ready to receive more chyme, the release of CCK decreases.

gastric emptying The process by which food leaves the stomach and enters the small intestine.

cholecystokinin (CCK) (CHO – le – cys – to – KI – nin) A hormone, produced by the small intestine, that stimulates the release of enzymes from the pancreas and contraction of the gallbladder.

***Helicobacter pylori* (*H. pylori*)** A bacterium residing in the GI tract that causes peptic ulcers.

gastroesophageal reflux disease (GERD) A condition caused by the weakening of the gastroesophageal sphincter, which enables gastric juices to reflux into the esophagus, causing irritation to the mucosal lining.

Millions of Americans describe the symptoms as "fire" in the belly—that burning sensation often mistaken for indigestion. However, peptic ulcers and gastroesophageal reflux disease (GERD) are more than just a little indigestion. These terms both refer to painful conditions that affect the upper GI tract.

Peptic Ulcers

Contrary to popular belief, the majority of peptic ulcers are not caused by stress or by eating spicy foods. Rather, most ulcers (80%) are caused by a small spiral-shaped bacterium called **Helicobacter pylori** (*H. pylori*).[3] This discovery revolutionized standard medical treatment that traditionally consisted of a diet of bland food, milk, stress reduction, and rest. However, the idea that a bacterium could cause ulcers was slow to gain acceptance in the medical community. It seemed highly doubtful that bacteria could survive the acidic environment of the stomach. This changed in 1982 when Drs. Barry Marshall and Robin Warren demonstrated that *H. pylori* could burrow into the thick protective mucus layer, exposing the sensitive underlying stomach layers to acidic gastric juice.

To prove their theory that *H. pylori* could survive the acidic environment of the stomach, Dr. Marshall willingly swallowed the bacterium.[4] Ten days later, he developed acute gastritis (inflammation of the stomach lining). The presence of *H. pylori* was later confirmed by examining a sample of his gastric mucosal lining. Because of Marshall and Warren's scientific conviction, the *H. pylori* bacterium is now widely accepted as the primary cause of ulcers. In 2005, the Nobel Prize in Physiology or Medicine was awarded to Marshall and Warren for their discovery that peptic ulcers and gastritis could be caused by a bacterium.

The most common symptom associated with ulcers is dull, gnawing pain in the stomach that is often relieved by eating. Other symptoms include intermittent pain in the abdominal region, weight loss, loss of appetite, and vomiting. Today, doctors treat most people diagnosed with ulcers with a combination of therapies, including antibiotics to address the underlying bacterial infection and acid-blocking medications to help promote healing. Although most ulcers are caused by *H. pylori*, irritants such as nonsteroidal anti-inflammatory agents (aspirin and ibuprofen) and alcohol can cause them as well.

Gastroesophageal Reflux Disease (GERD)

Gastroesophageal reflux disease (GERD) is a persistent disorder that affects a striking percentage of the U.S. population. According to the American College of Gastroenterology, more than 15 million Americans experience GERD symptoms daily. It occurs when there is reverse flow (reflux) of the stomach contents into the esophagus. Since the esophagus does not have a thick protective mucus layer, repeated exposure to acidic chyme can irritate its lining.

This causes a burning sensation in the upper chest, the most common symptom of GERD. Because of the location of the pain, people with GERD often complain of what is

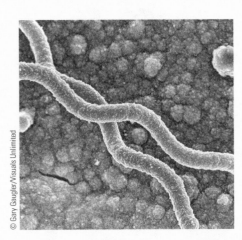

In 1982, Drs. Robin Warren (left) and Barry Marshall (right) proposed that this small spiral-shaped bacterium called *Helicobacter pylori* (*H. pylori*) was the cause of most peptic ulcers. In 2005, these researchers were awarded the Nobel Prize in Physiology or Medicine for their discovery.

(continued)

frequently called heartburn. Other physical complaints associated with GERD include a dry cough, asthma, and difficulty swallowing.

The primary cause of GERD is the inappropriate relaxation of the gastroesophageal sphincter, which normally prevents the stomach contents (chyme) from moving back into the esophagus. Dietary habits that contribute to this include eating large portions of foods and lying down soon after eating. GERD is also associated with consuming certain foods and beverages such as onions, chocolate, mint, high-fat foods, spicy foods, citrus juices, alcohol, and caffeinated beverages. Other factors associated with GERD include being overweight, smoking, and wearing tight-fitting clothes. Pregnancy-related hormonal and anatomical changes can also contribute to GERD.

Making the appropriate lifestyle changes is the first step in preventing and managing GERD. However, when lifestyle changes are not enough, over-the-counter or prescription medications may be necessary. Fortunately, many people are able to manage GERD effectively by making lifestyle changes such as eating smaller portions of foods, avoiding foods that trigger reflux, and remaining upright after eating.

GERD can lead to chronic inflammation of the esophagus that, left untreated, is a risk factor for esophageal cancer. A procedure called **endoscopy** is used to examine the lining of the esophagus for damage and early precancerous changes. For this reason, it is important for all people experiencing GERD to seek medical attention and treatment.

endoscopy A procedure used to examine the lining of the GI tract

duodenum (du – o – DE – num) The first segment of the small intestine.

jejunum (je – JU – num) The midsection of the small intestine, located between the duodenum and the ileum.

ileum (IL – e – um) The last segment of the small intestine that comes after the jejunum.

THE SMALL INTESTINE IS THE PRIMARY SITE OF CHEMICAL DIGESTION AND NUTRIENT ABSORPTION

The small intestine is the primary site of chemical digestion and nutrient absorption. This 20-foot-long narrow tube has a diameter of about 1 inch and is well-suited to carry out these functions. The small intestine consists of three regions—the **duodenum,** the **jejunum,** and the **ileum** (Figure 1.21). Chyme

FIGURE 1.21 Overview of the Small Intestine and Accessory Organs

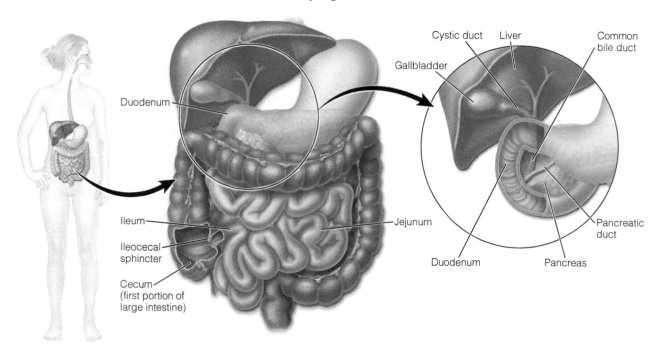

A. The small intestine is divided into 3 regions: the duodenum, jejunum, and ileum. The ileocecal sphincter regulates the flow of material from the ileum, the last segment of the small intestine, into the cecum, the first portion of the large intestine.

B. The duodenum receives secretions from the gallbladder via the common bile duct. The pancreas releases its secretions into the pancreatic duct, which also empties into the duodenum.

first passes into the duodenum, the receiving end of the small intestine. In addition to chyme, the duodenum receives secretions from the gallbladder (bile) and the pancreas (pancreatic juice). When the gallbladder contracts, bile is forced into the cystic duct, which ultimately joins the common bile duct and empties into the duodenum. Similarly, pancreatic juice is released into the pancreatic duct, which also empties into the duodenum.

The lining of the small intestine has a large surface area, which is aptly suited for the process of digestion and nutrient absorption. As illustrated in Figure 1.22, the inner lining of the small intestine (the mucosa and underlying submucosa) is arranged in large, circular folds called **plica circulares.** These circular folds face inward, toward the lumen of the small intestine, and

plica circulares Circular folds in the mucosal lining of the small intestine.

FIGURE 1.22 Absorptive Surface of the Small Intestine

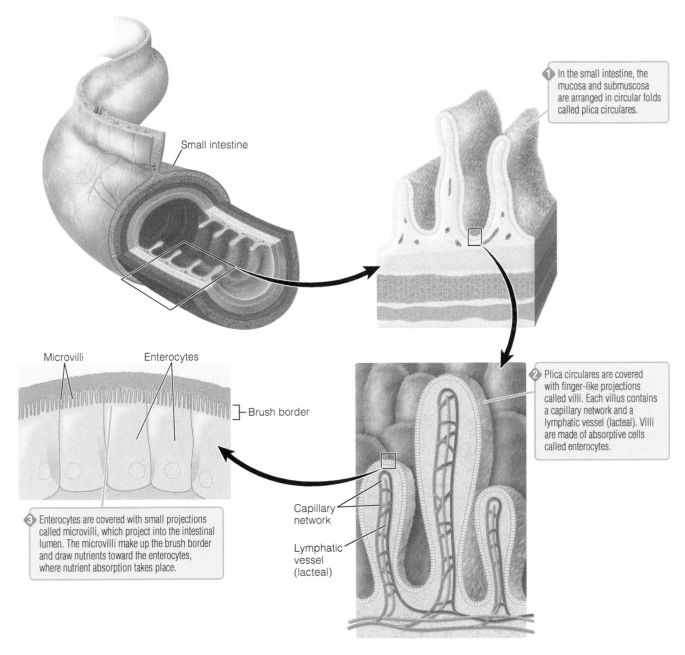

Small intestine

1. In the small intestine, the mucosa and submucosa are arranged in circular folds called plica circulares.

2. Plica circulares are covered with finger-like projections called villi. Each villus contains a capillary network and a lymphatic vessel (lacteal). Villi are made of absorptive cells called enterocytes.

Microvilli   Enterocytes

Brush border

3. Enterocytes are covered with small projections called microvilli, which project into the intestinal lumen. The microvilli make up the brush border and draw nutrients toward the enterocytes, where nutrient absorption takes place.

Capillary network

Lymphatic vessel (lacteal)

are covered with tiny finger-like projections called **villi** (plural form of *villus*). Another way to think of the inner lining of the small intestine is to imagine a looped-style bathroom rug folded like an accordion. The folds in the rug represent the plica circulares, whereas each tiny loop that covers the surface of the rug represents a villus.

Each villus is made up of hundreds of absorptive epithelial cells called **enterocytes.** The surface of the enterocyte that faces the intestinal lumen is covered with thousands of minute projections called **microvilli.** The microvilli comprise the absorptive surface of the small intestine, or what is called the **brush border.** This vast surface area is approximately the size of the playing surface on a standard tennis court. The brush border of the small intestine is where the final stages of nutrient digestion and absorption take place. Each villus contains a network of blood capillaries and a lymph-containing lymphatic vessel called a **lacteal,** both of which circulate absorbed nutrients away from the small intestine.

Digestion in the Small Intestine Is Regulated by Hormones In addition to CCK, the small intestine produces a hormone called **secretin.** With great precision, these hormones help regulate the process of digestion by coordinating the release of secretions from the pancreas and gallbladder, the relaxation of sphincters, and GI motility. The actions of secretin and CCK ensure that nutrient digestion and absorption in the small intestine are rapid and efficient. Indeed, within 30 minutes of the arrival of chyme in the small intestine, the final stages of digestion are complete.

Pancreatic Juice Protects the Small Intestine The pancreas plays an important role in protecting the small intestine from the acidity of chyme (Figure 1.23). Recall that chyme has a pH of approximately 2 and is potentially damaging to the unprotected lining of the small intestine. The arrival of chyme in the small intestine stimulates the release of the hormone secretin, which in turn signals

villi (plural of *villus*) (VI – li, VI – lus) Small, finger-like projections that cover the inner surface of the small intestine.

enterocytes (en – TER – o – cytes) Epithelial cells that make up the lumenal surface of each villus.

microvilli (MI – cro – vi – li) Hairlike projections on the luminal surface of enterocytes.

brush border The absorptive surface of the small intestine made up of thousands of microvilli that cover the luminal surface of enterocytes.

lacteal (LAC – te – al) A lymphatic vessel found in an intestinal villus.

secretin (se – CRE – tin) A hormone, secreted by the small intestine, that stimulates the release of sodium bicarbonate and enzymes from the pancreas.

FIGURE 1.23 The Pancreas The pancreas releases pancreatic juice into the pancreatic duct. The pancreatic juice helps neutralize chyme when it enters the small intestine and contains enzymes needed for nutrient digestion.

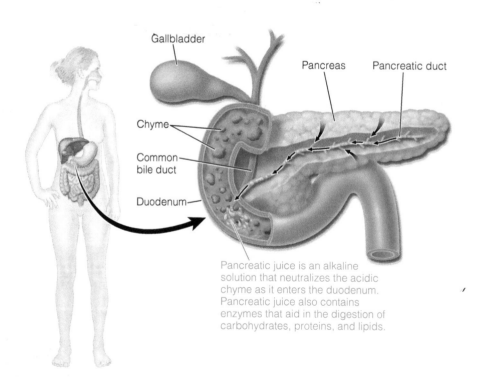

Gallbladder

Pancreas   Pancreatic duct

Chyme

Common bile duct

Duodenum

Pancreatic juice is an alkaline solution that neutralizes the acidic chyme as it enters the duodenum. Pancreatic juice also contains enzymes that aid in the digestion of carbohydrates, proteins, and lipids.

FIGURE 1.24 The Role of the Liver and Gallbladder in Digestion The liver produces bile, which is stored in the gallbladder. The gallbladder releases bile into the common bile duct, which empties into the small intestine. Bile plays an important role in lipid digestion.

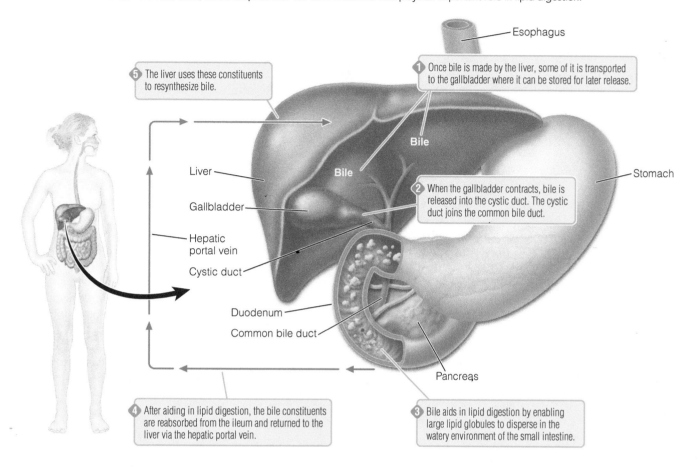

Esophagus

5 The liver uses these constituents to resynthesize bile.

1 Once bile is made by the liver, some of it is transported to the gallbladder where it can be stored for later release.

Bile

Bile

Stomach

Liver

Gallbladder

2 When the gallbladder contracts, bile is released into the cystic duct. The cystic duct joins the common bile duct.

Hepatic portal vein

Cystic duct

Duodenum

Common bile duct

Pancreas

4 After aiding in lipid digestion, the bile constituents are reabsorbed from the ileum and returned to the liver via the hepatic portal vein.

3 Bile aids in lipid digestion by enabling large lipid globules to disperse in the watery environment of the small intestine.

the pancreas to release **pancreatic juice.** This alkaline solution consists of water, sodium bicarbonate, and various enzymes needed for digestion. The sodium bicarbonate in pancreatic juice quickly neutralizes chyme as it enters the duodenum.

Bile Is Needed for Fat Digestion A substance called **bile** also plays an important role in digestion, especially when you consume fatty foods (Figure 1.24). Bile, which is made in the liver, is a watery solution that consists primarily of cholesterol, bile acids, and bilirubin—a pigment that gives bile its characteristic yellowish-green color. Once bile is formed, it is transported to the gallbladder, where some of it is stored and the rest is released into the small intestine.

Fats are not soluble in the watery environment of the small intestine and are therefore more difficult to digest and absorb than other food components. To counteract this, the presence of fat-containing chyme in the small intestine signals the release of CCK, which causes the gallbladder to contract, emptying its contents into the duodenum. The bile acids and cholesterol in bile act like detergents, dispersing large globules of fat into smaller droplets that are easier for the enzymes to digest. Once lipid digestion is complete, bile is reabsorbed through the ileum and returned to the liver via the **hepatic portal vein.** This process, called **enterohepatic circulation,** enables the liver to recycle many of the constituents that make up bile. In fact, only 5% of the bile escapes into the large intestine and is lost in the feces.

pancreatic juice Pancreatic secretions that contain bicarbonate and enzymes needed for digestion.

bile A fluid, made by the liver and stored in and released from the gallbladder, that contains bile salts, cholesterol, water, and bile pigments.

hepatic portal vein A blood vessel that circulates blood to the liver from the GI tract.

enterohepatic circulation (en – ter – O – he – PA – tic) Circulation between the small intestine and the liver used to recycle compounds such as bile.

Digestion in the Small Intestine Is Facilitated by Enzymes Both the small intestine and pancreas provide enzymes needed to chemically break down nutrients even further. Pancreatic enzymes are released into the duodenum in response to the hormone CCK. Intestinal enzymes are made in the brush border epithelial cells, where the final steps of enzymatic digestion take place.

Nutrient Absorption When the process of digestion is complete, nutrients are ready to be absorbed. The transfer of nutrients from the GI tract to the circulatory system, or what is referred to as **nutrient absorption,** takes place by passive and active transport mechanisms: simple diffusion, facilitated diffusion, carrier-mediated active transport, and to a lesser extent endocytosis. The stomach has a minor role in nutrient absorption—only water and alcohol are absorbed there to any significant extent. By far, the majority of nutrients are absorbed along the brush border surface of the small intestine. Once absorbed, nutrients enter either the blood or lymphatic circulatory systems.

The vast surface area and unique structure of the small intestine make nutrient absorption very efficient. The sweeping action of the microvilli traps and pulls nutrients toward the enterocytes. However, the transfer of nutrients from the lumen of the small intestine into the enterocyte is really only the first step in nutrient absorption. To enter the blood or lymph, nutrients must also cross the **basolateral membrane,** the cell membrane of the enterocyte that faces away from the lumen toward the submucosa. Thus, nutrient absorption includes both entry into and exit out of the enterocyte, as illustrated in Figure 1.25. Disease states that affect the absorptive surface of the small intestine, such as celiac disease, can lead to nutritional deficiencies. You can read more about celiac disease in the Focus on Clinical Applications feature.

⟨CONNECTIONS⟩ Recall that an autoimmune disease develops when the immune system produces antibodies that attack and destroy cells in the body.

nutrient absorption The transfer of nutrients from the lumen of the GI tract to the circulatory system.

basolateral membrane The cell membrane that faces away from the lumen of the GI tract and toward the submucosa.

FIGURE 1.25 Nutrient Absorption and Circulation Nutrient absorption includes both entry into and exit out of enterocytes. Once nutrients cross the basolateral membrane, they are circulated away from the intestine by either blood or lymph. Water-soluble nutrients are circulated in blood, whereas fat-soluble nutrients are circulated in lymph.

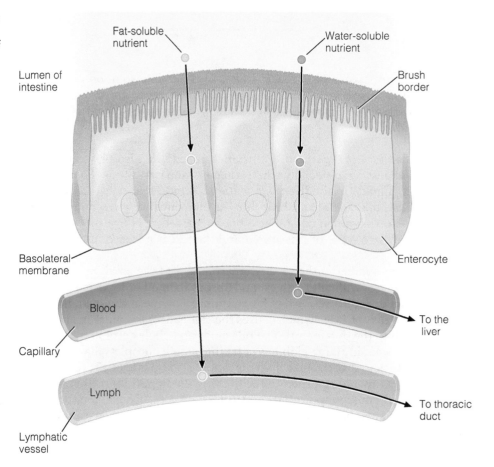

The extent to which a nutrient or other food component is absorbed is called its **bioavailability.** The bioavailability of a particular nutrient can be influenced by physiological conditions, other dietary components, and certain medications. For example, the body absorbs only the amount of iron it needs and excretes the excess in feces. This regulatory step helps protect us from iron toxicity. It is also well known that some nutrients can markedly affect the bioavailability of other nutrients. For example, vitamin C can enhance the absorption of certain forms of iron, and vitamin D can increase the bioavailability of calcium. This is why it is important to be aware of the impact that certain nutrients and drugs may have on nutrient bioavailability.

How Does the Body Circulate Nutrients and Eliminate Cellular Waste Products?

Once nutrient digestion and absorption are complete, the next task is to transport the nutrients throughout the body. This is accomplished by the body's extensive circulatory system, which is made up of veins, arteries, and lymphatic vessels. In addition to delivering nutrients and oxygen, the circulatory system aids in the elimination of cellular waste products.

NUTRIENTS ABSORBED FROM THE SMALL INTESTINE ARE CIRCULATED TO THE LIVER

Upon absorption, water-soluble nutrients enter the bloodstream through the capillaries contained within each villus. (As we will discuss later in this chapter, fat-soluble nutrients instead enter lymphatic vessels.) Once water-soluble nutrients enter the bloodstream, they circulate directly to the liver. This arrangement gives the liver first access to the nutrient-rich blood leaving the small intestine. Nutrients taken up by the liver can then be stored, undergo metabolic changes, or be released into the systemic circulation, which in turn delivers the nutrients to other parts of the body.

THE CARDIOVASCULAR SYSTEM CIRCULATES NUTRIENTS, OXYGEN, AND OTHER SUBSTANCES

In addition to having a continuous supply of essential nutrients and oxygen, cells must also have a way to rid themselves of metabolic waste products. The cardiovascular system, which consists of the heart and an elaborate vascular network, helps meet these needs. The cardiovascular system consists of two separate loops: (1) the systemic circulation and (2) the pulmonary circulation. As illustrated in Figure 1.26 on p. 37, each of these circulatory loops delivers blood to specific regions within the body.

The Systemic Circulation Delivers Blood to Body Organs Substances are transported to and from cells in the blood, which flows through vessels called **arteries** and **veins** in a continuous loop that begins and ends at the heart. This part of the cardiovascular system, called the **systemic circulation,** delivers blood to all the body's organs except the lungs. Oxygenated blood leaves the heart through the **aorta,** which then branches into an intricate maze of arteries. Arteries circulating blood away from the heart divide, subdivide, and eventually form beds of microscopic vessels called **capillaries** situated around organs and tissues. Capillaries have thin walls

bioavailability The extent to which nutrients are absorbed into the blood or lymphatic system.

artery A blood vessel that carries blood away from the heart.

vein A blood vessel that carries blood toward the heart.

systemic circulation The division of the cardiovascular system that begins and ends at the heart and delivers blood to all the organs except the lungs.

aorta (a – OR – ta) The main artery that initially carries blood from the heart to all areas of the body except the lungs.

capillaries (CAP – il – lar – ies) Blood vessels with thin walls, which allow for the exchange of materials between blood and tissues.

It has only been in the last 50 years that researchers have begun to understand celiac disease and how to treat it. In 1888, Dr. Samuel Gee recognized that, when celiac patients avoided foods that contained starch, they "suffered" far less.[5] Many years later, Dr. Gee's observations were confirmed. Indeed, **celiac disease** is an inflammatory response to a specific protein called **gluten** found in a variety of cereal grains such as wheat, rye, barley, and possibly oats. When people with celiac disease consume gluten-containing foods, the lining of the small intestine becomes damaged and eventually causes severe symptoms. Celiac disease is far more common than previously estimated, and its prevalence may be as high as 3 million—1 out of 133 Americans.[6]

Researchers now know that people with celiac disease (also called gluten-sensitive enteropathy) experience an immunological response to gluten. More specifically, the consumption of gluten-containing foods triggers the production of specific antibodies that attack the intestinal microvilli, causing them to flatten.[7] As a result, nutrient absorption is impaired. Like many other types of autoimmune diseases, celiac disease runs in families.

Because of the progressive damage to the absorptive surface of the small intestine, people with celiac disease experience diarrhea, weight loss, and malnutrition. In fact, children with celiac disease often experience slow growth and tend to be small for their age.[8] More recently, migraine headaches, osteoporosis, miscarriages, and infertility have also been attributed to celiac disease.[9] Celiac disease is often misdiagnosed because the signs and symptoms are similar to other common GI disorders. When celiac disease is suspected, a blood test may be performed to screen for the presence of antibodies made in response to gluten. Although antibody testing is important, a definitive diagnosis can only be made by biopsy, a procedure that requires taking a small piece of tissue from the intestinal lining.

Fortunately, once diagnosed, many people with celiac disease manage to live symptom-free by eliminating gluten from their diet. This was certainly the case with Emily, who shared her story in Chapter 2. Recall that it took several months before she was accurately diagnosed with celiac disease. Once she realized that her iron deficiency was due to celiac disease and not insufficient dietary iron, her health quickly returned to normal. However, given the numerous food products made with wheat and other cereal grains (such as breads, crackers, cookies, cakes, and pasta), adherence to a gluten-free diet is easier advised than done. In addition to occurring naturally in many cereal-based products, gluten is often added to many processed foods. People with celiac disease should be vigilant when reading food labels because gluten may be present in less obvious foods such as meats, soups, candies, soy sauce, malt beverages, and even in some medications. In addition, foods made with modified food starch, hydrolyzed vegetable protein, and binders may contain gluten. For this reason, foods claiming to be wheat free are not always gluten free.

For people with celiac disease, it is important to buy foods that are gluten free. A variety of gluten-free products are available in most grocery stores.

---

celiac disease (CE – li – ac) An autoimmune response to the protein gluten that damages the absorptive surface of the small intestine; also called gluten-sensitive enteropathy.

gluten (GLU – ten) A protein found in cereal grains such as wheat, rye, barley, and possibly oats.

interstitial fluid (in – ter – STI – tial) Fluid that surrounds cells.

and narrow diameters, making them well suited for their primary function—the exchange of materials, nutrients, and gases between the blood and tissues. Tiny pores in the capillary walls allow materials such as nutrients and oxygen to pass from the blood into the **interstitial fluid** that surrounds cells (Figure 1.27 on p. 38). At the same time, cellular waste products (such as carbon dioxide) can be taken up from the interstitial fluid and carried away by the blood.

A capillary network marks the end of the arterial blood flow to the cell and the beginning of the venous blood flow away from the cell and back to the

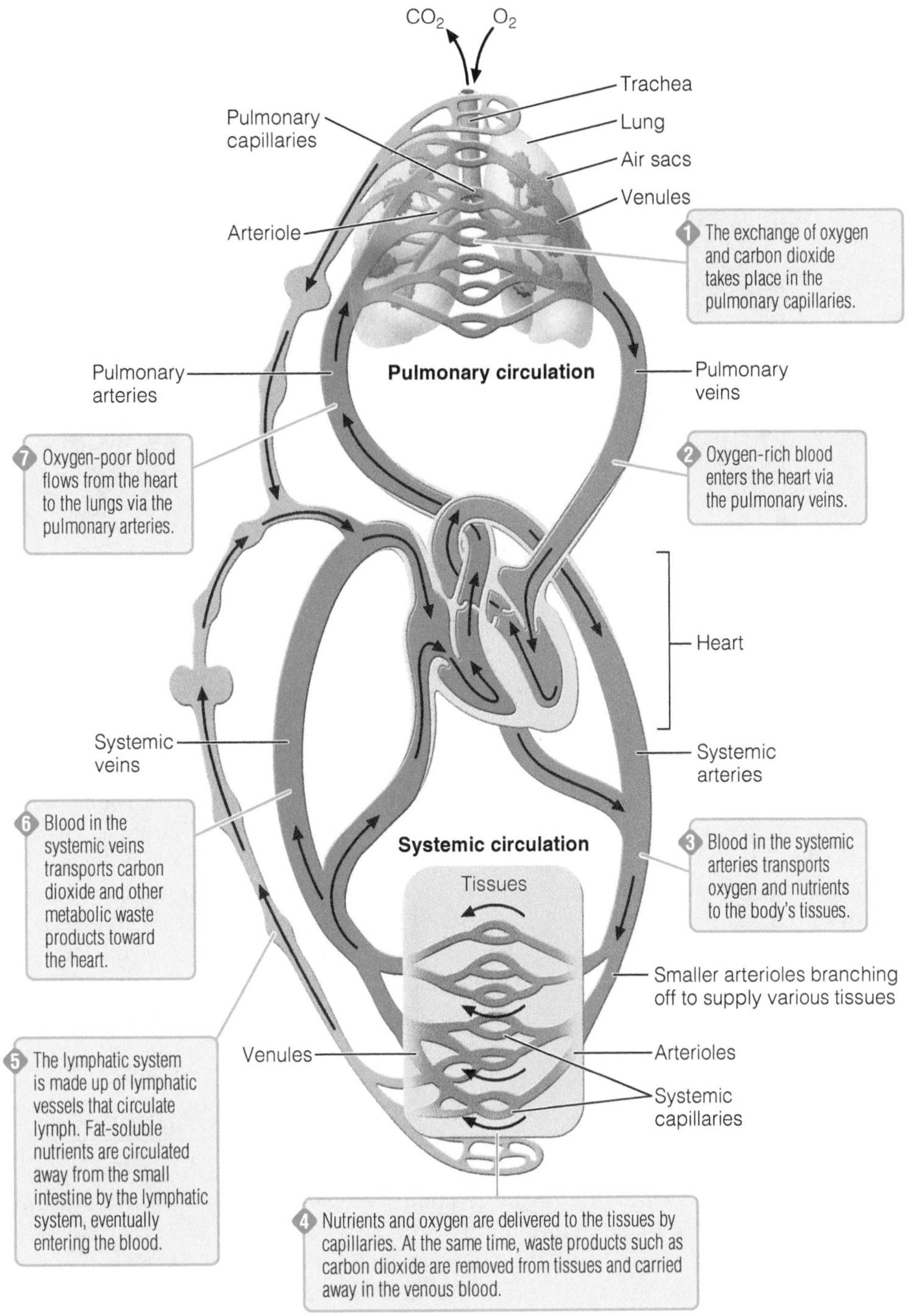

FIGURE 1.26 Systemic and Pulmonary Circulation Blood circulates from the heart to the body and back to the heart via the systemic circulation. Blood circulates from the heart to the lungs and back to the heart via the pulmonary circulation. Lymph circulates through the lymphatic system, which is made up of lymphatic vessels.

heart. Although the arterial and venous vascular systems have many similarities, they differ in several ways.

- Oxygen-rich arterial blood flows toward capillaries, whereas the oxygen-poor venous blood flows away from them.
- Arteries leading to the capillaries become progressively smaller (arteries → **arterioles** → capillaries), whereas veins leading away from capillaries become progressively larger (capillaries → **venules** → veins).
- Arterial circulation flows away from the heart, whereas the venous blood flows toward the heart.
- Arterial blood delivers nutrients and oxygen to cells, whereas the venous blood carries metabolic waste products away from cells.

arteriole Small blood vessel that branches off from arteries.

venule Small blood vessel that branches off from veins.

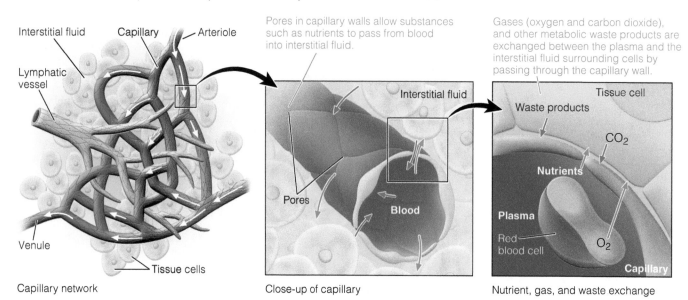

FIGURE 1.27 **Nutrient and Gas Exchange across the Capillary Wall** The exchange of nutrients, gases (oxygen and carbon dioxide), and other cellular waste products takes place between the plasma and the interstitial fluid.

Interstitial fluid

Capillary

Arteriole

Lymphatic vessel

Venule

Tissue cells

Capillary network

Pores in capillary walls allow substances such as nutrients to pass from blood into interstitial fluid.

Interstitial fluid

Pores

Blood

Close-up of capillary

Gases (oxygen and carbon dioxide), and other metabolic waste products are exchanged between the plasma and the interstitial fluid surrounding cells by passing through the capillary wall.

Tissue cell

Waste products

CO_2

Nutrients

Plasma

Red blood cell

O_2

Capillary

Nutrient, gas, and waste exchange

The Pulmonary Circulation Moves Blood between the Heart and Lungs Another component of the cardiovascular system involves the flow of blood between the heart and lungs. This circuit, referred to as the **pulmonary circulation,** begins with the arrival of partially deoxygenated (oxygen-poor, carbon dioxide–rich) venous blood to the heart. The **pulmonary arteries** transport blood from the right side of the heart to the lungs, where the exchange of carbon dioxide and oxygen takes place across the pulmonary (lung) capillaries. By exhaling air through our nose and mouth, carbon dioxide that has crossed out of the capillaries and into the air sacs of the lungs is eliminated from the body. Likewise, during inhalation, oxygen is taken into the lungs, where it crosses into the capillaries and enters the blood. The oxygen-rich blood returns to the heart through the **pulmonary veins** and is pumped out of the heart through the aorta to the rest of the body.

THE LYMPHATIC SYSTEM TRANSPORTS FAT-SOLUBLE NUTRIENTS AWAY FROM THE GI TRACT

Another major component of the circulatory system, the **lymphatic system,** plays an important role in circulating fat-soluble nutrients (mostly lipids and some vitamins) away from the GI tract and eventually delivering them to the cardiovascular system. Each villus contains a lacteal that connects to a network of lymphatic vessels that circulate a clear liquid called lymph. This circulatory route initially bypasses the liver, eventually emptying into the blood. At this point, fat-soluble nutrients can circulate in the bloodstream where they can be taken up and used by cells.

THE KIDNEYS PLAY AN IMPORTANT ROLE IN EXCRETING CELLULAR WASTE PRODUCTS

Nutrients taken up by cells undergo considerable metabolic change. That is, they are transformed, broken down, or used to synthesize other materials. These processes result in the formation of a variety of cellular waste products such as carbon dioxide, water, and urea—a nitrogen-containing compound resulting from the breakdown of protein. Because the accumulation of waste

pulmonary circulation (PUL – mo – nar – y) The division of the cardiovascular system that circulates deoxygenated blood from the heart to the lungs, and oxygenated blood from the lungs back to the heart.

pulmonary artery Blood vessel that transports oxygen-poor blood from the right side of the heart to the lungs.

pulmonary vein Blood vessel that transports oxygen-rich blood from the lungs to the heart.

lymphatic system (lym – PHAT – ic) A component of the circulatory system made up of lymphatic vessels and lymph that flows from organs and tissues, drains excess fluid from spaces that surround cells, and picks up dietary fats from the digestive tract.

FIGURE 1.28 Overview of the Urinary System The urinary system serves three important functions—filtration, reabsorption, and excretion.

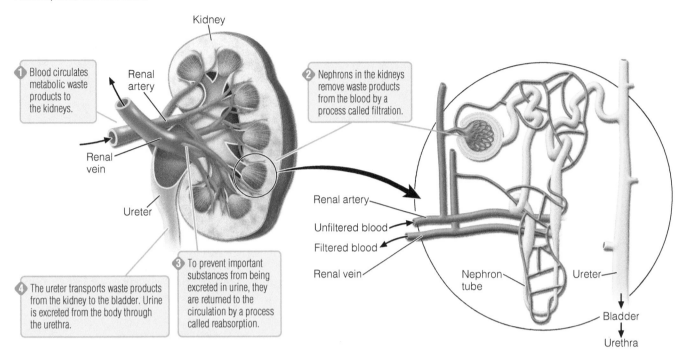

Kidney

1 Blood circulates metabolic waste products to the kidneys.

Renal artery

Renal vein

Ureter

4 The ureter transports waste products from the kidney to the bladder. Urine is excreted from the body through the urethra.

3 To prevent important substances from being excreted in urine, they are returned to the circulation by a process called reabsorption.

2 Nephrons in the kidneys remove waste products from the blood by a process called filtration.

Renal artery

Unfiltered blood

Filtered blood

Renal vein

Nephron tube

Ureter

Bladder

Urethra

products can be toxic, it is important that they be eliminated from the body. Whereas the respiratory system removes carbon dioxide, the kidneys have the responsibility of ridding the body of other cellular wastes, via the urine.

Urinary Excretion of Cellular Waste Products As shown in Figure 1.28, the kidneys play an important role in removing metabolic wastes from the plasma and delivering them for excretion in the urine. Blood flows to the kidneys at a rate of about 1,200 mL (approximately 1.5 quarts) per minute. **Nephrons,** which are the functional units of the kidney, perform the important functions of filtration and reabsorption. **Filtration** initially removes substances such as urea, excess water, electrolytes, salts, and minerals from the blood as it flows through the nephrons, most of which are then excreted from the body in the urine. Because the body can reuse some of the materials that have been filtered out of the blood, the kidneys carry out another important function called **reabsorption,** which means to "absorb again." Substances that are reabsorbed are returned to the blood, enabling the body to reclaim compounds such as amino acids, glucose, and other important nutrients that would otherwise be excreted in the urine.

The inability of our kidneys to perform these important functions can cause toxic waste products to accumulate in the blood. Although some forms of kidney disease are associated with hereditary factors, diabetes and high blood pressure are the two most common causes of impaired kidney function. People with impaired kidney function often need to adhere to strict dietary restrictions so that their kidneys do not become overburdened. However, if kidney function becomes severely impaired, a person may require hemodialysis. **Hemodialysis** is a medical procedure that uses a machine to filter waste products from the blood and to restore proper fluid balance.

The Formation and Excretion of Urine Urine contains a variety of substances including water, salts, and urea. In fact, the composition of urine is so well

nephron (NEPH – ron) Functional unit of the kidney that filters waste materials from the blood that are later excreted in the urine.

filtration The process of selective removal of metabolic waste products from the blood.

reabsorption The return of previously removed materials to the blood.

hemodialysis A medical procedure that uses a machine to filter waste products from the blood and to restore proper fluid balance.

defined that urine analysis can be used to diagnose certain diseases. For example, finding large amounts of glucose in the urine could indicate a condition called diabetes. Urine in the bladder is virtually sterile, meaning that no microorganisms such as bacteria are present. However, when certain microorganisms gain entry into the urinary system, a urinary tract infection may occur. You can read more about the impact of diet, specifically cranberry juice, on urinary tract infections in the Focus on Clinical Applications feature.

What Is the Role of the Large Intestine?

Not all of the food that we eat is completely broken down by the process of digestion. Rather, some food components are able to resist the powerful actions of digestive enzymes. The remaining undigested food residue continues to move through the GI tract, eventually approaching the last leg of its journey—the large intestine. Here it will spend another 10 to 24 hours before leaving the GI tract entirely. The major functions of the large intestine are (1) absorption and reabsorption of fluids and electrolytes, (2) microbial action, and (3) storage and elimination of solid waste (feces).

THE LARGE INTESTINE AIDS IN THE ELIMINATION OF SOLID WASTE PRODUCTS

cecum (CE – cum) The first portion of the large intestine.

The large intestine has four general regions: the cecum, colon, rectum, and anal canal (Figure 1.29). The **cecum**, the first portion of the large intestine, is a short, saclike structure with an attached appendage, consisting of

FIGURE 1.29 Overview of the Large Intestine The large intestine has four general regions: the cecum, colon, rectum, and anal canal.

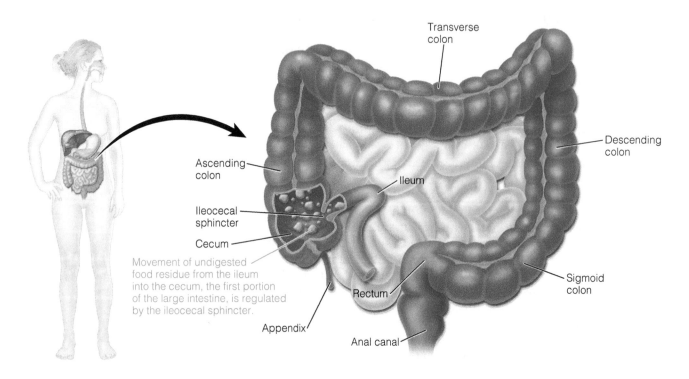

Transverse colon

Descending colon

Ascending colon

Ileum

Ileocecal sphincter

Cecum

Movement of undigested food residue from the ileum into the cecum, the first portion of the large intestine, is regulated by the ileocecal sphincter.

Sigmoid colon

Rectum

Appendix

Anal canal

Hype or reality: is there evidence to support the claim that cranberry juice helps prevent urinary tract infections (UTIs)? For years, the evidence was anecdotal, but several clinical intervention studies now provide some evidence that cranberry juice may protect against the recurrence of UTIs.[10] UTIs, which are more common in women than men, result from bacteria adhering to the lining of the bladder and urinary tract. The bacterium *Escherichia coli (E. coli)* is the most common cause of UTIs. More than 11 million prescriptions are written yearly for antibiotics to treat these infections.

Typically, a person with a UTI experiences urgent, frequent, and painful sensations associated with urination. How does drinking cranberry juice help? In a word—proanthocyanidins. It may be a mouthful to pronounce, but proanthocyanidins are phytochemicals, found in both cranberry and blueberry juices, which may help prevent UTIs. These compounds, which are the pigments responsible for the deep red or purple color of these juices, appear to prevent *E. coli* from adhering to the bladder lining. A clinical intervention study published in the *British Medical Journal* reported a significant reduction in the incidence of UTIs among women who consumed daily a concentrate consisting mainly of cranberry juice.[11] The rate of UTI recurrence was 16% for women given cranberry juice and 36% for women given a placebo. This and several other studies provide considerable support that cranberry juice helps prevent recurrent UTIs.[12] Although more studies are needed to clarify effective doses, as little as 10 ounces of cranberry juice daily may be enough to help ward off UTIs. It is important to emphasize, however, that cranberry juice is not a treatment for an established

UTI. Instead, drinking cranberry juice on a daily basis may offer protection from the recurrence of UTIs and is an example of how a food can offer both medicinal and nutritional benefits. This is one reason why cranberry juice, in addition to being nutritious, is considered by some to be a functional food.

Proanthocyanidins are phytochemicals found in cranberries that may help protect against the recurrence of urinary tract infections.

lymphatic tissue, called the appendix. On occasion, trapped material can cause the appendix to become inflamed, which can necessitate an appendectomy—its surgical removal. The function of the appendix remains unclear, and people do not experience ill effects related to its removal. The **ileocecal sphincter** regulates the intermittent flow of material from the ileum into the cecum.

The **colon,** which makes up most of the large intestine, is shaped like an inverted letter U (∩). The right side of the colon is called the ascending colon, whereas the portion spanning right to left across the abdomen is the transverse colon. From there, the descending colon continues downward, on the left side of the body. Following the colon is the **rectum,** which terminates at the anal canal, the segment of the large intestine that leads outside of the body. A thickening of smooth muscle around the anal canal forms the internal and external **anal sphincters.**

ileocecal sphincter (il – e – o – CE – cal) The sphincter that separates the ileum from the cecum and regulates the flow of material between the small and large intestines.

colon The portion of the large intestine that carries material from the cecum to the rectum.

rectum The lower portion of the large intestine between the sigmoid colon and the anal canal.

anal sphincters Internal and external sphincters that regulate the passage of feces through the anal canal.

FLUIDS AND ELECTROLYTES ARE ABSORBED AND REABSORBED IN THE LARGE INTESTINE

Material entering the large intestine consists mostly of undigested fibrous material from plants, water, bile, cellular debris, and electrolytes. In the large intestine, slow, churning segmentation movements called **haustral contractions** help expose the undigested food residue to the absorptive (mucosal) lining of the colon. Some water and electrolytes are absorbed for the first time from the colon. However, most are actually being reabsorbed because they were released into the colon as GI secretions. As material passes through the various regions of the colon, water and electrolytes are extracted and circulated away from the GI tract in the blood for reuse by the body. This is another example of the body's ability to reclaim its important resources. Between intervals of haustral contractions, peristalsis slowly propels the material forward for further processing.

The consistency of the remaining material, now called **feces,** reflects how much water was reabsorbed. For example, diarrhea, characterized by loose, watery fecal matter, can result when material moves too quickly through the colon, not allowing sufficient time for water removal. Prolonged diarrhea can result in excessive loss of fluids and electrolytes from the body, which can lead to serious complications such as dehydration. Conversely, slow colonic movements can cause too much water to be removed, resulting in hard, dry fecal matter, a condition known as constipation. Constipation can make elimination difficult and put excessive strain on the colonic muscles. Two other conditions that can cause intestinal discomfort are irritable bowel syndrome and inflammatory bowel disease, which are discussed further in the Focus on Clinical Applications feature.

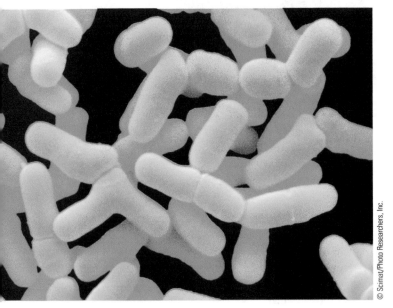

Bifidobacterium cells on colon epithelium.

© Scimat/Photo Researchers, Inc.

haustral contractions (HAU – stral) Slow muscular movements that move the colonic contents back and forth, helping to compact the feces.

feces Waste matter consisting mostly of undigested food residue that is eliminated from the body through the anus.

intestinal microbiota (MI – cro – bi – O – ta) Bacteria that reside in the large intestine (also called intestinal microbiome).

MICROBIAL ACTION IN THE LARGE INTESTINE BREAKS DOWN UNDIGESTED FOOD RESIDUE

Although bacteria reside throughout the GI tract, the large intestine provides the most suitable environment for them. The colon's optimal pH, sluggish haustral contractions, and lack of antimicrobial secretions present ideal conditions for bacteria to grow and flourish. The number and variety of bacteria residing in the large intestine is astronomical. In fact, more than 400 species of bacteria can be found in the large intestine, contributing to nearly one-third of the dry weight of feces. This natural microbial population, also referred to as the **intestinal microbiota** (or **microbiome**), is important for a healthy colonic ecosystem. First, these bacteria break down some of the undigested food residue, which consists mostly of fibrous plant material. Intestinal bacteria also produce vitamin K, limited amounts of certain B vitamins, and some lipids. Nutrients and other substances produced by the intestinal bacteria are absorbed into the blood. Perhaps most important, the natural microbiota help protect us from infection by competing with pathogenic bacteria for limited resources (nutrients and space) in the large intestine. You can read more about how to establish a healthy intestinal microbiota in the next Focus on Clinical Applications feature.

After we eat and enjoy our food, the GI tract dutifully takes over without us having to give it further thought. That is, unless something goes wrong. Health conditions that affect GI function can seriously impair the ability of the GI tract to digest food and absorb nutrients. This is certainly the case with **inflammatory bowel disease (IBD)** such as ulcerative colitis or Crohn's disease. Another GI disorder is irritable bowel syndrome (IBS), a poorly understood condition that affects up to 20% of Americans. While irritable bowel syndrome and inflammatory bowel disease (IBD) may sound similar, they are very different. Nonetheless, both these GI disturbances can have serious implications in terms of nutritional health.

Because **irritable bowel syndrome (IBS)** is not associated with any known structural abnormalities, it is considered to be a functional disorder in that normal bowel activity is disrupted. People with IBS experience bouts of abdominal discomfort such as cramping, bloating, diarrhea, and constipation. Although the underlying cause of IBS has yet to be determined, some clinicians believe that it is a psychological manifestation. While emotional stress itself is not a cause of IBS, it can be a contributing factor. There is also some speculation that IBS may occur when the colon overreacts to stimuli, causing it to spasm. Typically, the diagnosis of IBS is based on ruling out other, better-defined intestinal disorders such as ulcerative colitis or Crohn's disease. It is important for people with IBS to identify and avoid foods that trigger IBS episodes and seek out those that bring comfort and relief. IBS is sometimes treated with antispasmodic medication. Fortunately, IBS does not progress to other, more serious illnesses.

Ulcerative colitis and **Crohn's disease** are both considered forms of IBD, and are characterized by inflammation of the lining of the GI tract.[13] IBD is classified as an autoimmune disease, although this is somewhat speculative.

Many researchers believe that IBD can develop when the intestinal lining is exposed to a foreign protein, or what is called an antigen. Possible immune triggers (antigens) include proteins found in food and exposure to viruses or bacteria. In response to the antigen, a person's immune system produces antibodies, which triggers inflammation. Because the inflammatory response is often prolonged and excessive, the intestinal lining can become damaged. For reasons that are not clear, the occurrence of IBD has dramatically increased over that past 30 years.[14] Currently, it is estimated that more than 1.5 million Americans have IBD, 25–30% of whom are children or adolescents.[15]

Crohn's disease and ulcerative colitis share many of the same clinical signs and symptoms. However, Crohn's disease tends to affect the lower portion of the small intestine (ileum), although it can occur anywhere along the GI tract. In contrast, ulcerative colitis tends to occur along the inner lining of the colon. IBD is typically diagnosed by a procedure called colonoscopy, which involves the insertion of a small scope into the anus. This scope is threaded through the rectum, allowing the physician to inspect the intestinal wall for signs of inflammation or ulcers. A biopsy (tissue sample) can also be taken at this time. Because IBD increases a person's risk of developing colon cancer, it is important for people with this disease to have regular medical exams.

IBD flare-ups can cause diarrhea, fatigue, weight loss, abdominal pain, diminished appetite, and, on occasion, rectal bleeding. Although dietary practices do not cause IBD, nutritional support is important. Nutrient malabsorption and loss of appetite can cause significant weight loss and a variety of nutritional problems. Poor nutritional status can even make IBD worse because adequate nutrient and energy intake are needed to repair the damaged tissue. This is why the right assistance by a qualified team of health care professionals can help prevent further complications associated with IBD.

THE LARGE INTESTINE STORES AND ELIMINATES SOLID WASTE PRODUCTS FROM THE BODY

Egestion refers to the process whereby solid waste (feces) is eliminated from the body. As solid waste moves through the colon, it eventually collects in the rectum, which serves as a holding chamber for feces. The accumulation of

Critical Thinking: Paige's Story Now that you understand the function of the large intestine, think back to Paige's story. Paige was slowly losing function in her large intestine. As a result, she was experiencing weight loss, diarrhea, and severe abdominal pain. However, Paige's condition was not quickly diagnosed. In fact, it was first suggested that she was lactose intolerant or had celiac disease or irritable bowel syndrome. Why could any of these conditions cause the same signs and symptoms as Crohn's disease?

inflammatory bowel disease (IBD) Chronic conditions such as ulcerative colitis and Crohn's disease that cause inflammation of the lower GI tract.

irritable bowel syndrome (IBS) A condition that typically affects the lower GI tract, causing abdominal pain, muscle spasms, diarrhea, and constipation.

ulcerative colitis (co – LI – tis) A type of inflammatory bowel disease (IBD) that causes chronic inflammation of the colon.

Crohn's disease A chronic inflammatory condition that usually affects the ileum and first portion of the large intestine.

The role that intestinal bacteria play in GI function and disease prevention has only recently become appreciated. It has been estimated that several million bacteria reside in the human GI tract, a number far greater than previously believed. Bacteria begin to colonize the GI tract shortly after birth. Age-related changes in the gut, antibiotic therapy, and dietary choices can disrupt this balance. In fact, by 65 years of age, the number of "friendly" microbes residing in a human's colon declines a thousand fold.[16] For these and other reasons, there has been considerable interest in fully understanding factors that help establish and maintain a healthy microbiota.

There is now substantial evidence that certain types of food contribute to a healthy intestinal microbiota.[17] These foods are referred to as probiotic and prebiotic foods. **Probiotic foods** contain live bacterial cultures that colonize the colon and have health-promoting benefits. For example, microorganisms associated with probiotic foods help inhibit the colonization of pathogenic bacteria by adhering to the intestinal lining. Other potential health benefits include immune-enhancing effects and protection against cancer, allergies, and autoimmune diseases. Dietary supplements are one source of probiotic bacteria, as are some "cultured" dairy products such as yogurt, buttermilk, sour cream, and cottage cheese. Some yogurts produced in the United States are made by adding the probiotic bacteria *Streptococcus thermophilus* to milk. Foods with live bacteria in them are often labeled as such.

Scientists have also found that the addition of probiotic bacteria to soy milk improved the bioavailabilty of bioactive compounds (isoflavones) and calcium, effects that are beneficial to health.[18] Thus, probiotic foods may have the added advantage of enhancing the body's ability to absorb certain components in food. Studies are also investigating a potential therapeutic role of various probiotics in managing disease states such as IBS and IBD.[19] Researchers have also discovered that the millions of bacteria that populate our intestinal tract may play a role in body weight regulation.[20] Although information linking body weight and human microbiota is limited, it has been hypothesized that intestinal bacteria may use energy from food to maintain a stable microbial "community." Future treatments for weight management may someday include populating the intestine with certain bacteria through ingestion of probiotic foods. Although probiotic foods have been available for years, the American public is just beginning to recognize their potential as a natural approach to staying healthy. In fact, probiotics are considered part of the functional foods trend.

Another way to increase colonization of beneficial bacteria is to consume **prebiotic foods,** which are typically fiber-rich and selectively promote the growth of nonpathogenic bacteria. Because dietary fiber can resist digestion, it passes into the colon and provides a source of nourishment for the microbiota. Dietary fiber is found in whole grains, cereals, fruits, vegetables, and legumes and is discussed in detail in Chapter 8. Together, consumption of probiotic and prebiotic foods helps maintain a well-colonized microbial population in the large intestine, providing an important defense against pathogenic bacteria. In addition, the microbiota produce many substances that are likely beneficial to our health.

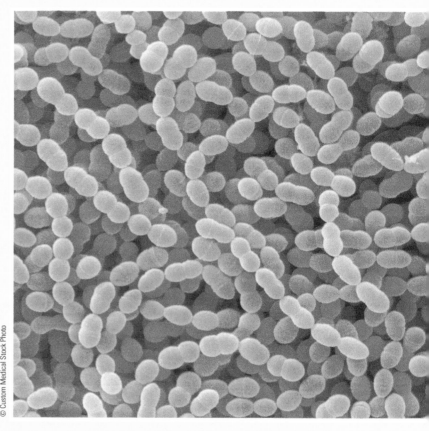

Streptococcus thermophilus is a bacteria used to produce yogurt. It is beneficial to the gastrointestinal tract, and therefore considered a probiotic bacteria.

Critical Thinking: Paige's Story As a result of Crohn's disease, Paige had a portion of her colon surgically removed. How might this impact normal physiological functions in the colon? Why do you think her physician recommended she take nutrient supplements and probiotics?

feces causes the walls of the rectum to stretch, signaling the need to **defecate.** However, the external anal sphincter, which is under conscious control, enables us to determine whether the time is right for waste elimination. By keeping the external anal sphincter contracted, a person can delay defecation. When the internal anal sphincter relaxes, the feces move into the anal canal and then are expelled from the body, via the process of defecation.

probiotic (PRO – bi – o – tic) Food or dietary supplement that contains beneficial live bacteria.

prebiotic food (PRE – bi – o – tic) Food that stimulates the growth of beneficial bacteria that naturally reside in the large intestine.

defecation The expulsion of feces from the body through the rectum and anal canal.

Notes

1. Spechler SJ. Clinical manifestations and esophageal complications of GERD. American Journal of Medical Sciences. 2003;326:279–84.

2. Meyer JH. Gastric emptying of ordinary food: effect of antrum on particle size. American Journal of Physiology. 1980;239:G133–5. Hunt JN. Mechanisms and disorders of gastric emptying. Annual Review of Medicine. 1983;34:219–29.

3. DeCross AJ, Marshall BJ. The role of *Helicobacter pylori* in acid-peptic disease. American Journal of Medical Sciences. 1993;306:381–92.

4. Blaser MJ. The bacteria behind ulcers. Scientific American. 1996;274:104–7. Meuler DA. Helicobacter pylori and the bacterial theory of ulcers. 2010 National Center for Case Study Teaching in Science. University of Buffalo. Available from: http://sciencecases.lib.buffalo.edu/cs/files/peptic_ulcer.pdf.

5. Impact. A publication of the University of Chicago Celiac Disease Center. A brief history of celiac disease. 2007;7:1–2. Available from: http://www.celiacdisease.net/assets/pdf/SU07CeliacCtr.News.pdf.

6. Fasano A, Berti I, Gerarduzzi T, Not T, Colletti RB, Drago S, Elitsur Y, Green P, Guandalini S, Hill ID, Pietzak M, Ventura A, Thorpe M, Kryszak D, Fornaroli F, Wasserman SS, Murray JA, Horvath K. Prevalence of Celiac Disease in At-Risk and Not-At-Risk Groups in the United States. A Large Multi-center Study. Archives of Internal Medicine. 2003;163:286–92.

7. Nelsen DA. Gluten-sensitive enteropathy (celiac disease): More common than you think. American Family Physician. 2002;66:2259–66.

8. Mearin ML. Celiac disease among children and adolescents. Current Problems in Pediatrics and Adolescent Health Care. 2007;37:86–105.

9. Hernandez L, Green PH. Extraintestinal manifestations of celiac disease. Current Gastroenterology Reports. 2006;8:383–9.

10. McMurdo ME, Bissett LY, Price RJ, Phillips G, Crombie IK. Does ingestion of cranberry juice reduce symptomatic urinary tract infections in older people in hospital? A double-blind, placebo-controlled trial. Age and Ageing. 2005;34:256–61. Howell AB. Cranberry proanthocyanidins and the maintenance of urinary tract health. Critical Reviews in Food Science and Nutrition. 2002;42:273–8.

11. Kontiokari T, Sundqvist K, Nuutinen M, Pokka T, Koskela M, Uhari M. Randomised trial of cranberry-lingonberry juice and Lactobacillus GG drink for the prevention of urinary tract infections in women. British Medical Journal. 2001;30;322:1571.

12. L Strothers. A randomized trial to evaluate effectiveness and cost effectiveness of naturopathic cranberry products as prophylaxis against urinary tract infection in women. Canadian Journal of Urology 2002;9:1558–62.

13. Xavier RJ, Podolsky DK. Unravelling the pathogenesis of inflammatory bowel disease. Nature. 2007;448:427–34.

14. Lakatos PL. World recent trends in the epidemiology of inflammatory bowel diseases: Up or down? Journal of Gastroenterology. 2006;12:6102–8.

15. Abraham C, Cho JH. Inflammatory bowel disease. New England Journal of Medicine. 2009;361:2066–78.

16. Kolida S, Saulnier DM, Gibson GR. Gastrointestinal microflora: Probiotics. Advances in Applied Microbiology. 2006;59:187–219.

17. Rastall RA. Bacteria in the gut: Friends and foes and how to alter the balance. Journal of Nutrition. 2004;134:2022S–6S.

18. Pham TT, Shah NP. Biotransformation of isoflavone glycosides by bifidobacterium animalis in soymilk supplemented with skim milk powder. Journal of Food Science. 2007;72:316–24.

19. Guslandi MJ. Probiotic agents in the treatment of irritable bowel syndrome. Journal of International Medical Research. 2007;35:583–9. Hedin C, Whelan K, Lindsay JO. Evidence for the use of probiotics and prebiotics in inflammatory bowel disease: A review of clinical trials. Proceedings of the Nutrition Society. 2007;66:307–15.

20. DiBaise JK, Zhang H, Crowell MD, Krajmalnik-Brown R, Decker GA, Rittmann BE. Gut microbiota and its possible relationship with obesity. Mayo Clinic Proceedings. 2008;83:460–9. Kalliomäki M, Collado MC, Salminen S, Isolauri E. Early differences in fecal microbiota composition in children may predict overweight. American Journal of Clinical Nutrition. 2008;87:534–8. Cani PD, Delzenne NM. Gut microflora as a target for energy and metabolic homeostasis. Current Opinion in Clinical Nutrition and Metabolic Care. 2007;10:729–34.

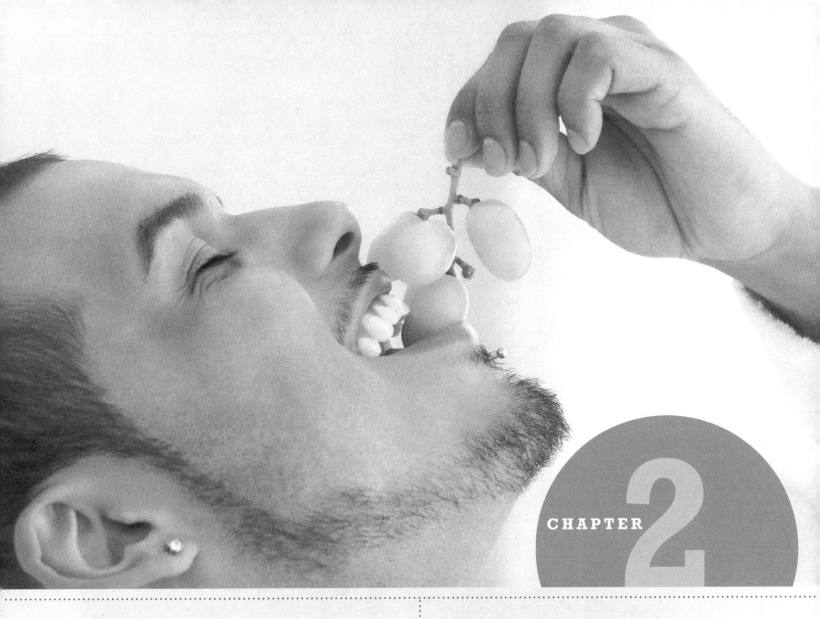

Key Nutrition Concepts and Terms

NUTRITION SCOREBOARD

1. Calories are a component of food. **True/False**

2. Nutrients are substances in food that are used by the body for growth and health. **True/False**

3. Inadequate intakes of vitamins and minerals can harm health, but high intakes do not. **True/False**

4. "Dietary Reference Intakes" (DRIs) provide science-based standards for nutrient intake. **True/False**

Answers can be found at the end of the chapter.

After completing Chapter 2
you will be able to:

- Explain the scope of nutrition
  as an area of study.

- Demonstrate a working
  knowledge of the meaning of
  the 10 nutrition concepts.

The Meaning of Nutrition

• Explain the scope of nutrition as an area of study.

What is nutrition? It can be explained by situations captured in photographs as well as by words. This introduction presents a photographic tour of real-life situations that depict aspects of the study of nutrition.

Before the tour begins, take a moment to make yourself comfortable and clear your mind of clutter. Take a careful look at the photographs shown below and on the next two pages, pausing to mentally describe in two or three sentences what each photograph shows.

Not everyone who thinks about the photographs will describe them in the same way. Reactions will vary somewhat due to personal experiences, interests, beliefs, and cultural background. An individual trying to gain weight will probably react differently to the photograph of the person on the scale than someone who is trying to lose weight. The photo of a dad measuring his son's growth progress may bring back memories of the "measuring wall" you knew as a child, and how you were encouraged to eat your vegetables to grow up strong and tall. Depending on your background, you may recognize the photo of ham hocks, greens, and beans as your favorite holiday meal. If the foods are unfamiliar, you may wonder what they are or why is there so much meat on the plate. The final photo, showing a crowd of children and adults at a soup kitchen, may have reminded you that food is essential for life.

Although knowledge about nutrition is generated by impersonal and objective methods, it can be a very personal subject.

Nutrition Defined

nutrition The study of foods, their nutrients and other chemical constituents, and the effects that foods and food constituents have on health.

In a nutshell, **nutrition** is the study of foods and health. It is a science that focuses on foods, their nutrient and other chemical constituents, and the effects of food and food constituents on body processes and health. The scope of nutrition extends from food choices to the effects of diet and specific food components on biological processes and health.

Gary Conner/PhotoEdit

Royalty free/Corbis

Jupiterimages/Photolibrary/Getty Images

Phil Schermeister/Encyclopedia/Corbis

Photodisc

Uschi Gerschner/Newscom

Photodisc

Budimir Jevtic/Shutterstock.com

Pat Shearman/Alamy

Christian Science Monitor/Getty Images

Nutrition Is a "Melting Pot" Science The broad scope of nutrition makes it an interdisciplinary science. Knowledge provided by the behavioral and social sciences, for example, is needed in studies that examine how food preferences develop and how they may be changed. Information generated by the biological, chemical, physical, and food sciences is required to propose and explain diet and disease relationships. The knowledge and skills of mathematicians and statisticians are needed to develop and implement appropriate research designs and analysis strategies that produce objective, reliable research results. The study of nutrition will bring you into contact with information from a variety of disciplines.

Nutrition Knowledge Is Applicable As you study the science of nutrition, you will discover answers to a number of questions about your own diet, health, and eating behaviors. Is obesity primarily due to eating habits, physical inactivity, the food environment, or your genes? How do you know whether new information you hear about nutrition is true? Can sugar harm more than your teeth? Can the right diet or supplement give you a competitive edge? What is a healthful diet and how do you know if you have one? If improvements seem warranted, what's the best way to go about changing your diet for the better? These are just a few of the questions that will be addressed during the course of your study of nutrition. You will take from this learning experience not only knowledge about nutrition and health, but also skills that will keep the information and insights working to your advantage for a long time to come.

Foundation Knowledge for Thinking about Nutrition

- **Demonstrate a working knowledge of the meaning of the 10 nutrition concepts.**

You don't have to be a bona fide nutritionist to think like one. What you need is a grasp of the language and basic concepts of the science. The purpose of this unit is to give you that background. The essential topics covered here are explored in greater depth in units to come, and they build on this foundation of knowledge. With a working knowledge of nutrition terms and concepts, you will have an uncommonly good sense of nutrition.

food security Access at all times to a sufficient supply of safe, nutritious foods.

food insecurity Limited or uncertain availability of safe, nutritious foods—or the ability to acquire them in socially acceptable ways.

NUTRITION CONCEPT #1

Food is a basic need of humans.

Humans need enough food to live, and they need the right amount and assortment of foods for optimal health. In the best of all worlds, the need for food is combined with the condition of **food security**. People who experience food security have access at all times to a sufficient supply of the safe, nutritious foods that are needed for an active and healthy life. They are able to acquire acceptable foods in socially acceptable ways; they do not have to scavenge or steal food in order to survive or to feed their families. It is important to note that food security emphasizes access to an adequate supply of *nutritious* food rather than simply sufficient food. **Food insecurity** exists whenever the availability of safe, nutritious foods—or the ability to acquire them in socially acceptable ways—is limited or uncertain (Illustration 2.1).[1]

Adults who live in food-insecure households are more likely to have poor-quality diets, to be obese, and to have hypertension, heart disease, or diabetes than adults who are poor but food secure.[2] The ready availability of inexpensive, high-calorie foods; poverty; the absence of local supermarkets; limited opportunities

Illustration 2.1 "It is possible to go an entire lifetime without knowing about people's experiences with hunger."
—*Meghan LeCates, Capital Area Food Bank*

for exercise; and lack of cooking facilities may be partly responsible for the higher rates of obesity and chronic disease among food-insecure adults.[3] Although many children living in food-insecure households are nourished adequately, successful in school, and develop high levels of social skills, as a group they are at higher risk of poor school performance and social and behavioral problems.[4]

Food insecurity exists in 14.3% of U.S. and 7.7% of Canadian households.[5,6] The rate in the United States is over twice as high as the target of 6% established as a national health goal.[7]

Who Are the Food Insecure?

Although unemployment, disabilities, and homelessness contribute to food insecurity, most food-insecure individuals live in households with one or more full-time worker.[8]

One in four of such households include an adult serving in the military.[9] College students are being increasingly recognized as a risk group for food insecurity. Over 120 U.S. colleges have established food pantries for students, and more are being developed as you read this.[10] Low wages and high food costs can limit the ability of households and college students to afford sufficient, nutritious food.

Food Security and Stainable Diets Climate change is affecting worldwide food security. Floods and drought precipitated by global warming can displace fertile farmlands and water supplies, and reduce crop yields and fish and seafood availability.[11] Food production, in turn, affects climate change by contributing greenhouse gases that trap heat in the atmosphere.[12] Fossil fuels, which are widely used in food production, processing, and transport, are the largest source of greenhouse gas emissions in the United States.[13]

The extent of production of greenhouse gases is generally assessed by measuring the amount of carbon dioxide released into the atmosphere as a result of food production and other processes. The result is referred to as the "**carbon footprint**."

Food security for future generations will be enhanced by environmentally friendly, sustainable food production methods that leave low carbon footprints. In general, plant-based and lower-calorie diets leave a lower carbon footprint than diets rich in animal products such as dairy and meat. Plant foods are being increasingly recognized as the most important component of a healthy and sustainable diet.[14] Sustainable diets are considered during development of national dietary intake recommendations.[15]

carbon footprint Generally defined as the amount of the greenhouse gas carbon dioxide emitted by farming, food production, a person's activities, or a product's manufacture and transport.

calorie A unit of measure of the amount of energy supplied by food. (Also known as a kilocalorie, abbreviated kcal, or the *large Calorie* with a capital C.)

nutrients Chemical substances in food that are used by the body for growth and health. The six categories of nutrients are carbohydrates, proteins, fats, vitamins, minerals, and water.

NUTRITION CONCEPT #2

Foods provide energy (calories), nutrients, and other substances needed for growth and health.

People eat foods for many different reasons. The most compelling reason is that we need the calories, nutrients, and other substances supplied by foods for growth and health.

Calories

A **calorie** is a unit of measure of the amount of energy in a food—and of how much energy will be transferred to the person who eats it. Although we often refer to the number of calories in this food or that one, calories are not a substance present in food. And, because calories are a unit of measure, they do not qualify as a nutrient.

Nutrients

Nutrients are chemical substances present in food that are used by the body to sustain growth and health (Illustration 2.2). Essentially everything that's in our body was once a component of the food we consumed.

Illustration 2.2 Foods provide nutrients. "Please pass the complex carbohydrates, thiamin, and niacin . . . I mean, the bread!"

Brand X Pictures/Getty Images

There are six categories of nutrients (Table 2.1). Each category (except water) consists of a number of different substances that are used by the body for growth and health. The carbohydrate category includes simple sugars and complex carbohydrates (starches and dietary fiber). The protein category includes 20 amino acids, the chemical units that serve as the "building blocks" for proteins. Several different types of fat are included in the fat category. Of primary concern are the saturated fats, unsaturated fats, essential fatty acids, *trans* fats, and cholesterol. The vitamin category consists of 14 vitamins, and the mineral category includes 15 minerals. Water makes up a nutrient category by itself.

Carbohydrates, proteins, and fats supply calories and are called the *energy nutrients.* Although each of these three types of nutrients performs a variety of functions, they share the property of being the body's only sources of fuel. Vitamins, minerals, and water are chemicals that the body needs for converting carbohydrates, proteins, and fats into energy and for building and maintaining muscles, blood components, bones, and other tissues in the body.

Other Substances in Food

Food also contains many other substances, some of which are biologically active in the body. One major type of such substances is the **phytochemicals**. There are thousands of them in plants. Illustration 2.3 presents examples of plant foods that are particularly rich sources of phytochemicals. Phytochemicals provide plants with color, give them flavor, foster their growth, and protect them from insects and diseases. In humans, consumption of certain phytochemicals in diets is strongly related to a reduced risk of developing certain types of cancer, heart disease, infections, and other disorders.[16]

Specific phytochemicals have names that may be hard to pronounce and difficult to remember. Nevertheless, here are a few examples. Plant pigments, such as lycopene (like-o-peen), which help make tomatoes red, anthocyanins (an-tho-sigh-an-ins), which give blueberries their characteristic blue color, and beta-carotene (bay-tah-kar-o-teen), which imparts an orange color to carrots, are phytochemicals that act as **antioxidants**.

They protect plant cells—and in some cases, human cells, too—from damage that can make them susceptible to disease. Various types of sulfur-containing phytochemicals are present in cabbage, broccoli, cauliflower, brussels sprouts, and other vegetables of the same family. These substances help prevent a number of different types of cancer in people with specific gene types.[17]

Some Nutrients Must Be Provided by the Diet Many nutrients are required for growth and health. The body can manufacture some of these from raw materials supplied by food, but others must come assembled. Nutrients that the body cannot generally produce, or cannot produce in sufficient quantity, are referred to as **essential nutrients**. Here *essential* means "required in the diet." Vitamin A, iron, and calcium are examples of essential nutrients. Table 2.2 lists all the known essential nutrients.

Nutrients used for growth and health that can be manufactured by the body from components of food in our diet are considered nonessential. Cholesterol, creatine, and

Table 2.1 The six categories of nutrients

| 1. Carbohydrate |
| 2. Protein |
| 3. Fat |
| 4. Vitamins |
| 5. Minerals |
| 6. Water |

phytochemicals   Chemical substances in plants (phyto = plant). Some phytochemicals perform important functions in the human body. They give plants color and flavor, participate in processes that enable plants to grow, and protect plants against insects and diseases. Also called phytonutrients.

antioxidants   Chemical substances that prevent or repair damage to cells caused by exposure to oxidizing agents such as environmental pollutants, smoke, ozone, and oxygen. Oxidation reactions are a normal part of cellular processes.

essential nutrients   Nutrients required for normal growth and health that the body can generally not produce, or produce in sufficient amounts. Essential nutrients must be obtained in the diet.

Photodisc

Illustration 2.3   Examples of good food sources of beneficial phytochemicals.

Table 2.2 Essential nutrients for humans: A reference table

Energy Nutrients	Vitamins	Minerals	Water
Carbohydrates	Biotin	Calcium	Water
Fats[a]	Folate	Chloride	
Proteins[b]	Niacin (B_3)	Chromium	
	Pantothenic acid	Copper	
	Riboflavin (B_2)	Fluoride	
	Thiamin (B_1)	Iodine	
	Vitamin A	Iron	
	Vitamin B_6 (pyroxidine)	Magnesium	
	Vitamin B_{12}	Manganese	
	Vitamin C (ascorbic acid)	Molybdenum	
	Vitamin D	Phosphorus	
	Vitamin E	Potassium	
	Vitamin K	Selenium	
	Choline[c]	Sodium	
		Zinc	

[a]Fats supply the essential nutrients linoleic and alpha-linolenic acids.
[b]Proteins are the source of nine "essential amino acids": histidine, isoleucine, leucine, lysine, methionine, phenylalanine, threonine, tryptophan, and valine. The other 11 amino acids are not a required part of our diet; they are considered "nonessential."
[c]A dietary source of choline may not be required during all stages of the life cycle.

nonessential nutrients Nutrients required for normal growth and health that the body can manufacture in sufficient quantities from other components of the diet. We do not require a dietary source of nonessential nutrients.

glucose are examples of **nonessential nutrients**. Nonessential nutrients are present in food and used by the body, but they are not required parts of our diet because we can produce them ourselves.

Both essential and nonessential nutrients are required for growth and health. The difference between them is whether or not we need to obtain the nutrient from a dietary source. A dietary deficiency of an essential nutrient will cause a specific deficiency disease, but a dietary lack of a nonessential nutrient will not. People develop scurvy (the vitamin C–deficiency disease), for example, if they do not consume enough vitamin C. But you could have zero cholesterol in your diet and not become "cholesterol deficient," because your liver produces cholesterol.

Our Requirements for Essential Nutrients The amount of essential nutrients humans need each day varies a great deal, from amounts measured in cups to micro-grams. (See Table 2.3 to get a notion of the amount represented by a gram, milligram, and other measure.) Generally speaking, adults need 11 to 15 cups of water from fluids and foods, 9 tablespoons of protein, one-fourth teaspoon of calcium, and only one-thousandth teaspoon (a 30-microgram speck) of vitamin B_{12} each day.

We all need the same nutrients, but not always in the same amounts. The amounts needed vary among people based on:

• Age

• Sex

• Growth status

• Body size

• Genetic traits

and the presence of conditions such as:

• Pregnancy

• Breastfeeding

Table 2.3 Units of measure commonly employed in nutrition

Measure	Abbreviation	Equivalents
Kilogram	kg	1 kg = 2.2 lb = 1,000 grams
Pound	lb	1 lb = 16 oz = 454 grams = 2 cups (liquid)
Ounce	oz	1 oz = 28 grams = 2 tablespoons (liquid)
Gram	g	1 g = 1/28 oz = 1,000 milligrams
Milligram	mg	1 mg = 1/28,000 oz = 1,000 micrograms
Microgram	mcg, µg	1 mcg = 1/28,000,000 oz

1 egg = 50 grams or 1³/₄ oz; 212 milligrams (0.2 grams) of cholesterol in yolk

1 slice of bread = 1 oz = 28 grams

1 nickel = 5 grams

1 teaspoon of sugar = 4 grams, 1 grain of sugar = 200 micrograms

- Illnesses

- Drug/medication use

- Exposure to environmental contaminants

Each of these factors, and others, can influence nutrient requirements. General diet recommendations usually make allowances for major factors that influence the level of nutrient need, but they cannot allow for all of the factors.

Nutrient Intake Standards Recommendations for daily levels of nutrient intake were first developed in the United States in 1943 and have been updated periodically since then. Called the *Recommended Dietary Allowances* (RDAs), these standards were established in response to the high rejection rate of World War II recruits, many of whom were underweight and had nutrient deficiencies. The recommended levels of nutrient intake provided are based on age, gender, and condition (pregnant or breastfeeding). Because the science underlying nutrient intake and health advances with time, these standards are periodically revised and expanded (Illustration 2.4).

Dietary Reference Intakes Nutrient intake standards now in place are referred to as *Dietary Reference Intakes* (DRIs); they include categories of nutrient intake in addition to the RDAs. The current RDAs referenced in the DRIs reflect nutrient intake levels that protect almost all healthy individuals from developing deficiency disease and that also reduce the risk of common chronic diseases. Table 2.4 provides examples of endpoints aimed at chronic disease prevention used by the DRI Committee to determine the RDAs. A category labeled *Adequate Intakes* (AIs) has been added to indicate "tentative RDAs" for a few nutrients such as vitamin K and fluoride, for which too few reliable scientific studies have been done to establish an RDA.

The DRI standards include a category called *Estimated Average Requirement* (EAR). This category represents nutrient intake levels that are estimated to meet the nutrient intake requirements of 50% of individuals within an age, sex, and condition (pregnant or breastfeeding) group.

DRI standards consider the effects of excessively high intake of nutrients, primarily from supplements and fortified foods, on health. These standards are labeled *Tolerable Upper Intake Levels*, abbreviated *ULs* for "upper levels." Table 2.5 graphically displays the

Illustration 2.4 The latest DRI report on calcium and vitamin D, 2010.

Table 2.4 Examples of primary endpoints used to estimate DRIs

Carbohydrate: Amount needed to supply optimal levels of energy to the brain.

Total Fiber: Amount shown to provide the greatest protection against heart disease.

Folate: Amount that maintains normal red blood cell folate concentration.

Iodine: Amount that corresponds to optimal functioning of the thyroid gland.

Selenium: Amount that maximizes its function in protecting cells from damage.

Table 2.5 Terms and abbreviations used in the DRIs and a graphic representation of their meaning

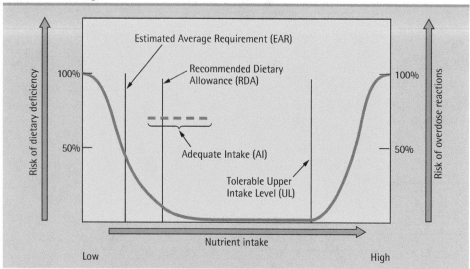

- **Dietary Reference Intakes (DRIs).** This is the general term used for nutrient intake standards for healthy people.

- **Recommended Dietary Allowances (RDAs).** These are levels of essential nutrient intake judged to be adequate to meet the known nutrient needs of practically all healthy persons while decreasing the risk of certain chronic diseases.

- **Adequate Intakes (AIs).** These are "tentative" RDAs. AIs are based on less conclusive scientific information than are the RDAs.

- **Estimated Average Requirements (EARs).** These are nutrient intake values that are estimated to meet the requirements of half the healthy individuals in a group. The EARs are used to assess adequacy of intakes of population groups.

- **Tolerable Upper Intake Levels (ULs).** These are upper limits of nutrient intake compatible with health. The ULs do not reflect desired levels of intake. Rather, they represent total daily levels of nutrient intake from food, fortified foods, and supplements that should not normally be exceeded.

relationships between nutrient intake level and the various categories of the DRI standards now in use, and presents definitions of the DRI nutrient intake categories.

Developed by nutrition scientists from the United States and Canada, the RDAs apply to 97 to 98% of all healthy people in both countries. The fundamental premise of the first RDAs—that nutrient intake should come primarily from foods—is maintained in the current nutrient intake standards.

DRI tables are shown on the inside front and back covers of this text. Check out these tables. They can be used to identify recommended daily levels of essential nutrient intake and levels of intake that should not normally be exceeded.

NUTRITION CONCEPT #3

Health problems related to nutrition originate within cells.

Cells are the main employers of nutrients (Illustration 2.5). All body processes required for growth and health take place within cells and the fluid that

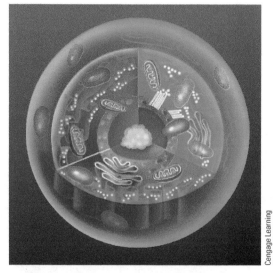

Illustration 2.5 Schematic representation of the structure of a human cell.

Cengage Learning

surrounds them. Humans contain over 100 trillion cells in body tissues (and around 100 times more than that number if bacteria and other microorganisms in the large intestine are included).[18] The functions of each cell are maintained by the nutrients it receives. Problems begin when a cell's need for nutrients differs from the available supply.[19]

Nutrient Functions at the Cellular Level Cells are the building blocks of tissues (such as muscles and bones), organs (such as the kidneys, heart, and liver), and systems (such as the respiratory, reproductive, circulatory, and nervous systems). Normal cell health and functioning are maintained when a nutritional and environmental utopia exists within and around the cells. Such circumstances allow **metabolism**—the chemical changes that take place within and outside of cells—to proceed flawlessly. Disruptions in the availability of nutrients—or the presence of harmful substances in the cell's environment—initiate diseases and disorders that eventually affect tissues, organs, and systems. Here are two examples of how cell functions can be disrupted by the presence of low or high concentrations of nutrients:

metabolism The chemical changes that take place in the body. The formation of energy from carbohydrates is an example of a metabolic process.

1. Folate, a B vitamin, is required for protein synthesis within cells. When too little folate is available, cells produce proteins with abnormal shapes and functions. Abnormalities in the shape of red blood cell proteins, for example, lead to functional changes that produce loss of appetite, weakness, and irritability.[20]

2. When too much iron is present in cells, the excess reacts with and damages cell components. If cellular levels of iron remain high, the damage spreads, impairing the functions of organs such as the liver, pancreas, and heart.[21]

Health problems in general begin with disruptions in the normal activity of cells. Humans are as healthy as their cells.

Poor nutrition can result from both inadequate and excessive levels of nutrient intake.

For each nutrient, every individual has a range of optimal intake that produces the best level for cell and body functions. On either side of the optimal range are levels of intake associated with impaired body functions.[22] This concept is presented in Illustration 2.6. Inadequate essential nutrient intake, if prolonged, results in obvious deficiency diseases. Marginally deficient nutrient intake generally produces subtle changes in behavior or physical condition. If the optimal intake range is exceeded, mild to severe changes in mental and physical functions occur, depending on the amount of the excess and the nutrient. Severe zinc deficiency, for example, is related to diarrhea, respiratory infection, and stunted growth. Mild zinc deficiency causes disturbances in the sense of taste and smell, and reduces appetite and food intake. Excessive intake of zinc is associated with vomiting and a decline in the body's ability to fight infections.[23] Nearly all cases of vitamin and mineral overdose result from excessive use of supplements or errors made in the level of nutrient fortification of food products. They are almost never caused by foods. For nutrients, "enough is as good as a feast."

Steps in the Development of Nutrient Deficiencies and Toxicities Poor nutrition due to inadequate diet generally develops in the stages outlined in Illustration 2.7. To help explain the stages, this illustration includes an example of how vitamin A deficiency develops.

After a period of deficient intake of an essential nutrient, the body's tissue reserves of the nutrient become depleted. Blood levels of the nutrient then decrease because there are no reserves left to replenish the blood supply. Without an adequate supply of the nutrient in the blood, cells get shortchanged. They no longer have the supply of nutrients needed to maintain normal function. If the dietary deficiency is prolonged, the malfunctioning cells cause

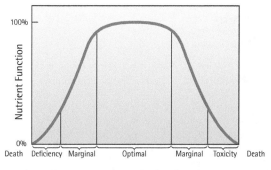

Illustration 2.6 For every nutrient, there is a range of optimal intake that corresponds to the optimal functioning of that nutrient in the body.

Illustration 2.7 The usual sequence of events in the development of a nutrient deficiency and an example of how vitamin A deficiency develops.[24,25]

| Inadequate dietary intake | → | Depletion of tissue stores of the nutrient | → | Decreased blood levels of the nutrient | → | Decreased nutrient available to cells |

EXAMPLE: Deficient vitamin A intake

Photodisc

EXAMPLE: Reduced liver stores of vitamin A

EXAMPLE: Reduced blood levels of vitamin A

Photodisc

EXAMPLE: Decreased vitamin A available to cells within eye

ICI Pharmaceuticals

| Long-term impairment of health | ← | Physical signs and symptoms of deficiency | ← | Impaired cellular functions |

EXAMPLE: Outer covering of the eyes dries out and thickens; vision is lost

Photodisc

EXAMPLE: Outer covering of the eyes dries out, thickens, and becomes susceptible to infection

EXAMPLE Impaired ability to see in dim light

sufficient impairment to produce physically obvious signs of a deficiency disease. Eventually, some of the problems produced by the deficiency may no longer be repairable, and permanent changes in health and function may occur. In most cases, the problems resulting from the deficiency can be reversed if the nutrient is supplied before this final stage occurs.

Excessively high intake of many nutrients such as vitamin A and selenium produces toxicity diseases. The vitamin A toxicity disease is called *hypervitaminosis A,* and the disease for selenium toxicity is called *selenosis.* Signs of the toxicity disease stem from increased levels of the nutrient in the blood and subsequent oversupply of the nutrient to the cells. The high nutrient load upsets the balance needed for normal cell function. The changes in cell functions lead to the signs and symptoms of the toxicity disease.

For both deficiency and toxicity diseases, the best time to correct the problem is usually at the level of dietary intake, before tissue stores are adversely affected. In that case, no harmful effects on health and cell function occur—they are prevented.[26]

Nutrient Deficiencies Are Often Multiple Most foods contain many nutrients, so poor diets will affect the intake level of more than one nutrient (Illustration 2.8). Inadequate diets generally produce a spectrum of signs and symptoms related to multiple nutrient deficiencies. For example, protein, vitamin B$_{12}$, iron, and zinc are packaged

Illustration 2.8 This woman has iron deficiency anemia. Chances are good that she has poor status of other nutrients in addition to iron.

Chad Johnston/Masterfile

together in many high-protein foods. The protein-deficient, starving children you may see in news reports are rarely deficient just in protein. They may also be deficient in iron, zinc, and vitamin B_{12}.

The "Ripple Effect" Dietary changes affect the level of intake of many nutrients. Switching from a high-fat to a low-fat diet, for instance, may result in a higher intake of protein, carbohydrate, or both; a higher intake of cholesterol and vitamin E; and increased intake of vitamin A, vitamin C, and iron. So dietary changes introduced for the purpose of improving the intake level of a particular nutrient produce a ripple effect on the intake of other nutrients.

NUTRITION CONCEPT #5

Humans have adaptive mechanisms for managing fluctuations in nutrient intake.

Healthy humans are equipped with a number of adaptive mechanisms that partially protect the body from poor health due to fluctuations in dietary intake. In the context of nutrition, adaptive mechanisms act to conserve nutrients when dietary supply is low and to eliminate them when they are present in excessively high amounts. Dietary surpluses of energy and some nutrients—such as vitamin A and vitamin B_{12}—can be stored within tissues for later use. In the case of iron, copper, and calcium, the body regulates the amounts absorbed in response to its need for them. The body is unable to store excess amino acids for very long. It uses excess amino acids consumed primarily as a source of energy.[27]

Here are some other examples of how the body adapts to changes in dietary intake:

- When calorie intake is reduced by fasting, starvation, or dieting, the body adapts to the decreased supply by lowering energy expenditure. Declines in body temperature and the capacity to do physical work also act to decrease the body's need for calories. When caloric intake exceeds the body's need for energy, the excess is stored as fat for energy needs in the future.

- The ability of the gastrointestinal tract to absorb dietary iron increases when the body's stores of iron are low. To help protect the body from iron overdose, the mechanisms that facilitate iron absorption in times of need shut down when enough iron has been stored.

- The body can protect itself from excessively high levels of intake of vitamin C from supplements by excreting the excess in the urine.

Although these built-in mechanisms do not protect humans from all the consequences of poor diets, they do provide an important buffer against the development of nutrient-related health problems.

NUTRITION CONCEPT #6

Malnutrition can result from poor diets and from disease states, genetic factors, or combinations of these factors.

Malnutrition means poor nutrition; it results from both inadequate and excessive availability of calories and nutrients in the body. Vitamin A toxicity, obesity, vitamin C deficiency (scurvy), and underweight are examples of malnutrition.

Malnutrition can result from poor diets and also from diseases that interfere with the body's ability to use the nutrients consumed. Diarrhea, alcoholism, cancer, bleeding ulcers, and HIV/AIDS, for example, may be primarily responsible for the development of malnutrition in people with these disorders.

In addition, a percentage of the population is susceptible to malnutrition and increased disease risk due to genetic factors. For example, people may be born with a

malnutrition Poor nutrition resulting from an excess or lack of calories or nutrients.

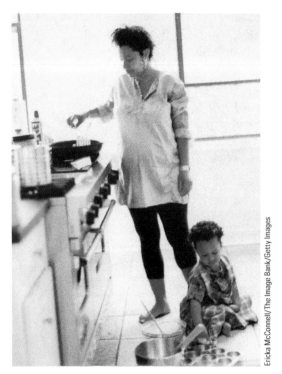

Illustration 2.9 Women who are pregnant or breast-feeding and infants are among the people who are at a higher risk of becoming inadequately nourished.

genetic tendency to produce excessive amounts of cholesterol, absorb high levels of iron, or use folate poorly. Some cases of obesity, diabetes, heart disease, and cancer are related to a combination of genetic and dietary factors.[28]

NUTRITION CONCEPT #7

Some groups of people are at higher risk of becoming inadequately nourished than others.

Women who are pregnant or breastfeeding, infants, growing children, the frail elderly, the ill, and those recovering from illness have a greater need for nutrients than other people. As a result, they are at higher risk than other people of becoming inadequately nourished (Illustration 2.9). The fetus during pregnancy and infants are developing rapidly and are particularly vulnerable to the adverse affects of poor nutrition. Poor nutrition experienced early in life can induce changes in gene function that adversely affect health status for a lifetime.[29] In cases of widespread food shortages, such as those induced by natural disasters or war, the health of these nutritionally vulnerable groups is compromised the soonest and the most.

Within the nutritionally vulnerable groups, certain people and families are at particularly high risk of malnutrition. These are people and families who are poor and least able to secure food, shelter, and high-quality medical services. The risk of malnutrition is not shared equally among all persons within a population.

NUTRITION CONCEPT #8

Poor nutrition can influence the development of certain chronic and other diseases.

chronic diseases Slow-developing, long-lasting diseases that are not contagious (e.g., heart disease, diabetes, and cancer). They can be treated but not always cured.

Poor nutrition does not result only in nutrient deficiency or toxicity diseases. Faulty diets play important roles in the development of **chronic diseases** such as hypertension, heart disease, cancer, and osteoporosis. Diets high in salt, for example, are related to the development of hypertension; those low in vegetables and fruit to cancer; low-calcium diets and poor vitamin D status to osteoporosis; and high-sugar diets to tooth decay. The harmful effects of poor dietary practices on chronic disease development generally accumulate over the course of years.[30]

NUTRITION CONCEPT #9

Adequacy, variety, and balance are key characteristics of healthy dietary patterns.

dietary pattern The quantities, proportions, variety, or combination of different foods, drinks, and nutrients in diets, and the frequency with which they are habitually consumed.

Healthy diets correspond to a **dietary pattern** associated with normal growth and development, a healthy body weight, health maintenance, and disease prevention. One such pattern is represented by the USDA's ChooseMyPlate food group intake guide (Illustration 2.10). Several other types of dietary patterns, including the Mediterranean-style dietary pattern and the DASH Eating Plan, have also been found to promote health and foster disease prevention.[15] Healthy dietary patterns are characterized by regular consumption of moderate portions of a variety of foods from each of the basic food groups. No specific foods or food preparation techniques are excluded in a healthy dietary pattern.

Healthy dietary patterns are plant-food based and include the regular consumption of vegetables, fruits, dried beans, fish and seafood, low-fat dairy products, poultry and lean

Illustration 2.10 ChooseMyPlate is the icon for the USDA's new food guidance system. It is intended to visually help people choose healthy meals.

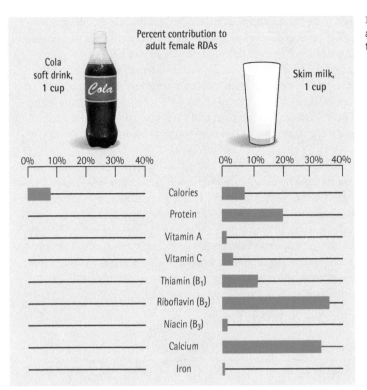

meats, nuts and seeds, and whole grains. Dietary patterns that include large or frequent servings of foods containing high amounts of *trans* fat, added sugars, salt, or alcohol miss the healthy dietary pattern mark. Healthy dietary patterns supply needed nutrients and beneficial phytochemicals through food rather than supplements or special dietary products.[31]

Energy and Nutrient Density Most Americans consume more calories than needed, become overweight as a result, *and* consume inadequate diets. This situation is partly due to over-consumption of **energy-dense foods** such as processed and high-fat meats, chips, candy, many desserts, and full-fat dairy products. Energy-dense foods have relatively high calorie values per unit weight of the food. Intake of energy-dense diets is related to the consumption of excess calories and to the development of overweight and diabetes.[32]

Many energy-dense foods are nutrient poor, meaning they contain low levels of nutrients given their caloric value. These foods are sometimes referred to as **empty-calorie foods**; these include products such as soft drinks, sherbet, hard candy, alcohol, and cheese twists. Excess intake of energy-dense and empty-calorie foods increases the likelihood that calorie needs will be met or exceeded before nutrients needs are met.[33] Diets most likely to meet nutrient requirements without exceeding calorie needs contain primarily **nutrient-dense foods**, foods with high levels of nutrients and relatively low calorie value. Nutrient-dense foods such as non-fat milk and yogurt, lean meat, dried beans, vegetables, and fruits provide relatively high amounts of nutrients compared to their calorie value.[32] Illustration 2.11 shows a comparison of the calorie and nutrient content of an empty-calorie and a nutrient-dense food.

energy-dense foods Foods that provide relatively high levels of calories per unit weight of the food. Fried chicken; cheeseburgers; a biscuit, egg, and sausage sandwich; and potato chips are energy-dense foods.

empty-calorie foods Foods that provide an excess of energy or calories in relation to nutrients. Soft drinks, candy, sugar, alcohol, and animal fats are considered empty-calorie foods.

nutrient-dense foods Foods that contain relatively high amounts of nutrients compared to their calorie value. Broccoli, collards, bread, cantaloupe, and lean meats are examples of nutrient-dense foods.

NUTRITION CONCEPT #10

There are no "good" or "bad" foods.

People tend to classify foods as being "good" or "bad," but such opinions over-simplify each food's potential contribution to a diet.[31] Typically hot dogs, ice cream, candy, bacon, and french fries are judged to be "bad," whereas vegetables, fruits, and whole-grain products are given the "good" stamp. Unless we're talking about spoiled stew, poison mushrooms, or something similar, however, no food can be firmly labeled as "good" or "bad." Ice cream can be a "good" food for physically active, normal-weight individuals with a high calorie need who have otherwise met their nutrient requirements by consuming nutrient-dense foods. Some people who eat only what they consider to be "good" foods such as broccoli, berries, brown rice, and tofu may still miss the healthful diet mark due to inadequate consumption of essential fatty acids and certain vitamins and minerals.

All foods can fit into a healthy dietary pattern as long as nutrient needs are met at calorie intake levels that maintain a healthy body weight.[31] If nutrient needs are not being met and calorie intake levels are too high, then the diet likely includes too many energy-dense or empty-calorie foods. Substituting nutrient-dense for energy-dense foods would help bring the diet back into balance.

The basic nutrition concepts presented here are listed in Table 2.6. It may help you to remember the concepts and to start thinking like a bona fide nutritionist if you go back over each concept and give several examples related to it. If you understand these concepts, you will have gained a good deal of insight into nutrition.

Table 2.6 **Nutrition concepts**

1. Food is a basic need of humans.
2. Foods provide energy (calories), nutrients, and other substances needed for growth and health.
3. Health problems related to nutrition originate within cells.
4. Poor nutrition can result from both inadequate and excessive levels of nutrient intake.
5. Humans have adaptive mechanisms for managing fluctuations in nutrient intake.
6. Malnutrition can result from poor diets and from disease states, genetic factors, or combinations of these factors.
7. Some groups of people are at higher risk of becoming inadequately nourished than others.
8. Poor nutrition can influence the development of certain chronic and other diseases.
9. Adequacy, variety, and balance are key characteristics of healthy dietary patterns.
10. There are no "good" or "bad" foods.

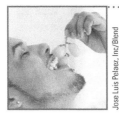

NUTRITION
up close

Nutrition Concepts Review

Focal Point: Nutrition concepts apply to diet and health relationships.

Write the number of the nutrition concept from Table 2.6 that most closely applies to the situation described. Use each concept and do not repeat concept numbers in your response.

Nutrition concept number	Situation
1. ____	The Irish potato famine caused thousands of deaths.
2. ____ and ____	Otis mistakenly thought that as long as he consumed enough calories from food along with vitamin and mineral supplements, he would stay healthy no matter what he ate.
3. ____	I feel guilty every time I eat potato chips. I wish they weren't bad for me.
4. ____	Phyllis was relieved to learn that her chronic diarrhea was due to the high level of vitamin C supplements she had been taking.
5. ____	A low amount of iron in Tawana's red blood cells was the reason for her loss of appetite and low energy level.
6. ____	Far more young children than soldiers died as a result of the 10-year civil war in Sudan.
7. ____	For the past 20 years, Don's idea of dinner was a big steak and potatoes. His recent heart attack changed his view of what's for dinner.
8. ____	During the two weeks they were backpacking in the Netherlands, Tomás and Ozzie ate very few vegetables and fruits. Their health remained robust, however.
9. ____	Zhang wasn't aware that he had the inherited condition hemochromatosis until he began taking iron supplements and developed iron overload symptoms.

Feedback on Nutrition Up Close is located in Appendix F.

REVIEW QUESTIONS

- **Explain the scope of nutrition as an area of study.**

1. Nutrition is defined as "the study of foods, their nutrients and other chemical constituents, and the effects that food constituents have on health."   True/False

2. ____ Cassandra is on her way to her first nutrition class and is thinking about what the course will cover. Listed below are her ideas. Three of the ideas correspond to the scope of the study of nutrition. Which idea is *not* a component of the scope of the study of nutrition?
   a. diet and disease relationships
   b. components of healthful diets
   c. magical powers of super foods for weight loss
   d. nutrient composition of foods

- **Demonstrate a working knowledge of the meaning of the 10 nutrition concepts.**

3. The word *nonessential* as in *nonessential nutrient* means that the nutrient is *not* required for growth and health.   True/False

4. Food insecurity is a problem in developing countries, but it is *not* a problem in the United States or Canada.   True/False

5. Nutrients are classified into five basic groups: carbohydrates, protein, fats, vitamins, and water.   True/False

6. The development of standards for nutrient intake levels was prompted in part by the high rejection rate of World War II recruits due to underweight and nutrient deficiencies.   True/False

7. Tissue stores of nutrients decline after blood levels of the nutrients decline.   True/False

8. To maintain health, all essential nutrients must be consumed at the recommended level daily.   True/False

9. An individual's genetic traits play a role in how nutrient intake affects disease risk.   True/False

10. ____ If you wanted to know the Recommended Dietary Allowance for protein for a 7-month-old infant, you would refer to tables on the:

    a. Dietary Guidelines for Americans
    b. Infant Nutritional Recommendations for Americans
    c. Recommended Daily Intakes
    d. Recommended Dietary Allowances

11. ____ The Tolerable Upper Intake Level for iron for a 65-year-old male in milligrams per day (mg/d) is:

    a. 1,100
    b. 45
    c. 350
    d. 3.5

12. Certain phytochemicals, or_____, such as lycopene and anthocyanins, act as _____.

13. Groups of people at higher risk than others of becoming inadequately nourished include _____ and _____.

For the questions below, match the term in column A with its definition in column B.

Column A   Column B

____14. Essential nutrients   a. Chemical substances in food that are used by the body for growth and health.

____15. Nutrients   b. Chemical substances that prevent or repair damage to cells caused by exposure to oxidizing agents such as environmental pollutants, smoke, ozone, and oxygen.

____16. Food insecurity   c. Foods that contain relatively high amounts of nutrients in relation to their calorie value.

____17. Antioxidants   d. The chemical changes that take place in the body.

____18. Calorie   e. Limited or uncertain availability of safe, nutritious foods—or the ability to acquire them in socially acceptable ways.

____19. Metabolism   f. A unit of measure of the amount of energy supplied by food.

____20. Malnutrition   g. Nutrients required for normal growth and health that the body can generally not produce, or produce in sufficient amounts.

____21. Nutrient dense   h. Poor nutrition resulting from an excess or lack of calories or nutrients.

Answers to these questions can be found in Appendix F.

NUTRITION SCOREBOARD ANSWERS

1. Calories are a measure of the amount of energy supplied by food. They're a property of food, not a substance present in food. **False**

2. That's the definition of nutrients. **True**

3. Excessive as well as inadequate intake levels of vitamins and minerals can be harmful to health. **False**

4. The DRIs are shown on the inside front and back covers of this text. **True**

Ariel Skelley/Blend Images/Jupiter Images

The Inside Story about Nutrition and Health

CHAPTER 3

NUTRITION SCOREBOARD

1 How long people live and how healthy they are primarily depends on four factors: lifestyle behaviors, the environments to which they are exposed, genetic factors, and access to quality health care. **True/False**

2 Diet is related to the top two causes of death in the United States. **True/False**

3 Biological processes of modern humans were designed over 40,000 years ago. **True/False**

4 Most economically developed countries regularly monitor levels of various contaminants and nutrients in foods and diets, but the United States does not. **True/False**

Answers can be found at the end of the chapter.

- Identify characteristics of diets related to the development of specific diseases.

- Explain how differences in diets of early versus modern humans may promote the development of certain diseases.

- List the types of food that are core components of healthful diets.

Nutrition in the Context of Overall Health

• **Identify characteristics of diets related to the development of specific diseases.**

Think of your body as a machine. How well this machine performs depends on a number of related factors: the quality of its design and construction, the appropriateness of the materials used to produce it, and how well it is maintained.

A machine designed to produce 10,000 copies a day will break down sooner if it is used to make 20,000 copies a day. The repair call will, in all probability, come earlier if the machine is overused *and* poorly maintained or if it has a part that doesn't work well. On the other hand, chances are good the copy machine will function at full capacity if it is free from design flaws, skillfully constructed from appropriate materials, properly used, and kept in good shape through regular maintenance.

Although much more complex and sophisticated, the human body is like a machine in some important ways. How well the body works and how long it lasts depend on a variety of interrelated factors. The health and fitness of the human machine depend on genetic traits (the design part of the machine), the quality of the materials used in its construction (your diet), and regular maintenance (your diet, other lifestyle factors, and health care).

Lifestyles exert the strongest overall influence on health and longevity (Illustration 3.1).[2] Behaviors that constitute our lifestyle—such as diet, smoking habits, illicit drug use or excessive drinking, level of physical activity or psychological stress, and the amount of sleep we get—largely determine whether we are promoting health or disease. Of the lifestyle factors that affect health, our diet is one of the most important.[3] In a sense, it is fortunate that diet is related to disease development and prevention. Unlike age, gender, and genetic makeup, our diets are within our control.

People have an intimate relationship with food—each year we put over a thousand pounds of it into our bodies! Food supplies the raw materials the body needs for growth and health; these, in turn, are affected by the types of food we usually eat. The diet we feed the human machine can hasten, delay, or prevent the onset of an impressive group of today's most common health problems.

KEY NUTRITION CONCEPTS

Unit 1 presents 10 key nutrition concepts that are fundamental to the science of nutrition. The content on diet and health covered in this unit directly relates to three of them:

1. Nutrition Concept #3: Health problems related to nutrition originate within cells.

Illustration 3.1 Conditions that contribute to death among adults under the age of 75 in the United States. Health care refers to access to quality care; environmental exposures include the safety of one's surroundings and the presence of toxins and disease-causing organisms in the environment; lifestyle factors include diet, exercise, obesity, smoking, genetic traits, and alcohol and drug use.[3,4]

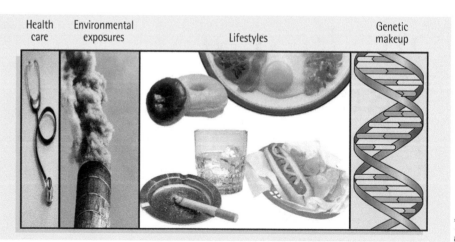

Photodisc

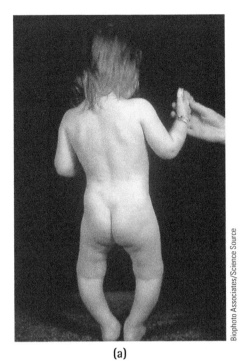

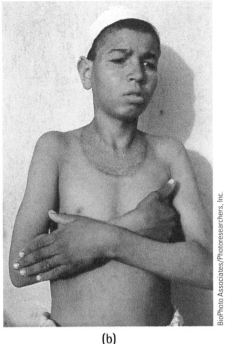

(a)

(b)

Illustration 3.2 Vitamin D deficiency (rickets shown on the left) and niacin deficiency (pellagra pictured on the right) were leading causes of hospitalization of children in the United States in the 1930s.

2. Nutrition Concept #6: Malnutrition can result from poor diets and from disease states, genetic factors, or combinations of these factors.

3. Nutrition Concept #8: Poor nutrition can influence the development of certain chronic diseases.

The Nutritional State of the Nation

Since early in the twentieth century, researchers have known that what we eat is related to the development of vitamin and mineral deficiency diseases, to compromised growth and impaired mental development in children, and to the body's ability to fight infectious diseases. Seventy years ago in the United States, widespread vitamin deficiency diseases filled children's hospital wards and contributed to serious illness and death in adults (Illustration 3.2). Now, however, dietary excesses are filling hospital beds and reducing the quality of life for millions of Americans.

Today, the major causes of death among Americans and in other developed countries are slow-developing, lifestyle-related **chronic diseases**.[6] Based on government survey data, 44% of Americans have a chronic condition such as **diabetes**, heart disease, cancer, **hypertension**, or high cholesterol levels; 13% have three or more of these conditions.[7]

The leading causes of death among Americans are heart disease and cancer (see Illustration 3.3). Together they account for 47% of all deaths. Western-type dietary patterns that are generally low in vegetables, fruits, whole grains, dried beans, poultry, nuts, and fish and relatively high in meat, refined grains, sugars, calories, and salt are linked to the development of a number of chronic diseases.[8] People who consume Western-type diets are at higher risk of developing obesity, diabetes, cancer, heart disease, and hypertension.[3,9] Examples of diseases and disorders associated with dietary intake are shown in Table 3.1.

chronic diseases Slow-developing, long-lasting diseases that are not contagious (e.g., heart disease, cancer, diabetes). They can be treated but not always cured.

diabetes Short for the term *diabetes mellitus*, a disease characterized by abnormal utilization of glucose by the body and elevated blood glucose levels. There are three main types of diabetes: type 1, type 2, and gestational diabetes. The word *diabetes* in this text refers to type 2 diabetes, by far the most common.

hypertension High blood pressure. It is defined as blood pressure exerted inside of blood vessel walls that typically exceeds 140/90 mm Hg (millimeters of mercury).

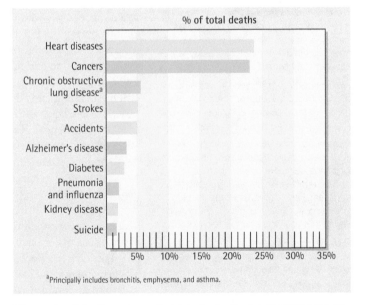

% of total deaths

Heart diseases
Cancers
Chronic obstructive lung disease[a]
Strokes
Accidents
Alzheimer's disease
Diabetes
Pneumonia and influenza
Kidney disease
Suicide

5% 10% 15% 20% 25% 30% 35%

[a]Principally includes bronchitis, emphysema, and asthma.

Illustration 3.3 Percentage of total deaths for the top 10 leading causes of death in the United States, 2012.[5]

Table 3.1 Examples of diseases and disorders linked to diet[8-14]

Disease or disorder	Dietary connections
Heart disease	Excessive body fat, high intake of *trans* fat, added sugar, and salt; low vegetable, fruit, fish, nuts, and whole-grain intake
Cancer	Low vegetable and fruit intake; excessive body fat and alcohol intake; regular consumption of processed meats
Stroke	Low vegetable and fruit intake; excessive alcohol intake; high animal-fat diets
Diabetes (type 2)	Excessive body fat; low vegetable, whole grain, and fruit intake; high added sugar intake
Cirrhosis of the liver	Excessive alcohol consumption; poor overall diet
Hypertension	Excessive sodium (salt) and low potassium intake; excess alcohol intake; low vegetable and fruit intake; excessive levels of body fat
Iron-deficiency anemia	Low iron intake
Tooth decay and gum disease	Excessive and frequent sugar consumption; inadequate fluoride intake
Osteoporosis	Inadequate calcium and vitamin D; low intake of vegetables and fruits
Obesity	Excessive calorie intake; overconsumption of energy-dense, nutrient-poor foods
Chronic inflammation and oxidative stress	Excessive calorie intake; excessive body fat; high animal-fat diets; low intake of whole grains, vegetables, fruit, and fish
Alzheimer's disease	Regular intake of high animal-fat products; low intake of olive oil, vegetables, fruits, fish, wine, and whole grains

chronic inflammation Low-grade inflammation that lasts weeks, months, or years. Inflammation is the first response of the body's immune system to infectious agents, toxins, or irritants. It triggers the release of biologically active substances that promote oxidation and other reactions to counteract the infection, toxin, or irritant. A side effect of chronic inflammation is that it may also damage lipids, cells, tissues, and certain body processes. Also called low-grade inflammation.

oxidative stress A condition that occurs when cells are exposed to more oxidizing molecules (such as free radicals) than to antioxidant molecules that neutralize them. Over time, oxidative stress causes damage to lipids, DNA, cells and tissues. It increases the risk of heart disease, type 2 diabetes, cancer, and other diseases.

osteoporosis A condition in which bones become fragile and susceptible to fracture due to a loss of calcium and other minerals.

free radicals Chemical substances (often oxygen-based) that are missing electrons. The absence of electrons makes the chemical substance reactive and prone to oxidizing nearby molecules by stealing electrons from them. Free radicals can damage lipids, proteins, cells, DNA, and eventually tissues by altering their chemical structure and functions.

antioxidants Chemical substances that prevent or repair damage to cells caused by oxidizing agents such as pollutants, ozone, smoke, and reactive oxygen. Oxidation reactions are a normal part of cellular processes. Vitamins C and E and certain phytochemicals function as antioxidants.

Shared Dietary Risk Factors A number of the diseases and disorders listed in Table 3.1 share the common risk factors of low intake of vegetables, fruits, and whole grains; excess calorie intake and body fat; and high animal-fat intake. These risk factors are associated with the development of **chronic inflammation** and **oxidative stress**, conditions that are strongly related to the development of heart disease, diabetes, **osteoporosis**, Alzheimer's disease, cancer, and other chronic diseases.[15]

Inflammation and Oxidative Stress Inflammation is an important part of the body's defense systems against cell and tissue damage due to the presence of infectious agents, chemical irritants, toxins, or physical injury. It can be classified as acute or chronic. Most of us are familiar with acute inflammation. That's a temporary reaction that occurs when you sprain your ankle or develop a fever. The ankle you injured or your head feels very warm to the touch and hurts, and your head may ache from the fever. Acute inflammation goes away with time as your body heals itself. Chronic, low-grade inflammation doesn't hurt; it occurs silently within the fluids, tissues, and cells inside your body; and it lasts weeks, months, or years. An important part of the body's inflammatory response is the production and release of oxidizing agents such as **free radicals** that destroy the offending substances. In the process, however, free radicals may also oxidize fats (lipids), cell membranes, and DNA inside of cells. In the short term, damage induced by oxidation reactions can generally be repaired by **antioxidants** produced by the body and consumed in vegetables, fruits, whole-grain products, and other plant foods.[16]

Inflammation is related to chronic disease development when it is present at a low level for a long time, or is *chronic*. Chronic inflammation and the resulting oxidative stress are sustained by irritants continually present in the body. Excess body fat and habitually high intake of foods high in *trans* fats and added sugar are examples of such irritants. If not countered by a sufficient supply of antioxidants, chronic inflammatory processes and oxidative stress impair the normal functioning of cells and tissues. Chronic inflammation represents a major pathway by which diet influences the development of chronic disease.[39]

Loss of excess body fat and a dietary pattern high in whole grains, vegetables, fruit, poultry, and fish and low in refined grains, added sugars, red and processed meat, high-fat dairy products, and sweetened beverages is associated with lower levels of inflammation compared to dietary patterns that do not.[17, 18] Table 3.2 summarizes the types of foods included in dietary patterns that tend to decrease or increase oxidative stress and inflammation.

Nutrient–Gene Interactions and Health Some diseases are promoted by interactions between nutrients and genes. One example of such an interaction involves gene types and the health effects of cruciferous vegetables such as broccoli, cauliflower, brussels sprouts, and cabbage. These vegetables contain isothiocyanates (pronounced ice-o-thee-o-sigh-ah-nates), which are involved in the prevention of cancer development. Individuals carrying certain types of genes that lead to the rapid breakdown of isothiocyanates and related compounds are more susceptible to cancer development than others who break down these substances slowly. Isothiocyanates appear to block mechanisms that promote tumor development.[19] Another example of nutrient–gene interactions that influence health relates to obesity. The causes of obesity are complex and include interactions between a number of gene types and environmental factors. One contributing genetic factor appears to be the form of the FTO gene present. This gene participates in processes that regulate appetite and food intake. People with the "high-risk" form of the FTO gene experience a 20% higher lifetime risk of becoming overweight or obese compared to those who have the "low-risk" form of the gene.[20]

Knowledge of nutrient–gene interactions in health and disease is expanding rapidly and is greatly enhancing our understanding of the relationship between diet and health. This knowledge will contribute to the development of personalized nutritional interventions targeted at an individual's genetic characteristics.[21]

The Importance of Food Choices

People are not born with an internal compass that directs them to select a healthy diet—and it shows. If given access to a food supply like that available in the United States, people show a marked tendency to choose a diet that is high in energy-dense, nutrient-poor foods[22] (Illustration 3.4). Such a diet tends to include processed foods high in saturated fat, salt, or sugar and low in whole grains, vegetables, fruits, and other basic foods. This type of diet poses the greatest risks to the health of Americans.[22]

Diet and Diseases of Western Civilization

- **Explain how differences in diets of early versus modern humans may promote the development of certain diseases.**

Why is the U.S. diet—a "Western" style of eating—hazardous to our health in so many ways? What is it about a diet that is high in animal fat, salt, and sugar and low in vegetables, fruits, and whole grains that promotes certain chronic diseases? A good deal of evidence indicates that the chronic diseases now prevalent in the United States and other Westernized countries have roots in dietary changes that have taken place over centuries.

Our Bodies Haven't Changed

The biological processes that control what the human body does with food were developed over the course of our evolution tens of thousands of years ago. These processes exist today because they are linked to the genetic makeup of humans and continue to influence how diet affects health.[23,24]

Then . . . For the first 200 centuries of their existence, humans survived by hunting and gathering (Illustration 3.5). They were constantly on the move, pursuing wild game or following the seasonal maturation of fruits and vegetables. Meat, berries, and many other plant products obtained from successful hunting and gathering journeys spoiled quickly, so they had to be consumed in a short time. Feasts would be followed by famines that lasted until the next successful hunt or harvest.[23]

. . . and Now The bodies of modern humans, adapted to exist on a diet of wild game, fish, fruits, nuts, seeds, roots, vegetables, and grubs; to survive periods of famine; and

Table 3.2 Types of food associated with decreased or increased inflammation, oxidative stress, or both[11,14,15]

Decreased
Colorful fruits and vegetables
Dried beans
Whole grains
Fish and seafood, fish oils
Red wines
Dark chocolates
Olive oil
Nuts
Coffee

Increased
Processed and high-fat meats
High-fat dairy products
Baked products, snack foods with *trans* fats
Soft drinks, other high-sugar beverages

Illustration 3.4   Lopsided, all-American food choices.

to sustain a physically demanding lifestyle are now exposed to a different set of circumstances. The foods we eat bear little resemblance to the foods available to our early ancestors (Illustration 3.6). Sugar, salt, alcohol, food additives, oils, margarine, dairy products, refined grain products, and processed foods were not a part of their diets. These ingredients and foods came with Western civilization.[23] Furthermore, we do not have to engage in strenuous physical activity to obtain food, and our feasts are no longer followed by famines.

The human body developed other survival mechanisms that are not the assets they used to be. Mechanisms that stimulate hunger in the presence of excess body fat stores, conserve the body's supply of sodium, and confer an innate preference for sweet-tasting foods—as well as a digestive system that functions best on a high-fiber diet—were advantages for early humans. They are not advantageous for modern humans, however, because our diets and lifestyles are now vastly different.

(a)   (b)

Illustration 3.5   Hunter-gatherers still exist in the world, but their numbers are diminishing. It is estimated that hunter-gatherers consume approximately 3,000 calories daily due to their physically demanding way of life.

Illustration 3.6   The disconnect between high animal fat, high salt, high sugar, and processed foods in Western-type diets (right) and wild plants and animal foods consumed by our early ancestors (left). Foods consumed by hunter-gatherers shown in the photograph include birds' eggs, wild cucumbers, roots, nuts, and berries. Not shown are grubs and other insects, which might be consumed as quickly as they are discovered.

Although the human body has a remarkable ability to adapt to changes in diet, the health problems of modern civilization such as heart disease, cancer, hypertension, and diabetes are thought to result, in part, from diets that are greatly different from those of our early ancestors. The human body was built to function best on a diet that is low in sugar and sodium, contains lean sources of protein, and is high in fiber, vegetables, and fruits.[25] Strong evidence for this conclusion is provided by studies that track how disease rates change as people adopt a Western style of eating.

Different Diets, Different Disease Rates

Many countries are adopting the Western diet and the pattern of disease that accompanies it. People in Japan, for example, live longer than almost anyone else in the world—until they move to the United States (Table 3.3). In Japan, the traditional diet consists mainly of rice, vegetables, fish, shellfish, broths, tofu, noodles, seaweed, eggs, tea, and small portions of meat

Table 3.3 **Life expectancy at birth for countries with high life expectancies, 2012[33]**

Country	Life expectancy (years)	Country	Life expectancy (years)
Japan	83	Germany	81
Switzerland	83	Ireland	81
France	83	Finland	81
Italy	83	Greece	81
Australia	82	Belgium	80
Spain	82	Portugal	80
United Kingdom	82	Denmark	80
Canada	81	Costa Rica	80
New Zealand	81	Cuba	79
Austria	81	United States	79

JTB MEDIA CREATION, Inc./Alamy

(Illustration 3.7).[26] When Japanese people move to the United States, their diets often change to include, on average, more fat, sugar, and calories and less fish and vegetables.[27] Japanese living in the United States are much more likely to develop diabetes (Illustration 3.8), heart disease, breast cancer, and colon cancer than people who remain in Japan.[29,31]

Dietary habits in Japan are rapidly becoming similar to those in the United States. Hamburgers, fries, steak, ice cream, and other high-fat foods are gaining in popularity. Rates of diabetes, heart disease, and cancer of the breast and colon are on the rise in Japan.[27,30] Similarly, the "diseases of Western civilization" are occurring at increasing rates in Russia, Greece, Israel, and other countries adopting the Western diet.[32] Obesity, diabetes, and heart disease rates tend to increase among some population groups after they immigrate to the United States. Latinos who move to the United States, for example, tend to consume poorer quality diets, are more likely to become obese, and are more likely to develop diabetes than people in their home countries.[34]

Today's food supply makes it a bit challenging to eat more like our early ancestors did. What types of foods would you choose if you wanted to shape a diet that is closer to that of hunter-gathers? Join Beth and Shandra in making these dietary decisions in the Reality Check.

The Power of Prevention

Heart disease, cancer, and other chronic diseases are not the inevitable consequence of Westernization. The types of diets that promote chronic disease can be avoided or changed. Although heart disease is still the leading cause of death in the United States, its rate has declined by over 50% in the last 30 years. About half of this decline is related to improvements in risk factors such as smoking, hypertension, and elevated blood cholesterol levels; the other half is primarily related to medical interventions for people with heart disease.[35] The American Heart Association concludes that future gains in heart health among Americans will stem primarily from improved dietary intakes, declines in rates of overweight and obesity, increased physical activity, and decreased smoking.[36]

Illustration 3.8   An increased rate of diabetes in Japanese men immigrating to Seattle corresponds to dietary changes.
Source: Tsunehara CH, et al. Diet of second-generation Japanese-American men with and without non-insulin-dependent diabetes. *Am J Clin Nutr.* 1990;52:731–38.

Table 3.4 Examples of the *Healthy People 2020* nutrition objectives for the nation[37]

Healthier food access
• Increase the proportion of schools that offer nutritious foods and beverages outside of school meals.
• Increase the proportion of Americans who have access to a food retail outlet that sells a variety of foods that are encouraged by the Dietary Guidelines.
Weight status
• Increase the proportion of adults who are at a healthy weight.
• Reduce the proportion of adults who are obese.
• Reduce the proportion of children and adolescents who are considered obese.
• Prevent inappropriate weight gain in youths and adults.
Food insecurity
• Eliminate very low food security among children.
• Reduce household food insecurity and in doing so reduce hunger.
Food and nutrient consumption
• Increase the variety and contribution of vegetables to diets.
• Increase the contribution of whole grains to diets.
• Reduce the consumption of calories from solid fats and added sugars.
• Reduce the consumption of saturated fat.
• Reduce the consumption of sodium.
• Increase the consumption of calcium.
• Reduce the consumption of sodium.
Iron deficiency
• Reduce iron deficiency among young children and females of childbearing age.
• Reduce iron deficiency among pregnant females.
Health care and worksite settings
• Increase the proportion of physician office visits that include counseling or education related to nutrition and weight.

Improving the American Diet

• **List the types of food that are core components of healthful diets.**

Many efforts are under way to improve the diet and health status of Americans. Like other countries with high rates of "Western" diseases, the United States sets national health goals and implements programs aimed at improving health. Goals and objectives for changes in health status in the United States are presented in the report *Healthy People 2020*. Examples of the *Healthy People 2020* objectives for nutrition are shown in Table 3.4. These objectives highlight the national emphasis on improving weight status and dietary intake of the population by the year 2020.

REALITY CHECK

Getting Back to the Basics—But How?

Beth and Shandra have been roommates for a year and usually shop for groceries together. For the next trip to the grocery store, they decide that each of them will make up a shopping list that includes foods that resemble those their early ancestors might have eaten. Here are the results.

Which list do you think comes closest to matching the basic foods consumed by our early ancestors?

Answers on page 74.

Beth: rice, yogurt, pork, honey, olive oil

Shandra: carrots, nuts, asparagus, fish, blueberries

ANSWERS TO REALITY CHECK
Getting Back to the Basics—But How?

Shandra's list contains unprocessed plant foods and fish, which would have been available to our early ancestors.

Beth: 👎

Shandra: 👍

Table 3.5 ChooseMyPlate.gov food guidance

Balancing calories

- Enjoy your food, but eat less.
- Avoid oversized portions.

Foods to increase

- Make half your plate fruits and vegetables.
- Make at least half your grains whole grains.
- Switch to fat-free or low-fat (1%) milk.

Foods to reduce

- Compare sodium in foods like soup, bread, and frozen meals—and choose the foods with lower numbers.
- Drink water instead of sugary drinks.

United States Department of Agriculture

What Should We Eat?

Evidence-based conclusions about nutrition and health are translated into food choices by the U.S. Department of Agriculture's (USDA) food guidance materials. Intended for public consumption, the materials specify food choices that build healthy dietary patterns and contribute to the prevention of common diseases such as diabetes, hypertension, and heart disease.

USDA's advice on healthy food choices changes as knowledge about diet and health relationships advances, and so do the names it uses to label the new food guidance materials. In the past nine decades, USDA's food choice guidance has changed from the "Basic Five Food Groups," to the "Basic Four Food Groups," to "MyPyramid." Beginning in 2011, food guidance materials became labeled "MyPlate" and "ChooseMyPlate" and reflect current concerns about food choices, nutrition, and health (Table 3.5).

Current food intake recommendations focus on basic, nutrient-dense foods such as whole-grain products, vegetables, fruits, lean meats, low-fat dairy products, dried beans, and fish. They call for reduced consumption of soft drinks and other sweetened beverages, high-fat meat and dairy products, and foods high in salt. Sweets, desserts, and packaged snacks are not excluded from the recommendations. They can be included in a healthy dietary pattern as long as food group recommendations are met and overall calorie needs are not exceeded. ChooseMyPlate interactive tools for planning and evaluating dietary intake are available at ChooseMyPlate.gov.

Nutrition Surveys: Tracking the American Diet

The food choices people make and the quality of the American diet and food supply are regularly evaluated by national surveys (Table 3.6). The first survey began in 1936. It was conducted in conjunction with the original national program aimed at reducing hunger, poor growth in children, and vitamin and mineral deficiency diseases. Results of nutrition surveys are used to identify problem areas within the food supply, characteristics of diets consumed by the public, and the prevalence of nutrition-related health disorders. The surveys provide information ranging from the amount of lead and pesticides in certain foods to the adequacy of diets of low-income families. Together with the results of studies conducted by university researchers and others, they provide the information needed to give direction to food and nutrition programs and to policies aimed at improving the availability and quality of the food supply. Many other countries, including Canada and Australia, have similar programs.

Table 3.6 Periodic national surveys of food, diet, and health in the United States

Survey	Purpose
1. National Health and Nutrition Examination Survey (NHANES)	Assesses dietary intake, health, and nutritional status in a sample of adults and children in the United States on a continual basis
2. Nationwide Food Consumption Survey (NFCS)	Performs regular surveys of food and nutrient intake and understanding of diet and health relationships among a national sample of individuals in the United States
3. Total Diet Study (sometimes called Market Basket Study)	Ongoing studies begun in 1961 that determine the levels of various contaminants and nutrients in foods and diets

Ariel Skelley/Blend Images/
Jupiter Images

NUTRITION
up close

Food Types for Healthful Diets

Focal Point: Separating the types of foods that fit into the ChooseMyPlate basic food groups from those that characterize Western-type diets.

The ChooseMyPlate food groups consist of foods that make up a healthful, disease-preventing dietary pattern. The typical Western-type dietary pattern, on the other hand, is related to the development of a number of diseases and includes many types of food that are not part of the ChooseMyPlate basic food groups. To which dietary pattern do the following types of food most appropriately belong?

Food types	Dietary pattern	
	ChooseMyPlate	Western
Mixed vegetables		
Cold cuts (ham, bologna, salami)		
Fish and seafood		
Whole-grain breads		
Fruit jams and jellies		
Potato, tortilla, and other snack chips		
Dried beans		
Poultry (chicken, turkey)		
Fruit		
Vegetables		
Gravy		
Ice cream		
Soft drinks		
Salad dressing, mayonnaise		
Skim milk		
Cake		

Feedback to the Nutrition Up Close is located in Appendix F.

REVIEW QUESTIONS

- **Identify characteristics of diets related to the development of specific diseases.**

1. Vitamin deficiency diseases remain a major health problem for children in the United States.   True/False

2. Diets are related to the top two causes of death.   True/False

3. Low intake of vegetables and fruits is related to the development of heart disease, cancer, hypertension, and osteoporosis.   True/False

4. The incidence of chronic diseases such as diabetes, cancer, and heart disease increases as countries adopt a Western style of eating.   True/False

- **Explain how differences in diets of early versus modern humans may promote the development of certain diseases.**

5. Our genetic makeup changes as our diets change.   True/False

6. List two major characteristics of the lifestyles of early humans cited in this unit that are rare in modern humans and don't involve types of foods available.

   a. _____

   b. _____

7. ___ Chronic inflammation and oxidative stress are sustained by irritants continually present in the body. Two examples of irritants that promote the presence of these conditions in the body are:

   a. physical inactivity and meal skipping
   b. weight loss and lack of sleep
   c. excess body fat and habitually high intake of added sugars
   d. high intake of water and excess dietary fiber

8. ___ Adverse effects of chronic inflammation and oxidative stress can be diminished by:

   a. loss of excess body fat
   b. increased water consumption
   c. decreased intake of high-fiber foods
   d. daily consumption of potatoes

9. Although individuals cannot change their genetic makeup, they can change their risk for chronic disease development by making healthful eating and lifestyle changes.   True/False

- **List the types of food that are core components of healthful diets.**

10. National surveys in the United States assess food intake, nutritional health of the population, and the safety of the food supply.   True/False

11. The *Basic Four Food Groups* is the most recently published guide to selection of a healthful diet.   True/False

12. ___ Which of the following statements about food choices is true?

   a. Given the availability of a wide assortment of foods, people tend to select and consume a healthful diet.
   b. People are *not* born with an internal compass that directs them to select and consume a healthy diet.
   c. Food choices usually change very little as people move from one country to another.
   d. Recommendations for healthy diets include a narrow range of food choices.

Questions 13–15 refer to the following case scenario:

Assume you are attending a family picnic and are staring at a table full of food. Your choices from the table are: fried chicken, cold-cut platter (bologna, salami), whole-grain rolls, shrimp, jell-o, spinach salad, fruit salad, baked beans, potato chips, zucchini squash, cake, and low-fat milk.

13. ___ From the food options available, which three foods would you put on your plate if you wanted to consume the basic foods?

   a. fried chicken, fruit salad, and baked beans
   b. zucchini squash, jell-o, and baked beans
   c. spinach salad, cold cuts, and shrimp
   d. spinach salad, zucchini squash, and shrimp

14. ___ Which of the following sets of foods would be considered creations of modern humans?

   a. potato chips and cold cuts
   b. cake and spinach salad
   c. jell-o and fruit salad
   d. fried chicken and zucchini squash

15. ___ You decide to have a glass of milk along with shrimp, fruit salad, baked beans, and spinach salad. Which basic food group is missing from your plate?

   a. vegetables
   b. fruits
   c. grains
   d. protein foods

Answers to these questions can be found in Appendix F.

NUTRITION SCOREBOARD ANSWERS

1. There are no secrets to a long, healthy life.   True

2. Diet is associated with the development of heart disease and cancer, which cause about half of all deaths in the United States.[1]   True

3. Hairstyles may be different, but our bodies are the same as they were 40,000 years ago.   True

4. The Total Diet Study has been ongoing in the U.S. since 1961.[38] False

Understanding Food and Nutrition Labels

CHAPTER 4

NUTRITION SCOREBOARD

1 Nutrition labels are required on all foods and dietary supplements sold in the United States. **True/False**

2 Nutrition labeling rules allow health claims to be made on the packages of certain food products. **True/False**

3 A food product labeled "less sugar" means the product contains very little sugar. **True/False**

4 Nutrition labels contain all of the information people need to make healthy decisions about what to eat. **True/False**

Answers can be found at the end of the chapter.

After completing Chapter 4 you will be able to:

- Apply knowledge about the four key elements of nutrition labeling to decisions about the nutritional value of foods.

- Evaluate nutrient content and health claims made on dietary supplement labels.

- Compare the characteristics of organically and conventionally produced food products.

- Identify strengths and weaknesses of various nutrition labeling systems on food packaging and calorie listings for food items.

Illustration 4.1 Nutrition information for produce, fish, and seafood can be presented on posters.[a]

Nutrition Labeling

- **Apply knowledge about the four key elements of nutrition labeling to decisions about the nutritional value of foods.**

Misleading messages, hazy health claims, and the slippery serving sizes that characterized food labels in the past led to a revolution in nutrition labeling. Consumers—especially those responsible for buying food for the family; people with weight concerns, food allergies, or diabetes; and the health-conscious—made it clear they wanted to end the mystery about what's in many foods. Passage of the 1990 Nutrition Labeling and Education Act by Congress indicated that their concerns had been heard. In 1993 the Food and Drug Administration (FDA) published rules for nutrition labeling; implementation and revision of the standards continue.[2]

KEY NUTRITION CONCEPTS

Content in this unit on nutrition labeling directly relates to three key nutrition concepts:

1. Nutrition Concept #2: Foods provide energy (calories), nutrients, and other substances needed for growth and health.

2. Nutrition Concept #4: Poor nutrition can result from both inadequate and excessive levels of nutrient intake.

3. Nutrition Concept #6: Malnutrition can result from poor diets and from disease states, genetic factors, or combinations of these factors.

Key Elements of Nutrition Labeling Standards

Nutrition labeling regulations cover four major areas:

- The Nutrition Facts panel
- Nutrient content claims
- Health claims
- Structure/function claims

The Nutrition Facts panel is required on most foods sold in grocery stores. Rules established for the other three areas must be followed when a claim about the product is made on the packaging.

The Nutrition Facts Panel With the exception of foods sold in very small packages or in stores like local bakeries and of alcohol-containing beverages such beer, wine, and liquor, foods containing more than one ingredient must display a Nutrition Facts panel. Single fresh foods, such as pears, a head of cabbage, or fresh shrimp, do not have to be labeled, but grocery stores are encouraged to present nutrition information on posters (Illustration 4.1).

Nutrition Facts panels provide specific information about a food's serving size, the calorie value and nutrient content of a serving, and ingredients. Illustration 4.2 shows a Nutrition Facts panel and provides explanations of what various components of the panel mean. The panel highlights the fat, saturated fat, *trans* fat, cholesterol, sodium, carbohydrate, dietary fiber, sugars, vitamins A and C, and calcium and iron content of a serving of food.

All labeled foods must provide the information shown on the Nutrition Facts panel in Illustration 4.2. Additional information

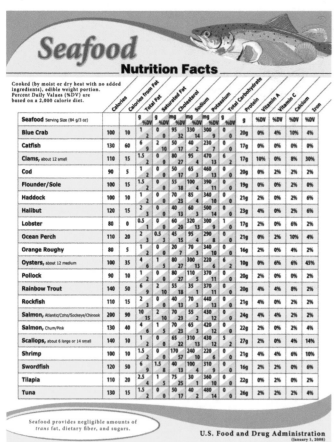

Seafood Nutrition Facts

Cooked (by moist or dry heat with no added ingredients, edible weight portion. Percent Daily Values (%DV) are based on a 2,000 calorie diet.

Seafood Serving Size (84 g/3 oz)	Calories	Calories from Fat	Total Fat (g / %DV)	Saturated Fat (g / %DV)	Cholesterol (mg / %DV)	Sodium (mg / %DV)	Potassium (mg / %DV)	Total Carbohydrate (g / %DV)	Protein (g)	Vitamin A %DV	Vitamin C %DV	Calcium %DV	Iron %DV
Blue Crab	100	10	1 / 2	0 / 0	95 / 32	330 / 14	300 / 9	0 / 0	20g	0%	4%	10%	4%
Catfish	130	60	6 / 9	2 / 10	50 / 17	40 / 2	230 / 7	0 / 0	17g	0%	0%	0%	0%
Clams, about 12 small	110	15	1.5 / 2	0 / 0	80 / 27	95 / 4	470 / 13	6 / 2	17g	10%	0%	8%	30%
Cod	90	5	1 / 2	0 / 0	50 / 17	65 / 3	460 / 13	0 / 0	20g	0%	2%	2%	2%
Flounder/Sole	100	15	1.5 / 2	0 / 0	55 / 18	100 / 4	390 / 11	0 / 0	19g	0%	0%	2%	0%
Haddock	100	10	1 / 2	0 / 0	70 / 23	85 / 4	340 / 10	0 / 0	21g	2%	0%	2%	6%
Halibut	120	15	2 / 3	0 / 0	40 / 13	60 / 3	500 / 14	0 / 0	23g	4%	0%	2%	6%
Lobster	80	0	0.5 / 1	0 / 0	60 / 20	320 / 13	300 / 9	1 / 0	17g	2%	0%	6%	2%
Ocean Perch	110	20	2 / 3	0.5 / 2	45 / 15	95 / 4	290 / 8	0 / 0	21g	0%	2%	10%	4%
Orange Roughy	80	5	1 / 2	0 / 0	20 / 7	70 / 3	340 / 10	0 / 0	16g	2%	0%	4%	2%
Oysters, about 12 medium	100	35	4 / 6	1 / 5	80 / 27	300 / 13	220 / 6	6 / 2	10g	0%	6%	6%	45%
Pollock	90	10	1 / 2	0 / 0	80 / 27	110 / 5	370 / 11	0 / 0	20g	2%	0%	0%	2%
Rainbow Trout	140	50	6 / 9	2 / 10	55 / 18	35 / 1	370 / 11	0 / 0	20g	4%	4%	8%	2%
Rockfish	110	15	2 / 3	0 / 0	40 / 13	70 / 3	440 / 13	0 / 0	21g	4%	0%	2%	2%
Salmon, Atlantic/Coho/Sockeye/Chinook	200	90	10 / 15	2 / 10	70 / 23	55 / 2	430 / 12	0 / 0	24g	4%	4%	2%	2%
Salmon, Chum/Pink	130	40	4 / 6	1 / 5	70 / 23	65 / 3	430 / 12	0 / 0	22g	2%	0%	2%	4%
Scallops, about 6 large or 14 small	140	10	1 / 2	0 / 0	65 / 22	310 / 13	430 / 12	5 / 2	27g	2%	0%	4%	14%
Shrimp	100	10	1.5 / 2	0 / 0	170 / 57	240 / 10	220 / 6	0 / 0	21g	4%	4%	6%	10%
Swordfish	120	50	6 / 9	1.5 / 8	40 / 13	100 / 4	310 / 9	0 / 0	16g	2%	2%	0%	6%
Tilapia	110	20	2.5 / 4	1 / 5	75 / 25	30 / 1	360 / 0	0 / 0	22g	0%	2%	0%	2%
Tuna	130	15	1.5 / 2	0 / 0	50 / 17	40 / 2	480 / 14	0 / 0	26g	2%	2%	2%	4%

Seafood provides negligible amounts of *trans* fat, dietary fiber, and sugars.

U.S. Food and Drug Administration
(January 1, 2008)

[a]Download posters from the FDA that show nutrition information for the twenty most frequently consumed raw fruits, vegetables, and fish in the United States by searching the term: Nutrition Information for Raw Fruits, Vegetables, and Fish.

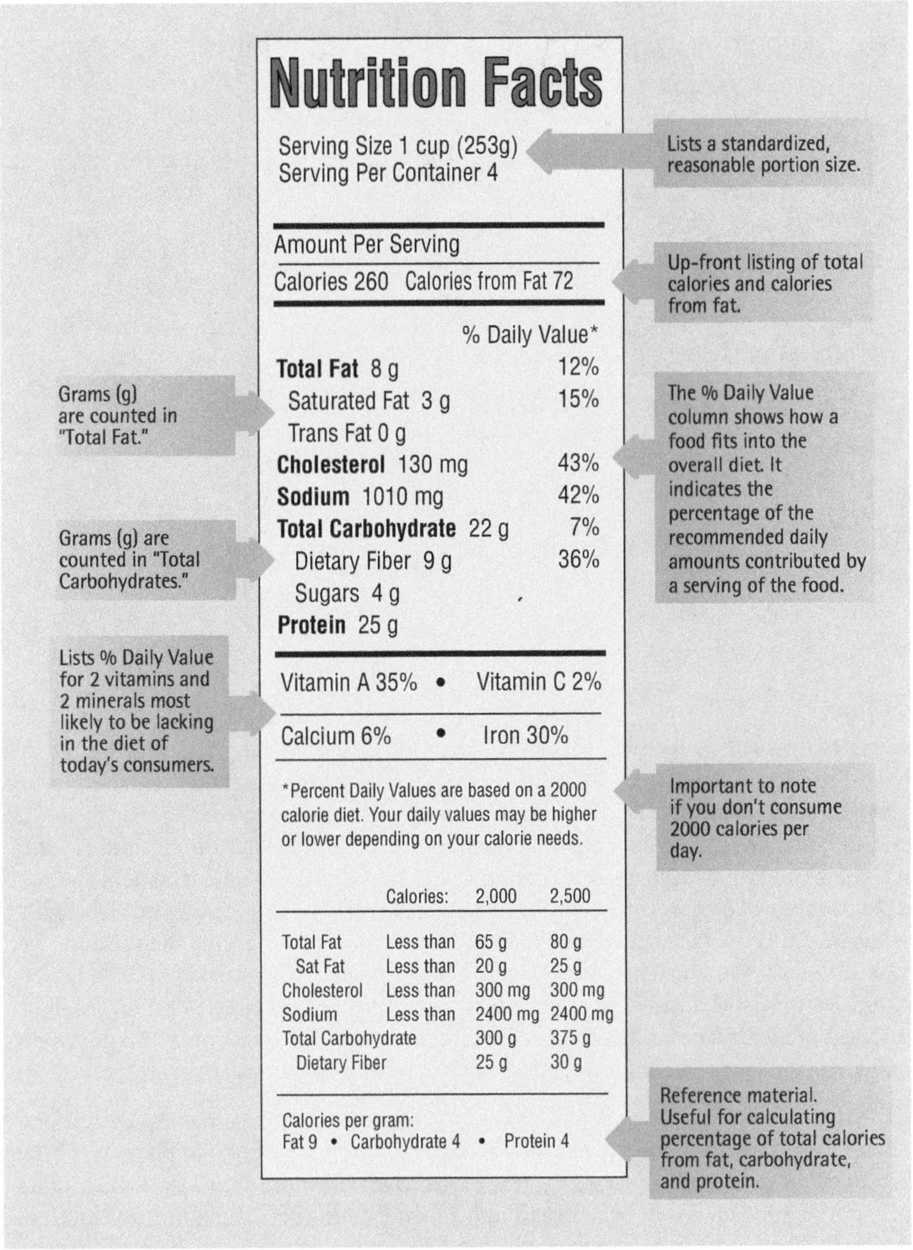

Illustration 4.2 Inside the Nutrition Facts panel.

Nutrition Facts

Serving Size 1 cup (253g)
Serving Per Container 4

> Lists a standardized, reasonable portion size.

Amount Per Serving

Calories 260   Calories from Fat 72

> Up-front listing of total calories and calories from fat.

% Daily Value*

Total Fat 8 g	12%
Saturated Fat 3 g	15%
Trans Fat 0 g	
Cholesterol 130 mg	43%
Sodium 1010 mg	42%
Total Carbohydrate 22 g	7%
Dietary Fiber 9 g	36%
Sugars 4 g	
Protein 25 g	

> Grams (g) are counted in "Total Fat."

> Grams (g) are counted in "Total Carbohydrates."

> The % Daily Value column shows how a food fits into the overall diet. It indicates the percentage of the recommended daily amounts contributed by a serving of the food.

Vitamin A 35%	•	Vitamin C 2%
Calcium 6%	•	Iron 30%

> Lists % Daily Value for 2 vitamins and 2 minerals most likely to be lacking in the diet of today's consumers.

*Percent Daily Values are based on a 2000 calorie diet. Your daily values may be higher or lower depending on your calorie needs.

> Important to note if you don't consume 2000 calories per day.

	Calories:	2,000	2,500
Total Fat	Less than	65 g	80 g
Sat Fat	Less than	20 g	25 g
Cholesterol	Less than	300 mg	300 mg
Sodium	Less than	2400 mg	2400 mg
Total Carbohydrate		300 g	375 g
Dietary Fiber		25 g	30 g

Calories per gram:
Fat 9 • Carbohydrate 4 • Protein 4

> Reference material. Useful for calculating percentage of total calories from fat, carbohydrate, and protein.

on specific nutrients such as vitamin D, vitamin E, folate, and potassium can be added to the panel on a voluntary basis. However, if the package makes a claim about the food's content of a particular nutrient that is not on the "must have" list, then information about that nutrient has to be added to the Nutrition Facts panel. Nutrition Facts panels also contain a column headed **% Daily Value (%DV)**. Figures given in this column are intended to help consumers answer such questions as "Does a serving of this macaroni and cheese provide more protein than the other brand?" and "How much fiber does this cereal provide compared to my daily need for it?"

Daily Values (DVs) are standard amounts of nutrients developed specifically for use on nutrition labels. They are based on an earlier edition of the Recommended Dietary Allowances.[3] The %DV figures listed on labels represent the percentages of the standard nutrient amounts obtained from one serving of the food product. Standard values for total fat, saturated fat, and carbohydrates are based on a daily intake of 2,000 calories. In general, % Daily Values of 5 or less are considered "low," Daily Values of 10 to 19% are considered "good," and those listed as 20% or more "high" sources of the nutrient listed on the label.[2]

% Daily Value (%DV) Daily Values are scientifically agreed-upon standards of daily intake of nutrients from the diet developed for use on nutrition labels. The "% Daily Values" listed in nutrition labels represent the percentages of the standards obtained from one serving of the food product.

Some Examples of What "Front of the Package" Nutrient-Content Claims Must Mean[4,5]

Term	Means that a serving of the product contains:
More	At least 10% more of the Daily Value for a vitamin, mineral, protein, dietary fiber, or potassium
Good source	From 10 to 19% of the Daily Value for a particular nutrient
High	20% or more of the Daily Value for a particular nutrient
Less	At least 25% less of a nutrient or of calories than appropriate reference foods (similar products)
Less sugar	At least 25% less sugar than appropriate reference foods (similar products)
Low calorie	40 calories or less
Low sodium	140 grams or less sodium
Low fat	3 grams or less of fat
Free	No, or negligible amounts of, fat, sugars, *trans* fat, or sodium
Gluten free	Less than 20 parts per million of gluten

Table 4.1 Examples of undefined claims used on food labels

Natural
All natural
Pure
Antibiotic-free
Raised without antibiotics
Additive-free
Pesticide-free
Hormone-free
Nutritionally improved
No cholesterol (on plant foods)
Free-range
Eco-friendly
Pasture-fed
Contains whole grains
Made with real fruit
Dairy-free
Probiotic
Vegan
Free of fructose
Real cane sugar
Agave nectar

Nutrient Content Remember seeing the labels "High Fiber," "Good Source," or "Gluten-Free" on food packages? These are examples of nutrient content claims, and they are usually placed on the front of food packages. The claims are used to characterize the level of calories and nutrients in a serving of a food product. (The Health Action feature in this unit contains additional information on nutrient claim criteria.) Nutrient content claims must be accurate and conform to specific criteria developed by the FDA. Foods labeled "low sugar," for example, must contain at least 25% less sugar per serving than similar products. Low-fat foods can be labeled with a "percent fat free" label, such as "98% fat-free" turkey. This label means that the product contains approximately 2% fat on a weight basis. Meat products labeled "lean" must contain fewer than 10 grams of fat, 4.5 grams of saturated fat and *trans* fat combined, and 95 milligrams of cholesterol per serving.

Evaluating Undefined Content Claims Other claims, such as "natural," "hormone-free," and "pure" are sometimes placed on food labels to imply a particular product is better than others not so labeled (Table 4.1). These terms have not been defined by the FDA, may or may not represent a health benefit, and may mislead consumers about the health value of products. The word *natural*, for example, can be found on labels for foods such as potato chips, butter-flavored microwave popcorn, and ice cream (see Illustration 4.3). About 60% of consumers look for the "natural" label on food packages and often assume that the label means the food has no artificial ingredients, pesticides, or **genetically modified organisms (GMOs)**. It is also commonly assumed that animals raised for meat and poultry and sold as "natural" were not given growth hormones or antibiotics. This is not the case.[6]

A label may declare that a product is "low-sugar, "made with real fruit," or "made with whole grains." Products labeled as low-sugar may be high in fat and calories, and the fruit or whole-grain content of foods labeled "made with real fruit" or "made with whole grains" may be as little as a gram of fruit or whole grains per serving. A look at the ingredients label will help you estimate how much sugar, fruit, or whole grains are in a product because ingredients are listed by weight. If fruit, for example, is listed as the first or second ingredient, the product likely contains a significant amount of fruit. If the fruit is listed further down, you know it probably doesn't have much fruit in it.

Health Claims In 1984 the Kellogg Company launched an ad campaign for All-Bran cereal that announced "eating the right foods may reduce your risk of some kinds of cancer." Sales of the high-fiber cereal increased 37% in one year, but then Kellogg had to withdraw the ads. The FDA ruled the All-Bran statement was equivalent to a claim for a drug. The campaign, however, started the nutrition and health claims revolution.[7,8]

On approval by the FDA, foods or food components with scientifically agreed-upon benefits for disease prevention can be labeled with a health claim (Table 4.2). However, the health claim must be based on the FDA's "model claim" statements. For instance, scientific consensus holds that diets high in fruits and vegetables may lower the risk of cancer, so a health claim to this effect is allowed. The FDA's model claim for labeling fruits and vegetables is "Low-fat diets rich in fruits and vegetables may reduce the risk of some types of cancer, a disease associated with many factors." The FDA approves health claims only for food products that are not high in fat, saturated fat, cholesterol, or sodium. Model health claims approved by the FDA are shown in Table 4.3.

Labeling Foods as Enriched or Fortified The vitamin and mineral content of foods can be increased by **enrichment** and **fortification**. Definitions for these terms were established more than 50 years ago. Enrichment pertains only to refined grain products, which lose vitamins and minerals when the germ and bran are removed during processing. Enrichment replaces the thiamin, riboflavin, niacin, and iron lost with the germ and bran. By law, producers of bread, cornmeal, pasta, crackers, white rice, and other products made from refined grains must use enriched flours. Beginning in 1998, federal regulations mandated that folate (a B vitamin) in the form of folic acid be added to refined grain products. This regulation was put into effect and is helping to reduce the incidence of a particular type of inborn, structural problem in children (neural tube defects) related to low blood levels of folate early in pregnancy.[9]

Any food product can be fortified with vitamins and minerals—and many are. One of the few regulations governing the fortification of foods is that the amount of vitamins and minerals added must be listed in the Nutrition Facts panel. Illustration 4.4 shows some examples of enriched and fortified foods.

Food enrichment and fortification began in the 1930s to help prevent deficiency diseases such as rickets (vitamin D deficiency), goiter (from iodine deficiency), pellagra

Photo by Judith Brown

Illustration 4.3 Here's the ingredient list for the ice cream labeled "Made with natural ingredients": skim milk, cream, sugar, corn syrup, whey, molasses, acacia gum, guar gum, ground vanilla beans, vanilla extract, carob bean gum, carrageenan, xanthan gum.

genetically modified organisms (GMOs) Food products that contain genetic material (usually from a bacteria) that has been transferred from the genetic material of another organism to give the food specific properties when produced.

enrichment The replacement of the thiamin, riboflavin, niacin, and iron lost when grains are refined.

fortification The addition of one or more vitamins and/or minerals to a food product.

Table 4.2 FDA-approved health claims[7]

1. Calcium, vitamin D, and osteoporosis
2. Dietary lipids (fats) and cancer
3. Dietary saturated fat and cholesterol and risk of coronary heart disease
4. Dietary non-cariogenic carbohydrate sweeteners and dental caries
5. Fiber-containing grain products, fruits, vegetables, and cancer
6. Folic acid and neural tube defects
7. Fruits, vegetables, and cancer
8. Fruits, vegetables, and grain products that contain fiber, particularly soluble fiber, and risk of coronary heart disease
9. Sodium and hypertension
10. Soluble fiber from certain foods and risk of coronary heart disease
11. Soy protein and risk of coronary heart disease
12. Stanols/sterols and risk of coronary heart disease

Scott Goodwin Photography

Table 4.3 Examples of model health claims approved by the FDA for labels of foods that qualify based on nutrient content; model claims are often abbreviated on food labels[a]

Food and related health issues	Model health claim
Whole-grain foods and heart disease, certain cancers	Diets rich in whole-grain foods and other plant foods and low in fat, saturated fat, and cholesterol may reduce the risk of heart disease and certain cancers.
Sugar alcohols and tooth decay	Frequent between-meal consumption of food high in sugars and starches promotes tooth decay. Xylitol, the sugar alcohol in this food, may reduce the risk of tooth decay.
Saturated fat, cholesterol, and heart disease	Development of heart disease depends on many factors. Eating a diet low in saturated fat and cholesterol and high in fruits, vegetables, and grain products that contain fiber may lower blood cholesterol levels and reduce your risk of heart disease.
Calcium, vitamin D, and osteoporosis	Regular exercise and a healthy diet with enough calcium and vitamin D help maintain good bone health and may reduce the risk of osteoporosis later in life.
Fruits and vegetables and cancer	Low-fat diets rich in fruits and vegetables may reduce the risk of some types of cancer, a disease associated with many factors.
Folate and neural tube defects	Women who consume adequate amounts of folate daily throughout their childbearing years may reduce their risk of having a child with a brain or spinal cord defect.

[a]Nutrition labeling regulations exist in many countries. If you travel to Europe, you may notice that the health claim "helps maintain normal bowel function" is allowed on prune juice.[16]

(niacin deficiency), and iron-deficiency anemia.[10–12] Today, foods are increasingly being fortified for the purpose of reducing the risk of chronic diseases such as osteoporosis, cancer, and heart disease. Other foods, such as snack bars, fruit drinks, and sweetened ready-to-eat cereals, are being fortified with an array of vitamins and minerals. Regular consumption of fortified foods increases the risk that people will exceed Tolerable Upper Intake Levels (ULs) of nutrients designated in the Recommended Dietary Allowances (RDAs). Regular use of multiple vitamin and mineral supplements, along with liberal intake of fortified foods, enhances the likelihood that excessive amounts of some nutrients will be consumed.[12]

The Ingredient Label Still more useful information about the composition of food products is listed on ingredient labels. The label of any food that contains more than one ingredient must list the ingredients in order of their contribution to the weight of the food (Table 4.4).

The FDA requires that ingredient labels include the presence of common food allergens in products. Potential food allergens include milk, eggs, fish, shellfish, tree nuts, wheat, peanuts,

Illustration 4.4 Examples of foods that are enriched (*left*) and fortified (*right*).

Table 4.4 Almost all foods with more than one ingredient must have an ingredient label. Here are the rules.

1. Ingredients must be listed in order of their contribution to the weight of the food, from highest to lowest.

2. Beverages that contain juice must list the percentage of juice on the ingredient label.

3. The terms *colors* and *color added* cannot be used. The name of the specific color ingredients must be given (for example, caramel color, turmeric).

4. Milk, eggs, fish, and five other foods to which some people are allergic must be listed on the label.

and soybeans. These foods, sometimes called the "Big Eight," account for 90% of food allergies. Precautionary statements such as "may contain," or "processed in a facility with" are allowed.[13]

Food Additives on the Label Specific information about **food additives** must be listed on the ingredient label. About 10,000 chemical substances may be added to food to enhance its flavor, color, texture, cooking properties, shelf life, or nutrient content. Food additives considered GRAS (*Generally Recognized As Safe*) by the FDA can be used in food without preapproval. Food dyes, however, must be approved by the FDA prior to their use in food.[14] Specific ingredients allowed on the GRAS list change with time as new evidence becomes available. Caffeine, *trans* fats, added sugar, and salt, currently on the GRAS list, are being considered for removal by the FDA due to safety concerns. It is possible that these ingredients will require premarketing approval prior to use or carry a warning label in the future.[17]

Table 4.5 provides examples of the functions of some additives used in food. The most common food additives are sugar and salt, but trace amounts of polysorbate, potassium benzoate, and many other additives that are not so familiar are also included in foods. Appendix D lists many of the most common additives and indicates their function in foods.

Trace amounts of substances such as pesticides; hormones and antibiotics given to livestock; fragments of packaging materials such as plastic, wax, aluminum, or tin; very small fragments of bone; and insects may end up in foods. These are considered "unintentional additives" and do not have to be included on the label.

Irradiated Foods Food irradiation is an odd example of a food additive. It is actually a process that doesn't add anything to foods. Irradiation uses X-rays, gamma rays, or electron beams to kill insects, bacteria, molds, and other microorganisms in food. Food irradiation

food additives Any substances added to food that become part of the food or affect the characteristics of the food. The term applies to substances added both intentionally and unintentionally to food.

Table 4.5 Solving the mystery of ingredient label terms[a]

Cake mix ingredient label		Additive	Function
	Ingredients: Sugar, enriched flour bleached (wheat flour, niacin), iron, thiamin mononitrate (vitamin B$_1$), riboflavin (vitamin B$_2$), vegetable shortening (contains partially hydrogenated soybean cottonseed oil), sodium aluminum phosphate, dextrose, leavening (baking soda, monocalcium phosphate, dicalcium phosphate, aluminum sulfate), wheat starch, propylene glycol monoesters, modified corn starch, salt, egg white, vanilla, dried corn syrup, polysorbate 60, nonfat milk, **mono-** and **diglycerides**, sodium citrate, xanthan gum, soy lecithin.	Sodium aluminum phosphate	Gives baked products a light texture
		Propylene glycol monoesters	Helps blend ingredients uniformly, enhances moisture content and texture
		Mono- and diglycerides	Maintains product softness after baking
		Xanthan gum	Thickening agent, helps hold product together after baking

Note: All foods with more than one ingredient must have an ingredient label.
[a]See Appendix D for more information on the definitions of ingredient terms.

Illustration 4.5 The "radura," as the symbol is called, must be displayed on irradiated foods. The words "treated by irradiation, do not irradiate again" or "treated with radiation, do not irradiate again" must accompany the symbol.

dietary supplement Any product intended to supplement the diet, including vitamins, minerals, proteins, enzymes, herbs, hormones, and organ tissues. Such products must be labeled "Dietary Supplement."

structure/function claim Statement appearing primarily on dietary supplement labels that describes the effect a supplement may have on the structure or function of the body. Such statements cannot claim to diagnose, cure, mitigate, treat, or prevent disease.

Nutrition Facts
Per 125mL (87g)

Amount	% Daily Value
Calories 80	
Fat 0.5 g	1%
Saturated 0 g + Trans 0 g	0%
Cholesterol 0 mg	
Sodium 0 mg	0%
Carbohydrate 18 g	6%
Fibre 2 g	8%
Sugars 2 g	
Protein 3 g	
Vitamin A 2%	Vitamin C 10%
Calcium 0%	Iron 2%

Scott Goodwin Photography

Illustration 4.7 Dietary supplement label.

enhances the shelf life of food products and decreases the risk of food-borne illness. Irradiation must be performed according to specific federal rules.[15]

Irradiated foods retain no radioactive particles. The process is like having your luggage X-rayed at the airport. Your luggage doesn't become radioactive, nor has anything in it changed. The process leaves no evidence of having occurred. Actually, this lack of change creates a challenge because inspection agencies can't determine whether a food has been irradiated or not. All irradiated foods—except spices that are added to processed foods—are required to display the international "radura" symbol and to indicate that the food has been irradiated (Illustration 4.5). Irradiation is approved for use on chicken, turkey, pork, beef, eggs, grains, fresh fruits and vegetables, and other foods in the United States.[15]

Genetically Modified Organisms (GMOs) GMOs contain selected, individual genes transferred from one organism, usually a bacteria, to another. It is generally done to improve a crop's resistance to diseases, increase production, improve drought resistance, or enhance the nutrient content of the GMO product. GMO foods have been consumed by humans since the mid-1990s. They have passed risk assessment tests and have not been found to pose risks to human health. Consequently, the FDA determined that the labeling of GMO foods is not warranted and GMO foods do not need to be labeled in the United States.[24,25] Other countries do require GMO foods to be labeled, based primarily on concerns such as crop biodiversity and the control of seed crops by a few agricultural companies (Illustration 4.6). Some people are fearful about consuming genes, but they are a component of most of the food we eat, are digested, and do not function as genes in our bodies.[24,26]

Dietary Supplement Labeling

- **Evaluate nutrient content and health claims made on dietary supplement labels.**

A **dietary supplement** is a product taken by mouth that contains a "dietary ingredient" intended to supplement the diet. The dietary ingredients in these products include vitamins, minerals, proteins, enzymes, herbs, hormones, and organ tissues. Nutrition labeling regulations place dietary supplements in a special category under the general umbrella of "foods," not drugs. Dietary supplements differ from drugs in that they do not have to undergo rigorous testing and obtain FDA approval before they are sold. In return, dietary supplement labels cannot claim that the products treat, cure, or prevent disease.[18]

According to FDA regulations, dietary supplements must be labeled as such and include a "Supplement Facts" panel that lists serving size, ingredients, and percent Daily Value (%DV) of essential nutrient ingredients (Illustration 4.7). Like foods, qualifying dietary supplements can be labeled with nutrient claims (e.g., "high in calcium"), and health claims (e.g., "diets low in saturated fat and cholesterol that include sufficient soluble fiber may reduce the risk of heart disease"). Dietary supplement labels can also include **structure/function claims.**

INGREDIENTS: SUGAR, WATER, MAIZE FLOUR (PRODUCED FROM GENETICALLY MODIFIED MAIZE), EGG, FLAVOURINGS.

Illustration 4.6 Example of GMO labeling in the UK for a corn (maize) product.[26]

Structure or Function Claims

Dietary supplements can be labeled with statements that describe effects the supplement may have on body structures or functions (shown in Illustration 4.8). Under this regulation, for example, certain supplements can be labeled with the following statements: "Promotes healthy heart and circulatory function," "Helps maintain mental health," or "Supports the immune system." If a structure/function claim is made on the label or package inserts, the label or insert must acknowledge that the FDA does not support the claim: *"This statement has not been evaluated by the FDA. This product is not intended to diagnose, treat, cure, or prevent any disease."* Dietary supplements labeled with misleading or untruthful information and those that are not safe can be taken off the market and the manufacturers fined.

Structure and function claims do not have to be approved before they can appear on product labels. The FDA and FTC have taken action against companies responsible for fraudulent claims.[18] The FDA has issued new guidelines that define the quality of scientific evidence needed to substantiate structure/function claims.[19]

The COOL Rule

USDA requires retailers to display a country-of-origin label (COOL) on certain products (Illustration 4.9). The rule is meant to expand informed consumer choices and to help track down food-borne illness outbreaks. The rule applies to meats, fish, seafood, fruits, vegetables, many nuts, and some herbs.[20]

Organic Foods

- **Compare the characteristics of organically and conventionally produced food products.**

For many years, consumers and producers of organic foods urged Congress to set criteria for the use of the term *organic* on food labels. Consumers wanted to be assured that foods were really organic, and producers wanted to keep the business honest. Standards were needed because consumers cannot distinguish organically produced foods from others by looking at them, tasting them, or reading their nutrient values. The U.S. Department of Agriculture developed and is implementing standards for organic foods.[15,21]

Labeling Organic Foods

Rules that qualify foods as organic are shown in Illustration 4.10. If organic growers and processors qualify according to USDA-approved certifying organizations, they can place the green-and-white USDA Organic seal on product labels (Illustration 4.11). The Canadian organic label, also shown in Illustration 4.11, is equivalent to the USDA

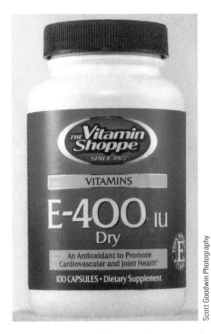

Illustration 4.8   Example of a structure/function claim.

Illustration 4.9   Country-of-origin labels are now required on certain foods.

Illustration 4.10   USDA rules for qualifying foods as organic.

1. Plants
- Must be grown in soils not treated with synthetic fertilizers, pesticides, and herbicides for at least three years
- Cannot be fertilized with sewer sludge
- Cannot be treated by irradiation
- Cannote be grown from genetically modified seeds or contain genetically modified ingredients

2. Animals
- Cannot be raised in "factory-like" confinement conditions
- Cannot be given antibiotics or hormones to prevent disease or promote growth
- Must be given feed products that are 100% organic

United States Department of Agriculture

Canada Organic Regime (COR), Canadian Food Inspection Agency (CFIA)

Illustration 4.11 Foods certified as organic by the USDA can display the "USDA Organic" seal on food packages. The approved seal for Canada is below it.

Organic label. The USDA and Canadian authorities can impose financial penalties on companies that use the seal inappropriately. Organically grown and produced foods can be labeled in four other ways:

1. "100% Organic" if they contain entirely organically produced ingredients.

2. "Organic" if they contain at least 95% organic ingredients.

3. "Made with organic ingredients" if they contain at least 70% organic ingredients.

4. "Some Organic Ingredients" if the product contains less than 70% organic ingredients.

Some people choose organic foods because they perceive them to be better for the environment and for health than conventionally grown foods.[22] Most organic foods, however, are not totally free of some of these ingredients. Due to "pesticide drift" from sprayed crops, past use of pesticides on farmland, and the leaching of chemicals used on crops into groundwater, organically grown plants may contain traces of pesticides. There are positive and negative characteristics of both organic and conventional foods, and so far neither type has been shown to be superior overall in nutrient content, safety, or effects on health.[22,23]

Other Nutrition Labeling Systems

- **Identify strengths and weaknesses of various nutrition labeling systems on food packaging and calorie listings for food items.**

If you are a nutrition label reader, you have probably noticed that a variety of food company–initiated labeling systems appear on the front of food packages. Three of these, "Nutrition at a Glance," "Smart Choices Made Easy," and the "Smart Choices Program," are shown in Illustration 4.12. The Smart Choices Program features a symbol that identifies more nutritious choices within specific food product categories, and lists the calories per serving and number of servings per package. This system has been selected for use by a number of food companies in Europe and the United States.[27] Other nutrition labeling systems, including those from the American Heart Association and various grocery store chains, are being used on food product packaging.

Industry-initiated labeling systems are voluntary and generally employ science-based criteria for judging the nutrient qualities of food products.[28] The labels supplement the Nutrition Facts panel of the food products. The primary weaknesses of the food industry–backed labeling systems are that they may cause confusion among consumers, criteria used to label foods vary, they may favor the company's products, and the labels only appear on selected food products.[27,29]

Labeling systems in place may not adequately emphasize information consumers need to know to help curb high-calorie intake and obesity. This concern is being addressed through increased emphasis on the labeling of calories in portions of food sold in chain restaurants.

Illustration 4.12 A few examples of food industry–initiated nutrition labeling systems used on food packages.

Scott Goodwin Photography

Calories on Display

Since 2012, several large cities and states have implemented legislation that requires chain restaurants with more than 20 locations to post the calorie value of a serving of each standardized menu item offered (Illustration 4.13). Details regarding the item's content of 11 other nutrients must be available on request. This calorie labeling system for chain restaurants was adopted for nationwide implementation beginning in 2012. The labeling requirement also extends to food offered in vending machines for establishments with 20 or more locations nationwide.[30]

Study results indicate that calorie labeling of restaurant menu items is associated with no change or somewhat lower-calorie food choices among diners, and increased availability of lower-calorie menu options in some fast food restaurants.[31,32]

Illustration 4.13 An example of the required calorie labeling of chain restaurant menu items.

Upcoming Nutrition Label Revisions

With the exception of the addition of *trans* fats to the label in 2006, few major changes have occurred in nutrition labeling in the United States. Science marches on, however, and knowledge about the behavioral science behind the design of effective communication tools has advanced. To bring nutrition labeling standards up to date, the FDA began the process of revising the labels in 2013. The new labels will reflect advances in nutrition science and national dietary intake recommendations, reality-based serving size standards, and an improved design.[3,4]

The major changes proposed for the new labels include:

1. Reality-based serving sizes. The serving size listed on the Nutrition Facts Panel would represent the average portion consumed by people over the age of 2 in the United States. A standard serving of ice cream, for example, would change from ½ cup to 1 cup.

2. Calorie content of a serving of the food would be highlighted in large type on the Nutrition Facts Panel.

3. *Added sugars* would be placed on the Nutrition Facts Panel under *Sugars*. It is proposed that added sugars be defined as "sugar or sugar-containing ingredient that is added during processing."

4. The vitamin D and potassium content of a serving of the food would be required on the Nutrition Facts Panel because intake of these two nutrients tends to be low. Vitamins A and C would no longer be required because intakes are rarely low.

5. Updated Daily Values (DVs) for vitamin D and sodium would be applied for calculating %DV of these nutrients. Establishing a DV for added sugars of less than 10% of total calories and reporting %DV, grams, and teaspoons of added sugar per serving on the label have also been proposed.[15]

REALITY CHECK
Nutrition Labeling

Foods labeled as "fat free" have few or no calories.

Who gets the thumbs up?

Answers on page 88.

Jared: Right. You take the fat out of food, and calories go away.

Ronald: Maybe, maybe not. "Fat-free" foods could still contain sugars, protein, and other ingredients that have calories.

Ronald got it. "Fat free" does not equal "calorie free." Fat-free Caesar salad dressing, for example, provides 40 calories in a 2-tablespoon serving.

Jared:

Ronald:

The new standards are expected to be fully implemented by 2017.[3] A sample of the Nutrition Facts Panel highlighting the proposed changes appears in Illustration 4.14.

Beyond Nutrition Labels

Even with the new labels, consumers can't be stupid.
—MAX BROWN

Illustration 4.14 The proposed Nutrition Facts Panel.

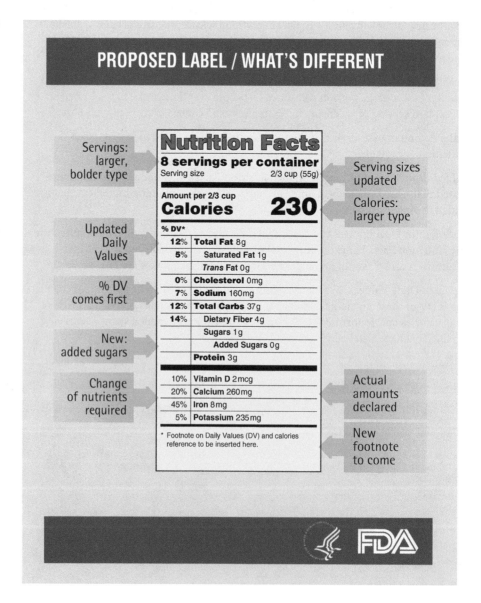

Understanding and applying the information on nutrition labels calls for more nutrition knowledge on the public's part. As highlighted in the Reality Check, people need to know about nutrition before they can understand nutrition labels and incorporate labeled foods appropriately into an overall diet. Good diets include more than foods with nutrition labels. The ice cream cone from the stand in the mall; the orange, potato, or fish we buy in the store; and the pizza delivered to the dorm are unlabeled parts of many diets. We need to know enough about the composition of unlabeled foods to fit them into a healthy diet. Nutrition labels often list a few vitamins and minerals, but many more are required for health. It's particularly important for people to know if their diet is varied enough to supply needed vitamins and minerals. In addition, nutrient and health claims made on food packaging do not address potentially negative aspects of food products, such as "low fiber," "high saturated fat," or "may promote tooth decay." Use of the MyPlate food group guide should go hand in hand with label reading for diet planning.

Not every food we eat has to have the "right" label profile. Serving mostly low-fat or low-calorie foods to children, for example, might have unintended unhealthy effects. Young children need calories and fat for growth and development. If diets are severely restricted, growth and development will be impaired.

Nutrition labels are an important tool for helping people make informed food-purchasing decisions. About 6 in 10 adults use them to guide their food purchasing decisions, and those who do use them tend to consume healthier diets than those who don't.[23] However, labels do not now—nor will they ever—provide all the information needed to make wise decisions about food. Only people who are well informed about nutrition can do that.

NUTRITION
up close

Understanding Food Labels

Focal Point: Understanding nutrition labels so you know what you are getting.

Are you ready to put your knowledge of nutrition labels to use? Examine the front-of-package, facts panel, and ingredient labels shown in the Illustrations below. Then answer the following questions:

Courtesy of Brown

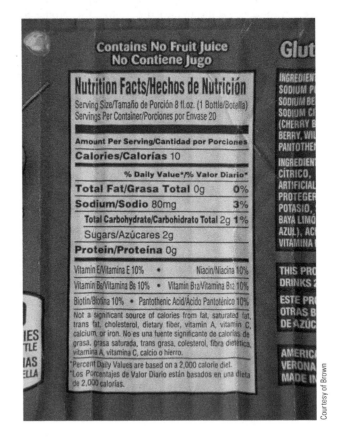

Courtesy of Brown

1. Fruit drinks are not part of a healthy dietary pattern. Does the lower sugar content and the addition of B vitamins and vitamin E to a fruit drink change that?

2. Does the product contain any fruit or fruit juice?

3. The product is fortified with biotin, pantothenic acid, niacin, vitamin B_6, and vitamin B_{12}. Which of these vitamins are included in the group of vitamins considered to be of public health significance?

4. The fruit drink is labeled gluten-free even though there is no reason to expect that a fruit drink would contain gluten. Why do you think this label was added?

5. If nutrition labels were required to label both the positive and negative characteristics of a product, which positive and negative characteristics would you attribute to this product?

Feedback to the Nutrition Up Close is located in Appendix F.

....

REVIEW QUESTIONS

- **Apply knowledge about the four key elements of nutrition labeling to decisions about the nutritional value of foods.**

1. Almost all multiple-ingredient foods must be labeled with nutrition information.   True/False

2. Food manufacturers can list any serving size they want on Nutrition Facts panels.   True/False

3. In general, %DV listed for nutrients in Nutrition Facts panels of 10% or more are considered "low," and those listed as 50% or more are considered "high."   True/False

4. An overriding principle of nutrition labeling regulations is that nutrient content and health claims made about a food on the packaging must be truthful.   True/False

5. The term *enriched* on a food label means that extra vitamins and minerals have been added to the food to bolster its nutritional value.   True/False

6. The ingredient that makes up the greatest portion of a food product's weight must be listed first on ingredients labels.   True/False

7. Food labeling regulations do *not* require food manufacturers to list the presence of major food allergens on ingredient labels.   True/False

8. ___ A tea manufacturer labels a green tea product as "high in the antioxidant vitamin C." After brewing, a serving of the green tea was found to contain 10 mg of vitamin C. Which of the following statements about the green tea nutrient content claim is true?
   a. The green tea qualifies for the "high in" nutrient content claim made.
   b. The green tea qualifies for a "natural" content claim.
   c. The green tea qualifies for a "healthy" nutrient content claim.
   d. The green tea qualifies for a "good source of" nutrient content claim.

9. ___ You see a yogurt product at the grocery store labeled "fat free." This nutrient content claim means that a standard serving of the yogurt contains:
   a. No, or negligible amounts of, fat.
   b. No, or negligible amounts of, fat, *trans* fat, and sodium.
   c. Fewer than 10 grams of fat and 4.5 grams of saturated fat.
   d. Three grams of fat or less.

10. ___ John eats three cookies for a bedtime snack while reading the Nutrition Facts panel on the cookie package. He notices that a three-cookie serving provides 2 grams of saturated fat and 0 grams of *trans* fat. The cookie qualifies for a nutrient content claim of:
   a. Low saturated fat.   c. Extra lean.
   b. *Trans* fat free.   d. None of the above.

Questions 11 through 14 refer to the following case scenario.

You have just been informed by your health care provider that your cholesterol level is "borderline high" and that you need to lose 10 pounds. The health care provider recommends that you reduce your calorie intake by 400 calories a day, cut back on saturated fats, and come back to the clinic in three months.

Here's your plan on how to decease calories and saturated fat: You will cut back on your intake of snack food. You have been eating three 1-ounce servings of potato chips, and two 1-ounce servings of cheese crackers nightly. You notice from the Nutrition Facts panels on the chips and crackers that each ounce of potato chips (15 chips) provides 160 calories, and each ounce of cheese crackers (25 crackers) provides 140. Also, one serving of potato chips provides 1 gram (5% Daily Value) of saturated fat, and one serving of cheese crackers has 1.5 grams (8% Daily Value) of saturated fat.

11. How many calories were you consuming from the chips and crackers daily? ___ calories

12. If you reduced your intake of both potato chips and cheese crackers to one serving per day, you would reduce your intake from these snack foods by ___ calories.

13. The amount of saturated fat in terms of grams and % Daily Value provided by three servings of potato chips and two servings of cheese crackers is ___ grams and ___ % Daily Value.

14. If you decrease your intake of potato chips and cheese crackers to one ounce of each a day, you will reduce your saturated fat intake from these snacks by ___ grams and by ___ % Daily Value.

- **Evaluate nutrient content and health claims made on dietary supplement labels.**

15. Dietary supplements, such as those for herbs and vitamins, are considered drugs and are not regulated by FDA's Nutrition Labeling rules.   True/False

16. Foods, but *not* dietary supplements, can be labeled with nutrient content and health claims.   True/False

17. ___ Assume a dietary supplement made from pomegranate is labeled with the claims "lowers plaque formation in the arteries" and "improves blood flow." It turns out that the FDA becomes aware of the claims and subsequently sues the maker of the dietary supplement. What would be the most likely reason for the FDA suit?
   a. There are no approved nutrient claims for "lowers plaque formation" and "improves blood flow."
   b. Dietary supplement labels cannot claim that a product treats a disease.
   c. There was insufficient research to support the claims.
   d. The FDA did not preapprove the claim before it was made for the juice.

- **Compare the characteristics of organically and conventionally produced food products.**

18. Animals providing meats labeled organic cannot be given antibiotics or hormones.   True/False

19. Foods bearing the USDA Organic seal are certified as organic by the U.S. Department of Agriculture.   True/False

- **Identify strengths and weaknesses of various nutrition labeling systems on food packaging and calorie listings for food items.**

20. Nutrition labels provide all the nutrition information we need to make healthful food choices.   True/False

Answers to these questions can be found in Appendix F.

NUTRITION SCOREBOARD ANSWERS

1. Labeling is required for almost all processed foods and for dietary supplements, but remains largely voluntary for fresh fruits, vegetables, and fish.[1,2] False

2. Health claims for food products are allowed on many food packages. The claims must be truthful and adhere to FDA standards. True

3. Less sugar doesn't necessarily mean *low-sugar*. The "less sugar" label means the labeled product has at least 25% less sugar per serving than comparable products. It still may contain a good deal of sugar.[2] False

4. Nutrition labels are necessarily short and can't tell the whole story about food and health. They help people make several key decisions about a food's composition. False

Nutrition, Attitudes, and Behavior

CHAPTER 5

NUTRITION SCOREBOARD

1 Food preferences are primarily determined by genetic factors. **True/False**

2 Food habits never change. **True/False**

3 Sweet, sour, salty, and bitter tastes can be sensed over all parts of the tongue. **True/False**

4 Children's acceptance of a wide variety of vegetables and fruits can be increased by frequently offering them an assortment of vegetables and fruits. **True/False**

Answers can be found at the end of the chapter.

- Identify factors that influence an individual's food choices and preferences.

- Apply the process for making healthful changes in food choices to a specific change in food intake.

- Differentiate between scientifically supported and unsupported conclusions about diet and behavior relationships.

Origins of Food Choices

- **Identify factors that influence an individual's food choices and preferences.**

Horse meat is a favorite food in a large area of north-central Asia. Pork, which is widely consumed in North and South America, Europe, and other areas, is rigidly avoided by many people in Islamic countries. Bone-marrow soup and sautéed snails are delicacies in France, while kidney pie is traditional in England. Dog is a popular food in Borneo, New Guinea, the Philippines, and other countries, whereas snake is a delicacy in China. Some people enjoy insects (Illustration 5.1). And then there are steamed clams and raw oysters—food passions for some, but absolutely disgusting to others.[3]

When did you first think "yecck!"? The food choices just described would elicit that response among people from a variety of cultures, but they would not necessarily be responding to the same foods.

Why do people eat what they do? People learn from their family and the society in which they live what animals and plants are considered food and which are not.[4] Once items are identified as food, they develop a legacy of strong symbolic, emotional, and cultural meanings. Comfort foods, health foods, junk foods, fun foods, soul foods, fattening foods, mood foods, and pig-out foods, for example, have been identified in the United States. All cultures have their "super food": in Russia and Ireland, it's potatoes; in Central America, it's corn and yucca (a starchy root, also called *manioc*); in Somalia, it's rice. The designation refers to the cultural significance of the food and not to its nutritional value.[5]

In countries such as the United States where a wide variety of foods are available and people have the luxury of selecting which foods they will eat, food choices are influenced by a range of factors (Illustration 5.2). Of these factors, food preference has the largest impact. Food preferences vary a good deal among individuals and lead to a wide array of specific food choices. Rather than being inborn, food preferences are primarily learned.[7]

Illustration 5.1 Grasshoppers are Mexican delicacies, served at Girasoles Restaurant in Mexico City.

Gavriel Jecan/Photodisc/Getty Images

KEY NUTRITION CONCEPTS

Our examination of the factors involved in food choices relates to three of the key nutrition concepts introduced in Chapter 2:

1. Nutrition Concept #1: Food is a basic need of humans.

2. Nutrition Concept #7: Some groups of people are at higher risk of becoming inadequately nourished than others.

3. Nutrition Concept #8: Poor nutrition can influence the development of certain chronic diseases.

We Don't Instinctively Know What to Eat

The food choices people make are not driven by a need for nutrients or guided by food selection genes. People deficient in iron, for example, do not seek out iron-rich foods. If we're overweight, no inner voice tells us to reject high-calorie foods. Women who are pregnant don't instinctively know what to eat to nourish their growing fetuses. No evidence indicates that young children, if offered a wide variety of foods, would select and ingest a well-balanced diet.[8]

Humans are born with mechanisms that help them decide when and how much to eat, however.[7] An inborn attraction to sweet-tasting foods, a dislike for the taste of bitter or sour, and the response of thirst when water is needed all influence food and fluid intake to an extent (Illustration 5.3).[6,9] There is evidence to suggest that people deficient in sodium experience an increased preference for salty foods.[7]

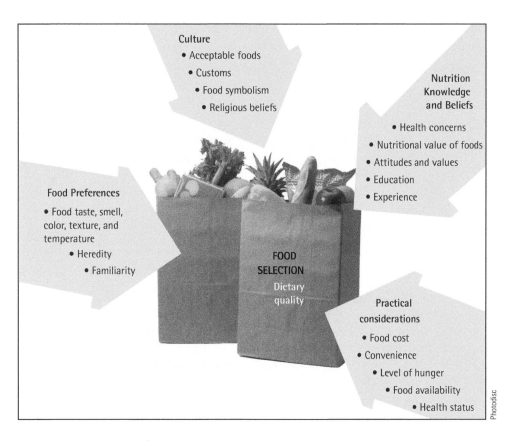

Illustration 5.2 Factors influencing food selection and dietary quality. Each of these sets of factors interacts with the others.[6]

Culture
- Acceptable foods
- Customs
- Food symbolism
- Religious beliefs

Nutrition Knowledge and Beliefs
- Health concerns
- Nutritional value of foods
- Attitudes and values
- Education
- Experience

Food Preferences
- Food taste, smell, color, texture, and temperature
- Heredity
- Familiarity

FOOD SELECTION
Dietary quality

Practical considerations
- Food cost
- Convenience
- Level of hunger
- Food availability
- Health status

Photodisc

Food Preferences

You're going to eat that!?! Do you have any idea what's in that hot dog?

Yeah, I do. There's barbecue in the backyard, ball games with my mom and dad, and the parties at my friend's house. The memories taste great!

The strong symbolic, emotional, and cultural meanings of food come to life in the form of food preferences. We choose foods that, based on our cultural background and other learning experiences, give us pleasure. Foods give us pleasure when they relieve our hunger pains, delight our taste buds, provide the right feeling of texture in our mouth, or give us comfort and a sense of security. We find foods pleasurable when they outwardly demonstrate our superior intelligence, our commitment to total fitness, or our pride in our ethnic heritage. We reject foods that bring us discomfort, guilt, and unpleasant memories, and those that run contrary to our values and beliefs.[10]

The Symbolic Meaning of Food Food symbolism, cultural influences, and emotional reasons for food choices are broad concepts that may become clearer with concrete examples. Here are a few examples to consider.

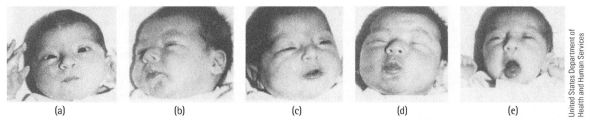

(a)   (b)   (c)   (d)   (e)

United States Department of Health and Human Services

Illustration 5.3   Newborn infants respond to different tastes: (a) baby at rest, (b) tasting distilled water, (c) tasting sugar, (d) tasting something sour, and (e) tasting something bitter.
Source: Taste-induced facial expressions of neonate infants from the classic studies of J. E. Steiner, in Taste and Development: *The Genesis of Sweet Preference*, ed. J. M. Weiffenbach, HHS Publication no. NIH 77-1068 (Bethesda, MD: U.S. Department of Health and Human Services, 1977), pp. 173–89, with permission of the author.

Status Foods Vance Packard, in his book *The Status Seekers*, provides a memorable example of the symbolic value of food:

> As a lad, this man had grown up in a poor family of Italian origin. He was raised on blood sausages, pizza, spaghetti, and red wine. After completing high school, he went to Minnesota and began working in logging camps, where—anxious to be accepted—he soon learned to prefer beef, beer, and beans, and he shunned "Italian" food. Later, he went to a Detroit industrial plant, and eventually became a promising young executive. . . . In his executive role he found himself cultivating the favorite foods and beverages of other executives: steak, whiskey, and seafood. Ultimately, he gained acceptance in the city's upper class. Now he began winning admiration from people in his elite social set by going back to his knowledge of Italian cooking, and serving them, with the aid of his manservant, authentic Italian treats such as blood sausage, spaghetti, and red wine![11]

Comfort Foods Ice cream, apple pie, chicken noodle soup, boxed chocolates, meat loaf and mashed potatoes: these are popular comfort foods in the United States. The feelings of security and love that came along with the tea and honey or chicken soup that your mother or father gave you when you had a cold, or with the ice cream and popsicles lovingly given to soothe a sore throat, are renewed with comfort foods. Some comfort foods can bring pleasure and reduce anxiety just by their image (Illustration 5.4).[12,14]

Once the symbolic value of a food is established as a comfort food, its nutritional value will remain secondary.[13] Food status is a strong determinant of food choices; and after all, as a noted nutritionist once said, "Life needs a little bit of cheesecake."[14]

"Discomfort Foods" Memories of bad experiences with food and expectations that certain foods will harm us in some way, each contribute to our learning about food and affect our food preferences. Eating a piece of blueberry pie right before an attack of the flu or overindulging in sweet pickles or olives, for example, may take these foods off your preferred list for a long time. Children who have had the experience of not being able to leave a table until they ate their green peas (or another food) may hold on to that "discomforting" memory for a lifetime.

Cultural Values Surrounding Food A team of scientists observed that the diet of certain groups in the Chin States of Upper Burma was seriously deficient in animal protein. After considerable study, a way was found to improve the situation by cross-breeding the small, local black pigs raised by the farmers with an improved strain to obtain progeny giving a greater yield of meat. The entire operation, however, completely failed to benefit the nutrition of the population because of one fact, which had been viewed as irrelevant. The cross-bred pigs were spotted. And it was firmly believed—as firmly as we believe that to eat, say, mice, would be disgusting—that spotted pigs were unfit to eat.[3]

Dietary change introduced into a culture for the purpose of improving health can be successful only if it is accepted by the culture. Cultural norms are not easily modified.[4]

Other Factors Influencing Food Choices and Preferences
Food preferences and selections are also affected by the desire to consume foods that are considered healthy. Reducing fat intake, eating more fruits and vegetables, and cutting down on sweets bring rewards and pleasures such as weight loss and maintenance, an end to constipation, lower blood cholesterol level, and a newly discovered preference for basic foods.[15]

Food Cost and Availability Food choices are affected by the cost of food. Researchers found that college students eating in dining halls have better diets when they prepay for their meals for the entire term rather than paying at each meal.[16] The cost of vegetables and fruits in grocery stores affects intake; the higher the price, the lower the intake. Intake of vegetables and fruits tends to increase when they become cheaper to buy.[17]

Jon Edwards Photograph/Age Fotostock

Illustration 5.4 Do you find comfort in this photo?

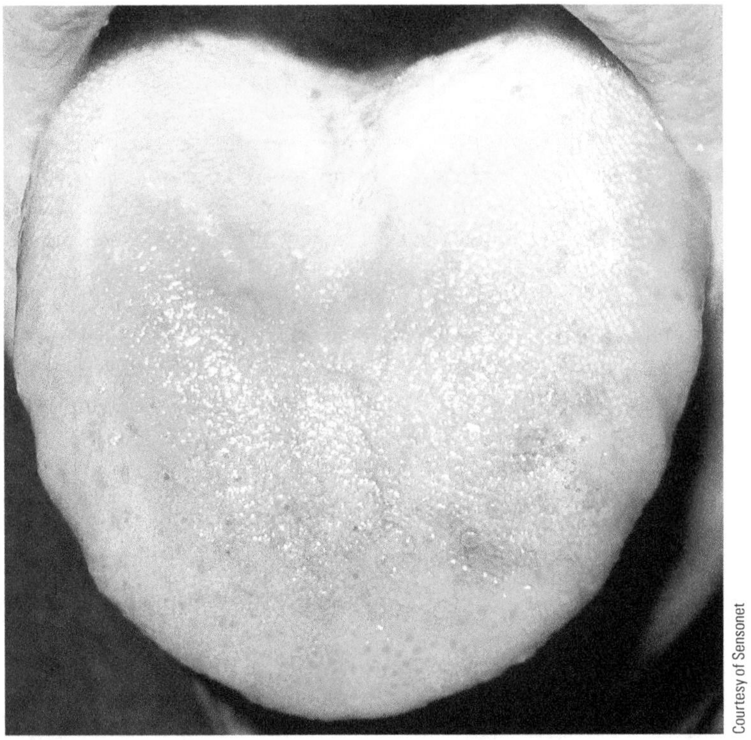

Illustration 5.5 This tongue is colored with blue food dye to show the small "bumps" on the tongue called papillae. Taste receptor cells are located in taste buds within the papillae.[1]

Taste Sensitivities and Food Preferences Taste is a major factor influencing food choices and appears to affect the reinforcing value of food and body weight. Sensitivity to the basic tastes varies among individuals. The differences appear to be strongly influenced by genetic factors.[18]

Five basic tastes have been identified: sweet, salty, bitter, sour, and umami (savory or meaty).[17] (Evidence that a sixth basic taste exists for fatty is growing.[19,20]) When food is consumed, taste receptor cells located primarily in tiny taste buds all over the tongue transmit to the brain information on the sensation and intensity of the basic tastes present (Illustration 5.5). The perceived intensity of the tastes will vary among individuals, and that is largely due to genetic differences. Some people, for example, are very sensitive to a bitter taste (supertasters), others will detect it to a limited extent (medium tasters), and still other people will not taste bitter at all (nontasters).[1] Supertasters tend not to eat vegetables such as brussels sprouts, cabbage, and broccoli, and to dislike bitter-tasting teas, wine, and tonic water.[21,22] About 25% of the U.S. population are supertasters, 50% are medium tasters, and 25% are nontasters.[23] Adults with above-average taste intensity scores appear to be more likely to gain weight over time than others with lower scores.[18]

Taste is closely aligned with smell, and genetic variations in the sense of smell also influence food preferences. A dislike of cilantro or arugula, for example, is related to a perceived adverse smell of the food. Most people immediately dislike and will decide not to eat a food that that tastes very bitter or sour, or that smells "bad."[24]

Food Preferences and Choices Do Change

- **Apply the process for making healthful changes in food choices to a specific change in food intake.**

Who says old dogs can't learn new tricks? Most Americans aren't eating the way they used to. Recent surveys indicate that food choices made in the United States are trending toward healthier dietary patterns. On average, people are consuming more vegetables, fish, chicken, whole grains, and berries and less sugar, soft drinks, salt, and *trans* fats than in the past. Caloric intake is on a downward trend, as are portion sizes and the frequency of dining at fast food restaurants.[26-28] The American consumer is increasingly interested in flavorful ethnic and spicy foods, fresh vegetables, whole grains such as quinoa, brown and wild rice, and nuts and seeds. Restaurants are taking this development seriously, with menu offerings of appetizing, lower-calorie, plant-centered dishes that appeal to health-conscious consumers (Illustration 5.6).[26,29,30]

Food choices are largely learned and do change with time as we learn more about foods and health. Individuals who did not like foods such as oysters, asparagus, beets, mushrooms, or sushi as children sometimes discover that they really like these foods later in life.[25] Children's acceptance of a wide variety of vegetables and fruits can be increased by frequently offering them an assortment of vegetables and fruits.[2] Perhaps your food choices have changed over time. How do the food choices you make now compare with the choices you made five years ago?

How Do Food Choices Change?

What are the ingredients for change in food choices? Why do some people succeed in improving their food choices while other people find that very hard to do? Nutrition knowledge, attitudes, and values have a lot to do with changing food choices for the better.

Nutrition Knowledge and Food Choices Sound knowledge about good nutrition necessarily precedes the selection of a healthful diet. But is knowledge enough to ensure that healthy changes in diet will be made? The answer is "yes" for some people and "no" for others.[6] Many people know far more about the components of a good diet than they put into practice, but between knowledge and practice lie multiple beliefs and experiences that act as barriers to change (Illustration 5.7). Change of any type is most likely to succeed when the benefits of making the change outweigh the disadvantages. This makes changes in food choices a very individual decision, with each person deciding whether a change is in his or her best interest.[6] But what sorts of circumstances, in addition to

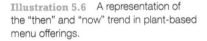
Illustration 5.6 A representation of the "then" and "now" trend in plant-based menu offerings.

©stockcreations/Shutterstock.com

| "I feel guilty about eating the foods I like." | "Eating right is too expensive." | "I tried eating better, but I didn't stick with it." | "I don't have the time to eat right." | "The vegetables I like aren't available." | "I'm healthy now... Why should I worry about my diet?" |

Photodisc

Illustration 5.7   Why knowledge about a good diet may not be enough to improve food choices.

increased knowledge, make changes in food choices worthwhile for individuals—and even highly desired?

Nutrition Attitudes, Beliefs, and Values   The value individuals place on diet and health is reflected in the food choices they make. A survey of restaurant patrons found that food choices varied according to the consumer's perceptions of the importance of diet to health.[31]

- "Unconcerned" consumers—people who are not concerned about the connection between diet and health and who tend to describe themselves as "meat and potato eaters"—select foods for reasons other than health.

- "Committed" consumers believe that a good diet plays a role in the prevention of illness. They tend to consume a diet consistent with their commitment to good nutrition.

- "Vacillating" consumers—people who describe themselves as concerned about diet and health but who do not consistently base food choices on this concern—tend to vary their food choices depending on the occasion. These consumers are likely to abandon diet and health concerns when eating out or on special occasions, but they generally adhere to a healthy diet.

Avoiding illness and curing or diminishing current health problems are likewise strong incentives for changing food choices (Table 5.1). In almost all instances, the key to lasting improvement in diet is to make changes you feel confident you can make, and can do without a lot of "will power." Changes that bring you more pleasure than inconvenience or discomfort have staying power, while those that require a lot of will power to maintain rarely last.[6]

Successful Changes in Food Choices

The primary reason efforts to improve food choices fail is that the changes attempted are too drastic. Improvements that last tend to be the smallest acceptable changes needed to do the job.[15]

The Process of Changing Food Choices   Assume you need to lose weight and want to keep it off by modifying your food choices. A promising plan to accomplish this goal would begin by identifying food choices you would like to change and lower-calorie food options you would be willing and able to eat (Table 5.2). Then you could make the plan more specific by identifying the changes that would be easiest to implement. For example, assume the low-calorie foods you like include frozen nonfat yogurt and oranges. You might decide to eat yogurt or an orange in place of your usual bedtime snack of ice cream. A specific dietary change such as this is much easier to implement than a broad notion,

Table 5.1 Factors that enhance food intake changes[6]

- Attitude that nutrition is important
- Belief that diet affects health
- Perceived susceptibility to diet-related health problems
- Perception that benefits of change outweigh barriers to change
- Confidence that the behavior change can be made

Table 5.2 Changing food choices for the better[6,15]

The process	An example
1. Identify a healthful change in your diet you'd like to make.	1. I'd like to lower my fat intake.
2. Identify two food choices you make that should change because they contribute to the need for the healthful change you identified.	2. I eat at fast food restaurants three times a week. I usually have a large order of fries and once a week I eat fried chicken.
3. Identify two or more specific, acceptable options for more healthful food choices than the ones identified under number 2.	3. Options identified: • Order tossed salad with low-calorie dressing instead of fries. • Eat a grilled chicken sandwich every other week instead of fried chicken. • Eat Mexican fast food more often.
4. Decide which option is easiest to accomplish and requires the smallest change to get the job done.	4. I love Mexican food. It would be easy to eat tacos instead of french fries or fried chicken.
5. Plan how to incorporate the change into your diet.	5. Mondays and Fridays, I'll eat tacos.
6. Implement the change. Be prepared for midcourse corrections.	6. Midcourse correction: On Fridays, when I'm with my friends, it's easier to eat at the restaurant they like. I'll order the grilled chicken sandwich and coleslaw.

such as "eat less." Although weight loss will take a while, such a small acceptable change has a much better chance of working than a drastic change in diet.[15]

Planning for Relapses When making a change in your diet, be prepared for relapses. Relapses happen for a number of reasons, and they don't mean the attempt has failed. People often return to old habits because the change they attempted was too drastic or because they tried to make too many changes at once. If the change undertaken doesn't work out, rethink your options and make a midcourse correction.

Strategies for Improving Diets in Groups of People Improvements in the diets of people in the United States are needed to reduce the risk of obesity and diet-related diseases. A number of behavioral changes that help people in general eat better are:

- Making healthy, delicious, affordable food available at home, in restaurants, schools, colleges, at work, and in military settings

- Decreasing fast food restaurant dining

- Increasing family meals and home cooking

- Learning to cook

- Self-monitoring of calorie intake and weight

- Maintaining the healthful eating patterns of a home country after immigrating to the United States

Improvements in farm-to-table transfer of locally produced foods and reliable access to healthy, affordable foods also tend to improve dietary intakes.[15]

Does Diet Affect Behavior?

- **Differentiate between scientifically supported and unsupported conclusions about diet and behavior relationships.**

Food affects behavior in some rather striking ways. Irritable, hungry, crying infants rapidly change into cooing, sleepy angels after they are fed. Low-on-sleep employees perk up after their morning coffee. A high-calorie lunch makes some people feel calm and sleepy, and eating favorite foods with family and friends can make people feel happy.[32] Not only

Table 5.3 Examples of ways in which diet may affect behavior

Dietary characteristic	Behavioral outcomes
Malnutrition, growth stunting in early childhood	Lower intellectual functioning and school performance, increased antisocial behavior in childhood[33]
Nutritional supplementation of malnourished young children	Improved growth and intellectual functioning in adulthood[34]
Very low carbohydrate intake (less than 20 grams per day)	Reduced short-term memory, slower reaction times, increased attention span[35]
Ingestion of certain color dye additives	Moderate increases in hyperactivity and other behaviors in susceptible children[36,37]
Lead consumption	Higher risk of violent and aggressive behavior, hyperactivity, and mental and behavioral problems[38]
Iron deficiency in young children	Long-term deficits in learning ability and social skills[39]

do our behaviors affect our diets, but also our diets can affect our behaviors. Examples of associations between dietary characteristics and behavior are listed in Table 5.3. One common belief about food and behavior is the subject of this unit's Reality Check.

Malnutrition and Mental Performance

Like growth and health, mental development and intellectual capacity can be affected by diet. The effects range from mild and short term to serious and lasting, depending on when the malnutrition occurs, how long it lasts, and how severe it is. The effects are most severe when malnutrition occurs while the brain is growing and developing.

Severe deficiency of protein, calories, or both early in life leads to growth retardation, low intelligence, poor memory, short attention span, and social passivity. When the nutritional insult is early and severe, some or all of these effects may be lasting (Illustration 5.8).[40] In Barbados, for example, children who experienced protein-calorie malnutrition in the first year of life did not fully recover even with nutritional rehabilitation. Growth improved but academic performance did not. Compared to well-nourished children, those experiencing protein and calorie deficits during infancy were more likely to drop out of school and were four times more likely to have symptoms of attention deficit hyperactivity disorder (ADHD).[41]

Protein-calorie malnutrition that occurs later in childhood, after the brain has developed, produces behavioral effects that can be corrected with nutritional rehabilitation.

REALITY CHECK

Can Food Be a Love Potion?

Toward the end of a friend's birthday party, Glenda and Cassell got into a heated debate about the existence of food aphrodisiacs. Part of the conversation went like this:

Is it all in Glenda's head?

Answers on page 102.

Glenda: Chocolate and vanilla are natural love potions, there's no doubt about it. If I eat a chocolate truffle and spritz on vanilla flavoring like perfume before a date, the guy always goes nuts for me!

Cassell: Are you kidding? No way! I've heard about oysters and this tree bark that are supposed to work miracles, too. It's all in your head.

Illustration 5.8 Malnutrition in early childhood has long-lasting effects. Some children never fully recover.

Illustration 5.9 There are many opportunities for overexposure to lead. Young children are especially vulnerable.

Correction of other deficits that often accompany malnutrition, such as lack of educational and emotional stimulation and harsh living conditions, hastens and enhances recovery.[40]

Protein-calorie malnutrition severe enough to cause permanent delays in mental development rarely occurs in the United States. When it does, malnutrition is usually due to neglect or inadequate care giving. More common dietary events that impair learning in U.S. children are skipping breakfast, fetal exposure to alcohol, iron deficiency, and lead toxicity.

Early Exposure to Alcohol Affects Mental Performance Mental development can be permanently delayed by exposure to alcohol during fetal growth. Although growth is also retarded, the most serious effects of fetal exposure to alcohol are permanent delays in mental development and behavioral problems associated with them. Women are advised not to drink if they are pregnant or may become pregnant.[42]

Iron Deficiency Impairs Learning Most cases of iron-deficiency anemia in children result from inadequate intake of dietary iron. Iron-deficiency anemia in children is a widespread problem in developed and developing countries and likely is the most common single nutrient deficiency.[43] The potential impact of iron-deficiency anemia on the functional capacity of humans represents staggering possibilities.

Until recently, it was thought that the effects of iron-deficiency anemia on intellectual performance were short term and could be reversed by treating the anemia. Although some of the effects can be lessened by treatment early in life, some are lasting. Five-year-old children in Costa Rica treated for iron-deficiency anemia during infancy scored lower on hand–eye coordination and other motor skill tests than similar children without a history of anemia. Studies in the United States have identified fearful behavior, shyness, lack of playfulness, and reduced problem-solving ability in iron-deficient children.[39,44]

Overexposure to Lead There are many opportunities for overexposure to lead. Approximately 84% of U.S. houses built before 1980 contain some lead-based paint.[45] Children living in or near these houses may eat the paint flakes (they taste sweet), or the old paint may contaminate the soil near the houses (Illustration 5.9). Lead also ends up in soil from industrial and agricultural chemicals, in water from lead-based pipes and solder, and in the air from the days when leaded gas was used.

Although the use of lead in cans, pipes, and gasoline has decreased dramatically, lead remains in the environment for long periods. Lead also stays in the body, stored principally in the bones, for a long time—20 years or more. It takes over a year of treatment to reduce blood lead levels. The effects of excessive exposure to lead in children include increased absenteeism from school, impaired reading skills, higher dropout rates, and increased aggressive behavior.[46] Occupational exposure to lead in adults is associated with poor kidney function and the risk of hypertension.[47] Blood lead levels in children have dropped substantially in recent decades. Despite this drop, 500,000 young children in the United States still have elevated blood lead levels.[48]

Food Additives, Sugar, and Hyperactivity The notion that certain food additives are related to hyperactivity in children has been popular since the related Feingold Hypothesis was announced in the mid-1970s. Since then, multiple studies have been undertaken to test whether the hypothesis is true. In a study involving a large group of healthy 3-, 8-, and 9-year-old children, intake of a beverage containing four food colorants (types of yellow, orange, and red color additives) and a preservative (sodium benzoate) was related to development of hyperactivity. Signs of hyperactivity detected in the children consuming the beverage containing the food dyes included overactivity, impulsiveness, and short attention spans. Not all children consuming the beverage demonstrated hyperactive behavior, however, and the same effects were observed in children with or without attention deficit hyperactivity disorder.[36] It appears that some children may be vulnerable to the effects of these food additives while others are not.[36,37]

Studies examining the effects of sugar intake on hyperactivity in children have not demonstrated that such a relationship exits.[49] The excitement that often accompanies high–sugar eating occasions such as Halloween and birthday parties—or the expectation that sugar causes hyperactivity—may be responsible for the reported effect.[50] (See Illustration 5.10.)

The Future of Diet and Behavior Research

Identifying the effects of nutrition on behavior is a tricky business. Many factors in addition to diet influence behavior, making it difficult to separate diet from social, economic, educational, and genetic factors. We still have much to learn, and many assumptions about diet and behavior must await confirmation through research.

Illustration 5.10 Do food additives or lots of sugary foods make children hyperactive?

9 Denis Rozhnovsky/ Shutterstock.com

NUTRITION
up close

Improving Food Choices

Focal Point: Developing a plan for healthier eating.

Identify a change in your diet that you would like to make. Then develop a plan for making the change by thinking through and responding to each element of the dietary change process listed. (Refer to Table 5.2 for examples of responses.)

Dietary Change Process

1. Identify a healthful change in your diet you'd like to make.

2. Identify two food choices you make that should change because they contribute to the need for the healthful change you identified.

3. Identify two specific, acceptable options for food choices more healthful than those identified under number 2.

4. Decide which option is easiest to accomplish and requires the smallest change to get the job done.

5. Plan how to incorporate the change into your diet.

Feedback to the Nutrition Up Close is located in Appendix F.

Your Response

1. _____

2. _____

3. _____

4. _____

5. _____

REVIEW QUESTIONS

- **Identify factors that influence an individual's food choices and preferences.**

1. Food preferences are universal—everyone likes the same foods.   True/False

2. Foods will be rejected by a population, no matter how nutritious they may be, if the foods don't fit into a culture's definition of what foods are appropriate to eat. True/False

3. Food choices are driven largely by a person's need for nutrients.   True/False

4. Potatoes, rice, corn, and yucca are examples of "super foods" in specific countries due to their cultural significance.   True/False

5. Nutrition knowledge is an important prerequisite for making healthful food choices.   True/False

6. _____ Emanuel is at a dinner party and the hostess asks, "Would you like to try the brussels sprouts? They're fresh from the garden and delicious!" Emanuel immediately responds, "No, thank you. I've tried brussels sprouts at least ten times before and they just don't taste good to me." Chances are good that Emanuel

   a. is being a picky eater.
   b. dislikes all vegetables.
   c. is upset about something.
   d. really does not like the taste of brussels sprouts.

7. _____ Last week Yu asked her neighbor where she should take her visiting parents to dinner. The neighbor recommends Al's Restaurant, saying it has "the best food in town." Later that week Yu spoke with her neighbor and the neighbor asked how her parents liked Al's Restaurant. Yu responds, "Well, actually, we thought it was awful. They used way too much garlic in everything." The neighbor is shocked. What's the most likely reason for this reaction?

   a. The neighbor assumed that because she likes the food at Al's, everyone will.
   b. Yu's family doesn't recognize good food.
   c. Yu's family members are genetically sensitive to the taste of garlic.
   d. The neighbor is unable to detect the presence of garlic in food.

- **Apply the process for making healthful changes in food choices to a specific change in food intake.**

8. Broad dietary changes, such as a decision to simply "eat less," are more likely to produce lasting behavioral changes than are small changes in diets, such as snacking on favorite fruits rather than candy.   True/False

9. Changes in dietary intake that are acceptable to an individual and easy to implement are the types of changes that are most likely to last.   True/False

10. Changing food choices for the better takes planning and includes individual decisions on specifically how the change in food choices will be implemented.   True/False

11. Individuals need to plan for modifying their approach to improving food choices because even the best-planned changes in food choices sometimes fail.   True/False

The next two questions refer to this scenario:

Assume your wife decides to increase her vegetable intake. She travels from place to place for work and eats out a lot. She loves the premade mixed greens salad and the single-serve yogurt dressing packet you can get at some gas station convenience stores. She decides she would happy to have that for lunch twice a week instead of her usual burger or pizza slices.

12. ____Does this plan have a chance of working?
    a. Yes, because she seems to be a person with great will power.
    b. Yes, she is planning a specific and acceptable change.
    c. No, the plan isn't likely to work because the change is too drastic.
    d. No, the plan won't work because she will grow tired of eating salads.

13. ____The salad-for-lunch plan worked so well that your wife decides she will lose 10 pounds in the next two months by only eating salads. Will this plan likely work?
    a. It has a good chance of working. She really loves salads.
    b. Yes, she won't be consuming very many calories.
    c. It's unlikely to work because it is too drastic a change in her diet.
    d. It probably won't work because she will start craving burgers and pizza.

- **Differentiate between scientifically supported and unsupported conclusions about diet and behavior relationships.**

14. Examples of ways in which dietary intake affects behavior include the relationship between sugar intake and hyperactivity in children.   True/False

15. Excessive exposure to lead during childhood is related to impaired reading skills and higher school dropout rates.   True/False

Answers to these questions can be found in Appendix F.

NUTRITION SCOREBOARD ANSWERS

1. Although genetics plays a role in food preferences, the predominant influences are environmentally determined.   **False**

2. The idea that food habits don't change is a myth.   **False**

3. Sweet, sour, salty, and bitter tastes are sensed by taste cells in all parts of the tongue, and not, as popularly thought, clustered by type of taste in different zones on the tongue. Taste receptor cells are also located in the back of the throat and nasal cavity.[1]   **True**

4. **True**[2]

Alistair Berg/Digital Vision/Getty Images

CHAPTER 6

Healthy Dietary Patterns, Dietary Guidelines, MyPlate, and More

NUTRITION SCOREBOARD

1 Healthy dietary patterns include the regular consumption of legumes, nuts, poultry, and whole-grain products. **True/False**

2 Over half of U.S. adults fail to consume three or more servings of vegetables and two or more servings of fruits daily. **True/False**

3 The basic food groups include a "healthy snack" group. **True/False**

4 "Value" fast food meals may provide double the calories of regular-sized fast food meals. **True/False**

Answers can be found at the end of the chapter.

- Apply the characteristics of healthy dietary patterns to the design of one.

- Identify characteristics of dietary patterns that promote health and those that do not.

- Utilize ChooseMyPlate.gov guidance materials and interactive tools for dietary planning and evaluation.

Healthy Eating: Achieving a Balance between Good Taste and Good for You

- **Apply the characteristics of healthy dietary patterns to the design of one.**

May I have your attention please? For a moment, think about the foods in Illustration 6.1. If your mouth is watering and you're ready to go out and buy some ripe peaches, you have found the balance between good taste and good for you. Who said foods that taste good aren't good for you?

KEY NUTRITION CONCEPTS

This unit explores what constitutes a healthy diet and ways of evaluating and assessing diets. The discussion directly relates to four key nutrition concepts:

1. Nutrition Concept #2: Foods provide energy (calories), nutrients, and other substances needed for growth and health.

2. Nutrition Concept #4: Poor nutrition can result from both inadequate and excessive levels of nutrient intake.

3. Nutrition Concept #8: Poor nutrition can influence the development of certain chronic and other diseases.

4. Nutrition Concept #9: Adequacy, variety, and balance are key characteristics of healthy dietary patterns.

Illustration 6.1 Can you smell it? Can you taste it? A plump, golden peach. It's so ripe that juice spurts from it and drips down your chin when you take a bite. A golden brown turkey just taken out of the oven. The wonderful smell fills the kitchen. A steaming loaf of homemade bread just set out to cool. A perfect ripe tomato just picked from the garden. It melts in your mouth.

Susie M. Eising/StockFood Creative/Getty Images

Photodisc

Photodisc

Wally Eberhart/Visuals Unlimited/Getty Images

Characteristics of Healthy Dietary Patterns

Healthy **dietary patterns** come in a variety of forms. They may include bread, olives, nuts, fruits, beans, vegetables, and lamb (as in Greece); rice, vegetables, fish, and seaweed (as in China); or black beans, rice, chicken, and tropical fruits (as in Cuba and Costa Rica). Illustration 6.2 gives examples of the diverse foods that can be part of a healthy dietary pattern. Although the types of foods that go into them can vary substantially, dietary guidelines support healthful dietary patterns that share four basic characteristics: adequacy, variety, balance, and health maintenance.[4,5]

Adequate diets include a wide variety of foods that together provide sufficient levels of calories and **essential nutrients**. What's considered sufficient in the United States and Canada? For calories, it's the number that maintains a healthy body weight. For essential nutrients, sufficiency corresponds to intakes that are in line with recommended intake levels represented by the *Recommended Dietary Allowances (RDAs)* and *Adequate Intakes (AIs)*. (Tables showing these levels appear on the inside covers of this book.) Recommended amounts of essential nutrients should be obtained from foods to reap the benefits offered by the variety of naturally occurring substances in foods that promote health.

Variety is a core characteristic of healthy diets because the nutrient content of food differs. Consumption of an assortment of foods from each of the basic food groups increases the probability that the diet will provide enough of them all. You could, for example, eat three servings of vegetables a day by eating only potatoes. But you would consume a much broader variety of vitamins, minerals, and beneficial substances in plants— such as antioxidants—if you consumed spinach and tomatoes along with the potatoes.

dietary patterns   The quantities, proportions, variety, or combination of different foods, drinks, and nutrients in diets, and the frequency with which they are habitually consumed. Also called "eating patterns".

adequate diet   A diet consisting of foods that together supply sufficient protein, vitamins, and minerals and enough calories to meet a person's need for energy.

essential nutrients   Substances the body requires for normal growth and health but cannot manufacture in sufficient amounts; they must be obtained in the diet.

variety   A diet consisting of many different foods from all of the food groups.

Illustration 6.2   Foods that contribute to healthy diets in different countries. (a) Dinner in Italy might be linguine primavera (pasta with vegetables). (b) Pad Thai, rice noodles and vegetables, is a favorite dish in Thailand. (c) Tamales are a celebration food in many Latin cultures. (d) Dal, curry dishes, vegetables, and chicken are popular parts of the cuisine of India.

a.

b.

c.

d.

Macronutrient	Carbohydrate	Added sugars*	Protein	Fat	Linoleic acid	Alpha-linolenic acid
AMDR	45–65%	≤10%	10–35%	20–35%	05–10%	0.6–1.2%
Average intake	52%	13%	15%	33%	6.3%	0.6%

*The 2015 Dietary Guidelines Advisory Committee concluded that total added sugar intake should be10% or less of total calories.[1]

Illustration 6.3 The match between Acceptable Macronutrient Distribution Ranges (AMDRs) and average intake by adults in the United States.[1,6]

A **balanced diet** provides calories, nutrients, and other components of food in the right proportion—neither too much nor too little. Diets that contain too much sodium or too little fiber, for example, are out of balance. Diets that provide more calories than needed to maintain a healthy body weight are also out of balance.

People need relatively large amounts of carbohydrates, proteins, fats, and water. These nutrients are collectively classified as **macronutrients**. Guidelines indicating the percentages of total caloric intake that should consist of carbohydrate, protein, and fat are listed in a DRI table labeled *Acceptable Macronutrient Distribution Ranges*, or *AMDRs*. AMDRs have also been set for linoleic acid and alpha-linolenic acid (the two essential fatty acids). Illustration 6.3 shows the acceptable ranges for carbohydrate, protein, fat, **added sugars**, and essential fatty acid intake, as well as the average intake levels of U.S. adults.

Another key attribute of healthy dietary patterns is health maintenance. As methods used to assess dietary patterns have advanced, researchers have been able to examine the relationship between complex dietary patterns and health.[5] Although the utility of dietary adequacy, moderation, and balance remains, it is important to consider the effects of overall long-term dietary pattern on health. People consume foods, not nutrients, and the nutrients and other components of food such as phytochemicals interact with each other in ways that can modify health outcomes over time. Several dietary patterns have been shown to reduce the risk of cardiovascular disease, obesity, hypertension, and type 2 diabetes compared to Western-style dietary patterns. Healthy dietary patterns provide levels of calories and nutrients needed for growth, health, and life. In particular, they provide adequate amounts of potassium, vitamin D, calcium, and fiber—nutrients that are most likely to be found in low amounts in the diets of people in the United States.[1] Only 14% of adults consume three or more servings of vegetables and two or more servings of fruits daily, and about 10% consume at least one serving of dark-green or colorful vegetable daily.[7]

Healthy Dietary Patterns Identified for the United States

Also called healthy "eating patterns," healthy dietary patterns most consistently associated with positive health and nutrient intake outcomes are the:

- Healthy Mediterranean-Style Dietary Pattern

- Healthy U.S. Dietary Pattern (represented by USDA's MyPlate recommendations)

- Healthy Vegetarian-Style Dietary Pattern

The foods represented in these dietary patterns, and their proportionate representation in pattern, vary somewhat. However, they all are associated with reduced the risk of obesity, type 2 diabetes, heart disease, and other diseases and disorders, and with improved nutrient intake compared to our existing Western-style dietary pattern.[1,4,5]

Healthy dietary patterns are anchored by plant foods. They are represented by the regular consumption of fruits, vegetables, **whole grains**, nuts, legumes, oils, low-fat dairy

balanced diet A diet that provides neither too much nor too little of nutrients and other components of food such as fat and fiber.

macronutrients The group name for carbohydrate, protein, fat, and water. They are called macronutrients because we need relatively large amounts of them in our daily diet.

added sugars Sugars that are either added during the processing of foods, or are packaged as such and include sugars (free, mono-, and disaccharides), syrups, naturally occurring sugars that are isolated from a whole food and concentrated, and other caloric sweeteners.

whole grains Cereal grains that consist of the intact, ground, cracked, or flaked kernel, which includes the bran, the germ, and the innermost part of the kernel (the endosperm).

products, poultry, lean meat, fish and seafood, and moderate alcohol consumption (by adults who choose to drink). There is room in these patterns for the occasional dessert or other treat, and individuals can decide which specific foods they will eat from the overall food types represented in the pattern. Table 6.1 summarizes the general characteristics of healthy dietary patterns. Two specific healthy dietary patterns, the Healthy U.S.-Style Dietary Pattern and Mediterranean-Style Dietary Pattern, are presented in Table 6.6. The table provides examples of these dietary patterns based on 2,000 calories. Recommendations based on 11 other calorie need levels are included in the 2015 Dietary Guidelines report.[30] The third healthy dietary pattern is the Vegetarian-Style Dietary Pattern.

National Guides for Healthful Diets

Due to the impact of food choices on the health of individuals and population groups, many countries establish recommendations for dietary intake. Such recommendations are periodically updated as new knowledge about diet and health emerges. Most of the guidelines include recommendations for physical activity and food safety as well.

National recommendations for diet and physical activity usually apply to children over the age of 2 and aren't appropriate for every individual in a population. General guidelines may not completely match the needs of individuals who, for example, are strict vegetarians, have food allergies, or have specific genetic traits that affect nutrient utilization. A basic premise of national dietary guidelines is that nutrient needs should be met through food consumption without reliance on dietary or nutrient supplements.[1,4]

Benefits of population-based dietary and physical activity recommendations are multiple. The information is science-based, free, and made widely available to the public on the Internet.

Table 6.1 Characteristics of healthy dietary patterns[1,4,5]

Healthy dietary patterns include the regular consumptions of:
• Fruits and vegetables
• Whole grains and whole-grain products, and other high-fiber foods
• Nuts, all types
• Oils (such as olive oil, seed and vegetable oils)
• Legumes (such as navy and pinto beans)
• Low-fat dairy products
• Poultry, lean meats
• Fish and seafood
• Alcohol in moderation (for adults who choose to drink).
Healthy dietary patterns are *not* characterized by the regular consumptions of:
• Foods and beverages high in added sugar such as sugar-sweetened soft drinks, fruit drinks, and sweetened teas
• Foods high in sodium such as some fast and processed foods
• Refined grain products such as white bread, rice cakes, pasta
• Processed meats such as salami and bologna
• Foods high in saturated fats, such as animal fat and tropical oils

Table 6.2 **Examples of dietary guidelines from around the world**[7,8]

Country	Example of dietary guidelines
Japan	Eat 30 or more different kinds of foods daily.
China	Eat clean and safe food.
Norway	FOOD + JOY = HEALTH.
United Kingdom	Encourage and support the production of lower saturated fat foods.
Mexico	Eat more dried beans and less food of animal origin.
South Africa	Enjoy a variety of foods. Be active.
Cuba	Fish and chicken are the healthiest meats.

Adherence to the information can help people stay healthy and lower their risk of developing disorders such as diabetes, heart disease, cancer, osteoporosis, and obesity.[5]

Some of the national guidelines give credit to the cultural and social importance of food. Guidelines for Japanese people, for example, include "Happy eating makes for happy family life; sit down and eat together and talk; treasure family taste and home cooking." Table 6.2 provides examples of dietary guidelines established for a variety of countries.

National guidance on healthful diets for some countries is accompanied by extensive information on how to select a healthful diet and achieve recommended levels of physical activity. In the United States, the national guidelines for diet are called the Dietary Guidelines for Americans, and the major how-to guide for implementing the guidelines is represented by MyPlate.

Dietary Guidelines for Americans

• **Identify characteristics of diets that promote health and those that do not.**

The Dietary Guidelines for Americans provide science-based recommendations to promote health and reduce the risk for major chronic diseases through diet and physical activity. Due to their credibility and focus on health promotion and disease prevention for the public, the Dietary Guidelines form the basis of federal food and nutrition programs and policies.[1]

By law, the Dietary Guidelines for Americans are updated every five years. The first edition of the Dietary Guidelines was published in 1980. The 2015 Dietary Guidelines highlight dietary patterns associated with decreased risk of obesity, heart disease, diabetes, and nutrient inadequacies. The Guidelines also encourage adequate physical activity. The 2015 Dietary Guidelines Advisory Committee, which consisted of scientific experts on nutrition and health, concluded that the health of the U.S. population could be improved, and common diseases and disorders prevented, if Americans consume a healthy dietary pattern (shown in Table 6.1) and exercise regularly. Key elements of the 2015 Dietary Guidelines related to food and nutrient intake, dietary pattern, and physical activity are listed in Table 6.3.

Most Americans do not currently consume diets that match the recommendations presented in the Dietary Guidelines.[1,10] There are many reasons for this, including access to affordable and nutritious foods, food preferences, opportunities for physical activity, and fast-paced lifestyles. Sometimes the reason can be that we simply don't think about the broad assortment of foods that can be included in a healthy dietary pattern. The Take Action feature is intended to highlight a variety of specific foods that can both be part of a healthy diet and match our food preferences.

Table 6.3 2015 Dietary Guidelines for Americans: Key Recommendations[30],[a] Guidelines that encourage a healthy eating pattern:

1. Follow a healthy eating pattern at an appropriate calorie level across the lifespan. Healthy eating patterns include:

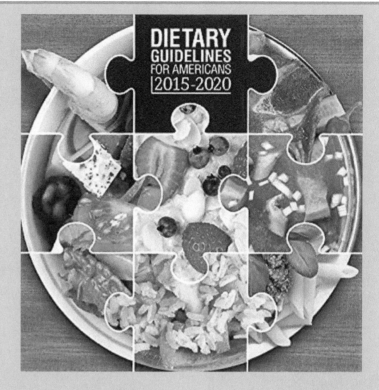

- Vegetables, including dark green, red, and orange colored vegetables

- Fruits, especially whole fruits

- Grains, at least half of which are whole grains

- Fat-free or low-fat fairy products, including milk, yogurt, cheese, and/or fortified soy beverages

- A variety of protein foods, including seafood, lean meats and poultry, eggs, legumes (beans and peas), and nuts and seeds.

- Oils

- Alcohol consumption by adults of drinking age who chose to consume it. Alcohol should be consumed in moderation—up to one drink a day for women and two for men.

- The variety of foods needed to fully meet people's need for nutrients.

Healthy eating patterns limit intake of:

- Saturated fat and added sugars to less than 10% of total calories

- Exclude *trans* fats, and limit sodium intake to less than 2,300 mg per day.

2. Shift to healthier food and beverage choices.

- Choose nutrient-dense foods and beverages across and within all food groups in place of less healthy choices.

3. Support healthy eating patterns for all.

- Everyone has a role in helping to create and support healthy eating patterns from home to school to work communities.

4. Physical Activity Recommendations
In addition to healthy eating patterns, adults should meet the Physical Activity Guidelines for Americans (at least 150 minutes of moderate intensity activity per week if physically able) to help promote health and reduce the risk of chronic disease.

[a]The Dietary Guidelines apply to the United States population over the age of 2 years.

Application of the Dietary Guidelines to Public Programs The Dietary Guidelines for Americans form the basis of federal food and nutrition programs and policies (Table 6.4). Food options served in school lunch programs, nutrition labeling standards, USDA's food guidance materials, and foods served to military personnel, for example, are based on the Dietary Guideline recommendations.

Interactive dietary assessment and planning tools for individuals and families based on the Dietary Guideline recommendations are becoming increasingly available. The most important and useful materials on how to implement the Dietary Guidelines are found at ChooseMyPlate.gov, USDA's site for the MyPlate food intake guidance materials.

Take Action

To bring variety to your diet by identifying the types of whole grain products, legumes, and fish and seafood you already like (but may not think about) or would be willing to try.

Please a √ in the appropriate column across from the food choice.

	Do not like	Never tried it	I like this food	I would try this one
Whole-Grain Foods				
popcorn	___	___	___	___
whole-grain crackers	___	___	___	___
cornmeal bread	___	___	___	___
brown rice	___	___	___	___
buckwheat pancakes	___	___	___	___
wild rice	___	___	___	___
barley	___	___	___	___
millet	___	___	___	___
hominy grits	___	___	___	___
polenta (whole grain type)	___	___	___	___
Legumes				
black beans	___	___	___	___
pinto beans	___	___	___	___
red beans	___	___	___	___
split peas	___	___	___	___
navy beans	___	___	___	___
black-eyed peas	___	___	___	___
chickpeas (garbanzos)	___	___	___	___
soy nuts	___	___	___	___
lentils	___	___	___	___
fava beans	___	___	___	___
lima beans	___	___	___	___
Fish and Seafood				
salmon	___	___	___	___
tuna	___	___	___	___
catfish	___	___	___	___
cod	___	___	___	___
tilapia	___	___	___	___
clams	___	___	___	___
shrimp	___	___	___	___
squid	___	___	___	___
anchovies	___	___	___	___
crab	___	___	___	___
scallops	___	___	___	___
lobster	___	___	___	___

Table 6.4 Examples of federal food and nutrition programs that utilize the U.S. Dietary Guidelines recommendations[9]

- Supplemental Food and Nutrition Assistance Program (SNAP, formerly known as Food Stamps)
- Head Start
- WIC (Special, Supplemental Nutrition Program for Women, Infants, and Children)
- National School Lunch Program
- Military food allowance program
- Nutrition labeling
- MyPlate educational materials
- Indian Health Service
- Healthy People 2020 (national objectives for improvements in weight status and diet)
- Older Americans Nutrition Program

MyPlate

- **Utilize ChooseMyPlate.gov guidance materials and interactive tools for dietary planning and evaluation.**

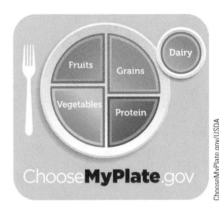

ChooseMyPlate.gov/USDA

Food group guides from the USDA have been available in the United States since 1916. Known by such names as the Basic Four Food Groups, and the Food Guide Pyramid, these guides are periodically updated to reflect advances in knowledge about foods, diets, and health. The latest revision of USDA's food guidance materials is called MyPlate and is represented by the MyPlate logo. Illustration 6.4 shows this logo and a plate of foods set up to match its messages. The MyPlate logo is intended to give consumers a visual reminder of the types and proportions of food that make up healthy meals. The logo shows a plate with four sections in different colors representing the proportion of vegetables, fruits, grains, and protein foods that should be on your plate. Next to the plate is a circle that represents a dairy product such as low-fat milk or other low-fat dairy product. The types of foods included on the plate are basic and nutrient dense. Information offered by ChooseMyPlate is available in English and Spanish at ChooseMyPlate.gov.

ChooseMyPlate.gov Healthy Eating Messages

ChooseMyPlate.gov supports the healthy eating messages in the Dietary Guidelines by offering the following key pieces of advice:

- Make at least half your plate fruits and vegetables.

- Enjoy your food but eat less.

- Make half your grains whole grains.

- Eat fewer foods that are high in saturated fat, added sugar, and sodium.

- Avoid oversized portions.

- Switch to fat-free or low-fat milk.

- Drink water instead of sugary drinks.

Judith Brown, 2011

Illustration 6.4 Foods shown on the plate consist of stewed turkey, barley, Swiss chard, mandarin oranges, and 1% milk.

Fruits	Vegetables	Grains	Protein Foods	Dairy
Focus on fruits.	**Vary your veggies.**	**Make at least half your grains whole.**	**Go lean with protein.**	**Get your calcium-rich foods.**
>> See Fruit Group	>> See Vegetable Group	>> See Grains Group	>> See Protein Foods Group	>> See Dairy Group

Illustration 6.5 USDA's ChooseMyPlate basic food groups and priority messages related to each group.

- Compare sodium in foods like soup, bread, and frozen meals and choose the foods with lower numbers.

The importance of consuming enough calories for growth and health while not eating extra calories and gaining weight is also stressed. Regular physical activity (60 minutes per day for children and adolescents and 2½ hours or more per week of moderate-intensity activity for adults weekly) is stressed because it contributes to weight control and provides many other health benefits.

Healthy U.S.-Style Dietary Pattern

This dietary pattern is based on the Dietary Approaches to Stop Hypertension Eating Plan ("The DASH Diet") and is used to formulate dietary recommendations for USDA's ChooseMyPlate educational materials (Illustration 6.5). Grains, vegetables, fruits, dairy, and protein foods are the designated groups. Interactive, educational material provided by ChooseMyPlate includes details about the types of foods that belong to each group.

Portion Sizes and Food Measure Equivalents Unlike previous food group guides, the current version does not recommend serving sizes or numbers of servings individuals in general should consume from the food groups. This information is provided if requested using a personalized interactive tool on menu planning such as the Daily Food Plan. The personalized information generated shows amounts of each food group to consume and food portion sizes that correspond to that amount.

Information on amounts of basic foods recommended for different levels of calorie need in the ChooseMyPlate materials are based on cup- and ounce equivalents. Table 6.5 provides a listing of cup- and ounce equivalents for the food groups. You can get additional examples of serving size equivalents by highlighting a food group shown on the Daily Food Plan output and clicking on "What counts as an ounce?" or "What counts as a cup?"

Sample Menus What does an eating pattern based on USDA's food group guidelines look like? Seven days of menus based on a 2,000-calorie diet that meet USDA's food group recommendations and nutrient needs are available from the "Sample Menus and Recipes" link on the ChooseMyPlate.gov homepage. Three days of the menus are shown in Illustration 6.6. The menus are intended to give consumers specific and general ideas about the types of foods to include in meals on a daily basis.

Table 6.5 How much food counts as a cup or an ounce?[36]

Vegetables and Fruits: 1 cup (c-eq) = 1 cup raw or cooked vegetables or fruit, 1 cup vegetable or fruit juice, or 2 cups leafy salad greens, ½ cup dried vegetable or fruit
Dairy: 1 cup equivalent (c-eq) = 1 cup milk, yogurt, or fortified soy milk, or 1½ ounces natural or 2 ounces processed cheese
Grains: 1 ounce equivalent (oz-eq) = 1 slice of bread, ½ cup cooked rice, cereal, or pasta; or 1 ounce ready-to-eat cereal (about 1 cup flaked cereal)
Protein: 1 ounce equivalent (oz-eq) = 1 ounce lean meat, poultry, or seafood; 1 egg; 1 Tbsp peanut butter; ½ ounce nuts or seeds; ¼ cup cooked dried beans or peas

Sample Menus for a 2000 Calorie Food Pattern

DAY 1	DAY 2	DAY 3
BREAKFAST Creamy oatmeal (cooked in milk): ½ cup uncooked oatmeal 1 cup fat-free milk 2 Tbsp raisins 2 tsp brown sugar Beverage: 1 cup orange juice **LUNCH** Taco salad: 2 ounces tortilla chips 2 ounces cooked ground turkey 2 tsp corn/canola oil (to cook turkey) ¼ cup kidney beans* ½ ounce low-fat cheddar cheese ½ cup chopped lettuce ⅓ cup avocado 1 tsp lime juice (on avocado) 2 Tbsp salsa Beverage: 1 cup water, coffee, or tea** **DINNER** Spinach lasagna roll-ups: 1 cup lasagna noodles(2 oz dry) ½ cup cooked spinach ½ cup ricotta cheese 1 ounce part-skim mozzarella cheese ½ cup tomato sauce* 1 ounce whole wheat roll 1 tsp tub margarine Beverage: 1 cup fat-free milk **SNACKS** 2 Tbsp raisins 1 ounce unsalted almonds	**BREAKFAST** Breakfast burrito: 1 flour tortilla (8" diameter) 1 scrambled egg ⅓ cup black beans* 2 Tbsp salsa ½ large grapefruit Beverage: 1 cup water, coffee, or tea** **LUNCH** Roast beef sandwich: 1 small whole grain hoagie bun 2 ounces lean roast beef 1 slice part-skim mozzarella cheese 2 slices tomato ¼ cup mushrooms 1 tsp corn/canola oil (to cook mushrooms) 1 tsp mustard Baked potato wedges: 1 cup potato wedges 1 tsp corn/canola oil (to cook potato) 1 Tbsp ketchup Beverage: 1 cup fat-free milk **DINNER** Baked salmon on beet greens: 4 ounce salmon filet 1 tsp olive oil 2 tsp lemon juice ⅓ cup cooked beet greens (sauteed in 2 tsp corn/canola oil) Quinoa with almonds: ½ cup quinoa ½ ounce slivered almonds Beverage: 1 cup fat-free milk **SNACKS** 1 cup cantaloupe balls	**BREAKFAST** Cold cereal: 1 cup ready-to-eat oat cereal 1 medium banana ½ cup fat-free milk 1 slice whole wheat toast 1 tsp tub margarine Beverage: 1 cup prune juice **LUNCH** Tuna salad sandwich: 2 slices rye bread 2 ounces tuna 1 Tbsp mayonnaise 1 Tbsp chopped celery ½ cup shredded lettuce 1 medium peach Beverage: 1 cup fat-free milk **DINNER** Roasted chicken: 3 ounces cooked chicken breast 1 large sweet potato, roasted ½ cup succotash (limas & corn) 1 tsp tub margarine 1 ounce whole wheat roll 1 tsp tub margarine Beverage: 1 cup water, coffee, or tea** **SNACKS** ¼ cup dried apricots 1 cup flavored yogurt (chocolate)

ChooseMyPlate.gov/USDA

Illustration 6.6 Sample menus for a 2,000 calorie food pattern.

USDA's Interactive Diet Planning Tools Do you want to lose weight? Want to learn about healthful foods for your preschooler or when you are pregnant? Do you need to look up information on the calorie value of different foods or keep track of your food intake? You can access this information and more from the ChooseMyPlate.gov site. To help lose weight, for example, access the Daily Food Plan interactive tool. This feature can be used to identify the amount of food from each group you should consume daily based on your age, sex, weight, height, and physical activity level (Illustration 6.7). An allowance for teaspoons of oil, and an "empty calorie" allowance for extra fats and sugars are included in the Daily Food Plan results. (This section of the Daily Food Plan will be updated and address added sugar intake when full implementation of the 2015 Dietary Guidelines takes place.)

ChooseMyPlate's SuperTracker can help you plan, analyze, and track your diet and physical activity. Access SuperTracker by searching the term Supertracker My Reports and you will see six interactive tools for diet, weight management, meal planning, and physical activity planning, assessment, and analysis (Illustration 6.8):

- **Food-A-Pedia** enables you to identify and compare the nutrient content of foods.

- **Physical Activity Tracker** can be used to compare your level of physical activity to the Physical Activity Guidelines for Americans. It can also be used to plan your activities and evaluate progress in meeting your physical activity goals.

- **My Weight Manager** provides tips for weight loss and helps you track progress in meeting weight-loss goals.

- **My Top 5 Goals** presents 19 options for goals related to weight management, physical activity, calorie intake, food group intake, and nutrient intake. You can select and track progress on meeting up to five of these goals. For example, you can set a weight goal and receive a calorie intake plan that will help you reach the goal. Graphs on changes in weight over time and tips for weight loss can be generated by this tool.

Illustration 6.7 Daily food plan results for a person with a 2,200 calorie need.

Source: ChooseMyPlate.gov. Food group recommendations presented are for a 28-year-old female, 5'6" tall, weighing 135 pounds, and exercising 30–60 minutes a day.

My Daily Food Plan

Based on the information you provided, this is your daily recommended amount for each food group.

GRAINS 7 ounces	VEGETABLES 3 cups	FRUITS 2 cups	DAIRY 3 cups	PROTEIN FOODS 6 ounces
Make half your grains whole Aim for at least **3 1/2 ounces** of whole grains a day	**Vary your veggies** Aim for these amounts each week: **Dark green veggies** = 2 cups **Red & orange veggies** = 6 cups **Beans & peas** = 2 cups **Starchy veggies** = 6 cups **Other veggies** = 5 cups	**Focus on fruits** Eat a variety of fruit Choose whole or cut-up fruits more often than fruit juice	**Get your calcium-rich foods** Drink fat-free or low-fat (1%) milk, for the same amount of calcium and other nutrients as whole milk, but less fat and Calories Select fat-free or low-fat yogurt and cheese, or try calcium-fortified soy products	**Go lean with protein** Twice a week, make seafood the protein on your plate Vary your protein routine—choose beans, peas, nuts, and seeds more often Keep meat and poultry portions small and lean

Find your balance between food and physical activity

Be physically active for at least **150 minutes** each week.

Know your limits on fats, sugars, and sodium

Your allowance for oils is **6 teaspoons** a day.
Limit Calories from solid fats and added sugars to **270 Calories** a day.
Reduce sodium intake to less than **2300 mg** a day.

Your results are based on a 2200 Calorie pattern.

Name: _____

This Calorie level is only an estimate of your needs. Monitor your body weight to see if you need to adjust your Calorie intake.

ChooseMyPlate.gov/USDA

Daily Food Plan
Get a personalized plan just for you

SuperTracker:
My foods. My fitness. My health.

Daily Food Plans for Preschoolers
Get your child's Plan today

SuperTracker and MyReports
Help you plan, analyze, and track your diet and physical activity.

Daily Food Plans for Moms
Start out right as a new mom or mom-to-be

ChooseMyPlate.gov/USDA

Illustration 6.8 Interactive tools available from SuperTracker.

- **My Reports** can be used to track changes in food group, calorie, and nutrient intake; and changes in physical activity level (www.supertracker.usda.gov/myReports.aspx). Over 40 different reports can be generated. The reports automatically compare your results to the appropriate recommendations.

Several other useful planning tools are also accessible from the ChooseMyPlate home page:

- **Healthy Eating on a Budget** features tips for saving money on food purchases and preparing healthy low-cost meals, and provides a two-week budget-friendly menu plan with recipes and a grocery-list builder.

- **MyPlate Kids' Place** offers fun educational materials for educators and parents. At this site you will find healthy eating–related games, activity sheets, videos, and songs that can be used as part of a curriculum or by parents.

- **What's Cooking?** An interactive tool that can help with. healthy meal planning, cooking, and grocery shopping, this site features a searchable database of healthy recipes, options to save recipes to a cookbook, print recipe cards, and share recipes via social media. All of the recipes previously on ChooseMyPlate.gov are being moved to this site.

  Stay tuned to MyPlate.gov. Additional useful tools are periodically added to this site.

Limitations of MyPlate Materials available from MyPlate are almost entirely made available on the Internet, making the information inaccessible to people who do not use computers or have access to the Internet. MyPlate does not provide specific recommendations for infants, individuals on therapeutic diets, or vegans.

Menus suggested by MyPlate interactive tools may not correspond to individual food preferences and contain relatively few ethnic foods. As with past food guides, planning and evaluating how mixed dishes (such as stews, soups, salads, and various types of pizza) fit into the food groups can still be perplexing.

Other Healthy Dietary Patterns Other types of dietary patterns have been shown to promote health and prevent disease. Two such patterns are the Dietary Approaches to Stop Hypertension (DASH) and the Healthy Mediterranean Dietary Pattern.

The DASH Eating Plan

Originally identified as a diet that helps control mild and moderate **hypertension**, the DASH Eating Plan has also been found to reduce the risk of certain types of cancer, osteoporosis, and heart disease. Improvements in blood pressure are generally seen within two weeks of starting this dietary pattern.[11,12]

The DASH dietary pattern emphasizes fruits, vegetables, low-fat dairy foods, whole-grain products, poultry, fish, and nuts. Only small amounts of fats, red meats, sweets, and sugar-containing beverages are included. This dietary pattern provides ample amounts of potassium, magnesium, calcium, fiber, and protein, and limited amounts of saturated and *trans* **fats**.[13,14] Although two calorie levels are shown in the table, DASH Eating Plans are available for 12 levels (1,600 to 3,200 calories) online.[13]

The Healthy Mediterranean Dietary Pattern

The traditional Mediterranean diet ranks with the USDA's Food Guide and the DASH Eating Plan when it comes to health promotion and chronic disease prevention.[5] The Mediterranean diet was originally based on foods consumed by people in Greece, Crete, southern Italy, and other Mediterranean areas where rates of chronic disease were low and life expectancy long.[15] The DASH Eating Plan and the Mediterranean Diet were used to formulate the "Healthy-U.S.-Style Dietary Pattern" and the "Healthy Mediterranean-Style Dietary Pattern" recommended in the Dietary Guidelines (Table 6.6).

hypertension High blood pressure. It is defined as blood pressure exerted inside blood vessel walls that typically exceeds 140/90 millimeters of mercury.

trans fats A type of unsaturated fatty acid produced by the addition of hydrogen to liquid vegetable oils to make them more solid. Small amounts of naturally occurring *trans* fat are found in milk and meat.

Table 6.6 Composition of the Healthy U.S. Style-and the Healthy Mediterranean-Style Eating Patterns based on a need for 2,000 calories daily

Food group		Healthy U.S.-Style Eating Pattern		Healthy Mediterraanean-Style Eating Pattern
Grains[a]		6 oz-eq		6 oz-eq (half or more should be whole grains)
Vegetables (include dark green, red, and orange vegetables)		2½ c-eq	2½ c-eq	2½ c-eq
				½ cup raw or cooked vegetables
				½ cup vegetable juice
Fruits		2 c-eq	2 c-eq	2½ c-eq
				½ cup fresh, frozen, or canned fruit
				½ cup fruit juice
Dairy		3 c-eq	3 c-eq	2 c-eq
Protein Foods		5½ oz-eq	5½ oz-eq	6½ oz-eq
Oils		6 tsp	6 tsp	6 tsp
Calories, other[a]		270/day	270/day	260/day

[a]Includes calories from added sugars, solid fats, alcohol, or additional basic foods

The Mediterranean dietary pattern, represented by the Mediterranean Diet Pyramid shown in Illustration 6.9, emphasizes plant foods, such as fruits, vegetables, grains (mostly whole), beans, and nuts. Fish and seafood is represented in the diet at least twice a week, poultry and eggs twice weekly or less, and cheese and yogurt one to seven times a week. Meats and sweets form the smallest part of the pyramid and are consumed infrequently. Wine in moderation is a traditional part of the Mediterranean diet, and water intake is encouraged. A number of studies have shown that this dietary pattern is associated with lower risks of heart disease, stroke, diabetes, several forms of cancer, and overall mortality.[16,17]

Realities of the Food Environment

Decisions people make about which foods to eat and how much of them to consume are affected by nutrition knowledge; food affordability, availability, and preferences; time availability; and other factors. The selection and consumption of a healthy dietary pattern is related to all of these factors, and largely takes place in an environment that promotes the overconsumption of relatively inexpensive, convenient, and highly palatable foods. In this environment, the larger the portion of foods that can be served that meets consumers' preferences, time constraints, and budgets, the higher the profit for those selling the foods. This situation produces a food environment that encourages the over-consumption of foods, especially when eating out.[3]

Portion Distortion

Do you know how much food you ate yesterday? If you had meat, what was the size of the serving of meat you consumed? Did you drink something with that? How much of that beverage did you consume?

Few people are aware of how much food they eat. Some people think a serving of food is the same as the portion of food they are served or eat. Portion sizes or "servings" of food today tend to exceed standard serving sizes developed by the USDA for use in planning healthful diets. In this era of growing portion and people sizes it's particularly important to become aware of how much food is eaten.[3]

Individual ideas of normal serving amounts are based on past experiences at family meals, the size of portions provided by restaurants, and packaged food and beverage sizes. Supersized meals at fast food restaurants, large portions served by other restaurants, large bakery products, and larger cups of soft drinks are contributing to the problem of portion distortion. Table 6.7 provides examples of how food portion sizes are expanding. Table 6.8 provides a guide for estimating food portion sizes.

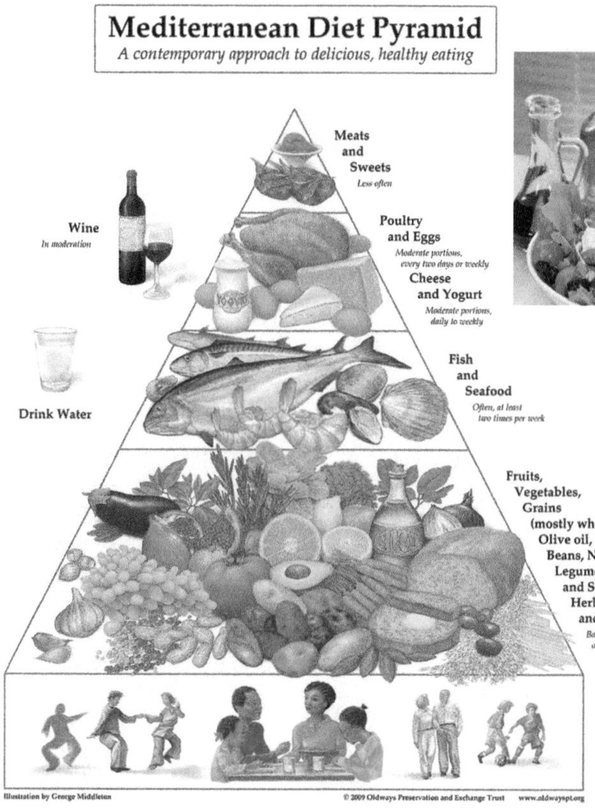

Are Supersized Portions Supersizing Americans? Supersized fast food meals or "value" meals can contain double the caloric content compared to their regular-sized counterparts. A single, supersized meal of a double quarter-pound cheeseburger, large fries, and thick shake provides more calories (about 2,200) than many people need in a day. Larger portions don't cost restaurants much more than smaller portions, they increase sales volume, and they encourage people to eat more. Many Americans are eating a good deal more food than they need, and it is appears that rising rates of obesity are partly related to large portion sizes.[3]

Illustration 6.9 The Mediterranean Diet Pyramid.

Table 6.7 Typical portion sizes and calorie content of foods in the marketplace versus calorie content and portion sizes 20 years ago[18,19]

Food	Portion size Calories 20 years ago	Marketplace portion size Calories now
Bagel	3-inch diameter	6-inch diameter
	140 calories	350 calories
Cheeseburger	4.3 ounces	7.1 ounces
	343 calories	535 calories
French fries	2.4 ounces	6.9 ounces
	210 calories	610 calories
Soft drink	6.5 ounces	20 ounces
	85 calories	820 calories
Muffin	1.5 ounce	6.5 ounce
	167 calories	724 calories

Table 6.8 Portion size estimators

1 cup = baseball	Skizer/Shutterstock.com
½ cup = tennis ball	Vlad09/Shutterstock.com
¼ cup = golf ball or extra-large egg	Cameramannz/Shutterstock.com
2 Tablespoons = ping-pong ball	Tomas1111/Dreamstime.com
1 teaspoon = fingertip	RunPhoto/StockbyteGetty Images
3 ounces of meat = deck of cards, palm of hand	Dedyukhin Dmitry/Shutterstock.com

For single-serve foods, check the weight given on the food package label.

nutrient-dense foods Foods that contain relatively high amounts of nutrients compared to their calorie value.

Children and adults tend to eat more when offered larger portions of foods than smaller portions. Children and adolescents have been reported to consume 5–12% more soft drinks, pizza, french fries, and salty snacks when offered large portions.[20] Among adults, a 50% increase in portion sizes of meals has been found to increase daily energy intake by 423 calories.[21] Frequent dining at fast food restaurants (three or more times per week) is associated with higher intakes of calories, sugar, and sodium and with a higher risk of overweight and obesity than less frequent dining at these restaurants.[22] On a positive note, some restaurants are offering smaller portions and healthier menu options than in the past.[23]

Can You Still Eat Right When Eating Out? The question about what to eat often boils down to choosing the right restaurant. According to USDA data, 41% of American adults eat out at least weekly.[24] In general, foods eaten away from home have lower nutrient content and are higher in fat than foods eaten at home (see Illustration 6.10).[25] In addition, children and teenagers who eat dinner with their families most days tend to have more healthful diets and eating patterns than those who never or only occasionally eat dinner with their family.[26]

Staying on Track While Eating Out You'll find it easier to stick to a healthy diet if you decide what to eat before you enter a restaurant and look over the menu (Illustration 6.11). You could make the decision to order soup and a salad, broiled meat, a half-portion of the entrée (or to split an entrée with someone else), or no dessert *before* entering the restaurant. "Impulse ordering" is a hazard that can throw diets out of balance. If you're going to a party or a business event where food will be served, decide before you go what types of food you will eat and what you will drink. If only high-calorie foods are offered, plan on taking a small portion and stopping there. Some people find it helps to have a healthy snack before going to a party or an event where high-calorie food will be served to avoid being really hungry when they get there.

Can Fast Foods Be Part of a Healthy Diet? As the information in Table 6.9 demonstrates, many of the foods served in fast food restaurants deserve their reputation of being high in calories, sodium, and sugar. In recognition of this fact, and in response to legislative requirements that chain restaurants label the caloric value of menu items, lower-calorie and more **nutrient-dense foods** are being added to menus. The addition of low-fat milk,

68/Ocean/Corbis

Envision/Encyclopedia/Corbis

Illustration 6.10 Hamburgers, french fries, and pizza are top-selling food items in U.S. restaurants.[27]

oatmeal, fruited yogurt, coleslaw, multigrain breads, apple slices, a variety of salads, and bean burritos to fast food and fast causal food menus makes it easier to eat well when eating out.

The Slow Food Movement An interesting trend in food preparation and consumption is making its way across the globe. The trend is away from fast and processed foods, and toward sustainable, eco-friendly agricultural practices and locally grown foods. The Slow Food movement represents some aspects of this trend. Part of an international group, Slow Food USA is an educational organization that supports ecologically sound food production; the revival of the kitchen and the table as centers of pleasure, culture, and community; and living a slower and more harmonious rhythm of life.[28] The trend is placing the topic of healthy eating in a new light for many individuals and communities and may help bring people closer to family, friends, and the environment.

What If You Don't Know How to Cook? With so many convenience foods available and time at a premium, there is growing concern that we're becoming a nation of cooking illiterates. Cooking at home gives you control over what you eat and how it's prepared. Some people immensely enjoy cooking and get a thrill out of making their specialties for friends and family. It's becoming a popular leisure-time activity: 43% of U.S. adults have taken it up for enjoyment.[29]

If you don't know how to cook, there are several ways to learn. You could start on your own by using the recipes on food packages like pasta, tomato sauce, or dried beans. You could search for recipes online or buy a basic cookbook and make dishes like salads, tacos, shish kebab, and lentil soup. You could even take a community education course. Illustration 6.12 shows some examples of good starter cookbooks. Read the sections on basic cooking skills and learn what types of equipment and utensils you need to prepare basic dishes. Get the foods and other supplies you need. Select the recipes that look doable and be sure they pass your taste and nutritional standards tests. Voilà! You're cooking.

Bon Appétit!

Dietary guidelines from some other countries contain one other rule of healthy eating that would serve Americans well: "Enjoy your meals." Eating a healthy diet should be enjoyable. If it's too much of a struggle, the healthy diet won't last. The best diets are those that keep us healthy and enhance our sense of enjoyment and wellbeing. The trick is to remember the broad array of nutritious foods we like that give us good taste and enjoyment when we eat them. And remember not to feel guilty when you occasionally eat a hamburger or some ice cream. Enjoy them to the utmost—as much as ripe oranges, papaya, homemade soups, roast turkey, hummus, and countless other nutritious delicacies.

Illustration 6.11 "No, no thank you. Extra cheese isn't part of what I planned to eat."

Illustration 6.12 Some good starter cookbooks and recipe sources.

REALITY CHECK

Portion Distortion

Mohammad and Kevin decided to eat at an Italian restaurant after soccer practice. Just for the fun of it, they agreed to guess how many cups of spaghetti they would consume if they ate the portion served to them.

Who has the better guess?

Answers on page 124.

Mohammad: Let me see . . . I'd guess it would be 3 cups.

Kevin: I bet it's about a cup. It looks like a normal serving to me.

ANSWERS TO REALITY CHECK
Portion Distortion

The average portion size of pasta served by restaurants is nearly 3 cups.[18] For people on a 2,000 calorie per day diet, that's a day's allotment for the foods from the grain group.

Mohammad: 👍

Kevin: 👎

Table 6.9 Calorie, sodium, and sugar content of some fast foods[a]

Food	Calories	Sodium, mg	Sugar, g
A.1® Ultimate bacon cheeseburger	850	1,480	8
Apple slices	15	0	3
BK® breakfast muffin sandwich with ham, egg, and cheese	430	1,140	3
Big Mac	530	960	9
Catfish fillet	460	1,140	0
Chicken burrito on flour tortilla with black beans, sour cream, guacamole	915	330	1
Chicken McNuggets, 6 pc	280	540	0
Chipotle BBQ snack wrap (grilled)	260	700	7
Chocolate frosted donut, 1	270	340	13
Dutch apple pie	340	310	25
Fruit and maple oatmeal	290	160	32
Hamburger	240	480	6
Iced coffee, large	20	15	0
Ketchup packet	10	90	2
McCafe blueberry pomegranate smoothie, medium	260	50	54
McCafe frappe mocha, medium	540	160	71
McFlurry with OREO® cookies, 12 oz.	510	280	64
Multigrain bagel, 1	350	450	8
Peach tea, medium	130	0	32
Premium grilled chicken ranch BLT sandwich	450	1,230	14
Premium McWrap Southwest (crispy)	670	1,470	12
Sausage breakfast burrito	310	820	2
Sausage biscuit	420	1,090	3
Spicy chicken thigh, 1	440	640	0
Spicy original chicken sandwich	550	1,310	5
Yoplait Go-Gurt	50	35	6

[a]Data obtained from www.mcdonalds.com/us/en/food/full_menu/full_menu_explorer.html, www.bk.com/pdfs/nutrition.pdf, popeyes.com/menu/nutrition, www.dunkindonuts.com/content/dunkindonuts/en/menu/nutrition/nutrition_catalog.html, accessed 11/10/14. Data on "sugar" content was listed as sugar, sugars, or total sugar on these sites.

Alistair Berg/Digital Vision/Getty Images

NUTRITION
up close

Focal Point: Fast foods can be a major source of calories, sodium, and sugar in diets—or not, depending on how often and which fast foods are consumed.

Meet Benson, a person with a busy life who usually gets her meals at fast food or fast casual restaurants on most days during the work week. The Tolerable Upper Intake Level for sodium that applies to Benson is 2,300 mg per day, and the recommended limit on sugar intake is less than 10% of total calories consumed daily. There are 4 calories in a gram of sugar.

1. Below is a list of the foods Benson consumed at fast food restaurants on one day. Using the information provided in Table 6.9, fill in the columns and complete the calculations indicated in sections a–d.

	Calories	Sodium, mg	Sugar, g
Breakfast:			
BK® breakfast muffin sandwich with ham, egg, and cheese	_____	_____	_____
Iced coffee, large	_____	_____	_____
Lunch			
Premium McWrap Southwest (crispy)	_____	_____	_____
McCafe blueberry pomegranate smoothie, medium	_____	_____	_____
Dinner			
Spicy, original chicken sandwich	_____	_____	_____
Dutch apple pie	_____	_____	_____
Peach tea, medium	_____	_____	_____
a. Totals	_____	_____	_____

b. Difference in sodium intake from 2,300 mg _____mg

c. Calories from sugar: total grams sugar × 4 = _____ calories

d. % calories from sugar: sugar calories/total calories × 100= _____%

Feedback to the Nutrition Up Close is located in Appendix F.

REVIEW QUESTIONS

- **Apply the characteristics of healthful diets to the design of one.**

1. Adequate diets are defined as those that provide sufficient calories to relieve hunger and maintain a person's body weight.   **True/False**

2. The diets of Americans tend to be out of balance in a number of ways.   **True/False**

3. _____ "Macronutrients" consist of:
   a. fiber, vitamins, carbohydrates, and minerals
   b. water, calories, fat, and fiber
   c. protein, minerals, vitamins, and fats
   d. carbohydrates, proteins, fat, and water

4. _____ Andre, a normal-weight man with little body fat, takes a multivitamin and mineral supplement each morning to make sure he gets all the vitamins and minerals needed daily. The rest of the day he eats whatever he wants, such as candy, donuts, pizza, burgers, and soft drinks. Is Andre consuming a healthful diet?
   a. Most likely yes, because he is getting all the vitamins and minerals he needs.
   b. Yes, because he is normal weight.

c. Probably not because his diet lacks balance and variety.

d. It depends on the composition of the multivitamin and mineral supplement.

- **Identify characteristics of diets that promote health and those that do not.**

5. The Healthy Mediterranean-Style Dietary Pattern is an example of a vegetarian diet plan that promotes weight loss. **True/False**

6. Healthy dietary patterns are represented by the Healthy U.S.-Style Dietary Pattern, Healthy Vegetarian-Style Dietary Pattern, and the Healthy Mediterranean-Style Dietary Pattern **True/False**

7. The DASH Eating Plan represents a dietary pattern that is intended for vegetarians. **True/False**

8. The 2015 Dietary Guidelines include a recommendation to lower total fat intake. **True/False**

9. The 2015 Dietary Guidelines include recommendations related to saturated fat intake and physical activity. **True/False**

10. _____ As the head of food service for Lincoln Elementary School you have been charged with making sure the new cafeteria lunch menus conform to the Dietary Guidelines recommendations. Which of the following sets of menu items would you *limit* offering in the lunch menu?

a. tossed salads and salad dressing
b. corn and tuna fish
c. hot dogs and fruit drinks
d. fruit salad and corn bread

11. Healthy dietary patterns share the basic characteristics of providing

a. adequate levels of nutrients
b. low amounts of fat
c. low amounts of carbohydrate
d. high amounts of protein

- **Utilize ChooseMyPlate.gov guidance materials and interactive tools for dietary planning and evaluation.**

12. ChooseMyPlate provides the major tools used in the United States to help people implement the Dietary Guidelines. **True/False**

13. _____ Jack wants to get into shape for baseball season and decides that adopting a healthy diet may help him do that. Which interactive tool from ChooseMyPlate should he use to best help him plan a healthful diet given his size and current physical activity level?

a. Food-A-Pedia
b. My Reports
c. Athlete's Food Guide
d. My Daily Food Plan

14. _____ Marta has recently been diagnosed as having "iron deficiency" and decides to assess the amount of iron in her usual diet. Which of the following ChooseMyPlate tools will best help her get this information?

a. What's Cooking?
b. SuperTracker
c. My Top 5 Goals
d. Daily Food Plan

15. _____ Maya lost 10 pounds and would like to keep the weight off. Tonight she's headed to a party where lots of delicious pastries will be served and she doesn't want to eat too many of them. What can she do to help with that?

a. Try to convince herself that she really doesn't like pastries.
b. Decide before she goes to the party to enjoy one pastry.
c. Skip the party. It would be impossible to resist the pastries.
d. Go to the party late because the pastries may be gone.

16. _____ Which of the following statements about large portion size is true?

a. Adults and children tend to eat more when offered large portions of foods rather than small portions.
b. People eat until they feel full and then stop eating, regardless of portion size.
c. Large portions tend to make adults and children eat less than if given smaller portions.
d. Children tend to eat more when offered more food but adults do not.

Answers to these questions can be found in Appendix F.

NUTRITION SCOREBOARD ANSWERS

1. Healthy dietary patterns also include fish, seafood, vegetables, fruits, seeds, low-fat dairy products, and a moderate amount of alcohol (for adults who choose to drink).[1] **True**

2. 86% of U.S. adults fail to meet this recommendation.[2] **True**

3. The basic food groups do not include a "healthy snack" food group. **False**

4. "Value" (aka supersized) fast food meals can pile on calories.[3] **True**

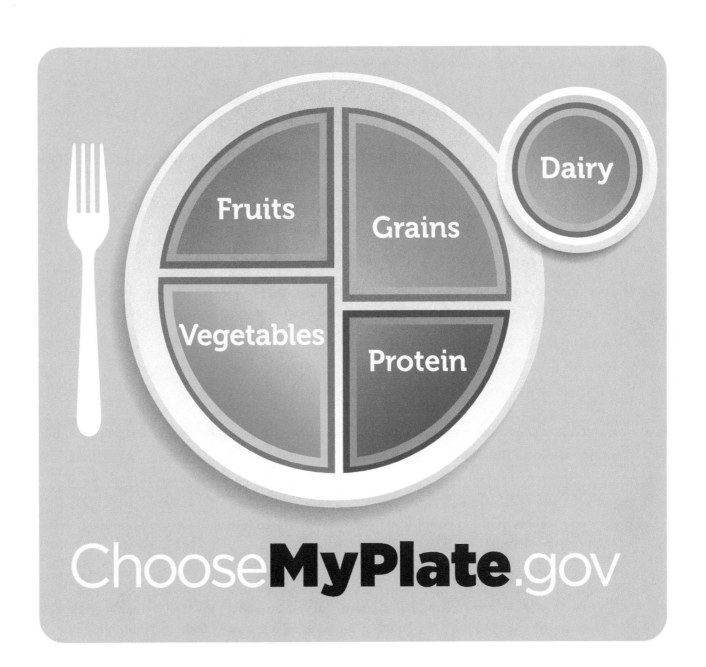

Fruits

Grains

Dairy

Vegetables

Protein

ChooseMyPlate.gov

ChooseMyPlate.gov

Dietary Guidelines for Americans 2010

The Dietary Guidelines for Americans 2010 provide evidence-based advice to help people attain and maintain a healthy weight, reduce the risk of chronic diseases, and promote overall health through diet and physical activity. Below are their key recommendations.

Balancing Calories to Manage Weight

- Prevent and/or reduce overweight and obesity through improved eating and physical activity behaviors.

- Control total calorie intake to manage body weight. For people who are overweight or obese, this will mean consuming fewer calories from foods and beverages.

- Increase physical activity and reduce time spent in sedentary behaviors.

- Maintain appropriate calorie balance during each stage of life—childhood, adolescence, adulthood, pregnancy and breastfeeding, and older age.

Foods and Food Components to Reduce

- Reduce daily sodium intake to less that 2,300 milligrams and further reduce intake to 1,500 milligrams among persons who are 51 and older and those of any age who are African American or have hypertension, diabetes, or chronic kidney disease. The 1,500 milligrams recommendation applies to about half of the U.S. population, including children and the majority of adults.

- Consume less than 10 percent of calories from saturated fatty acids by replacing them with monounsaturated and polyunsaturated fatty acids.

- Consume less than 300 milligrams per day of dietary cholesterol.

- Keep *trans* fatty acid consumption as low as possible by limiting foods that contain synthetic sources of *trans* fats, such as partially hydrogenated oils, and by limiting other solid fats.

- Reduce the intake of calories from solid fats and added sugars.

- Limit the consumption of foods that contain refined grains, especially refined grain foods that contain solid fats, added sugars, and sodium.

- If alcohol is consumed it should be consumed in moderation—up to one drink per day for women and two drinks per day for men—and only by adults of legal drinking age.

Dietary Guidelines for Americans 2010

continued from page 128

Foods and Nutrients to Increase

- Increase vegetable and fruit intake.

- Eat a variety of vegetables, especially dark-green and red and orange vegetables and beans and peas.

- Consume at least half of all grains as whole grains. Increase whole-grain intake by replacing refined grains with whole grains.

- Increase intake of fat-free or low-fat milk and milk products, such as milk, yogurt, cheese, or fortified soy beverages.

- Choose a variety of protein foods, which include seafood, lean meat and poultry, eggs, beans and peas, soy products, and unsalted nuts and seeds.

- Increase the amount and variety of seafood consumed by choosing seafood in place of some meat and poultry.

- Replace protein foods that are higher in solid fats with choices that are lower in solid fats and calories and/or are sources of oils.

- Use oils to replace solid fats where possible.

- Choose foods that provide more potassium, dietary fiber, calcium, and vitamin D, which are nutrients of concern in American diets. These foods include vegetables, fruits, whole grains, and milk and milk products.

Building Healthy Eating Patterns

- Select an eating pattern that meets nutrient needs over time at an appropriate calorie level.

- Account for all foods and beverages consumed and assess how they fit within a total healthy eating pattern.

- Follow food safety recommendations when preparing and eating foods to reduce the risk of foodborne illnesses.

NOTE: These guidelines are intended for adults and children ages 2 and older.

SOURCE: The *Dietary Guidelines for Americans,* www.dietaryguidelines.gov

Dietary Reference Intakes for Calcium and Vitamin D

In 2010, the Committee on Dietary Reference Intakes released new calcium and vitamin D recommendations, listed below.

Life Stage Group	Calcium			Vitamin D					
	Estimated Average Requirement	Recommended Dietary Allowance	Tolerable Upper Intake Level	Estimated Average Requirement		Recommended Dietary Allowance		Tolerable Upper Intake Level	
	(mg/day)	(mg/day)	(mg/day)	(IU/day)	(µg/day)	(IU/day)	(µg/day)	(IU/day)	(µg/day)
Infants 0 to 6 months	---	200	1,000	---	---	400	10	1,000	25
Infants 6 to 12 months	---	260	1,500	---	---	400	10	1,500	38
1–3 years old	500	700	2,500	400	10	600	15	2,500	63
4–8 years old	800	1,000	2,500	400	10	600	15	3,000	75
9–13 years old	1,100	1,300	3,000	400	10	600	15	4,000	100
14–18 years old	1,100	1,300	3,000	400	10	600	15	4,000	100
19–30 years old	800	1,000	2,500	400	10	600	15	4,000	100
31–50 years old	800	1,000	2,500	400	10	600	15	4,000	100
51–70 years old males	800	1,000	2,000	400	10	600	15	4,000	100
51–70 years old females	1,000	1,200	2,000	400	10	600	15	4,000	100
71+ years old	1,000	1,200	2,000	400	10	800	20	4,000	100
14–18 years old, pregnant/lactating	1,100	1,300	3,000	400	10	600	15	4,000	100
19–50 years old, pregnant/lactating	800	1,000	2,500	400	10	600	15	4,000	100

NOTE: Values for infants are Adequate Intakes (AI). Dashes indicate that values have not been determined.
SOURCE: Committee on Dietary Reference Intakes, *Dietary Reference Intakes for Calcium and Vitamin D* (Washington, DC: National Academies Press), 2011.

Healthy People 2020

The Healthy People program identifies national health priorities and guides policies to "increase the quality and years of healthy life" and "eliminate health disparities." The nutrition and weight status objectives for Healthy People 2020 are listed below.

Healthy People 2020 Nutrition and Weight Status Objectives

- Increase the proportion of adults who are at a healthy weight.

- Reduce the proportion of adults who are obese.

- Reduce iron deficiency among young children and females of childbearing age.

- Reduce iron deficiency among pregnant females.

- Reduce the proportion of children and adolescents who are overweight or obese.

- Increase the contribution of fruits to the diets of the population aged 2 years and older.

- Increase the variety and contribution of vegetables to the diets of the population aged 2 years and older.

- Increase the contribution of whole grains to the diets of the population aged 2 years and older.

- Reduce consumption of saturated fat in the population aged 2 years and older.

- Reduce consumption of sodium in the population aged 2 years and older.

- Increase consumption of calcium in the population aged 2 years and older.

- Increase the proportion of worksites that offer nutrition or weight management classes or counseling.

- Increase the proportion of physician office visits that include counseling or education related to nutrition or weight.

- Eliminate very low food security among children in U.S. households.

- Prevent inappropriate weight gain in youth and adults.

- Increase the proportion of primary care physicians who regularly measure the body mass index of their patients.

- Reduce consumption of calories from solid fats and added sugars in the population aged 2 years and older.

- Increase the number of states that have state-level policies that incentivize food retail outlets to provide foods that are encouraged by the *Dietary Guidelines*.

- Increase the number of states with nutrition standards for foods and beverages provided to preschool-aged children in childcare.

- Increase the percentage of schools that offer nutritious foods and beverages outside of school meals.

NOTE: "Nutrition and Weight Status" is one of 38 topic areas, each with numerous objectives.
SOURCE: Healthy People 2020, www.healthypeople.gov

Sample Menus for a 2000 Calorie Food Pattern

Use this 7-day menu as a motivational tool to help put a healthy eating pattern into practice, and to identify creative new ideas for healthy meals. Averaged over a week, this menu provides the recommended amounts of key nutrients and foods from each food group. The menus feature a large number of different foods to inspire ideas for adding variety to food choices. They are not intended to be followed day-by-day as a specific prescription for what to eat.

Spices and herbs can be used to taste. Try spices such as chili powder, cinnamon, cumin, curry powder, ginger, nutmeg, mustard, garlic powder, onion powder, or pepper. Try fresh or dried herbs such as basil, parsley, cilantro, chives, dill, mint, oregano, rosemary, thyme, or tarragon. Also try salt-free spice or herb blends.

While this 7-day menu provides the recommended amounts of foods and key nutrients, it does so at a moderate cost. Based on national average food costs, adjusted for inflation to March 2011 prices, the cost of this menu is less than the average amount spent for food, per person, in a 4-person family.

DAY 1

BREAKFAST
Creamy oatmeal (cooked in milk):
- *½ cup uncooked oatmeal*
- *1 cup fat-free milk*
- *2 Tbsp raisins*
- *2 tsp brown sugar*
- Beverage: 1 cup orange juice

LUNCH
Taco salad:
- *2 ounces tortilla chips*
- *2 ounces cooked ground turkey*
- *2 tsp corn/canola oil (to cook turkey)*
- *¼ cup kidney beans**
- *½ ounce low-fat cheddar cheese*
- *½ cup chopped lettuce*
- *½ cup avocado*
- *1 tsp lime juice (on avocado)*
- *2 Tbsp salsa*
- Beverage:
- 1 cup water, coffee, or tea**

DINNER
Spinach lasagna roll-ups:
- *1 cup lasagna noodles(2 oz dry)*
- *½ cup cooked spinach*
- *½ cup ricotta cheese*
- *1 ounce part-skim mozzarella cheese*
- *½ cup tomato sauce**
- 1 ounce whole wheat roll
- *1 tsp tub margarine*
- Beverage: 1 cup fat-free milk

SNACKS
- 2 Tbsp raisins
- 1 ounce unsalted almonds

DAY 2

BREAKFAST
Breakfast burrito:
- *1 flour tortilla (8" diameter)*
- *1 scrambled egg*
- *⅓ cup black beans**
- *2 Tbsp salsa*
- ½ large grapefruit
- Beverage:
- 1 cup water, coffee, or tea**

LUNCH
Roast beef sandwich:
- *1 small whole grain hoagie bun*
- *2 ounces lean roast beef*
- *1 slice part-skim mozzarella cheese*
- *2 slices tomato*
- *¼ cup mushrooms*
- *1 tsp corn/canola oil (to cook mushrooms)*
- *1 tsp mustard*
Baked potato wedges:
- *1 cup potato wedges*
- *1 tsp corn/canola oil (to cook potato)*
- *1 Tbsp ketchup*
- Beverage: 1 cup fat-free milk

DINNER
Baked salmon on beet greens:
- *4 ounce salmon filet*
- *1 tsp olive oil*
- *2 tsp lemon juice*
- *⅓ cup cooked beet greens (sauteed in 2 tsp corn/canola oil)*
Quinoa with almonds:
- *½ cup quinoa*
- *½ ounce slivered almonds*
- Beverage: 1 cup fat-free milk

SNACKS
- 1 cup cantaloupe balls

DAY 3

BREAKFAST
Cold cereal:
- *1 cup ready-to-eat oat cereal*
- *1 medium banana*
- *½ cup fat-free milk*
- 1 slice whole wheat toast
- *1 tsp tub margarine*
- Beverage: 1 cup prune juice

LUNCH
Tuna salad sandwich:
- *2 slices rye bread*
- *2 ounces tuna*
- *1 Tbsp mayonnaise*
- *1 Tbsp chopped celery*
- *½ cup shredded lettuce*
- 1 medium peach
- Beverage: 1 cup fat-free milk

DINNER
Roasted chicken:
- *3 ounces cooked chicken breast*
- 1 large sweet potato, roasted
- *½ cup succotash (limas & corn)*
- *1 tsp tub margarine*
- 1 ounce whole wheat roll
- *1 tsp tub margarine*
- Beverage:
- 1 cup water, coffee, or tea**

SNACKS
- ¼ cup dried apricots
- 1 cup flavored yogurt (chocolate)

Sample Menus for a 2000 Calorie Food Pattern (cont'd)

DAY 4

BREAKFAST
1 whole wheat English muffin
  1 Tbsp all-fruit preserves
1 hard-cooked egg
Beverage:
1 cup water, coffee, or tea**

LUNCH
White bean-vegetable soup:
  1 ¼ cup chunky vegetable soup
  with pasta
  *½ cup white beans**
6 saltine crackers*
½ cup celery sticks
Beverage: 1 cup fat-free milk

DINNER
Rigatoni with meat sauce:
  1 cup rigatoni pasta (2 oz dry)
  2 ounces cooked ground beef
  (95% lean)
  2 tsp corn/canola oil (to cook beef)
  *½ cup tomato sauce**
3 Tbsp grated parmesan cheese
Spinach salad:
  1 cup raw spinach leaves
  ½ cup tangerine sections
  ½ ounce chopped walnuts
  4 tsp oil and vinegar dressing
Beverage:
1 cup water, coffee, or tea**

SNACKS
1 cup nonfat fruit yogurt

DAY 5

BREAKFAST
Cold cereal:
  1 cup shredded wheat
  ½ cup sliced banana
  ½ cup fat-free milk
1 slice whole wheat toast
  2 tsp all-fruit preserves
Beverage:
1 cup fat-free chocolate milk

LUNCH
Turkey sandwich
  1 whole wheat pita bread (2 oz)
  3 ounces roasted turkey, sliced
  2 slices tomato
  ¼ cup shredded lettuce
  1 tsp mustard
  1 Tbsp mayonnaise
½ cup grapes
Beverage: 1 cup tomato juice*

DINNER
Steak and potatoes:
  4 ounces broiled beef steak
  ⅔ cup mashed potatoes made
  with milk and 2 tsp tub margarine
½ cup cooked green beans
  1 tsp tub margarine
  1 tsp honey
1 ounce whole wheat roll
  1 tsp tub margarine
Frozen yogurt and berries:
  ½ cup frozen yogurt (chocolate)
  ¼ cup sliced strawberries
Beverage: 1 cup fat-free milk

SNACKS
1 cup frozen yogurt (chocolate)

DAY 6

BREAKFAST
French toast:
  2 slices whole wheat bread
  3 Tbsp fat-free milk and
  ⅔ egg (in French toast)
  2 tsp tub margarine
  1 Tbsp pancake syrup
½ large grapefruit
Beverage: 1 cup fat-free milk

LUNCH
3-bean vegetarian chili on baked
potato:
  *¼ cup each cooked kidney beans,**
  *navy beans, * and black beans**
  *½ cup tomato sauce**
  ¼ cup chopped onion
  2 Tbsp chopped jalapeno peppers
  1 tsp corn/canola oil (to cook
  onion and peppers)
  ¼ cup cheese sauce
  1 large baked potato
½ cup cantaloupe
Beverage:
1 cup water, coffee, or tea**

DINNER
Hawaiian pizza
  2 slices cheese pizza, thin crust
  1 ounce lean ham
  ¼ cup pineapple
  ¼ cup mushrooms
  1 tsp safflower oil (to cook
  mushrooms)
Green salad:
  1 cup mixed salad greens
  4 tsp oil and vinegar dressing
Beverage: 1 cup fat-free milk

SNACKS
3 Tbsp hummus
5 whole wheat crackers*

DAY 7

BREAKFAST
Buckwheat pancakes with berries:
  2 large (7") pancakes
  1 Tbsp pancake syrup
  ¼ cup sliced strawberries
Beverage: 1 cup orange juice

LUNCH
New England clam chowder:
  3 ounces canned clams
  ½ small potato
  2 Tbsp chopped onion
  2 Tbsp chopped celery
  6 Tbsp evaporated milk
  ¼ cup fat-free milk
  1 slice bacon
  1 Tbsp white flour
10 whole wheat crackers*
1 medium orange
Beverage: 1 cup fat-free milk

DINNER
Tofu-vegetable stir-fry:
  4 ounces firm tofu
  ½ cup chopped Chinese cabbage
  ¼ cup sliced bamboo shoots
  2 Tbsp chopped sweet red peppers
  2 Tbsp chopped green peppers
  1 Tbsp corn/canola oil (to cook
  stir-fry)
1 cup cooked brown rice (2 ounces
raw)
Honeydew yogurt cup:
  ¾ cup honeydew melon
  ½ cup plain fat-free yogurt
Beverage:
1 cup water, coffee, or tea**

SNACKS
1 large banana spread with
  *2 Tbsp peanut butter**
1 cup nonfat fruit yogurt

Notes:

*Foods that are reduced sodium, low sodium, or no-salt added products. These foods can also be prepared from scratch with no added salt. All other foods are regular commercial products, which contain variable levels of sodium. Average sodium level of the 7-day menu assumes that no salt is added in cooking or at the table.

**Unless indicated, all beverages are unsweetened and without added cream or whitener.

Italicized foods are part of the dish or food that precedes it.

Sample Menus for a 2000 Calorie Food Pattern (cont'd)

Average amounts for weekly menu:

Food group	Daily average over 1 week
GRAINS	**6.2 oz eq**
Whole grains	3.8
Refined grains	2.4
VEGETABLES	**2.6 cups**
Vegetable subgroups (amount per week)	
Dark green	1.6 cups per week
Red/Orange	5.6
Starchy	5.1
Beans and Peas	1.6
Other Vegetables	4.1
FRUITS	**2.1 cups**
DAIRY	**3.1 cups**
PROTEIN FOODS	**5.7 oz eq**
Seafood	8.8 oz per week
OILS	**29 grams**
CALORIES FROM ADDED FATS AND SUGARS	245 calories

Nutrient	Daily average over 1 week
Calories	1975
Protein	96 g
Protein	19% kcal
Carbohydrate	275 g
Carbohydrate	56% kcal
Total fat	59 g
Total fat	27% kcal
Saturated fat	13.2 g
Saturated fat	6.0% kcal
Monounsaturated fat	25 g
Polyunsaturated fat	16 g
Linoleic Acid	13 g
Alpha-linolenic Acid	1.8 g
Cholesterol	201 mg
Total dietary fiber	30 g
Potassium	4701 mg
Sodium	1810 mg
Calcium	1436 mg
Magnesium	468 mg
Copper	2.0 mg
Iron	18 mg
Phosphorus	1885 mg
Zinc	14 mg
Thiamin	1.6 mg
Riboflavin	2.5 mg
Niacin Equivalents	24 mg
Vitamin B6	2.4 mg
Vitamin B12	12.3 mcg
Vitamin C	146 mg
Vitamin E	11.8 mg (AT)
Vitamin D	9.1 mcg
Vitamin A	1090 mcg (RAE)
Dietary Folate Equivalents	530 mcg
Choline	386 mg

Carbohydrates

Carbohydrates comprise a diverse group of compounds produced primarily by plants. The sugars in fruit, the fibers in celery, and the starch in potatoes are all different forms of carbohydrate. Although all carbohydrates are made of sugar molecules, they differ from each other in a variety of ways. For example, some carbohydrates contain only one or two sugar molecules, whereas others are made of hundreds of sugar molecules. As the number of sugar molecules in a carbohydrate increases, so does its size and complexity. Carbohydrates are plentiful in a variety of foods and serve many essential functions within the body. For example, carbohydrates provide cells with a vital source of energy and are components of ribonucleic and deoxyribonucleic acids (RNA and DNA, respectively). As such, they are an important part of a healthy, well-balanced diet. In this chapter, we will look at carbohydrate chemistry, dietary sources of carbohydrates, functions of carbohydrates in the body, and guidelines for carbohydrate intake.

What It Takes to Stay in the Race

After completing over 20 marathons, Laura was ready for her next challenge—the Ironman® competition. This ultra-endurance event is one of the most rigorous athletic competitions: a 2.4-mile swim, a 112-mile bike ride, and a marathon (26.2 miles). After six months of relentless training, Laura finally began the final stages of planning and preparation. As she packed her bag, she put the two most essential items in last—her insulin pump and her glucometer. Yes, Laura has type 1 diabetes. In addition to worrying about flat tires, bike crashes, and consuming enough energy to make it through the grueling 14-hour competition, Laura has an additional worry that most Ironman competitors do not have—monitoring and regulating her blood glucose levels throughout the competition. For Laura, plummeting blood glucose levels could easily mean the end of the race.

Like most competitive athletes, Laura maintains a strict training regimen. In addition to recording her distances and times, she also keeps a detailed food diary. Laura understands that rigorous exercise can cause unexpected highs and lows in blood glucose, and determining what and how much food to eat can be a real challenge.

For those with type 1 diabetes, testing blood glucose throughout the day using a glucometer becomes routine. This requires obtaining a small drop of blood using a lancet device, placing the blood onto the test strip, and then inserting the test strip into the glucometer. How to manage this while biking and running was an obvious challenge for Laura. After a few near-crashes on her bike she decided to try a continuous glucose-monitoring device. The tiny glucose sensor inserted under the skin allows Laura to more easily monitor and record her blood glucose levels throughout the day and night. The information from the sensor is transmitted to a wireless monitor. Continuous glucose monitoring is not intended for long-term use, but it enables Laura to determine the right combination and amount of food she needs to compete.

Training for the Ironman® competition is hard for anyone, but it is especially hard for someone with diabetes.

But Laura's motto has always been "Be ready for anything." This means being able to detect and respond to sudden and unexpected fluctuations in blood glucose. Knowing when to eat, what to eat, and when it is time for insulin is critical. Watches, alarms, needles, pumps, and carefully selected food keep Laura on pace and in the race. You may be wondering why someone with type 1 diabetes, or anyone for that matter, competes in such an intense and exhausting competition. In addition to family and work, much of Laura's inspiration comes from a nonprofit organization called Triabetes. This online network for triathletes with type 1 diabetes provides support and strategies to navigate these largely uncharted waters. As the finisher's medal was placed around Laura's neck, she thought to herself, "What an amazing experience."

Laura Benson

Critical Thinking: Laura's Story

Type 1 diabetes is a condition that is not easily ignored. In fact, a person with this disease must be vigilant when it comes to blood glucose control. What insights did you gain from Laura's story about what it is like to have type 1 diabetes? What do you think your biggest challenges would be if you had this condition?

What Are Simple Carbohydrates?

A carbohydrate is an organic compound made up of one or more sugar molecules (Figure 8.1). Most people think of sugar as a substance used to sweeten their foods. Although this is true, sugars encompass far more than this. For example, cells use a special type of sugar, glucose, for an important source of energy. **Carbohydrates** are abundant in a wide variety of foods, and there are many different types. A carbohydrate consisting of a single sugar molecule is called a **monosaccharide;** a carbohydrate made of two sugar molecules is a **disaccharide.** Because of their small size, monosaccharides and disaccharides are called **simple carbohydrates** or simple sugars.

MONOSACCHARIDES ARE SINGLE SUGAR MOLECULES

Monosaccharides are single-sugar molecules that are made up of carbon, hydrogen, and oxygen atoms in the ratio of 1:2:1. For example, if a sugar has 6 carbon atoms, it also has 12 hydrogen and 6 oxygen atoms (written $C_6H_{12}O_6$). Because the number of carbon atoms and the arrangement of atoms can vary, monosaccharides typically have different shapes and sizes. There are hundreds of different naturally occurring monosaccharides, but the three most plentiful in food are glucose, fructose, and galactose. Although the structures of these sugars differ, they all have one thing in common: each contains six carbon atoms. Therefore these monosaccharides are referred to as **hexose** (*hexa*, meaning six) sugars.

Although glucose, fructose, and galactose have the same molecular formula ($C_6H_{12}O_6$), each has a different arrangement of atoms (Figure 8.2). Notice that each monosaccharide exists as a ring structure and that each carbon atom in the ring is assigned a number. This numbering system provides a way to describe important structural differences in monosaccharides. As you will soon learn, it also provides the basis for describing different types of chemical bonds that attach one sugar molecule to the next.

Glucose Is the Most Abundant Sugar in Blood **Glucose,** the most abundant monosaccharide in the body, is produced when chlorophyll-containing plants combine carbon dioxide and water in the presence of sunlight (Figure 8.3). This process, called **photosynthesis,** provides an important energy source

carbohydrate Organic compound made of varying numbers of monosaccharides.

monosaccharide (mo – no – SAC – cha – ride) (*mono-*, one; -*saccharide*, sugar) Carbohydrate consisting of a single sugar.

disaccharide (di – SAC – cha – ride) (*di-*, two) Carbohydrate consisting of two monosaccharides bonded together.

simple carbohydrate, or simple sugar Category of carbohydrates consisting of mono- and disaccharides.

hexose Monosaccharide made of six carbon atoms.

glucose (GLU – kose) A six-carbon monosaccharide produced by photosynthesis in plants.

photosynthesis (*photo-*, light; -*synthesis*, product) Process whereby plants use energy from the sun to produce glucose from carbon dioxide and water.

FIGURE 8.1 Classification of Carbohydrates Carbohydrates are classified as simple or complex. These categories are further subdivided into mono-, di-, oligo-, and polysaccharides, depending on the number of monosaccharides (sugars) they contain.

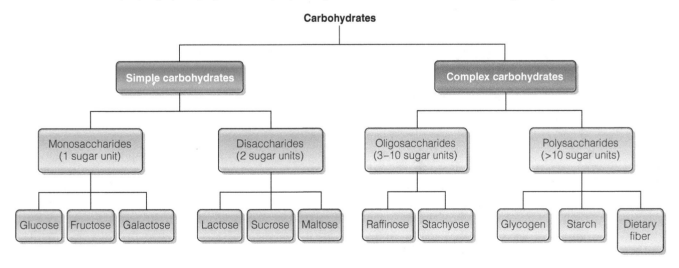

FIGURE 8.2 The Structure of Monosaccharides Glucose, galactose, and fructose are monosaccharides that contain six carbon atoms.

Monosaccharides that have 6 carbon atoms are called hexose sugars.
All hexose sugars have 6 carbon, 12 hydrogen, and 6 oxygen atoms (1:2:1).

Fructose has a 5-sided ring structure, whereas glucose and galactose have 6-sided ring structures.

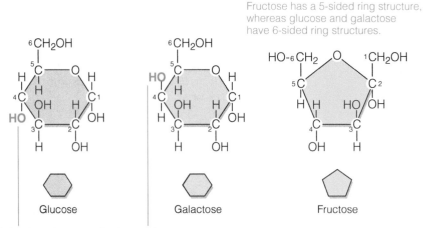

Glucose

Galactose

Fructose

Notice that glucose and galactose have a similar chemical structure with the exception of the hydroxyl groups (–OH), which face in opposite directions.

FIGURE 8.3 Photosynthesis The process of photosynthesis combines carbon dioxide and water in the presence of sunlight to produce glucose. When we consume plants, the glucose provides us with a source of energy.

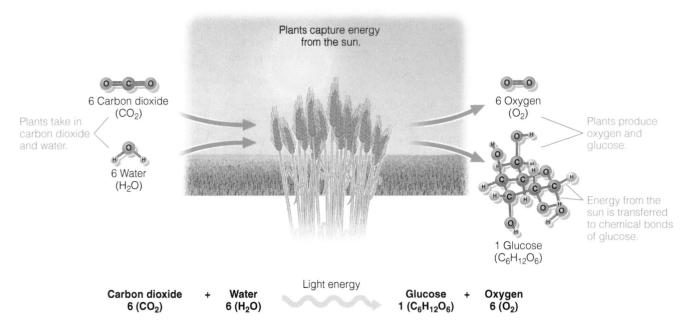

Plants capture energy from the sun.

6 Carbon dioxide
(CO_2)

Plants take in carbon dioxide and water.

6 Water
(H_2O)

6 Oxygen
(O_2)

Plants produce oxygen and glucose.

Energy from the sun is transferred to chemical bonds of glucose.

1 Glucose
($C_6H_{12}O_6$)

| Carbon dioxide | + | Water | Light energy | Glucose | + | Oxygen |
| 6 (CO_2) | | 6 (H_2O) | | 1 ($C_6H_{12}O_6$) | | 6 (O_2) |

⟨**CONNECTIONS**⟩ The nervous system is made up of the brain, spinal cord, and nerves (Chapter 1, page 16).

for plants. Plants can use glucose to form large, complex carbohydrates. When plant foods are consumed, the body breaks down (or in other words, digests) these large carbohydrates into glucose, which is subsequently used by cells as a source of energy. In this way, plants and animals are each part of the delicate balance that exists in nature.

The primary function of glucose in the body is to provide cells with a source of energy (ATP). While most cells use a combination of energy sources, glucose is the preferred energy source for the nervous system and the sole source of energy for red blood cells. Glucose is also used to synthesize other compounds in the body. For example, it can be converted to some amino acids (the building blocks of proteins) and fat for long-term energy storage. In addition, the body can store small amounts of glucose as a compound called glycogen.

Fructose Is the Most Abundant Sugar in Fruits and Vegetables **Fructose** is a naturally occurring monosaccharide found primarily in honey, fruits, and vegetables. Even though fructose has a five-sided ring structure, it is still classified as a hexose sugar because it contains a total of six carbons. While fructose is abundant in fruits and vegetables, the majority of fructose in the Western diet comes from foods made with high-fructose corn syrup. **High-fructose corn syrup** (HFCS) is a widely used sweetener found in soft drinks, fruit juice beverages, and a variety of other foods. Derived from corn, HFCS consists of almost equal amounts of fructose and glucose. It is used so extensively by food manufacturers that it now accounts for approximately 7% of total energy intake in the United States.[1] The U.S. Department of Agriculture estimates that the typical American consumes approximately 35 pounds of HFCS per year.[2] In fact, consumption of foods and beverages sweetened with HFCS now exceeds that of those sweetened with table sugar (Figure 8.4). Some scientists speculate that excess consumption of HFCS may be fueling the American obesity epidemic.[3] This controversy is discussed further in the Focus on Diet and Health feature.

Galactose Is Found in Dairy Products At first glance, the monosaccharides glucose and **galactose** look rather similar. However, notice that the hydroxyl (–OH) groups on carbon 4 actually face in opposite directions. Even this minor structural variation results in important differences in physiological functions of galactose and glucose. Few foods contain galactose in its free

fructose (FRUC – tose) A six-carbon monosaccharide found in fruits and vegetables; also called levulose.

high-fructose corn syrup A substance derived from corn that is used to sweeten foods and beverages.

galactose (ga – LAC – tose) A six-carbon monosaccharide found mainly bonded with glucose to form the milk sugar lactose.

FIGURE 8.4 Trends (1970–2010) in Consumption of Foods and Beverages Sweetened with High-Fructose Corn Syrup and Table Sugar

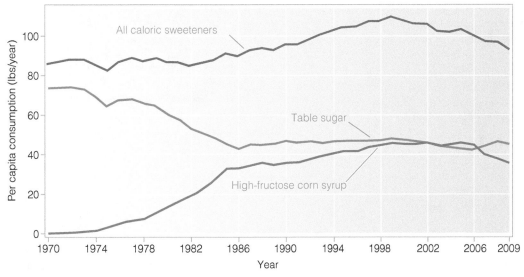

SOURCES: U.S. Department of Agriculture, Economic Research Service. 2011. Table 51—Refined cane and beet sugar: estimated number of per capita calories consumed daily, by calendar year. Table 52—High fructose corn syrup: estimated number of per capita calories consumed daily, by calendar year. Available from: http://www.ers.usda.gov/Briefing/Sugar/Data.htm.

Fructose is a naturally occurring sugar that makes fruits deliciously sweet. However, most of the fructose we consume in the United States does not come from fruit but instead comes from foods and beverages sweetened with high-fructose corn syrup (HFCS).

HFCS is produced when corn is treated with enzymes, resulting in a glucose-rich syrup. During this process, some of the glucose molecules are converted to fructose. While sucrose consists of glucose and fructose in roughly equal amounts, the ratio of fructose to glucose in HFCS can be considerably higher. For example, HFCS used in the production of soft drinks contains approximately 55% fructose and 45% glucose. According to the National Soft Drink Association, HFCS works better than refined sugar in beverage production because it blends more readily with liquids. In addition, the cost to make products with HFCS is substantially less than that of using sucrose.

Recently, scientists observed that an increase in HFCS consumption has paralleled an increasing rate of obesity.[3] Coincidence or not, some claim that HFCS has unique properties that may be cause for concern. Although researchers have raised questions about a link between HFCS consumption and obesity, as yet studies have not shown a definite relationship.[4] Clearly, no single cause is responsible for the obesity epidemic. And it should come as no surprise that weight gain is likely to ensue when any energy-dense foods and beverages are consumed in excess. So what is it about HFCS, aside from its caloric content, that has sounded the alarm?

Both sucrose and HFCS, when consumed in excess, can provide considerable amounts of fructose. However, the fact that HFCS has a high fructose-to-glucose ratio has led some researchers to believe that it may elicit a physiological response that can lead to unwanted weight gain. For example, some researchers speculate that fructose is more efficiently converted to fat than other types of sugars. Yet few studies have demonstrated metabolic differences in response to HFCS and sucrose ingestion.[5] Similarly, there is little or no credible evidence that HFCS and sucrose differ in terms of their effect on satiety, food intake, and subsequent energy intake.[6]

Until scientists learn more about the effects of HFCS on health, we can all agree that consuming too many calories in the form of sugar and HFCS can contribute to unwanted weight gain. Undoubtedly, the relationship between high intakes of HFCS and obesity is likely due to excess consumption of energy-dense foods and beverages and not HFCS *per se*.

High-fructose corn syrup is a widely used sweetener found in a variety of foods and beverages.

state. Rather, most of the galactose in our diet comes from the disaccharide lactose (comprised of galactose and glucose) in dairy products. The body uses galactose to make certain components of cell membranes and to synthesize lactose, an important sugar in breast milk. However, the majority of galactose in the body is converted to glucose and used as a source of energy (ATP).

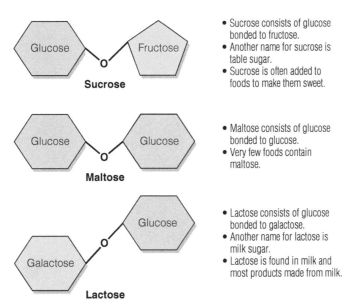

- Sucrose consists of glucose bonded to fructose.
- Another name for sucrose is table sugar.
- Sucrose is often added to foods to make them sweet.

Sucrose

- Maltose consists of glucose bonded to glucose.
- Very few foods contain maltose.

Maltose

- Lactose consists of glucose bonded to galactose.
- Another name for lactose is milk sugar.
- Lactose is found in milk and most products made from milk.

Lactose

FIGURE 8.5 Disaccharides The disaccharides sucrose, maltose, and lactose consist of two monosaccharides bonded together. Notice that each disaccharide has at least one glucose molecule.

DISACCHARIDES CONSIST OF TWO MONOSACCHARIDES

Disaccharides consist of two monosaccharides bonded together. The most common disaccharides are **lactose** (galactose and glucose), **maltose** (glucose and glucose), and **sucrose** (fructose and glucose). In all of these disaccharides, at least one of the monosaccharides in the pair is glucose (Figure 8.5). A condensation reaction chemically joins monosaccharides together by a glycosidic bond. As shown in Figure 8.6, a **glycosidic bond** is formed when the hydroxyl group (–OH) from one monosaccharide interacts with a hydrogen group (–H) from another monosaccharide, resulting in the loss of one molecule of water (H_2O). Disaccharides have a molecular formula of $C_{12}H_{22}O_{11}$, which reflects the loss of the water molecule ($2C_6H_{12}O_6 \rightarrow C_{12}H_{22}O_{11} + H_2O$).

In addition to consisting of different pairs of monosaccharides, disaccharides can also have different types of glycosidic bonds. Numbers are used to designate which carbon atoms form the glycosidic bond. For example, a

lactose Disaccharide consisting of glucose and galactose; produced by mammary glands.

maltose (MAL – tose) Disaccharide consisting of two glucose molecules bonded together; formed during the chemical breakdown of starch.

sucrose (SU – crose) Disaccharide consisting of glucose and fructose; found primarily in fruits and vegetables.

glycosidic bond (gly – co – SI – dic) A type of chemical bond that forms between two monosaccharides.

⟨CONNECTIONS⟩ A condensation reaction occurs when molecules are bonded together with the release of a molecule of water (Chapter 1, page 8).

FIGURE 8.6 Formation of Glycosidic Bonds A glycosidic bond is formed by a condensation reaction between two monosaccharides. In this example, two glucose molecules join to form water and the disaccharide maltose.

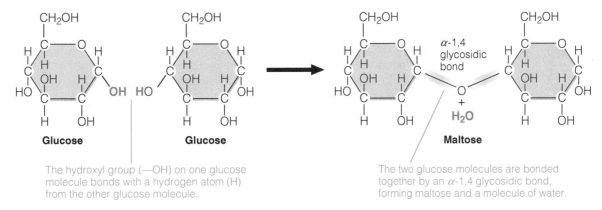

The hydroxyl group (—OH) on one glucose molecule bonds with a hydrogen atom (H) from the other glucose molecule.

The two glucose molecules are bonded together by an α-1,4 glycosidic bond, forming maltose and a molecule of water.

FIGURE 8.7 The Naming of Glycosidic Bonds A glycosidic bond joins two monosaccharides. The numbers in a glycosidic bond refer to the carbon atoms that form the bond. If the glycosidic bond is facing down, it is called an alpha (α) glycosidic bond. A glycosidic bond facing up is called a beta (β) glycosidic bond.

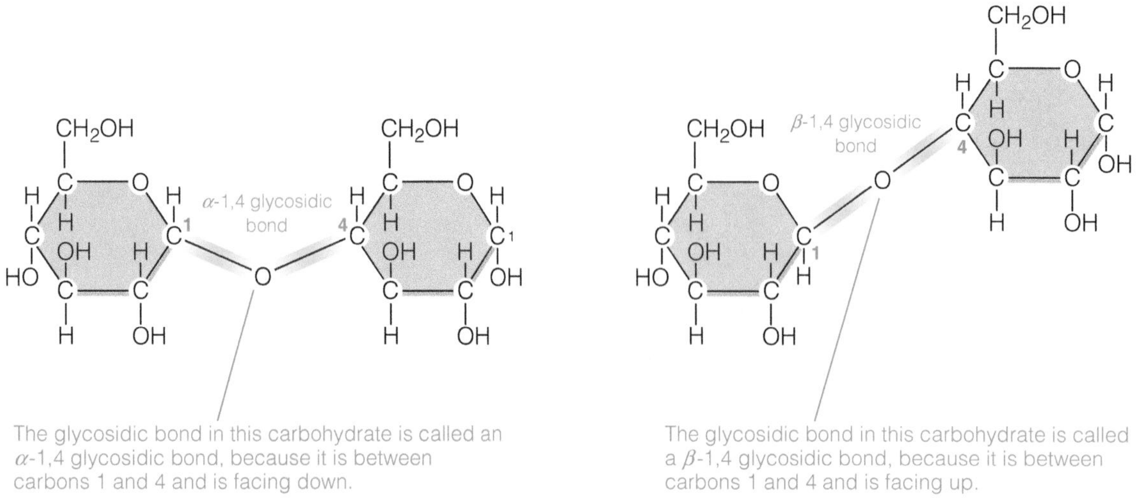

The glycosidic bond in this carbohydrate is called an α-1,4 glycosidic bond, because it is between carbons 1 and 4 and is facing down.

The glycosidic bond in this carbohydrate is called a β-1,4 glycosidic bond, because it is between carbons 1 and 4 and is facing up.

alpha (α) glycosidic bond A downward-facing type of glycosidic bond between two monosaccharides.

beta (β) glycosidic bond An upward-facing type of glycosidic bond between two monosaccharides.

glycosidic bond between carbon 1 on the glucose molecule and carbon 4 on the other monosaccharide is referred to as a 1,4 glycosidic bond. In addition, the bond may be an **alpha (α) bond,** which faces down, or a **beta (β) bond,** which faces up (Figure 8.7). This system offers a convenient way to describe important structural features of carbohydrates. Understanding these basic differences between glycosidic bonds is also important because they can determine whether a carbohydrate is digestible or indigestible.

Lactose, Maltose, and Sucrose Are Disaccharides The disaccharides lactose, maltose, and sucrose are found in a wide variety of foods. **Lactose** is the most abundant carbohydrate in milk, and is the only disaccharide that has a beta glycosidic bond between the monosaccharides. During lactation, enzymes in the mammary glands (breasts) combine glucose and galactose to produce lactose. Thus, milk and milk products (such as yogurt, cheese, and ice cream) contain lactose, although its concentrations can vary.

The disaccharide maltose is formed during starch digestion. Maltose is not found in many foods, but the breakdown of starch to maltose is an important step in the production of beer. During the brewing process, enzymes convert cereal starches such as those found in barley to maltose. Maltose is then fermented by bacteria to produce the alcohol found in beer.

Sucrose is found in many plants, and is especially abundant in sugar cane and sugar beets. These plants can be crushed to produce a juice that is processed to make a brown liquid called molasses. Further treatment and purification forms pure crystallized sucrose, otherwise known as refined table sugar. Because most people enjoy the intense sweetness of sucrose, it is often added to processed foods.

Naturally Occurring Sugars and Added Sugars Foods like fruits, vegetables, and milk contain naturally occurring sugars, while many processed foods contain added sugar or syrups. On food labels, the term "sugar" usually refers to both added sugar and those that occur naturally. As you can see in Table 8.1, many types of sugars and syrups can be added to foods, including white sugar, brown sugar, corn syrup, honey, and molasses. Sweetness is not the only reason sugar is added to foods. For example, sugars and syrups can be used to thicken and to alter the texture of foods.

Although added sugars are chemically identical to naturally occurring sugars, foods with naturally occurring sugars are usually nutrient dense, providing high amounts of nutrients relative to the amount of calories. For example, besides sucrose, fruits are naturally rich in vitamins, minerals, and fiber. In contrast, foods with large amounts of added sugars, such as soft drinks, cakes, and candy, often have little nutritional value beyond the calories they contain. Some people believe that sweeteners such as honey are healthier alternatives to sweeteners such as refined white sugar. However, like refined sugars, natural sweeteners also have limited nutritional value other than energy.

In recent years, consumption of added sugars has increased throughout the United States. Today, the average American consumes about 89 grams (22 teaspoons) of added sugars—a total of 355 kcal—every day.[7] Although this may not sound like much, the calories can add up quickly. Nearly half (46%) of the added sugars in the American diet are from sugar-sweetened soft drinks.[8] Moderate evidence suggests that greater consumption of sugar-sweetened beverages is associated with increased body weight in adults.[9] This is not surprising when you consider that the average 12-oz can of sweetened soft drink contains 8 teaspoons of added sugar—the equivalent of 130 kcal. To lower the caloric content of foods, food manufacturers often use nonsugar, low-calorie sweeteners. Because sugar-laden foods are associated with tooth decay, nonsugar substitutes may provide a healthier sweet alternative. You can learn more about these products in the Focus on Food feature.

TABLE 8.1 Selected Terms Used on Food Labels for Added Sweeteners

Brown rice syrup
Brown sugar
Concentrated fruit juice sweetener
Confectioner's sugar
Corn syrup
Dextrose
Fructose
Glucose
Granulated sugar
High-fructose corn syrup
Honey
Invert sugar
Lactose
Levulose
Maltose
Maple sugar
Molasses
Natural sweeteners
Raw sugar
Sucrose
Turbinado sugar
White sugar

What Are Complex Carbohydrates?

In contrast to the simple carbohydrates that contain one or two monosaccharides, **complex carbohydrates** consist of many monosaccharides bonded together in a variety of arrangements. Complex carbohydrates include **oligosaccharides,** made of 3 to 10 monosaccharides, and **polysaccharides,** which consist of more than 10 monosaccharides. However, most polysaccharides are made of hundreds of monosaccharides bonded together. You can refer back to Figure 8.1 to see how oligosaccharides and polysaccharides fit in the overall scheme of carbohydrate classification.

OLIGOSACCHARIDES ARE COMPONENTS OF CELL MEMBRANES

Oligosaccharides are present in a variety of foods, including dried beans, soybeans, peas, and lentils. Raffinose and stachyose are the two most common oligosaccharides. However, because humans lack the enzymes needed to digest these two carbohydrates, they pass undigested into the large intestine, where bacteria break them down. As a result, some people experience abdominal discomfort (bloating and cramps) and flatulence (gas). Commercial products such as Beano® supply enzymes needed to break down oligosaccharides, making them more digestible and therefore less available to intestinal bacteria.

In the body, oligosaccharides are components of cell membranes, and allow cells to recognize and interact with one another. Oligosaccharides are

complex carbohydrates Category of carbohydrate that includes oligosaccharides and polysaccharides.

oligosaccharide (o – li – go – SAC – cha – ride) (*oligo-*, few; *-saccharide*, sweet) Carbohydrate made of relatively few (3 to 10) monosaccharides.

polysaccharide (po – li – SAC – cha – ride) (*poly-*, many; *-saccharide*, sweet) Complex carbohydrate made of many monosaccharides.

Nutritive sweeteners are naturally occurring, digestible carbohydrates such as sucrose. However, food manufacturers can also use nonsugar alternatives to sweeten foods, including saccharin, aspartame, acesulfame K, and sugar alcohols. In addition to these alternative sweeteners, a sweet-tasting herbal dietary product called *stevia* has recently been approved by the Food and Drug Administration (FDA). This long-awaited decision means that food companies can now use stevia to sweeten foods and beverages.

Choosing which nonsugar sweetener to use is not as simple as you might think. For example, some lose their sweetness when heated and therefore are not recommended for baking or cooking. Others are not chemically stable and become bitter over time. In addition, ongoing debates questioning the safety of some nonsugar substitutes have raised health concerns among consumers.

- **Saccharin.** Saccharin was one of the first nonsugar *artificial* sweeteners to be widely used in the United States. During World War I, the use of saccharin increased because of sugar rationing. Although there was concern that saccharin might cause cancer, recent studies support its safety for human consumption. Most experts agree that the use of saccharin in moderation poses no health risks.[10] Saccharin is extremely sweet, very stable, and inexpensive to produce. Commercial products with saccharin include Sweet'N Low® and Sugar Twin®.

- **Aspartame.** Aspartame, another nonsugar sweetener, sold by its trade names NutraSweet® and Equal®, consists of two amino acids (phenylalanine and aspartic acid) bonded together. Although aspartame has the same energy content as sucrose (4 kcal per gram), it is almost 200 times as sweet. The food industry uses aspartame in sugar-free beverages. However, it is not heat stable and therefore cannot be used in products that require cooking. Beverages made with aspartame must be clearly labeled because of a potential risk to people with a genetic condition called phenylketonuria (PKU). Individuals with PKU cannot metabolize the amino acid phenylalanine, a component of this sweetener. Although the FDA has judged aspartame safe, some people claim it has adverse effects, including headaches, dizziness, nausea, and seizures. While several studies have reported that feeding laboratory animals aspartame increased the occurrence of cancer, hundreds of studies have failed to find similar effects in humans.[11]

- **Acesulfame K.** Another nonsugar sweetener, called acesulfame K, is actually a salt that contains potassium (K is the symbol for potassium). Acesulfame K has been used extensively in Europe and was approved in 1998 for use in the United States, where it is sold under the trade name of Sweet One®. Unlike aspartame, this artificial sweetener is heat stable and can be used in a wide variety of commercial products. Acesulfame K has been approved for use in refrigerated and frozen desserts, yogurt, dry dessert mixes, candies, gum, syrups, and alcoholic beverages.

- **Sucralose.** Sucralose (trade name Splenda®), another low-calorie sweetener, was approved by the FDA in 1990. The production of sucralose starts with sucrose, which then undergoes considerable processing. The end result is a sugar molecule with chlorine atoms attached. Sucralose is about 600 times sweeter than sucrose but provides minimal calories because it is difficult for the body to digest and absorb. Because it is water soluble and stable, sucralose is used in a broad range of foods and beverages.

- **Sugar Alcohols.** Sugar alcohols, another group of alternative sweeteners, are neither sugars nor alcohols. Rather, they are "polyols," meaning that the sugar molecule has multiple alcohol groups attached. Sugar alcohols occur naturally in plants, particularly fruits, and have half the sweetness and calories of sucrose (only 2 to 3 kcal per gram). Sorbitol, mannitol, and xylitol, the most common sugar alcohols, are often found in "sugar-free" products such as chewing gums, breath mints, candies, toothpastes, mouthwashes, and cough syrups. One advantage of sugar alcohols is that, unlike sucrose, they do not readily promote tooth decay. However, when eaten in excessive amounts, sugar alcohols can have a laxative effect, leading to diarrhea.

- **Stevia.** Stevia is sometimes referred to as the "natural sugar substitute" because it is derived from sunflowers that grow freely in tropical regions of Central and South America. The native people of these regions have long known about the sweet properties of stevia. Stevia is essentially calorie free and is considerably sweeter than table sugar. Now that the FDA no longer objects to the use of stevia, product development by several large food manufacturers is underway. Stevia has multiple trade names including Only Sweet™, PureVia™, Reb-A™, Rebiana™, SweetLeaf™, and Truvia®.

Although foods sweetened with nonsugar alternatives may appear to offer sweet indulgences with little caloric cost, this is not always the case. It is important to understand that foods sweetened with low-energy sweeteners are not necessarily healthy food choices. Many of these foods offer little in terms of nutritional value. Furthermore, replacing caloric sweeteners with low-energy sugar substitutes is a common practice among those trying to lose weight. Substituting nonsugar sweeteners for sugar sweeteners alone will not facilitate weight loss.[12] As obesity rates continue to climb, so does the use of low-energy sugar substitutes.

also made in the breasts, where they are incorporated into human milk. These oligosaccharides are part of a complex system that helps protect infants from disease-causing pathogens, and they are one of the many reasons why mothers are encouraged to breastfeed.

POLYSACCHARIDES DIFFER IN THE TYPES AND ARRANGEMENTS OF SUGAR MOLECULES

Polysaccharides are made of many monosaccharides bonded together by glycosidic bonds. The types and arrangements of the sugar molecules determine the shape and form of the polysaccharide. For example, some polysaccharides have an orderly linear appearance, whereas others are shaped like branches on a tree. The three most common polysaccharides—starch, glycogen, and dietary fiber—are discussed next.

Amylose and Amylopectin Are Types of Starch Recall that plants synthesize glucose by the process of photosynthesis. To store this important source of energy, plants convert the glucose to starch. There are two forms of starch, **amylose** and **amylopectin,** both of which consist entirely of glucose molecules. However, what distinguishes amylose from amylopectin is the

amylose (A – my – lose) A type of starch consisting of a linear (unbranched) chain of glucose molecules.

amylopectin (a – my – lo – PEK – tin) A type of starch consisting of a highly branched arrangement of glucose molecules.

FIGURE 8.8 Structure of Starch Plants store glucose in the form of starch. Both forms of starch (amylose and amylopectin) consist of glucose molecules bonded together.

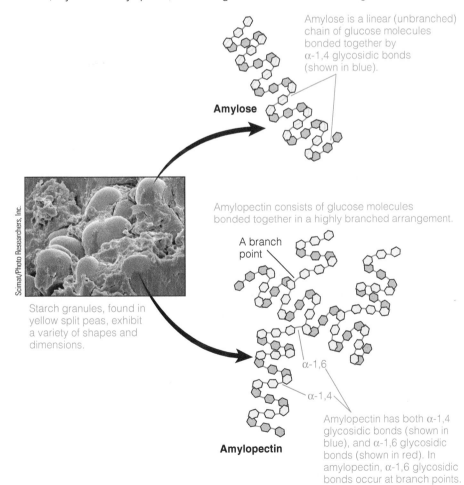

Amylose is a linear (unbranched) chain of glucose molecules bonded together by α-1,4 glycosidic bonds (shown in blue).

Amylose

Amylopectin consists of glucose molecules bonded together in a highly branched arrangement.

A branch point

α-1,6

α-1,4

Amylopectin has both α-1,4 glycosidic bonds (shown in blue), and α-1,6 glycosidic bonds (shown in red). In amylopectin, α-1,6 glycosidic bonds occur at branch points.

Amylopectin

Starch granules, found in yellow split peas, exhibit a variety of shapes and dimensions.

Scimat/Photo Researchers, Inc.

arrangement of glucose molecules (Figure 8.8). Whereas amylose is a linear (unbranched) chain of glucose molecules held together entirely by α-1,4 glycosidic bonds, amylopectin is a highly branched arrangement of glucose molecules. The linear portions of amylopectin contain α-1,4 glycosidic bonds, whereas α-1,6 glycosidic bonds occur at branch points.

Plants typically contain a mixture of both types of starch—amylose and amylopectin. Examples of starchy foods include grains (such as corn, rice, and wheat), products made from them (such as pasta and bread), and legumes (such as lentils and split peas). Potatoes and winter (hard) squashes are also sources of starch. Starch is often added to food to enhance its texture and stability. For instance, cornstarch forms a thick gel when heated, which is why it is used to thicken sauces and gravies. Food scientists have also developed ways to chemically modify starch to improve its functionality. This is why modified food starch is an ingredient commonly used in food production.

The Body Stores Glucose as Glycogen The body stores small amounts of glucose in the form of **glycogen**. Although many tissues store small amounts of glycogen, the majority is found in liver and skeletal muscles. Like amylopectin, glycogen is a highly branched arrangement of glucose molecules, consisting of both α-1,4 (linear portions) and α-1,6 (branch points) glycosidic bonds (Figure 8.9). The numerous branch points in glycogen provide a physiological advantage because enzymes can hydrolyze multiple glycosidic bonds simultaneously. As a result, glycogen can be broken down quickly when energy is needed.

glycogen (GLY – co – gen) Polysaccharide consisting of a highly branched arrangement of glucose molecules; found primarily in liver and skeletal muscle.

FIGURE 8.9 Structure of Glycogen Animals store glucose in the form of glycogen, a polysaccharide consisting of many glucose molecules.

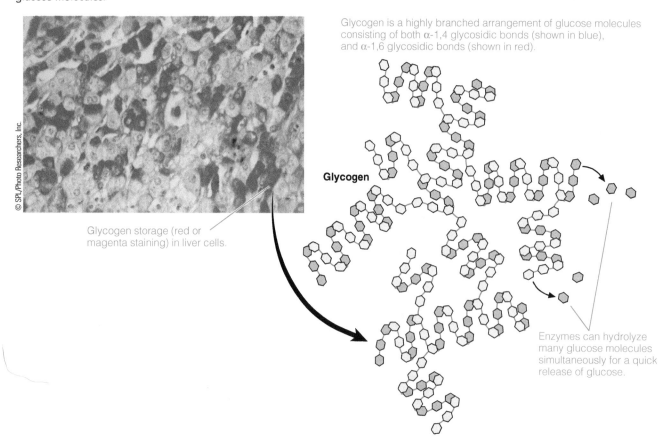

Glycogen is a highly branched arrangement of glucose molecules consisting of both α-1,4 glycosidic bonds (shown in blue), and α-1,6 glycosidic bonds (shown in red).

Glycogen

© SPL/Photo Researchers, Inc.

Glycogen storage (red or magenta staining) in liver cells.

Enzymes can hydrolyze many glucose molecules simultaneously for a quick release of glucose.

FIGURE 8.10 Humans Are Unable to Digest Dietary Fiber Unlike starch (which can be fully broken down into glucose molecules), dietary fiber remains intact in the gastrointestinal tract. This is because humans lack digestive enzymes that can break the glycosidic bonds in fiber.

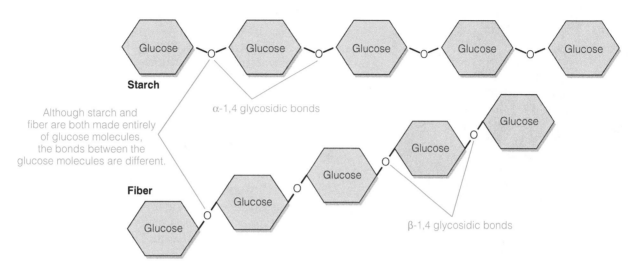

Because glucose is an important energy source for cells, the body turns to glycogen when glucose availability is low, such as during fasting and strenuous or prolonged exercise. When liver glycogen is broken down, the glucose can be released directly into the blood. Therefore, liver glycogen plays an important role in blood glucose regulation. However, muscle glycogen serves a somewhat different role. Unlike the liver, muscle lacks the enzyme needed to release glucose into the blood. In this case, the glucose that results from the breakdown of muscle glycogen is used to fuel physical activity. Some athletes try to increase their muscle glycogen stores by a technique called **carbohydrate loading.**

Humans Are Unable to Digest Fiber The term **fiber** generally refers to a diverse group of plant polysaccharides that, unlike starch, cannot be digested in the human small intestine. This is because fiber contains β-glycosidic bonds that are resistant to digestive enzymes (Figure 8.10). Undigested fiber passes from the small to the large intestine relatively intact. Intestinal bacteria then begin to break down the fiber, producing gas, lipids, and other substances. Although gas production (flatulence) by intestinal bacteria may be an annoyance, other substances made by these microorganisms serve useful purposes such as nourishing cells that line the colon. In addition, dietary fiber promotes the selective growth of beneficial intestinal bacteria, which in turn help inhibit the growth of other, disease-causing (pathogenic) bacteria.[13]

Classification of Fiber Types Our diets contain many kinds of fiber. **Dietary fiber** occurs naturally in plants, whereas **functional fiber** refers to fiber added to food as an ingredient. Functional fiber is typically derived from natural fibrous plant sources and used to manufacture foods or other types of products. It can also be synthetically manufactured. For a fiber to be called functional fiber, it must have demonstrated beneficial physiological effects. Functional fiber can increase the total fiber content of certain foods, which in turn provides health benefits. Thus, the term **total fiber** refers to the combination of dietary fiber that exists naturally in a food plus any functional fiber that is added during manufacturing.

carbohydrate loading A technique used to increase the body's glycogen stores.

fiber Polysaccharide found in plants that is not digested or absorbed in the human small intestine.

dietary fiber Fiber that naturally occurs in plants.

functional fiber Fiber that is added to food to provide beneficial physiological effects.

total fiber The combination of dietary fiber and functional fiber.

Fiber is an important part of a healthy diet. Soluble fiber retains water, and can help lower blood cholesterol. Insoluble fiber adds bulk, and can aid in weight management. Most fiber-rich foods such as dried beans and peanuts provide a combination of both soluble and insoluble fiber.

⟨**CONNECTIONS**⟩ Recall that prebiotics promote the growth of healthy bacteria in the colon (Chapter 1, page 44).

soluble fiber Dietary fiber that dissolves in water.

insoluble fiber Dietary fiber that does not dissolve in water.

Dietary fiber is found in a variety of plant foods such as whole grains, legumes, vegetables, and fruits. Different foods contain different types of dietary fiber, which is commonly classified on the basis of physical properties. For example, fiber is often categorized according to its solubility in water. **Soluble dietary fiber** tends to dissolve or swell in water, while **insoluble dietary fiber** remains relatively unchanged. Gel-forming soluble fibers are readily broken down (fermentable) by bacteria residing in the colon, as opposed to insoluble fibers, which are largely nonfermentable.

Health Benefits Associated with Fiber While fiber is not a required dietary component, it is an important part of a healthful diet. Studies consistently show that fiber from whole foods protects against cardiovascular disease, obesity, and type 2 diabetes. Furthermore, a fiber-rich diet is essential for digestive health. The evidence supporting the health benefits of fiber-rich foods is so substantive that the FDA has approved several fiber-related health claims (Table 8.2).[14] Though fiber is not a magic bullet, its credentials are impressive.

Foods such as oats, barley, legumes, rice, bran, psyllium seeds, soy, and some fruits contain mostly soluble, viscous dietary fibers that "absorb" water and swell. Examples of fibers with these physical properties include pectin, gums, and β-glucan. The sponge-like effect of these types of fiber can help soften fecal matter, reducing strain and making elimination easier. Consumption of soluble, viscous fiber can also help reduce blood cholesterol levels in some people.[15] Research indicates that the viscous fiber may bind with dietary fat and cholesterol in the GI tract, making it less likely to be absorbed. Similarly, eating foods with soluble, viscous fiber can delay gastric emptying, which can help promote satiety. Delayed gastric emptying may also help lower blood glucose levels.[16] In addition to these health-promoting benefits, soluble, fermentable fibers can also function as a prebiotic, promoting the growth of "friendly" bacteria in the colon.

Cellulose is the most abundant insoluble dietary fiber in food. Examples of foods high in cellulose include whole-grain flour, wheat bran, whole-grain breakfast cereals, seeds, and many vegetables including carrots, broccoli, celery, peppers, and cabbage. In addition to cellulose, insoluble dietary

TABLE 8.2 List of FDA-Approved Health Claims Concerning Fiber

- "Diets low in fat and rich in high-fiber foods may reduce the risk of certain cancers."
- "Diets low in fat and rich in soluble fiber may reduce risk of heart disease."
- "Diets low in fat and rich in fruits and vegetables may reduce the risk of certain cancers."
- "Diets low in fat and rich in whole oats and psyllium seed husk can help reduce the risk of heart disease."
- "Diets high in whole-grain foods and other plant foods and low in total fat, saturated fat, and cholesterol may help reduce the risk of heart disease and certain cancers."

SOURCE: U.S. Food and Drug Administration Health Claims Meeting Significant Scientific Agreement (SSA). Available from http://www.fda.gov/Food/LabelingNutrition/LabelClaims/HealthClaimsMeetingSignificantScientificAgreementSSA/default.htm

TABLE 8.3 Common Dietary Fibers, Food Sources, and Potential Physiological Effects

Dietary Fiber	Description	Food Sources	Potential Physiological Effects
Cellulose	Insoluble fiber consisting of glucose molecules with β–glycosidic bonds; the main structural component of plant cell walls.	Whole grains Bran Cereals Broccoli	Increases stool weight; may decrease transit time
Hemicellulose	Insoluble fiber consisting of a variety of different monosaccharide molecules (e.g., glucose, arabinose, mannose, and xylose).	Cabbage Legumes Apples Root vegetables	
Pectin	Soluble fiber found in the skin of ripe fruits; consists of a variety of monosaccharide molecules; is used commercially to make jams and jellies.	Apples Citrus fruits Strawberries Raspberries	Some evidence exists that it lowers cholesterol by increased excretion of bile acids and cholesterol; significantly reduces glycemic response
β-Glucan	A nonstarch polysaccharide composed of branched chains of glucose molecules.	Mushrooms Barley Oats	Oat bran increases stool weight because of viscosity and fermentation by bacteria; reduces cholesterol; may reduce blood glucose levels
Gums	Highly soluble and viscous nonstarch polysaccharides used to thicken foods.	Oats Legumes	Reduces blood cholesterol and blood glucose
Psyllium	Insoluble nonpolysaccharide dietary fiber, consisting of numerous alcohol units, found within the woody portion of plants.	Berries Wheat	May decrease lipid absorption

SOURCE: Institute of Medicine. Dietary reference intakes for energy, carbohydrate, fiber, fat, fatty acids, cholesterol, protein, and amino acids. Washington, DC: National Academies Press; 2005.

fiber includes hemicellulose and lignin, both of which are found in wheat and some green, leafy vegetables (Table 8.3). Because insoluble fibers do not readily dissolve in water, they do not form viscous gels. Nor do bacteria in the colon readily ferment them. Instead, insoluble fiber passes intact through the GI tract, which helps to increase fecal weight and volume. In general, large amounts of fecal mass move through the colon quickly by stimulating peristaltic contractions in the colon, propelling the material forward. When consumed with sufficient amounts of fluid, insoluble fiber can help prevent and alleviate constipation.

The production of hard, dry feces, characteristic of insufficient fiber, not only makes elimination more difficult but may also contribute to a condition called **diverticular disease** (Figure 8.11). When it is chronic, constipation can cause areas along the colon wall to become weak. As a result, protruding pouches, called diverticula, can form. It is not uncommon for older adults, especially those with low intakes of fibrous foods, to develop diverticula. By 70 years of age, approximately 50% of adults have diverticulosis, 10 to 20% of whom develop complications such as diverticulitis.[17] **Diverticulitis** occurs when the diverticula become infected or inflamed (note that the suffix –itis refers to inflammation). Symptoms include cramping, diarrhea, fever, and on occasion bleeding from the rectum. Scientists think that dietary fiber helps prevent the formation of diverticula by increasing fecal mass, making bowel movements easier. Therefore, a diet high in fiber with plenty of fluids may help protect against diverticular disease.

Whole-Grain Foods There are many ways to ensure adequate fiber intake. In addition to eating a variety of fruits and vegetables every day, it is important to select foods made from whole grains. The nutritional value of grains is greatest when all three components of the grain—bran, germ, and endosperm—are

diverticular disease, or diverticulosis (di – ver – TI – cu – lar) (di – ver – ti – cu – LO – sis) Condition in the large intestine; characterized by the presence of pouches that form along the intestinal wall.

diverticulitis (di – ver – ti – cu – LI – tis) Inflammation of diverticula (pouches) in the lining of the large intestine.

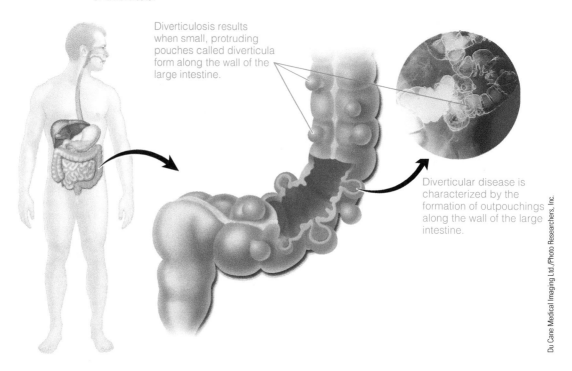

FIGURE 8.11 **Diverticular Disease** Diverticular disease is most common in older adults who consume low intakes of fibrous foods. Diverticulitis is the inflammation of diverticula.

Diverticulosis results when small, protruding pouches called diverticula form along the wall of the large intestine.

Diverticular disease is characterized by the formation of outpouchings along the wall of the large intestine.

Du Cane Medical Imaging Ltd./Photo Researchers, Inc.

present (Figure 8.12). Whereas the **bran** contains most of the fiber, the **germ** supplies much of the vitamins and minerals. The **endosperm** is mostly starch. Because milling removes the bran and germ, products made with refined flour often contain very little fiber. Although some of the lost nutritive value is restored when food manufacturers fortify their products with vitamins and minerals, many important nutrients, as well as fiber, may still be lacking.

When reading food labels, look for the words "whole-grain cereal" and "whole-wheat flour." Foods made with "wheat flour" are not necessarily high in fiber. To make it easier for consumers to select whole-grain products, the FDA has published a tentative definition: **whole-grain foods** should contain

bran The outer layer of a grain; contains most of the fiber.

germ The portion of a grain that contains most of its vitamins and minerals.

endosperm The portion of a grain that contains mostly starch.

whole-grain foods Cereal grains that contain bran, endosperm, and germ in the same relative proportion as they exist naturally.

FIGURE 8.12 **Anatomy of a Wheat Kernel** Foods made with whole-wheat flour are typically more nutritious than foods made with refined wheat flour because the components of the wheat kernel—the bran, the germ, and the endosperm—have not been removed. Each component contributes important nutrients needed for good health.

Germ contains vitamins and minerals.

Endosperm is mostly starch.

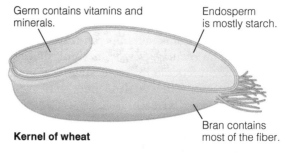

Kernel of wheat

Bran contains most of the fiber.

The processing of grains often removes the germ and bran portions, which contain the majority of the vitamins, minerals, and fiber.

TABLE 8.4 The Dietary Fiber Content of Selected Foods

Food	Serving Size	Insoluble Fiber (g)	Soluble Fiber (g)	Total Dietary Fiber (g)
Fruits				
Apple	1 medium	2.0	0.9	2.9
Orange	1 medium	0.7	1.3	2.0
Banana	1 medium	1.4	0.6	2.0
Vegetables				
Broccoli	1 stalk	1.4	1.3	2.7
Carrot	1 large	1.6	1.3	2.9
Tomato	1 small	0.7	0.1	0.8
Potato	1 medium	0.8	1.0	1.8
Corn	$^2/_3$ cup	1.4	0.2	1.6
Grains				
All-Bran® cereal	$^1/_2$ cup	7.6	1.4	9.0
Oat bran	$^1/_2$ cup	2.2	2.2	4.4
Cornflakes® cereal	1 cup	0.5	0	0.5
Rolled oats	$^3/_4$ cup	1.7	1.3	3.0
Whole-wheat bread	1 slice	1.1	0.3	1.4
White bread	1 slice	0.1	0.3	0.4
Macaroni	1 cup cooked	0.3	0.5	0.8
Legumes				
Green peas	$^2/_3$ cup	3.3	0.6	3.9
Kidney beans	$^1/_2$ cup	4.9	1.6	6.5
Pinto beans	$^1/_2$ cup	4.7	1.2	5.9
Lentils	$^2/_3$ cup	3.9	0.6	4.5

SOURCE: Anderson JW, Bridges SR. Dietary fiber content of selected foods. American Journal of Clinical Nutrition. 1988; 47: 440–7.

the three key ingredients of cereal grains—bran, endosperm, and germ—in the same relative proportion as the naturally-occurring grain.[18]

Based on USDA food-labeling guidelines, a food labeled as an "excellent source of fiber" must contain 20% or more of the Daily Value of fiber or at least 5 grams of fiber per serving. Foods with labels claiming to be "good" sources of fiber have 2.5 to 4.9 g of fiber per serving. The fiber content of selected foods is shown in Table 8.4.

How Are Carbohydrates Digested, Absorbed, and Circulated in the Body?

Because carbohydrates are structurally diverse, the steps involved in the digestive process vary. During digestion, carbohydrates undergo extensive chemical transformations as they move through the GI tract. With the help of digestive enzymes, the glycosidic bonds that hold disaccharides and starches together are broken. The ultimate goal of carbohydrate digestion is to break down large, complex molecules such as starches into small, absorbable monosaccharides. This process requires a series of enzymes produced in the salivary glands, pancreas, and small intestine.

The nutritional value is greater in whole-grain foods than in refined-grain foods because all three components of the grain—bran, germ, and endosperm—are present.

⟨CONNECTIONS⟩ The brush border is the absorptive surface of the small intestine that is made up of minute projections called microvilli, which line enterocytes. (Chapter 1, page 32).

⟨CONNECTIONS⟩ Hydrolytic reactions break chemical bonds by the addition of water (Chapter 1, page 8).

salivary α-amylase (A – my – lase) Enzyme, produced by the salivary glands, which digests starch by hydrolyzing α-1,4 glycosidic bonds.

dextrin A partial breakdown product formed during starch digestion, consisting of varying numbers of glucose units.

pancreatic α-amylase (pan – cre – A – tic Al – pha A – my – lase) Enzyme, produced by the pancreas, which digests starch by hydrolyzing α-1,4 glycosidic bonds.

maltase (MAL – tase) Brush border enzyme that hydrolyzes maltose into two glucose molecules.

brush border enzyme Enzyme, produced by enterocytes, which aids in the final steps of digestion.

limit dextrin (DEX – trin) A partial breakdown product formed during amylopectin digestion that contains three to four glucose molecules and an α-1,6 glycosidic bond.

α-dextrinase (DEX – stri – nase) Brush border enzyme that hydrolyzes α-1,6 glycosidic bonds.

disaccharidase (di – SAC – cha – ri – dase) Brush border enzyme that hydrolyzes glycosidic bonds in disaccharides.

sucrase Brush border enzyme that hydrolyzes sucrose into glucose and fructose.

lactase Brush border enzyme that hydrolyzes lactose into glucose and galactose.

lactose intolerance Inability to digest the milk sugar lactose; caused by a lack of the enzyme lactase.

STARCH DIGESTION BEGINS IN THE MOUTH

Recall that the two basic forms of starch, amylose and amylopectin, are both made entirely of glucose molecules, although the types of bonds are slightly different. Amylose has only α-1,4 glycosidic bonds, while amylopectin contains both α-1,4 and α-1,6 glycosidic bonds. Nonetheless, many of the same enzymes are involved in digesting both amylose and amylopectin (Figure 8.13).

The chemical digestion of starch begins in the mouth when the salivary glands release the enzyme salivary α-amylase. **Salivary α-amylase** hydrolyzes the α-1,4 glycosidic bonds in both amylose and amylopectin, resulting in shorter polysaccharide chains of varying lengths called **dextrins.**

However, because food stays in the mouth only a short time, very little starch digestion actually takes place there. Once dextrins enter the stomach, the acidic environment stops the enzymatic activity of salivary α-amylase. Dextrins then pass unchanged from the stomach into the small intestine, where they encounter pancreatic α-amylase, an enzyme produced by the pancreas. Like salivary α-amylase, **pancreatic α-amylase** hydrolyzes α-1,4 glycosidic bonds, transforming dextrins into the disaccharide maltose. Last, **maltase**, a **brush border enzyme** produced by enterocytes, finishes the job of starch digestion by hydrolyzing the last remaining chemical bond in maltose, resulting in two free (unbound) glucose molecules.

The combined efforts of salivary α-amylase, pancreatic α-amylase, and maltase complete the chemical transformation of amylose into multiple glucose molecules. However, an additional step, the hydrolysis of α-1,6 glycosidic bonds, is needed to complete the digestion of amylopectin. Because salivary and pancreatic α-amylases only hydrolyze α-1,4 glycosidic bonds, partial breakdown products called limit dextrins form during amylopectin digestion. **Limit dextrins** consist of three to four glucose molecules and contain α-1,6 glycosidic bonds that were located at branch points in the original amylopectin molecules. The enzyme **α-dextrinase,** also a brush border enzyme, accomplishes the hydrolysis of α-1,6 glycosidic bonds, completing the digestion of amylopectin. Thus, amylose and amylopectin digestion results in the production of numerous glucose molecules that are now ready to be transported into the enterocytes.

DISACCHARIDES ARE DIGESTED IN THE SMALL INTESTINE

The digestion of disaccharides (maltose, sucrose, and lactose) takes place entirely in the small intestine (Figure 8.14). The enterocytes provide the enzymes needed for disaccharide digestion, and are collectively referred to as the **disaccharidases**. Each disaccharide requires a specific disaccharidase. For example, **sucrase** hydrolyzes sucrose into glucose and fructose, maltase hydrolyzes maltose into two glucose molecules, and **lactase** hydrolyzes lactose into glucose and galactose. Once disaccharides have been digested into their component monosaccharides, they can be transported into the enterocytes.

Lactose Intolerance Although most babies produce enough lactase to digest the high amounts of lactose found in milk, some people produce very little or none of this enzyme, so they have difficulty digesting lactose—a condition called **lactose intolerance**. When people with lactose intolerance consume lactose-containing foods, much of the lactose enters the large intestine undigested. Bacteria in the large intestine break down (ferment) the lactose, producing several by-products, including gas. Symptoms such as abdominal cramping,

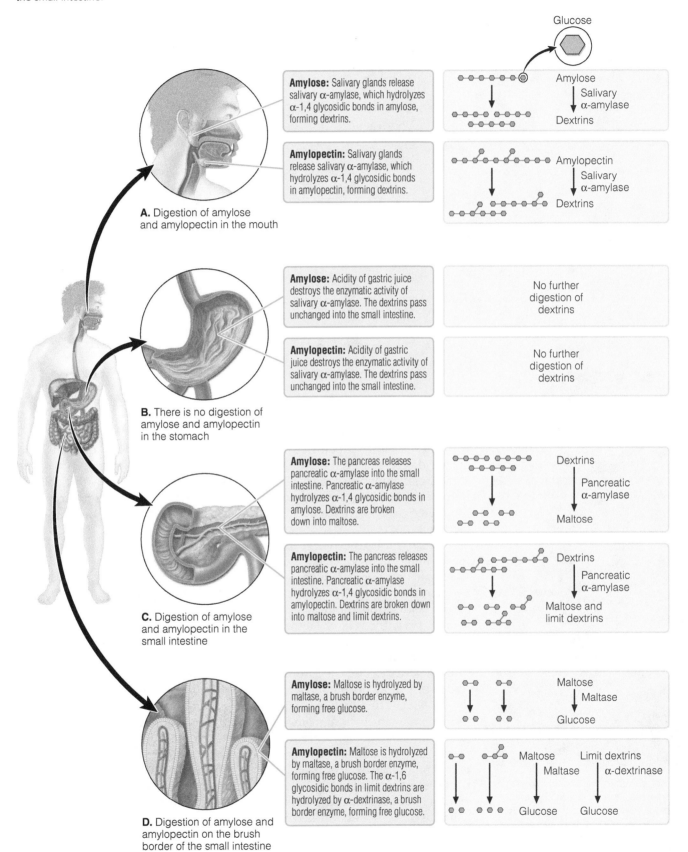

FIGURE 8.13 **Digestion of Amylose and Amylopectin** Amylose and amylopectin digestion require many of the same enzymes. Through the process of digestion, both amylose and amylopectin are broken down into molecules of glucose. Most starch digestion takes place in the small intestine.

Glucose

Amylose: Salivary glands release salivary α-amylase, which hydrolyzes α-1,4 glycosidic bonds in amylose, forming dextrins.

Amylose
Salivary
α-amylase
Dextrins

Amylopectin: Salivary glands release salivary α-amylase, which hydrolyzes α-1,4 glycosidic bonds in amylopectin, forming dextrins.

Amylopectin
Salivary
α-amylase
Dextrins

A. Digestion of amylose and amylopectin in the mouth

Amylose: Acidity of gastric juice destroys the enzymatic activity of salivary α-amylase. The dextrins pass unchanged into the small intestine.

No further digestion of dextrins

Amylopectin: Acidity of gastric juice destroys the enzymatic activity of salivary α-amylase. The dextrins pass unchanged into the small intestine.

No further digestion of dextrins

B. There is no digestion of amylose and amylopectin in the stomach

Amylose: The pancreas releases pancreatic α-amylase into the small intestine. Pancreatic α-amylase hydrolyzes α-1,4 glycosidic bonds in amylose. Dextrins are broken down into maltose.

Dextrins
Pancreatic
α-amylase
Maltose

Amylopectin: The pancreas releases pancreatic α-amylase into the small intestine. Pancreatic α-amylase hydrolyzes α-1,4 glycosidic bonds in amylopectin. Dextrins are broken down into maltose and limit dextrins.

Dextrins
Pancreatic
α-amylase
Maltose and limit dextrins

C. Digestion of amylose and amylopectin in the small intestine

Amylose: Maltose is hydrolyzed by maltase, a brush border enzyme, forming free glucose.

Maltose
Maltase
Glucose

Amylopectin: Maltose is hydrolyzed by maltase, a brush border enzyme, forming free glucose. The α-1,6 glycosidic bonds in limit dextrins are hydrolyzed by α-dextrinase, a brush border enzyme, forming free glucose.

Maltose   Limit dextrins
Maltase   α-dextrinase
Glucose   Glucose

D. Digestion of amylose and amylopectin on the brush border of the small intestine

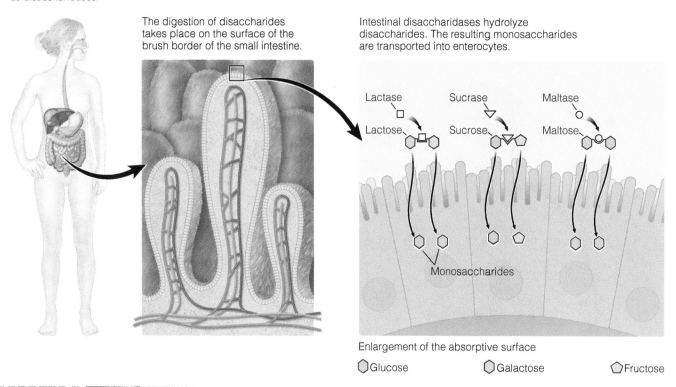

The digestion of disaccharides takes place on the surface of the brush border of the small intestine.

Intestinal disaccharidases hydrolyze disaccharides. The resulting monosaccharides are transported into enterocytes.

Lactase Sucrase Maltase

Lactose Sucrose Maltose

Monosaccharides

Enlargement of the absorptive surface

⬡ Glucose   ⬡ Galactose   ⬠ Fructose

Lactose-reduced milk and other dairy products are widely available at most grocery stores. These dairy products contain all of the nutrients found in regular milk products, with the exception of lactose.

© Scott Bauer/USDA

bloating, flatulence, and diarrhea can occur within 30 to 60 minutes. The severity of symptoms depends on how much lactose is consumed and how little lactase the person produces. Most people with lactose intolerance can consume small amounts of dairy products without experiencing any discomfort. Some dairy products such as yogurt and cheese are easier to tolerate because the bacteria used to make these dairy products convert some of the lactose to lactic acid.

Lactose intolerance is especially common among certain ethnic groups; more than half of Asian Americans, Native Americans, and African Americans are reportedly lactose intolerant.[19] In comparison, only 12% of people of northern European descent are lactose intolerant. Certain diseases, medications, and surgery that damage the intestinal mucosa can also increase the risk of lactose intolerance, although this is usually temporary; lactase production returns once the underlying condition is no longer present.

In most cases, lactose intolerance does not pose a serious health threat. Although it may be annoying and sometimes inconvenient, people can easily live with it. Products such as lactose-free milk and over-the-counter lactase enzyme tablets have made it easier for lactose-intolerant people to enjoy dairy products and consume sufficient calcium. These lactose-reduced dairy products contain all of the nutrients found in regular milk products, with the exception of lactose. However, if these calcium-rich, lactose-reduced alternatives are not available or acceptable, it is important for lactose-intolerant individuals to choose alternate sources of calcium such as fortified soy products, fish with edible bones, fortified orange juice, and calcium-rich vegetables such as Swiss chard and rhubarb. They may also need calcium supplements to ensure the body is receiving adequate amounts of this important mineral.

FIGURE 8.15 Absorption and Circulation of Monosaccharides Glucose and galactose are absorbed into the enterocytes by carrier-dependent, energy-requiring active transport. Fructose is absorbed by facilitated diffusion. Once absorbed, monosaccharides are circulated to the liver via the hepatic portal system.

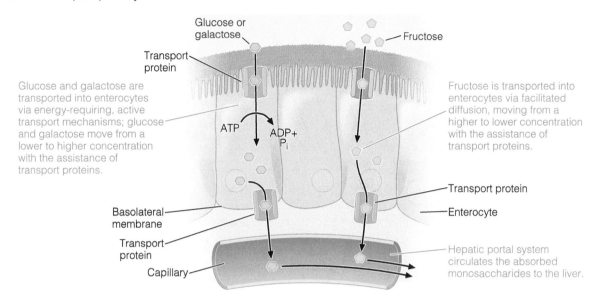

Glucose or galactose

Fructose

Transport protein

Glucose and galactose are transported into enterocytes via energy-requiring, active transport mechanisms; glucose and galactose move from a lower to higher concentration with the assistance of transport proteins.

ATP   ADP+ Pi

Fructose is transported into enterocytes via facilitated diffusion, moving from a higher to lower concentration with the assistance of transport proteins.

Transport protein

Enterocyte

Basolateral membrane

Transport protein

Capillary

Hepatic portal system circulates the absorbed monosaccharides to the liver.

MONOSACCHARIDES ARE READILY ABSORBED FROM THE SMALL INTESTINE

Once disaccharide and starch digestion is complete, the resulting monosaccharides (glucose, galactose, and fructose) are readily absorbed from the small intestine into the blood. As shown in Figure 8.15, glucose and galactose are absorbed across the lumenal membrane of the enterocytes via carrier-mediated active transport, whereas fructose is absorbed through facilitated diffusion. After crossing the basolateral membrane (via facilitated diffusion), monosaccharides then circulate in the blood directly to the liver via the hepatic portal system.

The Glycemic Response A rise in blood glucose levels can be detected shortly after we eat carbohydrate-rich foods. However, not all carbohydrates have the same effect on blood glucose levels. Some foods cause blood glucose levels to rise quickly and remain elevated, while others elicit a more gradual increase. The change in blood glucose following the ingestion of a specific food is called the **glycemic response** (Figure 8.16).

Scientists have long believed that simple carbohydrates elicit higher glycemic responses than complex carbohydrates. However, this is not always the case. In fact, blood glucose response to starchy foods—such as potatoes, refined cereal products, white bread, some whole-grain breads, and white rice—are sometimes higher than foods rich in simple sugars, such as soft drinks. Scientists now recognize that, although carbohydrate complexity may influence the glycemic response elicited by a food, other factors such as the presence of fat or protein, viscosity, processing, ripeness, and food additives are also important.

The **glycemic index (GI)** is a rating system based on a scale of 0 to 100 that can be used to compare the glycemic responses of various foods. In this system, blood glucose response to the consumption of 50 grams of pure glucose is used as a reference point represented by the highest GI value, 100. The GI value of a food is determined experimentally: subjects consume an amount of food that contains 50 grams of carbohydrate, and their blood

⟨**CONNECTIONS**⟩ Carrier-mediated active transport requires a transport protein and energy (ATP) to transport material across cell membranes moving from a lower to a higher concentration. Facilitated diffusion requires only a transport protein moving from a higher to a lower concentration (Chapter 1, page 11).

glycemic response (glyc-, sugar) The change in blood glucose following the ingestion of a specific food.

glycemic index (GI) A rating system used to categorize foods according to the relative glycemic responses they elicit.

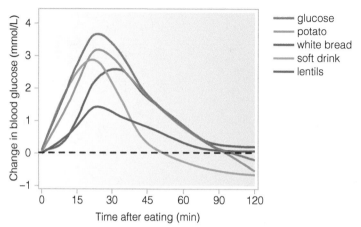

FIGURE 8.16 Glycemic Response Some foods cause blood glucose levels to rise quickly and remain elevated (high glycemic response), while others elicit a more gradual increase (low glycemic response).

SOURCE: Brand-Miller J, McMillan-Price J, Steinbeck K, Carterson I. Dietary glycemic index: Health implications. Journal of the American College of Nutrition. 2009;28:446S–9S.

⟨**CONNECTIONS**⟩ The hepatic portal system is the vascular connection from the small intestine to the liver that circulates absorbed nutrients (Chapter 1, page 33).

glucose response is measured. Foods that elicit glycemic responses similar to that of pure glucose are considered high-GI foods (GI ≥ 70), whereas those that cause a lower or more gradual rise in blood glucose are considered low-GI foods (GI ≤ 55).

One limitation of using GI values to compare the effects of foods on glycemic response is that the amount of carbohydrate found in a typical serving of a food is not taken into account. Rather, GI values are based on a standard *amount* of carbohydrate (50 grams), which may or may not represent the *amount* a person would normally eat. For example, to consume 50 g of carbohydrates from carrots, a person would need to eat more than a pound of carrots, which is unrealistic. To counter this problem, another rating system is sometimes used to evaluate the glycemic response to various foods. In this system, foods are assigned a value called its **glycemic load (GL).** The GL of a food is different from its GI in that the GL takes into account the typical portion of food consumed.

Glycemic load (GL) is calculated by dividing the glycemic index (GI) of a food by 100 and then multiplying by the grams of carbohydrate in one serving. For example, the GL of carrots that have a GI of 16 and contain approximately 8 g carbohydrates per serving is 1.3 (16 ÷ 100 × 8). By comparison, the GL of a 12-oz soft drink that has a GI of 63 and 39 g carbohydrate is approximately 25 (63 ÷ 100 × 39). These values are probably closer to what you may have predicted. You can compare the GI and GL values of selected foods in Table 8.5.

The impact of glycemic response on long-term health remains controversial. While some studies have demonstrated health benefits associated with low-GL diets, others are less clear.[20] Until we know more, it is important to make healthy food choices by increasing our consumption of whole grains, fruits, and vegetables, while decreasing our consumption of foods made with refined flour and added sugar.

MONOSACCHARIDES HAVE SEVERAL FUNCTIONS IN THE BODY

Once absorbed, monosaccharides circulate directly to the liver via the hepatic portal system, where the majority of galactose and fructose is converted into other compounds—most notably glucose. However, some

glycemic load (GL) A rating system used to categorize the body's glycemic response to foods that takes into account the glycemic index as well as the amount of carbohydrate typically found in a single serving of the food.

TABLE 8.5 Glycemic Index (GI) and Glycemic Load (GL) of Selected Foods Based on Glucose as a Reference Food

Value	Glycemic Index (GI)	Glycemic Load (GL)
High	≥70	≥20
Medium	56–69	11–19
Low	≤55	≤10

Food (serving size)	GI	Carbohydrates per Serving (g)	GL	Food	GI	Carbohydrates per Serving (g)	GL
Pastas, Grains, Legumes, Breads, Starchy Vegetables, and Cereals							
Baked potato (150 g)	85	30	26	Corn Chex™ (30 g)	83	30	25
Waffles (35 g)	76	13	10	Corn (80 g)	48	16	8
French fries (150 g)	75	29	22	Popcorn (20 g)	54	11	6
Bagel (70 g)	72	35	25	Cracked wheat (150 g)	48	26	12
Oat bran bread (30 g)	44	18	8	Pancakes (80 g)	67	58	39
White rice (150 g)	56	42	24	Apple muffins (60 g)	44	29	13
Angel food cake (50 g)	67	29	19	Lentils (50 g)	30	40	12
Raisin bran cereal (30 g)	61	19	12				
Fruits, Beverages, and Snack Foods							
Apple (22 g)	38	22	8	Grapes (120 g)	43	17	7
Raisins (60 g)	64	44	28	Apple juice (250 mL)	39	25	10
Banana (120 g)	48	25	12	Tomato juice (250 mL)	38	9	3
Potato chips (28 g)	54	15	8	Plums (120 g)	24	14	3
Jelly beans (28 g)	80	26	21	Chocolate (28 g)	49	18	9
Cherries (120 g)	22	12	3	Sucrose (5 g)	65	5	3

NOTES: Glycemic load (GL) of a food is the glycemic index (GI) divided by 100 and multiplied by the amount of carbohydrate in a single serving (in grams). Therefore, a food's GL is numerically lower than its GI.

SOURCE: Foster-Powell K, Holt SH, Brand-Miller JC. International table of glycemic index and glycemic load values: 2002. American Journal of Clinical Nutrition. 2002; 76: 5–56.

monosaccharides are converted to ribose, a constituent of many vital compounds including ATP, RNA, and DNA. While monosaccharides serve numerous functions within the body, the ability to transform the energy contained in glucose into ATP is probably their most important role.

How Do Hormones Regulate Blood Glucose and Energy Storage?

The concentration of glucose in your blood fluctuates throughout the day. After you have gone several hours without eating, your blood glucose decreases. Conversely, blood glucose increases after you eat a meal rich in carbohydrates. Blood glucose levels are the lowest in the morning after an overnight fast, returning to normal shortly after breakfast. Because our cells need energy 24 hours a day, the pancreatic hormones insulin and glucagon work vigilantly to maintain blood glucose levels within an acceptable range at all times.

THE HORMONES INSULIN AND GLUCAGON ARE PRODUCED BY THE PANCREAS

The pancreas, which is partially composed of hormone-secreting cells, plays a major role in glucose homeostasis. These endocrine cells, collectively known as the islets of Langerhans, consist mainly of beta (β) cells that produce the hormone **insulin** and alpha (α) cells that produce the hormone **glucagon** (Figure 8.17). The pancreas, which has a rich blood supply, releases these hormones directly into the circulation in response to fluctuations in blood glucose.

Insulin and glucagon assist in blood glucose regulation and energy storage. When blood glucose levels increase, the pancreas takes action by releasing more insulin. Insulin in turn lowers blood glucose by facilitating the uptake of glucose into many kinds of cells. In addition, when meals provide more glucose than we require, insulin stimulates its storage as glycogen. Once muscles and the liver reach their glycogen storage capacity, excess glucose is converted to fat, which is stored primarily in adipose tissue. The hormonal balance shifts toward glucagon when blood glucose levels decrease. To increase glucose availability, glucagon stimulates the breakdown of glycogen stores in the liver. It is important to recognize that the release of insulin and glucagon is not an all-or-nothing situation. That is, the relative concentrations of these two hormones in the blood determine the shift between energy storage and mobilization (Figure 8.18). We will now take a closer look at these two important hormones and how each contributes to blood glucose regulation.

⟨**CONNECTIONS**⟩ Homeostasis is a state of balance or equilibrium (Chapter 1, page 15).

insulin Hormone secreted by the pancreatic β-cells in response to increased blood glucose.

glucagon (GLU – ca – gon) Hormone secreted by the pancreatic α-cells in response to decreased blood glucose.

FIGURE 8.17 Release of Insulin and Glucagon from the Pancreas Insulin is made and released by the pancreatic β-cells, and glucagon is made and released by the pancreatic α-cells. Both hormones play an important role in blood glucose regulation.

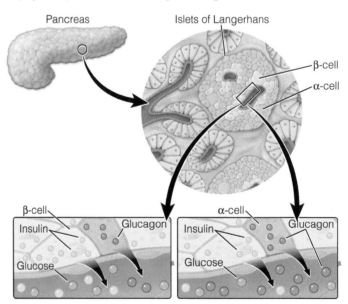

In response to elevated blood glucose levels, the pancreas increases its release of the hormone insulin.

In response to low blood glucose levels, the pancreas increases its release of the hormone glucagon.

FIGURE 8.18 Hormonal Regulation of Blood Glucose
Insulin increases when the level of glucose in the blood increases. Glucagon increases when the level of glucose in the blood decreases.

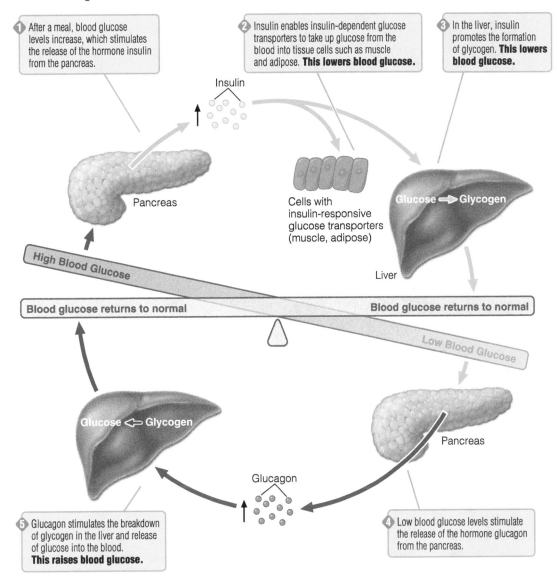

1 After a meal, blood glucose levels increase, which stimulates the release of the hormone insulin from the pancreas.

2 Insulin enables insulin-dependent glucose transporters to take up glucose from the blood into tissue cells such as muscle and adipose. **This lowers blood glucose.**

3 In the liver, insulin promotes the formation of glycogen. **This lowers blood glucose.**

Insulin

Pancreas

Cells with insulin-responsive glucose transporters (muscle, adipose)

Glucose ⟹ Glycogen

Liver

High Blood Glucose

Blood glucose returns to normal

Blood glucose returns to normal

Low Blood Glucose

Glucose ⟸ Glycogen

Glucagon

Pancreas

5 Glucagon stimulates the breakdown of glycogen in the liver and release of glucose into the blood. **This raises blood glucose.**

4 Low blood glucose levels stimulate the release of the hormone glucagon from the pancreas.

INSULIN LOWERS BLOOD GLUCOSE AND PROMOTES ENERGY STORAGE

Glucose enters cells via facilitated diffusion, mediated by carrier proteins known as **glucose transporters** (Figure 8.19). Some glucose transporters require insulin to transport glucose across the cell membrane, whereas others do not. In fact, it is the type of glucose transporter in a cell that determines if insulin is needed for the uptake of glucose. Examples of tissues that have **insulin-responsive glucose transporters** are skeletal muscle and adipose tissue. Brain and liver tissue have glucose transporters that do not require insulin for the uptake of glucose.

After a person eats carbohydrate-containing foods, blood glucose levels quickly rise, causing the pancreas to increase its release of insulin. When insulin encounters cells with insulin-responsive glucose transporters, it binds

glucose transporters Proteins that assist in the transport of glucose molecules across cell membranes.

insulin-responsive glucose transporters Glucose transporters that require insulin to function.

FIGURE 8.19 Role of Insulin in Cellular Uptake of Glucose Binding of insulin to its receptor causes insulin-responsive glucose transporters to relocate from the cytoplasm to the cell membrane, facilitating the movement of glucose molecules across the cell membrane.

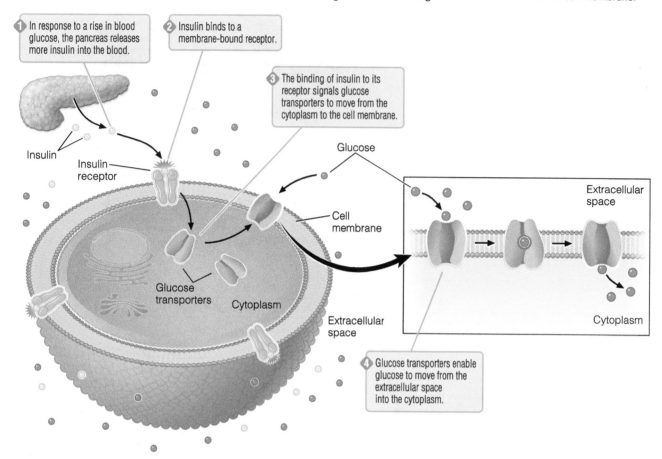

1 In response to a rise in blood glucose, the pancreas releases more insulin into the blood.

2 Insulin binds to a membrane-bound receptor.

3 The binding of insulin to its receptor signals glucose transporters to move from the cytoplasm to the cell membrane.

4 Glucose transporters enable glucose to move from the extracellular space into the cytoplasm.

Insulin

Insulin receptor

Glucose transporters

Cytoplasm

Glucose

Cell membrane

Extracellular space

Extracellular space

Cytoplasm

to **insulin receptors** on the surface of cell membranes. The binding of insulin to its receptor signals glucose transporters to relocate from the cytoplasm to the cell membrane, allowing glucose to enter the cell. During exercise, skeletal muscle cells can temporarily take up glucose without insulin. This is because muscle contractions, like insulin, can activate the movement of glucose transporters from the cytoplasm to the cell membrane, allowing for glucose uptake. The insulin-responsive glucose transporters return to their insulin-requiring state within a few hours after exercise stops.

Sometimes insulin-responsive glucose transporters have difficulty transporting glucose across cell membranes, causing glucose to accumulate in the blood, a condition called **hyperglycemia.** When this occurs, a person is said to have **impaired glucose regulation** or, in more serious cases, diabetes. Impaired blood glucose regulation and diabetes are serious health conditions that can lead to long-term complications. Diabetes prevention and management are fully addressed in the Nutrition Matters following this chapter.

Insulin Promotes Energy Storage After a meal more glucose may be available than is needed. When this occurs, the body stores the excess energy contained in glucose for later use. The hormone insulin promotes the storage of excess energy from glucose in the form of glycogen and body fat (Figure 8.20). In addition, insulin stimulates protein synthesis and inhibits the breakdown of muscle.

insulin receptors Proteins, located on the surface of certain cell membranes, that bind insulin.

hyperglycemia (hi – per – gly – CE – mi – a) (*hyper-*, excessive) Abnormally high level of glucose in the blood.

impaired glucose regulation Condition characterized by elevated levels of glucose in the blood.

FIGURE 8.20 Insulin Promotes Energy Storage The pancreas increases its release of the hormone insulin in response to high blood glucose. Insulin stimulates glucose transport into cells and promotes energy storage.

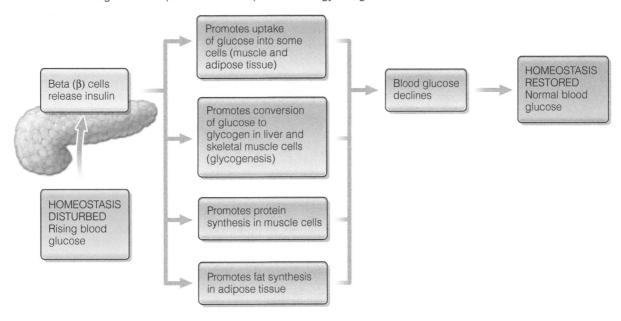

To store excess glucose as glycogen, insulin stimulates a process called **glycogenesis,** which occurs mainly in the liver and skeletal muscle. However, the body can only store a limited amount of glycogen. Once the limit is reached, insulin promotes the uptake of excess glucose by adipose tissue, where the glucose is converted to fatty acids (a type of lipid). Unlike what occurs with glycogen storage, the body has a seemingly endless capacity to store excess energy from glucose as body fat. It is important to note that the conversion of glucose into a fatty acid is irreversible. That is, once glucose is transformed into a fatty acid, the fatty acid cannot be converted back into glucose. This is very different from the reversible metabolic transformation between glucose and glycogen.

Critical Thinking: Laura's Story Think back to Laura's story about her experience with type 1 diabetes, a condition characterized by the body's inability to produce insulin. Based on what you have read about the role of insulin in blood glucose regulation and energy storage, why is it so important for Laura to monitor her blood glucose while exercising? How might low insulin and high blood glucose levels, common in people with diabetes, influence Laura's ability to train for and compete in the Ironman® competition?

GLUCAGON HELPS INCREASE BLOOD GLUCOSE

Clearly insulin is instrumental in helping the body use glucose after a meal and to promote energy storage. However, to maintain homeostasis, the body also requires another hormone that predominates during periods of low blood glucose. Several hours after a person eats, blood glucose levels begin to fall. Unless glucose is replenished by food, the body must begin to break down liver glycogen to maintain adequate levels of blood glucose. The hormone glucagon provides the signal for this to occur.

Because the brain and other components of the nervous system cannot store glucose, they are dependent on circulating glucose for energy. The brain is very sensitive to low blood glucose levels, and even a relatively small decrease

glycogenesis (gly – co – GE – ne – sis) (-*genesis,* coming into being) Formation of glycogen.

in blood glucose, a condition referred to as **hypoglycemia,** can make some people feel nauseated, dizzy, anxious, lethargic, and irritable. This is one reason why it is hard to concentrate when you have not eaten for a long time. Hypoglycemia has many causes. For example, a person with diabetes can experience hypoglycemia if too much insulin is injected. In addition, blood glucose levels can fall in response to prolonged or vigorous exercise. In nondiabetic individuals, there are two main types of hypoglycemia—reactive hypoglycemia and fasting hypoglycemia.

Reactive and Fasting Hypoglycemia **Reactive hypoglycemia** (now more commonly referred to as idiopathic postprandial hypoglycemia) occurs when the pancreas "over-responds" to high blood glucose levels following the consumption of food. The release of too much insulin ultimately results in extraordinarily low blood glucose. Researchers are not sure what causes idiopathic postprandial (reactive) hypoglycemia and even if it really exists, but speculate that some people may be especially sensitive to increases in blood glucose levels after a meal. People who experience symptoms associated with hypoglycemia are encouraged to eat small, frequent meals throughout the day. Other dietary recommendations include avoiding foods with caffeine, alcohol, and foods that cause a rapid and/or elevated glycemic response. Until blood glucose levels stabilize, keeping a food diary can help identify foods that trigger reactive hypoglycemia.

Unlike reactive hypoglycemia, **fasting hypoglycemia** is not associated with eating. Instead, it occurs when the pancreas releases too much insulin, even when food has not been consumed. This form of hypoglycemia is typically caused by pancreatic tumors, medications, hormonal imbalances, and certain illnesses, and it requires medical attention.

The Breakdown of Glycogen Glucagon stimulates the breakdown of glycogen in the liver and the release of glucose into the blood. This metabolic process, called **glycogenolysis,** literally means the breakdown of glycogen. Liver glycogen can supply glucose for approximately 24 hours before being depleted. Although glycogenolysis also occurs in skeletal muscle, the process differs in two important ways from that in the liver. First, muscle tissue is not responsive to the hormone glucagon; in other words, it is glucagon insensitive. Instead, the hormone **epinephrine,** which is released from the adrenal glands, is an important regulator of glycogenolysis in skeletal muscle. Another important difference is that muscles lack the enzyme needed to release glucose garnered from glycogenolysis into the blood. Rather, glucose released from glycogen during muscle glycogenolysis is used by muscle cells themselves and is not made available to other tissues. Therefore, muscle glycogen does not play a role in blood glucose regulation. The role of glucagon in blood glucose regulation and in the breakdown of energy stores is summarized in Figure 8.21.

Gluconeogenesis Increases Glucose Availability The breakdown of liver glycogen is an effective short-term solution for providing cells with glucose. However, because this energy reserve can be quickly depleted, the body must find an alternative source of glucose. Because fatty acids cannot be converted to glucose, the body uses amino acids from muscle protein to make glucose when glycogen stores are no longer available. The synthesis of glucose from noncarbohydrate sources such as amino acids is called **gluconeogenesis.** The same hormones that stimulate glycogenolysis—glucagon and epinephrine—also stimulate gluconeogenesis. Not surprisingly, insulin inhibits gluconeogenesis.

hypoglycemia (*hypo-*, under or below) Abnormally low level of glucose in the blood.

reactive hypoglycemia (also called idiopathic postprandial hypoglycemia) Low blood glucose that occurs when the pancreas releases too much insulin in response to eating carbohydrate-rich foods.

fasting hypoglycemia Low blood glucose that occurs when the pancreas releases excess insulin during periods of low food intake.

glycogenolysis (gly – co – ge – NO – ly – sis) (-*lysis*, to break apart) The breakdown of liver and muscle glycogen into glucose.

epinephrine (e – pi – NEPH – rine) Hormone released from the adrenal glands in response to stress; helps increase blood glucose levels by promoting glycogenolysis.

gluconeogenesis (glu – co – ne – o – GE – ne – sis) (*neo-*, new; -*genesis*, bringing forth) Synthesis of glucose from noncarbohydrate sources.

FIGURE 8.21 Glucagon Promotes Mobilization of Stored Energy The pancreas increases its release of the hormone glucagon in response to low blood glucose. Glucagon promotes the breakdown of liver glycogen, fat (adipose tissue), and protein (muscle). Glucagon also promotes the synthesis of ketones and the production of glucose from noncarbohydrate sources.

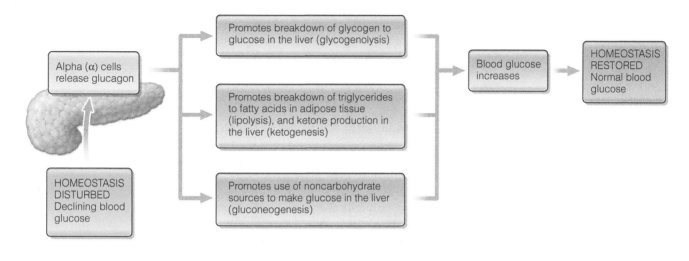

FIGHT-OR-FLIGHT RESPONSE PROVIDES AN IMMEDIATE ENERGY SOURCE

Whereas glucagon is involved in the day-to-day homeostatic regulation of blood glucose, other hormones dominate when there is a need for immediate energy. For example, the adrenal glands release the hormones epinephrine and **cortisol** when the body needs a sudden surge in blood glucose. These hormones have a variety of metabolic effects. For example, epinephrine increases glucose availability by stimulating glycogenolysis and gluconeogenesis. In addition to stimulating gluconeogenesis, cortisol also decreases glucose uptake by tissues other than the brain. These hormonal responses to the body's need for immediate energy, components of what is sometimes called the fight-or-flight response, ensure glucose availability under extreme circumstances.

KETONES ARE THE BODY'S ALTERNATIVE ENERGY SOURCE

Although gluconeogenesis increases glucose availability, it can also have negative consequences. This is because the breakdown of muscle protein to provide a source of glucose can have damaging effects, especially if it involves organs such as the heart. To minimize the loss of muscle, the body reduces its dependency on glucose by using an alternative energy source called **ketones.** These organic compounds form when fatty acids are broken down in a relative absence of glucose. This process, called **ketogenesis,** occurs in the liver and is stimulated by the hormone glucagon. Once formed, ketones are released into the blood and used for energy by the brain, heart, skeletal muscle, and kidneys. This glucose-sparing response helps minimize the loss of muscle protein by lessening the body's demand for glucose.

Ketone synthesis is not without its own consequences, however. A condition called **ketosis** occurs when ketone production exceeds the rate of ketone use, resulting in the accumulation of ketones in the blood. This happens when

⟨**CONNECTIONS**⟩ Dietary Reference Intakes (DRIs) are reference values for nutrient intake. When adequate information is available, Recommended Dietary Allowances (RDAs) are established, but when less information is available, Adequate Intake (AI) values are provided. Tolerable Upper Intake Levels (ULs) are intake levels that should not be exceeded.

cortisol (COR – ti – sol) Hormone secreted by the adrenal glands in response to stress; helps increase blood glucose availability via gluconeogenesis and glycogenolysis.

ketone (KE – tone) Organic compound used as an energy source during starvation, fasting, low-carbohydrate diets, or uncontrolled diabetes.

ketogenesis (ke – to – GE – ne – sis) Metabolic pathway that leads to the production of ketones.

ketosis (ke – TO – sis) Condition resulting from excessive ketones in the blood.

energy intake is very low or when the diet provides insufficient amounts of carbohydrate. Ketosis causes a variety of complications, including loss of appetite. In fact, many of the currently popular low-carbohydrate diets promote ketosis as the "fat-burning key." These diets are discussed in more detail in Chapter 11.

How Much Carbohydrate Do We Require?

Technically speaking, carbohydrates are not essential nutrients. They are, nonetheless, an important part of your diet because they provide energy and dietary fiber. Some tissues, such as the brain, typically rely almost exclusively on glucose for energy. Furthermore, there is evidence that certain carbohydrates may help prevent chronic diseases. Thus, to ensure that individuals make "healthy" carbohydrates a part of their eating pattern, the 2010 Dietary Guidelines for Americans emphasize the importance of eating a variety of brightly colored vegetables (dark green, red, and orange) and legumes. Also emphasized is the importance of replacing at least half our intake of refined grains with whole-grain foods. This recommendation is particularly important because refined-grain foods often contain solid fats, added sugars, and excess sodium. By making this one dietary change, overall health can be improved in many ways. The recommendations discussed in the next section—developed to minimize risk for chronic disease and promote optimal health—point out that it is important to pay attention to the amount as well as to the source of dietary carbohydrates.

DIETARY REFERENCE INTAKES FOR CARBOHYDRATES

The Institute of Medicine's Dietary Reference Intakes (DRIs) established a Recommended Dietary Allowance (RDA) for carbohydrate of 130 g per day for adults; this represents the minimum amount of glucose utilized by the brain. However, during pregnancy and lactation, glucose requirements increase to 175 g and 210 g per day, respectively. The Institute of Medicine does not provide any special recommendations about carbohydrate intake for athletes. However, people who are physically active and work out regularly are often advised to consume more carbohydrate to prevent fatigue and replenish glycogen stores. DRI values for carbohydrates are provided inside the front cover of this book.

There is no Tolerable Upper Intake Level (UL) for any individual carbohydrate class or for total carbohydrates. However, the DRI committee recognizes that there is considerable evidence that overconsumption of added sugars is associated with various adverse health effects, such as dental caries and obesity. The Institute of Medicine recommends that added sugars contribute no more than 25% of total calories.

To meet the body's need for energy and decrease the risk for chronic diseases, the Institute of Medicine suggests an Acceptable Macronutrient Distribution Range (AMDR) of 45 to 65% of total energy from carbohydrates. As there are 4 kcal per gram of carbohydrate, this means a person needing 2,000 kcal daily should consume between 225 and 325 g of carbohydrate (900 and 1,300 kcal, respectively) each day. You can easily determine how much carbohydrate is recommended for you by following the steps outlined in Table 8.6.

Recommendations for Dietary Fiber Intake Based on the DRI recommendations, the 2010 Dietary Guidelines for Americans recommend that adults consume

TABLE 8.6 Calculating Recommended Total Carbohydrate Intake

Total Energy Requirement (kcal/day)	Recommended % kcal from Carbohydrate (AMDR)	Calculation	Recommended Total Carbohydrate Intake (g/day)
1,500	45–65	1,500 × 0.45/4 1,500 × 0.65/4	169–244
2,000	45–65	2,000 × 0.45/4 2,000 × 0.65/4	225–325
2,500	45–65	2,500 × 0.45/4 2,500 × 0.65/4	281–406

NOTE: Intake goals are based on the Acceptable Macronutrient Distribution Ranges (AMDRs).

14 g of dietary fiber per 1,000 kcal. Adequate Intake (AI) values for dietary fiber range from 21 to 38 g per day for adults. The average daily intake of fiber in the United States is about half this amount (approximately 15 g).[21] One reason for this is the rise in popularity of low-carbohydrate weight-loss diet plans, some of which recommend minimal intakes of fruits, vegetables, and grains. Encouraging people to consume more fruits, vegetables, and whole-grain foods is a crucial step in helping increase fiber consumption.

Although UL values for dietary fiber and functional fiber have not been established, some people may experience indigestion and other gastrointestinal disturbances when they consume large amounts of fiber. While some fibers have a tendency to bind minerals such as calcium, zinc, iron, and magnesium, it is doubtful that consuming dietary fiber in recommended amounts affects mineral status in healthy adults. Most experts agree that high-fiber diets are beneficial and have minimal adverse effects.

DIETARY GUIDELINES AND MYPLATE EMPHASIZE WHOLE GRAINS, FRUITS, AND VEGETABLES

When it comes to carbohydrate-rich foods, there are many from which to choose. While the choices can be overwhelming, health experts agree that the best strategy is to maximize your intake of foods that are nutrient dense and consume plenty of fiber, vitamins, and minerals. Conversely, it is important to minimize your intake of foods high in fat and sugar. Following the 2010 Dietary Guidelines for Americans can help eliminate much of the guesswork when it comes to determining which carbohydrate-rich foods to choose. One general guideline is to consume more fruits, vegetables, and whole-grain foods; these foods provide naturally occurring sugars, fiber, and plenty of vitamins and minerals. The MyPlate food guidance system, which emphasizes whole grains, fruit, and vegetables as the foundation of a healthy diet, is another useful tool for evaluating your carbohydrate food choices. The current recommendation for adults is to consume an equivalent of 5 to 10 ounces of grain-based foods daily, half of which should be whole grain.

Another general guideline is to minimize highly processed foods that contain added sugars. Although there is no consensus as to how much total added sugar to consume, health experts generally agree that one's consumption of foods with high amounts of added sugar (such as cookies, soda, sugary cereals, and heavy syrups) should be minimized. Table 8.7 lists the average amounts of added sugar in selected foods and beverages. This chapter's Food Matters feature provides suggestions for reducing your intake of added sugars.

Working Toward the Goal: Focus on Reducing Added Sugars

A healthy eating pattern means selecting foods that are nutrient dense—high in nutrients relative to the amount of calories. To achieve this goal, the 2010 Dietary Guidelines for Americans suggest that we reduce the amount of added sugars in our diet. In many cases, this can be easily achieved by replacing highly processed foods with fresh fruits, vegetables, and whole grains. The following suggestions can help you reduce the amount of added sugars in your diet.

- The Nutrition Facts panel on food labels lists both total carbohydrates and total sugars. In highly processed foods, most of the sugar is likely to be added sugar. Compare the number of grams of sugar with the number of grams of carbohydrate. If the difference in these two numbers is small, the food is likely to be high in added sugar and should only be consumed in moderation.

- Limit intake of soft drinks, fruit punch, and other sweetened beverages. Water is the best way to quench your thirst. If you do drink these beverages, avoid large sizes.

- When you want something sweet, choose fresh fruit rather than candy or cookies. Also, when buying canned fruit, choose brands that are packed in natural juices rather than sweetened varieties.

- Avoid presweetened breakfast cereals. Small amounts of sweeteners can be added to oatmeal and other cereals if desired.

- Add fruit to plain yogurt rather than buying sweetened varieties with fruit added.

TABLE 8.7 Added Sugars in Your Diet

Food	Serving Size	Added Sugar (g)*
Soft drink	12 oz	43
Milkshake	10 oz	36
Fruit punch	8 oz	38
Chocolate	1.5 oz	24
Sweetened breakfast cereal (flakes)	1 cup	15
Yogurt with fruit	1 cup	33
Ice cream (vanilla)	1 cup	28
Cake with frosting	1 slice	28
Cookies (chocolate chip)	2 (medium)	14
Jam or jelly	1 tbsp	2

*1 teaspoon equals 4.75 grams

SOURCE: United States Department of Agriculture Database for the Added Sugars Content of Selected Foods, Release 1 (2006). Available from www.nal.usda.gov/fnic/foodcomp/Data/add_sug/addsug01.pdf. Krebs-Smith SM. Choose beverages and foods to moderate your intake of sugars: Measurement requires quantification. The Journal of Nutrition. 2001;131:527S–5S. Available from: http://jn.nutrition.org/content/131/2/527S.full.

Laura Benson

Critical Thinking: Laura's Story Think back to Laura's story about her experience with type 1 diabetes. Based on what you have read about how foods affect blood glucose, are there foods that Laura should avoid or limit? Review the recommendations and guidelines for carbohydrate intake and identify some ways these might be different for an athlete with type 1 diabetes compared to an athlete without diabetes.

Diet Analysis **PLUS** ✚ Activity

PART A. Total Carbohydrates

1. Using your "Profile DRI goals" report from your Diet Analysis Plus printouts, how many total kilocalories (kcals) should you be consuming per day? _____

2. Next to your recommendation for total carbohydrates in italics, what percentage range of total kcals is recommended to come from carbohydrates? _____

3. Using this information, calculate the recommended number of grams (g) of carbohydrates you should be eating.

   Lower end of range:

   AMDR total kcals × 0.45 = kcal of carbohydrates ÷ 4 kcal/gram = grams of carbohydrates

   _____ × 0.45 = _____ kcal
   ÷ 4 kcal/g = _____ g

   Upper end of range:

   _____ × 0.65 = _____ kcals
   ÷ 4 kcal/g = _____ g

4. Compare your "Profile DRI goals" with your printed and calculated DRI range for total carbohydrates.

   DRI recommended range for total carbohdyrates

   _____

   Calculated range for number of grams of carbohydrates _____

   Compare these values to determine if your diet is providing the recommended amounts (g) and recommended range of total kilocalories for carbohydrates.

PART B. Fiber

1. Refer to your "Intake and DRI Goals Compared" report. How much fiber did you actually consume? _____ g

2. Does the amount you actually consumed meet your AI for fiber? _____

3. Look at your individual days' "Intake Spreadsheets." List the one food consumed each day that contains the most fiber.

4. If you answered no to Question 2, what foods could you eat more of to meet your AI for fiber?

PART C. Added Sugar

1. The DRI suggests that added sugars should account for no more than 25% of the day's total energy intake. To determine the percentage of sugar calories you actually consumed, fill in the blanks below. Use your "Intake and DRI Goals Compared" report to locate the information you need.

   Grams of sugar consumed _____ × 4 kcal/g

   = _____ kcal of sugar consumed

   _____ (kcal from sugar) ÷ _____ (total kcal consumed) × 100 = _____% of calories as sugars

2. What is your assessment of your sugar intake?

3. Looking at your individual days' "Intake Spreadsheets," identify the food that contributed the most sugar each day.

Conclusions

Do you need to improve or change your overall carbohydrate intake? If so, what changes need to be made?

Notes

1. Bray GA, Nielsen SJ, Popkin BM. Consumption of high-fructose corn syrup in beverages may play a role in the epidemic of obesity. American Journal of Clinical Nutrition. 2004;9:537–43. Gaby AR. Adverse effects of dietary fructose. Alternative Medicine Review. 2005;10:294–306.

2. U.S. Department of Agriculture, Economic Research Service. 2011. Table 51—Refined cane and beet sugar: estimated number of per capita calories consumed daily, by calendar year. Table 52—High fructose corn syrup: estimated number of per capita calories consumed daily, by calendar year. Available from: http://www.ers.usda.gov/Briefing/Sugar/Data.htm.

3. Jacobson MF. High-fructose corn syrup and the obesity epidemic. American Journal of Clinical Nutrition. 2004;60:1081–2.

4. Forshee RA, Storey ML, Allison DB, Glinsmann WH, Hein GL, Lineback DR, Miller SA, Nicklas TA, Weaver GA, White JS. A critical examination of the evidence relating high fructose corn syrup and weight gain. Critical Reviews in Food Science and Nutrition. 2007;47:561–82.

5. Melanson KJ, Zukley L, Lowndes J, Nguyen V, Angelopoulos TJ, Rippe JM. Effects of high-fructose corn syrup and sucrose consumption on circulating glucose, insulin, leptin, and ghrelin and on appetite in normal-weight women. Nutrition. 2007;23:103–12.

6. Monsivais P, Perrigue MM, Drewnowski A. Sugars and satiety: Does the type of sweetener make a difference? American Journal of Clinical Nutrition. 2007;86:116–23.

7. Johnson R, Appel LJ, Brands M, Howard BV, Lefevre M, Lustig RH, Sacks F, Steffen LM, Wylie-Rosett J. Dietary sugars intake and cardiovascular health: A scientific statement from the American Heart Association. Circulation. 2009;120:1011–20.

8. National Cancer Institute. Sources of added sugars in the diets of the U.S. population ages 2 years and older, NHANES 2005–2006. Risk Factor Monitoring and Methods. Cancer Control and Population Sciences. http://riskfactor.cancer.gov/diet/foodsources/added_sugars/table5a. html.

9. U.S. Department of Agriculture and U.S. Department of Health and Human Services. Dietary Guidelines for Americans, 2010, 7th edition. Washington, DC: U.S. Government Printing Office, December 2010.

10. Whitehouse CR, Boullata J, McCauley LA. The potential toxicity of artificial sweeteners. American Association of Occupational Health Nurses. 2008;56:251–9.

11. Soffritti M, Belpoggi F, Tibaldi E, Esposti DD, Lauriola M. Life-span exposure to low doses of aspartame beginning during prenatal life increases cancer effects in rats. Environmental Health Perspectives. 2007;11:1293–7. Magnuson BA, Burdock GA, Doull J, Kroes RM, Marsh GM, Pariza MW, Spencer PS, Waddell WJ, Walker R, Williams GM. Aspartame: A safety evaluation based on current use levels, regulations, and toxicological and epidemiological studies. Critical Reviews in Toxicology. 2007;37:629–727.

12. Swithers SE, Davidson TL. A role for sweet taste: calorie predictive relations in energy regulation by rats. Behavioral Neuroscience. 2008;122:161–73.

13. Slavin J. Why whole grains are protective: Biological mechanisms. Proceedings of the Nutrition Society. 2003;62:129–34.

14. Marquart L, Wiemer KL, Jones JM, Jacob B. Whole grains health claims in the USA and other efforts to increase whole-grain consumption. Proceedings of the Nutrition Society. 2003;62:151–60.

15. Jenkins DJ, Kendall CW, Axelsen M, Augustin LS, Vuksan V. Viscous and nonviscous fibres, nonabsorbable and low glycaemic index carbohydrates, blood lipids and coronary heart disease. Current Opinion in Lipidology. 2000;11:49–56.

16. Anderson JW. Whole grains protect against atherosclerotic cardiovascular disease. Proceedings of the Nutrition Society. 2003;62:135–42.

17. Stollman N, Raskin JB. Diverticular disease of the colon. Lancet. 2004;363:631–9.

18. U.S. Department of Health and Human Services, Food and Drug Administration. Whole grain label statements. 2006 Available from: http://www.cfsan.fda.gov/-dms/flgragui.html.

19. Vesa TH, Marteau P, Korpela R. Lactose intolerance. Journal of the American College of Nutrition. 2000;19:165S–75S.

20. Foster-Powell K, Holt SH, Brand-Miller JC. International table of glycemic index and glycemic load values: 2002. American Journal of Clinical Nutrition. 2002;76:5–56.

21. Lang R, Jebb SA. Who consumes whole grains, and how much? Proceedings of the Nutrition Society. 2003;62:123–7.

Nutrition and Diabetes

Eat a healthy diet, exercise, and *watch your weight* are all important health recommendations that you have undoubtedly heard before. Yet there is a convincing reason for us to take this advice seriously—in a word, *diabetes*. In the past 40 years, the prevalence of diabetes has dramatically increased. In fact, diabetes has now reached epidemic proportions. According to the Centers for Disease Control and Prevention (CDC), 11% of the U.S. adult population (aged 20 years or older) has diabetes (18.8 million diagnosed and 7 million undiagnosed), which adds up to approximately 25.8 million Americans. Furthermore, an additional 79 million Americans have a prediabetic condition characterized by blood glucose levels that are elevated but not high enough to be considered diabetes.[1] These numbers are staggering—but not surprising to many health experts.

Although some types of diabetes cannot be prevented, many professionals believe the most common form of diabetes, type 2, is largely attributable to lifestyle. While genetics plays a role, physical inactivity, obesity, and unhealthy dietary practices have paved the way for this modern epidemic. The good news is that diet and other lifestyle practices can play an equally powerful role in managing and preventing type 2 diabetes.[2] Although some aspects of diabetes are not within our control, action is the key to diabetes prevention, treatment, and management.

What Is Diabetes?

Diabetes mellitus is a metabolic disorder characterized by elevated levels of glucose in the blood, otherwise known as hyperglycemia. Diabetes was first described more than 2,000 years ago as an affliction that caused excessive thirst, weight loss, and honey-sweet urine that attracted ants and flies. By the 16th century, physicians began treating diabetic patients with special diets. For example, some physicians recommended diets consisting of milk, barley, water, and bread to replace the sugar lost in the urine. Physicians later tried diets of meat and fat. Regardless of diet, patients died within a few months of diagnosis. With nothing else to offer, physicians attempted to keep patients alive by restricting their food intake. Until the discovery of insulin, there was little hope for people with diabetes.

THE DISCOVERY OF INSULIN

More than 100 years ago, two physiologists made the unexpected discovery that the surgical removal of the pancreas in dogs caused diabetes. This medical finding captured the interest of Frederick Banting, who was certain it would lead to a cure for diabetes. Banting—along with associates Charles Best, John Macleod, and James Collip—were the first to successfully control blood glucose in diabetic dogs by injecting them with secretions from the pancreas. Shortly thereafter, Leonard Thompson, a 14-year-old boy who was dying of diabetes, was the first human to be injected with Banting and Best's pancreatic extract. Amazingly, the young boy gained weight and seemed to recover from the debilitating effects of diabetes. The pancreatic extract, which was later given the name insulin, was obtained from the pancreas of cattle and pigs. Today, human insulin is produced by genetic engineering, and is used successfully by millions of people to manage their diabetes.

DIABETES IS CLASSIFIED BY ITS UNDERLYING CAUSE

Since the discovery of insulin, major advances have been made in understanding diabetes. By 1960, it became clear that there were different types of diabetes, which over the years were classified in a variety of ways. For example, diabetes was once categorized according to typical age of onset: juvenile-onset diabetes versus adult-onset diabetes. This classification scheme was

---

diabetes mellitus (di – a – BE – tes MEL – lit – tus) (*diabetes*, to siphon; *mellitus*, sugar) Medical condition characterized by a lack of insulin or impaired insulin utilization that results in elevated blood glucose levels.

Billy Leroy, one of the first patients to receive insulin. Pictures were taken of Billy as an infant before diabetes, with diabetes, and after several months of treatment with insulin.

What Is Type 1 Diabetes?

Type 1 diabetes occurs when the pancreas is no longer able to produce insulin. Without insulin, most cells cannot take up glucose, causing blood glucose levels to become dangerously high (Figure 1). This is the reason why individuals with type 1 diabetes require daily insulin injections to control blood glucose. It is also the reason why it was previously called insulin-dependent diabetes mellitus. Approximately 5 to 10% of all people with diabetes have type 1.[4] Although type 1 diabetes can develop at any age, it most often occurs during childhood and early adolescence, typically around 12 to 14 years of age. This is why type 1 diabetes was once called juvenile-onset diabetes.

confusing because the various forms of diabetes can occur at any age. Diabetes was later classified on the basis of whether insulin was required as a treatment: insulin-dependent diabetes versus non–insulin-dependent diabetes. The problem with this classification system was that some people diagnosed with non–insulin-dependent diabetes require insulin to control hyperglycemia. In 1997 the American Diabetes Association developed a new system of diabetes classification based on etiology, or underlying cause.[3] Today diabetes is categorized as type 1 diabetes, type 2 diabetes, gestational diabetes, or secondary diabetes. These are summarized in Table 1.

type 1 diabetes Previously known as juvenile-onset diabetes and as insulin-dependent diabetes mellitus, this form of diabetes results when the pancreas is no longer able to produce insulin due to a loss of insulin-producing β-cells.

Courtesy of Clendening History of Medicine Library, University of Kansas

TABLE 1 Classification of Types of Diabetes

Category	Typical Age of Onset	Underlying Cause	Description
Type 1 diabetes; formerly called insulin-dependent diabetes mellitus (IDDM) or juvenile-onset diabetes	Childhood and adolescence; can develop in adults	Lack of insulin production by the pancreatic ß-cells	An autoimmune disorder that destroys the insulin-producing ß-cells of the pancreas, resulting in little or no insulin production. Person requires insulin delivered via daily injections or insulin pump.
Type 2 diabetes; formerly called non–insulin-dependent diabetes mellitus (NIDDM) or adult-onset diabetes mellitus	Middle-aged and older adults; increasing incidence in childhood and adolescence	Insulin resistance	Skeletal muscle and adipose tissue develop insulin resistance, resulting in blood glucose levels that are above normal. Genetics, obesity, and physical inactivity play major roles in insulin resistance. Can often be managed by diet, weight loss, and exercise; may require glucose-lowering medication or insulin injections.
Gestational diabetes	Occurs in pregnant women	Insulin resistance	A temporary form of diabetes that develops in 4 to 7% of pregnant women. Characterized by insulin resistance brought on by hormonal changes that take place during pregnancy. Women who develop gestational diabetes are at increased risk for developing type 2 diabetes.
Secondary diabetes	Varies	Varies	Brought on by other diseases, medical conditions, and medications.

FIGURE 1 Type 1 Diabetes Blood glucose regulation depends on the release of insulin from the pancreas in response to elevated blood glucose. In the case of type 1 diabetes, the pancreas is not able to produce insulin.

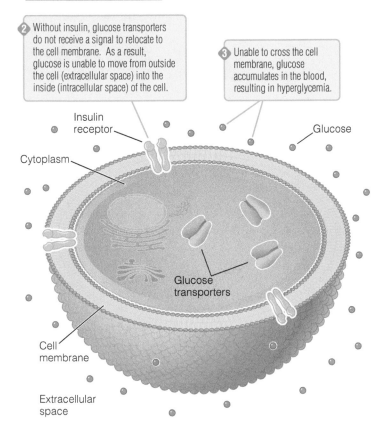

1. The insulin producing β-cells of the pancreas have been destroyed, and are unable to produce insulin.

2. Without insulin, glucose transporters do not receive a signal to relocate to the cell membrane. As a result, glucose is unable to move from outside the cell (extracellular space) into the inside (intracellular space) of the cell.

3. Unable to cross the cell membrane, glucose accumulates in the blood, resulting in hyperglycemia.

Insulin receptor

Cytoplasm

Glucose

Glucose transporters

Cell membrane

Extracellular space

TYPE 1 DIABETES IS CAUSED BY A LACK OF INSULIN PRODUCTION

Type 1 diabetes is caused by complex interactions among genetics, environmental factors, and the immune system that lead to an inability of the pancreas to produce insulin. Scientists have long recognized that type 1 diabetes seems to be more common in some families and in certain ethnic groups, suggesting a genetic component.[5] Type 1 diabetes, which is classified as an autoimmune disease, is triggered by something in the environment. Recall that autoimmune diseases are disorders caused by the production of antibodies that attack and destroy tissues in the body. Many researchers believe that, in genetically susceptible people, a viral infection stimulates the immune system to produce antibodies.[6] Type 1 diabetes results when the antibodies attack and destroy the insulin-producing cells (β-cells) of the pancreas.[7] Why this occurs is not fully understood.

The destruction of pancreatic β-cells is a gradual process. With little or no insulin-producing ability left, the pancreas cannot keep up with the body's need for insulin, resulting in severe hyperglycemia (Figure 2). Symptoms tend to develop rapidly and, because of their severity, are not easily overlooked.

METABOLIC DISTURBANCES RESULT FROM TYPE 1 DIABETES

The lack of insulin causes the rapid breakdown of fat stores and muscle. This is why people with untreated type 1 diabetes experience profound weight loss. Furthermore, without insulin, most cells cannot take up glucose, an energy source on which they normally depend.

Cells compensate for the lack of glucose by metabolizing fat and protein (muscle) for energy. However, this can have serious consequences. Using large amounts of fat for energy with a relative absence of glucose results in ketone formation. The accumulation of acidic ketones in the blood brings on flu-like symptoms such as abdominal pain, nausea, and vomiting. If the situation is not corrected, a life-threatening condition called **diabetic ketoacidosis** can occur. Diabetic ketoacidosis can lead to coma or death and requires immediate medical attention.

In addition to rapid weight loss, other symptoms associated with type 1 diabetes include extreme thirst, hunger, and frequent urination (Table 2). These symptoms are caused by high levels of blood glucose. For example, the kidneys are not able to remove the excess glucose from the blood, resulting in the "spilling" of glucose into the urine. However, before the glucose can be excreted in the

⟨**CONNECTIONS**⟩ Recall that autoimmune disorders occur when the immune system produces antibodies that attack and destroy tissues in the body. Antibodies are proteins produced by the immune system and typically respond to the presence of foreign proteins in the body.

diabetic ketoacidosis (ke – to – a – ci – DO – sis) Severe metabolic condition resulting from the accumulation of ketones in the blood.

FIGURE 2 **Cause of Type 1 Diabetes** Type 1 diabetes is caused by an autoimmune disorder in which antibodies destroy the insulin-producing cells of the pancreas.

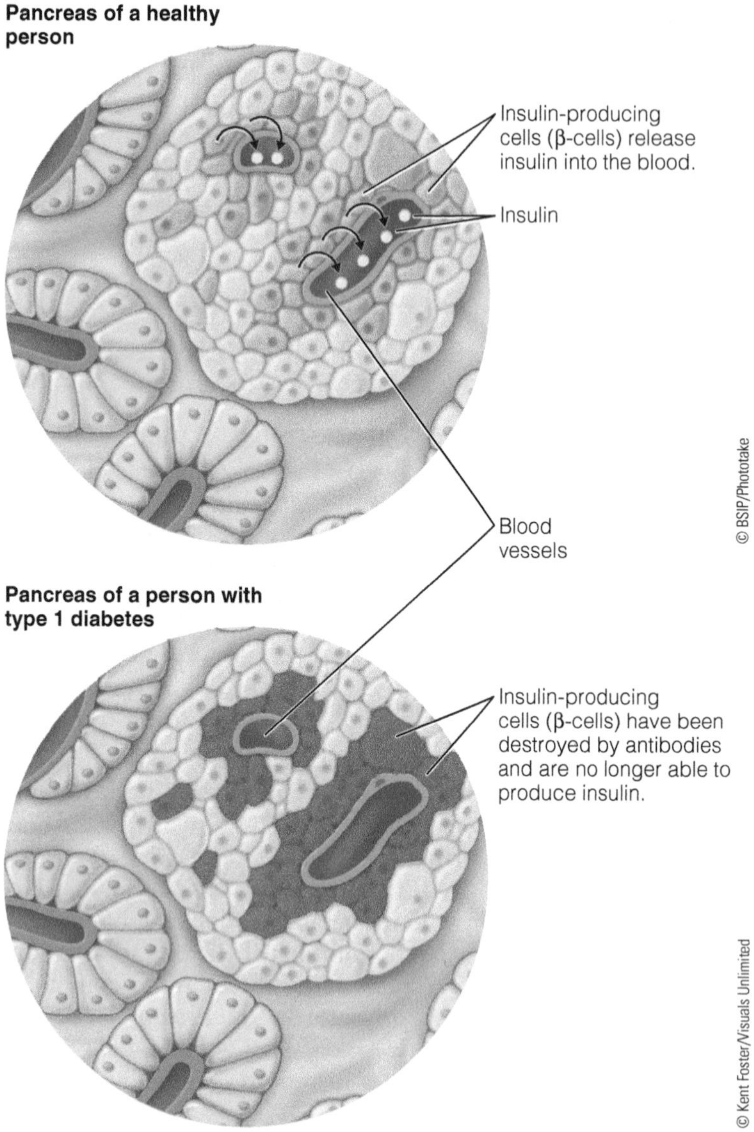

Pancreas of a healthy person

Insulin-producing cells (β-cells) release insulin into the blood.

Insulin

Blood vessels

Pancreas of a person with type 1 diabetes

Insulin-producing cells (β-cells) have been destroyed by antibodies and are no longer able to produce insulin.

© BSIP/Phototake

© Kent Foster/Visuals Unlimited

TABLE 2 Symptoms Associated with Type 1 and Type 2 Diabetes

Type 1 Diabetes	Type 2 Diabetes
Frequent urination	Frequent urination
Excessive thirst	Excessive thirst
Fatigue	Fatigue
Unusual weight loss	Frequent infections
Extreme hunger	Blurred vision
Ketosis	Vaginal itching
	Cuts or bruises that are slow to heal
	Tingling or numbness in hands or feet
	Frequent urinary tract infections

urine, it must first be diluted. This is accomplished by drawing water out of the blood. As a result, people urinate frequently, resulting in dehydration and thirst. Type 1 diabetes also causes a person to feel hungry and weak because cells are starved for energy. In fact, diabetes is often described as "starvation in the midst of plenty."

TYPE 1 DIABETES REQUIRES INSULIN INJECTIONS OR A PUMP

Insulin cannot be taken orally because it is a protein and therefore would be destroyed by enzymes in the digestive tract. Therefore, people with type 1 diabetes need multiple daily injections of insulin to control blood glucose. However, some people prefer to use an insulin pump, which looks like a small pager and is worn on a belt or on the waistband of clothing. Work is also under way to develop an artificial pancreas that can monitor blood glucose levels and automatically dispense the right amount of insulin into the blood. Hopefully this new technology will make diabetes care and blood glucose management easier for those who currently depend on a daily, labor-intensive regimen to maintain control of blood glucose.

To know how much insulin to administer, it is important for people with type 1 diabetes to monitor their blood glucose using a medical device called a **glucometer.** Glucometers are able to measure the approximate concentration of glucose in the blood using only a small drop of blood. Most glucometers have memory chips that can display a readout of average blood glucose measures over time.

The most successful care plans for type 1 diabetes focus on individual needs, preferences, and cultural practices. There is no such thing as "one-plan-fits-all" when it comes to treating and managing type 1 diabetes. However, the primary focus for people with type 1 diabetes is to balance insulin injections with a healthy diet and physical activity.

glucometer A medical device used to measure the concentration of glucose in the blood.

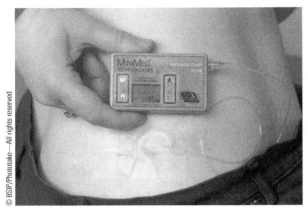

Individuals with type 1 diabetes must either inject insulin or use an insulin pump. Insulin pumps deliver insulin directly under the skin through narrow tubing.

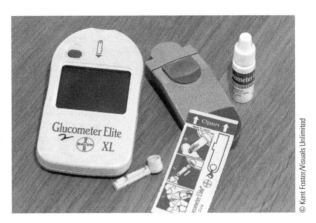

Glucometers are an important part of diabetes management. It is important for people who have diabetes to monitor blood glucose levels several times daily.

Type 1 Diabetes Can Lead to Serious Health Complications It is important for people to know that type 1 diabetes can lead to other serious health problems. For example, diabetes increases the risk for having heart attacks and strokes. It can also lead to blindness, limb amputation, impaired kidney function, and a loss of feeling in the feet and hands. These long-term complications are largely attributed to the harmful effects of excess glucose on blood vessels and nerves.

For many years, it was unclear if the occurrence of long-term complications associated with type 1 diabetes could be delayed or reduced by controlling blood glucose levels as close to normal as possible. However, a large clinical study called the Diabetes Control and Complications Trial (DCCT) was conducted to answer this important question.[8] This study showed that maintaining blood glucose levels as close to normal as possible can prevent many of the long-term complications associated with type 1 diabetes.

What Is Type 2 Diabetes?

Type 2 diabetes is by far the most common form of diabetes, with 90 to 95% of people with diabetes falling into this category.[1] In the United States, type 2 diabetes has become so widespread that an estimated one out of four people have or will have this disease or have a family member with it.[9] Because of the high prevalence of type 2 diabetes, the American Diabetes Association recommends that all adults be considered for screening beginning at 45 years of age.[10] Individuals who have one or more risk factors for type 2 diabetes (Table 3) should be screened at a younger age. Screening for type 2 diabetes involves

an overnight fast and a blood draw. This screening method, called a fasting plasma glucose, is relatively simple and inexpensive.

Although type 2 diabetes can occur at any age, it most frequently develops in middle-aged and older adults. For this reason, type 2 diabetes was once called adult-onset diabetes. However, the growing number of children and teens diagnosed with type 2 diabetes is alarming.[11] Nearly 2 million adults were newly diagnosed with type 2 diabetes in 2010. Perhaps more alarming is the fact that between 2002 and 2005, approximately 3,500 children and adolescents were newly diagnosed with type 2 diabetes annually.[1] Therefore, type 2 diabetes can no longer be thought of as a condition that affects adults only.

TABLE 3 Risk Factors for Type 2 Diabetes

- Family history of diabetes (parent, brother, or sister with diabetes)
- History of gestational diabetes or delivery of a baby weighing more than nine pounds at birth
- Sedentary lifestyle (exercise fewer than three times per week)
- Overweight or obese (BMI ≥ 25 kg/m²)
- History of vascular disease
- Being over 45 years of age
- Being of African, Hispanic, Native American, or Pacific Island descent
- Having polycystic ovary syndrome (females)
- Having high blood pressure (≥140/90 mm Hg)
- Having low HDL cholesterol (≤35 mg/dL)
- Having high triglycerides (≥250 mg/dL)

SOURCE: Adapted from American Diabetes Association. Position statement from the American Diabetes Association: Screening for type 2 diabetes. Diabetes Care. 2004;27:s11–s14. Available from: http://care.diabetesjournals.org/content/27/suppl_1/s11.full.

TYPE 2 DIABETES IS CAUSED BY INSULIN RESISTANCE

Unlike type 1 diabetes, most people with type 2 diabetes have normal or even elevated levels of insulin in their blood. Because type 2 diabetes is often managed without insulin treatments, it was once called non–insulin-dependent diabetes mellitus (NIDDM). **Type 2 diabetes** is caused by **insulin resistance,** which means that insulin receptors have difficulty recognizing or responding to insulin (Figure 3).[12] As discussed in Chapter 8, insulin acts like a cellular key by binding to insulin receptors and enabling glucose to enter the cell. Because insulin is not able to do its job, the transport of glucose across the cell membrane is impaired, which results in the accumulation of glucose in the blood.

Elevated blood glucose levels cause the pancreas to release even more insulin into the blood. In many cases, this response keeps blood glucose levels within a relatively normal range. It is not clear why some people with insulin resistance develop type 2 diabetes whereas others do not. Perhaps after many years of working overtime to produce extra insulin, the pancreas becomes worn out and can no longer produce insulin in amounts needed to lower blood glucose. When blood glucose levels remain elevated over time, a person begins to experience symptoms associated with type 2 diabetes. Blood glucose values used to diagnose diabetes are shown in Table 4.

GENETIC AND LIFESTYLE FACTORS INCREASE THE RISK OF DEVELOPING TYPE 2 DIABETES

Many risk factors are associated with the development of type 2 diabetes (see Table 3, shown previously). For example, the prevalence of type 2

FIGURE 3 Type 2 Diabetes Blood glucose regulation depends on the release of insulin from the pancreas in response to elevated blood glucose. Insulin binds to insulin receptors, allowing glucose to be taken up by the cell. Type 2 diabetes occurs when insulin receptors are unable to respond to insulin, resulting in hyperglycemia. This is called "insulin resistance."

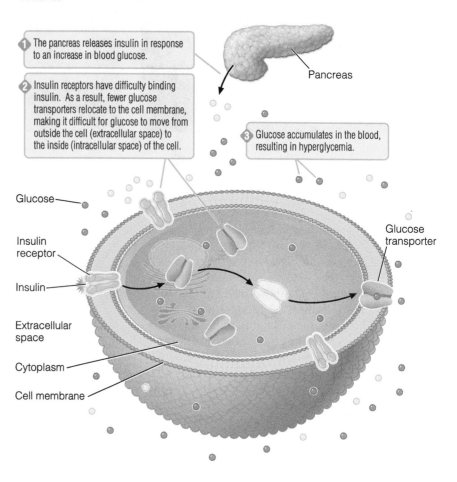

1. The pancreas releases insulin in response to an increase in blood glucose.

2. Insulin receptors have difficulty binding insulin. As a result, fewer glucose transporters relocate to the cell membrane, making it difficult for glucose to move from outside the cell (extracellular space) to the inside (intracellular space) of the cell.

3. Glucose accumulates in the blood, resulting in hyperglycemia.

Pancreas

Glucose

Insulin receptor

Insulin

Extracellular space

Cytoplasm

Cell membrane

Glucose transporter

TABLE 4 Diagnostic Values for Diabetes

Diagnosis	Fasting Plasma Glucose[a]	Two-Hour Oral Glucose-Tolerance Test[b]
Normal	<100 mg/dL	<140 mg/dL
Diabetes	≥126 mg/dL	≥200 mg/dL

[a] Person must fast for at least 8 hours prior to testing.
[b] Person must fast for at least 8 hours prior to testing. After an initial fasting blood sample is taken, a 75-g glucose beverage is consumed. A second blood sample is taken two hours later.

SOURCE: American Diabetes Association. Standards of medical care in diabetes—2011. Diabetes Care. 2011;34:s11–s61.

type 2 diabetes Previously known as adult-onset diabetes and non–insulin-dependent diabetes mellitus, this form of diabetes results when insulin-requiring cells have difficulty responding to insulin.

insulin resistance Condition characterized by the inability of insulin receptors to respond to the hormone insulin.

diabetes among the Caucasian population is approximately 10%.[1] However, in the African American population, it increases to about 19%.[1] The Pima Indians have the highest rates of type 2 diabetes in the United States. Aside from ethnicity and genetics, other risk factors associated with type 2 diabetes are obesity and the nutritional environment *in utero* and during infancy.

Obesity and Type 2 Diabetes The risk of developing type 2 diabetes is influenced, in part, by genetics.[13] However, this risk dramatically increases when accompanied by obesity. Overweight adults are more likely to develop type 2 diabetes than are lean individuals.[14] Approximately 80% of people diagnosed with type 2 diabetes are overweight. In addition, having a particular distribution of body fat can influence risk for type 2 diabetes. Body fat stored in the abdominal region of the body poses a greater risk than does that stored in the lower body regions of the body.[15] As the prevalence of obesity in the United States continues to climb, so does the prevalence of type 2 diabetes (Figure 4). If these trends continue, researchers estimate that 18 million Americans will have type 2 diabetes by the year 2020.[16]

The mechanisms that link obesity and insulin resistance remain unclear. However, a recent discovery that adipose tissue secretes a variety of hormones that trigger insulin resistance has generated new insights.[17] Although there is more to learn about these hormones, the link between obesity and insulin resistance is indisputable.

Birth Weight, Early Growth, and Type 2 Diabetes In addition to genetics and obesity, there is evidence that a poor nutritional environment *in utero* and during infancy can influence a person's risk for developing type 2 diabetes later in life.[18] A U-shaped relationship between birth weight and type 2 diabetes indicates that both accelerated (high birth weight) and slow (low birth weight) fetal growth can increase a person's risk of type 2 diabetes later in life (Figure 5).[19] This relationship is most evident when low birth weight is followed by rapid postnatal weight gain. Although the mechanisms remain unclear, scientists believe that epigenetic modification of DNA may be the means whereby adverse nutritional conditions early in life can increase health risks as adults.[20] You will learn more about epigenetics in Chapter 9.

SIGNS AND SYMPTOMS OF TYPE 2 DIABETES ARE OFTEN IGNORED

Because symptoms associated with type 2 diabetes tend to develop gradually, they are easily ignored. This is why type 2 diabetes can go undiagnosed

FIGURE 4 2008 Age-Adjusted Estimates of the Percentages of Adults with Diagnosed Diabetes

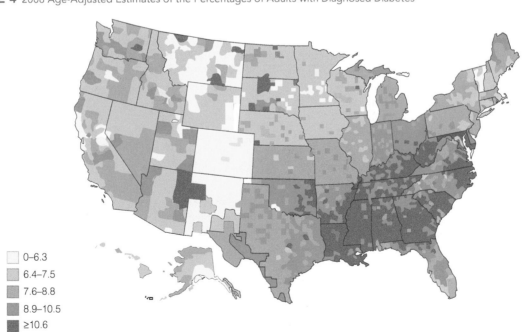

- 0–6.3
- 6.4–7.5
- 7.6–8.8
- 8.9–10.5
- ≥10.6

SOURCE: Centers for Disease Control and Prevention, Behavioral Risk Factor Surveillance System (BRFSS).

FIGURE 5 Relationship Between Birth Weight and Relative Risk of Developing Type 2 Diabetes Later in Life The U-shaped curve indicates that both low (<5.5 lb) and high birth weight (>9.9 lb) are associated with a higher occurrence of type 2 diabetes later in life.

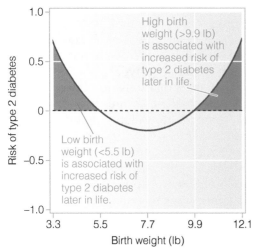

High birth weight (>9.9 lb) is associated with increased risk of type 2 diabetes later in life.

Low birth weight (<5.5 lb) is associated with increased risk of type 2 diabetes later in life.

SOURCE: Harder T, Rodekamp E, Schellong K, Dudenhausen JW, Plageman A. Birth weight and subsequent risk of type 2 diabetes: A Meta-Analysis. American Journal of Epidemiology. 2007;165:849–57.

for many years. Some of the early symptoms associated with type 2 diabetes include fatigue, frequent urination, and excessive thirst (see Table 2, shown previously). However, if type 2 diabetes is left untreated, it eventually causes more noticeable symptoms, such as blurred vision, frequent urinary tract infections, slow wound healing, and vaginal itching.

MANAGING TYPE 2 DIABETES CAN HELP PREVENT LONG-TERM COMPLICATIONS

Like type 1 diabetes, people with type 2 diabetes are at increased risk of developing long-term complications such as heart disease, blindness, and impaired kidney function. However, the good news is that maintaining blood glucose within near-normal ranges can often prevent these complications. In the case of type 2 diabetes, this can often be accomplished through weight management, regular exercise, and a healthy diet. For example, losing as little as 10 to 15 pounds or taking a 30-minute walk daily has been shown to lower blood glucose. Similar to the Diabetes Control and Complications Trial, the United Kingdom Prospective Diabetes Study revealed that controlling blood glucose within near-normal ranges can help

reduce long-term complications associated with type 2 diabetes.[21]

While many people with type 2 diabetes are able to control blood glucose by making lifestyle changes, others may require medication that lowers blood glucose by making cells more responsive to insulin. Even though the pancreas produces insulin, some people with type 2 diabetes need additional insulin to lower blood glucose. This is why some people with type 2 diabetes require insulin injections. However, even if medication is used to manage blood glucose, diet remains a critical component of diabetes management.

Diet and Type 2 Diabetes Nutritional guidelines developed by the American Diabetes Association stress the importance of diet in the management of type 2 diabetes.[22] Most of these guidelines, which are summarized in Table 5, can be applied by anyone who wants to eat a healthy diet. To help control type 2 diabetes, the American Diabetes Association recommends consuming carbohydrates mainly from fruits, vegetables, whole grains, legumes, and low-fat milk. Strategies such as the the diabetic exchange system and carbohydrate counting can be used to help monitor total carbohydrate intake (see Appendix D). Although people with diabetes are often told to avoid foods high in sugar, there is no need to totally eliminate sugars from the diet. Still, whole grains, fruits, and vegetables are preferred over high-sugar foods because they are good sources of micronutrients, phytochemicals, and dietary fiber.

Other dietary interventions for controlling type 2 diabetes and other health-related complications include reducing energy intake in order to reduce body weight and limiting foods high in saturated and *trans* fatty acids, cholesterol, and sodium. Consuming two or more servings of fish per week is also recommended to ensure adequate intakes of omega-3 fatty acids, which are important for good health. At this time there is insufficient evidence to recommend the use of vitamin or mineral supplements to treat diabetes. Rather, nutritionists recommend that a well-balanced diet be the source of these important nutrients.

Physical Activity and Type 2 Diabetes For people with type 2 diabetes, the benefits of regular exercise and physical activity are substantial. When combined with an energy-controlled meal plan, exercise can facilitate weight loss and can help in

TABLE 5 American Diabetes Association Dietary Recommendations for Type 2 Diabetes

- Overweight or obese individuals are encouraged to lose weight by decreasing energy intake and increasing physical activity.
- Carbohydrates from fruits, vegetables, whole grains, legumes, and low-fat milk should be emphasized.
- A diet that provides a variety of fiber-containing foods is encouraged.
- There is no need to totally eliminate sugars from the diet, but whole grains, fruits, and vegetables are recommended over high-sugar foods.
- Limit dietary cholesterol to <200 mg/day, saturated fats to <7% of total calories, and minimize consumption of *trans* fatty acids.
- Two or more servings of fish per week should be consumed to provide omega-3 fatty acids.
- There is not sufficient evidence to recommend the use of vitamin and mineral supplements to prevent or treat diabetes.
- Moderate alcohol consumption, defined as one drink/day for adult women and two drinks/day for adult men, poses no significant health concern for people with diabetes.

SOURCE: American Diabetes Association. Standards of Medical Care in Diabetes. Diabetes Care. 2011;34:S11–S61.

long-term maintenance of weight loss.[23] Regular exercise has many other health benefits including improved insulin sensitivity and blood glucose regulation. In addition, exercise can also help reduce blood pressure, improve cardiovascular fitness, reduce stress, and decrease lipid levels in the blood. For these reasons, exercise and physical activity should always be encouraged. Exercise ought to be thought of as a vital component of a type 2 diabetes management plan, and as little as 30 minutes of moderate physical activity on most days of the week can benefit health.

LIFESTYLE PRACTICES CAN INFLUENCE RISK OF DEVELOPING TYPE 2 DIABETES

We cannot do much about our genetics; however, there is much we can do in terms of lifestyle choices to help prevent type 2 diabetes. Studies show that even individuals who are at high risk for developing type 2 diabetes can significantly reduce their risk by eating a variety of healthy foods, staying physically fit, and maintaining a recommended body weight.[24] The story of the Pima Indians, presented in the Focus on the Process of Science feature, provides an example of how alterations in diet and physical activity can help prevent type 2 diabetes, even in a genetically susceptible population.

Dietary Recommendations for Preventing Type 2 Diabetes Individuals at high risk for developing type 2 diabetes are encouraged to eat a healthy diet that balances energy intake with energy expenditure.

Diets that provide whole grains and adequate amounts of dietary fiber have been shown to be beneficial in preventing diabetes.[25] Currently, there is not enough evidence to suggest that eating foods with low-glycemic loads can help prevent diabetes.[26] Although moderate alcohol consumption poses no significant health concern for people with type 2 diabetes and some studies suggest that moderate alcohol intake may actually reduce the risk for diabetes, the American Diabetes Association does not recommend alcohol consumption for diabetes prevention.[27] To help prevent type 2 diabetes in children and adolescents, a healthy, well-balanced diet that adequately supports normal growth and development is especially important.

What Are Secondary Diabetes and Gestational Diabetes?

The majority of people with diabetes have type 1 or type 2. However, a small percentage of people have **secondary diabetes,** which is hyperglycemia that develops as a result of certain diseases, medical conditions, or medications. For example, some medications may affect how the body uses or produces insulin. Once the underlying cause of secondary diabetes is treated, normal blood glucose control is often restored. Another type of diabetes, called gestational diabetes, is brought on by pregnancy.

secondary diabetes Diabetes that results from other diseases, medical conditions, or medication.

The nature-versus-nurture issue is an ongoing and hotly debated topic. Are chronic diseases determined by our genes or by our lifestyle? Is it a combination of both, and, if so, does one have more influence than the other? The Pima Indians have provided researchers with a unique opportunity to study these important questions.[28]

For thousands of years the Pima Indians, originally from the Sierra Madre Mountains in Mexico, farmed, hunted, and fished as a way of life. The Pima, living in a hot, dry climate, developed ingenious farming practices and an elaborate irrigation system that allowed their crops to flourish. For some Pima, this way of life changed about 100 years ago, when new settlers to the region began farming upstream from the Pima, diverting the water supply for their own crops. Unable to farm, many Pima migrated north and relocated to the state of Arizona. For these people, the traditional lifestyle was gone. Not only did their diet change, but so did their way of life. The physical demands of farming gave way to a sedentary lifestyle. Instead of traditional foods, they consumed diets rich in sugar, white flour, and lard.

For the past 30 years researchers from the National Institutes of Health have been studying the Pima Indians.[29] The high rates of obesity and type 2 diabetes among this group have been of particular concern. This is because the prevalence of type 2 diabetes is an astonishing 34% of men and 41% of women.[30] Many Pima develop type 2 diabetes at a remarkably young age, and 95% of those with type 2 diabetes are obese. To complete their investigation, the research team also studied a group of Pima Indians still living in a remote area in Mexico. The diet and lifestyle of these traditional Mexican Pima were similar to those of 100 years ago. Although the two groups of Pima Indians (those living in Arizona and those living in Mexico) were genetically similar, the Mexican Pima had very low rates of obesity and type 2 diabetes.

These divergent lifestyles of the U.S. and Mexican Pima clearly show that lifestyle can influence the development of obesity and type 2 diabetes. Although both groups of Pima have similar genetics, only those living in a permissive environment of abundant food and reduced physical activity developed diabetes. Further studies have shown that lifestyle changes, such as increased exercise and improved diets, can significantly reduce the high rates of obesity and type 2 diabetes among the Pima living in Arizona.[31] The good news for all of us is that lifestyle practices can sometimes override the influence of genetics on development of obesity and type 2 diabetes.

National Photo Company Collection, Prints & Photographs Division, Library of Congress, LC-USZ62-126404

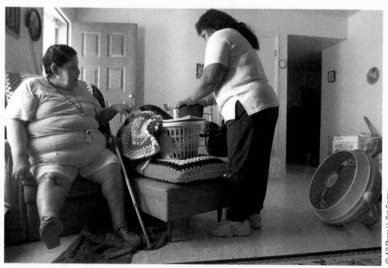
© AP Photo/J. Pat Carter

Genetics and environment both play a role in the development of type 2 diabetes. Research conducted on the Pima Indians has provided evidence that obesity is a major risk factor in the development of diabetes. The prevalence of diabetes is only 8% among Pima Indians living a traditional lifestyle in Mexico, while it is greater than 34% among Pima Indians living in the United States.

SOME PREGNANT WOMEN DEVELOP GESTATIONAL DIABETES

During pregnancy, most women experience insulin resistance to some extent. This normal and healthy response to pregnancy helps make glucose more available to the fetus. However, approximately 18% of all pregnant women develop a more severe form of insulin resistance called **gestational diabetes.**[1] This temporary form of diabetes typically develops after the 24th week of pregnancy, disappearing within 6 weeks after delivery. Gestational diabetes is more common in obese women and those who have a family history of type 2 diabetes.[32] Risk factors associated with gestational diabetes are listed in Table 6.

Effects of Gestational Diabetes Although maternal insulin does not cross the placenta, glucose does. As a result, the fetus of a woman with gestational diabetes is exposed to high levels of glucose. This causes the fetus's pancreas to increase its production of insulin, which can lead to increased glucose uptake and the deposit of extra fat in the growing baby. For this reason, infants born to women with gestational diabetes tend to be large, which can make delivery difficult. Babies born to mothers with poorly controlled gestational diabetes can weigh as much as 10 to 12 pounds.

Women who develop gestational diabetes tend to have a high recurrence of this condition in subsequent pregnancies. In addition, having gestational diabetes increases a woman's risk for developing type 2 diabetes by up to 60% in the next 10–20 years.[1] Infants born to women with gestational diabetes may also be at increased risk for developing type 2 diabetes, especially if there is a family history of diabetes.[33]

TABLE 6 Risk Factors Associated with Gestational Diabetes

- Overweight or very overweight
- History of abnormal glucose tolerance
- Over 25 years of age
- Being of African, Hispanic, Native American, or Pacific Islander descent
- Family history of type 2 diabetes or gestational diabetes
- Previous delivery of a stillborn or a very large baby
- Having gestational diabetes with previous pregnancy

SOURCE: U.S. Department of Health and Human Services, National Institutes of Health, and the National Institute of Child Health and Human Development. NIH Pub. No. 00–4818; 2005. Available from: http://www.nichd.nih.gov/publications/pubs/upload/gest_diabetes_risk_2005.pdf.

Diagnosing and Managing Gestational Diabetes Unless a woman is considered low risk for gestational diabetes (younger than 25 years with a body mass index less than 27 kg/m²), the American Diabetes Association recommends routine screening for gestational diabetes.[34] This is typically done around the 24th week of pregnancy and involves measuring blood glucose levels. To help achieve and maintain optimal blood glucose control, it is important for women with gestational diabetes to receive counseling on diet and exercise. The American Diabetes Association recommends a meal plan that provides adequate nutrients and energy to meet the needs of pregnancy but restricts carbohydrate intake to 35 to 40% of daily calories.[35] Another important component of gestational diabetes management is exercise. A physically active lifestyle is an important part of a healthy pregnancy, and this can help lower maternal glucose concentrations, reduce stress, and prevent excess weight gain. However, every pregnant woman should first consult with her physician before beginning an exercise program.

Managing Diabetes Today Can Help Prevent Health Problems Tomorrow

While the prevalence of diabetes continues to increase in the United States and around the world, so do the related medical expenditures. Indeed, having diabetes can be costly both emotionally and monetarily. It has been estimated that people with diabetes, on average, have medical expenditures that are approximately 2.3 times higher than those without diabetes.[36] Making dietary and other lifestyle changes is not easy and often requires assistance from specially trained health care professionals. The day-to-day management of these medical conditions can be demanding, and the emotional side of this task is often overlooked. It is not uncommon for people with diabetes to follow "all the rules" and still have difficulties controlling blood glucose, creating a feeling of frustration. However, people who participate actively as members of their health care team by asking questions and making good choices can avoid complications and achieve positive effects on their health and well-being.

gestational diabetes (ges – TA – tion – al) Type of diabetes characterized by insulin resistance that develops in pregnancy.

Key Points

What Is Diabetes?

- The four types of diabetes are categorized by the underlying cause.

What Is Type 1 Diabetes?

- Type 1 diabetes is classified as an autoimmune disease and is caused by an interaction among genetics, environment, and the immune system.

- In the case of type 1 diabetes, antibodies produced in response to something in the environment attack and destroy the insulin-producing cells of the pancreas.

- Without insulin, cells are unable to take up glucose, resulting in hyperglycemia, rapid weight loss, accumulation of ketones in the blood, extreme thirst, hunger, frequent urination, and fatigue.

- Treatment of type 1 diabetes focuses on balancing insulin injections with a healthy diet and physical activity.

- Controlling blood glucose reduces the complications associated with diabetes.

- Chronic hyperglycemia damages blood vessels and nerves, leading to heart attacks, strokes, blindness, limb amputation, impaired kidney function, and loss of feeling in the feet and hands.

What Is Type 2 Diabetes?

- Type 2 diabetes is caused by insulin resistance, meaning that insulin-requiring cells do not respond to insulin in a normal way.

- Most people with diabetes have type 2, which is strongly associated with obesity.

- Symptoms associated with type 2 diabetes include fatigue, frequent urination, excessive thirst, chronic hyperglycemia, blurred vision, frequent urinary tract infections, slow wound healing, and vaginal itching.

- The treatment focus of type 2 diabetes is on maintaining near-normal blood glucose through a healthy diet, weight management, and physical activity.

What Are Secondary Diabetes and Gestational Diabetes?

- Diabetes that develops as a result of other diseases, medical conditions, or medications is called secondary diabetes.

- Gestational diabetes is a form of diabetes that can develop during the later stages of pregnancy. It is more common in obese women and those with a family history of type 2 diabetes.

- Infants born to mothers with gestational diabetes are often large because the fetal pancreas increases its production of insulin in response to being exposed to high blood glucose levels from the mother.

- Women who develop gestational diabetes are at high risk for recurrence, and both the infant and mother are at increased risk for developing type 2 diabetes later in life.

Managing Diabetes Today Can Help Prevent Health Problems Tomorrow

- A team made up of health care professionals is important for management of diabetes and metabolic syndrome.

- A healthy lifestyle is a critical component of all management plans.

Notes

1. Centers for Disease Control and Prevention. National diabetes fact sheet: national estimates and general information on diabetes and prediabetes in the United States, 2011. Atlanta, GA: U.S. Department of Health and Human Services, Centers for Disease Control and Prevention, 2011. Available from: http://www.cdc.gov/diabetes/pubs/pdf/ndfs_2011.pdf.

2. Lindstrom J, Peltonen M, Tuomilehto J. Lifestyle strategies for weight control: Experience from the Finnish diabetes prevention study. Proceedings of the Nutrition Society. 2005;64:81–8.

3. Expert Committee on the Diagnosis and Classification of Diabetes Mellitus. Report of the Expert Committee on the diagnosis and classification of diabetes mellitus. Diabetes Care. 1997;20:1183–97.

4. Sperling MA, editor. Type 1 diabetes: Etiology and treatment. Totowa, NJ: Human Press; 2003.

5. Eisenbarth GS. Type 1 diabetes: Molecular, cellular and clinical immunology. Advances in Experimental Medicine and Biology. 2004;552:306–10. Hirschhorn JN. Genetic epidemiology of type 1 diabetes. Pediatrics and Diabetes. 2003;4:87–100.

6. Hintermann E, Christen U. Viral infection—a cure for type 1 diabetes? Current Medicinal Chemistry. 2007;14:2048–52.

7. Knip M. Environmental triggers and determinants of beta-cell autoimmunity and type 1 diabetes. Endocrine and Metabolic Disorders. 2003;4:213–23.

8. Diabetes Control and Complications Trial Research Group. The effect of intensive treatment of diabetes on the development and progression of long-term complications in insulin-dependent diabetes mellitus. New England Journal of Medicine. 1993;329:977–86.

9. American Diabetes Association. Position statement from the American Diabetes Association: The prevention or delay of type 2 diabetes. Diabetes Care. 2002;25:742–9.

10. American Diabetes Association. Position statement from the American Diabetes Association: Screening for type 2 diabetes. Diabetes Care. 2004;27:s11–s14.

11. Draznin MB. Type 2 diabetes. Adolescence Medicine State of the Art Review. 2008;3:498–506.

12. Leahy JL. Pathogenesis of type 2 diabetes mellitus. Archives of Medical Research. 2005;36:197–209.

13. Malecki MT. Genetics of type 2 diabetes mellitus. Diabetes Research and Clinical Practice. 2005;68:S10–21. Tusie Luna MT. Genes and type 2 diabetes mellitus. Archives of Medical Research. 2005;36:210–22.

14. van Dam RM, Rimm EB, Willett WC, Stampfer MJ, Hu FB. Dietary patterns and risk for type 2 diabetes mellitus in U.S. men. Annals of Internal Medicine. 2002;136:201–9.

15. Solomon CG, Manson JE. Obesity and mortality: A review of the epidemiologic data. American Journal of Clinical Nutrition. 1997;66:1044S–50S.

16. Green A, Christian Hirsch N, Pramming SK. The changing world demography of type 2 diabetes. Diabetes/Metabolism Research and Reviews. 2003;19:3–7.

17. Fischer-Posovszky P, Wabitsch M, Hochberg Z. Endocrinology of adipose tissue—an update. Hormone and Metabolic Research. 2007;39:314–21.

18. Mathers JC. Early nutrition: Impact on epigenetics. Forum of Nutrition. 2007;60:42–8.

19. Dunger DB, Salgin B, Ong KK. Session 7: Early nutrition and later health early developmental pathways of obesity and diabetes risk. The Proceedings of the Nutrition Society. 2007;66:451–7.

20. Nobili V, Alisi A, Panera N, Agostoni C. Low birth weight and catch-up-growth associated with metabolic syndrome: a ten year systematic review. Pediatric Endocrinology Reviews. 2008;2:241–7.

21. UK Prospective Diabetes Study (UKPDS) Group. Intensive blood-glucose control with sulphonylureas or insulin compared with conventional treatment and risk of complications in patients with type 2 diabetes (UKPDS 33). Lancet. 1998;352:837–53. Anderson JW, Kendall CW, Jenkins DJA. Importance of weight management in type 2 diabetes: Review with meta-analysis of clinical studies. Journal of the American College of Nutrition. 2003;22:331–9.

22. American Diabetes Association. Position statement from the American Diabetes Association: Nutrition recommendations and interventions for diabetes. Diabetes Care. 2008;31:S61–78.

23. Sigal RJ, Kenny GP, Wasserman DH, Castaneda-Sceppa C. Physical activity, exercise, and type 2 diabetes. Diabetes Care. 2005;27:2518–39.

24. Wareham NJ. Epidemiological studies of physical activity and diabetes risk, and implications for diabetes prevention. Applied Physiology, Nutrition, and Metabolism. 2007;32:778–82.

25. Kelley DE. Sugars and starch in the nutritional management of diabetes mellitus. American Journal of Clinical Nutrition. 2003;78:858S–64S.

26. Thomas D, Elliott EJ. Low glycaemic index, or low glycaemic load, diets for diabetes mellitus.Cochrane Database Systematic Reviews. 2009;1:CD006296.

27. Katsilambros N, Liatis S, Makrilakis K. Critical review of the international guidelines: What is agreed upon—what is not? Nestle Nutrition Workshop Series. Clinical Performance and Programme. 2006;11:207–18.

28. Bennett PH. Type 2 diabetes among the Pima Indians of Arizona: An epidemic attributable to environmental change? Nutrition Reviews. 1999;57:S51–4.

29. Baier LJ, Hanson RL. Genetic studies of the etiology of type 2 diabetes in Pima Indians: Hunting for pieces to a complicated puzzle. Diabetes. 2004;53:1181–6.

30. Schulz LO, Bennett PH, Ravussin E, Kidd JR. Effects of traditional and western environments on prevalence of Type 2 diabetes in Pima Indians in Mexico and the U.S. Diabetes Care. 2006;29:1866–71.

31. Pavkov ME, Hanson RL, Knowler WC, Bennett PH, Krakoff J, Nelson RG. Changing patterns of type 2 diabetes incidence among Pima Indians. Diabetes Care. 2007;30:1758–63.

32. American Diabetes Association. Position statement from the American Diabetes Association: Gestational diabetes mellitus. Diabetes Care. 2008;31:s12–s54.

33. Stocker CJ, Arch JR, Cawthorne MA. Fetal origins of insulin resistance and obesity. Proceedings of the Nutrition Society. 2005;64:143–51.

34. The American College of Obstetricians and Gynecologists. Screening for gestational diabetes mellitus: Recommendations and rationale. Obstetrics & Gynecology. 2003;101:393–5.

35. American Diabetes Association. Position statement from the American Diabetes Association: Standards of medical care in diabetes—2011. Diabetes Care. 2011;34:s11–s61.

36. Economic costs of diabetes in the U.S. in 2007. Diabetes Care. 2008;31:596–615.

Protein

In Chapter 8, you learned about carbohydrates, which are a critical source of energy and serve as building blocks for complex molecules such as deoxyribonucleic acid (DNA) and ribonucleic acid (RNA). However, the importance of dietary carbohydrates to health is arguably rivaled by the next class of macronutrient: protein. The term *protein* was derived more than 170 years ago from the Greek *prota*, meaning "of first importance."[1] There is no debate that the proteins your body makes (or does not make) can be critical to your health. And because you need to eat protein to make protein, obtaining enough of the right kinds of protein from your diet is vitally important.

Proteins constitute the most abundant organic substance in your body, making up at least 50% of your dry weight.[2] But why exactly do you need these proteins, and how can the proteins you eat influence the proteins you make? The answers to these questions are somewhat complex, and understanding some fundamental concepts related to protein nutrition will help you answer them.

In this chapter, you will learn what proteins are, what foods are good sources of them, how you digest and absorb dietary proteins, and how your body uses these substances to maintain health. You will also learn about vegetarianism; how certain proteins in foods cause allergic reactions; and how dietary choices, genetic makeup, and environment can interact to influence your risk of various diseases. With this information, you can make more informed decisions concerning which protein sources to choose and how much is needed to optimize long-term health and well-being.

Living with Peanut Allergy

Tyler is an adorable 10-year-old boy who loves to camp and play ice hockey, and you might think he does not have a care in the world. However, that assumption is far from the truth. When Tyler was two years old, he and his mother, Christie, were camping. After eating a peanut butter sandwich, Tyler began to feel sick and complained that his stomach hurt. When he would not stop crying, his mother took him to the emergency room. At that point, life dramatically changed, as Tyler was diagnosed with having a severe allergic response to peanuts. In fact, he was allergic to all legumes.

Nothing is easy when you have a food allergy. For example, Tyler's mom once gave him a bowl of vegetable soup that contained a couple of lima beans, a type of legume. She was horrified when his tongue swelled and he began having difficulty breathing. Once again, a trip to the emergency room was necessary. Clearly, living with food allergies is extremely scary for Tyler and his family, and Christie told us that she lives with the fear that she "will lose Tyler because someone just was not being careful."

When Tyler was a toddler, it was not difficult to limit him to "safe" foods. However, as he got older, this became increasingly difficult. "Peanut-Free Zone" signs were posted in his classroom and at a special table in the lunchroom. Notes were sent home requesting that only peanut-free foods be brought for birthdays and other celebrations. It was also very important for Tyler to be aware of which foods were "safe" and "unsafe." But even these precautions were not enough. One day Tyler traded a bag of carrots for a chocolate cookie with a friend at lunch. Both kids thought the cookie was safe, but it had been in a bag with a peanut butter sandwich. When hives developed on Tyler's neck and his breathing became labored, the school called Christie, and she once again rushed him to the emergency room.

Along with Christie, Tyler's teachers and school nurse continue to be vigilant in keeping him safe. The school nurse told us that "handling these sometimes life-threatening conditions can be very confusing, but the best advice is continued education and awareness." Although Christie says her goal has always been to prevent peanut allergy from affecting Tyler's social life, striking a balance between being overprotective and not hurting someone's feelings is difficult. For example, Tyler is not allowed to eat with many of his friends at school—instead, he must eat at the designated "peanut-free table."

Clearly, living with peanut allergy is challenging. We asked Tyler what it is like, and he told us the following: "It's really scary. If I could have one wish in the world, it would be to be like my friends and not have peanut allergy."

© Shelley McGuire

Critical Thinking: Tyler's Story

Do you or anyone you know have a food allergy? If so, how has this affected your (or his or her) life choices? How might having a food allergy be especially difficult in college, and what might you do differently if you learned that your roommate had a severe allergy to some type of food?

What Are Proteins?

Perhaps the first questions you might ask when beginning your study of proteins are "*What are proteins, how does one protein differ from another, and what makes proteins different from other nutrients?*" **Proteins** (also called **peptides**) are complex molecules made from smaller subunits called **amino acids** that are joined together by **peptide bonds.** The formation of a peptide bond is an example of a condensation reaction in which a hydroxyl group ($-OH$) from one amino acid is joined with a hydrogen atom ($-H$) from another amino acid, releasing a molecule of water (H_2O) in the process. Proteins contain appreciable amounts of nitrogen, since each amino acid contains one or more nitrogen atoms. This makes proteins chemically distinct from the other macronutrients (carbohydrates and lipids).

Protein size is quite variable, depending on the number of amino acids present. Whereas some are very simple, containing only a few amino acids, others contain thousands of these building blocks. Most proteins, however, have 250 to 300 amino acids. This is why proteins are often classified on the basis of their number of amino acids: dipeptides have two amino acids, tripeptides have three, and so forth. Those with 2 to 12 amino acids are collectively called oligopeptides, and proteins that have more than 12 amino acids are called **polypeptides.**

AMINO ACIDS ARE THE BUILDING BLOCKS OF PROTEINS

The numerous proteins in the body are amazingly diverse. But what makes one protein different from another? The key to this diversity lies in the number and types of amino acids they contain as well as the sequence in which they are linked together. To understand protein diversity, you must first understand the basic components of the amino acids and what makes each one unique.

Most amino acids have three common parts: (1) a central carbon atom bonded to a hydrogen atom, (2) a nitrogen-containing **amino group** ($-NH_2$), and (3) a carboxylic acid group ($-COOH$). In the body, the amino and carboxylic acid groups almost always exist in "charged" states. Specifically, the amino group has a positive charge, and the carboxylic acid has a negative charge. Each amino acid also contains a side-chain, called an **R-group.** The structure of the R-group makes each amino acid uniquely different from the others. The R-group can be as simple as a hydrogen atom or as complex as a ring structure. The chemical and physical properties of amino acids depend on subtle differences in the R-groups. For example, some R-groups are negatively charged, some are positively charged, and some have no charge at all. The three charged components of amino acids—the amino group, the carboxylic acid group, and the R-group—cause proteins to bend and take on unique and complex shapes important for their functions. A "generic" amino acid and some examples of R-groups are shown in Figure 9.1.

AMINO ACIDS ARE CLASSIFIED AS ESSENTIAL, NONESSENTIAL, OR CONDITIONALLY ESSENTIAL

The body needs 20 different amino acids to make all the proteins it requires, and these amino acids can be categorized as essential, nonessential, or conditionally essential (Table 9.1).[*3] The nine essential amino acids are those you

*Essential amino acids are also called indispensable amino acids, nonessential amino acids are called dispensable amino acids, and conditionally essential amino acids are called conditionally indispensable amino acids.

protein (peptide) Nitrogen-containing macronutrient made from amino acids.

amino acid Nutrient composed of a central carbon bonded to an amino group, carboxylic acid group, and a side-chain group (R-group).

peptide bond A chemical bond that joins amino acids.

polypeptide A string of more than 12 amino acids held together via peptide bonds.

amino group ($-NH_2$) The nitrogen-containing component of an amino acid.

R-group The portion of an amino acid's structure that distinguishes it from other amino acids.

The structure of the R-group makes one amino acid different from another.

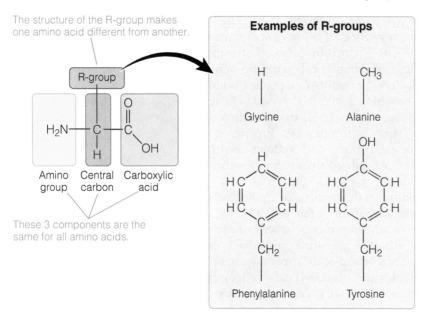

This nitrogen is positively charged, because it has gained a hydrogen atom.

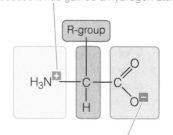

These 3 components are the same for all amino acids.

Examples of R-groups

This oxygen is negatively charged, because it has lost a hydrogen atom.

must consume in your diet because the body cannot make them or cannot make them in required amounts. The remaining 11 amino acids are nutritionally nonessential because the body can make them from other compounds. To synthesize a nonessential amino acid, the body transfers an amino group from one amino acid to another compound called an **α-keto acid,** which is basically an amino acid without its amino group. This process, called **transamination,** results in the synthesis of a new amino acid.

Under some conditions, however, the body may be unable to synthesize one or more of the nonessential amino acids. For example, some infants born prematurely cannot make several of the traditionally nonessential amino acids. When this happens, some nonessential amino acids become conditionally essential because they must be obtained from the diet until the baby matures.[4] Fortunately, infants fed human milk and/or infant formula receive sufficient amounts of these conditionally essential amino acids. For premature infants, however, it is sometimes necessary to fortify milk with conditionally essential amino acids to ensure optimal growth and development.[5]

α-keto acid A compound that accepts an amino group from an amino acid in the process of transamination.

transamination The process whereby an amino group is formed via the transfer of an amino group from one amino acid to another organic compound (an α-keto acid).

TABLE 9.1 Essential, Nonessential, and Conditionally Essential Amino Acids

Essential	Nonessential	Conditionally Essential
Histidine	Alanine	Arginine
Isoleucine	Asparagine	Cysteine
Leucine	Aspartic acid	Glutamine
Lysine	Glutamic acid	Glycine
Methionine	Serine	Proline
Phenylalanine		Tyrosine
Threonine		
Tryptophan		
Valine		

SOURCE: Institute of Medicine. Dietary Reference Intakes for energy, carbohydrate, fiber, fat, fatty acids, cholesterol, protein, and amino acids. Washington, DC: National Academies Press; 2005.

Diseases, such as **phenylketonuria (PKU),** can also cause traditionally nonessential amino acids to become conditionally essential. People with PKU do not make one of the enzymes required to convert the essential amino acid phenylalanine to the normally nonessential amino acid tyrosine. Thus, tyrosine becomes conditionally essential. In all, six amino acids are considered to be conditionally essential, and these are listed in Table 9.1.

Are All Food Proteins Equal?

It is not surprising that some foods contain more protein than others. For example, meat, poultry, fish, eggs, dairy products, and nuts provide more protein (per gram) than grains, fruits, and vegetables. An exception to this is legumes such as dried beans, lentils, peas, and peanuts. Leguminous plants are unique in that their roots are associated with bacteria that can take nitrogen from the air and incorporate it into amino acids. These amino acids are then used by the associated plant. This is why legumes tend to have higher protein content than most other plants.

Moreover, foods with the same amount of total protein can contain different combinations of amino acids. For example, both a cup of cottage cheese and a cup of cooked lima beans provide about 15 g of protein. However, the amino acids in these two foods are quite different. Some foods provide balances of amino acids that are more nutritionally useful for the body than others. This is somewhat analogous to having two piggy banks—one containing a pound of coins made up of ½ pennies and ½ nickels and the other containing a pound of coins made up of ⅓ each of nickels, dimes, and quarters. In this analogy, you can think about the pennies and nickels as nonessential amino acids and the dimes and quarters as essential amino acids. Although both of these banks contain 1 pound of coins, that containing the dimes and quarters is more valuable. Similarly, some food proteins are "worth" more than others, with animal foods generally being more "valuable" than plant foods. Take our comparison of cottage cheese versus lima beans. Foods of animal origin like cottage cheese tend to be more "valuable" than do plant-derived foods like lima beans because the former generally have greater amounts of *essential* amino acids than do the latter.

phenylketonuria (PKU) An inherited disease in which the body cannot convert phenylalanine into tyrosine.

complete protein source A food that contains all the essential amino acids in relative amounts needed by the body.

incomplete protein source A food that lacks or contains very low amounts of one or more essential amino acids.

limiting amino acid The essential amino acid in the lowest concentration in an incomplete protein source.

COMPLETE AND INCOMPLETE PROTEINS

Food proteins can be categorized based on the balance of amino acids they contain. Those containing adequate amounts of all essential amino acids are said to be **complete protein sources,** whereas those supplying low amounts of one or more of the essential amino acids are **incomplete protein sources.** The amino acids that are missing or in low amounts in an incomplete protein source are called **limiting amino acids.** When a limiting amino acid is missing from the diet, we cannot make proteins that contain that particular amino acid, even if we have all of the other essential amino acids.

In general, meat, poultry, eggs, and dairy products are complete protein sources, whereas plant products are incomplete protein sources. For example, corn protein is generally low in lysine and tryptophan—these are considered two of its limiting amino acids. Another incomplete protein source is wheat, which provides only small amounts of lysine.

© Ulrich Kerth/StockFood/Getty Images

Animal-derived foods such as meat, milk, fish, and eggs are excellent sources of complete protein.

Protein complementation involves consumption of a variety of incomplete protein sources so that all essential amino acids are consumed.

PROTEIN COMPLEMENTATION

You may be wondering how people who eat only plant foods (vegetarians) get all of their essential amino acids if these types of foods only contain incomplete proteins. The answer to this question is that combining a variety of foods with incomplete proteins can provide adequate amounts of all the essential amino acids. This dietary practice, called **protein complementation,** is customary around the world, especially in regions that traditionally rely heavily on plant sources for protein.[6] Examples of commonly consumed foods whose proteins "complement" each other are rice and beans as well as corn and beans. Both rice and corn have several limiting amino acids (for example, lysine) but generally provide adequate methionine. Beans and other legumes tend to be limiting in methionine but provide adequate lysine. In general, protein complementation allows diets containing a variety of plant protein sources to provide all of the needed essential amino acids.

PROTEIN QUALITY

The "quality" of a food protein, however, is more complex than simply which amino acids it contains. Factors that determine whether food proteins are good sources of essential amino acids include whether the protein is complete or incomplete and the body's ability to absorb the amino acids the protein contains (i.e., its bioavailability).[7] If a food is a complete protein source and it provides easily absorbed amino acids (has high bioavailability), it is a **high-quality protein source.** In general, animal-derived foods are sources of high-quality protein, and foods containing incomplete proteins and/or those in which the protein has low bioavailability are **low-quality protein sources.**

For example, protein from processed (polished) rice is of low quality because it has low amounts of an essential amino acid (lysine) and has low bioavailability. Thus, people who rely solely on rice for their amino acid requirements become lysine deficient unless they consume extremely large amounts of rice. A person with lysine deficiency cannot make enough of *any* of the proteins that require this amino acid. Thus, lysine deficiency negatively affects many aspects of health, including oxygen transport (insufficient hemoglobin), muscle synthesis (inadequate muscle proteins), and digestive processes (lack of digestive enzymes).

Traditionally, the protein quality of a plant was something that we could not influence, making protein complementation vital in cultures consuming little or no animal-derived food. However, scientists are now able to alter the amino acid composition of some plants with the ultimate goal of transforming low-quality protein sources into high-quality protein sources. You can read more about this in the Focus on Food feature.

How Are Proteins Made?

You now know the fundamental concepts related to what makes up a protein, how proteins differ from each other, and why some foods are considered better sources of protein than others. But to really understand why proteins are essential in the diet, you must also understand what your body does with protein once you eat it. In other words, how does your body convert the proteins you eat into the exact and functional proteins you need? And how do these new proteins function? To understand this you must first know how the body makes its own proteins. This is because proteins acquire their distinct structures and functions as they are made in your cells, and it is the shape

protein complementation Combining incomplete protein sources to provide all of the essential amino acids in relatively adequate amounts.

high-quality protein source A complete protein source with high amino acid bioavailability.

low-quality protein source A food that is either an incomplete protein source or one that has low amino acid bioavailability.

Which specific proteins a particular plant or animal produces is determined by its genetic code. Therefore, alterations in the genetic material (DNA) can influence the proteins a plant makes and ultimately affect the protein quality of foods that we make from the plant. Although purposefully altering the DNA of an organism was in the realm of science fiction only decades ago, this is no longer the case. Indeed, the DNA and protein quality of a food can now be enhanced by the production of a genetically modified plant or animal (also called a **genetically modified organism,** or GMO).[8] Genetic modification involves manipulating the genetic material of an organism to "force" it to produce different or altered proteins. Sometimes this is done by simply modifying the DNA that the organism has. Other times, DNA from another organism is inserted in the nucleus. In this way, a plant with one or more limiting amino acids can be enriched with those amino acids and, as a result, be transformed into a complete protein source.

An example of a GMO is the Opaque-2 corn plant.[9] To produce this modified corn, scientists altered the corn's genetic code so that it produces proteins with more lysine and tryptophan—the two limiting amino acids in unmodified corn. The availability of this genetically modified corn may be especially important for people living in areas of the world that rely heavily on corn for their protein intake, because its consumption decreases the risk of amino acid deficiency in these at-risk populations.

However, the development and use of GMOs is somewhat controversial. For example, some people worry that GMOs may increase risk of food allergies and antibiotic resistance and result in unintended transfer of genetic materials to other organisms. Others have expressed concern that it is unethical to tamper with nature by mixing genes among species. This is currently an area of active debate. In response, the U.S. Food and Drug Administration (FDA) has developed policies and guidelines that must be followed in developing new foods made with GMOs. For example, genetic modification must not negatively alter the nutrient quality of the food. The FDA does not require foods made with GMOs to be so labeled, although those that do not contain GMOs can be labeled as such. You can find out more about GMOs and the FDA's rules and regulations related to them on its website (http://www.fda.gov).

that determines a protein's function. The process of protein synthesis, which involves three basic steps—cell signaling, transcription, and translation—is discussed next and illustrated in Figure 9.2.

STEP 1: CELL SIGNALING INITIATES PROTEIN SYNTHESIS

Nearly all cells in the body make proteins. However, different cells need different proteins, and the amounts needed are highly variable. As a result, protein synthesis within a cell is not random; instead, it is regulated by what the individual cell and the entire body need. To initiate this process, the cell must first be told to make a particular protein. This process, called **cell signaling,** is initiated by proteins (like hormones) or other substances (like vitamins) associated with the cell. Cell signaling communicates environmental conditions or cellular needs to the nucleus of the cell, just as an "indoor/outdoor" thermometer can be used to monitor the outside temperature from inside a house.

The "turning on" of protein synthesis by cell signaling is called **up-regulation.** Conversely, sometimes cells are instructed to "turn off" the synthesis of a certain protein. This type of cell signaling is called **down-regulation.** Nutrients are involved in both of these processes. The body's ability to orchestrate the up- and down-regulation of protein synthesis via cell signaling allows it to produce only the proteins it needs—and to stop producing proteins that are not needed. This involves a cell's genes: the genetic material found in its nucleus. For example, cell signaling of the genes needed for calcium absorption in your small intestine is up-regulated when your blood calcium level is low and down-regulated when it is elevated. In this way, blood calcium level is maintained in its optimal range.

genetically modified organism (GMO) An organism (plant or animal) made by genetic engineering.

cell signaling The first step in protein synthesis, in which the cell receives a signal to produce a protein. Note that this term is also used for a variety of other processes (aside from protein synthesis) within the cell.

up-regulation In the context of protein synthesis, increased expression of a gene.

down-regulation In the context of protein synthesis, decreased expression of a gene.

FIGURE 9.2 The Steps of Protein Synthesis Protein synthesis involves three basic steps: cell signaling, transcription, and translation.

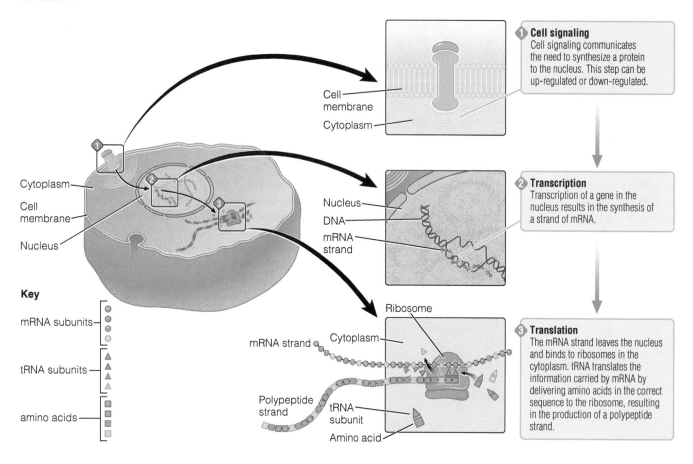

Key

mRNA subunits

tRNA subunits

amino acids

1 Cell signaling
Cell signaling communicates the need to synthesize a protein to the nucleus. This step can be up-regulated or down-regulated.

2 Transcription
Transcription of a gene in the nucleus results in the synthesis of a strand of mRNA.

3 Translation
The mRNA strand leaves the nucleus and binds to ribosomes in the cytoplasm. tRNA translates the information carried by mRNA by delivering amino acids in the correct sequence to the ribosome, resulting in the production of a polypeptide strand.

STEP 2: TRANSCRIPTION TRANSFERS GENETIC INFORMATION TO mRNA

Cell signaling (or more specifically, the up-regulation of protein synthesis) initiates the second step of protein synthesis, **transcription.** One way to think of transcription is to compare it to reading an instruction manual describing how to put together a newly purchased product. In protein synthesis, the "written" instruction manual is in the nucleus of each cell—coded in strands of deoxyribonucleic acid, or DNA. Found in a cell's nucleus, coiled strands of DNA combine with special proteins to form **chromosomes.** Each chromosome is subdivided into thousands of units called **genes.** Each gene provides the instructions needed to make a peptide. In other words, a gene tells a cell which amino acids are needed and in what order they must be arranged to synthesize a particular protein (or portion thereof).

However, proteins are not made directly from the genes comprising your chromosomes. Rather, the gene's DNA code must first be communicated to protein-synthesizing organelles found outside the cell's nucleus. This is accomplished by enzymes that "unzip" the DNA. Next, the genetic information contained in the DNA is "decoded" or transcribed by a molecule called **messenger ribonucleic acid (mRNA).** In the nucleus, a series of mRNA subunits bind to the up-regulated gene. These mRNA subunits then join, forming a strand of mRNA. The newly formed mRNA strand separates from the DNA, exits the nucleus, and enters the cytoplasm, where the mRNA participates in the next step of protein synthesis, called translation. As such, the information

transcription The process whereby mRNA is made using DNA as a template.

chromosome A strand of DNA and associated proteins in a cell's nucleus.

gene A portion of a chromosome that codes for the primary structure of a protein.

messenger ribonucleic acid (mRNA) A form of RNA involved in gene transcription.

originally contained in the DNA sequence of a gene is transcribed into an mRNA sequence, which can be understood by the protein-synthesizing organelles housed in the cytoplasm.

STEP 3: TRANSLATION PRODUCES A NEW PEPTIDE CHAIN

The third step of protein synthesis, called **translation,** begins when the mRNA strand binds to ribosomes. **Ribosomes** are found in the cytoplasm and assist in translating the information contained in mRNA into a peptide. This process requires yet another type of RNA called **transfer ribonucleic acid (tRNA).** The function of tRNA is to transport amino acids to the ribosomes for protein synthesis. For translation to proceed, the ribosome moves along the mRNA strand "reading its sequence." The sequence of the mRNA, in turn, instructs specific tRNAs to transfer the amino acids they are carrying to the ribosome. One by one, amino acids join together via peptide bonds to form a growing peptide chain. When translation is complete, the newly formed peptide separates from the ribosome. However, this is not the final step in the formation of a new protein. The final structure and shape of the protein is yet to be determined, and how this happens is described next.

How Do Proteins Get Their Shapes?

When genetic material has been transcribed and translated, a peptide chain is released from the ribosome. However, to form the complex shape of the final protein the peptide must fold and, in some cases, combine with other peptide chains or substances. Because the final shape must be exact for the protein to function correctly, it is important to consider the many levels of protein structure and what can happen when something goes wrong.

PRIMARY STRUCTURE DICTATES A PROTEIN'S BASIC IDENTITY

There are several levels of protein structure, but the most basic is determined by the number and sequence of amino acids in a single peptide chain. This is called a protein's **primary structure** or primary sequence (Figure 9.3) and is determined by the DNA code. Each peptide chain has a unique primary structure and is therefore a unique molecule. Although understanding the concept of primary structure is relatively simple, take a moment to contemplate the enormous number of primary structures that can be made from just 20 amino acids. Consider, as an analogy, the English alphabet, which has a somewhat similar number of letters—26 in all. As you know, the number and variety of words that can be made with only 26 letters is astonishing. Some are short; some are long. Some contain just a few different letters; others contain many different letters. The same holds true for proteins: some are short, some are long. Some contain a handful of different amino acids, whereas others may contain all 20 amino acids. The possibilities are seemingly endless.

A Protein's Primary Structure Is Critical The primary structure of a protein is critical to its function because it determines the protein's most basic chemical and physical characteristics. In other words, the primary structure represents the basic identity of the protein. Changes in a protein's primary structure can profoundly affect the ability of the protein to do its job. Some

translation The process whereby amino acids are linked together via peptide bonds on ribosomes, using mRNA and tRNA.

ribosome An organelle, associated with the endoplasmic reticulum in the cytoplasm, involved in gene translation.

transfer ribonucleic acid (tRNA) A form of RNA in the cytoplasm involved in gene translation.

primary structure The sequence of amino acids that make up a single peptide chain.

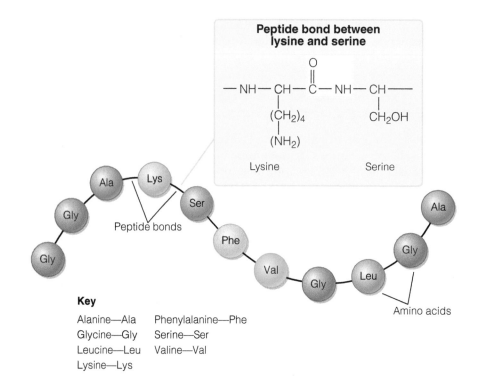

FIGURE 9.3 The Primary Structure of a Protein The primary structure of a protein is the sequence and number of the amino acids in its peptide chain.

Peptide bond between lysine and serine

Lysine   Serine

Peptide bonds

Amino acids

Key

Alanine—Ala	Phenylalanine—Phe
Glycine—Gly	Serine—Ser
Leucine—Leu	Valine—Val
Lysine—Lys	

alterations in a protein's primary structure can be caused by inherited genetic variations in the DNA sequence. **Sickle cell anemia** (also called sickle cell disease) is an example of a disease caused by an inherited genetic variation. The complications associated with sickle cell anemia, such as fatigue and increased risk for infection, are due to a small "error" in the DNA that ultimately results in the production of defective molecules of the protein hemoglobin within red blood cells.[10] Because hemoglobin is responsible for circulating oxygen and carbon dioxide in the blood, these complications can sometimes be serious. Additional detail on sickle cell anemia is provided in the Focus on Clinical Applications feature.

SECONDARY STRUCTURE FOLDS AND TWISTS A PEPTIDE CHAIN

In addition to each peptide's unique primary structure, most proteins have a three-dimensional shape that is more complex than a linear chain of amino acids. Recall that the backbone of the peptide chain is made of a series of positively charged amino acids and negatively charged carboxylic acid groups. These charges attract and repel each other like magnets, folding portions of the peptide into organized and predictable patterns. These interactions form weak bonds[†] that twist and fold the primary structure into a three-dimensional pattern. This level of folding is called the **secondary structure** of a protein. The two most common folding patterns are the **α-helix** and **β-folded sheets,** illustrated in Figure 9.4. You can think of an α-helix as being like a spiral staircase and a β-folded sheet as being similar to a folded paper fan.

TERTIARY STRUCTURE ADDS COMPLEXITY

The next level of protein complexity is called **tertiary structure,** which is additional folding due to interactions among the R-groups. This causes the

[†] Technically, these are hydrogen bonds.

---

sickle cell anemia A disease in which a small change in the amino acid sequence of hemoglobin causes red blood cells to become misshapen and decreases the ability of the blood to carry oxygen and carbon dioxide.

secondary structure Folding of a protein because of weak bonds that form between elements of the amino acid backbone (not R-groups).

α-helix (AL – pha – he – lix) A common configuration that makes up many proteins' secondary structures.

β-folded sheet (BAY – ta – fold – ed) A common configuration that makes up many proteins' secondary structures.

tertiary structure (TER – ti – a – ry) Folding of a polypeptide chain because of interactions among the R-groups of the amino acids.

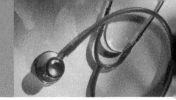

Sickle cell anemia is a disease caused by a single error in the amino acid sequence of the protein hemoglobin. More specifically, people with sickle cell anemia have an alteration, or mutation, in the gene that codes for one of the polypeptides in hemoglobin. Hemoglobin is a large and complex protein, found in red blood cells, that carries oxygen from the lungs to the body's tissues and carbon dioxide back to the lungs for removal. The sickle cell mutation results in the insertion of an incorrect amino acid (valine) for the correct amino acid (glutamic acid) during translation. As a result of this seemingly small error in its primary structure, the shape of the hemoglobin molecule is completely altered, causing it to function improperly.

Red blood cells with normal hemoglobin are smooth and glide through blood vessels. Red blood cells with "sickle cell" hemoglobin are rigid, sticky, and shaped like a sickle or crescent. This alteration in shape causes the cells to form clumps and get stuck in blood vessels. The clumps of sickle-shaped cells can block blood flow in the vessels that lead to the limbs and organs, resulting in pain, serious infections, and organ damage. Signs and symptoms of this disease include anemia (the inability of the blood to carry oxygen); pain in the chest, abdomen, and joints; severe fatigue; swollen hands and feet; frequent infections; stunted growth; and vision problems.

Scientists hypothesize that, hundreds of years ago, the genetic alteration responsible for sickle cell anemia somehow protected people from the serious and sometimes deadly disease malaria.[11] During this time, a malaria epidemic killed tens of thousands of people in parts of Africa, the Mediterranean, the Middle East, and India. People with the sickle cell mutation, however, are thought to have survived this outbreak better than those who did not have the mutated gene. As a result, survivors were able to reproduce and pass on their genetic code to their offspring. Today, millions of people all over the world have the gene for sickle cell anemia, especially those with African, Mediterranean, Middle Eastern, or Indian ancestry. In the United States, sickle cell anemia is most common in people of black African heritage; 1 in 500 African Americans has the disease.[11] Note that in order for a person to have

sickle cell anemia he or she must have received a faulty gene from *both* parents. If a sickle cell gene is inherited from only one parent, the person will not have serious signs or symptoms of sickle cell anemia but could pass the gene on to the next generation. Such individuals are said to have the sickle cell "trait." Approximately 1 in 12 African Americans has sickle cell trait.[11]

Although there is currently no cure for sickle cell anemia, scientists hope that they might soon have the technology to "correct" the defective DNA.[12] This technology, called **gene therapy,** may be able to cure this and other genetically based diseases. In the meantime, complications of this disease are treated with analgesic drugs to reduce pain, antibiotics to treat infection, blood transfusions, supplemental oxygen, folic acid supplementation, and bone marrow transplants.

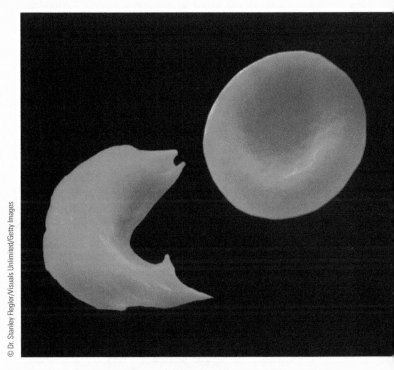

© Dr. Stanley Flegler/Visuals Unlimited/Getty Images

Normal red blood cells (right) are disc shaped, whereas those of people with sickle cell anemia (left) are crescent shaped.

entire protein to have an even more complex, three-dimensional structure. Imagine, for example, what would happen to a folded paper fan if you were to gently crumple it in your hand. The fan's original folds (analogous to a protein's secondary structure) would remain, but they might be further bent and twisted in some regions (analogous to a protein's tertiary structure). This is what happens in a protein to form its tertiary structure.

gene therapy The use of altered genes to enhance health.

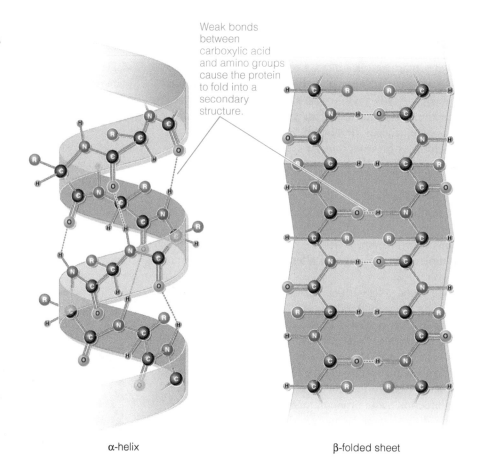

FIGURE 9.4 The Secondary Structure of a Protein Weak chemical bonds fold proteins into α-helix and β-folded sheet patterns, resulting in a secondary structure.

Weak bonds between carboxylic acid and amino groups cause the protein to fold into a secondary structure.

α-helix

β-folded sheet

Some of the strongest interactions between R-groups occur between amino acids that contain sulfur atoms (for example, cysteine), which react with other sulfur-containing amino acids to form sulfur–sulfur (disulfide) bonds. These bonds are particularly stable, and anything that disturbs them can severely disrupt the protein's tertiary structure. In fact, it is the number and arrangement of disulfide bonds in hair that determine if it is straight or wavy. Breaking and reforming these bonds is the basis for some "permanent wave" treatments used in hairstyling.

SOME PROTEINS HAVE QUATERNARY STRUCTURE AND PROSTHETIC GROUPS

The fourth level of protein structure is called **quaternary structure.** Quaternary structure exists when two or more peptide chains come together, as shown in Figure 9.5. This level of complexity would be somewhat like putting two or three crumpled paper fans together. Not all proteins have a quaternary structure—only those made from more than one peptide chain.

To function properly, some proteins have precisely positioned nonprotein components called **prosthetic groups,** which often contain minerals. Hemoglobin is an example of a protein with quaternary structure and prosthetic groups. Specifically, it is made from four separate polypeptide chains (hemoglobin's quaternary structure), each of which combines with an iron-containing heme (hemoglobin's prosthetic group). Heme is the portion of hemoglobin that actually transports the oxygen and carbon dioxide gases in the blood.

quaternary structure (quat – ER – nar – y) The combining of peptide chains with other peptide chains in a protein.

prosthetic group A nonprotein component of a protein that is part of the quaternary structure.

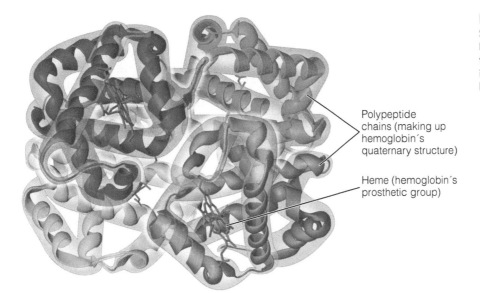

FIGURE 9.5 **The Quaternary Structure and Prosthetic Groups of Hemoglobin** Hemoglobin is made from four polypeptide chains and four iron-containing prosthetic groups called heme.

Polypeptide chains (making up hemoglobin's quaternary structure)

Heme (hemoglobin's prosthetic group)

DENATURING AGENTS ALTER A PROTEIN'S SHAPE AND FUNCTION

A protein's primary, secondary, tertiary, and quaternary structures determine its final shape. This shape is critical to allowing the protein to carry out its function. However, there are many conditions that can disrupt a protein's shape. One example is **denaturation.** Denaturation occurs when a protein unfolds. Note, however, that only the secondary, tertiary, and quaternary structures are affected by denaturation; the primary structure remains intact. This would be akin to flattening out one of the pieces of paper making up your folded paper fan "protein." Flattening out the fan would result in a fan that does not work, just like denaturation can cause a protein to lose its function. Compounds and conditions that cause denaturation are called denaturing agents and include physical agitation (e.g., shaking), heat, detergents, acids, bases, salts, alcohol, and heavy metals (e.g., lead and mercury).

A familiar example of denaturation occurs when an egg white is heated; the proteins unfold, and the egg white changes from a thick, clear liquid to an opaque solid. Another example of a denaturing agent is mercury, which can disrupt disulfide bonds and thus tertiary structure. Because of its denaturing effects on proteins involved in neural function, high levels of mercury exposure can cause numbness, hearing loss, visual problems, difficulty walking, and severe emotional and cognitive difficulties. In fact, we now know that a condition termed "mad hatter's disease" that was widespread in the 19th century was due to mercury poisoning.[13] During that period, hat makers used large amounts of mercury-containing compounds to treat felt and fur and suffered the consequences in that they went "mad."

Because of the denaturing effects of heavy metals such as lead and mercury, efforts are being made to decrease the concentrations of these compounds in the environment. For example, there are concerns that certain types of fish contain dangerously high amounts of mercury. In fact, although most fish are safe

© Lisa Romerein/Taxi/Getty Images

It is important to limit our consumption to certain types of fish because some contain mercury, which can cause protein denaturation in the body.

denaturation The alteration of a protein's three-dimensional structure by heat, acid, chemicals, enzymes, or agitation.

to consume, the U.S. Food and Drug Administration (FDA) and the Environmental Protection Agency (EPA) have made the following recommendations for women who might become pregnant, women who are pregnant or lactating, and young children.[14] These recommendations are meant to decrease mercury exposure while encouraging consumption of low-mercury fish in these at-risk populations.

- Do not eat shark, swordfish, king mackerel, or tilefish.
- Eat up to 12 ounces (two average meals) a week of a variety of fish and shellfish that are lower in mercury: shrimp, canned light tuna, salmon, pollock, and catfish.
- Check local health departments about the safety of fish caught in local lakes, rivers, and coastal areas. If no advice is available, eat up to 6 ounces per week of fish you catch from local waters, but do not consume any other fish during that week.

As previously discussed, protein synthesis is a complex process that begins when cell signaling initiates transcription of a gene. This ultimately results in the body producing all of the many proteins that it needs. But have you ever considered why some cells make different proteins than other cells, and how your nutritional status may affect this process? We will now look at how genes interact with nutrition to determine the particular proteins you make, in turn influencing your health and well-being.

Genetics, Epigenetics, Nutrition, and Nutrigenomics

The genetic material (DNA) in a cell's nucleus provides the cell with instructions to make all the proteins it needs, and the exact DNA sequences individuals have are determined by what DNA their parents passed on to them. When a sperm cell fertilizes an egg cell, the chromosomes in each combine to become the genetic makeup of the offspring. This is how an infant inherits his or her parents' **genetic makeup** or genotype. Except for identical twins, no two individuals have the same genotype, which in part explains the vast diversity among humans.

Although the overall sequences of genes coding for specific proteins are often identical among people, some genes have modifications that are unique. These genetic differences give us each our individual physical characteristics, such as eye color and hair color. These characteristics do not really affect our health. However, alterations in other genes can have more important consequences.

GENETIC ALTERATIONS: MUTATIONS AND POLYMORPHISMS

Sometimes a modification in a gene results in a protein with an altered amino acid sequence. When this is due to a chance genetic modification, it is called a **mutation.** While some mutations have no measurable effect on health, others do. For example, you have learned that sickle cell anemia (which negatively impacts the function of red blood cells) is a sometimes serious disease caused by a mutation. Other mutations, such as the one involved in PKU, can influence metabolism. Still others alter protein synthesis, causing cells to experience uncontrollable growth, which, in turn, can lead to cancer. Alterations (mutations) in the DNA of egg and sperm cells can be passed on to the offspring and are therefore inherited. PKU and sickle cell anemia are the result of inherited DNA mutations.

genetic makeup (genotype) The particular DNA contained in a person's cells.

mutation The alteration of a gene.

Some mutations are more common than others, and when a particular genetic mutation is present in at least 1% of the population, it is called a **polymorphism.** Polymorphisms can impact a person's health and risk for disease. They can also influence nutrient requirements. Thus, nutritional scientists are very interested in understanding common genetic polymorphisms and how they interact with nutrition to influence health.

EXPERTS BELIEVE THAT NUTRITION MAY BE RELATED TO EPIGENETICS

Although a person's genetic makeup (genotype) has long been thought to reflect only the *sequence* of the DNA making up his or her chromosomes, scientists have recently learned that the connection between genes and physiology is actually much more complex. For example, the DNA and its accompanying organelles can be altered in a variety of ways (not related to DNA sequence) that regulate whether a particular gene is expressed or not. The term **epigenetics** (the prefix *epi-* is derived from the Greek word meaning "over" or "above") refers to these types of modifications. As such, two people can have exactly the same genetic sequences in their DNA but produce different amounts and combinations of proteins. Like mutations in DNA, some of these epigenetic differences can be passed on to the next generation; in other words, they are inheritable.

Epigenetic modifications are responsible for establishing and maintaining the diverse patterns of gene expression that distinguish different cell and tissue types.[15] For example, although bone and nerve cells contain exactly the same DNA, they synthesize very different proteins in large part because of different epigenetics. Scientists now think that epigenetic modifications may also play important roles in the development of many chronic degenerative diseases such as cancer, type 2 diabetes, and cardiovascular disease.[16] For instance, epigenetic changes can influence kidney function and therefore have an effect on blood pressure regulation.

Of great importance to the field of nutrition is growing evidence that nutritional status in very early life may affect long-term health via epigenetic modifications. For example, babies who are malnourished during fetal life but then experience accelerated growth in infancy or childhood may be at increased risk for cardiovascular disease and type 2 diabetes later in life, partly due to differences in epigenetic patterns.[17] Whether later alterations in a person's environment, such as better nutrition, can reverse this epigenetic effect remains to be discovered. Clearly epigenetics is an exciting new area of nutrition research.

THE HUMAN GENOME PROJECT HAS OPENED THE DOOR TO NUTRIGENOMICS

Until about 20 years ago, scientists knew very little about the genes that make up the human genome. However, the Human Genome Project was carried out in the 1990s to describe all the genes within our chromosomes. The Human Genome Project, described in more detail in the Focus on the Process of Science feature, has allowed scientists to better understand how genetic variations influence protein synthesis, overall health, and risk for disease. In addition, it has opened up the field of nutrition research referred to as nutrigenomics.[18] **Nutrigenomics** is the study of how nutrition and genetics interact to influence health. Nutritional scientists and other health professionals hope that, by understanding your individual nutritional needs, nutritional status, *and* genetics, they will someday be able to optimize your health and

polymorphism An alternation in a gene that is present in at least 1% of the population.

epigenetics (epi – ge – NE – tics) Alterations in gene expression that do not involve changes in the DNA sequence.

nutrigenomics (nu – tri – gen – O – mics) The science of how genetics and nutrition together influence health.

Scientists, philosophers, and psychologists have long been interested in what makes one person different from the next. Although this question may represent the ultimate mystery, researchers continue to try to understand the physiological and chemical reasons for why we all differ. In the 1800s, Gregor Mendel discovered that many physical characteristics could be passed on to offspring, launching the field of modern genetics. Genetics has, however, progressed significantly since that time.

In 1990, the U.S. government and an international team of scientists initiated the Human Genome Project, which was designed to decode all the genes making up human chromosomes. This information has become invaluable for our understanding of genetic diversity and how it influences health and well-being.

Nutritional scientists can now directly test what they have long thought: nutrition interacts with genetics to influence health. For example, it is now much easier to study how dietary factors influence whether a gene is turned on or turned off. Additionally, emerging studies suggest that nutrition likely determines some epigenetic alterations.

Indeed, many scientists and health professionals hope that someday each of us will be able to inexpensively and noninvasively find out what type of diet is best for our personal genetic makeup. This may be a relatively simple process only requiring that you provide a swab taken from the inside of your cheek. The sample could then be mailed to a laboratory, which would analyze it and make dietary recommendations about specific amounts of foods, nutrients, supplements, and exercise to best help prevent illness.

For example, if your genes suggest that your risk for heart disease is high, your list of recommendations would likely stress a diet low in fat, saturated fat, *trans* fatty acids, and sodium but high in cardioprotective foods. In other words, you would receive a *personalized* dietary prescription. Today a handful of commercial laboratories claim to be able to make these personal diet recommendations for anywhere from $500 to $2,000. However, their validity is the subject of *significant* debate. Nonetheless, this amazing concept is no longer science fiction but is clearly the direction in which modern nutritional science is moving.

well-being by precisely matching dietary recommendations with your unique genetic—and perhaps epigenetic—makeup.

You now know how proteins are made in the body and how alterations in genetics and epigenetics can influence this process. However, to make proteins, the body must first obtain their basic building blocks—the amino acids—from the diet. As such, dietary proteins must be digested, absorbed, and circulated to the cells that need them. How the body completes these tasks is described next.

How Are Dietary Proteins Digested, Absorbed, and Circulated?

The process of digestion disassembles food proteins into amino acids that are then absorbed and circulated to cells where the amino acids are reassembled into the proteins that the body needs. This is somewhat like disassembling someone else's house and then using the materials to rebuild another house that perfectly fits your own exact needs. In addition to using *dietary* proteins, the body efficiently and systematically breaks down and recycles its own proteins when they become old and nonfunctional. In this way, intestinal enzymes, digestive secretions, and degraded cells can contribute more than once to the amino acids available for protein synthesis. In fact, you can think of the body as having its own "protein recycling center." The stages of protein digestion, absorption, and circulation are shown in Figure 9.6 and described next.

⟨**CONNECTIONS**⟩ Recall that hydrochloric acid (HCl) is released from parietal cells, whereas pepsinogen is released from chief cells (Chapter 1, page 25).

PROTEIN DIGESTION BEGINS IN THE STOMACH

Proteins in the foods you eat must be broken down into their component amino acids; only then can they be used by the body. Although a small amount of

FIGURE 9.6 Overview of Protein Digestion Protein digestion occurs in both the stomach and small intestine.

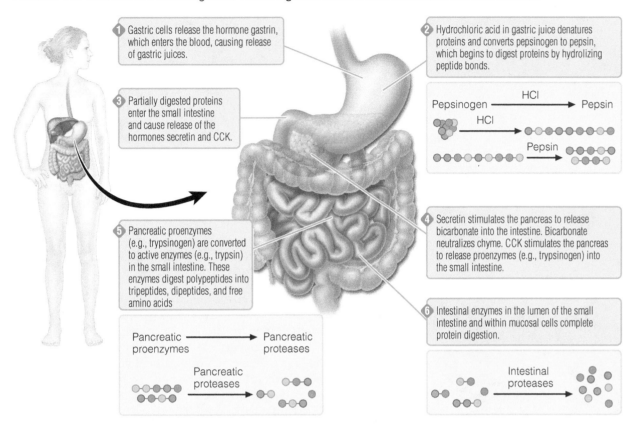

1. Gastric cells release the hormone gastrin, which enters the blood, causing release of gastric juices.

2. Hydrochloric acid in gastric juice denatures proteins and converts pepsinogen to pepsin, which begins to digest proteins by hydrolizing peptide bonds.

Pepsinogen —HCl→ Pepsin

—HCl→

—Pepsin→

3. Partially digested proteins enter the small intestine and cause release of the hormones secretin and CCK.

4. Secretin stimulates the pancreas to release bicarbonate into the intestine. Bicarbonate neutralizes chyme. CCK stimulates the pancreas to release proenzymes (e.g., trypsinogen) into the small intestine.

5. Pancreatic proenzymes (e.g., trypsinogen) are converted to active enzymes (e.g., trypsin) in the small intestine. These enzymes digest polypeptides into tripeptides, dipeptides, and free amino acids

6. Intestinal enzymes in the lumen of the small intestine and within mucosal cells complete protein digestion.

Pancreatic proenzymes → Pancreatic proteases

Pancreatic proteases →

Intestinal proteases →

mechanical digestion of food protein occurs as you chew, chemical digestion of protein begins when it enters the stomach. The presence of food in your stomach causes some gastric cells to release a hormone called **gastrin** which is released from endocrine cells found deep in the gastric pits. Gastrin triggers the release of hydrochloric acid (HCl), **pepsinogen,** mucus, and substances from other stomach cells that make up the gastric pits. Remember these substances are collectively referred to as *gastric juices*. Note that pepsinogen is an example of a **proenzyme** (also called a zymogen), which is an inactive form of an enzyme.

It is important to understand how all these substances (HCl and enzymes) work together to contribute to protein digestion. First, HCl disrupts the chemical bonds responsible for the protein's secondary, tertiary, and quaternary structures. This process of unfolding, or denaturation, straightens out the complex protein structure so that the peptide bonds can be exposed to digestive enzymes. Next, HCl converts pepsinogen (the proenzyme form) into its active form, called **pepsin.** Pepsin is an example of a **protease.** Protease enzymes hydrolyze or break peptide bonds between amino acids. Note that the stomach does not produce the active protease enzyme—in this case, pepsin. Instead, it produces and stores the inactive or "safe" proenzyme (pepsinogen), thus protecting itself from the enzyme's protein-digesting function until it is needed.

As a result of the actions of the gastric juices, proteins that you eat are partially digested to shorter peptides and some free amino acids. The partially broken down proteins are then ready to leave the stomach and enter the small intestine to be digested further.

PROTEIN DIGESTION CONTINUES IN THE SMALL INTESTINE

Protein digestion in the small intestine takes place both in the lumen and within the enterocytes that line it. Initiating this cascade of digestive events, amino

⟨**CONNECTIONS**⟩ Recall that hydrolysis is a chemical reaction in which chemical bonds are split by the addition of a water molecule (Chapter 1, page 8).

gastrin (GAS – trin) A hormone, secreted by endocrine cells in the stomach, which stimulates the production and release of gastric juice.

pepsinogen (pep – SIN – o – gen) The inactive form (proenzyme) of pepsin, produced in the stomach.

proenzyme (zymogen) An inactive precursor of an enzyme.

pepsin (PEP – sin) (*peptein*, to digest) An enzyme needed for protein digestion.

protease An enzyme that cleaves peptide bonds.

acids and smaller polypeptides coming from the stomach stimulate the release of the hormones secretin and cholecystokinin (CCK) from intestinal cells into the blood. Secretin and CCK in turn signal the pancreas to release bicarbonate into the lumen of the duodenum, neutralizing the acid from the stomach and inactivating pepsin. These hormones also cause the release of **trypsinogen, chymotrypsinogen, proelastase,** and **procarboxypeptidase,** all of which are proenzymes made in the pancreas. These inactive proenzymes are converted in the small intestine to their active protease forms: **trypsin, chymotrypsin, elastase,** and **carboxypeptidase.** Each of these enzymes recognizes specific amino acids in polypeptide chains, breaking the peptide bonds holding them together and forming tripeptides, dipeptides, and free amino acids. Finally, the di- and tripeptides are further broken down by enzymes produced in the brush border enterocytes. In some cases, the enzymes are released into the lumen of the small intestine. However, most di- and tripeptides are first transported into the enterocytes, where their digestion is completed.

AMINO ACIDS ARE ABSORBED IN THE SMALL INTESTINE AND CIRCULATED IN THE BLOOD

People who have peanut allergy must be very careful to not eat foods, like peanut butter, that contain peanuts.

When protein digestion is complete, some amino acids are already in the enterocytes. However, those remaining in the intestinal lumen must be transported into the brush border cells. This occurs via both passive and active transport mechanisms. Because amino acids with chemically similar R-groups are often transported by the same carrier molecules, such amino acids can compete with each other for transport into the blood. This is one reason why some experts recommend that you avoid taking large quantities of certain amino acid supplements. Most amino acids are absorbed in your duodenum, where they enter your blood and circulate to the liver via the hepatic portal system.

What Are Food Allergies and Intolerances? The breakdown of proteins into amino acids is usually quite complete and typically results in the absorption of amino acids (not proteins) into the circulation. Sometimes, however, larger peptide chains are absorbed. When this happens, the body's immune system may respond as if these peptides were dangerous. In such cases, the person is said to have an "allergic response," or what is more commonly called a **food allergy.**[19] The majority of food allergies are caused by proteins present in eggs, milk, peanuts, soy, and wheat. Researchers estimate that approximately 2% of adults and 5% of infants and young children in the United States have food allergies. Recall Tyler, the boy with a severe food allergy featured at the beginning of this chapter. Tyler is allergic to peanuts and other legumes. This is because relatively large portions of intact legume proteins are absorbed into his bloodstream, and his immune system responds as if they were dangerous—mounting a strong and sometimes dangerous inflammatory response.

Note, however, that not all adverse reactions to foods are true food allergies. A nonimmunological reaction to a substance in a food is called a **food intolerance** (or food sensitivity). An example of a food intolerance is lactose intolerance, which was discussed in Chapter 8.

trypsinogen, chymotrypsinogen, proelastase, and **procarboxypeptidase** Inactive proenzymes produced in the pancreas and released into the small intestine in response to CCK.

trypsin, chymotrypsin, elastase, and **carboxypeptidase** Active enzymes (proteases) involved in protein digestion in the small intestine.

food allergy A condition in which the body's immune system reacts against a protein in food.

food intolerance A condition in which the body reacts negatively to a food or food component but does not mount an immune response.

Signs and Symptoms of Food Allergies For some people, an allergic reaction to a particular food protein may cause mild physical reactions such as skin rashes or gastrointestinal (GI) distress. For others, like Tyler, an allergic food reaction can be frightening and even life threatening. Signs and symptoms of a food allergy usually develop within a few minutes to an hour after eating the food and are dependent upon what type of food allergy a person has and how his or her immune system reacts to it. The most common signs and symptoms are listed here.

- Tingling in the mouth
- Hives, itching, or eczema
- Swelling of the lips, face, tongue, and throat or other parts of the body
- Wheezing, nasal congestion, or trouble breathing
- Abdominal pain, diarrhea, nausea, or vomiting
- Dizziness, lightheadedness, or fainting

In a severe allergic reaction to food, a person may have more extreme symptoms. For example, **anaphylaxis** might result. Anaphylaxis is a rapid immune response that causes a sudden drop in blood pressure, rapid pulse, dizziness, and a narrowing of the airways. This can, almost immediately, block normal breathing. When this occurs emergency treatment is critical. Anaphylaxis due to food allergies is responsible for thousands of emergency room visits and as many as 200 deaths in the United States each year.[20]

What to Do if You Have a Food Allergy If you have a food allergy, the best way to prevent an allergic reaction is to know which foods to avoid. It is especially important to read food labels carefully, and wearing a medical alert bracelet or necklace may be advantageous. The U.S. Food and Drug Administration (FDA) requires that all foods containing the most common allergens be labeled as such. In the case of children, parents should talk with members of their families as well as friends, child care providers, and school personnel so that they can help avoid exposing the child to offending foods when he or she is under their supervision. It is also important to make sure that the child knows what foods to avoid and to ask for help if needed.

Now that you understand better what causes food allergies and how to avoid them, you may want to reconsider the questions posed at the beginning of this chapter concerning the issue of food allergies on a college campus. For example, what actions can you take to help protect a roommate or friend who has a food allergy?

© Shelley McGuire

Critical Thinking: Tyler's Story Explain, at a basic physiologic level, what causes the allergic responses Tyler experiences when he eats peanuts and other legumes. Specifically, how is this related to protein nutrition, digestion, and absorption? Can you imagine ways that food manufacturers might treat or process peanut-containing foods to make them safe for people with peanut allergy?

What Are the Major Functions of Proteins and Amino Acids in the Body?

Once amino acids are circulated away from the GI tract, the body uses them to make the thousands of proteins it needs via the protein synthetic reactions previously described. Using the previous analogy, this is the stage at which you would use all of the disassembled materials from someone else's house to build one of your own. However, there are many different types of things you can build even from a single material like wood, such as walls, cabinets, stairways,

anaphylaxis A severe and potentially life-threatening allergic reaction.

TABLE 9.2 The Major Functions of Proteins in the Body

Function	Description	Example(s)
Structure	Proteins making up the basic structure of tissues such as bones, teeth, skin	• Hydroxyapatite in bones • Collagen in skin, teeth, ligaments, and tendons • Keratin in hair and fingernails
Catalysis	Enzymes	• Lingual lipase digests lipids in mouth • Pancreatic amylase digests carbohydrates in intestine • Pepsin digests proteins in intestine
Movement	Proteins found in muscles, ligaments, and tendons	• Actin and myosin in muscle
Transport	Proteins involved in the movement of substances across cell membranes and within the circulatory system	• Glucose and sodium transporters in cell membranes • Lipoproteins and hormone transport proteins in blood
Communication	Protein hormones and cell-signaling proteins	• Insulin and glucagon regulate blood glucose • CCK helps regulate digestion in the small intestine • Cell-signaling proteins initiate protein synthesis
Protection	Skin proteins and immune proteins	• Collagen in skin • Fibrinogen helps blood clot • Antibodies fight off infection
Regulation of fluid balance	Proteins that—via the process of osmosis—regulate the distribution of fluid in the body's various compartments	• Albumin is a major regulator of fluid balance in the circulatory system
Regulation of pH	Proteins that readily take up and release hydrogen ions (H^+) to maintain pH of the body	• Hemoglobin is an important regulator of blood pH

and furniture. Each of these household items has its own function. Similarly, the proteins the body makes can be classified into general categories related to their functions (Table 9.2). For example, some proteins, like those in the muscles, are needed for movement. Others, like the hormone insulin, are used to communicate blood glucose levels. Proteins can also be broken down for energy (ATP), and some amino acids can be used to make glucose. In this section, you will learn more about the various types of proteins you need as well as other ways your body uses proteins and amino acids to function and stay healthy.

PROTEINS PROVIDE STRUCTURE

Proteins provide most of the structural materials in the body, being constituents of the muscles, skin, bones, hair, and fingernails. For instance, collagen is a structural protein that forms a supporting matrix in bones, teeth, ligaments, and tendons. Proteins are also important structural components of cell membranes and organelles. The synthesis of structural proteins is especially important during periods of active growth and development such as infancy and adolescence.

ENZYMES ARE PROTEINS THAT CATALYZE CHEMICAL REACTIONS

Molecules called "enzymes" (most of which are proteins) function as biological catalysts, driving the myriad chemical reactions that occur in the body. A **catalyst** is a substance that speeds up a chemical reaction but is not consumed or altered in the process. Without the catalytic functions of protein enzymes, the thousands of chemical reactions needed by the body would simply not occur or, at best, occur at very slow rates.

As an analogy, consider everything it takes to prepare a meal. Although this clearly requires the availability of all the ingredients, just having them in your kitchen will never result in their being prepared and served. Instead, there must also be a person who is willing and able to slice, dice, cook, and season the foods appropriately. In the same way, chemical reactions (like meals) will not occur readily without enzymes (like chefs) to arrange the molecules in the correct positions and supply the needed expertise to facilitate the appropriate chemical changes. Examples of protein enzymes you have already learned about are amylase and pepsin, which catalyze reactions needed to digest carbohydrates and proteins, respectively.

Proteins are essential for movement.

MUSCLE PROTEINS FACILITATE MOVEMENT

Protein is also needed for movement, which results from the contraction and relaxation of the many muscles in the body. There are three distinct types of muscle: skeletal muscle, which is responsible for all voluntary movements; cardiac muscle, which enables the heart to contract and relax; and smooth muscle, which lines many of our tissues and organs such as the stomach, intestine, and blood vessels. For example, as you read this book, you are experiencing contraction of all three types of muscle, allowing your body to simultaneously sit in your chair, turn the pages of the book, breath, and perhaps even drink a cup of coffee or eat a snack. Nearly half of the body's protein is present in skeletal muscle, and adequate protein intake is required to form and maintain muscle mass and function throughout life. Although there are many proteins related to movement, perhaps the most important are actin and myosin, which make up much of the machinery needed to contract and relax your muscles. This is why protein deficiency can cause muscle wasting and weakness.

SOME PROTEINS PROVIDE A TRANSPORT SERVICE

You also need amino acids to make transport proteins, which are responsible for escorting substances into and around the body as well as across cell membranes. For example, absorption of many nutrients (such as calcium) requires transport proteins to help the nutrients cross the cell membranes of enterocytes. Protein deficiency can decrease the body's production of intestinal transport proteins, resulting in secondary malnutrition of a variety of nutrients. Other transport proteins that you have previously learned about are the glucose transporters, which move glucose from the blood into tissues. In addition, many nutrients and other substances are circulated in the blood bound to even more transport proteins. Examples of circulating transport proteins include hemoglobin, which transports gases (oxygen and carbon dioxide), and

catalyst A substance that increases the rate by which a chemical reaction occurs, without being consumed in the process.

a variety of "binding proteins," which carry hormones and fat-soluble vitamins in the blood. If the diet does not provide adequate amounts of the essential amino acids, the synthesis of these proteins will decrease, and health consequences (such as anemia and vitamin deficiencies) can result.

HORMONES AND CELL-SIGNALING PROTEINS ARE CRITICAL COMMUNICATORS

Tissues and organs have a variety of ways to communicate with each other, and most of these involve proteins. Although not all hormones are proteins, most are—for example, secretin, gastrin, insulin, and glucagon.[‡] There are also proteins embedded in cell membranes that communicate information about the extracellular environment to the intracellular space. Some of these proteins are involved in the cell-signaling process that initiates protein synthesis itself. Others regulate cellular metabolism. Together, hormones and cell-signaling proteins make up part of the body's critical communication network. Thus, protein deficiency can have profound effects on the body's ability to coordinate its myriad functions.

PROTEINS PROTECT THE BODY

Perhaps one of the most basic functions of the proteins in the body is protecting it from physical danger and infection. For instance, skin is made mainly of proteins that form a barrier between the outside world and the internal environment. In addition to making up the skin, proteins provide other forms of vital protection. For example, if you cut yourself, blood clots (made possible by the presence of specialized proteins) close off this possible entry point for infection. In this way, blood clotting acts as a second level of defense. However, if a bacterium or other foreign substance does enter the body, the immune system fights back by producing proteins called **antibodies** or immunoglobulins. Antibodies bind foreign substances so they can be destroyed. Protein deficiency can make it difficult for the body to prevent and fight certain diseases, because its natural defense systems become weakened.

FLUID BALANCE IS REGULATED IN PART BY PROTEINS

Another function of proteins is regulating how fluids are distributed in the body. Most of the body consists of water, which is found both inside of cells (intracellular space) and outside of cells (extracellular space). In addition, the extracellular space can be subdivided into fluid found in blood and lymph vessels (intravascular fluid) and fluid found between cells (interstitial fluid). The amount of fluid in these spaces is highly regulated in a variety of ways, some of which involve proteins. For example, **albumin,** a protein present in the blood in relatively high concentrations, plays such a role. As the heart contracts, blood is propelled through blood vessels that become increasingly narrower. As the intravascular pressure builds, the fluid portion of the blood is forced out of the tiny capillaries and into the interstitial space. Albumin, which remains in the blood vessels, becomes more concentrated as more fluid accumulates in the interstitial space sourrounding tissues. The high intravascular concentration of albumin creates a powerful force that, like a sponge, draws the fluid from the interstitial space back into the blood vessel (Figure 9.7). Severe protein deficiency can impair albumin synthesis, resulting in low levels of this important protein in the blood and, in turn, accumulation

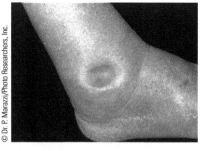

Because protein is needed for fluid balance, severe protein deficiency can cause edema. This photo depicts what happens when finger pressure is applied to an ankle with severe edema—a condition called pitting edema.

antibody A protein, produced by the immune system, that helps fight infection.

albumin A protein important in regulating fluid balance between intravascular and interstitial spaces.

[‡] The non-protein hormones, many of which are steroid hormones, include some of the reproductive hormones such as estrogen and testosterone.

FIGURE 9.7 Regulation of Fluid Balance by Albumin Albumin is a protein in the blood that helps regulate fluid balance between the intravascular space and the interstitial space.

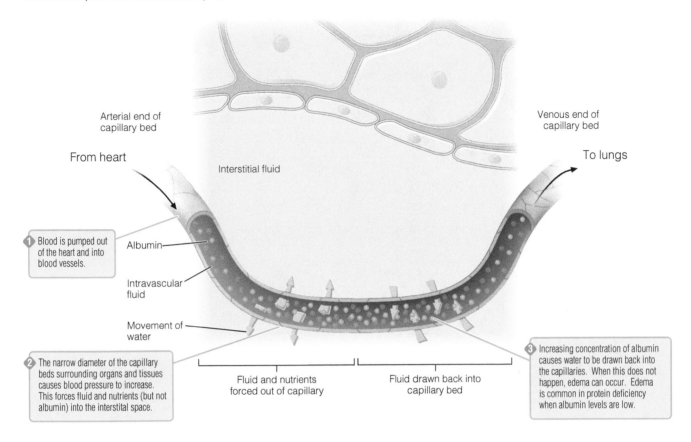

Arterial end of capillary bed

From heart

Interstitial fluid

Venous end of capillary bed

To lungs

1 Blood is pumped out of the heart and into blood vessels.

Albumin

Intravascular fluid

Movement of water

2 The narrow diameter of the capillary beds surrounding organs and tissues causes blood pressure to increase. This forces fluid and nutrients (but not albumin) into the interstitial space.

Fluid and nutrients forced out of capillary

Fluid drawn back into capillary bed

3 Increasing concentration of albumin causes water to be drawn back into the capillaries. When this does not happen, edema can occur. Edema is common in protein deficiency when albumin levels are low.

of fluid in the interstitial space. This condition is called **edema** and sometimes can be seen as swelling in the hands, feet, and abdominal cavity in severely malnourished individuals.

PROTEINS HELP REGULATE pH

Proteins also regulate how acidic or basic the body fluids are. Recall that the body must maintain a certain pH for optimal function. As you learned in Chapter 1, the pH of a fluid is determined by its hydrogen ion (H^+) concentration, and the body's fluids are kept in specific pH ranges. For example, the pH range for stomach fluid is 1.5 to 3.5, while the pH of blood is 7.3 to 7.5. One way that blood's pH is maintained is through the buffering action of proteins such as hemoglobin. Recall that the components of amino acids often have a net positive or negative charge. This is because they can readily accept and donate charged hydrogen ions. When the hydrogen ion concentration in the blood is too high (acidic), negatively charged proteins can bind excess hydrogen ions, restoring the blood to its proper pH. Conversely, proteins can release hydrogens into the blood when the hydrogen concentration is too low (basic). As such, the body can have difficulty maintaining optimal pH balance during periods of severe protein deficiency.

PROTEINS ARE SOURCES OF GLUCOSE AND ENERGY (ATP)

Aside from their role in protein structure, the body can use some amino acids for glucose synthesis and energy (ATP) production. Excess protein can also be converted to fats for more long-term energy storage. Together, these processes

edema (e – DE – ma) The buildup of fluid in the interstitial spaces.

FIGURE 9.8 Protein and Energy Metabolism The fate of dietary protein depends on the body's need for glucose and energy (ATP).

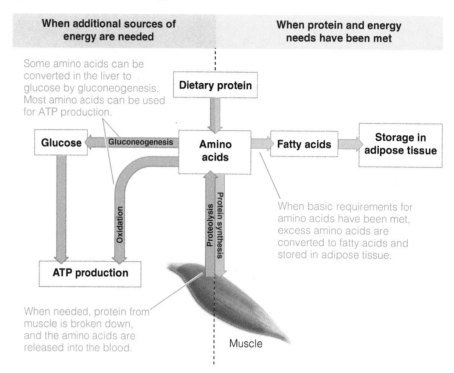

When additional sources of energy are needed	When protein and energy needs have been met

Some amino acids can be converted in the liver to glucose by gluconeogenesis. Most amino acids can be used for ATP production.

Dietary protein

Glucose ← Gluconeogenesis ← Amino acids → Fatty acids → Storage in adipose tissue

Oxidation

Protein synthesis

Proteolysis

ATP production

When basic requirements for amino acids have been met, excess amino acids are converted to fatty acids and stored in adipose tissue.

When needed, protein from muscle is broken down, and the amino acids are released into the blood.

Muscle

help the body (1) maintain blood glucose at appropriate levels, (2) generate energy (ATP) to power chemical reactions even when glucose and fat availability is limited, and (3) store excess energy when dietary protein intake is more than adequate (Figure 9.8).

Some Amino Acids Can Be Converted to Glucose When energy (ATP) availability is low, the body first turns to stored glycogen for a source of glucose. However, when glycogen stores are depleted, the body turns to protein (amino acids) as an alternate source of glucose. To convert amino acids to glucose, muscle tissue is broken down, and the liver takes up the amino acids. Gluconeogenesis is the process whereby glucose-yielding amino acids, called **glucogenic amino acids,** are converted to glucose. As it implies, the term **gluconeogenesis** refers to metabolic pathways that make glucose from noncarbohydrate sources. To do this, the nitrogen-containing amino group of the glucogenic amino acid is removed in the liver—a process called **deamination.** The remaining carbon skeleton is converted to glucose, which can then be metabolized to produce ATP. In addition, many cells can harvest the energy stored in amino acids by oxidizing them directly. Like carbohydrates, oxidation of 1 g of protein yields approximately 4 kcal of energy. Thus, consuming 10 g of dietary protein is equivalent to consuming 40 kcal.

Excess Amino Acids Are Converted to Fat Although some people may think that eating large amounts of protein or taking amino acid supplements will increase their muscle mass, this is generally not the case. In fact, eating extra protein during times of glucose and energy sufficiency generally contributes to more fat storage, not muscle growth. This is because, during times of glucose and energy excess, the body redirects the flow of amino acids away from gluconeogenesis and ATP-producing pathways and instead converts

glucogenic amino acid An amino acid that can be converted to glucose via gluconeogenesis.

gluconeogenesis (glu – co – ne – o – GE – ne – sis) (*neo-*, new; *-genesis*, bringing forth) Synthesis of glucose from noncarbohydrate sources.

deamination The removal of an amino group from an amino acid.

them to lipids. For this to happen, the nitrogen-containing amino group of each amino acid must first be removed via deamination. However, instead of being converted to glucose as would happen in gluconeogenic pathways, the remaining carbon skeletons are diverted to lipid-producing pathways. The resulting lipids can subsequently be stored as body fat for later use.

AMINO ACIDS SERVE MANY ADDITIONAL PURPOSES

In addition to their role as protein components, amino acids themselves serve many other purposes in the body. Some regulate protein breakdown, others are involved in cell signaling, and still others are converted to neurotransmitters, which function as messengers in your nervous system. Some amino acids can also stimulate or inhibit the activity of enzymes involved in metabolism and provide nitrogen for the synthesis of many important nonprotein, nitrogen-containing compounds such as DNA and RNA. Thus, you need amino acids not only for protein synthesis and as an energy source, but for a multitude of other functions as well.

Protein Turnover, Urea Excretion, and Nitrogen Balance

Proteins are the body's "workhorses" and like any working livestock, how long they are useful is finite. Fortunately, the body can recycle and reuse most of the amino acids from these "retired" proteins to synthesize new ones. The process of continuously breaking down and resynthesizing protein is known as **protein turnover.** By regulating protein turnover, the body can adapt to periods of growth and development during childhood and maintain relatively stable amounts of protein during adulthood without requiring enormous amounts of dietary protein.

PROTEIN TURNOVER HELPS MAINTAIN AN ADEQUATE SUPPLY OF AMINO ACIDS

In addition to amino acids from foods, amino acids from recycled proteins can also be used by the body for protein synthesis. In fact, about half of the amino acids the body uses each day come from worn-out proteins that have "served their time."[21] Protein degradation (proteolysis), which is catalyzed by special protein-cleaving enzymes, releases amino acids into what is called the body's **labile amino acid pool.** Dietary amino acids also contribute to the body's labile amino acid pool. As such, new proteins are produced from a mixture of newly obtained and recycled amino acids.

Protein turnover is regulated mainly by hormones, which coordinate the appropriate balance between protein degradation and synthesis. For example, after you eat a meal, high concentrations of the hormone insulin inhibit the breakdown of your body's protein and stimulate overall protein synthesis. In this way, insulin promotes protein accumulation when energy is abundant. In contrast, when you experience various types of stress, thyroid hormone and cortisol are released, stimulating protein degradation and inhibiting protein synthesis. This results in increased protein turnover. As such, these stress-related hormones help your body maintain an immediate supply of labile, or "free" amino acids, which can be used for gluconeogenesis and ATP production.

protein turnover The cycle involving both protein synthesis and protein degradation in the body.

labile amino acid pool Amino acids that are immediately available to cells for protein synthesis and other purposes.

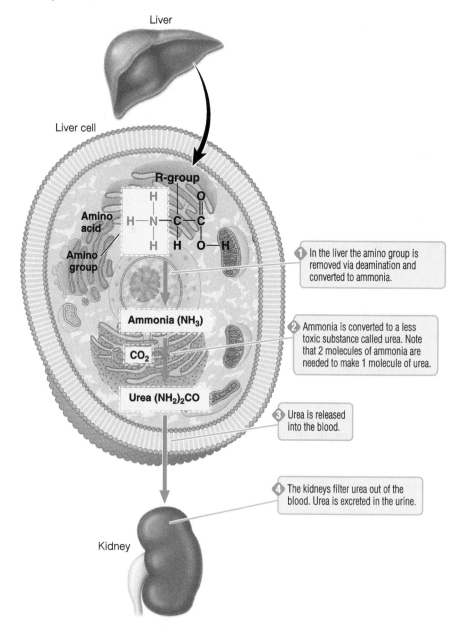

FIGURE 9.9 Urea Synthesis and Excretion Urea is synthesized in the liver and excreted by the kidneys in urine.

Liver

Liver cell

R-group

Amino acid

Amino group

1 In the liver the amino group is removed via deamination and converted to ammonia.

Ammonia (NH$_3$)

CO$_2$

2 Ammonia is converted to a less toxic substance called urea. Note that 2 molecules of ammonia are needed to make 1 molecule of urea.

Urea (NH$_2$)$_2$CO

3 Urea is released into the blood.

4 The kidneys filter urea out of the blood. Urea is excreted in the urine.

Kidney

NITROGEN IS EXCRETED AS UREA

As previously stated, amino acids must first be deaminated before they can be converted to glucose or used as a source of energy. This process produces ammonia (NH$_3$), which is toxic to cells. In response, your liver quickly converts ammonia to a less toxic substance called **urea.** As shown in Figure 9.9, urea is released into the blood, filtered out by the kidneys, and excreted in the urine. This metabolic transformation requires both energy (ATP) and carbon dioxide.

WHAT IS NITROGEN BALANCE?

Protein turnover results in a somewhat complex flux of amino acids, or "remodeling," in the body and measuring protein turnover can tell a health

urea (u – RE – a) A relatively nontoxic, nitrogen-containing compound that is produced from ammonia.

professional important information about overall protein status. More specifically, protein status can be assessed by comparing protein intake with the amount of nitrogen lost in body secretions such as urine, sweat, and feces.[22] When protein (or nitrogen) loss equals protein intake, a person is in **nitrogen balance. Negative nitrogen balance** occurs when nitrogen loss exceeds intake; this can occur during starvation, illness, or stress. When nitrogen intake exceeds the loss of nitrogen from the body, as occurs during childhood or recovery from illness, a person is in **positive nitrogen balance.** Knowing whether an individual is in positive or negative nitrogen balance can help clinicians diagnose and treat certain diseases and physiologic conditions. For example, people with kidney failure who are on dialysis often experience negative nitrogen balance (nitrogen loss > nitrogen intake) and therefore require specialized nutritional support. Conversely, growing children should be in a state of positive nitrogen balance (nitrogen intake > nitrogen loss). If this is not the case, protein intake may not be adequate to support growth.

Healthy growing children are in a state of positive nitrogen balance.

You have now learned important details concerning how the body adjusts nitrogen turnover to support nitrogen balance and optimal health. This process is largely regulated to supply constant and adequate amounts of all the amino acids needed to synthesize the many proteins you need. Dietary recommendations for proteins and amino acids therefore have been developed to take into account both nitrogen balance and the varying requirements for specific amino acids during different periods of the lifespan.

How Much Protein Do You Need?

You need to consume dietary protein for two major reasons: (1) to supply adequate amounts of the essential amino acids; and (2) for the additional nitrogen needed to make the nonessential amino acids and other nonprotein, nitrogen-containing compounds such as DNA. As such, recommendations for dietary amino acids as well as overall protein consumption reflect these needs. These recommendations are described next.

DIETARY REFERENCE INTAKES (DRIs) FOR AMINO ACIDS

To begin with, consider how much of each essential amino acid you should eat every day. Currently, the best estimates can be obtained from the Institute of Medicine's Recommended Dietary Allowances (RDAs), which are shown in Figure 9.10.[2] Note that these values have the unit of "milligrams per kilogram per day" (mg/kg/day) because they represent requirements of the essential amino acids *relative* to body size; the larger you are, the more essential amino acids you need. For example, because the RDA for the essential amino acid valine in adults is 4 mg/kg/day, a woman weighing 140 lb (~64 kg) would require 256 milligrams (4 × 64) of this amino acid in her diet each day. A person weighing 200 lb (~91 kg) would require 364 milligrams (4 × 91) of valine each day.

Because the DRI committee concluded that there is no compelling evidence that high intakes of any of the essential amino acids pose known health risks, the Institute of Medicine did not establish Tolerable Upper Intake Levels (ULs) for them.

nitrogen balance The condition in which protein (nitrogen) intake equals protein (nitrogen) loss by the body.

negative nitrogen balance The condition in which protein (nitrogen) intake is less than protein (nitrogen) loss by the body.

positive nitrogen balance The condition in which protein (nitrogen) intake is greater than protein (nitrogen) loss by the body.

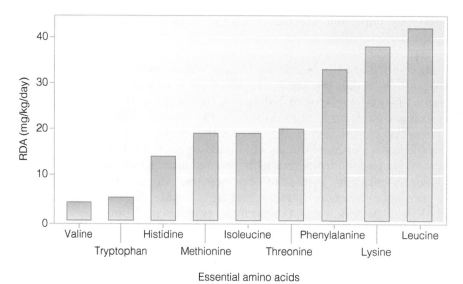

FIGURE 9.10 The RDAs for the Essential Amino Acids in Adults

SOURCE: Institute of Medicine. Dietary reference intakes for energy, carbohydrate, fiber, fat, fatty acids, cholesterol, protein, and amino acids. Washington, DC: National Academies Press; 2005.

DIETARY REFERENCE INTAKES (DRIs) FOR PROTEINS

⟨**CONNECTIONS**⟩ Dietary Reference Intakes (DRIs) are reference values for nutrient intake. Recommended Dietary Allowances (RDAs) are set so that 97% of people have their needs met. Adequate Intake (AI) values are provided when RDAs cannot be established. Tolerable Upper Intake Levels (ULs) indicate intake levels that should not be exceeded.

DRIs for *overall* protein intake have also been published. These are listed inside the cover of this book. Although not all protein sources are created equal, researchers generally agree that most diets in affluent countries such as the United States provide a balanced mix of all the essential amino acids. Therefore, the dietary recommendations for protein intake do not distinguish between people who consume high-quality proteins and those who do not.

The RDA values for protein are provided in two ways—the first is expressed as grams per day (g/day) and is considered to reflect requirements for a "typical" person in a particular life-stage group. These recommended protein intakes increase with age and are somewhat higher for males than females because, in general, males are larger than females and have more muscle mass. Using this set of values, a typical college-age male needs 56 g/day protein, whereas a comparable female needs 46 g/day.

The second type of RDA values for protein is expressed as grams per kilogram body weight per day (g/kg/day); like those for the essential amino acids, these recommendations adjust for body weight. The Institute of Medicine recommends that healthy adults consume 0.8 g/kg/day of protein. For example, adults weighing 140 lb (~64 kg) would require about 51 g of protein (0.8 × 64) each day regardless of whether they are male or female. You would easily get this much protein by eating a bowl of wheat flake cereal and low-fat milk (12 g protein) for breakfast, a hamburger (24 g protein) for lunch, and a bean burrito (15 g protein) for supper.

During infancy, the most rapid phase of growth in the life cycle, protein requirements (when adjusted for body weight) are relatively high. Protein requirements also increase during pregnancy and lactation, because additional protein is needed to support growth and milk production.[22] Note that these RDA values apply only to healthy individuals. For a variety of reasons, people recovering from trauma (such as burns) or illness may require more protein.[23] People who are healthy show little evidence of harmful effects of high protein intake, and therefore no UL values are set for this macronutrient.

EXPERTS DEBATE WHETHER ATHLETES NEED MORE PROTEIN

Although many people believe that athletes have higher protein requirements than nonathletes, this is a topic of active debate. The DRI committee that established the recommendations for amino acid and protein intake carefully considered this question. The committee concluded that physically active people likely require similar amounts of protein *on a body-weight basis*, and adult athletes can generally estimate their protein requirement the same way as other adults by using the same mathematical formula of 0.8 g/kg/day. On the other hand, in a position statement published in 2007, the International Society of Sports Nutrition concluded that protein intakes of 1.4 to 2.0 g/kg/day for physically active individuals are not only safe but may improve the training adaptations to exercise.[24] The American College of Sports Medicine likewise recommends that endurance and strength-trained athletes consume protein in the range of 1.2–1.7 g/kg/day.[25] Thus, scientists continue to grapple with this issue. You can read more about the use of protein supplements by athletes in the Focus on Sports Nutrition feature.

ADDITIONAL RECOMMENDATIONS FOR PROTEIN INTAKE

Aside from the RDA values, several other sets of recommendations for protein intake are also available. For example, the Institute of Medicine's Acceptable Macronutrient Distribution Ranges (AMDRs) recommend that you consume 10 to 35% of your energy as protein. Using this recommendation, consider a moderately active college student with an energy requirement of 2,000 kcal/day. How much protein should this student consume? To answer this question, you must first determine that 200 to 700 kcal (0.10 × 2,000 and 0.35 × 2,000) should come from protein. This translates to 50 to 175 g of protein (200 ÷ 4 and 700 ÷ 4).[§] Note that 1 medium hamburger patty and 1 cup of skim milk provide approximately 25 and 10 g, respectively, of protein to the diet, making the recommended amount of protein quite easy to obtain—especially at the lower end of the range.

The 2010 Dietary Guidelines for Americans and the accompanying MyPlate food guidance system provide additional recommendations concerning intake of high-protein foods such as dairy products, meat, and dried beans and peas (legumes). Aside from supporting the AMDR for protein (10 to 35% of calories from protein), the Guidelines specifically encourage a range of intakes of fat-free or low-fat milk, lean meats, seafood, eggs, nuts, seeds, and soy products to support optimal health. These food groups represent nutrient-dense, high-protein foods. More specifically, the Dietary Guidelines for Americans, Food Patterns, and MyPlate recommend 1.5 to 5 ounces of lean meat, poultry, or eggs; 0.5 to 1.5 ounces of seafood; 0.14 to 0.7 ounces of nuts, seeds, or soy products; and 2 to 3 cups of fat-free or low-fat dairy products daily, depending on caloric needs. Periodic consumption of beans and peas is also encouraged. Note that 3 ounces of lean meat is generally equivalent to a small steak, lean hamburger patty, chicken breast, or piece of fish, and that portions of meat served in restaurants and cafeterias are often much larger than this. To determine precisely how many servings are recommended for you, visit the MyPlate website (http://www.choosemyplate.gov). You can learn about including a balance of healthful, high-protein foods in your daily food choices in the Food Matters section.

Athletes with increased muscle mass, such as football players and body builders, can probably estimate their protein requirements using the same equations used by nonathletes; they need about 0.8 grams protein for each kilogram of body weight daily. However, this is an area of active debate among scientists.

[§] Remember that 1 g of protein supplies 4 kcal of energy.

FOCUS ON SPORTS NUTRITION
Do Protein and Amino Acid Supplements Enhance Athletic Performance?

Optimizing performance is the ultimate goal for most athletes. Indeed, coaches and trainers have long sought training regimens and dietary plans that increase strength, speed, and agility. Methods and products used to optimize athletic performance are called ergogenic aids. One example of an ergogenic aid that has gained significant popularity is the use of protein and amino acid supplements—especially those containing branched-chain amino acids (BCAA). Of the essential amino acids, leucine, isoleucine, and valine are classified as BCAA because they each contain a chemically "branched" R-group. BCAA are naturally present in all protein-containing foods, but the best sources are red meat and dairy products.

Because protein is the major constituent of muscle, one could argue that increased protein or amino acid intake should result in greater muscle mass and, ultimately, increased strength and performance. Furthermore, BCAA appear to be preferentially used for ATP synthesis by muscle tissue during strenuous exercise.[26] These relatively simple facts have motivated the nutritional supplement industry to produce and sell a variety of protein and amino acid powders and drinks, especially those that are high in BCAA. However, is there evidence that protein or amino acid supplementation increases protein synthesis, decreases the breakdown of muscle, or provides especially usable energy?

The answers to these questions are not entirely clear. In fact, supporting evidence—at least for the impact of protein supplementation on protein synthesis—is weak at best. Keep in mind that, before muscle actually grows, the rate of protein synthesis must increase, and the synthesis of specific proteins is highly regulated, requiring the initial step of cell signaling. In other words, something must first initiate the complex processes resulting in DNA transcription and translation before a new protein is made. Although some of the amino acids might be involved in initiating protein transcription, increased dietary protein or amino acid intake does not, *by itself*, signal this process. In fact, contrary to long-held belief, research now suggests that physical activity, and especially resistance exercise, may actually decrease a person's dietary protein requirements.[27] This is because physical activity may trigger the body to become more efficient in its use of amino acids and proteins, resulting in

decreased protein turnover and ultimately a decreased requirement for dietary protein. Clearly, muscle growth and maintenance are complex processes, and more research is needed before we fully understand how exercise affects the protein requirements of athletes.

There are, however, a handful of convincing studies that suggest that amino acid supplementation—especially BCAA—may help inhibit muscle breakdown during intensive training.[28] In response to this, the International Society of Sports Nutrition recently published a position paper on protein and exercise stating that, under certain circumstances, supplementation with BCAA may improve exercise performance and recovery from exercise.[24]

Can protein or amino acid supplements give you that winning edge? Right now, there is little evidence to support this notion, although BCAA may play a special role in preventing muscle breakdown during strenuous activities.[29] The bottom line: (1) eat a well-balanced diet providing sufficient energy with an appropriate mix of carbohydrates, fats, and protein, and (2) train long and hard.

© David Young-Wolff/PhotoEdit

It is important for you to understand basic recommendations concerning how much of each essential amino acid you should consume. In addition, you can calculate your total protein requirement and can refer to the MyPlate food guidance system for suggestions that will help you meet these goals. But what if you were to decide not to eat meat or other high-quality

Working Toward the Goal: Obtaining Sufficient Protein While Minimizing Fats

The 2010 Dietary Guidelines recommend that we meet our protein requirements by choosing and preparing a balance of lean meat and poultry, eggs, seafood, legumes, nuts, and seeds in our meals, along with low-fat or fat-free milk or milk products. This approach helps ensure adequate intake of amino acids and many other micronutrients while minimizing calories, total fat, saturated fat, and cholesterol. The following selection and preparation tips will help you meet your nutrient requirements with high-protein foods while avoiding unnecessary fat consumption.

- Try to consume significant amounts of protein from at least three different food groups daily. This is especially true for vegetarians, who need to consider protein complementation.

- When selecting high-protein dairy foods such as ice cream and yogurt, choose those that are labeled "reduced fat," "low fat," or "fat free," when possible.

- When it is time for a snack, instead of foods with lower nutrient densities such as chips, choose reasonable amounts of high-protein items such as nuts and seeds.

- When comparing similar foods, choose higher protein, lower fat options by comparing Daily Values found on Nutrition Facts labels.

- Add slivered almonds or other nuts to steamed vegetables and fresh salads.

- Experiment with ethnic cuisines—such as Indian or Mexican—that frequently utilize a variety of "pulses" (legumes). These foods are great sources of protein, while being nutrient dense and low in fat.

- To maximize the nutrient density of meat products, trim excess fat prior to cooking.

protein sources? How would this affect your health, and what might you do to make sure your diet was adequate? The answers to these questions are discussed next.

Vegetarian Diets: Healthier Than Other Dietary Patterns?

People have many different reasons for deciding which foods they will and will not eat. This seems especially true about meat and other animal products. For example, some religious groups avoid some or all meats and animal products. Economic considerations and personal preference can also determine whether people eat meat and/or which meats they choose. Because animal products provide high-quality protein in our diets as well as a multitude of other essential nutrients, it is important to consider the effect of animal food consumption (or lack thereof) on issues related to nutritional status. Some issues you might want to consider if you ever think about becoming a vegetarian are discussed next.

Vegetarian diets can provide all the essential nutrients, but care must be taken to make sure sufficient protein, iron, calcium, zinc, and vitamin B_{12} are consumed.

THERE ARE SEVERAL FORMS OF VEGETARIANISM

The term **vegetarian** (from the Latin *vegetus*, meaning "whole," "sound," "fresh," "lively") was first used in 1847 by the Vegetarian Society of the United Kingdom to refer to a person who does not eat any meat, poultry, or fish or their related products, such as milk and eggs. Today most "vegetarians"

vegetarian Someone who does not consume any or selected foods and beverages made from animal products.

consume dairy products and eggs and are called **lacto-ovo-vegetarians.**[28] Alternatively, **lactovegetarians** include dairy products, but not eggs, in their diets. Vegetarians who avoid all animal products are referred to as **vegans.** Thus, you might want to ask a vegetarian what type he or she is.

VEGETARIAN DIETS SOMETIMES REQUIRE THOUGHTFUL CHOICES

But do vegetarians run any special nutritional risks? The answer to this question depends on what kind of "vegetarian" a person is. In general, a well-balanced lacto-ovo- or lactovegetarian diet can easily provide adequate protein, energy, and micronutrients. Dairy products and eggs are convenient sources of high-quality protein and many vitamins and minerals. However, because meat is often the primary source of bioavailable iron, eliminating it can lead to iron deficiency. Furthermore, vegans may be at increased risk of being deficient in several micronutrients, including calcium, zinc, iron, and vitamin B_{12}.[29] This risk increases further during pregnancy, lactation, and periods of growth and development such as infancy and adolescence.[30] Thus, it is especially important that vegetarians consume sufficient amounts of plant-based foods rich in these micronutrients.

Special Dietary Recommendations for Vegetarians Because some types of vegetarian diets pose certain nutritional risks, it is important to follow special dietary strategies if you decide to make this dietary choice. The MyPlate food guidance system specifically recognizes protein, iron, calcium, zinc, and vitamin B_{12} as nutrients for vegetarians to focus on, making specific recommendations as to how to get adequate amounts. In addition, it is pointed out that some meat replacements, such as cheese, can be very high in calories, saturated fat, and cholesterol. Thus, lower-fat versions should be chosen, and they should be consumed in moderation. The following comments and suggestions are offered to help ensure optimal health in individuals who choose—for whatever reason—to become vegetarians. Note that these recommendations are especially pertinent to vegans.

- Select sources of protein that are naturally low in fat, such as skim milk and legumes.
- Minimize the amount of high-fat cheese used as meat replacements.
- If you do not consume dairy products, consider drinking calcium-fortified, soy-based beverages. These can provide calcium in amounts similar to milk and are usually low in fat and do not contain cholesterol.
- Add vegetarian meat substitutes (such as tofu) to soups and stews to boost protein without adding saturated fat and cholesterol.
- Recognize that most restaurants can accommodate vegetarian modifications to menu items by substituting meatless sauces, omitting meat from stir-fries, and adding vegetables or pastas in place of meat.
- Consider eating out at Asian or Indian restaurants, as they often offer a varied selection of high-protein vegetarian dishes.
- Be mindful of getting enough vitamin B_{12}; because this vitamin is naturally found only in foods that come from animals, fortified foods or dietary supplements may be necessary for vegans.

The key to a healthy vegetarian diet, like any diet, is to enjoy a wide assortment of foods and to consume them in moderation. Because no single food provides all the nutrients the body needs, eating a variety of foods can help ensure that vegetarians get the necessary nutrients and other substances that promote good health.

lacto-ovo-vegetarian A type of vegetarian who consumes dairy products and eggs in an otherwise plant-based diet.

lactovegetarian A type of vegetarian who consumes dairy products (but not eggs) in an otherwise plant-based diet.

vegan (VE – gan) A type of vegetarian who consumes no animal products.

What Are the Consequences of Protein Deficiency?

Although generally not a concern in industrialized countries (even in people consuming a vegetarian diet), protein deficiency is common in regions where the amount and variety of foods is limited. Children are especially likely to be affected because their protein requirements are higher (per unit body weight) than those of adults. Protein deficiency is also seen in adults with some debilitating conditions such as cancer. Because of the importance of proteins and amino acids for optimal health, protein deficiency can have significant health implications, some of which are described next.

PROTEIN DEFICIENCY IS MOST COMMON IN EARLY LIFE

Even in economically poor countries, protein deficiency is rare during the first months of life when infants are consuming human milk or infant formula. However, once weaned from these high-quality protein sources to foods that lack adequate protein, infants become at greater risk for protein deficiency. Because protein-deficient diets are almost always also lacking in energy, protein deficiency is often referred to as **protein-energy malnutrition (PEM).** Children with PEM are typically deficient in one or more micronutrients, as well. Thus, PEM is a condition of overall malnutrition and has many implications for child health. For example, children with PEM are at great risk for infection and illness. Recall that protein is needed to make several components of your immune system, as well as skin and membranes that keep pathogens from entering the body. The World Health Organization (WHO) estimates that PEM plays an important role in at least one-third of all child deaths each year, many of which are complicated by infection.[31]

Severe PEM actually encompasses a spectrum of malnutrition: at the extremes are two distinct types of severe PEM, and between them are conditions that combine features of both.[32] At one end of the spectrum is a condition called **marasmus,** which results from severe, chronic, overall malnutrition. In marasmus, fat and muscle tissue are depleted, and the skin hangs in loose folds, with the bones clearly visible beneath the skin. Children with marasmus tend at first to be alert and ravenously hungry, although with increasing severity they become apathetic and lose their appetites. Clinicians often say that marasmus represents the body's survival response to long-term, chronic dietary insufficiency.

The other extreme type of PEM, called **kwashiorkor,** is often distinguished from marasmus by the presence of severe edema. Note that edema sometimes is present in children with marasmus, but those with kwashiorkor usually have more extensive edema, which typically starts in the legs but often involves the entire body. When fluid accumulates in the abdominal cavity, it is referred to as **ascites.** Recall that one of the roles of protein in the body is regulation of fluid balance. Children with kwashiorkor sometimes have large, distended abdomens due to ascites. Because malnourished children often have intestinal parasites, the presence of worms sometimes contributes to this abdominal distension as well. Children with kwashiorkor often are apathetic and have cracked and peeling skin, enlarged, fatty livers, and sparse, unnaturally blond or red hair. Although many characteristics of kwashiorkor were once thought simply to be caused by protein deficiency, this does not appear to be the case.[33] Researchers now believe that many of the signs and symptoms of kwashiorkor are the result of micronutrient deficiencies, for example, vitamin A deficiency, in combination with infection or other environmental stressors.

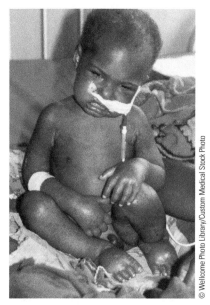

Children with kwashiorkor often have swollen abdomens (ascites), edema in their hands and feet, cracked and peeling skin, and an apathetic nature. These children are at increased risk for infection.

protein-energy malnutrition (PEM) Protein deficiency accompanied by inadequate intake of protein and often of other essential nutrients as well.

marasmus (ma – RAS – mus) (*marainein*, to waste away) A form of PEM characterized by extreme wasting of muscle and adipose tissue.

kwashiorkor (kwa – she – OR – kor) (Kwa, a language of Ghana; refers to "what happens to the first child when the next is born") A form of PEM often characterized by edema in the extremities (hands, feet).

ascites (a – SCI – tes) Abnormal accumulation of fluid in the abdominal cavity.

PROTEIN DEFICIENCY IN ADULTS

PEM can also occur in adults. Unlike children, however, adults with PEM rarely experience kwashiorkor. Instead, adult PEM generally takes the form of marasmus. There are many causes of PEM in adulthood, including inadequate dietary intake (such as occurs in alcoholics and those with eating disorders), protein malabsorption (such as occurs with some gastrointestinal disorders such as celiac disease), excessive and chronic blood loss, cancer, infection, and injury (especially burns).[34]

Adults with PEM can experience extreme muscle loss because the body's muscles are broken down to provide glucose and energy. In addition, fatty liver and edema are common. Adults with severe PEM experience decreased function of many vital physiological systems, including the cardiovascular system, renal system (kidneys), digestive system (gastrointestinal tract and accessory organs), and endocrine and immune systems. There are many causes of PEM in adults, and treatment is often long and difficult. For example, if the cause is infection, treatment may involve both dietary intervention and use of antibiotics. In contrast, if protein deficiency is a result of an eating disorder, psychological counseling becomes a key component of the health care plan. Regardless of its etiology, effective treatment of adult PEM presents a special challenge to any medical team.

Protein Excess: Is There Cause for Concern?

Protein deficiency can result in serious health concerns. But what about protein excess? Can this also be unhealthy or even life-threatening? Contrary to popular belief, high-protein diets do not cause adverse health outcomes such as osteoporosis, kidney problems, heart disease, obesity, and cancer in most people. This conclusion has been confirmed by the DRI committee, which carefully considered the peer-reviewed literature related to the potential health consequences of high-protein diets. This panel of experts concluded that, although epidemiologic studies offer limited evidence that high protein intake is associated with some adverse health outcomes, experimental data do not support this contention. In other words, the association between high protein consumption and poor health is likely not a causal relationship. In fact, the upper limit of the AMDR for protein intake (35% of energy from protein) was developed not because there was evidence that additional protein might pose a health risk, but solely to complement the recommendations for carbohydrate and fat intakes.

Nonetheless, high intakes of protein are often accompanied by high intakes of fat, saturated fat, and cholesterol. Because these dietary components are risk factors for heart disease, it is important to choose a variety of lean and low-fat protein foods, such as those recommended by the MyPlate food guidance system.

⟨CONNECTIONS⟩ Recall that causal relationships are typically determined from intervention studies, whereas epidemiologic studies are powerful in showing relationships or associations within a population.

HIGH RED MEAT CONSUMPTION MAY BE RELATED TO INCREASED RISK FOR CANCER

Growing evidence suggests that high, chronic intake of red meat (beef, lamb, and pork) or processed meats (bacon, sausage, hot dogs, ham, and cold cuts) is associated with increased risk for colorectal cancer.[35] As a result, the World Cancer Research Fund and the American Institute for Cancer Research in 2007 recommended that we limit our intake of red meat to no more than 500 g (18 oz) each week and eat very little processed meat.[36] On average, this would be about 70 g (2.6 ounces) of meat each day—an amount less than that recommended by the 2010 Dietary Guidelines and MyPlate food guidance system. Importantly, the panel of experts who made this recommendation emphasized that they do not recommend avoiding all meat or foods of animal origin. Clearly, these foods can be a valuable source of many essential nutrients and should be considered part of a healthy diet. Like several other issues related to protein nutrition, this topic continues to be one of active debate.

Critical Thinking: Tyler's Story Consider again Tyler, the little boy who has to avoid legumes, which are generally considered excellent sources of dietary protein. Given the severity of peanut allergy for children like Tyler, do you think a food manufacturer that markets a product with peanuts without proper labeling should be held legally responsible? What about restaurants that serve a peanut-containing food without revealing this fact to the consumer? In other words, who is responsible?

Diet Analysis **PLUS** ✚ Activity

The current DRI recommendation for protein intake is 0.8 grams of protein per kilogram of body weight. However, Americans generally consume more protein than is recommended. Some college students may have misconceptions about protein intake and think they need to eat a diet higher in protein than is recommended. This activity provides students with an opportunity to critically appraise their protein intake and compare it to both the RDA and the AMDR.

First, they will calculate their personal RDA for protein based on their healthy body weight. Another important part of this activity demonstrates the importance of using a healthy weight range when determining an RDA for protein. Students are directed to determine the weight range for a healthy body mass index (BMI), between 18.5 and 24.9 kg/m². They will then determine the protein RDA based on that range. This may be particularly enlightening to underweight and overweight students. They will then compare their RDA for protein to the International Society of Sports Nutrition (ISSN) recommendations. They will also determine an adequate protein intake for their calorie intake using the AMDR recommendations. Then, they will compare their RDA and the ISSN recommendations to the AMDR range. Finally, they will determine where their average intake, as determined using Diet Analysis Plus (DA+), fits within that range.

1. Since calculations for protein recommendations are based on body weight, start by calculating your healthy weight range and BMI range. Healthy BMI range is 18.5–24.9 kg/m².
   - lowest healthy weight (lb) = 18.5 × your height in inches²/703

- highest healthy weight (lb) = 24.9 × your height in inches²/703
- Healthy weight range: _____ (low) to _____ (high) lb.

2. Determine a range of protein intake based on your healthy weight range.
   - Lowest weight (lb)/2.2 × 0.8 g = lowest recommended protein in grams for your height.
   - Highest weight (lb)/2.2 × 0.8 g = highest recommended protein in grams for your height.
   - Protein intake range: _____ g to _____ g.

3. From your Profile DRI Goals report, indicate below how many grams of protein are recommended for you. Use the value that represents the daily requirement based on grams per kilogram of body weight (g/kg/day): _____ g. This will fall in the range that you calculated above.

4. Using your current weight, and to reconfirm the DA+ program's calculation, solve the following equation by filling in the blanks.
   a. Convert your body weight in pounds to kilograms.
      _____ lb ÷ 2.2 = _____ kg.
   b. Using kilograms of body weight, calculate your daily protein requirement in grams using your RDA (g/kg/day).
      _____ kg body weight × _____ (g/kg/day) based on sex/age and life stage = _____ g/day.

5. Another way to determine an acceptable protein intake is using the AMDR (Acceptable Macronutrient Distribution Range). To do this, find your recommended calorie

intake in your DA+ program. Then multiply it by the % total energy intake recommended in the AMDRs (10–35% of total energy should come from protein).

- Recommended energy intake × 0.10 = kilocalories from protein. Kilocalories from protein/4 = grams of protein.
- Recommended energy intake × 0.35 = kilocalories from protein. Kilocalories from protein/4 = grams of protein.
- Example: 2,000 kilocalories × 0.10 = 200 kilocalories 200 kcal/4 = 50 grams of protein. 2,000 kilocalories × 0.35 = 700 kilocalories. 700 kcal/4 = 175 grams of protein.
- AMDR for protein for 2,000 kilocalories is 50 to 175 grams.

6. The International Society of Sports Nutrition (ISSN) has concluded that "protein intake of 1.4 to 2.0 g/kg/day for physically active individuals is safe and could improve the training adaptation to exercise." Assuming that you are physically active, calculate your recommended protein intake using this range of protein intakes.

- Your body weight (lb) weight/2.2 = _____ kg body weight × 1.4 g/kg/day = _____ g protein/day.
- Your body (lb) weight/2.2 = _____ kg body weight × 2.0 g/kg/day = _____ g protein/day.

Recommended range of intake for physically active individual = _____ to _____

7. How does that intake range compare with the RDA and the AMDR calculation? _____

8. How does your protein intake as determined in your DA+ activity compare with these protein intake ranges? _____

9. Using the DA+ program, find foods that contribute protein in your diet.

Name of Food	Standard Serving Size	Grams of Protein per Serving

10. Based on your findings, use DA+ to plan a diet that meets the recommended protein intake range of your choosing. Use only fruits, vegetables, dairy products, grains and meat, fish, eggs, legumes, nuts, and seeds in your meal plan. Do not include special drinks and supplements.

11. Use DA+ and protein complementation principles to plan one day's vegetarian meals that meet the recommended protein intake.

Fill in this chart describing the digestion, absorption, and circulation of dietary proteins (blue areas only).

Organs	Secretions: Enzymes, Hormones, and Other	Digestion of Protein	Absorption of Amino Acids	Circulation of Amino Acids
Mouth	Saliva	Chewing and mixing with saliva		
Stomach	HCl, pepsinogen to pepsin	Proteins are denatured by HCl, and then peptide bonds are hydrolyzed. Polypeptide chains are broken down into smaller peptide units.		
Small intestine			Transport into the small intestine cells requires both passive and active transport mechanisms.	Amino acids enter the capillaries where they travel to the liver.
Accessory organs: Pancreas, liver, and gallbladder	Pancreatic enzymes specific for protein digestion: trypsin, chymotrypsin elastase, and carboxypeptidase. Also bicarbonate, to neutralize chyme.			

Notes

1. Carpenter KJ. Short history of nutritional science: Part 1 (1785–1885). Journal of Nutrition. 2003;133:638–45.

2. Institute of Medicine. Dietary reference intakes for energy, carbohydrate, fiber, fat, fatty acids, cholesterol, protein, and amino acids. Washington, DC: National Academies Press; 2005.

3. Furst P, Stehle P. What are the essential elements needed for the determination of amino acid requirements in humans? Journal of Nutrition. 2004;134:1558S–65S.

4. Erlandsen H, Patch MG, Gamez A, Straub M, Stevens RC. Structural studies on phenylalanine hydroxylase and implications toward understanding and treating phenylketonuria. Pediatrics. 2003;112:1557–65.

5. American Academy of Pediatrics. Pediatric nutrition handbook, 6th ed. Kleinman RE, ed. Elk Grove Village, IL: American Academy of Pediatrics; 2008.

6. Reeds PJ, Garlick PJ. Protein and amino acid requirements and the composition of complementary foods. Journal of Nutrition. 2003;133:2953S–61S.

7. Schaafsma, G. The protein digestibility-corrected amino acid score. Journal of Nutrition. 2000;130:1865S–7S.

8. Santerre CR. Food biotechnology, 9th ed. Bowman BA, Russell RM, editors. Washington, DC: ILSI Press; 2006.

9. Huang S, Adams WR, Zhou Q, Malloy KP, Voyles DA, Anthony J, Kriz AL, Luethy MH. Improving nutritional quality of maize proteins by expressing sense and antisense zein genes. Journal of Agricultural and Food Chemistry. 2004;52:1958–64. Zarkadas CG, Hamilton RI, Yu ZR, Choi VK, Khanizadeh S, Rose NGW, Pattison PL. Assessment of the protein quality of 15 new northern adapted cultivars of quality protein maize using amino acid analysis. Journal of Agricultural and Food Chemistry. 2000;48:5351–61.

10. Schnog JB, Duits AJ, Muskeit FAJ, ten Cate H, Rojer RA, Brandjes DPM. Sickle cell disease: A general overview. Journal of Medicine. 2004;62:364–74.

11 National Institutes of Health. National Heart, Lung, and Blood Institute. NIH Publication No. 96-4057. 1996.

12. Bank A. On the road to gene therapy for beta-thalassemia and sickle cell anemia. Pediatric Hematology and Oncology. 2008;25:1–4. Lebensburger J, Persons DA. Progress toward safe and effective gene therapy for beta-thalassemia and sickle cell disease. Current Opinion in Drug Discovery and Development. 2008;11:225–32.

13 Taber KH, Hurley RA. Mercury exposure: effects across the lifespan. Journal of Neuropsychiatry and Clinical Neuroscience. 2008;20:iv–389.

14. U.S. Department of Health and Human Services and U.S. Environmental Protection Agency. What you need to know about mercury in fish and shellfish. 2004 EPA and FDA advice for women who might become pregnant, women who are pregnant, nursing mothers, young children. Available from: http://www.cfsan.fda.gov/~dms/admehg3.html.

15 Ballestar E. An introduction to epigenetics. Advances in Experimental Medicine and Biology. 2011;711:1–11.

16 Martin-Subero JI, Esteller M. Profiling epigenetic alterations in disease. Advances in Experimental Medicine and Biology. 2011;711:162–77. Wierda RJ, Geutskens SB, Jukema JW, Quax PH, van den Elsen PJ. Epigenetics in atherosclerosis and inflammation. Journal of Cellular and Molecular Medicine. 2010;15(6A):1225–50.

17. Burdge GC, Hanson MA, Slater-Jefferies JL, Lillycrop KA. Epigenetic regulation of transcription: A mechanism for inducing variations in phenotype (fetal programming) by differences in nutrition during early life? British Journal of Nutrition. 2007;97(6):1036–46. Mathers JC. Early nutrition: Impact on epigenetics. Forum in Nutrition. 2007;60:42–8.

18. DeBusk R. The role of nutritional genomics in developing an optimal diet for humans. Nutrition in Clinical Practice. 2010;25:627–33. Stover PJ. Nutritional genomics. Physiological Genomics. 2004;16:161–5.

19. Taylor SL, Hefle SL. Food allergy. In: Present knowledge in nutrition, 9th ed. Bowman BA, Russell RM, editors. Washington, DC: ILSI Press; 2006.

20. National Institute of Allergy and Infectious Diseases. Food allergy. Report of the NIH expert panel on food allergy research. 2006. Available at: http://www.niaid.nih.gov/topics/foodallergy/research/pages/reportfoodallergy.aspx.

21. Fuller MF, Reeds PJ. Nitrogen cycling in the gut. Annual Review of Nutrition. 1998;18:385–411. Rand WM, Pellet PL, Young VR. Meta-analysis of nitrogen balance studies for estimating protein requirements in healthy adults. American Journal of Clinical Nutrition. 2003;77:109–27.

22. Dewey KG. Energy and protein requirements during lactation. Annual Review of Nutrition. 1997;17:19–36.

23. Dickerson RN. Estimating energy and protein requirements of thermally injured patients: Art or science? Nutrition. 2002;18:439–42. Gudaviciene D, Rimdeika R, Adamonis K. Nutrition of burned patients. Medicina. 2004;40:1–8. Wilmore DW.

24. Campbell B, Kreider RB, Zeigenfuss T, La Bounty P, Roberts M, Burke D, Landis J, Lopez H, Antonio J. International Society of Sports Nutrition position stand: Protein and exercise. Journal of the International Society of Sports Nutrition. 2007;4:8.

25. American Dietetic Association; Dietitians of Canada; American College of Sports Medicine, Rodriguez NR, DiMarco NM, Langley S. American College of Sports Medicine position stand. Nutrition and athletic performance. Medicine and Science in Sports Exercise. 2009;41:709–31.

26. Mero A. Leucine supplementation and intensive training. Sports Medicine. 1999;27(6):347–58.

27. Phillips SM. Protein requirements and supplementation in strength sports. Nutrition. 2004;20:689–95. Wilson J, Wilson GJ. Contemporary issues in protein requirements and consumption for resistance trained athletes. Journal of the International Society for Sports Nutrition. 2006;5:7–27.

28. Bedford JL, Barr SI. Diets and selected lifestyle practices of self-defined adult vegetarians from a population-based sample suggest they are more 'health conscious.' International Journal of Behavior, Nutrition, and Physical Activity. 2005;2:4. Haddad EH, Tanzman JS. What do vegetarians in the United States eat? American Journal of Clinical Nutrition. 2003;78:626S–32S.

29. Antony AC. Vegetarianism and vitamin B-12 (cobalamin) deficiency. American Journal of Clinical Nutrition. 2003;78:3–6. Hunt JR. Bioavailability of iron, zinc, and other trace minerals from vegetarian diets. American Journal of Clinical Nutrition. 2003;78:633S–9S.

30. Mangels AR, Messina V. Considerations in planning vegan diets: Infants. Journal of the American Dietetic Association. 2001;101:670–7. Messina V, Mangels AR. Considerations in planning vegan diets: Children. Journal of the American Dietetic Association. 2001;101:661–9.

31. United Nations Children's Fund. Tracking progress on child and maternal nutrition. 2009. UNICEF, New York, NY.

32. Jeejeebhoy KN. Protein nutrition in clinical practice. British Medical Bulletin. 1981;37:11–17. Waterlow JC. Classification and definition of protein-calorie malnutrition. British Medical Journal. 1972;3:566–9.

33. Golden M. The development of concepts of malnutrition. Journal of Nutrition. 2002;132:2117S–22S.

34. Hansen RD, Raja C, Allen BJ. Total body protein in chronic diseases and in aging. Annals of the New York Academy of Sciences. 2000;904:345–52.

35. Chao A, Thun MJ, Connell CJ, McCullough ML, Jacobs EJ, Flanders D, Rodriguez C, Sinha R, Calle EE. Meat consumption and risk of colorectal cancer. JAMA (Journal of the American Medical Association). 2005;293:172–82.

36. World Cancer Research Fund/American Institute for Cancer Research. Food, nutrition, physical activity, and the prevention of cancer: A global perspective. Washington, DC: American Institute for Cancer Research; 2007.

Food Safety

© AP Photo/L. G. Patterson

Most of us are fortunate to have an abundant supply of healthful food. Yet there are times when the food we eat causes serious illness. Clearly, it is important to avoid eating food that is spoiled, unclean, or stored improperly. However, food can appear, smell, and taste safe to eat but still harbor dangerous disease-causing agents.

Although every effort is made to ensure that our food is safe, food safety remains an important public health concern. The U.S. Centers for Disease Control and Prevention (CDC) estimate that each year 1 out of 6 Americans (48 million people) gets sick, 128,000 are hospitalized, and 3,000 die from foodborne diseases.[1] Understanding foodborne illness requires a basic knowledge of how foods come to contain disease-causing agents and how the body reacts to them. In this Nutrition Matters, you will gain an understanding of food risks and how to avoid them so that you can prevent foodborne illness for years to come.

What Causes Foodborne Illness?

We are exposed to thousands of microscopic organisms (microbes) each day. These microbes populate the world we live in, frequently serving useful purposes. For example, helpful microbes reside in your GI tract; these assist in food digestion and prevention of some diseases. But other microbes are pathogenic (disease-causing) and make us sick. Consuming pathogenic microbes in foods and beverages is the main cause of foodborne illness.[2] A **foodborne illness** is a disease caused by ingesting unsafe food and is sometimes referred to as "food poisoning." There are many forms and causes of foodborne illness, and you will learn about them next.

FOODBORNE ILLNESSES ARE CAUSED BY INFECTIOUS AND NONINFECTIOUS AGENTS

You can get a foodborne illness by ingesting either "infectious" or "noninfectious" substances. **Infectious agents,** or pathogens, include living microorganisms such as bacteria, viruses, molds, fungi, and parasites. Some people also consider prions, which are not living but are the cause of mad cow disease, infectious agents of foodborne illness. **Noninfectious agents** that cause foodborne illness include nonbacterial toxins; chemical residues from processing, pesticides, and antibiotics; and physical hazards such as glass and plastic.

DIFFERENT STRAINS OF A MICROORGANISM ARE CALLED SEROTYPES

Although most foodborne illnesses are caused by microbes, not all microbes are harmful. In fact, even related pathogenic microbes can cause very different signs and symptoms. To really understand how to prevent foodborne illness, you must first have a basic knowledge of microbiology and understand how various organisms can make you sick.

Each general group of closely related microorganisms can have several genetic strains or types called **serotypes.** Some serotypes are harmless,

---

foodborne illness A disease caused by ingesting unsafe food.

infectious agent of foodborne illness A pathogen in food that causes illness and can be passed or transmitted from one infected animal or person to another.

noninfectious agent of foodborne illness An inert (nonliving) substance in food that causes illness.

serotype A specific strain of a larger class of organism.

such as the ones living in our GI tracts, whereas others are pathogenic (disease-causing). For example, some of the various serotypes of the bacterium *Escherichia coli* (*E. coli*) live without any risk to us in our GI tract, while others (such as *E. coli* O157:H7) cause foodborne illness.[3] Furthermore, different serotypes of a single type of pathogenic bacterium can cause illness in different ways. For instance, some serotypes of pathogenic *E. coli* cause mild intestinal discomfort within one to three days, while others result in more severe symptoms, sometimes taking up to eight days to develop. The time elapsed between when a person eats an infected food and when he or she becomes sick is called the **incubation period.** Different pathogens have different incubation periods, and the length of the incubation period is frequently used by health care providers to help determine which pathogen is likely involved. Some of the most common infectious agents (including their incubation periods) are summarized in Table 1, and the ways in which they cause illness are described next.

incubation period The time between when infection occurs and signs or symptoms begin.

TABLE 1 Infectious Agents of Foodborne Illness, Food Sources, and Symptoms of Infection

Organism	Incubation Period	Duration of Illness	Commonly Associated Foods	Signs and Symptoms
Bacteria				
Campylobacter jejuni	2–5 days	2–10 days	Raw or undercooked poultry, untreated water, unpasteurized milk	Diarrhea (often bloody), abdominal cramping, nausea, vomiting, fever, fatigue
Clostridium botulinum	12–72 hours	From days to months	Home-canned foods with low acid content, improperly canned commercial foods, herb-infused oils	Vomiting, diarrhea, blurred vision, drooping eyelids, slurred speech, dry mouth, difficulty swallowing, weak muscles
Clostridium perfringes	8–16 hours	24–48 hours	Raw or undercooked meats, gravy, dried foods	Abdominal pain, watery diarrhea, vomiting, nausea
Escherichia coli O157:H7	1–8 days	5–10 days	Raw or undercooked meat, raw fruits and vegetables, unpasteurized milk and juice, contaminated water	Nausea, abdominal cramps, severe diarrhea (often bloody)
Escherichia coli (enterotoxigenic)	1–3 days	Variable	Water or food contaminated with human feces	Watery diarrhea, abdominal cramps, vomiting
Listeria monocytogenes	9–48 hours for GI symptoms, 2–6 weeks for invasive disease	Variable	Raw or inadequately pasteurized dairy products; ready-to-eat luncheon meats and frankfurters	Fever, muscle aches, nausea, diarrhea, premature delivery, miscarriage, or stillbirths
Salmonella	1–3 days	4–7 days	Raw poultry, eggs, and beef; fruit and alfalfa sprouts; unpasteurized milk	Diarrhea, fever, abdominal cramps, severe headaches
Shigella	24–48 hours	4–7 days	Raw or undercooked foods or water contaminated with human fecal material	Fever, fatigue, watery or bloody diarrhea, abdominal pain
Staphylococcus aureus	1–6 hours	24–48 hours	Improperly refrigerated meats, potato and egg salads, cream pastries	Severe nausea and vomiting, diarrhea, abdominal cramps
Vibrio cholerae	1–7 days	2–8 days	Contaminated water; undercooked foods; shellfish	Watery diarrhea, vomiting

TABLE 1 *(continued)*

Organism	Incubation Period	Duration of Illness	Commonly Associated Foods	Signs and Symptoms
Viruses				
Hepatitis A virus	15–50 days	2 weeks–3 months	Mollusks (oysters, clams, mussels, scallops, and cockles)	Jaundice, fatigue, abdominal pain, loss of appetite, nausea, diarrhea, fever
Norovirus	12–48 hours	12–60 hours	Raw or undercooked shellfish, contaminated water	Nausea, vomiting, diarrhea, abdominal pain, headache, fever
Parasites				
Trichinella (worm)	1–2 days for initial symptoms; others begin 2–8 weeks after infection	Months	Raw or undercooked pork or meats of carnivorous animals	Acute nausea, diarrhea, vomiting, fatigue, fever, and abdominal pain
Giardia intestinalis (protozoan)	1–2 weeks	Days to weeks	Contaminated water, any uncooked food	Diarrhea, flatulence, stomach cramps
Molds				
Aspergillus flavus	Days to weeks	Weeks to months	Wheat, flour, peanuts, soybeans	Liver damage

Adapted from Centers for Disease Control and Prevention. Diagnosis and management of foodborne illnesses: A primer for physicians and other health care professionals. Morbidity and Mortality Weekly Reports. 2004; 53:1–33. Available at: http://www.cdc.gov/mmwr/PDF/rr/rr5304.pdf. Murano PS. Understanding food science and technology. Thomson/Wadsworth, 2003.

SOME ORGANISMS MAKE TOXINS BEFORE WE EAT THEM

Some pathogenic organisms produce toxic substances while they are growing in foods. These toxins are called **preformed toxins,** because they have already contaminated the foods before we eat them. When we consume these foods, the toxins cause serious and rapid (one to six hours) reactions such as nausea, vomiting, diarrhea, and sometimes neurological damage.

Staphylococcus aureus (S. aureus) One bacterium that produces preformed toxins is *Staphylococcus aureus* (*S. aureus*), which causes nearly 250,000 foodborne illnesses annually in the United States. Foods commonly infected with *S. aureus* include raw or undercooked meat and poultry, cream-filled pastries, and unpasteurized dairy products. Symptoms of *S. aureus* infection include sudden onset of severe nausea and vomiting, diarrhea, and abdominal cramps, typically occurring within one to six hours of consuming the contaminated food. Note that, because the toxin produced by *S. aureus* is quite heat-stable, it is not easily destroyed by cooking.

Methicillin-Resistant *S. aureus* (MRSA)—An Emerging Concern You may have heard about an antibiotic-resistant strain of *S. aureus* called **methicillin-resistant *Staphylococcus aureus* (MRSA),** which has received considerable attention recently by public health officials. MRSA was identified more than 40 years ago and was thought to be spread only by direct contact with an infected person—usually in a hospital setting. However, it is now recognized that MRSA can also be "community acquired," meaning the infection is not linked to contact with a health care facility.[4] Most cases of community-acquired MRSA are attributed to sharing towels and equipment in athletic and school facilities. In addition, it now appears that community-acquired infection can also occur via consumption of MRSA-infected foods.

To date, there has been only one documented outbreak of foodborne illness caused by MRSA in the United States. In this case, it is thought that an infected food preparer transmitted the bacterium to coleslaw. Three family members later consumed the coleslaw and became ill within three to four hours, having severe nausea, vomiting, and stomach cramps.[5] Because MRSA cannot be treated

preformed toxin Poisonous substance produced by microbes while they are in a food (prior to ingestion).

methicillin-resistant *Staphylococcus aureus* (MRSA) A type of *S. aureus* that is resistant to most antibiotics.

effectively by antibiotics, public health officials are carefully watching this bacterium, especially as it relates to foodborne illness. To help prevent foodborne MRSA infections, proper hand washing is essential.

Clostridium botulinum (*C. botulinum*) Another bacterium that produces preformed toxins, *Clostridium botulinum* (*C. botulinum*), is found mainly in inadequately processed, low-acid, home-canned foods such as green beans. You can also become infected with this microbe by eating improperly canned commercial foods. Unlike *S. aureus*, however, cooking contaminated foods at sufficiently high temperatures can destroy the toxin produced by *C. botulinum*.

Infection with *C. botulinum* causes a disease called **botulism.** Mild cases result in vomiting and diarrhea, whereas symptoms of severe botulism include double vision, blurred vision, drooping eyelids, slurred speech, difficulty swallowing, dry mouth, and muscle weakness. In severe cases, botulism can cause paralysis, respiratory failure, and death.

In 2007, the CDC reported an outbreak of eight cases of *C. botulinum* infection in Indiana, Texas, and Ohio.[6] All infected persons had consumed a particular brand of hot dog chili sauce that was quickly recalled by the manufacturer. Although all infected persons recovered, they endured several days of painful GI symptoms. Because the botulism toxin is destroyed by high temperatures, some experts recommend that home-canned foods be boiled for 10 minutes before they are consumed.

Aspergillus While most molds that grow on foods such as cheese and bread are not dangerous to eat, others can produce dangerous preformed toxins. An example is **aflatoxin,** which is a toxin produced by the *Aspergillus* mold found on some agricultural crops such as peanuts. If the crop is not dried properly before storage, *Aspergillus* can continue to grow and produce toxic levels of aflatoxin. Consuming food contaminated with aflatoxin is of great concern because it can cause liver damage, cancer, and is often fatal.[7] Although many agricultural practices (such as sufficient drying) in the United States are used to help minimize contamination of our food with aflatoxin, it remains a significant public health issue in many other regions of the world.

SOME ORGANISMS MAKE ENTERIC (INTESTINAL) TOXINS AFTER WE EAT THEM

In contrast to organisms that produce preformed toxins in a food *before* it is consumed, others produce harmful toxins *after* they enter the GI tract. These toxins are called **enteric** or **intestinal toxins.** Enteric toxins draw water into the intestinal lumen, resulting in diarrhea. Although the symptoms are variable, the incubation period is generally one to five days, substantially longer than that for most preformed toxins.

Noroviruses **Noroviruses*** are examples of pathogens (viruses) that produce enteric toxins. Symptoms of norovirus infection usually include nausea, vomiting, diarrhea, and stomach cramping. Sometimes people also develop low-grade fevers, chills, headaches, muscle aches, and a general sense of tiredness. Symptoms usually begin one to two days after ingestion of the contaminated food.

An example of a norovirus outbreak occurred in 2008 on three college campuses in California, Michigan, and Wisconsin.[8] This outbreak resulted in approximately 1,000 cases of reported illness,

* Previously called Norwalk and Norwalk-like Virus.

Improper canning of low-acid foods can result in the finished product containing live *Clostridium botulinum*, which can cause severe illness.

iStockphoto.com/fcutrara

botulism The foodborne illness caused by *Clostridium botulinum*.

aflatoxin (a – fla – TOX – in) A toxic compound produced by certain molds, such as *Aspergillus*, that grow on peanuts, some grains, and soybeans.

enteric (intestinal) toxin A toxic agent produced by an organism after it enters the gastrointestinal tract.

norovirus A type of infectious pathogen (virus) that often causes foodborne illness.

including 10 hospitalizations, and prompted closure of one of the three campuses. Although it was never determined what caused these related outbreaks and if they were due to a common food, college campuses are at particularly high risk for norovirus outbreaks because of the extensive opportunities for transmission created by shared living and dining areas. Many evacuees from Hurricane Katrina also developed norovirus infections, and drinking contaminated water was the likely source.[9]

Unlike bacteria, viruses—such as the norovirus—cannot be treated with antibiotics. The only way to avoid norovirus infection is to follow the food safety guidelines—such as frequent hand washing and decontamination of food preparation utensils—outlined later in this section.

Some Serotypes of *E. coli* Although most forms of *E. coli* are harmless or even beneficial, some (called **enterotoxigenic** *E. coli*) produce enteric toxins, and consumption of food or water contaminated with these bacteria can cause severe GI upset. Symptoms typically occur within 6 to 48 hours after food consumption and include diarrhea, abdominal cramps, and nausea. Foods and beverages typically contaminated with these forms of *E. coli* include uncooked vegetables, fruits, raw or undercooked meats and seafood, unpasteurized dairy products, and untreated tap water. This type of *E. coli* is the primary cause of what is often referred to as "traveler's diarrhea," which is a clinical syndrome resulting from consuming microbially contaminated food or water while traveling. If you have experienced severe diarrhea and abdominal cramps during or shortly after traveling, you might have had this type of foodborne illness.

SOME ORGANISMS INVADE INTESTINAL CELLS

Some pathogens invade the cells of the intestine, seriously irritating the mucosal lining and causing severe abdominal discomfort and bloody diarrhea (dysentery); fever is also common. These types of pathogens are called **enterohemorrhagic** and include *Salmonella* and two especially dangerous serotypes of *E. coli* called *E. coli* O157:H7 and *E. coli* O104:H4. Incubation periods for enterohemorrhagic pathogens are generally several days.

Salmonella *Salmonella* is one of the most common causes of foodborne illness in the United States,

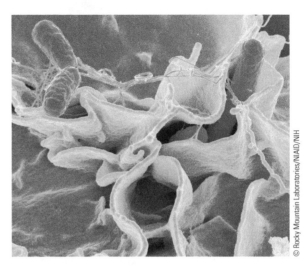

Salmonella is one of the most common causes of foodborne illness.

being typically found in raw poultry, eggs, beef, improperly washed fruit, alfalfa sprouts, and unpasteurized milk. The incubation period for this organism is one to three days, and symptoms include severe GI upset and headaches.

In 2007, *Salmonella*-infected frozen potpies caused at least 272 cases of foodborne illness in 35 states.[10] Unfortunately, the original source of the contamination was never determined. In 2010, over 1,900 *Salmonella*-related illnesses were reported in 11 states in the United States. This outbreak occurred in response to consumption of contaminated eggs, and resulted in a massive nationwide recall of eggs produced by two distributors located in Iowa.[11] Numerous *Salmonella* infections have also been reported from consuming raw alfalfa and mung bean sprouts. In fact, because they are particularly prone to a variety of bacterial infections, the FDA recommends raw sprouts not be consumed—especially by those with compromised immune systems such as children, the elderly, and persons with autoimmune conditions.[12]

***E. coli* O157:H7 and *E. coli* O104:H4** As previously mentioned, *E. coli* O157:H7 is an enterohemorrhagic serotype of *E. coli*. Since it was first identified as a human pathogen in 1982, many outbreaks of infection have been reported. Unpasteurized milk, apple juice, and apple cider can harbor this

---

enterotoxigenic Producing a toxin while in the GI tract.
enterohemorrhagic Causing bloody diarrhea.

pathogenic organism, as can improperly prepared meat (including poultry). Infection with this serotype of *E. coli* has an incubation period of one to eight days and results in nausea, abdominal cramps, and severe diarrhea that is often bloody.

In 2006, 183 persons were infected with *E. coli* O157:H7 in 26 states.[13] This outbreak rapidly gained national attention, as affected individuals had to be hospitalized, and at least three people eventually died. Fresh spinach was identified as the source of contamination, and the FDA quickly advised consumers not to eat bagged fresh spinach or fresh spinach–containing products unless they were cooked at 160°F for 15 seconds. Partly because of this outbreak, the FDA has issued new guidance to the food industry to minimize microbial contamination of fresh-cut fruits and vegetables.[14] In 2011, an outbreak of a different pathogenic serotype of *E. coli* (O104:H4) resulted in hundreds—if not thousands—of people becoming ill and many deaths throughout the European Union. Several cases (most of whom had recently traveled to Germany) were also reported in the United States. Authorities identified raw sprouts grown in Germany as the likely source of contamination.[15]

PROTOZOA AND WORMS ARE TYPES OF PARASITES

Parasites are complex one-celled or multicellular organisms that rely on other organisms to survive and are typically dangerous to the host. Not surprisingly, consumption of parasite-infested foods can cause foodborne illness. Because parasites often take up long-term residence in the body, their incubation periods are generally quite long. Several types of parasites can cause foodborne illness, the most common being protozoa and worms.

Protozoa Among the most ordinary of parasites are the **protozoa**, which are one-celled organisms that can live as parasites in the intestinal tracts of animals and humans. As part of their reproductive cycle, protozoa form **cysts** that are excreted in the feces. If cyst-containing feces come in contact with plants or animals, these food products can be contaminated as well. A foodborne illness can develop if foods or beverages contaminated with protozoan cysts are consumed.

One such parasitic protozoan, *Giardia intestinalis*, causes diarrhea, abdominal discomfort, and cramping; symptoms typically begin one to two weeks after infection. *Giardia intestinalis*

Drinking untreated water is not advised because of the possibility of it being contaminated with *Giardia intestinalis*.

can be found in chemically untreated swimming pools and hot tubs, and in rivers, ponds, and streams that have been contaminated with feces of an infected animal or person. You may already be aware that this is why chlorine (which kills this organism) is typically added to water in public swimming and bathing areas. This is also why boiling, chemically treating, or filtering water from ponds, streams, and lakes is recommended before drinking.

Worms Consuming foods that contain worms can also cause foodborne illness. Like protozoa, worms form cysts as part of their life cycles. Once ingested, cysts mature into worms that can cross the intestinal lining, travel through the blood, and eventually settle in various locations in the body, including muscles, eyes, and brain. An example of a worm is *Trichinella*, which is a roundworm that can invade a variety of animals, including pigs and some fish. Eating undercooked *Trichinella*-contaminated pork and seafood can result in the worm entering the body, causing muscle pain,

---

parasite An organism that, during part of its life cycle, must live within or on another organism without benefiting its host.

protozoa Very small (single-cell) organisms that are sometimes parasites.

cyst A stage of the life cycle of some parasites.

swollen eyelids, and fever. Although *Trichinella* infection is rare in the United States, it is still seen in areas of the world where wild game is consumed more frequently.

Anisakis simplex is another type of roundworm that causes foodborne illness. Within hours after ingestion of an *Anisakis simplex*–contaminated food, violent abdominal pain, nausea, and vomiting may occur. Some studies show that raw fish (sushi and sashimi) can contain this type of roundworm, as well as other parasites.[16] Consequently, the FDA advises that you always cook seafood thoroughly.[17] Because freezing kills many parasites, choosing to consume only sushi that has been made with previously frozen fish may be prudent.

PRIONS ARE INERT, NONLIVING PROTEINS THAT *MAY BE* INFECTIOUS

Although **prions** are not living organisms—in fact, they are not organisms at all—they may pose food safety concerns. Prions are altered proteins created when the secondary structure of a normal protein is disrupted, transforming α-helix coils into β-folded sheets. The resulting deformed protein is called a prion. Prions can cause other normal proteins to unravel, setting off a cascade of similar reactions converting hundreds of normal proteins into abnormal prions. When high levels of prions build up in a cell, the cell ruptures and releases the prions into the surrounding area, where they destroy other cells. Eventually, this kills the surrounding cells and gives the infected tissue a spongy appearance. Unfortunately, prions are extremely resilient and retain their ability to infect other cells even after exposure to extreme heat or acids. Therefore, cooking foods does not destroy prions. Furthermore, prions are not destroyed by the acidic conditions of the stomach, and if a prion-containing food is consumed, the prions can be absorbed into the bloodstream and begin the process of transforming normal proteins into prions in their new host. This is why some experts consider some prion-related diseases to be infectious.

Mad Cow Disease Several diseases in farm animals are known to be caused by prions, including scrapie in sheep and **bovine spongiform encephalopathy (BSE),** or **mad cow disease,** in bovine species.[18] Mad cow disease is characterized by loss of motor control, confusion, paralysis, wasting, and eventually death.

Creutzfelt-Jakob Disease and Variant Creutzfeldt-Jakob Disease A deadly human disease called **Creutzfeldt-Jakob disease,** caused by prions, is very rare and occurs in only one in a million people each year.[19] This disease has usually been attributed to direct infection by prions from contaminated medical equipment—for example, during surgery. However, during a BSE outbreak in the United Kingdom that occurred in the 1980s, researchers discovered a new form of Creutzfeldt-Jakob disease, which they called **variant Creutzfeldt-Jakob disease.** This disease has a relatively long incubation period (years) and is also fatal.

Considerable evidence links variant Creutzfeldt-Jakob disease to consumption of BSE-contaminated products.[20] To date, nearly 200 human cases have been reported worldwide, three of which were in the United States. No treatment exists for either the "classic" form of Creutzfeldt-Jakob disease or the variant form, and nothing can slow the progression of either disease.

Prions and Public Policy Because prions are found mainly in nerves and brains, the World Heath Organization (WHO) recommends that all governments prohibit the feeding of highly innervated tissue (such as the brain and spinal cord) from slaughtered cattle to other animals. Similarly, both Canada and the United States have banned the use of such products in human food, including dietary supplements, and in cosmetics. In 2006, Canada also banned inclusion of cattle tissues capable of transmitting BSE in all animal feeds, pet foods, and fertilizers. Partly in response to the concern about BSE, the U.S. Department of Agriculture (USDA) does not allow the slaughter of non-ambulatory ("downer") cattle for food purposes.[21] These measures, along with those established by the USDA, aim to provide a uniform national BSE policy with the goal of ensuring the safety of human food.

---

prion A misshapen protein that causes other proteins to also become distorted, damaging nervous tissue.

bovine spongiform encephalopathy (BSE; mad cow disease) A fatal disease in cattle caused by ingesting prions.

Creutzfeldt-Jakob disease A fatal disease in humans caused by a genetic mutation or surgical contamination with prions.

variant Creutzfeldt-Jakob disease A form of Creutzfeldt-Jakob disease that may be caused by consuming BSE-contaminated foods.

"Red tides" occur when marine algae produce brightly colored pigments; algae can also produce toxins that cause shellfish poisoning.

An example of shellfish poisoning is a phenomenon referred to as **red tide** that occurs when a particular marine alga begins to grow quickly and produce brightly colored pigments. These colorful "blooms" make the surrounding water appear red or brown. During this period of rapid growth, the algae produce potent toxins called **brevetoxins** which, when consumed by humans, causes classic shellfish poisoning. Every year, approximately 30 cases of shellfish poisoning are reported in the United States, typically in the coastal regions of the Atlantic Northeast and Pacific Northwest.[22]

How Can Noninfectious Substances Cause Foodborne Illness?

Consuming foods containing infectious pathogens poses the greatest risk of foodborne illness. However, noninfectious agents in food can also make you sick. These inert (nonliving) compounds include physical contaminants, such as glass and plastic, and other substances such as toxins, heavy metals and pesticides.

ALGAE TOXINS CAN MAKE SOME FISH AND SHELLFISH POISONOUS

One type of noninfectious foodborne illness, collectively called **shellfish poisoning,** can result from eating certain types of contaminated fish and shellfish (for example, clams and oysters). This is because these marine animals consume large amounts of algae that sometimes produce poisonous compounds called **marine toxins.** Consuming marine toxin–contaminated shellfish or fish can cause neurologic symptoms, including tingling, burning, numbness, drowsiness, and difficulty breathing. People with shellfish poisoning may also experience a strange phenomenon called hot–cold inversion, in which cold is perceived as hot and vice versa.

SOME PESTICIDES, HERBICIDES, ANTIBIOTICS, AND HORMONES ARE DANGEROUS

In addition to marine toxins, other noninfectious agents of foodborne illness can make their way into the food chain, including pesticides, herbicides, antibiotics, and hormones. Because exposure to some of these substances can make you ill, many federal and international agencies work together to ensure that their presence in foods is negligible or poses no risk to the consumer and the environment. These agencies include the Food and Agriculture Organization (FAO) of the United Nations, the U.S. Environmental Protection Agency (EPA), the FDA, and the USDA.

An example of such a compound is dichloro-diphenyl-trichloroethane (DDT), a pesticide once

---

shellfish poisoning A group of foodborne illnesses caused by consuming shellfish that contain marine toxins.

marine toxin A poison produced by ocean algae.

red tide Phenomenon in which certain ocean algae grow profusely, causing reddish discoloration of the surrounding water.

brevetoxin The toxin produced by red tide–causing algae, which, when consumed by humans, causes shellfish poisoning.

used to kill mosquitoes and increase crop yields. DDT was banned in the United States in 1972 when it was found to damage wildlife. However, the use of DDT to control malaria continues in many developing countries. More recently, **bovine somatotropin (bST),** otherwise known as bovine growth hormone, has attracted much public attention. This hormone is used in the dairy industry to increase milk production. Although some people are concerned about the safety of bST, a substantial amount of research suggests that it is safe for both cow and consumer.[23] Thus, the FDA allows its use, and there are no laws requiring milk produced by cows treated with bST be labeled as such. Even after a compound such as bST has been approved, however, the FDA continues to evaluate its safety.

FOOD ALLERGIES AND SENSITIVITIES CAN ALSO CAUSE FOODBORNE ILLNESS

Some noninfectious compounds in foods can cause illness in small segments of the population, but not because they are really toxic or poisonous in the way we have already discussed. Instead, they cause illness in a small percentage of people who are especially sensitive or even allergic to them. Because the percentage of individuals who have an adverse reaction to these compounds is small, their presence in food is not prohibited—but must be disclosed on the label. For example, monosodium glutamate (MSG), which is used as a flavor enhancer, causes severe headaches, facial flushing, and a generalized burning sensation in some people.[24] Because MSG is used extensively in Asian cuisines, this reaction is often referred to as "Chinese restaurant syndrome." The presence of MSG must be stated on the label of any food to which it has been added.

Similarly, **sulfites** are added to some foods (such as wine and dried fruits) to enhance color and prevent spoilage. Unfortunately, consuming them causes breathing difficulties in sulfite-sensitive people, especially in those with asthma. In response, the FDA requires food manufacturers to label all foods containing at least 10 parts per million (ppm) sulfites. You can check to see if a food has added sulfites by looking for the following terms: sodium bisulfite, sodium metabisulfite, sodium sulfite, potassium bisulfite, and potassium metabisulfite.

Exposure to some proteins naturally present in foods may also cause an allergic reaction in some people. For example, people who are allergic to peanuts and other legumes can experience a life-threatening allergic response when they eat these foods. You can read more about foodborne allergies (such as peanut allergy) in Chapter 9.

NEW FOOD SAFETY CONCERNS ARE ALWAYS EMERGING

Scientists and public health officials are continually trying to identify compounds in foods that may cause illness. Often these investigations receive attention in the popular press, causing public concern. Generally this is good, because public concern results in scientific scrutiny. In this section, you will learn about four foodborne substances that have garnered significant public health attention in the last few years.

Acrylamide One recently emergent food safety concern relates to a substance called **acrylamide.** Although acrylamide has probably always been in the food supply, concerns regarding its presence in foods were first reported in 2002.[25] Acrylamide is used to make polyacrylamide, found in some food-packaging materials, and small amounts of it may be present in food. It can also form in starchy foods exposed to very high temperatures, such as french fries and potato chips. Because very high levels of acrylamide cause cancer in laboratory animals, there is concern that dietary acrylamide may be harmful to humans as well.[26] However, most evidence suggests that humans would need to eat an unreasonably large amount of fried foods to consume a cancer-causing dose of acrylamide.[27] Nonetheless, both the FDA and researchers continue to study this issue.

Melamine Globalization of the world's food supply has led to increasing concerns that lower standards for food production in other countries may lead to unsafe food being imported to the United States. For example, in 2007 it was determined that pet food contaminated with **melamine** had caused

bovine somatotropin (bST; bovine growth hormone) A protein hormone produced by cattle and used in the dairy industry to enhance milk production.

sulfite A naturally occurring compound that is sometimes used as a food additive to prevent discoloration and bacterial growth.

acrylamide (a – CRYL – a – mide) A compound that is formed in starchy foods (such as potatoes) when heated to high temperatures.

melamine A nitrogen-containing chemical used to make lightweight plastic objects.

illness in and death of numerous dogs and cats.[28] Melamine is a nitrogen-containing chemical typically used to make lightweight plastic objects such as dishes. Investigations by the FDA determined that this compound had been added to gluten, a protein derived from wheat, which is used to produce animal foods. Because melamine is rich in nitrogen, some Chinese gluten manufacturers added it to their products to make them appear higher in protein than they actually were. The melamine-tainted gluten was then used to make pet foods in the United States.

Public attention quickly shifted from concern about pets to the possibility that melamine could enter the human food supply as well. The FDA and USDA originally concluded that consuming meat and milk from animals fed melamine-tainted feed posed very low risk to human health.[29] However, this conclusion was quickly revised when melamine-contaminated milk products caused nearly 60,000 infants and children in China to become sick in 2008 and 2009.[30] The FDA now advises consumers worldwide to avoid using any infant formula products made in China, as well as any milk products or products with milk-derived ingredients made in China. Unfortunately, this experience highlights the importance of a strong national food security monitoring system.

Bisphenol A Another somewhat recent concern receiving considerable attention relates to the chemical **bisphenol A (BPA),** which is found in some plastic food and beverage containers including baby bottles. This is because researchers have shown that BPA may influence long-term risk for cancer and reproductive abnormalities in laboratory animals.[31] Thus, scientists and public health officials are currently studying whether consuming foods and beverages from these types of containers poses health risks to humans as well.[32] In 2010, after reviewing the growing literature on BPA, the FDA announced that it did have some concern about the potential effects of BPA on the brain, behavior, and development of fetuses, infants, and young children. In response, the FDA has begun supporting efforts to decrease BPA exposure in the United States and is shifting to a more robust regulatory framework for its oversight.[32] Other countries, such as Canada, have prohibited use of some BPA-containing products such as plastic baby bottles. Individuals wishing to decrease their personal exposure to BPA can choose glass or metal containers over polycarbonate ones, not put plastics in the microwave, and wash plastic containers by hand instead of

in the dishwasher. In addition, parents of infants are advised by the U.S. Department of Health and Human Services to discard scratched plastic baby bottles and infant feeding cups, and not put boiling or very hot water, infant formula, or other liquids into bottles containing BPA while preparing them for use.[32] Without a doubt, you can be assured that public interest in compounds such as BPA will be intense for years to come.

How Do Food Manufacturers Prevent Contamination?

Many pathogens and inert compounds cause food-borne illness, and it is beyond the scope of this book to describe each one in detail. However, it is very important for you to understand how disease-causing agents are transmitted from food to food, also known as **cross-contamination,** and from person to person. It is also important for you to understand how food manufacturers strive to keep food safe.

CAREFUL FOOD-HANDLING TECHNIQUES HELP KEEP FOOD SAFE

To prevent foodborne illness, it is critical that people who handle food do so safely. This is because pathogens can be easily transmitted to almost any food by an infected person. In this way, a "perfectly clean" food can be made unsafe. Therefore, it is important that food handlers, including those who harvest, process, and prepare foods, avoid coughing or sneezing on foods. In addition, because many pathogens can pass through the GI tract and be excreted in the feces, fecal contamination of foods can cause foodborne illness. Consequently, people who process and prepare food are generally required to wear gloves while handling food and thoroughly wash their hands after using the toilet.

PROPER FOOD PRODUCTION, PRESERVATION, AND PACKAGING CAN PREVENT ILLNESS

In addition to safe food-handling techniques, the food-processing industry adheres to additional practices that help to keep your food safe. The

bisphenol A (BPA) A chemical used in the production of many plastic items including baby bottles.

cross-contamination The transfer of microorganisms from one food to another or from one surface or utensil to another.

goal is to produce a product that is as pathogen-free as possible and then to store it in a way that does not allow pathogenic growth. In 2011, the FDA Food Safety Modernization Act was signed into law to help ensure that the U.S. food supply is safe. These new regulations shifted the focus of federal regulators from *responding* to contamination to *preventing* it. For example, the FDA now has the authority to recall food products—a process that previously was only done voluntarily by food manufacturers.[33]

The USDA and the FDA have long had guidelines for food processors and handlers. These guidelines, called the **Hazard Analysis Critical Control Points (HACCP) system,** were developed to identify the critical points in food-processing where contamination is most likely. The USDA also requires that meat and poultry be inspected before sale and that guidelines for safe handling appear on packages. However, this inspection does not guarantee that the meat is pathogen free. Although we have no sure way to prevent a food from harboring dangerous pathogens or toxins, many techniques help keep food safe. These include drying, salting, smoking, fermentation, heating, freezing, and irradiation—all of which are described next.

Drying, Salting, Smoking, and Fermentation Meat has long been preserved by salting, smoking, and/or drying because these techniques—when done correctly—inhibit bacterial growth. Another technique called **fermentation** involves adding selected microorganisms to foods and is used in making sauerkraut, yogurt, pickles, and wine. Fermentation promotes the growth of nonpathogenic organisms, which minimize the growth of pathogenic organisms.

Heat Treatment: Cooking, Canning, and Pasteurization Because most foodborne pathogens prefer living in an environment between 40 to 140°F, cooling or heating foods below and above this range helps inhibit microbial growth. This is why food manufacturers often heat-treat their products and why you should also avoid this temperature range (called the **danger zone**) when storing and serving food. Table 2 provides temperature guidelines for cooking, serving, and reheating foods. As you can see, to be kept safe, different foods sometimes require different temperatures. Note that the USDA revised its recommended cooking temperature for pork, steaks, roasts, and chops to 145°F in 2011. The USDA now recommends cooking *all whole cuts of meat* (excluding poultry) to 145°F as measured

with a food thermometer placed in the thickest part of the meat, then allowing the meat to rest for three minutes before carving or consuming.

Heating food can also preserve it for later consumption. For example, during the canning process, foods are packaged (or "canned") in sanitized jars or cans and then heated at high temperatures. Pathogens are killed, and a vacuum is created within the jar. The air-free environment helps preserve the food, because most foodborne illness-causing organisms require oxygen to grow.

TABLE 2 Guidelines for Cooking, Serving, and Reheating Foods to Prevent Foodborne Illness

Note that internal temperatures should be measured with a thermometer.

Cooking	Beef and Pork
	• Cook beef roasts and steaks to a minimum of 145°F.
	• Cook ground beef and pork to at least 165°F.
	• Cook pork roasts and chops to a minimum of 145°F.
	Poultry
	• All poultry should be cooked to a minimum internal temperature of 165°F.
	Eggs
	• Cook eggs until the yolks and whites are firm.
	• Don't use recipes in which eggs remain raw or only partially cooked.
	Fish
	• Cook fish to 145°F or until the flesh is opaque and separates easily with a fork.
	• Avoid eating uncooked oysters and other shellfish. People with liver disorders or weakened immune systems are especially at risk for getting sick.
Serving	• Keep foods at 140°F or higher until served.
	• Keep foods hot with chafing dishes, slow cookers, and warming trays.
Reheating	• Reheat leftovers to at least 165°F.
	• When using a microwave oven, make sure food is evenly heated.

SOURCE: Adapted from the Partnership for Food Safety Education and FightBAC!® Available from: http://www.fightbac.org.

Hazard Analysis Critical Control Points (HACCP) system A USDA food safety protocol used to decrease contamination of foods during processing.

fermentation Metabolism, by bacteria, which occurs under relatively anaerobic conditions.

danger zone The temperature range between 40 and 140°F in which pathogenic organisms grow most readily.

The strawberries on the left were irradiated after they were picked, whereas those on the right were not.

FIGURE 1 The Radura Symbol Irradiated foods must include this symbol on their food label.

Recall, however, that some organisms, such as *Clostridium botulinum*, can survive in low-acid, anaerobic conditions. Making sure that home-canned foods have the proper acidity and are heated sufficiently helps keep them safe. You can find out more about canning foods safely on the USDA website at http://www.fsis.usda.gov/Help/FAQs_Hotline_Preparation/index.asp.

Pasteurization Perhaps the most common form of heat treatment used in food preservation is **pasteurization.** This process was named for the French scientist Louis Pasteur, who discovered in the 1860s that is was possible to prevent spoilage in wine and beer by exposing them to intense heat for a short period of time. Indeed, pasteurization is still commonly used today. Foods that are typically pasteurized include dairy products, juice, and spices. Pasteurization is especially good at killing some forms of *E. coli, Salmonella, Campylobacter jejuni*, and *Listeria monocytogenes*. Several outbreaks of *Salmonella* poisoning from unpasteurized apple cider prompted laws that now require commercially available apple cider to be pasteurized or labeled as being nonpasteurized, and the FDA recommends that homemade cider be heated for 30 minutes at 155°F or 15 seconds at 180°F.

Cold Treatment: Cooling and Freezing Foods also spoil less quickly if kept cold. To slow or halt the growth of microorganisms, you should always refrigerate foods at 40°F or freeze them soon after they are prepared. This helps prevent foods from staying in the "danger zone" for an extended period of time.

Irradiation In the 1950s, the National Aeronautics and Space Administration (NASA) first used irradiation to preserve food for space travel. Shortly thereafter, the FDA approved a form of food processing called **irradiation,** sometimes called "cold pasteurization," as a form of food preservation. Today, irradiation is approved for meat, poultry, shellfish, eggs, and other foods such as fresh fruits, vegetables, and spices. During irradiation, foods are exposed to radiant energy that damages or kills bacteria. It is important to understand that irradiation neither damages nutrients nor makes the foods radioactive—just as the use of X-rays to inspect luggage does not make it radioactive. Irradiation makes foods safer to eat and can dramatically increase their shelf life. For example, compared with nonirradiated strawberries, which have a brief shelf life, irradiated strawberries last for several weeks without spoiling. Irradiated foods must be labeled with the radura symbol (Figure 1).

What Steps Can You Take to Reduce Foodborne Illness?

There are many ways you can reduce your risk of foodborne illness. First, you should familiarize yourself with recommendations by regulatory groups and keep abreast of food safety alerts and recalls. Second, it is helpful to understand the basic concepts related to safe food handling. In this section, you will learn about a variety of consumer advisory bulletins that are available from the FDA and USDA. You will also learn about a national campaign, called FightBac!®, developed by the U.S.

pasteurization (pas – ter – i – ZA – tion) A food preservation process that subjects foods to heat to kill bacteria, yeasts, and molds.

irradiation A food preservation process that applies radiant energy to foods to kill bacteria.

government to provide useful tips for consumers on how to avoid foodborne illness.

CHECK CONSUMER ADVISORY BULLETINS

Both the USDA and FDA maintain user-friendly websites containing information about current food safety recommendations. In addition, there are toll-free phone numbers that can be used to find out important food safety information. Some of these are listed here.

- **General information**: www.foodsafety.gov
- **News and safety alerts**: www.foodsafety.gov/~fsg/fsgnews.html
- **Recalls**: www.fsis.usda.gov/FSIS_Recalls/index.asp
- **To report a foodborne illness**: http://www.fsis.usda.gov/FSIS_Recalls/problems_with_food_products/index.asp
- **USDA's Meat and Poultry Hotline**: 1-888-MPHotline
- **FDA's Food Safety Information Hotline**: 1-888-SAFEFOOD

THE FIGHTBAC!® CAMPAIGN PROVIDES BASIC FOOD SAFETY ADVICE

Although issues related to specific risk for various foodborne illnesses may change over time, consumers are encouraged to follow some basic rules. The USDA and the Partnership for Food Safety Education have developed a set of food safety guidelines, called **FightBAC!®**. The Fight-BAC!® campaign is a public education program focused on reducing foodborne illness. It specifically addresses avoidance of infectious agents of foodborne illness.

Clean Hands, Surfaces, and Cooking Utensils To remove and/or kill pathogens on hands, surfaces, and cooking utensils, you should frequently and adequately wash them with hot, soapy water. It is also recommended that you clean cutting boards thoroughly after each use and wash kitchen towels, cloths, and sponges frequently. Sponges can be best decontaminated by heating in a microwave for 1 minute.[34] Currently, there is no definitive answer as to whether plastic or wood cutting boards are best when it comes to food safety, although some studies suggest that wood might be preferable.[35] The USDA recommends that you periodically sanitize counters, equipment, utensils, and cutting boards with a solution of 1 tablespoon of unscented, liquid chlorine bleach in 1 gallon of water.

To prevent contamination of foods with pathogens found in fecal materials, you should always wash your hands after using the bathroom, changing diapers, or handling pets. To be effective, wash hands vigorously with soap for at least 20 seconds and rinse them thoroughly under clean, running warm water. Dry your hands using clean paper or cloth towels. Using antimicrobial gels can also help ensure that hands are pathogen free, although you should not consider this a substitute for proper hand washing.

Wash Fresh Fruits and Vegetables—Not Meat Washing fresh fruits and vegetables provides an important protection from foodborne pathogens and noninfectious agents such as pesticides. To do this effectively, you should wash produce under clean, cool, running water and, if possible, scrub with a clean brush. Fruits and vegetables should then be dried with a clean paper or cloth towel. There is some evidence that commercial fruit and vegetable cleaners may help remove *E. coli* 0157:H7 and *Salmonella* from produce,[36] but the USDA has not taken a stand on the effectiveness of these products. However, it is recommended that you *do not* wash raw meat, poultry, and fish, because doing so increases the danger of cross-contaminating otherwise noninfected surfaces and foods.

Separate Foods to Prevent Cross-Contamination To prevent transferring disease-causing pathogens from food to food (cross-contamination), you should separate raw meat and seafood from other foods in your grocery cart, in your refrigerator, and while preparing a meal. For example, put raw meats in separate, sealed plastic bags in your grocery cart and use separate cutting boards when preparing fresh produce and raw meat. Also, you should not put cooked meat back on the same plate that was used to hold the raw meat unless you have washed the plate thoroughly with hot, soapy water.

Cook Foods to the Proper Temperature As previously stated, heat kills most dangerous organisms. It can also alter the chemical composition of some preformed toxins, making them less dangerous. The temperatures and cooking times required to kill pathogens depend on the particular food and the organism or toxin. It is recommended that you measure the *internal temperature* of a cooked food and thoroughly clean meat thermometers between uses.

Also, just because a food has been previously cooked does not mean it is pathogen free. Before

FightBAC!® A public education program developed to reduce foodborne bacterial illness.

eating, you should reheat foods to their appropriate internal temperatures. This step is especially important for pregnant women, infants, older adults, and people with impaired immunity such as those with HIV or cancer, because these individuals are more likely than others to become seriously ill when exposed to pathogens. In fact, a *Listeria monocytogenes* outbreak in the late 1990s that resulted in at least six deaths and two miscarriages has prompted the recommendation that lunch meats and frankfurters be heated before eating.[37]

Keep Foods Cold, Chill Them Quickly, and *If in Doubt, Throw It Out* Because cold temperatures can slow the growth of microorganisms, you should always keep perishable foods refrigerated at 40°F or colder. Also, do not marinate or thaw foods at room temperature. Instead, marinate them in the refrigerator. Similarly, frozen foods should not be allowed to thaw at room temperature but instead should thaw in the refrigerator or microwave. It is also safe to thaw frozen foods in cold water in an airtight plastic wrapper or bag, changing the water every 30 minutes or so until the food is thawed. After a meal, you should refrigerate foods *as quickly as possible*, separating large amounts of foods into small, shallow containers to allow the food to cool rapidly.

Even properly chilled foods can become sources of foodborne illness, and for this reason you should consume leftovers within three to four days. Indeed, "If in doubt, throw it out!" You can see some of the recommended storage times for refrigerated foods by visiting the USDA website (http://www.fsis.usda.gov/PDF/Refrigeration_and_Food_Safety.pdf).

BE ESPECIALLY CAREFUL WHEN EATING OUT

Making sure your food is safe to eat can be especially difficult when you eat at restaurants, picnics, and buffets. However, you can decrease your risk of foodborne illness in these situations. In general, ask yourself whether the four basic concepts of the Fight Bac!® program have been followed.

- How likely was it that the people handling and preparing your foods used sanitary practices?
- What evidence is there that raw foods were kept separate from cooked foods?
- Do you think that foods were properly cooked and kept out of the thermic "danger zone"?
- Is there sufficient evidence that cold foods were kept cold?

Unfortunately, it is common at picnics and social gatherings to neglect these basic rules of food safety, resulting in foodborne illness. Therefore, use good judgment and eat only foods you know to be safe. Better safe than sorry!

What About Avoiding Foodborne Illness While Traveling or Camping?

When traveling or camping, there are added concerns related to food safety. To keep current on the latest information, visit the CDC website (http://wwwn.cdc.gov/travel/). In general, water and fresh produce pose the greatest risks to you, especially when you travel to foreign countries, as these foods often are the causes of traveler's diarrhea. You may not be able to completely prevent foodborne illness while traveling, but you can greatly lower your risk by being vigilant in choosing which foods and beverages to consume—and which to avoid. Finally, if a traveler does get traveler's diarrhea, it is important to replace lost fluids and electrolytes as soon as symptoms begin to develop. Clear liquids are routinely recommended for adults. Travelers who develop three or more loose stools in an eight-hour period may benefit from antibiotic therapy.

DRINK ONLY PURIFIED OR TREATED WATER

In general, when traveling out of the country (especially in less industrialized regions of the world) or camping in remote regions of the United States, it is advisable to drink bottled water and avoid using ice. Before drinking bottled beverages, be sure all containers have fully sealed caps. If seals are not intact, the bottles may have been refilled. If bottled water is not available, you should boil your water for one minute or, if at high altitudes (>2,000 m), three minutes. In addition, water can be chemically treated to kill pathogens that may be present, and portable water filters can also be used to remove some pathogens.

AVOID OR CAREFULLY WASH FRESH FRUIT AND VEGETABLES

Although fresh produce contaminated with bacteria may not cause illness for local people, it can cause serious illness (usually diarrhea) in visitors. Thus, when traveling to foreign locations, avoid or carefully wash fresh fruit (e.g., grapes) and

Traveling abroad requires special precautions regarding food safety.

vegetables (e.g., peppers) that are eaten without peeling them. Of course, only water known to be clean should be used.

TRAVELING IN AREAS WITH VARIANT CREUTZFELT-JAKOB DISEASE

When traveling in Europe or other areas that have reported cases of BSE in cows, the CDC recommends that concerned travelers consider either avoiding beef and beef products altogether or selecting only beef products composed of solid pieces of muscle meat. These considerations, however, should be balanced with the knowledge that the risk of disease transmission is very low. Milk and milk products from cows are not believed to pose any risk for transmitting BSE to consumers— even in areas of the world with emerging variant Creutzfelt-Jakob disease incidence.

What Are Some Emerging Issues of Food Biosecurity?

Recent worldwide events have raised concerns about the safety of our food. Indeed, terrorists and other malicious people could certainly do serious and widespread harm by contaminating the nation's food supply. As a result, **food biosecurity**— the prevention of terrorist attacks on our food supply—has gained intense national attention, and Congress authorized the "Public Health Security and Bioterrorism Preparedness and Response Act" (also called the **Bioterrorism Act**) in 2002.[38] Since then, additional regulations have been approved to help ensure that foods grown and produced both in the United States and abroad cannot be tampered with. Of particular interest in this regard is the possibility that *Clostridium botulinum* could be used by terrorists as a widespread disease-causing agent. As such, a supply of antitoxin against botulism is maintained by CDC in the event such an attack occurs. Clearly, this focus of deep individual and governmental interest will remain strong in coming years.

Changes in food production and distribution patterns both nationally and internationally may influence national food safety and risk of foodborne illness. For example, a contaminated food produced in Chile can easily find its way into the American food chain, be distributed to dozens of states, and perhaps appear on your table. In response, the origin of all foods must now be included on the food's label. As was clearly shown in the recent outbreak of melamine contamination of our pet food supply, the U.S. government has little power over whether optimal and honest food production policies are adhered to in other locations. Although the full impact of these changes in international commerce is not known, some experts warn that fewer restrictions on food importation into the United States may increase our risk for foodborne illness. For these changes to be beneficial, the health benefits imparted by the globalization of our food supply must greatly outweigh the associated risks.

---

food biosecurity Measures aimed at preventing the food supply from falling victim to planned contamination.

Bioterrorism Act Federal legislation aimed to ensure the continued safety of the U.S. food supply from intentional harm by terrorists.

Key Points

What Causes Foodborne Illness?

- Infectious foodborne illness can be caused by pathogenic organisms such as bacteria, viruses, parasites, molds, and fungi.

- Some organisms produce toxins while they grow in foods (preformed toxins), others produce toxins inside the intestinal tract (enteric toxins), and others invade intestinal cells.

How Can Noninfectious Substances Cause Foodborne Illness?

- Inert (noninfectious) compounds in foods can also cause foodborne illness.

- Shellfish poisoning is caused when seafood is contaminated with toxins produced by algae.

- Some people are especially sensitive or allergic to selected components of foods.

- Emerging concerns related to noninfectious agents include presence of acrylamide, melamine, and bisphenol A (BPA) in our food supply.

How Do Food Manufacturers Prevent Contamination?

- Salting, drying, and smoking remove water from foods, inhibiting pathogenic growth; fermentation involves adding nonpathogenic organisms that inhibit the growth of dangerous ones.

- Methods that slow the growth of or kill pathogens include heat and cold treatment as well as irradiation.

What Steps Can You Take to Reduce Foodborne Illness?

- You can help prevent foodborne illness by ensuring that you use proper cleaning, handling, heating, chilling, and storing techniques.

- Many consumer food safety recommendations are even more important for people who eat out often, attend picnics frequently, and travel.

What Are Some Emerging Issues of Food Biosecurity?

- Concern that terrorists can cause serious outbreaks of foodborne illness has prompted governmental programs targeted at ensuring food biosecurity in the United States.

Notes

1. Centers for Disease Control and Prevention. Estimates of foodborne illness in the United States. December 15, 2010. Available at http://www.cdc.gov/foodborneburden/index.html.

2. Centers for Disease Control and Prevention. Surveillance for foodborne disease outbreaks—1993–1997. Morbidity and Mortality Weekly Review. 2000;49:1–51. Available from: http://www.cdc.gov/mmwr/PDF/ss/ss4901.pdf.

3. Trabulsi LR, Keller R, Tardelli Gomes TA. Typical and atypical enteropathogenic Escherichia coli. Emerging Infectious Diseases. 2002;8:508–13.

4. Centers for Disease Control and Prevention. Methicillin-resistant Staphylococcus aureus (MRSA) infections. Available from: http://www.cdc.gov/mrsa/index.html.

5. Jones TF, Kellum ME, Porter SS, Bell M, Schaffner W. An outbreak of community-acquired foodborne illness caused by methicillin-resistant *Staphylococcus aureus*. Emerging Infectious Diseases. 2002;8:82–4.

6. Centers for Disease Control and Prevention. Botulism associated with canned chili sauce, July–August 2007. Updated August 24, 2007. Available from: http://www.cdc.gov/botulism/botulism.htm.

7. Abnet CC. Carcinogenic food contaminants. Cancer Investigation. 2007;25:189–96.

8. U.S. Centers for Disease Control and Prevention. Norovirus outbreaks on three college campuses—California, Michigan, and Wisconsin, 2008. Morbidity and Mortality Weekly Reports. 58:1095–1100, 2009. Available from: http://www.cdc.gov/mmwr/preview/mmwrhtml/mm5839a2.htm.

9. Centers for Disease Control and Prevention. Norovirus outbreak among evacuees from hurricane Katrina—Houston, Texas, September 2005. Morbidity and Mortality Weekly Report. 2005;54;1016–8. Available from: www.cdc.gov/mmwr/preview/mmwrhtml/mm5440a3.htm.

10. Centers for Disease Control and Prevention. Investigation of outbreak of human infections caused by Salmonella I 4,[5],12:i:-. Available from: http://www.cdc.gov/salmonella/4512eyeminus.html.

11. U.S. Centers for Disease Control and Prevention. Investigation update: Multistate outbreak of human *Salmonella* enteritidis infections associated with shell eggs. December 2, 2010. Available at http://www.cdc.gov/salmonella/enteritidis/.

12. US Food and Drug Administration. Safe handling of raw produce and fresh-squeezed fruit and vegetable juices. Available from: http://www.fda.gov/downloads/Food/ResourcesForYou/Consumers/UCM174142.pdf.

13. Centers for Disease Control and Prevention. Ongoing multistate outbreak of *Escherichia coli* serotype O157: H7 infections associated with consumption of fresh spinach—Unites States, September 2006. Morbidity and Mortality Weekly Reports. 2006;55:1–2. Available

from: http://www.cdc.gov/mmwr/preview/mmwrhtml/mm5538a4.htm.

14. US Food and Drug Administration. Guidance for industry: guide to minimize microbial food safety hazards of fresh-cut fruits and vegetables. February 2008. Available from: http://www.fda.gov/Food/GuidanceComplianceRegulatoryInformation/GuidanceDocuments/ProduceandPlanProducts/ucm064458.htm.

15. Federal Institute for Risk Assessment, Federal Office of Consumer Protection and Food Safety, and Robert Koch Institute. Information update on EHEC outbreak. June 10, 2011. Available from: http://www.rki.de/cln_144/nn_217400/EN/Home/PM082011.html. U.S. Centers for Disease Control and Prevention. Investigation update: outbreak of shiga toxin—producing *E. coli* O104 (STEC O104:H4) infections associated with travel to Germany. June 15, 2011. Available from: http://www.cdc.gov/print.do?url=http://www.cdc.gov/ecoli/2011/ecoliO104/.

16. Sakanari JA, McKerrow JH. Anisakiasis. Clinical Microbiology Reviews. 1989;2:278–84.

17. US Food and Drug Administration. 2005 Food code. Available from: http://www.cfsan.fda.gov/~dms/fc05-toc.html.

18. Mostl K. Bovine spongiform encephalopathy (BSE): The importance of the food and feed chain. Forum in Nutrition. 2003;56:394–6.

19. Ryou C. Prions and prion diseases: Fundamentals and mechanistic details. Journal of Microbiology and Biotechnology. 2007;17:1059–70.

20. Belay ED, Schonberger LB. The public health impact of prion diseases. Annual Review of Public Health. 2005;26:191–212. Centers for Disease Control and Prevention. vCJD (variant Creutzfeldt-Jakob disease). Available from: http://www.cdc.gov/ncidod/dvrd/vcjd/risk_travelers.htm. Roma AA, Prayson RA. Bovine spongiform encephalopathy and variant Creutzfeldt-Jakob disease: How safe is eating beef? Cleveland Clinic Journal of Medicine. 2005;72:185–94.

21. US Department of Agriculture. Food Safety and Inspection Service. Prohibition of the use of specified risk materials for human food and requirements for the disposition of non-ambulatory disabled cattle. Federal Register: January 12, 2004 (Volume 69, Number 7). Available from http://www.fsis.usda.gov/Frame/FrameRedirect.asp?main=http://www.fsis.usda.gov/OPPDE/rdad/FRPubs/03-025IF.htm.

22. National Center for Zoonotic, Vector-Borne, and Enteric Diseases. Marine toxins: general information. Last updated July 20, 2010. Available at http://www.cdc.gov/nczved/divisions/dfbmd/diseases/marine_toxins/.

23. Anonymous. Bovine somatotropin and the safety of cows' milk: National Institutes of Health technology assessment conference statement. Nutrition Reviews. 1991;49:227–32. Etherton TD, Kris-Etherton PM, Mills EW. Recombinant bovine and porcine somatotropin: Safety and benefits of these biotechnologies. Journal of the American Dietetic Association. 1993;93:177–80.

24. Walker R, Lupien JR. The safety evaluation of monosodium glutamate. Journal of Nutrition. 2000;130:1049S–52S.

25. Stadler RH, Blank E, Varga N, Robert F, Hau J, Guy PA, Robert MC, Riediker S. Food chemistry: Acrylamide from Maillard reaction products. Nature. 2002;419:449–50.

26. Bull RJ, Robinson M, Laurie RD, Stoner GD, Greisiger E, Meier RJ, Stober J. Carcinogenic effects of acrylamide in Sencar and A/J mice. Cancer Research. 1984;44:107–11.

27. Mucci LA, Dickman PW, Steineck G, Adami H-O, Augustsson K. Dietary acrylamide and cancer of the large bowel, kidney, and bladder: Absence of an association in a population-based study in Sweden. British Journal of Cancer. 2003;88:84–9.

28. US Food and Drug Administration. Melamine pet food recall of 2007. Available from: http://www.fda.gov/animal-veterinary/safetyhealth/recallswithdrawals/ucm129575.htm.

29. US Food and Drug Administration. FDA/USDA joint news release: Scientists conclude very low risk to humans from food containing melamine. Available from: http://www.fda.gov/NewsEvents/Newsroom/Press-Announcements/2007/default.htm.

30. US Food and Drug Administration. Melamine contamination in China. (Updated: January 5, 2009) Available from: http://www.fda.gov/NewsEvents/Public-HealthFocus/ucm179005.htm.

31. Durando M, Kass L, Piva J, Sonnenschein C, Soto AM, Luque EH, Muñoz-de-Toro M. Prenatal bisphenol A exposure induces preneoplastic lesions in the mammary gland in Wistar rats. Environmental Health Perspectives. 2007;115:80–6. Newbold RR, Jefferson WN, Padilla-Banks E. Long-term adverse effects of neonatal exposure to bisphenol A on the murine female reproductive tract. Reproductive Toxicology. 2007;24:253–8.

32. U.S. Food and Drug Administration. Bisphenol A (BPA). Update on bisphenol A(BPA) for use in food: January 2010. Available from: http://www.fda.gov/NewsEvents/PublicHealthFocus/ucm064437.htm.

33. U.S. Food and Drug Administration. FDA Food Safety Modernization Act. Public Law 111-353, Jan. 4. 2011. Available at http://www.gpo.gov/fdsys/pkg/PLAW-111publ353/pdf/PLAW-111publ353.pdf.

34. Sharma, M, Eastridge, J, Mudd, C. Effective disinfection methods of kitchen sponges [abstract]. Institute of Food Technologists. 2007;Control No. 3310.

35. Cliver DO. Cutting boards and Salmonella cross-contamination. Journal of AOAC International. 89:538–42, 2006.

36. Kenney SJ, Beuchat LR. Comparison of aqueous commercial cleaners for effectiveness in removing Escherichia coli O157:H7 and Salmonella meunchen from the surface of apples. International Journal of Food Microbiology. 2002;25:47–55.

37. Centers for Disease Control and Prevention. Update: Multistate outbreak of Listeriosis—United States, 1998–1999. Morbidity and Mortality Weekly Report. 1999;47:1117–8. Available from: http://www.cdc.gov/mmwr/preview/mmwrhtml/00056169.htm.

38. United States Congress. Public health security and bioterrorism preparedness and response act of 2002. June 12, 2002. H.R. 3448. Available from: http://www.gpo.gov/fdsys/pkg/BILLS-107hr3448enr/pdf/BILLS-107hr3448enr.pdf.

Lipids

Lipids are required for hundreds, if not thousands, of physiological functions in the body. For example, body fat (which contains lipids) protects vital organs, and lipids that surround cells provide a protective yet highly functional barrier. In addition, lipids make many foods flavorful—just think of butter, cream, olive oil, and well-marbled beef. However, for many people the thought of fatty foods simply conjures up images of unhealthy living. We often shop for "fat-free" foods and try to avoid fats altogether. Food manufacturers have even developed "fat substitutes" to replace the naturally-occurring fats found in our foods. Nonetheless, it is important to recognize that, although diets high in fat can lead to health complications such as obesity and heart disease, getting enough of the right *types* of fat is just as essential for optimal health as avoiding excess fat and the wrong kinds of fat. In this chapter, you will learn about the variety of fats and oils in foods and how the body uses them to make other substances vital for health. Guidelines for lipid intake that promotes optimal health are also discussed.

Gallbladder Surgery—When Things Do Not Go Smoothly

Nancy had noticed for quite some time that eating rich, high-fat meals frequently caused her to experience cramping and diarrhea. But these unpleasant effects did not always occur and were easily avoided if she chose her foods wisely. So when Nancy learned that she had mild gallbladder disease, she was not completely surprised by this news, and it did not cause much concern. Although Nancy's doctor advised her to have her gallbladder surgically removed, she decided to wait. After all, she was a working mother of three active sons and had very little free time. Nancy did not feel as if she could afford the amount of time required to have and recover from gallbladder surgery.

At first Nancy's plan seemed to work. She watched her diet carefully and avoided fatty foods. However, 10 years after her initial diagnosis Nancy woke up one morning with severe abdominal pain. The decision to not have surgery was no longer an option. Her gallbladder was inflamed, and after consultation with her doctor Nancy decided it was time to have it removed. This normally simple outpatient procedure, however, turned out to be anything but routine.

At first the surgery seemed to have gone smoothly, and although Nancy's abdomen was tender, she was discharged from the hospital five hours later. But the next morning she was feeling even worse. Not even strong narcotic medications could relieve the relentless abdominal pain she was experiencing. After calling her doctor, Nancy returned to the emergency room. Upon examination the problem became clear; whereas most people have one duct transporting bile from the liver to the gallbladder, Nancy had multiple ducts. Because her doctor did not know this when he performed the surgery, he had inadvertently severed these "accessory" ducts without closing them off. As a result, bile leaked from Nancy's liver into her abdomen—a very serious condition. An emergency surgery was quickly performed to repair the damage, and Nancy remained in the hospital for an additional two weeks.

The road to recovery was long, requiring a slow transition from a simple diet (e.g., rice, apples, bananas), to one containing small amounts of fats and oils. However, within two months Nancy was able to eat normally and once again enjoy an active life with her family. We asked her if she had garnered any wisdom from her experience that she would like to share, and she told us, "*I highly recommend that people with gallbladder disease do not put off getting treatment and be very careful to report abnormal signs and symptoms after their surgery if it is needed.*"

Courtesy of Shelley McGuire

Critical Thinking: Nancy's Story

Why do you think that people with mild gallbladder disease, which causes indigestion and diarrhea, often mistake it for other conditions? If you found out that you had gallbladder disease that could be handled effectively with dietary changes, would you agree to have your gallbladder removed? What factors might influence your decision?

What Are Lipids?

Lipids are important macronutrients because they provide a major source of energy and are critical for optimal health in many ways. Lipids come in many varieties, and it is important to understand how they differ both chemically and physiologically. In the following section, you will learn the fundamental concepts needed to grasp these basics.

FATS AND OILS ARE TYPES OF LIPIDS

It is helpful to begin with some basic lipid-related terminology and definitions. For example, fats and oils are examples of what chemists call **lipids,** which are relatively water-insoluble, organic molecules consisting mostly of carbon, hydrogen, and oxygen atoms. In other words, lipids are hydrophobic ("water fearing"). Lipids that are liquid at room temperature are called **oils,** and those that are solid at room temperature are called **fats.** The major lipids include fatty acids, triglycerides, phospholipids, sterols, and fat-soluble vitamins. Fatty acids, triglycerides, phospholipids, and sterols are discussed in detail in this chapter.

Oils and fats are both lipids. Whereas oils (like corn oil) are liquid at room temperature, fats (like margarine) are solid.

FATTY ACIDS ARE THE MOST COMMON TYPE OF LIPID

The most abundant lipids in your body and in the foods you eat are **fatty acids,** which are made entirely of carbon, hydrogen, and oxygen atoms (Figure 10.1). A chain of carbon atoms forms the backbone of each fatty acid. One end of this carbon chain, called the **alpha (α) end,** contains a carboxylic acid group (–COOH); the other end, called the **omega (ω) end,** contains a methyl group (–CH₃). Most fatty acids do not exist in their "free" (unbound) form. Instead, they are components of larger molecules, such as triglycerides and phospholipids (Figure 10.2). Fatty acids are also bound to cholesterol, forming cholesteryl esters. You will learn more about triglycerides, phospholipids, and cholesteryl esters later in this chapter.

There are hundreds of unique fatty acids, and they differ in the number of carbons they contain as well as the types and locations of the chemical bonds holding the carbon atoms together. These variations influence the physical properties of the fatty acids and the roles they play in foods and in the body.

lipid (*lipos*, fat) Organic molecule that is relatively insoluble in water and soluble in organic solvents.

oil (*oleum*, olive) A lipid that is liquid at room temperature.

fat A type of lipid that is solid at room temperature.

fatty acid A type of lipid consisting of a chain of carbons with a methyl (–CH₃) group on one end and a carboxylic acid group (–COOH) on the other.

alpha (α) end (*alpha*, the first letter in the Greek alphabet) The end of a fatty acid, which consists of a carboxylic acid (–COOH) group.

omega (ω) end (*omega*, the final letter in the Greek alphabet) The end of a fatty acid, which consists of a methyl (–CH₃) group.

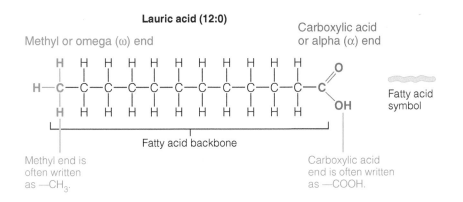

Lauric acid (12:0)

FIGURE 10.1 Fatty Acid Structure All fatty acids have three components: a carboxylic acid or alpha (α) end (–COOH), a methyl or omega (ω) end (–CH₃), and a fatty acid backbone.

FIGURE 10.2 Fatty Acids Can Take Several Forms In the body, fatty acids are found in several forms including being "free" (unbound) molecules and components of larger molecules such as triglycerides, phospholipids, and cholesteryl esters.

"Free" fatty acid symbol Triglyceride symbol Phospholipid symbol Cholesteryl ester symbol

Number of Carbons (Chain Length) The number of carbon atoms in the backbone of a fatty acid determines its **chain length,** as shown in Figure 10.3. In fact, it is helpful to think of a fatty acid as if it were a chain, with each link representing a carbon atom. Most naturally occurring fatty acids have an even number of links or carbons—usually 12 to 22—although some may be as short as 4 or as long as 26 carbons. Fatty acids with fewer than 8 carbons are called **short-chain fatty acids;** those with 8 to 12 carbons are **medium-chain fatty acids;** those with more than 12 carbons are **long-chain fatty acids.**

The chain length of a fatty acid affects its chemical properties and physiological functions. For example, it influences the temperature at which the fatty acid melts (melting point). Fatty acids with low melting points, such as short-chain fatty acids, typically take less heat to liquify (melt) compared with fatty acids with longer chain lengths. Lipids made predominantly of short-chain fatty acids are therefore likely to be oils or even gases. Conversely, lipids containing mostly longer-chain fatty acids can have high melting points and tend to be solids (fats) at room temperature. Chain length also affects the extent to which a fatty acid is soluble in water, with short-chain fatty acids generally being more water-soluble than long-chain fatty acids. Because humans are mostly water, short-chain fatty acids are more easily absorbed and transported in the body than are long-chain fatty acids.

Number and Positions of Double Bonds Aside from the *number* of carbon atoms they contain, fatty acids also differ in the *types* of chemical bonds between their carbon atoms (Figure 10.4). These carbon–carbon bonds can be either single bonds or double bonds. If a fatty acid contains all single carbon–carbon bonds, it is a **saturated fatty acid (SFA);** those containing one or more double bonds are **unsaturated fatty acids.** Fatty acids with one double bond are **monounsaturated fatty acids (MUFAs);** those with two or more double bonds are **polyunsaturated fatty acids (PUFAs).** Note that,

chain length The number of carbons in a fatty acid's backbone.

short-chain fatty acid A fatty acid having <8 carbon atoms in its backbone.

medium-chain fatty acid A fatty acid having 8–12 carbon atoms in its backbone.

long-chain fatty acid A fatty acid having >12 carbon atoms in its backbone.

saturated fatty acid (SFA) (*saturare,* to fill or satisfy) A fatty acid that contains only carbon–carbon single bonds in its backbone.

unsaturated fatty acid A fatty acid that contains at least one carbon–carbon double bond in its backbone.

monounsaturated fatty acid (MUFA) A fatty acid that contains one carbon–carbon double bond in its backbone.

polyunsaturated fatty acid (PUFA) A fatty acid that contains more than one carbon–carbon double bond in its backbone.

FIGURE 10.3 Fatty Acids Can Have Different Chain Lengths The number of carbon atoms making up the backbone of a fatty acid determines whether it is a short-, medium-, or long-chain fatty acid.

Medium-chain fatty acid

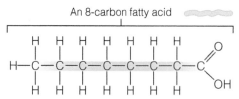

An 8-carbon fatty acid

Long-chain fatty acid

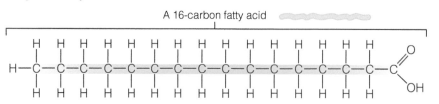

A 16-carbon fatty acid

FIGURE 10.4 Saturated and Unsaturated Fatty Acids The types of carbon–carbon bonds in a fatty acid determine whether it is saturated or unsaturated.

**Saturated fatty acid
(stearic acid; 18:0)**

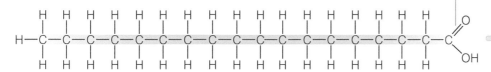

Note that this carbon–oxygen double bond does not make the fatty acid "unsaturated."

**Monounsaturated fatty acid
(oleic acid; *cis* 9–18:1)**

The presence of a double bond bends the fatty acid backbone.

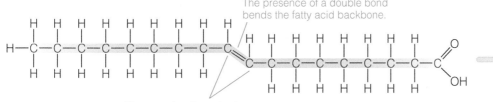

There are two fewer hydrogens for each double bond.

**Polyunsaturated fatty acid
(linoleic acid; *cis* 9, *cis* 12–18:2)**

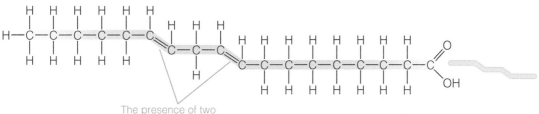

The presence of two double bonds causes two bends in the fatty acid backbone.

for each carbon–carbon double bond, two hydrogen atoms are lost from the fatty acid backbone. This is because carbon atoms can only have four chemical bonds. A double bond "counts" as two chemical bonds, so one hydrogen atom must be given up from each of the carbon atoms in the carbon–carbon double bond.

Like chain length, the number of double bonds can influence the physical nature of the fatty acid. As illustrated in Figure 10.4, each carbon atom in an SFA is surrounded (or "saturated") by hydrogen atoms. Being "saturated" with hydrogen atoms prevents the fatty acid from bending. Because of this rigidity, SFAs are highly organized and dense, making them solid at room temperature. Foods containing large amounts of SFAs (such as butter or coconut "oil"), therefore, tend to be solid fats at room temperature.

Compared with SFAs of similar chain length, unsaturated fatty acids (in other words, those with double bonds) have fewer hydrogen atoms and can bend. In fact, whenever there is a carbon–carbon double bond, there is a kink or a bend in the fatty acid backbone. These bends cause unsaturated fatty acids to become disorganized, preventing them from becoming

Some animal products, such as beef, contain a relatively high amount of saturated fatty acids.

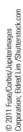

Saturated fatty acids tend to be straight (like uncooked spaghetti), whereas unsaturated fatty acids have bends (like cooked spaghetti).

densely packed. Imagine the difference between the organization of uncooked and cooked spaghetti noodles. The straight uncooked spaghetti noodles are neatly organized, whereas cooked spaghetti noodles are disorganized. This is similar to the difference between SFAs (like uncooked spaghetti) and PUFAs (like cooked spaghetti noodles). In general, organized molecules such as SFAs are solids (fats) at room temperature, and disorganized molecules like PUFAs are liquids (oils). MUFAs have chemical characteristics that lie between those of SFAs and PUFAs, being thick liquids or soft solids at room temperature. You may have noticed that olive oil, which is high in MUFAs, is a thick oil at room temperature.

Understanding *Cis* versus *Trans* Fatty Acids Unsaturated fatty acids can be further categorized depending on how the hydrogen atoms are arranged around the carbon–carbon double bonds. Most naturally occurring fatty acids have the hydrogen atoms positioned on the same side of the double bond, resulting in a **cis double bond** (Figure 10.5). When the hydrogen atoms are on opposite sides of the double bond, it is called a **trans double bond.** Unlike *cis* double bonds, *trans* double bonds do not cause bending. Fatty acids containing at least one *trans* double bond are called **trans fatty acids** and have fewer bends in their backbones than their *cis* counterparts. For this reason, *trans* fatty acids are also more likely to be solid (fats) at room temperature than fatty acids containing only *cis* double bonds.

Trans Fatty Acids in Food *Trans* fatty acids are found naturally in some foods such as dairy and beef products. However, most dietary *trans* fatty acids are

cis double bond (*cis*, on this side of) A carbon–carbon double bond in which the hydrogen atoms are positioned on the same side of the double bond.

trans double bond (*trans*, across) A carbon–carbon double bond in which the hydrogen atoms are positioned on opposite sides of the double bond.

trans fatty acid A fatty acid containing at least one *trans* double bond.

FIGURE 10.5 *Cis* versus *Trans* Fatty Acids Unsaturated fatty acids can differ by whether they have *cis* or *trans* carbon–carbon double bonds. *Cis* bonds cause the fatty acid to bend, whereas *trans* bonds do not.

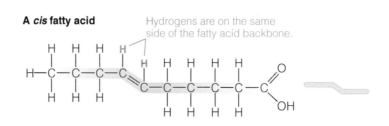

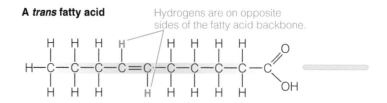

produced commercially via a process called **partial hydrogenation.** Partial hydrogenation converts the majority of carbon–carbon double bonds into carbon–carbon single bonds, causing oils (such as corn oil) to become fats (such as margarine or shortening). This is done by the chemical addition of hydrogen atoms. Aside from decreasing the number of double bonds and increasing the number of single bonds, the process of partial hydrogenation converts some of the *cis* double bonds to *trans* double bonds. As a result, the lipid is high in both saturated and *trans* fatty acids.

Partially hydrogenated lipids are often used in food manufacturing because they impart desirable food texture and reduce spoilage. Crackers, pastries, bakery products, shortening, and margarine are the main sources of the *trans* fatty acids in our diets.[1] However, the current focus on decreasing *trans* fatty acid intake has resulted in new food preparation and processing methods that decrease or eliminate *trans* fatty acids in many foods. For example, some fast-food chains have switched from frying their foods in high–*trans* fatty acid shortening to *trans* fatty acid–free vegetable oils.

Many public health agencies, including the Institute of Medicine and the U.S. Department of Agriculture, suggest that we limit our intake of *trans* fatty acids. These recommendations (which are described in more detail later) for lowering *trans* fatty acid intake have been made because some *trans* fatty acids may increase the risk for cardiovascular disease.[2] As of 2006, food manufacturers have been required to state the *trans* fatty acid content on their Nutrition Facts panels.* Because of the considerable concern about *trans* fatty acids in the diet, a number of cities in the United States, have become "*trans* fat–free zones." You can read more about this in the Focus on Diet and Health feature.

FATTY ACIDS ARE NAMED FOR THEIR STRUCTURES

There are several methods used to name or describe fatty acids. In general, these methods are based on the number of carbons, the number and types of double bonds, and the positions of the double bonds. Some fatty acids also have "common names."

Alpha (α) Naming System The "alpha (α)" naming system for fatty acids is based on the positions and types of double bonds relative to the carboxylic acid (α) end of the fatty acid. As an example, consider a fatty acid with the following characteristics:

- 18 carbons in length
- 2 *cis* double bonds
- first *cis* double bond between the 9th and 10th carbons from the carboxylic acid end
- second *cis* double bond between the 12th and 13th carbons from the carboxylic acid end.

The fatty acid's name is constructed beginning with an "18," signifying that there are 18 carbons. A "2" is added to form "18:2," signifying that there are two double bonds. Next, where the double bonds are located is designated as "9,12"–18:2, with the locations determined by counting *from the carboxylic acid (α) end*. Finally, because both double bonds are in the *cis* configuration,

---

*Because they are thought to not cause health problems, most naturally occurring *trans* fatty acids (such as those found in dairy products) are not included in this value.

partial hydrogenation A process whereby some carbon–carbon double bonds found in PUFAs are converted to carbon–carbon single bonds, resulting in the production of a lipid containing saturated and *trans* fatty acids.

Societies function best when they establish laws and regulations that limit dangerous behavior. For example, we have laws against driving under the influence of alcohol and smoking in public places. This concept sometimes carries over to nutrition, as well—such as the U.S. Food and Drug Administration's (FDA's) numerous regulations concerning accurate food labeling and safety of our food supply. The notion, however, of imposing laws concerning what kind of foods can and cannot be served in restaurants is relatively new. Such is the case for New York City, whose Board of Health voted in 2006 to ban the use of *trans* fats in the metropolis's restaurants. Specifically, restaurants must not serve foods containing *trans* fatty acids produced via partial hydrogenation. This ban, however, contains some exceptions. For instance, it still allows restaurants to serve *trans* fatty acid–containing foods that come in the manufacturer's original packaging. Thus, not all food sold must be *trans* fatty acid free. A similar rule was passed in Seattle, and one has been considered in Chicago, although some residents of this city have ridiculed it as unnecessary government meddling.[3] This trend is not confined to just a few large U.S. cities. For example, Tiburon (a small town in California), claims to have become the first "*trans* fat–free city" in 2004, and in 2003 Denmark issued national regulations limiting the amount of *trans* fat that could be used in processed foods.

Of course, these actions come on the heels of significant scientific evidence that consumption of *trans* fatty acids found in partially hydrogenated oils can increase risk for heart disease.[2] Still, many people argue that local municipalities have no business "outlawing" foods that the FDA has deemed safe enough to allow in the public food system. Clearly, this is an area of justifiable debate, and only time will tell whether this type of local food regulation will catch on in other cities—or, more important, improve the health of the local citizens.

the name is modified to *cis*9,*cis*12–18:2. This fatty acid (*cis*9,*cis*12–18:2) is shown here.

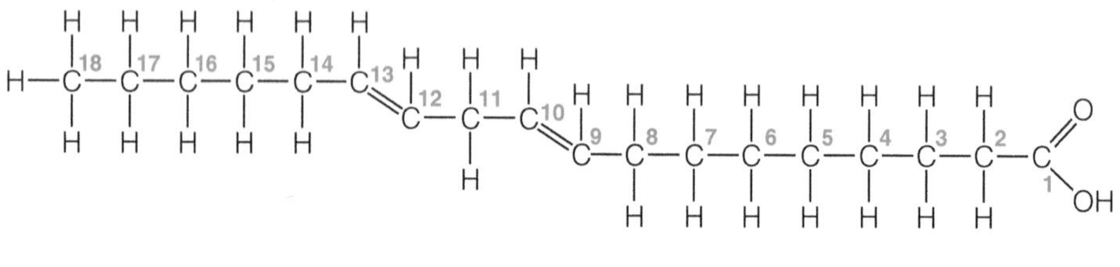

Omega (α) Naming System An alternate system for naming a fatty acid is the omega (ω) system. In this system, the numbers of carbons and double bonds are again distinguished (for example, 18:2). However, in the omega (ω) naming system, fatty acids are described on the basis of where the first double bond is located *relative to the methyl (ω) end* of the molecule. If the first double bond is between the third and fourth carbons from the ω end, the fatty acid is an **omega-3 (ω-3) fatty acid.** If the first double bond is between the sixth and seventh carbons, it is an **omega-6 (ω-6) fatty acid.** There are also ω-7 and ω-9 fatty acids. Unlike alpha nomenclature, omega nomenclature does not usually identify whether the double bonds are in the *cis* or *trans* configuration or the location of the other double bonds in the molecule.

Common Names Sometimes, fatty acids are referred to by their common names, which often reflect prominent food sources where they are found. For example, *palm*itic acid (16:0) is found in *palm* oil, and *arach*idonic acid

omega-3 (ω-3) fatty acid A fatty acid in which the first double bond is located between the third and fourth carbons from the methyl or omega (ω) end.

omega-6 (ω-6) fatty acid A fatty acid in which the first double bond is located between the sixth and seventh carbons from the methyl or omega (ω) end.

TABLE 10.1 Names and Food Sources of Some Important Fatty Acids in the Body

Alpha (α) Nomenclature	Omega (ω) Family[a]	Common Name	Food Sources
Saturated Fatty Acids			
12:0	—	Lauric acid	Coconut and palm oils
14:0	—	Myristic acid	Coconut and palm oils; most animal and plant fats
16:0	—	Palmitic acid	Animal and plant fats
18:0	—	Stearic acid	Animal fats, some plant fats
20:0	—	Arachidic acid	Peanut oil
Unsaturated Fatty Acids			
*cis*9–16:1	ω-7	Palmitoleic acid	Marine animal oils
*cis*9–18:1	ω-9	Oleic acid	Plant and animal fats, olive oil
*cis*9,*cis*12–18:2	ω-6	Linoleic acid	Nuts, corn, safflower, soybean, cottonseed, sunflower seeds, and peanut oil
*cis*9,*cis*12,*cis*15–18:3	ω-3	Linolenic acid (α-linolenic acid)	Canola, soybean, flaxseed, and other seed oils
*cis*5,*cis*8,*cis*11,*cis*14–20:4	ω-6	Arachidonic acid	Small amounts in plant and animal oils
*cis*5,*cis*8,*cis*11,*cis*14,*cis*17–20:5	ω-3	Eicosapentaenoic acid (EPA)	Marine algae, fish oils
*cis*4,*cis*7,*cis*10,*cis*13,*cis*16,*cis*19–22:6	ω-3	Docosahexaenoic acid (DHA)	Animal fats as phospholipid component, fish oils

[a]The omega (ω) nomenclature only applies to unsaturated fatty acids, because it refers to the position of the first carbon–carbon double bond in relation to the methyl (ω) end of the fatty acid.

SOURCE: Adapted from Gropper SS, Smith JL, Groff JL. Advanced nutrition and human metabolism, 5th ed. Belmont, CA: Thomson/Wadsworth; 2007.

(*cis*5,*cis*8,*cis*11,*cis*14–20:4; from *arachis,* meaning legume or peanut) is found in peanut butter. Some common names of fatty acids are listed in Table 10.1.

Which Fatty Acids Do We Need, and Where Do They Come From?

There are literally hundreds of different fatty acids—each with its own distinct structure, chemical properties, and name. Although we need many of them for optimal health, there are only two fatty acids that adults cannot synthesize, and are therefore essential to get from food. In this section, you will learn about several fatty acids—including the essential fatty acids—as well as their sources and functions.

⟨**CONNECTIONS**⟩ Essential nutrients are needed by the body but cannot be synthesized in sufficient quantities to meet our needs.

THERE ARE TWO ESSENTIAL FATTY ACIDS: LINOLEIC ACID AND LINOLENIC ACID

Although there is great diversity in the fatty acids found in foods, only two are dietary essentials: **linoleic acid** and **linolenic acid** (also called α-linolenic acid). Linoleic acid has 18 carbons, two *cis* double bonds, and is an ω-6 fatty acid. Linolenic acid has 18 carbons, three *cis* double bonds, and is an ω-3 fatty acid. Linoleic acid and linolenic acid are essential nutrients, because the body cannot create double bonds in the ω-3 and ω-6 positions.

The functions of the essential fatty acids are numerous. For example, they both serve as precursors of hormone-like compounds called eicosanoids (described below). Linolenic acid also is an important component of biological membranes, particularly in nerve tissue and the retina. In order for this to occur, these fatty acids must first be modified as described next.

linoleic acid An essential ω-6 fatty acid with 18 carbons and 2 double bonds.

linolenic acid An essential ω-3 fatty acid with 18 carbons and 3 double bonds.

Linoleic acid

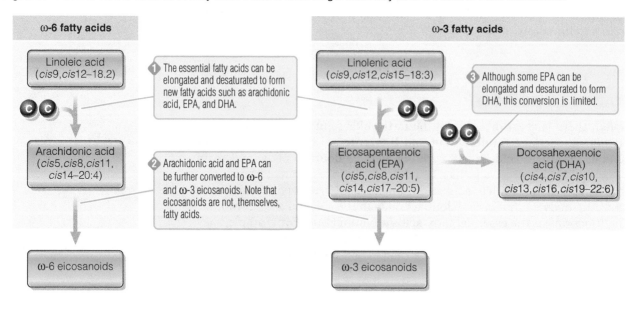

Linolenic acid

Converting Linoleic and Linolenic Acids to Longer-Chain Fatty Acids Dietary linoleic acid and linolenic acid provide the basic building blocks needed to make longer-chain ω-3 and ω-6 fatty acids. The synthesis of these newly formed (longer-chain) fatty acids is accomplished by increasing the number of carbon atoms (**elongation**) and the number of double bonds (**desaturation**). As shown in Figure 10.6, the essential fatty acid linoleic acid is used to make **arachidonic acid** (a 20-carbon, ω-6 fatty acid). This is accomplished by adding two carbon atoms via elongation and two double bonds via desaturation. Similarly, linolenic acid can be elongated and desaturated to **eicosapentaenoic acid** (**EPA**; a 20-carbon, ω-3 fatty acid with 4 double bonds), which can subsequently be elongated and desaturated further to **docosahexaenoic acid** (**DHA**; a 22-carbon, ω-3 fatty acid with 6 double bonds). These long-chain PUFAs have many functions in the body. For example, they are necessary for normal epithelial cell function and also assist in regulating gene expression. Because the conversion of linolenic acid to DHA is much lower than previously thought, dietary sources of this fatty acid may be especially important.[4]

Metabolism of the Essential Fatty Acids to Eicosanoids As you have just learned, the essential fatty acids can be elongated and desaturated to a variety of other

elongation The process whereby carbon atoms are added to a fatty acid, increasing its chain length.

desaturation The process whereby carbon–carbon single bonds are transformed into carbon–carbon double bonds in a fatty acid.

arachidonic acid (a – rach – i – DON – ic) A long-chain, polyunsaturated ω-6 fatty acid produced from linoleic acid.

eicosapentaenoic acid (EPA) (ei – co – sa – pen – ta – NO – ic) A long-chain, polyunsaturated ω-3 fatty acid produced from linolenic acid.

docosahexaenoic acid (DHA) (do – cos – a – hex – a – NO – ic) A long-chain, polyunsaturated ω-3 fatty acid produced from eicosapentaenoic acid (EPA).

FIGURE 10.6 Metabolism of the Essential Fatty Acids Linoleic Acid and Linolenic Acid Linoleic acid and linolenic acid can be elongated and desaturated to make other fatty acids. Some of these longer-chain fatty acids are used to make eicosanoids.

important fatty acids. Moreover, the essential fatty acids and some of their longer-chain metabolites can also be converted to other important compounds that are not, themselves, fatty acids but are lipid-like substances vital for health. One example is the **eicosanoids,** which act as chemical messengers that direct myriad physiologic functions. For example, eicosanoids play a particularly important role in regulating the immune and cardiovascular systems.[5] As shown in Figure 10.6, linoleic acid (an ω-6 fatty acid) is converted to ω-6 eicosanoids, whereas linolenic acid (an ω-3 fatty acid) is converted to ω-3 eicosanoids. An example of eicosanoids are the **prostaglandins,** which control dilation and constriction of blood vessels and therefore are important regulators of blood pressure.

The body produces both ω-3 and ω-6 eicosanoids, which have somewhat opposing actions. For instance, the ω-6 eicosanoids tend to cause inflammation and constriction of blood vessels, whereas the ω-3 eicosanoids are anti-inflammatory and stimulate dilation (or relaxation) of blood vessel walls. Also, ω-3 eicosanoids inhibit blood clotting. Both ω-3 and ω-6 eicosanoids are important for health, and the body can shift its relative production in response to its needs.

Your diet can also influence the amount and types of eicosanoids that you make. For example, Alaska natives consuming high amounts of ω-3 fatty acids from fish and marine mammals (e.g., seals) have enhanced physiologic responses stimulated by ω-3 eicosanoids. Consequently, Alaska natives consuming traditional diets tend to form blood clots more slowly than people who consume lesser amounts of ω-3 fatty acids.[6] Many studies suggest that alterations in the balance of ω-3 to ω-6 eicosanoids may influence a person's risk for conditions related to inflammation such as heart disease and cancer.[7] This is why experts recommend that we consume fish regularly, and these recommendations will be described in more detail later in this chapter.

Essential Fatty Acid Deficiency Essential fatty acid deficiencies are rare because of the almost endless supply of linoleic and linolenic acids stored in adipose tissue. Indeed, primary essential fatty acid deficiency generally occurs only in hospitalized patients receiving poor nutritional care. Secondary fatty acid deficiencies can occur with diseases that disrupt lipid absorption or utilization such as cystic fibrosis. Within two to four weeks, individuals with essential fatty acid deficiencies develop irritated and flaky skin, gastrointestinal problems, and impaired immune function. As a result, infections are common and wound healing may be slow. Children with essential fatty acid deficiencies also exhibit slow growth.

SOME FATTY ACIDS ARE CONDITIONALLY ESSENTIAL

In addition to linoleic and linolenic acids, other fatty acids may be conditionally essential during infancy. These include arachidonic acid and DHA. As described previously, in adults these fatty acids are made from linoleic acid and linolenic acid, respectively. However, babies may not be able to readily make them because they cannot produce the needed enzymes in sufficient amounts. The conditional essentiality of arachidonic acid and DHA during infancy is described in more detail in the Focus on Life Cycle Nutrition feature.

DIETARY SOURCES OF DIFFERENT TYPES OF FATTY ACIDS

Most foods contain a mixture of fatty acids. However, some foods are especially good sources of a particular fatty acid. For example, nuts (such as walnuts), seeds, and certain oils (such as those made from soybean, safflower, or corn) are generally abundant in linoleic acid. Linolenic acid is also found in

eicosanoids (ei – COS – a – noids) Biologically active compounds synthesized from arachidonic acid and eicosapentaenoic acid (EPA).

prostaglandins A group of eicosanoids involved in regulation of blood pressure; there are both ω-6 and ω-3 prostaglandins, having somewhat opposite effects.

Most young infants rely solely on either human milk or infant formula for all their nutritional needs. Although manufacturers strive to produce formulas that are similar to human milk, the lipids provided by these two infant foods are sometimes quite dissimilar. For example, human milk contains at least 47 different fatty acids, most of which are not found in infant formulas.[8] Scientists do not know precisely which of these fatty acids promote optimal growth and development during this time. However, some of the long-chain PUFAs found in human milk may be conditionally essential nutrients during infancy. These fatty acids, arachidonic acid and docosahexaenoic acid (DHA), are likely produced in sufficient amounts from linoleic and linolenic acids, respectively, in older children and adults. Therefore, they are not considered to be essential nutrients during these periods of the life cycle. However, infants (especially those born prematurely) have very low stores of arachidonic acid and DHA and may not be able to synthesize them in adequate amounts.[9] These fatty acids are thought to be important for growth and development of the eyes, nervous system, and brain.[10] Thus, many scientists believe that infants should consume adequate amounts of them during early life to achieve optimal growth and development.

Until relatively recently, only breastfed babies received these fatty acids, because infant formulas were not fortified with long-chain PUFAs. Because some research suggests that fortifying infant formula with long-chain PUFAs may be advantageous, many companies now produce infant formula fortified with arachidonic acid and

DHA, and these have been marketed in the United States since 2002.[11] Because little is known about the long-term effects of these long-chain PUFA-fortified formulas on human health, the FDA has asked manufacturers to closely monitor these products in the marketplace. Somewhat surprisingly, the American Academy of Pediatrics has no official position on the use of these fortified formulas in infant feeding.

Scientists are interested in the importance of other dietary lipids during infancy as well. An example is cholesterol. Although researchers have long known that human milk contains high amounts of cholesterol, infant formula generally lacks this lipid.[12] Whether early exposure to dietary cholesterol or the multitude of "nonessential" fatty acids and other lipids in human milk is important for optimal growth and development is unknown but continues to interest the medical and scientific communities.[13]

© 2011 Blend Images/JupiterImages Corporation

some oils (such as those made from canola, soybean, or flaxseed) as well as some nuts (such as walnuts). Some foods, such as soybean oil and walnuts, are good sources of both essential fatty acids. Because many of these foods and oils are common in the American diet, getting adequate amounts of linoleic acid and linolenic acid is easy.

Many experts also believe that it is important for adults to consume adequate amounts of the longer-chain ω-3 fatty acids such as EPA and DHA because higher intakes are related to lower risk for heart disease and stroke. This may be due to the potent anti-inflammatory effects of these compounds in the body. You can read more about the relationship between longer-chain ω-3 fatty and cardiovascular disease in the Nutrition Matters at the end of this chapter. EPA and DHA are plentiful in fatty fish and other seafood; smaller amounts are found in meat and eggs. Longer-chain ω-6 fatty acids are also found in meat, poultry, and eggs. Omega-6 fatty acids are typically more plentiful in our diets than are ω-3 fatty acids.

In general, animal foods contribute the majority of dietary SFAs (saturated fatty acids), whereas plant-derived foods supply the majority of PUFAs (polyunsaturated fatty acids). MUFAs (monounsaturated fatty acids) come from both plant and animal foods. However, some tropical oils, such as coconut and palm oils, contain relatively high amounts of SFAs, and many oily fish

© Michael Mahovlich/Masterfile

Many fish are excellent sources of ω-3 fatty acids.

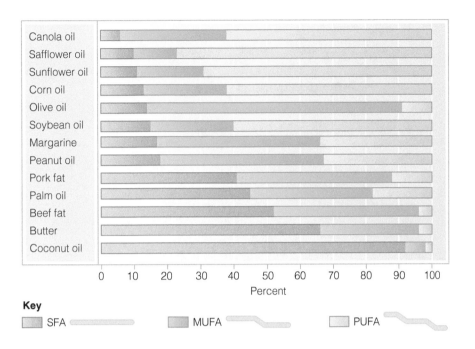

have high levels of PUFAs. The relative amounts of SFAs, MUFAs, and PUFAs in commonly consumed fats and oils are summarized in Figure 10.7.

Mono-, Di-, and Triglycerides: What's the Difference?

As previously mentioned, most fatty acids do not exist in their free (unbound) form in foods or in the body. Instead, they are part of larger, more complex molecules such as triglycerides, diglycerides, and monoglycerides. The number of fatty acids attached to the glycerol backbone determines whether the molecule is a mono-, di-, or triglyceride.[†] A **triglyceride** has three fatty acids, a **diglyceride** consists of two fatty acids, and a **monoglyceride** contains only one fatty acid. In all cases, the fatty acid molecules are attached to the glycerol backbone via "ester" linkages (Figure 10.8). These fatty acids can be saturated, monounsaturated, polyunsaturated, or a mixture of fatty acid types. Regardless of which types of fatty acids are attached to the glycerol molecule, most mono-, di-, and triglycerides are relatively hydrophobic.

TRIGLYCERIDES PLAY MANY ROLES IN THE BODY

Triglycerides—or, more accurately, their associated fatty acids—serve many purposes in the body. Perhaps most importantly, they are sources of the essential fatty acids needed for the body to function. Triglycerides are also vitally important for meeting both immediate and long-term energy needs.

Triglycerides as an Energy Source Compared with the other energy-yielding macronutrients, triglycerides represent the body's richest source of energy. The complete breakdown of 1 g of fatty acids yields approximately 9 kcal of energy, which is more than twice the yield from 1 g of carbohydrate or protein (4 kcal). Therefore, gram for gram, high-fat foods contain more calories than do other foods.

[†]Triglycerides, diglycerides, and monoglycerides are also called triacylglycerols, diacylglycerols, and monoacylglycerols, respectively.

triglyceride (also called triacylglycerol) (*tri*, three) A lipid composed of a glycerol molecule bonded to three fatty acids.

diglyceride (also called diacylglycerol) (*di*, two) A lipid made of a glycerol molecule bonded to two fatty acids.

monoglyceride (also called monoacylglycerol) (*monos*, single) A lipid made of a glycerol molecule bonded to a single fatty acid.

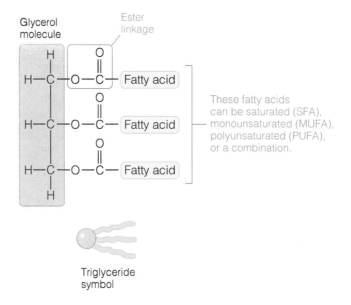

FIGURE 10.8 A Triglyceride Molecule A triglyceride molecule consists of a glycerol molecule bonded to three fatty acids via "ester" linkages.

Glycerol molecule

Ester linkage

Fatty acid

Fatty acid

Fatty acid

These fatty acids can be saturated (SFA), monounsaturated (MUFA), polyunsaturated (PUFA), or a combination.

Triglyceride symbol

Regardless of where they originate (from diet or adipose tissue), for triglycerides to be used as a source of energy, they must first be disassembled into glycerol and fatty acids. This process, called **lipolysis,** is catalyzed by enzymes called **lipases.** When triglycerides circulate in the blood, the form of lipase that removes the fatty acids from a triglyceride's backbone is called **lipoprotein lipase.** This is described in more detail in a later section of this chapter. When triglycerides within adipose tissue are being broken down for energy, the form of lipase that cleaves the fatty acids from their glycerol backbone is called **hormone-sensitive lipase.** To facilitate lipolysis in adipose tissue during times of starvation, low insulin-to-glucagon ratios stimulate hormone-sensitive lipase activity. Recall that insulin levels decrease and glucagon levels increase within a few hours of eating. Lipolysis via hormone-sensitive lipase is also stimulated by exercise and other forms of physiological stress.

Fatty Acids Can Also Be Converted to Energy-Yielding Ketones In addition to using fatty acids directly as an energy source, the body can convert them to other energy-yielding compounds called ketones. Recall from Chapter 8 that the production of ketones from fatty acids, a process called ketogenesis, occurs when glucose availability is low. Ketogenesis is important because some tissues such as brain, heart, skeletal muscle, and kidneys can use ketones for energy (ATP). In this way, ketones produced from fatty acids can serve as an important source of energy during times of severe glucose insufficiency (such as starvation) and can reduce the need for amino acids to be converted to glucose via the process of gluconeogenesis. This helps protect the body from having to use protein from its muscles and other tissues for energy.

Storage of Excess Energy as Triglycerides in Adipose Tissue But what happens when the energy available to the body exceeds its energy needs? During these times, excess fatty acids are stored as triglycerides in adipose tissue and, to a lesser extent, skeletal muscle. Adipose tissue consists of specialized cells called **adipocytes,** which can accumulate large amounts of lipid. Adipose tissue is found in many parts of the body, including under the skin (**subcutaneous adipose tissue**) and around the vital organs in the abdomen (**visceral adipose tissue** also called intra-abdominal fat). Many of the body's organs and tissues (such as the kidneys and breasts) have visceral adipose tissue associated with them, giving them ready access to fatty acids for their immediate energy

⟨**CONNECTIONS**⟩ Remember that gluconeogenesis is the synthesis of glucose from noncarbohydrate substances (Chapter 8, page 164).

lipolysis The breakdown of triglycerides into fatty acids and glycerol.

lipases Enzymes that cleave fatty acids from the glycerol backbones of triglycerides, phospholipids, and cholesteryl esters.

lipoprotein lipase An enzyme that hydrolyzes the ester linkage between a fatty acid and glycerol in a triglyceride, diglyceride, and monoglyceride molecule as they circulate in the bloodstream.

hormone-sensitive lipase An enzyme that catalyzes the hydrolysis of ester linkages that attach fatty acids to the glycerol molecule; mobilizes fatty acids stored in adipose tissue.

adipocyte (a – DIP – o – cyte) A specialized cell that makes up the majority of adipose tissue.

subcutaneous adipose tissue Adipose tissue found directly under the skin.

visceral adipose tissue Adipose tissue surrounding the vital organs.

needs. You will learn more about subcutaneous and visceral adipose tissue and how their distribution may influence health in Chapter 11.

Insulin Stimulates Lipogenesis and Triglyceride Storage The hormone insulin stimulates storage of triglycerides during times of energy excess—for instance, after a high-calorie meal. Insulin causes adipocytes, and to a lesser extent skeletal muscle cells, to take up glucose and fatty acids. It also stimulates the conversion of excess glucose to fatty acids. In turn, these newly formed fatty acids and excess dietary fatty acids can be incorporated into triglycerides. The synthesis of fatty acids and triglycerides is called **lipogenesis**. Because insulin inhibits the action of hormone-sensitive lipase, high levels of insulin also inhibit lipolysis, or the breakdown of lipids. Together, increased lipogenesis and decreased lipolysis after a meal help direct excess fatty acids to adipose tissue, where they are deposited for later use.

Energy Storage as Lipid Has Its Advantages Compared to glycogen, storage of excess energy as triglyceride has several advantages. First, because triglycerides are not stored with water (a bulky molecule), large amounts can fit in small spaces. Consequently, the body can store about six times more energy in one pound of adipose tissue than in one pound of liver glycogen. Indeed, the body has a seemingly infinite ability to store excess energy in adipose tissue, whereas its capacity to store glycogen is limited. In addition, gram for gram, lipids store more than twice the energy than glycogen.

Triglycerides Needed for Insulation and Protection Aside from their important role as an energy source, triglycerides stored in adipose tissue also protect your internal organs from injury. We also rely on adipose tissue for insulation, which keeps us warm. People with very little body fat can have difficulty regulating body temperature. In fact, one physiological response to being severely underweight is that very fine hair covering the body can develop. This hair, called *lanugo*, partially makes up for the absence of subcutaneous adipose tissue by providing a layer of external insulation. The presence of lanugo is common in people who have the eating disorder anorexia nervosa.[14]

What Are Phospholipids and Sterols?

In addition to triglycerides, the two other major lipid categories are phospholipids and sterols, both of which are essential components of cell membranes and are involved in the transport of lipids in the bloodstream. Because the body can synthesize all that it needs, there are no dietary requirements for either of these lipid classes. Nonetheless, they are commonly found in food.

PHOSPHOLIPIDS ARE CONSIDERED "AMPHIPATHIC"

Phospholipids are similar to triglycerides in that they contain a glycerol molecule bonded to fatty acids (Figure 10.9). However, instead of having three fatty acids, a phospholipid has only two fatty acids. Replacing the third fatty acid is a phosphate-containing **polar head group**. There are many different types of polar head groups, but the most common are choline, ethanolamine, inositol, and serine.

Phospholipids contain both polar (hydrophilic) and nonpolar (hydrophobic) regions. **Polar** substances are those with an unequal charge distribution. In other words, one portion of a polar molecule is positively charged, whereas another is negatively charged. Polar molecules, such as water and sodium chloride (NaCl; table salt), are attracted to other polar molecules. This

lipogenesis (*lipos*, fat; *genus*, birth) The metabolic processes that result in fatty acid and, ultimately, triglyceride synthesis.

phospholipid A type of lipid composed of a glycerol bonded to two fatty acids and a polar head group.

polar head group A phosphate-containing charged chemical structure that is a component of a phospholipid.

polar molecule A molecule (such as water) that has both positively and negatively charged portions.

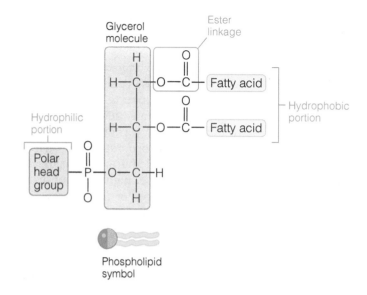

FIGURE 10.9 A Phospholipid Molecule A phospholipid consists of a glycerol molecule, a polar head group, and two fatty acids.

is why table salt dissolves readily in water. Because polar compounds dissolve in water, they are referred to as **hydrophilic** (or "water-loving"). Conversely, **nonpolar** compounds (such as triglycerides) have even charge distributions and do not dissolve easily in polar compounds. This is why oil and water do not mix, and why nonpolar compounds are often referred to as **hydrophobic** (or "water-fearing").

Having both nonpolar and polar portions makes *phospholipids* **amphipathic** compounds. The prefix "*amphi-*" means "both" and is the same as the prefix in the word *amphibian*, which refers to an animal capable of living in two environments, land and water. Similarly, amphipathic substances typically straddle two environments—in this case, lipids and water. In terms of a phospholipid, the two fatty acids are hydrophobic and therefore drawn to lipids, while the polar head group is hydrophilic and drawn to water. These opposing attractions allow phospholipids to act as major components of cell membranes and to play roles in the digestion, absorption, and transport of lipids in the body.

Most Foods Contain Phospholipids Because they are components of cell membranes (animals) and cell walls (plants), phospholipids are found naturally in most foods. They are also used as food additives. An example is **phosphatidylcholine** (the phospholipid with a choline polar head group), which is often added to foods such as mayonnaise to prevent them from separating. You can check whether a food has phosphatidylcholine by looking for the word "lecithin" (a common name for this compound) on the food label.

PHOSPHOLIPIDS ARE CRITICAL FOR CELL MEMBRANES AND LIPID TRANSPORT

As previously mentioned, phospholipids are the main structural component of membranes surrounding cells and organelles (Figure 10.10). For example, cell membranes consist of two layers (a bilayer) of phospholipids with the hydrophilic polar head groups pointing to the extra- and intracellular spaces, both of which are predominantly water. The hydrophobic tails face toward the interior of the membrane, forming a water-free lipid-rich zone. If the cell membrane were completely hydrophilic, it would dissolve and not create a barrier. If the cell membrane were completely hydrophobic, there would be no

hydrophilic substance (hy – dro – PHIL – ic) One that dissolves in or mixes with water.

nonpolar molecule One that does not have differently charged portions.

hydrophobic substance (hy – dro – PHO – ic) One that does not dissolve in or mix with water.

amphipathic (amphi-, on both sides) Having both nonpolar (noncharged) and polar (charged) portions.

phosphatidylcholine (also called lecithin) (PHOS – pha – tid – yl – CHO – line) A phospholipid that contains choline as its polar head group; commonly added to foods as an emulsifying agent.

FIGURE 10.10 A Cell Membrane Consists of a Phospholipid Bilayer, Proteins, and Cholesterol The amphipathic nature of phospholipids allows cell membranes to carry out their functions.

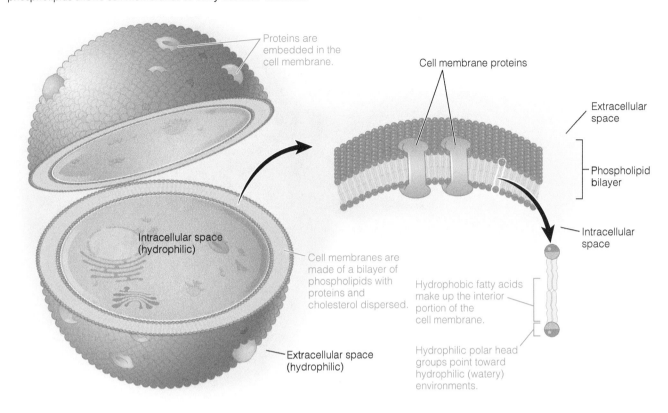

Proteins are embedded in the cell membrane.

Cell membrane proteins

Extracellular space

Phospholipid bilayer

Intracellular space (hydrophilic)

Intracellular space

Cell membranes are made of a bilayer of phospholipids with proteins and cholesterol dispersed.

Hydrophobic fatty acids make up the interior portion of the cell membrane.

Extracellular space (hydrophilic)

Hydrophilic polar head groups point toward hydrophilic (watery) environments.

chance of communication between the watery extra- and intracellular compartments. Thus, the incorporation of two layers of amphipathic phospholipids allows cell membranes to effectively carry out their functions.

Phospholipids also supply fatty acids for cellular metabolism and can act as biologically active compounds. For example, some activate enzymes important for energy metabolism, blood clotting, and cell turnover. Phospholipids also donate their fatty acids for eicosanoid production and can act as carriers of hydrophobic substances in the body. One example is the incorporation of phospholipids in the outer surface of a group of particles called lipoproteins. Lipoproteins transport lipids in the blood and are described later in this chapter.

STEROLS AND STEROL ESTERS ARE LIPIDS WITH RING STRUCTURES

Sterols are structurally different from other lipids in that they consist of multi-ring structures (Figure 10.11). A sterol can either be free (not bonded to another molecule) or attached to a fatty acid via an ester linkage. The latter is called a **sterol ester**.

Functions of Cholesterol and Cholesteryl Esters There are many types of sterols, but the most abundant and widely discussed is **cholesterol.** Some free cholesterol is found in the body, although most is bonded to a fatty acid (recall Figure 10.2). This cholesterol–fatty acid complex is called a **cholesteryl ester** and is an example of a sterol ester. Because they contain fatty acids,

sterol A type of lipid with a distinctive multi-ring structure; a common example is cholesterol.

sterol ester A chemical compound consisting of a sterol molecule bonded to a fatty acid via an ester linkage.

cholesterol (*kholikos*, bile; *stereos*, hard or solid) A sterol found in animal foods and made in the body; required for bile acid and steroid hormone synthesis.

cholesteryl ester A sterol ester made of a cholesterol molecule bonded to a fatty acid via an ester linkage.

FIGURE 10.11 Structures of a Sterol, Cholesterol, and a Cholesteryl Ester

A sterol

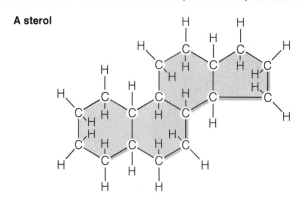

Cholesterol

The highlighted areas make this sterol a cholesterol molecule.

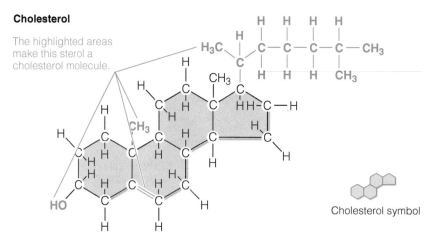

Cholesterol symbol

A cholesteryl ester

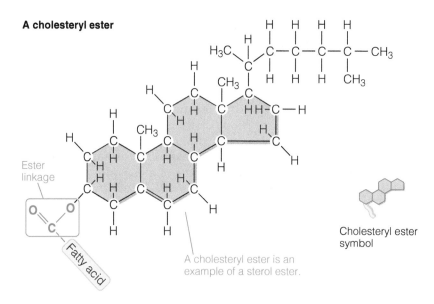

Ester linkage

Fatty acid

A cholesteryl ester is an example of a sterol ester.

Cholesteryl ester symbol

cholesteryl esters are more hydrophobic than is free cholesterol, which is weakly polar. The unhealthy relationship between cholesterol and heart disease is widely discussed, and you will learn more about this later in the chapter. You may not, however, be as familiar with the many essential roles cholesterol plays in the body.

For instance, cholesterol is needed to synthesize bile acids, which are important for the digestion and absorption of lipids. A **bile acid** consists of a

bile acid Amphipathic substance made from cholesterol in the liver; a component of bile important for lipid digestion and absorption.

cholesterol molecule attached to a very hydrophilic subunit, making it amphipathic, much like a phospholipid. Similar to phospholipids, cholesterol and cholesteryl esters are also components of membranes, such as those surrounding cells, where they help maintain fluidity. In addition, cholesterol is needed for the synthesis of the **steroid hormones,** which are important for reproduction, energy metabolism, calcium homeostasis and electrolyte balance. Cholesteryl esters also serve as crucial carrier molecules for fatty acids in the blood.

Sources of Cholesterol in the Body: Synthesis and Diet Almost every tissue in the body, especially the liver, can make cholesterol from glucose and fatty acids. Many dietary factors, however, influence how much cholesterol you make. For example, eating a low-calorie or low-carbohydrate diet can decrease cholesterol synthesis in some people.[15] This is not always the case, however, because dietary factors can interact with genetics to influence how much cholesterol a person makes. In other words, carbohydrate intake may only affect cholesterol synthesis in people with certain genetic variations. When dietary intervention is not effective in lowering blood cholesterol levels, medications are sometimes recommended. This is often the case for people at risk for heart disease. For example, the statin drugs Lipitor® (atorvastatin calcium) and Zocor® (simvastatin) decrease blood cholesterol by inhibiting one of the liver enzymes needed for its synthesis.

In addition to being made by the body, cholesterol is also obtained from animal-derived foods, such as shellfish, meat, butter, eggs, and liver (Figure 10.12). Plants do not produce substantial amounts of cholesterol, so exclusively plant-based foods are relatively low in cholesterol. However, because cholesterol is made by the body, vegans who eat no animal products are not at risk of cholesterol deficiency.

Animal-derived foods provide the vast majority of cholesterol to the diet.

steroid hormone (*stereos*, hard or solid; *hormon*, to urge on) A hormone made from cholesterol.

FIGURE 10.12 Cholesterol Content of Selected Foods

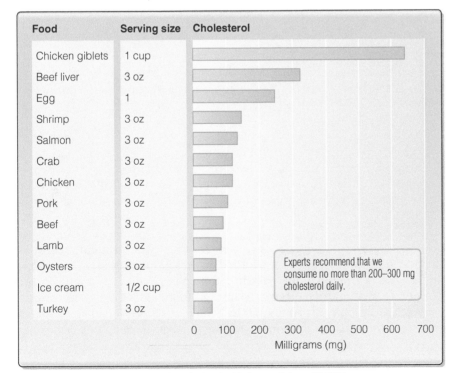

Food	Serving size	Cholesterol
Chicken giblets	1 cup	
Beef liver	3 oz	
Egg	1	
Shrimp	3 oz	
Salmon	3 oz	
Crab	3 oz	
Chicken	3 oz	
Pork	3 oz	
Beef	3 oz	
Lamb	3 oz	
Oysters	3 oz	
Ice cream	1/2 cup	
Turkey	3 oz	

Experts recommend that we consume no more than 200–300 mg cholesterol daily.

Milligrams (mg)

SOURCE: USDA National Nutrition Database for Standard Reference, Release 17, 2005.

Phytosterols and Phytostanols Are Sterol-Like Plant Compounds Although plants make very small amounts of cholesterol, some contain relatively large amounts of sterol-like compounds called **phytosterols** and **phytostanols.** One interesting group of phytosterols is found naturally in corn, wheat, rye, and other plants. Similar compounds are also produced commercially and are marketed under various names, such as Benecol®. These products are often found on the grocery shelf in butter substitutes, yogurt drinks, salad dressings, and even dietary supplements. Some studies suggest that consuming products containing phytosterols or stanols may decrease blood cholesterol, lowering the risk for cardiovascular disease.[16] As such, the FDA has approved the following health claim.

> *"Diets low in saturated fat and cholesterol that include two servings of foods that provide a daily total of at least 3.4 grams of plant sterols/stanols in two meals may reduce the risk of heart disease."*[17]

Because a typical serving of a plant sterol–fortified table spread contains about 1.1 gram of the sterol, you would need to consume about three servings daily to reach this goal. The mechanisms by which plant sterols/stanols decrease blood cholesterol are not well understood.[16] However, because they are not readily absorbed and appear to bind cholesterol in the intestine, consumption of sterols and stanols may increase cholesterol elimination in the feces.

How Are Dietary Lipids Digested?

Once lipids have been ingested, they must be digested, absorbed, and circulated away from the small intestine. As you will learn, digestion of the various types of dietary lipids is accomplished by enzymes and other secretions produced by the gastrointestinal tract and accessory organs.

DIGESTION OF TRIGLYCERIDES REQUIRES BILE AND LIPASES

The basic goal of triglyceride digestion is to cleave two of the fatty acids from the glycerol backbone. This is accomplished in your mouth, stomach, and small intestine when you eat a meal containing fats or oils. An overview of lipid digestion is illustrated in Figure 10.13 and described next.

A Small Portion of Triglyceride Digestion Occurs in Your Mouth The first stage of triglyceride digestion begins in the mouth. As chewing breaks apart food, **lingual lipase** (an enzyme produced by your salivary glands) begins to hydrolyze fatty acids from glycerol molecules. After the food is swallowed, lingual lipase accompanies the bolus into your stomach, where this enzyme continues to function.

Triglyceride Digestion Continues in Your Stomach The second stage of triglyceride digestion begins when food enters your stomach, stimulating the release of the hormone gastrin from specialized cells found in the gastric pits. Gastrin in turn circulates in the blood, where it quickly stimulates the release of the enzyme **gastric lipase,** also produced in stomach cells. Gastric lipase is a component of the "gastric juices" and continues where lingual lipase left off, breaking the bonds that attach fatty acids to glycerol molecules.

⟨**CONNECTIONS**⟩ Remember that hydrolysis is the breaking of chemical bonds by the addition of water. As a result, larger compounds are broken down into smaller subunits (Chapter 1, page 8).

phytosterol (*phuto*, plant) Sterol made by plants.

phytostanol Sterol-like compound made by plants.

lingual lipase (*lingua*, tongue; *lipos*, fat) An enzyme, produced in the salivary glands, that hydrolyzes ester linkages between fatty acids and glycerol molecules.

gastric lipase (*gaster*, belly; *lipos*, fat) An enzyme, produced in the stomach, that hydrolyzes ester linkages between fatty acids and glycerol molecules.

FIGURE 10.13 Overview of Triglyceride Digestion

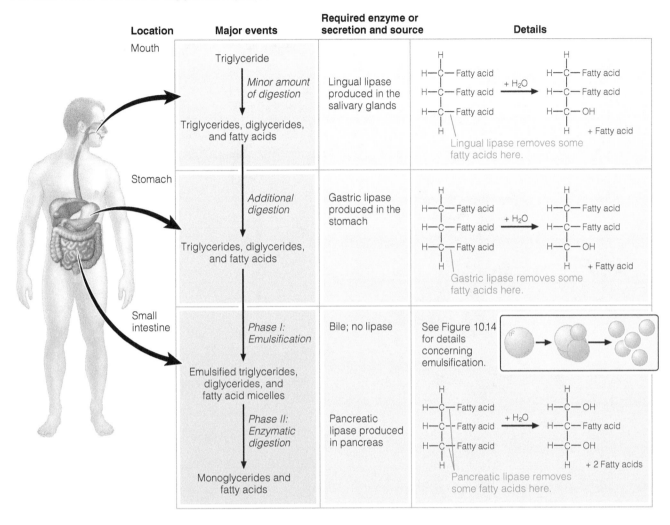

Location	Major events	Required enzyme or secretion and source	Details
Mouth	Triglyceride → *Minor amount of digestion* → Triglycerides, diglycerides, and fatty acids	Lingual lipase produced in the salivary glands	Lingual lipase removes some fatty acids here.
Stomach	*Additional digestion* → Triglycerides, diglycerides, and fatty acids	Gastric lipase produced in the stomach	Gastric lipase removes some fatty acids here.
Small intestine	*Phase I: Emulsification* → Emulsified triglycerides, diglycerides, and fatty acid micelles	Bile; no lipase	See Figure 10.14 for details concerning emulsification.
	Phase II: Enzymatic digestion → Monoglycerides and fatty acids	Pancreatic lipase produced in pancreas	Pancreatic lipase removes some fatty acids here.

Triglyceride Digestion Is Completed in Your Small Intestine Although some fatty acids are removed from the glycerol backbones in your stomach via gastric lipase, triglyceride digestion in the stomach is incomplete. This is in part because the watery environment of your GI tract causes lipids to clump together in large lipid globules, which are difficult to digest. To overcome this problem, the final stage of triglyceride digestion in your small intestine occurs in two complementary and consecutive phases.

Phase I: Emulsification of Lipids by Bile—Micelle Formation Phase I of intestinal lipid digestion begins with the arrival of lipids into the small intestine (Figure 10.14). This stimulates the release of the enteric hormone cholecystokinin (CCK), which in turn signals the gallbladder to contract and release bile. Recall that bile consists of a mixture of bile acids, cholesterol, and phospholipids. When bile acids and phospholipids are released into the duodenum, their hydrophobic portions are drawn toward the lipid globules, while their hydrophilic portions pull in the opposite direction, toward the surrounding water. These opposing forces disperse the large lipid globules into smaller droplets, a process called **emulsification.** Emulsification makes the ester linkages more accessible to the digestive enzymes in the small intestine.

emulsification The process whereby large lipid globules are broken down and stabilized into smaller lipid droplets.

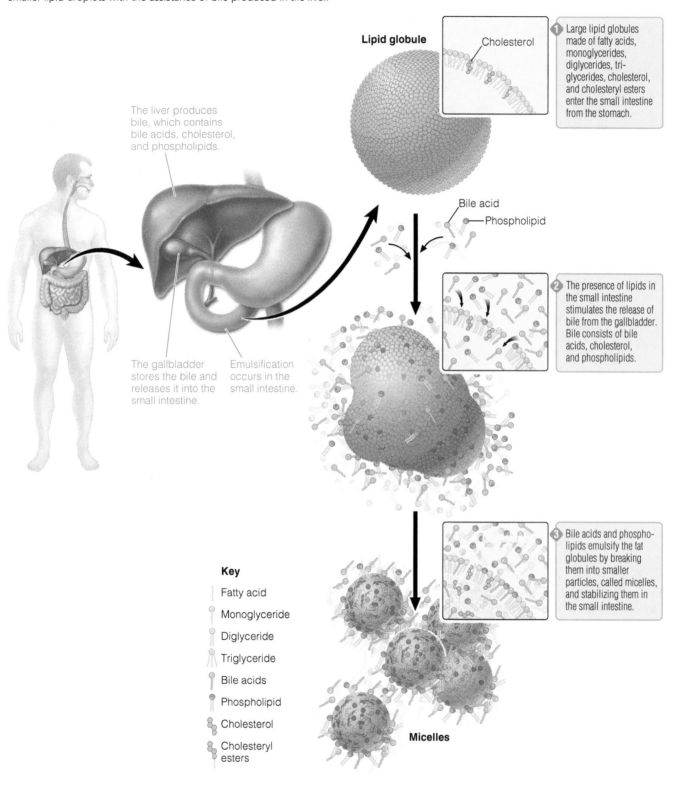

FIGURE 10.14 Emulsification of Lipids in the Small Intestine When lipid globules enter the small intestine, they are emulsified into smaller lipid droplets with the assistance of bile produced in the liver.

The liver produces bile, which contains bile acids, cholesterol, and phospholipids.

The gallbladder stores the bile and releases it into the small intestine.

Emulsification occurs in the small intestine.

Lipid globule

Cholesterol

Bile acid

Phospholipid

1 Large lipid globules made of fatty acids, monoglycerides, diglycerides, tri-glycerides, cholesterol, and cholesteryl esters enter the small intestine from the stomach.

2 The presence of lipids in the small intestine stimulates the release of bile from the gallbladder. Bile consists of bile acids, cholesterol, and phospholipids.

3 Bile acids and phospho-lipids emulsify the fat globules by breaking them into smaller particles, called micelles, and stabilizing them in the small intestine.

Key

Fatty acid

Monoglyceride

Diglyceride

Triglyceride

Bile acids

Phospholipid

Cholesterol

Cholesteryl esters

Micelles

micelle A water-soluble, spherical structure formed in the small intestine via emulsification.

Bile also disperses phospholipids and cholesteryl esters in much the same way. Bile acids and phospholipids then stay with the newly formed drop-lets, which are now referred to as **micelles.** Because they are coated with amphipathic substances, micelles do not gel together to reform larger lipid globules.

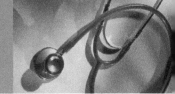

FOCUS ON CLINICAL APPLICATIONS
Gallblader Disease and Gallstones

Gallbladder disease can result when the gallbladder becomes inflamed or when bile contains an excess amount of cholesterol in relation to its other components. The accumulation of calcium and cellular debris around the cholesterol results in the formation of gallstones, which can range in size from 5 mm to more than 25 mm in diameter. For a point of reference, 1 millimeter is roughly the thickness of a dime. Some people with gallstones have no symptoms, while others can experience tenderness or extreme pain. Gallstones can also become lodged in the cystic or common bile ducts leading from the gallbladder to the small intestine, obstructing the flow of bile and pancreatic secretions into the intestine. Surgical removal of the gallbladder is the most common treatment for persistent gallstone-related problems. Because some people may have difficulty digesting fat after the gallbladder is removed, doctors often recommend that people initially avoid high-fat meals. However, the liver continues to supply bile to the small intestine even after the gallbladder is removed, so most people do not have to adjust their diet for very long.

Gallbladder disease is more common in women than men, and its prevalence increases with age.[18] Other risk factors include obesity, rapid weight loss, and pregnancy. It remains unclear if particular foods or dietary practices influence the formation of gallstones, but some studies report the prevalence to be lower in vegetarians than nonvegetarians.[19] In addition, moderate alcohol consumption (1–2 drinks/day), exercise, and aspirin use appear to have protective roles.

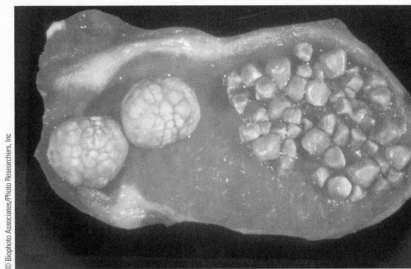

Gallstones are composed of calcium, cellular debris, and cholesterol.

Another way to think about lipid emulsification is to consider what happens when you shake water and cooking oil together with and without added soap. If you do not add soap, the water and oil initially appear to mix, but the oil eventually separates from the water. This is because oil is nonpolar (not charged), whereas water is polar (charged), and nonpolar and polar compounds do not willingly mix. However, if you add a few drops of dish soap and shake the mixture again, the oil becomes suspended as small droplets within the water. This is because soap is an amphipathic emulsifier, breaking up large lipid droplets into smaller ones and allowing them to disperse. Although most people never need to think about the importance of bile in lipid digestion, those with gallbladder disease fully understand. The Focus on Clinical Applications feature describes this disease in more detail.

The emulsification of lipids in the small intestine by bile acids is similar to the way dish soap makes grease combine with water.

Critical Thinking: Nancy's Story Recall Nancy, the woman featured at the beginning of the chapter. Now that you understand the importance of bile in the process of lipid digestion, can you explain why eating high-fat foods caused her to experience indigestion and abdominal pain? After her gallbladder surgery, why was it so important that she add back lipids *very slowly* to her diet?

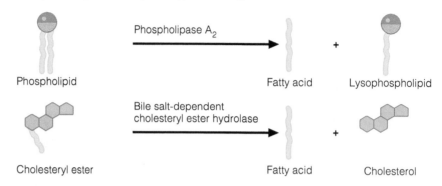

FIGURE 10.15 Phospholipid and Cholesteryl Ester Digestion Both phospholipids and cholesteryl esters rely on pancreatic enzymes (phospholipase A₂ and bile salt-dependent cholesteryl ester hydrolase, respectively) for their digestion in the small intestine.

Phospholipase A₂

Phospholipid

Fatty acid

Lysophospholipid

Bile salt-dependent cholesteryl ester hydrolase

Cholesteryl ester

Fatty acid

Cholesterol

Phase II: Digestion of Triglycerides by Pancreatic Lipase In response to lipid-containing chyme entering the duodenum, the small intestine releases the hormones secretin and cholecystokinin (CCK). These enteric hormones signal the pancreas to release pancreatic juices containing the enzyme **pancreatic lipase.** Pancreatic lipase completes triglyceride digestion by hydrolyzing additional fatty acids from glycerol molecules in the micelles. In general, two of the three fatty acids are removed from the triglyceride molecules, resulting in the release of a monoglyceride and two free (unbound) fatty acids.

DIGESTION OF PHOSPHOLIPIDS AND CHOLESTERYL ESTERS ALSO REQUIRES PANCREATIC ENZYMES

As is the case with triglycerides, very little digestion of the other major fatty acid-containing lipids (phospholipids and cholesteryl esters) occurs in the mouth and stomach. Instead, their digestion begins in the small intestine, where—like the triglycerides—phospholipids and cholesteryl esters contained in lipid globules are emulsified into micelles by bile. Phospholipids are then digested by the enzyme **phospholipase A₂,** produced by the pancreas and released in response to the intestinal hormone secretin. The products of digestion of a single phospholipid molecule are one fatty acid and a compound called a **lysophospholipid,** consisting of a glycerol molecule bonded to a fatty acid and a polar head group (Figure 10.15).

Although free cholesterol (not bonded to a fatty acid) does not need to be digested prior to absorption, cholesteryl esters must be broken down into cholesterol and fatty acids. This is accomplished by yet another pancreatic enzyme: **bile salt–dependent cholesteryl ester hydrolase.**

As you have learned, the pancreas is vitally important for the body to digest the many lipids found in foods. But for these lipids to be used by the body, they must be absorbed from the gastrointestinal tract and circulated to other tissues. These processes are described next.

pancreatic lipase An enzyme, produced in the pancreas, that hydrolyzes ester linkages between fatty acids and glycerol molecules.

phospholipase A₂ An enzyme, produced in the pancreas, that hydrolyzes fatty acids from phospholipids.

lysophospholipid A lipid composed of a glycerol bonded to a polar head group and a fatty acid; one of the final products of phospholipid digestion.

bile salt–dependent cholesteryl ester hydrolase An enzyme, produced in the pancreas, that cleaves fatty acids from cholesteryl esters.

How Are Dietary Lipids Absorbed and Circulated in the Body?

The products of lipid digestion are absorbed into the enterocytes and circulated away from the small intestine. This requires special handling because many of the products of digestion are hydrophobic, while both the interior of the enterocyte and the circulatory system are hydrophilic. It is the amphipathic properties of phospholipids that make both lipid absorption and circulation possible.

DIETARY LIPIDS ARE ABSORBED IN THE SMALL INTESTINE

Lipid absorption is accomplished in two ways, depending on how water-soluble (hydrophilic) the lipid is (Figure 10.16). Because they are relatively water soluble, short- and medium-chain fatty acids can be transported into the enterocytes unassisted. However, more hydrophobic compounds such as long-chain fatty acids, monoglycerides, and cholesterol must first be packaged into a form of micelle that can deliver them to the enterocytes. Once this micelle comes into contact with the brush border surface, its contents are released and transported into the enterocytes.

DIETARY LIPIDS ARE CIRCULATED AWAY FROM THE SMALL INTESTINE IN TWO WAYS

The process of lipid circulation in the body is complicated because these hydrophobic substances must somehow be transported in an overwhelmingly watery environment. Depending on how hydrophobic they are, lipids are initially circulated away from the small intestine in either the blood or the lymph.

Circulation of Relatively Hydrophilic Lipids in Your Blood The simplest case of dietary lipid circulation involves the short- and medium-chain fatty acids, which are somewhat water soluble (hydrophilic). As such, these fatty acids

FIGURE 10.16 Absorption and Circulation of Lipids in the Small Intestine How various lipids are absorbed and circulated depends on how hydrophilic they are. In general, it is easier for the body to absorb and circulate more hydrophilic lipids (such as short- and medium-chain fatty acids) than more hydrophobic lipids (such as long-chain fatty acids and monoglycerides).

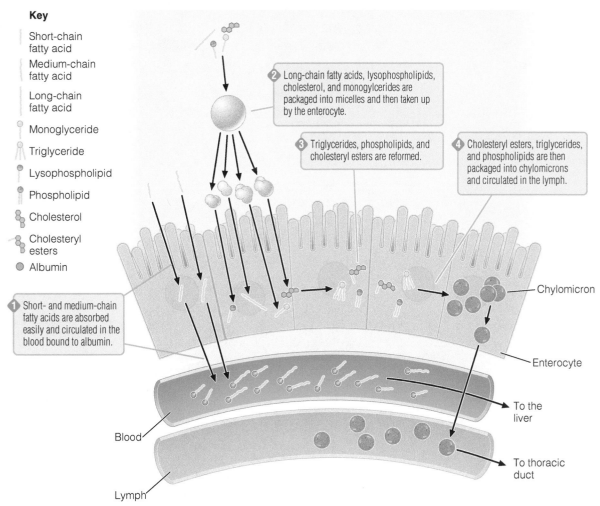

Key

- | Short-chain fatty acid
- | Medium-chain fatty acid
- | Long-chain fatty acid
- Monoglyceride
- Triglyceride
- Lysophospholipid
- Phospholipid
- Cholesterol
- Cholesteryl esters
- Albumin

2 Long-chain fatty acids, lysophospholipids, cholesterol, and monoglycerides are packaged into micelles and then taken up by the enterocyte.

3 Triglycerides, phospholipids, and cholesteryl esters are reformed.

4 Cholesteryl esters, triglycerides, and phospholipids are then packaged into chylomicrons and circulated in the lymph.

1 Short- and medium-chain fatty acids are absorbed easily and circulated in the blood bound to albumin.

Chylomicron

Enterocyte

To the liver

To thoracic duct

Blood

Lymph

can be circulated away from your small intestine in the blood. However, they are first bound to the protein albumin. Fatty acid–albumin complexes circulate in the blood from the small intestine to the liver, where they are either metabolized or rerouted for delivery to other cells in the body.

Circulation of More Hydrophobic Lipids in Lymph via Chylomicrons The circulation of larger lipids away from the GI tract is more involved. Long-chain fatty acids, monoglycerides, and lysophospholipids (products of phospholipid digestion) first enter the enterocyte, where they are reassembled into triglycerides and phospholipids. These large lipids, along with cholesterol and cholesteryl esters, are then incorporated into particles called **chylomicrons** (also called chylomicra), which are released into the lymph for initial circulation. Chylomicrons package their hydrophobic lipids (such as the triglycerides) within a hydrophilic exterior or "shell" formed mainly by the hydrophilic head groups of the phospholipids (Figure 10.17). The lymph carrying the

⟨**CONNECTIONS**⟩ Lymph is the fluid circulating in the lymphatic system (Chapter 1, page 17).

chylomicron A lipoprotein, made in the enterocyte, that transports large lipids away from the small intestine in the lymph.

FIGURE 10.17 The Lipoproteins The ratio of lipids to proteins determines a lipoprotein's density and, for most, its name.

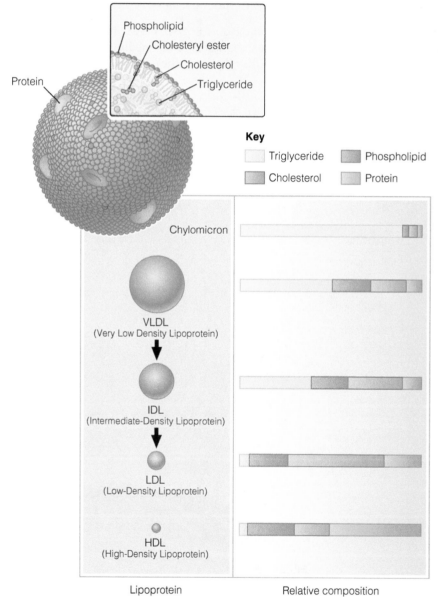

Adapted from: Skipski VP. Blood lipids and lipoproteins. Quantitation, composition, and metabolism. Nelson GJ, editor. 1972. Wiley-Interscience, New York; Christie W. American Oil Chemist Society Lipid Library. Lipoproteins. Available from: http://lipidlibrary.aocs.org/lipids/lipoprot/index.htm.

chylomicrons eventually mixes with the blood via the thoracic duct in the neck region. In this way, chylomicrons gradually enter the bloodstream, where they travel to cells that take up their contents. Note that chylomicrons are an example of a lipoprotein, which the body makes to transport dietary lipids. You will learn more about lipoproteins in the next section.

Chylomicrons in the blood deliver dietary fatty acids to cells via the enzyme lipoprotein lipase. Lipoprotein lipase is produced in many tissues, especially adipose and muscle. After it is produced in the cell's cytoplasm, this enzyme is relocated out of the cell and into the lumen (inside) of the neighboring capillary blood vessels. As the chylomicrons circulate in the blood, they are "attacked" by lipoprotein lipase. This enzymatic action releases fatty acids from the chylomicrons' triglycerides, allowing their uptake into surrounding cells. After delivering dietary fatty acids to cells, the chylomicron fragments remaining in the blood are called **chylomicron remnants.** These triglyceride-depleted particles are taken up by the liver, where they are broken down and their contents reused or recycled.

What Is the Role of Other Lipoproteins in Lipid Transport and Delivery?

As just described, small- and medium-chain dietary lipids are circulated in the blood away from the small intestine to the liver. The larger, more hydrophobic lipids are packaged into chylomicrons, circulating first in the lymph and then in the bloodstream and ultimately delivering dietary fatty acids to all of the tissues that need them. Eventually, fatty acid-depleted chylomicron remnants return to the liver, where they are broken down. As such, your liver serves as both the "central command center" and "recycling center" for lipid metabolism, receiving dietary lipids, as well as synthesizing and metabolizing other lipids as needed. To deliver these newly synthesized lipids as well as some dietary lipids to the body, the liver makes a series of lipoproteins that circulate in the blood. A summary of the origins and functions of the various lipoproteins (including the chylomicrons) is presented in Figure 10.18 and described next.

LIPOPROTEINS CONTAIN LIPIDS IN THEIR CORES

Because most lipids are very hydrophobic, their transport in the hydrophilic blood is somewhat complex. To aid in this process, the liver produces particles called **lipoproteins,** whose job is to transport lipids in the blood. Lipoproteins are complex globular structures containing varying amounts of triglycerides, phospholipids, cholesteryl esters, cholesterol, and proteins. The proteins embedded within the outer shell of the lipoproteins are called **apoproteins** or apolipoproteins. Like chylomicrons made in the small intestine, the lipoproteins made by the liver are also constructed so that their hydrophilic components (such as the apoproteins and the polar head groups of phospholipids) are situated on the outer surface, and their hydrophobic components (such as triglycerides) are facing inward (recall Figure 10.17).

You have already learned about one type of lipoprotein—the chylomicron, which is the largest and least dense member of the lipoprotein family. Whereas chylomicrons exclusively transport *dietary* lipids away from the small intestine, other lipoproteins carry lipids originating in various tissues and organs. For example, lipoproteins carry lipids that have been synthesized in the liver and lipids released from storage in adipose tissue.

chylomicron remnant The lipoprotein particle that remains after a chylomicron has lost most of its fatty acids.

lipoprotein A spherical particle made of varying amounts of triglycerides, cholesterol, cholesteryl esters, phospholipids, and proteins.

apoproteins Proteins embedded in the surface of lipoproteins.

FIGURE 10.18 The Origins and Major Functions of Lipoproteins
Both the liver and the small intestine make lipoproteins that circulate lipids in the body.

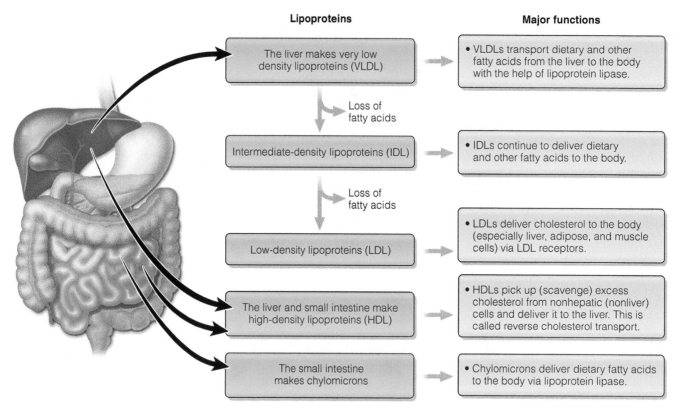

Lipoproteins

Major functions

The liver makes very low density lipoproteins (VLDL)
- VLDLs transport dietary and other fatty acids from the liver to the body with the help of lipoprotein lipase.

Loss of fatty acids

Intermediate-density lipoproteins (IDL)
- IDLs continue to deliver dietary and other fatty acids to the body.

Loss of fatty acids

Low-density lipoproteins (LDL)
- LDLs deliver cholesterol to the body (especially liver, adipose, and muscle cells) via LDL receptors.

The liver and small intestine make high-density lipoproteins (HDL)
- HDLs pick up (scavenge) excess cholesterol from nonhepatic (nonliver) cells and deliver it to the liver. This is called reverse cholesterol transport.

The small intestine makes chylomicrons
- Chylomicrons deliver dietary fatty acids to the body via lipoprotein lipase.

Because lipid is less dense than protein, the densities of the lipoproteins depend on their relative amounts (or percentages) of lipids and apoproteins. Lipoproteins with relatively more lipid than protein have lower densities than those with more protein and less lipid. With the exception of chylomicrons, lipoproteins are named according to their densities.

Fatty Acid Delivery by Very Low Density Lipoproteins (VLDLs) Whereas the small intestine produces chylomicrons and small amounts of other lipoproteins, the liver produces most of the lipoproteins responsible for lipid delivery in the body. One such lipoprotein is called a **very low density lipoprotein (VLDL).** VLDLs are similar to chylomicrons in that they contain triglycerides and cholesteryl esters in their cores surrounded by phospholipids, free cholesterol, and apoproteins on their surfaces. However, VLDLs have a lower lipid-to-protein ratio than do chylomicrons, making them smaller and denser. Like chylomicrons, the primary function of VLDLs is to deliver fatty acids to cells via the enzymatic action of lipoprotein lipase. However, unlike chylomicrons, which only deliver *dietary* fatty acids to the body, VLDLs also deliver fatty acids derived from liver and adipose tissue. When VLDLs in the blood circulate past cells that produce lipoprotein lipase, the enzyme cleaves fatty acids, which in turn are taken up by the surrounding cells.

Intermediate-Density Lipoproteins (IDLs) and Low-Density Lipoproteins (LDLs)
As VLDLs lose fatty acids via the action of lipoprotein lipase, they become denser and are called **intermediate-density lipoproteins (IDLs).** Some IDLs

very low density lipoprotein (VLDL) A lipoprotein, made by the liver, that contains a large amount of triglyceride; its major function is to deliver fatty acids to cells.

intermediate-density lipoprotein (IDL) A lipoprotein that results from the loss of fatty acids from a VLDL; many IDLs are ultimately converted to LDLs.

are taken up by the liver, whereas others remain in the circulation, where they continue to lose additional fatty acids. Eventually, most IDLs become cholesterol-rich **low-density lipoproteins (LDLs).** Specialized proteins called **LDL receptors** on cell membranes—especially those of liver, adipose tissue, and muscle—bind to the LDLs' apoproteins allowing the LDLs to be taken up and broken down by the cell. In this way, cholesterol is delivered to many tissues that use it for structural and metabolic purposes.

LDLs can also be taken up and degraded by white blood cells (macrophages) that have been drawn to a major blood vessel due to some sort of injury or infection as part of the body's inflammatory response. Some uptake of LDLs into macrophages is necessary, because the cells use the LDL's contents to synthesize important substances such as the eicosanoids and immune factors. However, uptake of too much LDL can result in buildup of a fatty substance called **plaque** within the vessel wall, slowing or even blocking blood flow. Epidemiologic studies suggest that high levels of LDL cholesterol (LDL-C) in the blood are related to increased risk for cardiovascular disease.[20] Thus, LDL-C has been deemed "bad cholesterol." More detail concerning LDL-C, plaque formation, and cardiovascular disease is presented in the Nutrition Matters at the end of this chapter.

Effect of Diet on LDL Cholesterol Concentrations Consuming high amounts of certain SFAs or *trans* fatty acids (particularly those in partially hydrogenated oils) can increase circulating LDL-C concentration in some people.[21] The exact mechanism by which this occurs is not known, although it may be related to variations in apoprotein production.[22] In some cases, high cholesterol intake is also associated with high LDL-C concentration.[23] The relationship between cholesterol intake and LDL-C is complex, however, because some people show no effect of high cholesterol intake on LDL-C levels. This is most likely because of genetic (or epigenetic) variation related to cholesterol absorption, metabolism, and other dietary and lifestyle factors (such as smoking) that also influence circulating LDL-C levels.[24] Conversely, diets high in PUFAs, ω-3 fatty acids, or dietary fiber can lower LDL-C levels in some people.[25] Clearly, many factors influence the concentration of LDL-C in a person's blood, with different diets having different effects on various people.

Cholesterol Uptake by High-Density Lipoproteins (HDLs) The liver and, to a lesser extent, the small intestine make another series of lipoproteins called **high-density lipoproteins (HDLs).** Compared with other lipoproteins, HDLs have the lowest lipid-to-protein ratio; thus, they have the highest densities. HDLs salvage excess cholesterol from cells, transporting it back to the liver. This transfer of cholesterol from nonhepatic (nonliver) cells back to the liver is called **reverse cholesterol transport.**[26] It is well established that high levels of HDL cholesterol (HDL-C) in blood are generally associated with lower risk for cardiovascular disease.[27] This is why HDL-C is often referred to as the "good cholesterol."

There are several types of HDL, and not all forms are equally effective in removing excess cholesterol. More specifically, different HDLs have different apoproteins, resulting in somewhat different functions. The presence of particular apoproteins makes some HDLs less efficient at reverse cholesterol transport than others.

Effects of Diet on HDL Cholesterol Concentrations Although it has long been thought that diets high in carbohydrates offer protection from cardiovascular disease, a considerable amount of research now suggests that these diets

low-density lipoprotein (LDL) A lipoprotein that delivers cholesterol to cells.

LDL receptor Membrane-bound protein that binds LDLs, causing them to be taken up and dismantled.

plaque A complex of cholesterol, fatty acids, cells, cellular debris, and calcium that can form inside blood vessels and within vessel walls.

high-density lipoprotein (HDL) A lipoprotein, made primarily by the liver, that circulates in the blood to collect excess cholesterol from cells.

reverse cholesterol transport Process whereby HDLs remove cholesterol from nonhepatic (nonliver) tissues for transport to the liver.

can sometimes lower HDL-C levels.[28] In other words, very high carbohydrate diets may actually increase risk for heart disease in some people. Conversely, research suggests that high MUFA intake or moderate alcohol consumption (1–2 drinks/day) can raise HDL-C levels in some people, lowering risk for cardiovascular disease.[29] Like LDL-C, many lifestyle and genetic factors influence HDL-C concentrations, and researchers continue to study these important interactions.

What Is the Relationship between Lipid Intake and Health?

After lipids have been digested, absorbed, and circulated, they can be used by the body for myriad purposes. As described previously, fatty acids and their metabolites regulate metabolic processes within cells and orchestrate a variety of physiological responses. Phospholipids are vital components of membranes, aid in lipid digestion and absorption, and also contribute fatty acids for cellular use. Cholesterol is incorporated into cell membranes, acts as a precursor for some hormones, and is involved in lipid digestion and absorption via its role in bile.

Although many lipids are vital for good health, certain types of lipids can be associated with poor health. For example, high dietary intake of some lipids is associated with increased risks for cardiovascular disease and certain cancers. These topics are discussed briefly here, but you will also learn more about them elsewhere in this book.

EXCESS LIPID INTAKE CAN LEAD TO OBESITY

Although consumption of a high-fat diet can increase risk for obesity, there are many other factors (such as genetics and exercise) that contribute equally to this health problem.

Obesity is defined as the overabundance of body fat, and its causes and consequences are described in detail in Chapter 11. Although obesity is complex, excess energy intake is a major factor. Because fatty acids provide more than twice as many calories per gram as carbohydrates and proteins, high fat intake is likely an important piece of the obesity puzzle. Whatever its causes, obesity is a major public health concern worldwide and is associated with increased risk for many diseases such as cardiovascular disease, type 2 diabetes, and some forms of cancer. In order to decrease the risk for obesity, experts recommend that we limit our energy intake. In response to consumer demand, many food manufacturers produce low-fat and fat-free items in addition to foods that contain fat substitutes. You can read more about fat substitutes in the following Focus on Food feature.

DIETARY LIPIDS MAY BE RELATED TO RISK OF CARDIOVASCULAR DISEASE

Like obesity, cardiovascular disease is caused by a complex web of factors, including genetics, physical inactivity, and poor diet. Cardiovascular disease risk is not only influenced by *total* dietary lipid intake (and obesity), but may also be affected by specific *types* of dietary lipids. As noted, high intakes of SFAs, *trans* fatty acids, and cholesterol can increase the risk for disease in some people, whereas MUFAs may have the opposite effect. However, a person's genetic makeup often interacts with his or her diet to influence health, and this is certainly true for lipids and cardiovascular disease.

Because many people want to consume "low-fat" diets, some food manufacturers have developed substances that have desired characteristics of lipids but with fewer or no calories. These "fat substitutes" are diverse in structure, some made from complex carbohydrates, some made from proteins, and some made from blends of carbohydrates and fatty acids.

Olestra is an example of a fat substitute made from sucrose (table sugar) bonded to six to eight fatty acids. Olestra cannot be digested by human lipases or colonic bacteria, and therefore provides no usable energy to the body. In fact, it passes through the small intestine relatively intact. In 1996, the FDA approved the use of olestra in savory snacks such as potato chips, cheese puffs, and crackers. Because olestra interferes with the absorption of the fat-soluble vitamins in the GI tract, the FDA requires that food manufacturers add vitamins A, D, E, and K to olestra-containing foods. There was initial concern that consuming large quantities of olestra would cause intestinal distress such as gas and diarrhea. However, review of many human studies led the FDA to conclude that, when people consume reasonable portions of olestra-containing foods, they are likely to experience only infrequent and mild gastrointestinal upset.[30] Thus, in most situations, consuming olestra poses no concern. However, as with any food, one should consume those containing olestra or other fat substitutes in moderation.

Other fat substitutes are made from carbohydrates or protein. Examples commonly used by food manufacturers include cellulose, vegetable gum fibers (for example, guar gum), Maltrin®, and Stellar®. These substances are used in many low-fat or fat-free foods such as salad dressings, table spreads, and frozen desserts. Some of these ingredients are not heat-stable, and therefore they cannot be used in foods that must be baked or fried. Note that, although cellulose and vegetable gums are considered dietary fiber and therefore do not contribute calories to the diet, Maltrin® and Stellar® are glucose polymers, providing 4 kcal per gram. Another fat substitute is Simplesse®, which is made from milk protein. Like Maltrin® and Stellar®, this substance contains 4 kcal per gram. Thus, foods containing these products may be "fat free," but they are certainly not "calorie free."

Fat-free ice cream often contains fat substitutes such as gum fiber.

© 2011 J&L Images/Jupiterimages Corporation

Many studies have shown that consuming diets high in cholesterol increases blood cholesterol and risk for cardiovascular disease.[21] However, some people can eat very high amounts of cholesterol-rich foods without experiencing this effect. This difference may arise because some people absorb dietary cholesterol more efficiently than others.[31] Similarly, genetic variation can influence a person's HDL and LDL levels and functionality. For example, the ability of cells to take up LDL particles (and thus form plaque) can be influenced by variations in the genes that code for LDL receptor proteins.[32] The Nobel Prize in Physiology or Medicine was awarded in 1985 to Michael Brown and Joseph Goldstein for this discovery.[33] As scientists learn more about how genetics interacts with diet to influence health (the field of nutrigenomics), health professionals may someday be able to "prescribe" the most heart-healthful diet given a person's individual genetic makeup.

THE RELATIONSHIP BETWEEN DIETARY LIPIDS AND CANCER IS UNCLEAR

Some studies show a link between high-fat diets and risk for cancer, but the data are inconclusive.[34] Obesity, however, is a risk factor for several types of cancer, including breast and colorectal cancers.[35] As obesity is often associated with consumption of high-fat diets, dietary lipids may play an indirect role. In 2007 the World Cancer Research Fund and the American Institute for Cancer Research jointly put forth their overall recommendations concerning diet and cancer prevention.[36] These experts advised that, to decrease risk of cancer, we should be within the normal range of body weight and avoid weight gain and increases in waist circumference as we age. They also recommended that we limit consumption of energy-dense foods and consume high-fat "fast foods" sparingly, if at all.

What Are the Dietary Recommendations for Lipids?

© Rita Maas/Getty Images

Even high-fat foods can be part of a healthy diet if eaten in moderation.

Although we have dietary requirements for only the essential fatty acids, we rely on additional fatty acids as important sources of energy. However, consuming too much lipid—or the wrong kind of lipid—can be associated with health problems. It is therefore important to maintain an optimal balance of lipid intake throughout life. Recommendations concerning how to best choose dietary lipids to help promote optimal health are discussed next.

CONSUME ADEQUATE AMOUNTS OF THE ESSENTIAL FATTY ACIDS

It is important to consume adequate amounts of the essential fatty acids, and the Dietary Reference Intakes (DRIs) for them are presented inside the cover of this book. Adequate Intake levels (AIs) for linoleic acid are 17 and 12 g/day for adult males and females, respectively. As a reference, this is the amount contained in approximately 1¾ tablespoons of corn oil. For linolenic acid, the AIs are 1.6 and 1.1 g/day for adult males and females, respectively. This is the amount contained in about 1½ tablespoons of soybean oil.

PAY SPECIAL ATTENTION TO THE LONG-CHAIN OMEGA-3 FATTY ACIDS

To help lower risk for cardiovascular disease, many organizations urge consumption of additional omega-3 fatty acids in our diet. For instance, the American Heart Association recommends eating fish (particularly fatty fish) at least two times a week. Each serving should be at least 3.5 ounces of cooked fish or about 3/4 cup of flaked fish. Higher amounts are recommended for individuals who already have cardiovascular disease. The Dietary Guidelines for Americans recommend weekly fish consumption, partly because of the benefits of omega-3 fatty acids. The Dietary Guidelines also recommend that women who are pregnant or breastfeeding consume at least 8 and up to 12 ounces (2–3 servings) of a variety of seafood per week to ensure that intakes of the omega-3 fatty acids are adequate for fetal growth and development, as well as

〈**CONNECTIONS**〉 Dietary Reference Intakes (DRIs) are reference values for nutrient intake that meets the needs of about 97% of the population. When adequate information is available, Recommended Dietary Allowances (RDAs) are established, but when less information is available, Adequate Intake (AI) values are provided. Tolerable Upper Intake Levels (ULs) indicate intake levels that should not be exceeded.

for development during early infancy and childhood. Special care, however, should be taken so that only fish known to be low in mercury are consumed.

DIETARY GUIDELINES AND INSTITUTE OF MEDICINE RECOMMEND LIMITING SATURATED FATTY ACIDS

Scientists have long known that SFA intake is positively related to risk for cardiovascular disease for some people. As such, many dietary guidelines are aimed at decreasing SFA intake. For example, the Institute of Medicine recommends that "intake of SFAs should be minimized while consuming a nutritionally adequate diet." In addition, the Dietary Guidelines suggest that SFA should constitute no more than 10% of total calories, and the American Heart Association recommends that SFA intake represent 7% or less of total calories.[37] Reading food labels such as those shown in Figure 10.19 can help you keep track of your SFA intake. You can read more about general suggestions for decreasing SFA intake in the Food Matters feature, and the relationship between SFA and heart disease is discussed in the Nutrition Matters section following this chapter.

FIGURE 10.19 Using Nutrition Facts Panels to Determine Fat Intake You can tell how much total fat, saturated fat, and *trans* fat is in a food by checking its Nutrition Facts panel.

Working Toward the Goal: Getting the Right Lipids in Your Diet

The 2010 Dietary Guidelines recommend that we limit our intake of saturated fatty acids, *trans* fatty acids, and cholesterol. The following selection and preparation tips will help you reach this goal.

Saturated Fats

- Limit intake of animal fats (e.g., beef and butter fat) and tropical oils (e.g., coconut and palm oils).

Trans Fats

- Decrease intake of cookies, crackers, cakes, pastries, doughnuts, and french fries made with partially hydrogenated shortening.

- Use oils instead of partially hydrogenated shortening.

- Use tub or "*trans* fat–free" margarine instead of stick margarine—or use butter.

Cholesterol

- Limit intake of high-cholesterol foods such as liver, eggs, cheesecake, and custards.

- Replace high-fat animal products with lower-fat products, such as lean cuts of meat and fat-free or low-fat milk.

TRANS FATTY ACIDS AND CHOLESTEROL SHOULD BE MINIMIZED

Although there are no absolute guidelines as to how much *trans* fatty acids we should (or should not) consume, the Institute of Medicine recommends that people "minimize their intakes of *trans* fatty acids." Similarly, the Dietary Guidelines recommend that *trans* fatty acid intake should be "as low as possible while consuming a nutritionally adequate diet." These guidelines apply to commercially produced *trans* fatty acids, not to naturally occurring ones.

No DRIs are established for cholesterol. However, the 2010 Dietary Guidelines and American Heart Association recommend that we consume no more than 200–300 mg of cholesterol daily. This is the amount found in one to two eggs or two servings of beef. The Institute of Medicine recommends that "cholesterol intake should be minimized," and the American Heart Association recommends it be limited to <300 milligrams/day.

National Cholesterol Education Program Recommends the Therapeutic Lifestyle Change (TLC) Diet The National Cholesterol Education Program (NCEP), which is part of the National Institutes of Health, promotes the **Therapeutic Lifestyle Change (TLC) diet** for individuals with increased risk for cardiovascular disease. This diet recommends that less than 7% of your day's total calories come from SFA and that you consume no more than 200 mg of cholesterol daily. Note that this advice for cholesterol intake is similar to that put forth by the Dietary Guidelines. You can find out more about the TLC diet by visiting its interactive website at http://www.nhlbi.nih.gov/cgi-bin/chd/step2intro.cgi.

GUIDELINES SET FOR TOTAL LIPID CONSUMPTION

Therapeutic Lifestyle Changes (TLC) diet A set of heart-healthy diet recommendations put forth by the National Cholesterol Education Program.

The Institute of Medicine has not established RDA or AI values for total lipid intake, except during infancy when AIs are set at approximately 30 g per day. However, the AMDRs recommend that healthy adults consume

20 to 35% of their energy from lipid. Based on a caloric requirement of 2,000 kcal per day, a person should consume between 400 to 700 kcal from dietary lipids. Considering that 1 g of lipid contains about 9 kcal of energy, the daily lipid intake for most adults should be between 44 and 78 g. This amount is easy to obtain. For example, a typical day's menu—including three servings of low-fat milk, a bagel with cream cheese, a peanut butter sandwich, a serving of spaghetti with meatballs, and a salad with ranch dressing—contains about 56 g of lipid.

Critical Thinking: Nancy's Story What would you do if you suspected that you had gallbladder disease at this stage of your life? Who would you call at your university to discuss your symptoms and get the care you would need? What dietary changes could you make to decrease the complications of gallbladder disease until you could be seen by a medical professional?

Diet Analysis PLUS ✚ Activity

How Healthy Is Your Fat Intake?

Part A: Total Fat

1. Your AMDR for total fat can be stated in many ways, which can often be confusing. The primary way that it can be represented is as a **percentage of calories.** Referring to your *Profile DRI Goals* report, you will see in italics the recommended percentage of 20–35% of total calories. To manually calculate your percentage, fill in the blanks below:

   _____ g × 9 kcal/gram = _____ kcal of total fat you consumed *(Grams of fat you consumed)*

   _____ kcals of total fat you consumed ÷ _____ kcals × 100 = _____ % *(Total kcals you consumed)*

2. Does your percentage for *total fat* fall between the AMDR for total fat of 20–35%? _____

3. Look at your individual day's *Intake Spreadsheets*. List the 10 foods having the highest content of total fat.

Part B: Types of Fat

Refer to your *Fat as Percentage of Total Calories* report to answer the first two questions.

1. Add up the percentages of all of the types of fat you consumed. What is the value? _____%

2. Does the percent you calculated in #1 equal the total fat percent on your Energy Nutrient Intake and AMDR Goals Ranges Compared report? _____

3. Using Chapter 10 in your text and for *each* TYPE of fat you consumed, list two significant food sources.

Part C: Saturated Fat

The Institute of Medicine recommends that less than 10% of your total calories come from saturated fat.

1. Referring to your *Intake and DRI Goals Compared* report, how many GRAMS of saturated fat did you consume? _____ g

2. Calculate the percent of saturated fat calories using the formula below:

   _____ g saturated fat × 9 kcal/g = _____ kcal saturated fat

   _____ kcal of saturated fat ÷ _____ × 100 = _____ % saturated fat *Total kilocalories consumed*

3. Compare your answer in #2 for saturated fat with that on *Fat as Percentage of Total Calories* report. Are they the same? _____

4. Is the percent of saturated fat in your diet less than 10% of calories? _____

5. Looking at your individual days' Intake Spreadsheets, identify the one food that contributed the most saturated fat for that day below.

Part D: Summary and Conclusions

1. What is your assessment of your fat intake? Is there anything you may need to change according to the AMDR guidelines? Please be specific in your answer.

Notes

1. Elias SL, Innis SM. Bakery foods are the major dietary source of trans-fatty acids among pregnant women with diets providing 30 percent energy from fat. Journal of the American Dietetic Association. 2002;102:46–51. Remig V, Franklin B, Margolis S, Kostas G, Nece T, Street JC. Trans fats in America: a review of their use, consumption, health implications, and regulation. Journal of the American Dietetic Association. 2010;110:585–92.

2. Bendesen NT, Christensen R, Bartels EM, Astrup A. Consumption of industrial and ruminant trans fatty acids and risk of coronary heart disease: a systematic review and meta-analysis of cohort studies. European Journal of Clinical Nutrition. 2011; March 23 (Epub ahead of print). Judd JT, Clevidence BA, Muesing RA, Wittes J, Sunkin ME, Podczasy JJ. Dietary trans fatty acids: Effects on plasma lipids and lipoproteins of healthy men and women. American Journal of Clinical Nutrition. 1994;59:861–8. Stender S, Dyerberg J. Influence of trans fatty acids on health. Annals of Nutrition and Metabolism. 2004;48:61–6.

3. Davey M. Chicago weighs new prohibition: bad-for-you fats. The New York Times. July 18, 2006.

4. Brenna JT, Salem N Jr, Sinclair AJ, Cunnane SC; International Society for the Study of Fatty Acids and Lipids, ISSFAL alpha-Linolenic acid supplementation and conversion to n-3 long-chain polyunsaturated fatty acids in humans. Prostaglandins, Leukotrienes, and Essential Fatty Acids. 2009;80:85–91. Saldanha LG, Salem N Jr, Brenna JT. Workshop on DHA as a required nutrient: overview. Prostaglandins, Leukotrienes, and Essential Fatty Acids. 2009;81:233–6.

5. Calder PC. Polyunsaturated fatty acids and inflammation. Biochemical Society Transactions. 2005;33:423–7. Hansen SN, Harris WS. New evidence for the cardiovascular benefits of long chain omega-3 fatty acids. Current Atherosclerosis Reports. 2007;9:434–40.

6. Vanschoonbeek K, de Maat MP, Heemskerk JW. Fish oil consumption and reduction of arterial disease. Journal of Nutrition. 2003;133:657–60. Wood DA, Kotseva K, Connolly S, Jennings C, Mead A, Jones J, Holden A, De Bacquer D, Collier T, De Backer G, Faergeman O, EUROACTION Study Group. Nurse-coordinated multidisciplinary, family-based cardiovascular disease prevention programme (EUROACTION) for patients with coronary heart disease and asymptomatic individuals at high risk of cardiovascular disease: A paired, cluster-randomised controlled trial. Lancet. 2008;371:1999–2012.

7. Mori TA, Beilin LJ. Omega-3 fatty acids and inflammation. Current Atherosclerosis Reports. 2004;6:461–7. Shahidi F, Miraliakbari H. Omega-3 (n-3) fatty acids in health and disease. Part 1: cardiovascular disease and cancer. Journal of Medicinal Foods. 2004; 7:387–401. Wijendran V, Hayes KC. Dietary n-6 and n-3 fatty acid balance and cardiovascular health. Annual Review of Nutrition. 2004;24:597–615. Zampelas A. Eicosapentaenoic acid (EPA) from highly concentrated n-3 fatty acid ethyl esters is incorporated into advanced atherosclerotic plaques and higher plaque EPA is associated with decreased plaque inflammation and increased stability. Atherosclerosis. 2010;212:34–5.

8. Jensen RG. Handbook of Milk Composition. New York: Academic Press; 1995.

9. Dangour AD, Uauy R. N-3 long-chain polyunsaturated fatty acids for optimal function during brain development and ageing. Asia Pacific Journal of Clinical Nutrition. 2008;17:185–8.

10. Alessandri JM, Guisnet P, Vancassel S, Astorg P, Denis I, Langelier B, Aid S, Poumes-Ballihaut C, Champeil-Potokar G, Lavialle M. Polyunsaturated fatty acids in the central nervous system: Evolution of concepts and nutritional implications throughout life. Reproduction, Nutrition, and Development. 2004;44:509–38. Heird WC. Infant feeding and vision. American Journal of Clinical Nutrition. 2008;87:1120. Heird WC, Lapillone A. The role of essential fatty acids in development. Annual Review of Nutrition. 2005;25:549–71.

11. Afleith M, Clandinin MT. Dietary PUFA for preterm and term infants: Review of clinical studies. Critical Review of Food Science and Nutrition. 2005;45:205–29. Smithers LG, Gibson RA, McPhee A, Makrides M. Effect of long-chain polyunsaturated fatty acid supplementation of preterm infants on disease risk and neurodevelopment: A systematic review of randomized controlled trials. American Journal of Clinical Nutrition. 2008;87:912–20.

12. Picciano MF, Guthrie HA, Sheehe DM. The cholesterol content of human milk. A variable constituent among women and within the same woman. Clinical Pediatrics. 1978;17:359–62. Shahin AM, McGuire MK, Anderson N, Williams J, McGuire MA. Effects of margarine and butter consumption on distribution of trans-18:1 fatty acid isomers and conjugated linoleic acid in major serum lipid classes in lactating women. Lipids. 2006;41:141–7.

13. German JB. Dietary lipids from an evolutionary perspective: sources, structures and functions. Maternal and Child Nutrition. 2011;7:2–16. Pond WG, Mersmann HJ, Su D, McGlone JJ, Wheeler MB, Smith EO. Neonatal dietary cholesterol and alleles of cholesterol 7-alpha hydroxylase affect piglet cerebrum weight, cholesterol concentration, and behavior. Journal of Nutrition. 2008;138:282–6.

14. Strumia R. Dermatologic signs in patients with eating disorders. American Journal of Clinical Dermatology. 2005;6:165–73.

15. Vidon C, Boucher P, Cachefo A, Peroni O, Diraison F, Beylot M. Effects of isoenergetic high-carbohydrate compared with high-fat diets on human cholesterol synthesis and expression of key regulatory genes of cholesterol metabolism. American Journal of Clinical Nutrition. 2001;73:878–84.

16. Gupta AK, Savopoulos CG, Ahuia J, Hatzitolios AI. Role of phytosterols in lipid-lowering: current perspectives. QMJ: monthly journal of the Association of Physicians. 2011;104:301–8.

17. U.S. Food and Drug Administration. A food labeling guide—Appendix C: health claims. October 2009. Available from http://www.fda.gov/Food/GuidanceComplianceRegulatory Information/GuidanceDocuments/FoodLabelingNutrition/ FoodLabelingGuide/ucm064919.htm.

18. National Institutes of Health. Gallstones. Medline Plus. Updated 2009. Available from: http://www.nlm.nih.gov/ medlineplus/ency/article/000273.htm.

19. Walcher T, Haenle MM, Mason RA, Koenig W, Imhof A, Kratzer W; EMIL Study Group. The effect of alcohol, tobacco and caffeine consumption and vegetarian diet on gallstone prevalence. European Journal of Gastroenterology and Hepatology. 2010;22:1345–51. Leitzmann C. Vegetarian diets: What are the advantages? Forum in Nutrition. 2005;57:147–56.

20. Adiels M, Olofsson SO, Taskinen MR, Borén J. Over-production of very low-density lipoproteins is the hallmark of the dyslipidemia in the metabolic syndrome. Arteriosclerosis Thrombosis Vascular Biology. 2008;28:1225–36.

Holvoet P. Oxidized LDL and coronary heart disease. Acta Cardiologica. 2004;59:479–84. Knopp RH, Paramsothy P, Atkinson B, Dowdy A. Comprehensive lipid management versus aggressive low-density lipoprotein lowering to reduce cardiovascular risk. American Journal of Cardiology. 2008;101:48B–57B.

21. Mozaffarian D, Willett WC. Trans fatty acids and cardio-vascular risk: A unique cardiometabolic imprint? Current Atherosclerosis Reports. 2007;9:486–93. Stender S, Dyerberg J. Influence of trans fatty acids on health. Annals of Nutrition and Metabolism. 2004;48:61–6.

22. Lands B. A critique of paradoxes in current advice on dietary lipids. Progress in Lipid Research. 2008;47:77–106. Yang Y, Ruiz-Narvaez E, Kraft P, Campos H. Effect of apolipoprotein E genotype and saturated fat intake on plasma lipids and myocardial infarction in the Central Valley of Costa Rica. Human Biology. 2007;79:637–47.

23. U.S. Department of Agriculture and U.S. department of Health and Human Services Dietary Guidelines Advisory Committee. Report of the Dietary Guidelines Advisory Committee on the Dietary Guidelines for Americans. 2010. National Technical Information Service, Springfield VA. Available at: http://www.cnpp.usda.gov/Publications/DietaryGuidelines/2010/DGAC/Report/2010DGACReport-camera-ready-Jan11-11.pdf.

24. Kathiresan S, Musunuru K, Orho-Melander M. Defining the spectrum of alleles that contribute to blood lipid concentrations in humans. Current Opinions in Lipidology. 2008;19:122–7. Wilson PW. Assessing coronary heart disease risk with tradi-tional and novel risk factors. Clinical Cardiology. 2004;27:7–11.

25. Anderson JW, Randles KM, Kendall CW, Jenkins DJ. Car-bohydrate and fiber recommendations for individuals with diabetes: A quantitative assessment and meta-analysis of the evidence. Journal of the American College of Nutrition. 2004;23:5–17. Christensen JH, Christensen MS, Dyerberg J, Schmidt EB. Heart rate variability and fatty acid content of blood cell membranes: A dose–response study with n-3 fatty acids. American Journal of Clinical Nutrition. 1999;70:331–7. Vanschoonbeek K, de Maat MP, Heemskerk JW. Fish oil con-sumption and reduction of arterial disease. Journal of Nutri-tion. 2003;133:657–60.

26. Cavigiolio G, Shao B, Geier EG, Ren G, Heinecke JW, Oda MN. The interplay between size, morphology, stability, and functionality of high-density lipoprotein subclasses. Biochemistry. 2008;47:4770–9. Florentin M, Liberopoulos EN, Wierzbicki AS, Mikhailidis DP. Multiple actions of high-density lipoprotein. Current Opinions in Cardiology. 2008;23:370–8.

27. Chiesa G, Parolini C, Sirtori CR. Acute effects of high-density lipoproteins: Biochemical basis and clinical find-ings. Current Opinions in Cardiology. 2008;23:379–85. Fernandez ML, Webb D. The LDL to HDL cholesterol ratio as a valuable tool to evaluate coronary heart disease risk. Journal of the American College of Nutrition. 2008;27:1–5.

28. Nettleton JA, Volcik KA, Hoogeveen RC, Boerwinkle E. Carbohydrate intake modifies associations between ANGPTL4[E40K] genotype and HDL-cholesterol concen-trations in white men from the Atherosclerosis Risk in Communities (ARIC) study. Atherosclerosis. 2009;203:214–20. Samaha FF, Foster GD, Makris AP. Low-carbohydrate diets, obesity, and metabolic risk factors for cardiovascular disease. Current Atherosclerosis Reports. 2007;9:441–7. Tay J, Brinkworth GD, Noakes M, Keogh J, Clifton PM. Metabolic effects of weight loss on a very-low-carbohydrate diet compared with an isocaloric high-carbohydrate diet in abdominally obese subjects. Journal of the American College of Cardiology. 2008;51:59–67.

29. Fan JG, Cai XB, Li L, Li XJ, Dai F, Zhu J. Alcohol consump-tion and metabolic syndrome among Shanghai adults: A randomized multistage stratified cluster sampling investiga-tion. World Journal of Gastroenterology. 2008;14:2418–24. Joosten MM, Beulens JW, Kersten S, Hendriks HF. Moderate alcohol consumption increases insulin sensitivity and ADIPOQ expression in postmenopausal women: A randomised, crossover trial. Diabetologia. 2008;51:1375–81. Paniagua JA, de la Sacristana AG, Sánchez E, Romero I, Vidal-Puig A, Berral FJ, Escribano A, Moyano MJ, Pérez-Martinez P, López-Miranda J, Pérez-Jiménez F. A MUFA-rich diet improves posprandial glucose, lipid and GLP-1 responses in insulin-resistant subjects. Journal of the American College of Nutrition. 2007;26:434–44.

30. Department of Health and Human Services. Food and Drug Administration. Food additives permitted for direct addition to food for human consumption; olestra; final rules. Federal Register 68:46363–402, 2003. Available at http://www.fda.gov/OHRMS/DOCKETS/98fr/03-19508.pdf.

31. Lammert F, Wang DQ. New insights into the genetic regula-tion of intestinal cholesterol absorption. Gastroenterology. 2005;129:718–34. Yang Y, Ruiz-Narvaez E, Kraft P, Campos H. Effect of apolipoprotein E genotype and saturated fat intake on plasma lipids and myocardial infarction in the Central Valley of Costa Rica. Wu K, Bowman R, Welch AA, Luben RN, Wareham N, Khaw KT, Bingham SA. Apolipo-protein E polymorphisms, dietary fat and fibre, and serum lipids: The EPIC Norfolk study. European Heart Journal. 2007;28:2930–6.

32. Brown MS, Goldstein JL. How LDL receptors influence cholesterol and atherosclerosis. Scientific American. 1984;251:52–60. Dedoussis GV, Schmidt H, Genschel J. LDL-receptor mutations in Europe. Human Mutation. 2004;443–59.

33. The Nobel Assembly at the Karolinska Institute. The 1985 Nobel Prize in Physiology or Medicine press release. Avail-able from: http://nobelprize.org/medicine/laureates/1985/press.html.

34. Prentice RL. Women's health initiative studies of postmeno-pausal breast cancer. Advances in Experimental Medicine and Biology. 2008;617:151–60. Van Horn L, Manson JE. The Women's Health Initiative: Implications for clinicians. Cleve-land Clinic Journal of Medicine. 2008;75:385–90. Wang J, John EM, Horn-Ross PL, Ingles SA. Dietary fat, cooking fat, and breast cancer risk in a multiethnic population. Nutrition and Cancer. 2008;60:492–504.

35. Al-Serag HB. Obesity and disease of the esophagus and colon. Gastroenterology Clinics of North America. 2005;34:63–82. Key TJ, Schatzkin A, Willett WC, Allen NE, Spencer EA, Travis RC. Diet, nutrition and the prevention of cancer. Public Health Nutrition. 2004;7:187–200. McTiernan A, Yang XR, Chang-Claude J, Goode EL, et al. Associations of breast cancer risk factors with tumor subtypes: a pooled analysis from the Breast Cancer Association Consortium studies. Journal of the National Cancer Institute. 2011;103:250–63. Obesity and cancer: The risks, science, and potential manage-ment strategies. Oncology. 2005;19:871–81.

36. World Cancer Research Fund/American Institute for Cancer Research. Food, nutrition, physical activity, and the prevention of cancer: A global perspective. Washington, DC: AICR; 2007.

37. Lichtenstein AH et al., Diet and lifestyle recommendations revision 2006. A scientific statement from the American Heart Association Nutrition Committee. Circulation. 2006;114:82–96.

Nutrition and Cardiovascular Health

The steadfast function of the cardiovascular system is likely one of the body's most impressive physiologic systems. Indeed, without you having to give it a single thought, your heart beats about 100,000 times a day and 40 million times a year. The resultant movement of blood keeps your body supplied with oxygen and nutrients while removing harmful waste products via 60,000 miles of blood vessels. To maintain your health and vigor, it is important to keep your cardiovascular system in top condition. However, as we enjoy longer lives, chronic degenerative diseases are becoming more common. Of these, heart disease and stroke are the first- and third-leading causes of death in the United States.[1] Heart disease and stroke are forms of **cardiovascular disease** (CVD), a term used to describe a variety of diseases of the heart and blood vessels. In this section, you will learn about the pathophysiology of CVD and the importance of nutrition in its prevention and treatment.

How Does Cardiovascular Disease Develop?

Cardiovascular disease is generally caused by a slowing or complete obstruction of blood flow to the heart or other parts of the body, including the brain (Figure 1). Remember that blood flows through blood vessels delivering oxygen and nutrients to cells. When blood flow is restricted, cells do not receive adequate oxygen and nutrients, ultimately causing cell death. Restriction of blood flow results from a condition called **atherosclerosis,** characterized by a narrowing and hardening of the blood vessels. Atherosclerosis can reduce the blood supply to the heart muscle (coronary arteries), brain (cerebral arteries), and other parts of the body. When it occurs in coronary arteries, it can cause **heart disease,** also called coronary heart disease, whereas atherosclerosis in cerebral arteries can cause a **stroke.**

Atherosclerosis can lead to other complications, which can also contribute to heart disease or stroke. For example, a blood vessel can become weak and distended, forming an **aneurysm.** Having an aneurysm is dangerous because the arterial walls become stretched and can rupture. When an aneurysm in a major blood vessel ruptures, blood pours into the body cavity, resulting in a rapid drop in blood pressure, which can deprive tissues of oxygen and nutrients. **Blood clots,** or thromboses, are another complication associated with atherosclerosis. When small pieces of clotted blood become lodged in an artery, blood flow to the target tissue can be reduced or cut off. The risks of a blood clot are even greater when the artery has been narrowed by atherosclerosis. Regardless of whether it occurs via atherosclerosis, aneurysm, blood clot, or a combination of factors, CVD can be life threatening.

ATHEROSCLEROSIS CAN LEAD TO CARDIOVASCULAR DISEASE

Atherosclerosis, a slowly developing, chronic, degenerative disease, is one of the most important risk factors associated with heart attacks and strokes. Atheroclerosis develops when fatty deposits called **plaque** accumulate within the walls of arteries. Plaques contain fatty acids, cholesterol, lipid-filled immune cells (called **foam cells**),

---

cardiovascular disease (car – di – o – VAS – cu – lar) (*kardia*, heart; *vascellum*, vessel) A disease of the heart or vascular system.

atherosclerosis (a – ther – o – scler – O – sis) (*atheroma*, cyst full of pus; *sklerosis*, hardening) The hardening and narrowing of blood vessels caused by buildup of fatty deposits (plaques) and inflammation in the vessel walls.

heart disease (also called coronary heart disease) A condition that occurs when the heart muscle does not receive enough blood.

stroke A condition that occurs when a portion of the brain does not receive enough blood.

aneurysm (AN – eu – rysm) (*aneurusma*, dilation) The outward bulging of a blood vessel.

blood clot (also called thrombosis) A small, insoluble particle made of blood cells and clotting factors.

plaque Fatty deposit that accumulates within walls of blood vessels, sometimes leading to atherosclerosis.

foam cell A type of cell—usually an immune cell—that contains large amounts of lipid.

FIGURE 1 Causes of Cardiovascular Disease Atherosclerosis, blood clots, and aneurysms can all reduce or stop blood flow, causing cardiovascular disease.

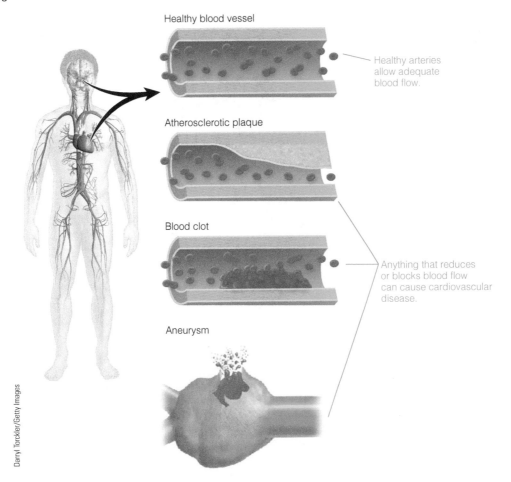

Healthy blood vessel

Healthy arteries allow adequate blood flow.

Atherosclerotic plaque

Blood clot

Anything that reduces or blocks blood flow can cause cardiovascular disease.

Aneurysm

Darryl Torckler/Getty Images

cellular waste products, calcium, and a variety of other substances. The accumulation of plaque within the arterial wall reduces blood flow. Plaque deposits can also dislodge (break off) from the arterial wall and enter the blood. When this happens, more white blood cells migrate to the injured area, in turn promoting inflammation that might be especially important in the etiology of CVD.

Chronic Inflammation Although scientists do not know what triggers the process of atherosclerosis, emerging research suggests that **chronic inflammation** may play a critical role.[2] The term *inflammation* refers to the body's response to a noxious (poisonous) stimulus, injury, or infection. Inflammation causes dilation of blood vessels, movement of white blood cells into the injured tissue, redness, and pain. Most experts now believe that many people with CVD may be in a state of chronic inflammation. This is likely one reason why taking low doses of aspirin (an anti-inflammatory drug) reduces the risk of heart disease and stroke.[3]

One potentially useful biological marker for inflammation is **C-reactive protein (CRP),** which is released into the blood by the liver, adipose tissue, and smooth muscle as part of the inflammatory response to injury or infection. High circulating concentration of CRP is related to increased risk for CVD.[4] The American Heart Association recommends the use of CRP screening only for individuals at intermediate risk for CVD.[5] This is because those at low risk are not likely to need intervention, and those at high risk should seek treatment regardless of their CRP levels. You can determine your risk for CVD by visiting their website at http://www.americanheart.org.

chronic inflammation A response to cellular injury that is characterized by chronic capillary dilation, white blood cell infiltration, release of immune factors, redness, heat, and pain.

C-reactive protein (CRP) A protein produced in the liver, adipose tissue, and smooth muscles in response to injury or infection that, when elevated, can indicate risk for cardiovascular disease.

FIGURE 2 Diagnosing Heart Disease Electrocardiograms, echocardiograms, and angiograms are all useful in diagnosing heart disease.

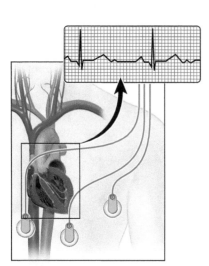

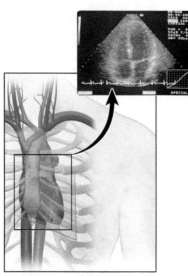

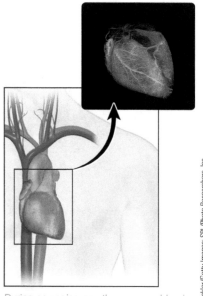

An electrocardiogram assesses heart function by recording electrical activity.

An echocardiogram assesses the structure and function of the heart using sound waves that create a moving image.

During an angiogram, the coronary blood vessels are visualized with the use of dyes.

Darryl Torckler/Getty Images; SPL/Photo Researchers, Inc.

HEART DISEASE IS A TYPE OF CARDIOVASCULAR DISEASE

In its less severe state, heart disease causes chest pain and discomfort called **angina pectoris,** or simply angina. Angina occurs when the heart tissue is still receiving some blood via the coronary arteries but not as much as it needs. Because the heart muscle does not receive adequate amounts of oxygen, the heart muscle can spasm, causing chest pain and shortness of breath. Although angina is not usually life threatening, it could indicate something more serious, such as an impending heart attack.

Heart attacks occur when the blood supply to the heart muscle is severely reduced or stopped. The medical term for a heart attack is "myocardial infarction." Because cardiac tissue can survive only for a few minutes without oxygen, heart attacks can permanently damage the heart muscle. If the damage is severe, the results can be fatal. Warning signs and symptoms of a heart attack include chest discomfort; pain radiating down one or both arms, back, neck, jaw, or stomach; shortness of breath; cold sweats; nausea; and lightheadedness. Women tend to experience somewhat different symptoms including unusual fatigue, sleep disturbances, shortness of breath, indigestion, and anxiety. It is important that anyone experiencing these symptoms contact emergency medical personnel immediately, as heart attacks are the leading cause of death in the United States.

Diagnosing Heart Disease When assessing a patient's risk for heart disease, a doctor will likely perform a complete physical examination and medical history. If heart disease is suspected, certain tests can help determine whether the heart is working normally and, if it is not, where the problem lies. These tests might include an **electrocardiogram** (EKG or ECG), which measures the electrical impulses in the heart; an **echocardiogram,** which uses sound waves to examine the heart's structure and motion; or an **angiogram,** in which the coronary arteries are visualized with the help of an injected dye. Angiograms enable physicians to determine if and where there are blockages in coronary arteries. These procedures are illustrated in Figure 2. Some doctors may also order blood tests such as that for C-reactive protein.

angina pectoris (an – GI – na pec – TOR – is) (*ankhone*, to strangle; *pectos*, breast) Pain in the region of the heart, caused by a portion of the heart muscle receiving inadequate amounts of blood.

heart attack (also called **myocardial infarction**) An often life-threatening condition in which blood flow to some or all of the heart muscle is completely blocked.

electrocardiogram A procedure during which the heart's electrical activity is recorded.

echocardiogram A visual image, produced using ultrasound waves, of the heart's structure and movement.

angiogram A procedure in which dye is injected into the blood, allowing the flow of blood through cardiac arteries to be visualized.

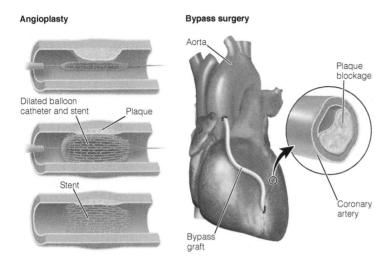

Angioplasty

Dilated balloon
catheter and stent

Plaque

Stent

Bypass surgery

Aorta

Plaque
blockage

Coronary
artery

Bypass
graft

FIGURE 3 Surgically Treating Heart Disease Surgeries such as angioplasty (the placing of a stent) and coronary bypass surgery are often life-saving for heart-disease patients.

Surgical Treatment of Heart Disease Sometimes heart disease is treated by performing surgical procedures such as placing a **stent** (wire mesh tube) in the affected artery during a procedure called **angioplasty** (or "balloon angioplasty"). During an angioplasty, the stent is initially collapsed to a small diameter and put over a ballooning catheter, which is inserted into the area of the blockage. When the balloon is inflated, the stent expands, locks in place, and forms a scaffold. The stent stays in the artery permanently, holding it open. An alternative to angioplasty is **coronary artery bypass surgery,** sometimes called CABG ("cabbage"). This type of surgery reroutes, or "bypasses," blood around clogged arteries to improve blood flow and aid in the delivery of oxygen to the heart. This is done by the removal of a segment of a blood vessel from another part of the body. This "graft" is then used to bypass blood flow around the blocked coronary arteries. These procedures are illustrated in Figure 3.

STROKE IS ANOTHER FORM OF CARDIOVASCULAR DISEASE

A stroke occurs when a portion of the brain is deprived of oxygen and critical nutrients. The extent of the resulting brain damage depends on the magnitude, duration, and area of the brain affected. Strokes often result in speech impairment or partial paralysis on one side of the body. However, if the stroke affects a large or critical part of the brain, it can be life threatening. One important warning sign for stroke is the occurrence of **transient ischemic attacks (TIAs),** sometimes referred to as "ministrokes." TIAs can occur when blood flow to the brain is temporarily disrupted.

Signs and symptoms of a TIA or stroke include sudden numbness or weakness (especially on one side of the body), confusion, slurred speech, dizziness, loss of balance, or a severe headache. When a TIA or stroke is suspected, a person should seek immediate medical assistance.

Once a stroke has been diagnosed, treatment depends on its type and location. Sometimes, medications that break up blood clots can be given. Other times, surgery is needed to place a stent in the affected artery. In all cases, time is of the essence, as quick treatment not only improves chances of survival but may also reduce the amount of resulting disability.

What Are the Risk Factors for Cardiovascular Disease?

Having any of the biological, lifestyle, and environmental risk factors associated with the development of atherosclerosis increases a person's chances of having a heart attack or stroke. The American Heart Association has categorized these factors as being either major risk factors or contributing risk

stent A device made of rigid wire mesh that is threaded into an atherosclerotic blood vessel to expand and provide support for a damaged artery.

angioplasty A procedure used to widen the heart's blood vessels by inserting a stent.

coronary bypass surgery A procedure in which a healthy blood vessel obtained from the leg, arm, chest, or abdomen is used to bypass blood from a diseased or blocked coronary artery to a healthy one.

transient ischemic attack (TIA) A "ministroke" that is caused by a temporary decrease in blood flow to the brain.

factors, depending on the availability of supporting evidence. Furthermore, they have subcategorized these risk factors into those that cannot be changed (nonmodifiable or biological risk factors) and those that can be modified, treated, or controlled by lifestyle choices or medication (modifiable risk factors).[6] The American Heart Association recommends that all adults—young and old—know their risk for CVD, and as previously mentioned, you can determine your risk by visiting their website at http://www.americanheart.org.

NONMODIFIABLE RISK FACTORS

The major nonmodifiable risk factors for CVD include age, sex, genetics (including race), prior stroke or heart attack, and having been born with low birth weight. For instance, CVD can strike at any age, but the older you get the more likely you are to develop it. In addition, men have a greater risk of CVD than women, and tend to develop it earlier in life. Further, compared with Caucasians, heart disease and stroke are more common in Mexican Americans, American Indians, native Hawaiians, and African Americans. This is thought to partly be due to higher rates of obesity and diabetes in these groups, although other genetic factors are likely important. People who were born with low birth weights also have increased risk for CVD later in life.[6] Remember that these factors cannot easily be controlled or treated. However, if you fall into any of these high-risk categories, it is even more important to take action.

MODIFIABLE RISK FACTORS

Modifiable or lifestyle risk factors include smoking, high blood pressure, elevated blood lipids, physical inactivity and obesity, diabetes, stress, and excessive alcohol consumption. Although many of these factors depend somewhat on biological influences, there are ways we can modify them. Some of these, especially those related to nutrition, are briefly described next.

Hypertension High blood pressure, also called hypertension, increases a person's risk for CVD partly because it puts additional demands on the heart muscle and damages blood vessels. High blood pressure can also cause plaque to break away from arterial walls. Dislodged plaque can both initiate inflammation and act like a blood clot, restricting blood flow even more. Biological and lifestyle factors can both contribute to high blood pressure.

For example, smoking and obesity are major causes of high blood pressure.[7] Choosing not to smoke and maintaining a healthy body weight can help prevent high blood pressure and thus CVD.

Elevated Blood Lipid Levels Are a Major Risk Factor
A high level of lipid in the blood, or **hyperlipidemia**, is also a risk factor for CVD.[8] Excess triglycerides (**hypertriglyceridemia**) and high levels of cholesterol (**hypercholesterolemia**) are of specific concern. Both forms of hyperlipidemia have genetic and lifestyle components and are often treated with a combination of medication (to address genetic factors) and lifestyle changes. Table 1 provides a summary of healthy and unhealthy blood lipid values.

hyperlipidemia (hy – per – li – pid – EM – i – a) Elevated levels of lipids in the blood.

hypertriglyceridemia (hy – per – tri – gly – cer – EM – i – a) Elevated levels of triglycerides in the blood.

hypercholesterolemia (hy – per – chol – est – er – ol – EM – i – a) Elevated levels of cholesterol in the blood.

TABLE 1 Reference Values for Blood Lipids

Blood Lipid Category and Value	Interpretation
Total Cholesterol	
≤200 mg/dL	Desirable
200–239 mg/dL	Borderline high
≥240 mg/dLs	High
Low-Density Lipoprotein (LDL)	
≤100 mg/dL	Optimal
100–129 mg/dL	Near optimal/above optimal
130–159 mg/dL	Borderline high
160–189 mg/dL	High
≥190 mg/dL	Very high
High-Density Lipoprotein (HDL)	
≤40 mg/dL	Low
≥60 mg/dL	High
Triglyceride	
≤150 mg/dL	Normal
150–199 mg/dL	Borderline high
200–499 mg/dL	High
≥500 mg/dL	Very high

SOURCE: U.S. Department of Health and Human Services. National Cholesterol Education Program. ATPIII guidelines at-a-glance quick desk reference, 2001. NIH Publication No. 01-3305. Available from: http://www.nhlbi.nih.gov/guidelines/cholesterol/atglance.pdf. Note that these reference values are endorsed by the American Heart Association.

Review of Lipid Transport in the Blood To better understand how blood lipid levels are related to CVD, it is important to briefly review how lipids (particularly cholesterol) are circulated in the blood. Recall that blood lipids are generally packaged as components of lipoproteins and that cholesterol is circulated mainly by two lipoproteins: low-density lipoproteins (LDLs) and high-density lipoproteins (HDLs). LDLs deliver cholesterol to cells, whereas HDLs pick up excess cholesterol for transport back to the liver. Many studies show that high levels of LDL cholesterol (LDL-C) are related to increased risk for plaque formation and CVD.[9] Because of this, LDL-C is called "bad cholesterol." Conversely, high levels of HDL cholesterol (HDL-C) protect people from CVD, and HDL-C is called "good cholesterol."[10]

Studies suggest that *oxidized* LDLs are much more atherogenic than *nonoxidized* LDLs.[11] Oxidized LDLs form when they are subjected to the damaging effects of free radicals. Experts believe that oxidized LDLs are especially atherogenic because foam cells (lipid-containing immune cells) are more likely to engulf them than their nonoxidized counterparts.[12] As such, it is the lipids contained in *oxidized* LDL that tend to form the core of the plaque that leads to atherosclerosis. This is one reason why dietary antioxidants, abundant in many fruits and vegetables, that decrease oxidation of LDL are likely important in preventing CVD. Because of differences in genetic make-up, some people are more susceptible than others to LDL oxidation and uptake of oxidized LDL into arterial walls.

What Is a Cholesterol Ratio? Of particular importance in terms of blood lipids is the distribution of cholesterol among the various lipoproteins—especially HDL (HDL-C) and LDL (LDL-C). As such, medical professionals sometimes use what is called a cholesterol ratio to assess a person's risk for CVD. The **cholesterol ratio** is calculated by dividing total cholesterol by HDL-C. The higher the ratio, the greater percentage of the total cholesterol is present in LDL relative to HDL. As an example, a person with a total cholesterol level of 200 mg/dL and an HDL-C of 50 mg/dL would have a cholesterol ratio of 4:1. Doctors recommend that this ratio be below 5:1 and optimally be 3.5:1 or less.

Obesity and Physical Inactivity—Related Risk Factors Of all the modifiable risk factors for CVD, likely the most important is being overweight or obese.[13] This is especially true if the excess body weight is stored in the waist area—in other words, if the person is apple-shaped. People with excess body weight are at higher risk of heart disease even if they have no other risk factors. Related to this is physical inactivity, which is an independent risk factor for CVD but also contributes to other risk factors such as obesity, hypertension, elevated blood lipids, and type 2 diabetes.

Unfortunately, there is no quick fix or magic bullet to help people shed unwanted pounds. However, being physically active and consuming a diet relatively low in fat and high in nutrient-dense foods are important in maintaining a healthy body weight. Even small reductions in weight can lower blood lipid levels and blood pressure, in turn lowering CVD risk.[14] Thus, doctors recommend that all people watch their caloric intake, maintain a healthy body weight, and engage in regular physical activity. The American Heart Association recommends at least 150 minutes per week of moderate exercise or 75 minutes per week of vigorous exercise (or a combination of moderate and vigorous activity). In other words, a goal of 30 minutes a day, five times a week is recommended.[15]

Diabetes—Another Modifiable Risk Factor Having either type 1 or type 2 diabetes is also a major risk factor for CVD. As previously discussed, diabetes is a condition in which the body can no longer effectively regulate blood glucose. Although genetics can play a critical role in determining who gets diabetes, other factors are also involved. For example, being overweight and physically inactive are risk factors for type 2 diabetes. Keeping these important health parameters in check can help prevent type 2 diabetes and, in turn, can reduce the risk for CVD. Furthermore, poorly treated diabetes can result in elevated blood glucose levels, which can damage blood vessels and lead to CVD. Controlling blood glucose levels with diet, exercise, and medication is therefore important in preventing heart disease and stroke.

cholesterol ratio The mathematical ratio of total blood cholesterol to high-density lipoprotein cholesterol (HDL-C).

How Does Nutrition Influence Cardiovascular Risk?

Consuming a varied diet in moderation can go a long way in preventing and treating CVD. This is because you need many micro- and macronutrients to maintain a healthy heart and vascular system. In this section, you will learn what scientists currently know about how diet influences several of the major risk factors for heart disease and stroke. You will also learn about how phytochemicals and functional foods are beginning to take center stage in the quest for optimal cardiovascular health.

HYPERTENSION CAN BE PARTIALLY CONTROLLED WITH DIET

Your dietary choices can make a big difference in determining whether you develop high blood pressure. For example, heavy alcohol consumption can cause blood pressure to rise, and people with hypertension are typically advised to limit their intake.[16] Further, some people's blood pressure is highly sensitive to sodium, or salt.[17] In other words, high sodium intakes can cause high blood pressure in salt-sensitive individuals. This is why most people diagnosed with high blood pressure are advised to consume relatively low-sodium diets. Consuming a low-sodium diet has become easier as food manufacturers have developed sodium-free and low-sodium food products, and recommendations concerning how to reduce dietary sodium are provided in Table 2.

The DASH Diet May Be Beneficial Many essential dietary minerals such as potassium, calcium, and magnesium are needed for maintaining healthy blood pressure levels. Because these minerals help lower blood pressure, it is important that people with increased risk for CVD consume enough foods that provide them. For example, legumes, seafood, dairy products, and many fruits and vegetables tend to be rich sources of these minerals and should be included in a heart-healthy diet.

One way to ensure sufficient intake of all these important nutrients is to follow what is called the **DASH (Dietary Approaches to Stop Hypertension) diet**, which emphasizes fruits, vegetables, and low-fat dairy products. This diet plan has been shown to lower blood pressure, especially in salt-sensitive people.[18] Although it is not clear whether a single nutrient (such as calcium or potassium) in the DASH diet is responsible for its healthy effects, the diet appears to work by increasing sodium excretion in the urine. Many organizations, such as the American Heart Association and the U.S. Department of Agriculture (USDA) recommend the DASH diet, especially for people predisposed to hypertension, and its components are summarized in Table 3.

CONTROLLING BLOOD LIPID LEVELS WITH DIET

Several dietary factors can influence circulating triglyceride and cholesterol levels, although not all people are equally affected. For example, diets high in soluble fiber (such as in oat or rice bran, oatmeal, legumes, barley, citrus fruits, and strawberries) may help lower LDL cholesterol.[19] Indeed, choosing wisely when it comes to what macronutrients we eat is important to a healthy heart.

Low-Carbohydrate Diets—Is There a Risk? Although people have long thought high-carbohydrate diets are heart-healthy, more recent studies show that very low total carbohydrate diets may be more beneficial.[20] However, because these types of diets are generally high in protein and fat, some professional organizations have cautioned against their use. The concern is that some people might restrict foods that provide essential nutrients and therefore not get the variety of foods needed to meet nutritional needs.[21] In addition, these dietary patterns tend to have excessive amounts of animal fats that might increase blood lipid levels. A recent review of the literature, however, concluded that low-carbohydrate diets can be used safely and effectively for short-term weight loss without adversely affecting cardiovascular risk factors.[22] In fact, most studies conducted in free-living populations show they are beneficial.[23] Unfortunately, little is known about the

---

TABLE 2 Strategies for Reducing Salt Intake

- Avoid adding salt to foods during cooking or dining.
- Choose fresh, frozen, or canned foods without salt.
- When dining out, ask for foods to be prepared without salt or with half the usual amount of salt.
- Cut down on highly salted snack foods, such as chips.
- Choose low-salt prepared foods when possible.
- Do not drink "sports drinks" unless specifically needed.

DASH (Dietary Approaches to Stop Hypertension) diet A dietary pattern emphasizing fruits, vegetables, and low-fat dairy products designed to lower blood pressure.

TABLE 3 Dietary Approaches to Stop Hypertension (DASH) Diet Basics

Food Group	Daily Servings	Examples of Foods
Grains*	6–8	1 slice bread 1 oz dry cereal† ½ cup cooked rice, pasta, or cereal
Vegetables	4–5	1 cup raw leafy vegetable ½ cup cut-up raw or cooked vegetable ½ cup vegetable juice
Fruits	4–5	1 medium fruit ¼ cup dried fruit ½ cup fresh, frozen, or canned fruit ½ cup fruit juice
Fat-free or low-fat milk and milk products	2–3	1 cup milk or yogurt 1½ oz cheese
Lean meats, poultry, and fish	6 or less	1 oz cooked meats, poultry, or fish 1 egg
Nuts, seeds, and legumes	4–5 per week	⅓ cup or 1½ oz nuts 2 Tbsp peanut butter 2 Tbsp or ½ oz seeds ½ cup cooked legumes (dry beans and peas)
Fats and oils	2–3	Use sparingly
Sweets and added sugars	5 or less per week	1 Tbsp sugar 1 Tbsp jelly or jam ½ cup sorbet, gelatin 1 cup lemonade

*Whole grains are recommended for most grain servings as a good source of fiber and nutrients.

†Serving sizes vary between ½ cup and 1¼ cups, depending on cereal type. Check the product's Nutrition Facts label.

SOURCE: Department of Health and Human Services and U.S. Department of Agriculture. Facts about the DASH eating plan. Washington, DC: U.S. Government Printing Office; 2006; available at http://www.nhlbi.nih.gov/health/public/heart/hbp/dash/new_dash.pdf.

long-term effects of these diets on the cardiovascular system, and a recently conducted clinical trial suggests that consumption of low-carbohydrate diets for one year increases total cholesterol and LDL cholesterol in persons trying to lose weight.[24] Clearly, more studies are needed.

Trans Fatty Acids, Saturated Fatty Acids, and Cholesterol Various dietary lipids can also influence blood lipid levels. For some people, high intake of dietary cholesterol can raise blood cholesterol and triglyceride levels.[25] Consuming high levels of saturated fatty acids and *trans* fatty acids can cause a rise as well.[26] This is why food labels list the content of these types of fatty acids. Although saturated fatty acids used to be considered the "worst" type of dietary lipid for maintaining a healthy cardiovascular system, mounting evidence suggests that *trans* fatty acids are equally bad if not more detrimental to health.[27] Studies show that saturated and *trans* fatty acids increase risk for heart disease, at least in part, by increasing circulating LDL concentrations.[28] *Trans* fatty acids do additional harm, however, because they not only decrease HDL but also promote inflammation.[29] However, the harmful effects of *trans* fatty acids on blood lipids may be true only for industrially produced *trans* fatty acids.[30] Those found naturally in foods—usually beef and dairy products—may actually be heart-healthy.[31]

Monounsaturated and ω-3 Fatty Acids Compared to saturated and *trans* fatty acids, which may *increase* blood lipids, some monounsaturated fatty acids (such as those in olive oil) and polyunsaturated fatty acids may help *lower* levels.[32] Studies suggest that the beneficial effects of the ω-3 fatty acids, such as those in fish and flaxseed oils, may be particularly important as they appear to decrease VLDL and triglyceride synthesis in the liver.[33] In addition, increased ω-3 fatty acid consumption may decrease blood pressure in some people,[34] reduce the risk of atherosclerotic plaque rupture,[35] and have anti-inflammatory properties.[36] However, the optimal level of ω-3 fatty acids (especially when also considering ω-6 fatty acids) is an area of active debate among scientists.[37]

Alcohol—Friend *and* Foe Consuming alcoholic beverages can have both detrimental and beneficial effects on blood lipid concentrations. Specifically, alcohol consumption tends to increase triglyceride levels (a negative effect), whereas moderate alcohol consumption (one to two drinks per day) can raise HDL cholesterol levels (a beneficial effect).[38] Therefore, if you drink, do so in moderation. It is never recommended, however, that nondrinkers begin to drink for health benefits.

OTHER DIETARY FACTORS AND PATTERNS ARE ASSOCIATED WITH LOWER RISK

In addition to dietary components *known* to influence cardiovascular risks such as saturated and *trans* fatty acids, other nutrients may be important as well. For example, the B vitamin folate (folic acid), vitamin B_6, and vitamin B_{12} help maintain

Garlic is a functional food because it contains sulfur-containing compounds thought to lower LDL cholesterol (LDL-C).

Consumption of a Mediterranean-type diet is associated with a lower risk for cardiovascular disease.

a healthy cardiovascular system. A deficiency in any of these vitamins can cause a compound called homocysteine to accumulate in the blood. This is important, because raised levels of homocysteine are strongly associated with increased risk for CVD.[39] Studies designed to determine how these vitamins are related to cardiovascular risk have not been conclusive, and more research is needed.[40]

As previously mentioned, dietary antioxidants may help decrease chronic inflammation. Indeed, emerging research suggests that dietary antioxidants may help inhibit the oxidation of LDLs, thus decreasing risk for CVD.[41] Examples of antioxidant nutrients include vitamins C and E, β-carotene (a precursor of vitamin A), zinc, and selenium. However, controlled clinical studies have not shown that increasing consumption of antioxidant supplements decreases risk for this disease, although higher vitamin E intake has been associated with lower circulating concentration of C-reactive protein (CRP).[42]

Functional Foods, Phytochemicals, and Zoonutrients Are Also Important It is important to recognize that the health benefits of foods extend beyond their classic micro- and macronutrient contents. Indeed, foods have other biologically active components thought to influence health. You may recall that these foods are referred to as functional foods because they contain phytochemicals or zoonutrients, and some of these are listed in Table 4. Phytochemicals that may influence heart health include plant sterols and stanols, isoflavones found mainly in soy products, compounds

TABLE 4 Examples of Heart-Healthy Functional Foods and Their Biologically Active Components

Functional Food	Possible Biologically Active Component
Black tea	Polyphenols
Blueberries	Anthocyanin
Cocoa	Flavanols
Dairy foods	Proteins, calcium, potassium, conjugated linoleic acid
Fish	ω-3 fatty acids
Flaxseed	ω-3 fatty acids
Garlic	Sulfur-containing compounds
Nuts	Unsaturated fatty acids, vitamin E, selenium
Olive oil	Monounsaturated fatty acids, phenolic compounds
Psyllium	Soluble fiber
Red wine and grapes	Resveratrol, quercetin
Soy	Proteins and flavonoids
Stanol-sterol-fortified foods	Plant stanols and sterols
Tomatoes	Lycopene, lutein
Whole grains	Soluble fiber, folate, antioxidants
Whole oats	β-glucan, soluble fiber

SOURCE: Adapted from Hasler CM, Bloch AS, Thomson CA, Enrione E, Manning C. Position of the American Dietetic Association: Functional foods. Journal of the American Dietetic Association. 2004;104:814–26.

found in red wine and grapes, and a variety of sulfur-containing compounds abundant in garlic, onions, and leeks.[43] Animal-derived foods that may provide protection from CVD are fatty fish and milk, which appear to contain fatty acids and proteins that are especially heart-healthy.[44]

However, not enough data are available to determine which components of these foods are beneficial or whether higher intakes of these compounds reduce the risk of heart attacks or strokes in all people. In the meantime, consuming a variety of foods—especially fruits, vegetables, whole-grain cereals, low-fat fish, and dairy products—is recommended to ensure adequate intake of these dietary factors.

HEART-HEALTHY DIETARY *PATTERNS*

In addition to knowing which nutrients, phytochemicals, or zoonutrients are important for heart health, it is also useful to consider which overall *types* of diets are associated with decreased risk for CVD. Dietary patterns that may help lower risk for CVD include the DASH diet, vegetarian diets, and the Mediterranean diet. As previously described, the DASH diet emphasizes fruits, vegetables, and low-fat dairy products. Vegetarian diets are typically those that exclude meat products but include milk, eggs, and sometimes fish. In addition, there is much interest in whether low- or high-carbohydrate diets are best for people with elevated risk for CVD.

Mediterranean Diets **Mediterranean diets** are not "diets" *per se,* but rather eating patterns common to traditional cuisines of those living in countries surrounding the Mediterranean Sea, such as Spain, Italy, and Greece. These diets emphasize fruits, fish, vegetables, whole grains, nuts, and seeds as well as olive oil and red wine; meats are consumed in moderation. Epidemiologic studies show that people consuming a Mediterranean-type diet have lower rates of heart disease.[45] However, it is difficult to determine which dietary components are protective and whether the reduced risk for disease is due to diet alone. More likely, a variety of confounding factors are involved, such as low rates of obesity and high rates of physical activity.

What Are the General Nutrition Guidelines for Healthy Hearts?

The multifaceted and complex relationship between diet and cardiovascular risk is yet another example of how dietary variety, moderation, and balance are essential to health and well-being. Clearly, many dietary components are related to risk for CVD. In response, agencies such as the American Heart Association, the U.S. Department of Agriculture, and the National Institutes of Health have set forth guidelines regarding dietary intakes that help lower risk for these diseases. These are outlined as follows.

BE MINDFUL OF ENERGY INTAKE AND MACRONUTRIENT BALANCE

The primary goal of energy consumption recommendations is to consume only enough calories to maintain a healthy body weight. People who need to lose weight are advised to decrease their energy intake and increase their energy expenditure to meet their body weight goals. In addition, there are several general recommendations concerning intake of lipids and other nutrients, some of which are listed here.

- **Total fat**—limit to 20 to 35% of total calories.
- **Saturated fat**—intake should be ≤10% of total calories; people who already have heart disease should consume <7% of total calories as saturated fats.
- ***Trans* fat**—intake should be minimal.
- **Omega-3 (ω-3) fatty acids**—recommended that everyone consume fish at least twice weekly; people with heart disease or elevated blood triglyceride should consume more.
- **Cholesterol**—limit to 300 mg/day; people with heart disease should consume ≤200 mg/day;

Mediterranean diet A dietary pattern, originating from the region surrounding the Mediterranean Sea, which is related to lower risk for cardiovascular disease.

Mixed salads provide many nutrients needed to maintain heart health.

those with very high levels of blood cholesterol should consider reducing intake even more.

- **Complex carbohydrates**—consume ≥3 servings of whole-grain products daily; emphasize complex carbohydrates such as those found in vegetables, fruits, and whole grains rather than refined carbohydrates, such as table sugar.

VITAMINS AND MINERALS ALSO MATTER

Although less is known about the relationship between micronutrients and CVD, many of these compounds are likely important for maintaining good health. Some recommendations concerning intakes of these substances are provided below.

- **Sodium**—intake should not exceed about 2,300 mg/day—the amount equivalent to 1 teaspoon of salt.
- **Calcium**—consume at least 1,000–1,200 mg/day; low-fat dairy products should be emphasized.
- **Potassium**—choose potassium-rich foods such as legumes, potatoes, seafood, and some fruits (such as bananas) regularly.
- **B vitamins**—select foods high in folate, vitamin B_6, and vitamin B_{12}; eating a variety of both animal- and plant-based foods on a regular basis is critical to meeting this goal.

Key Points

How Does Cardiovascular Disease Develop?

- Impaired blood flow to the heart or brain can cause heart attacks or strokes, respectively.
- Atherosclerosis, aneurysms, and blood clots are the major causes of heart disease and strokes.
- Chronic inflammation appears to be an important factor in CVD.

What Are the Risk Factors for Cardiovascular Disease?

- High levels of LDL—especially oxidized LDL—and low levels of HDL can increase risk for atherosclerosis.
- When medication and changes in lifestyle choices (e.g., nutrition) are not sufficient to lower risk for CVD, there are several surgical interventions that can help.
- Nonmodifiable factors for CVD include age, sex, genetics, and low birth weight.
- Modifiable risk factors include smoking, high blood pressure, elevated blood lipid levels, diabetes, obesity, and poor diet.
- Diets high in saturated fats or *trans* fats can increase risk for CVD, whereas those high in monounsaturated fatty acids decrease risk.

- Prevention and treatment of both obesity and diabetes involve restriction of caloric intake and increased activity levels.

How Does Nutrition Influence Cardiovascular Risk?

- Some dietary patterns such as the DASH diet, vegetarian diet, and Mediterranean diet are associated with decreased risk for CVD.

What Are the General Nutrition Guidelines for Healthy Hearts?

- Experts recommend that we consume a diet low in fat (<35% of calories), saturated fatty acids (≤10% of calories), *trans* fatty acids (minimal), cholesterol (<300 mg/d), and salt (<1 teaspoon per day).
- Fish and other foods rich in omega-3 fatty acids should be consumed on a regular basis as well as whole-grain foods and those containing soluble fiber.
- Consuming adequate amounts of some vitamins and minerals, such as calcium, potassium, and certain B vitamins, can reduce risk for CVD.

Notes

1. American Heart Association. Heart disease and stroke statistics—2011 update. Circulation. 2011;123:e18-e209.
2. Steinberg D, Witztum JL. Oxidized low-density lipoprotein and atherosclerosis. Atherosclerosis and Thrombosis in Vascular Biology. 2010;30:2311–6. Stocker R, Keaney JF, Jr. Role of oxidative modifications in atherosclerosis. Physiological Reviews. 2004;84:1381–478.

3. Berger JS, Brown DL, Becker RC. Low-dose aspirin in patients with stable cardiovascular disease: A meta-analysis. American Journal of Medicine. 2008;121:43–9.

4. Ridker PM, Rifai N, Rose L, Buring JE, Cook NR. Comparison of C-reactive protein and low-density lipoprotein cholesterol levels in the prediction of first cardiovascular events. New England Journal of Medicine. 2002;347:1557–65. Sakkinen P, Abbott RD, Curb JD, Rodriguez BL, Yano K, Tracy RP. C-reactive protein and myocardial infarction. Journal of Clinical Epidemiology. 2002;55:445–51. Alizadeh Dehnavi R, de Roos A, Rabelink TJ, van Pelt J, Wensink MJ, Romijn JA, Tamsma JT. Elevated CRP levels are associated with increased carotid atherosclerosis independent of visceral obesity. Atherosclerosis. 2008;200:417–23.

5. Pearson AP, Mensah GA, Alexander RW, Anderson JL, Cannon RO, Criqui M, Fadl YY, Fortmann SP, Hong Y, Myers GL, Rifai N, Smith SC, Taubert K, Tracy RP, Vinicor. Markers of inflammation and cardiovascular disease. Application to clinical and public health practice. A statement for healthcare professionals from the Centers for Disease Control and Prevention and the American Heart Association. Circulation. 2003;107:499–511.

6. American Heart Association. Risk factors and coronary heart disease. Available from: http://www.americanheart.org/presenter.jhtml?identifier=4726.

7. Katcher HI, Gillies PJ, Kris-Etherton PM. Atherosclerotic cardiovascular disease. In: Present knowledge in nutrition, 9th ed. Bowman BA, Russell RM, editors. Washington, DC: ILSI Press; 2006.

8. Hokanson JE, Austin MA. Plasma triglyceride level is a risk factor for cardiovascular disease independent of high-density lipoprotein cholesterol level: A meta-analysis of population-based prospective studies. Journal of Cardiovascular Risk. 1996;3:213–9. Krauss RM. Triglycerides and atherogenic lipoproteins: Rationale for lipid management. American Journal of Medicine. 1998;105:S58–S62.

9. Hawkins MA. Markers of increased cardiovascular risk. Obesity Research. 2004;12:107S–14S. Holvoet P. Oxidized LDL and coronary heart disease. Acta Cardiologica. 2004;59:479–84.

10. Gordon T, Castelli WP, Hjotland MC, Kannel WB, Dawber TR. High density lipoprotein as a protective factor against coronary heart disease. The Framingham study. American Journal of Medicine. 1977;62:707–14. Watson AD, Berliner JA, Hama SY, LaDu BN, Faull KF, Fogelman AM, Navab M. Protective effect of high density lipoprotein associated paraoxonase. Inhibition of the biological activity of minimally oxidized low density lipoprotein. Journal of Clinical Investigation. 1995;96:2882–91.

11. Hofnagel O, Luechtenborg B, Weissen-Plenz G, Robenek H. Statins and foam cell formation: Impact on LDL oxidation and uptake of oxidized lipoproteins via scavenger receptors. Biochimica Biophysica Acta. 2007;1771:1117–24. Pennathur S, Heinecke JW. Oxidative stress and endothelial dysfunction in vascular disease. Current Diabetes Reports. 2007;7:257–64.

12. Pereira MA, O'Reilly E, Augustsson K, Fraser GE, Goldbourt U, Heitmann BL, Hallmans G, Knekt P, Liu S, Pietinen P, Spiegelman D, Stevens J, Virtamo J, Willett WC, Ascherio A. Dietary fiber and risk of coronary heart disease. A pooled analysis of cohort studies. Archives of Internal Medicine. 2004;164: 370–6.

13. Clarke R, Smulders Y, Fowler B, Stehouwer CD. Homocysteine, B-vitamins, and the risk of cardiovascular disease. Seminars in Vascular Medicine. 2005;5:75–6. Eckel RH. Obesity and heart disease. A statement for health care professionals from the Nutrition Committee. American Heart Association. Circulation. 1997;96:3248–50.

14. Aucott L, Gray D, Rothnie H, Thapa M, Waweru C. Effects of lifestyle interventions and long-term weight loss on lipid outcomes—a systematic review. Obesity Reviews. 2011;12:e412–25.

15. American Heart Association. Guidelines for physical activity. Available at http://www.heart.org/HEARTORG/GettingHealthy/PhysicalActivity/GettingActive/American-Heart-Association-Guidelines_UCM_307976_Article.jsp. Updated January 19, 2011.

16. Lenz TL, Monaghan MS. Lifestyle modifications for patients with hypertension. Journal of the American Pharmacological Association. 2008;48:e92-9.

17. Weinberger MH. Sodium and blood pressure 2003. Current Opinions in Cardiology. 2004;19:353–6.

18. Miller ER 3rd, Erlinger TP, Appel LJ. The effects of macronutrients on blood pressure and lipids: an overview of the DASH and OmniHeart trials. Current Atherosclerosis Reports. 2006;8:460–5.

19. Mattson FH, Erickson BA, Kligman AM. Effect of dietary cholesterol on serum cholesterol in man. American Journal of Clinical Nutrition. 1972;25:589–94.

20. Mozaffarian D, Katan MB, Ascherio A, Stampfer MJ, Willett WC. Trans fatty acids and cardiovascular disease. New England Journal of Medicine. 2006;354:1601–13. Woodside JV, McKinley MC, Young IS. Saturated and trans fatty acids and coronary heart disease. Current Atherosclerosis Reports. 2008;10:460–6.

21. StJeor ST, Howard BV, Prewitt TE, Bovee V, Bazzarre T, Eckel RH. Dietary protein and weight reduction: A statement for healthcare professionals from the Nutrition Committee of the Council on Nutrition, Physical Activity, and Metabolism of the American Heart Association. Circulation. 2001;104:1869–74.

22. Katcher HI, Gillies PJ, Kris-Etherton PM. Atherosclerotic cardiovascular disease. In: Present knowledge in nutrition, 9th ed. Bowman BA, Russell RM, editors. Washington, DC: ILSI Press; 2006.

23. Kerksick CM, Wismann-Bunn J, Fogt D, et al. Changes in weight loss, body composition and cardiovascular disease risk after altering macronutrient distributions during a regular exercise program in obese women. Nutrition Journal. 2010;9:59–64. Larsen TM, Dalskov SM, van Baak M, et al. Diets with high or

low protein content and glycemic index for weight-loss maintenance. New England Journal of Medicine. 2010;363:2102–13.

24. Brinkworth GD, Noakes M, Buckley JD, Keogh JB, Clifton PM. Long-term effects of a very-low-carbohydrate weight loss diet compared with an isocaloric low-fat diet after 12 mo. American Journal of Clinical Nutrition. 2009; 90:23–32.

25. Katcher HI, Hill AM, Lanford JL, Yoo JS, Kris-Etherton PM. Lifestyle approaches and dietary strategies to lower LDL-cholesterol and triglycerides and raise HDL-cholesterol. Endocrinology and Metabolism Clinics North America. 2009;38:45–78.

26. Zaloga GP, Harvey KA, Stillwell W, Siddiqui R. Trans fatty acids and coronary heart disease. Nutrition in Clinical Practice. 2006;21:505–12. Katcher HI, Hill AM, Lanford JL, Yoo JS, Kris-Etherton PM. Lifestyle approaches and dietary strategies to lower LDL-cholesterol and triglycerides and raise HDL-cholesterol. Endocrinology and Metabolism Clinics North America. 2009;38:45–78.

27. Ascherio A, Katan MB, Zock PL, Stampfer MJ, Willett WC. Trans fatty acids and coronary heart disease. New England Journal of Medicine. 1999;340:1994–8.

28. Brouwer IA, Wanders AJ, Katan MB. Effect of animal and industrial trans fatty acids on HDL and LDL cholesterol levels in humans—a quantitative review. PLoS One. 2010;5:e9434. Remig V, Franklin B, Margolis S, Kostas G, Nece T, Street JC. Trans fats in America: a review of their use, consumption, health implications, and regulation. Journal of the American Dietetic Association. 2010;110:585–92.

29. Ascherio A. Trans fatty acids and blood lipids. Atherosclerosis Supplements. 2006;7:25–7. Mozaffarian D. Trans fatty acids—effects on systemic inflammation and endothelial function. Atherosclerosis Supplements. 2006;7(2):29–32.

30. Gebauer SK, Psota TL, Kris-Etherton PM. The diversity of health effects of individual trans fatty acid isomers. Lipids. 2007;42:787–99. Mensink RP. Metabolic and health effects of isomeric fatty acids. Current Opinions in Lipidology. 2005;16:27–30.

31. Pfeuffer M, Schrezenmeir J. Bioactive substances in milk with properties decreasing risk of cardiovascular diseases. British Journal of Nutrition. 2000;84:S155–9.

32. Moorandian AD, Haas MJ, Wong NCW. The effect of select nutrients on serum high density lipoprotein cholesterol and apolipoprotein A-I levels. Endocrine Reviews. 2006;27:2–16.

33. Riediger ND, Othman RA, Suh M, Moghadasian MH. A systemic review of the roles of n-3 fatty acids in health and disease. Journal of the American Dietetic Assocociation. 2009;109:668–79.

34. Geleijnse JM, Giltay EF, Grobbee DE, Donders AR, Kok FJ. Blood pressure response to fish oil supplementation: Metaregression analysis of randomized trials. Journal of Hypertension. 2002;20:1493–9.

35. Galli C, Risé P. Fish consumption, omega 3 fatty acids and cardiovascular disease. The science and the clinical trials. Nutrition and Health. 2009;20:11–20.

36. Mori TA, Beilin LJ. Omega-3 fatty acids and inflammation. Current Atherosclerosis Reports. 2004;5:461–7. Zhao G, Etherton TD, Martin KR, et al. Anti-inflammatory effects of polyunsaturated fatty acids in THP-1 cells. Biochemical and Biophysical Research Communications. 2005;336:909–17. Zampelas A. Eicosapentaenoic acid (EPA) from highly concentrated n-3 fatty acid ethyl esters is incorporated into advanced atheroscleroptic plaques and higher plaque EPA is associated with decreased plaque inflammation and increased stability. Altherosclerosis. 2010;212:34–5.

37. Nettleton JA, Koletzko B, Hornstra G. ISSFAL 2010 dinner debate: Healthy fats for healthy hearts—annotated report of a scientific discussion. Annals of Nutrition and Metabolism. 2011;58:59–65.

38. Pearson TA. Alcohol and heart disease. Circulation. 1996;94:3023–25. Renaud S, de Logeril M. Wine, alcohol, platelets and the French paradox for coronary heart disease. Lancet. 1992;339:1523–6.

39. McCully KS. Homocysteine, vitamins, and vascular disease prevention. American Journal of Clinical Nutrition. 2007;86:1563S–8S. Wang X, Qin X, Demirtas H, Li J, Mao G, Huo Y, Sun N, Liu L, Xu X. Efficacy of folic acid supplementation in stroke prevention: A meta-analysis. Lancet. 2007;369:1876–82.

40. Ciaccio M, Bellia C. Hyperhomocysteinemia and cardiovascular risk: effect of vitamin supplementation in risk reduction. Current Clinical Pharmacology. 2010;5:30–6.

41. Knekt P, Ritz J, Pereira MA, O'Reilly EJ, Augustsson K, Fraser GE, Goldbourt U, Heitmann BL, Hallmans G, Liu, S, Pietinen P, Spiegelman D, Stevens J, Virtamo J, Willett WC, Rimm EB, Ascherio A. Antioxidant vitamins and coronary heart disease risk: A pooled analysis of 9 cohorts. American Journal of Clinical Nutrition. 2004;80:1508–20. Wenger NK. Do diet, folic acid, and vitamins matter? What did we learn from the Women's Health Initiative, the Women's Health Study, the Women's Antioxidant and Folic Acid Cardiovascular Study, and other clinical trials? Cardiology Reviews. 2007;15:288–90.

42. Liepa GU, Basu H. C-reactive proteins and chronic disease: What role does nutrition play? Nutrition in Clinical Practice. 2003;18:227–33. Singh U, Devaraj S, Jialal I. Vitamin E, oxidative stress, and inflammation. Annual Review of Nutrition. 2005;25:151–74.

43. Badimon L, Vilahur G, Padro T. Nutraceuticals and atherosclerosis: human trails. Cardiovascular Therapies. 2010;28:202–15. Castro IA, Barroso LP, Sinnecker P. Functional foods for coronary heart disease risk reduction: A meta-analysis using a multivariate approach. American Journal of Clinical Nutrition. 2005;82:32–40. Gylling H, Miettinen TA. The effect of plant stanol- and sterol-enriched foods on lipid metabolism, serum lipids and coronary heart disease. Annals of Clinical Biochemistry. 2005;42:254–63.

44. American Heart Association Nutrition Committee. Diet and lifestyle recommendations revision 2006: A scientific statement from the American Heart Association Nutrition Committee. Circulation.

2006;114:82–96. Wang L, Manson JE, Buring JE, Lee IM, Sesso HD. Dietary intake of dairy products, calcium, and vitamin D and the risk of hypertension in middle-aged and older women. Hypertension. 2008;51:1073–9.

45. Bendinelli B, Masala G, Saieva C, Salvini S, Calonico C, Sacerdote C, Agnoli C, Grioni S, Frasca G, Mattiello A, Chiodini P, Tumino R, Vineis P, Palli D, Panico S. Fruit, vegetables, and olive oil and risk of coronary heart disease in Italian women: the EPICOR Study. American Journal of Clinical Nutrition. 2011;93:275–83. Bhupathiraju SN, Tucker KL. Coronary heart disease prevention: Nutrients, foods, and dietary patterns. Clinica Chimica Acta. 2011;412:1493-514. Oliveira A, Rodríguez-Artalejo F, Gaio R, Santos AC, Ramos E, Lopes C. Major habitual dietary patterns are associated with acute myocardial infarction and cardiovascular risk markers in a southern European population. Journal of the American Dietetic Association. 2011;111:241–50.

Energy Balance and Body Weight Regulation

Some count them, others curse them—but what role do calories actually play in body weight regulation? Moreover, while most people struggle to lose weight, some actually find it difficult to gain weight. If you have ever pondered these issues, you are not alone. In this chapter, you will learn about the relationships among energy intake, energy expenditure, and body weight. You will also understand how the foods you eat and the activities you engage in influence body weight. Factors related to body weight regulation, the growing prevalence of obesity in the United States, and weight loss and weight maintenance are addressed as well.

The Decision to Have Gastric Bypass Surgery

As far back as she can remember, August always had a weight problem. She certainly was larger and heavier than most of her peers. Although she tried to control her eating, it seemed the more August restricted her food intake, the hungrier she became. When the hunger became overwhelming, August felt out of control, and the cycle of food restriction followed by overeating would start again. Looking back, she recognizes that she never felt full. Hunger was a continuous force that drove August to eat.

As an adult, August felt tremendous guilt about her inability to lose weight. The drive to eat was overpowering. By the time she was 30, she weighed 275 pounds, making her feel self-conscious and sad. She found it difficult to look for a job, knowing that she might be judged on the basis of her weight, rather than her skill and ability.

In spite of these problems, August and her husband started a family, and had two beautiful children. August enjoyed staying home and raising them, but her weight continued to make life challenging. The simple act of sitting on the floor with her children was becoming increasingly difficult. Even walking them to school made her knees ache and her heart pound. Her health was deteriorating because of the excess weight. Diet after diet, nothing seemed to work.

Feeling hopeless, August finally found the help she needed. After a routine health exam, she and her physician discussed her long struggle with weight, and he suggested something she never considered before—gastric bypass surgery. He cautioned her that this was not the solution for everyone and gave her educational material to read. She soon learned that gastric bypass surgery would help her to feel full and eat less by reducing the size of her stomach. Food would also be rerouted through her digestive tract so that it bypassed a portion of her small intestine. Although there were clear benefits to gastric bypass surgery, like any surgical procedure, it was not without risks.

The following year, August was approved for gastric bypass surgery. She was fully aware that this radical procedure would change her life forever. It would mean eating differently and having to be careful that her nutrient needs were met. She was also told that she might experience recurring bouts of diarrhea and vomiting. Nonetheless, August decided this was her best hope.

It took several weeks for August to recover from the surgery and even longer to adjust to the physiological changes to her digestive system. Most importantly, however, August experienced something she had never experienced before—she did not feel the persistent sensation of hunger. In fact, she felt satisfied after each meal. Soon, the weight started coming off—50 pounds, 75 pounds, and eventually 125 pounds. Her physical and mental health improved. Today, August is back in school and studying to be a nurse. We asked her if she had any thoughts about obesity she wanted to share with others. In August's words:

"Gastric bypass surgery is a last resort, but for some people it is the lifeline we need. My quality of life has improved dramatically since having this procedure. Also, I want you to know that obesity is very complicated, and it is not something I would wish on anyone. Please don't judge people or think that because someone is obese they simply lack willpower. If I had one wish in life, it would be to not battle my body anymore."

© Ryan McVay/Lifesize/Getty Images

Critical Thinking: August's Story

What insights did you gain from reading August's story about what it is like to be obese? If you heard that someone you know was having gastric bypass surgery, what would you think?

What Is Energy Balance?

What determines whether the energy in the foods you eat is used to fuel your body or stored for later use? It all comes down to balance—the balance between energy intake and energy expenditure. In other words, the amount of energy you consume should be in balance with the amount of energy your body requires. Energy intake and energy expenditure are two important components of energy balance, with the third being energy stored (or mobilized). This relationship is often expressed as Energy stored = Energy consumed − Energy expended. Note that "Energy stored" can be either positive or negative depending on whether the amount of energy consumed is equal to, less than, or greater than that expended. When energy intake equals energy expenditure, a person is in a state of **energy balance** (energy intake = energy expenditure). When a person is in energy balance, body weight tends to be relatively stable. However, an **energy imbalance** can arise when the amount of energy consumed does not equal the amount of energy the body uses. Consuming more energy than the body needs puts you in a state of **positive energy balance,** which results in weight gain. Conversely, someone is in a state of **negative energy balance** when energy intake is less than energy expenditure, which results in weight loss. A change in body weight is a useful indicator of whether you are in positive or negative energy balance.

ENERGY BALANCE AFFECTS BODY WEIGHT

Body weight increases during periods of positive energy balance and decreases in response to negative energy balance. This relationship is illustrated in Figure 11.1. When a person is in positive energy balance, the amount of muscle or adipose tissue—or both—increases because excess energy is stored as protein (lean tissue) and/or fat (adipose tissue). This is desirable during periods of growth such as infancy, childhood, adolescence, and pregnancy. However, when increased body weight is primarily associated with increased body fat, positive energy balance often unhealthy.

Negative energy balance is a result of insufficient energy intake, excessive energy expenditure, or both. Under this condition, stored energy reserves are broken down, resulting in decreased body weight. Adipose tissue is the body's primary energy reserve, and is broken down during negative energy balance. One pound of adipose tissue, including supporting lean tissue, is equivalent to approximately 3,500 kcal. However, it is important to recognize that weight loss associated with negative energy balance is not always due entirely to loss of body fat. Water and muscle loss can also contribute to weight loss.

A Closer Look at Adipose Tissue There is a tendency to think of adipose tissue as a passive reservoir that buffers imbalances between energy intake and expenditure. However, adipose tissue is anything but passive. The source of several hormones, it plays an active role in regulating energy balance. The discovery that adipocytes are a source of hormones and hormone-like substances collectively called **adipokines,** has led many researchers to consider adipose tissue to be the largest endocrine organ in the body.[1] Adipose-derived hormones and adipokines provide an important communication link between adipocytes and other tissues and organs. In addition, many adipokines regulate immune and inflammatory processes, specifically those thought to be involved in weight-related diseases, such as type 2 diabetes.[2]

Adipose tissue is comprised largely of specialized cells called **adipocytes,** which contain a lipid-filled core that consists primarily of triglycerides. The number and size of adipocytes determine the amount of adipose tissue in

Zia Soleil/Iconica/Getty Images

Weighing yourself over time is a good way to determine if you are in positive or negative energy balance.

energy balance A state in which energy intake equals energy expenditure.

energy imbalance A state in which the amount of energy consumed does not equal the amount of energy used by the body.

positive energy balance A state in which energy intake is greater than energy expenditure.

negative energy balance A state in which energy intake is less than energy expenditure.

adipokines Hormone-like substances produced and released by adipocytes.

adipocytes Cells found in adipose tissue and used mainly for fat storage.

FIGURE 11.1 The Three States of Energy Balance
A change in body weight can indicate if a person is in positive or negative energy balance.

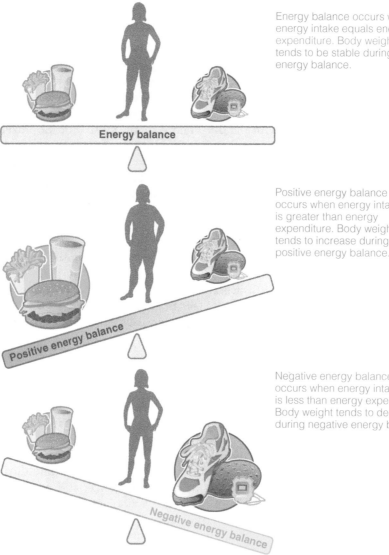

Energy balance occurs when energy intake equals energy expenditure. Body weight tends to be stable during energy balance.

Positive energy balance occurs when energy intake is greater than energy expenditure. Body weight tends to increase during positive energy balance.

Negative energy balance occurs when energy intake is less than energy expenditure. Body weight tends to decrease during negative energy balance.

© Susumu Nishinaga/Photo Researchers, Inc.

Adipose tissue consists of cells called adipocytes.

hypertrophic growth Growth associated with an increase in cell size.

hyperplastic growth Growth associated with an increase in cell number.

visceral adipose tissue (VAT) Adipose tissue deposited between the internal organs in the abdominal area.

subcutaneous adipose tissue (SCAT) Adipose tissue found directly beneath the skin.

the body. As an adipocyte fills with triglycerides, its size increases, a process called **hypertrophic growth.** To accommodate large amounts of lipid, the diameter of an adipocyte can increase 20-fold. When existing cells are full, new adipocytes are formed, a process called **hyperplastic growth.** When a person loses body fat, enlarged adipocytes return to normal size; however, the number of adipocytes remains constant.[3]

Some scientists believe that the number and size of adipocytes may influence our ability to maintain a healthy weight. For example, people with fewer, larger adipocytes may have less difficulty maintaining weight loss than those with a greater number of smaller adipocytes.[4] Figure 11.2 shows how adipocyte number and size change in response to weight gain and loss.

Visceral Adipose Tissue vs. Subcutaneous Adipose Tissue Although adipose tissue is found throughout the body, **visceral adipose tissue (VAT),** also called intra-abdominal fat, refers to adipose tissue surrounding the internal organs in the torso (Figure 11.3). However, much of our body fat reserve, which is called **subcutaneous adipose tissue (SCAT),** is found directly beneath the skin. SCAT is found throughout the body but is most predominant in the thighs, hips, and buttocks.

FIGURE 11.2 Hypertrophic and Hyperplastic Growth of Adipose Tissue The amount of adipose tissue a person has depends on the number and size of his or her adipose cells (adipocytes).

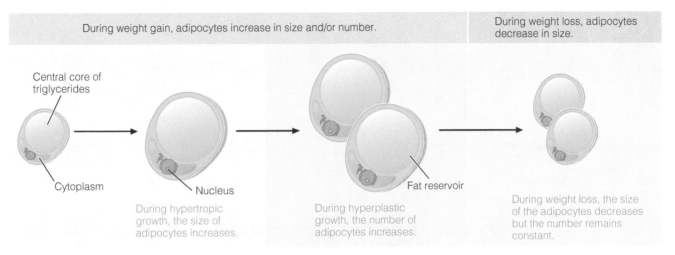

During weight gain, adipocytes increase in size and/or number.

During weight loss, adipocytes decrease in size.

Central core of triglycerides

Cytoplasm

Nucleus

Fat reservoir

During hypertropic growth, the size of adipocytes increases.

During hyperplastic growth, the number of adipocytes increases.

During weight loss, the size of the adipocytes decreases but the number remains constant.

What Determines Energy Intake?

Although it seems simple, understanding energy balance is more complicated than one might think. Your body gets energy from the foods you eat, but understanding what makes you eat and why you make certain choices is complex. Your food choices are related to many facets of your life, and the choices you make serve numerous purposes beyond providing your body with nourishment. It is also important to recognize that foods we enjoy as adults may be the very foods we avoided as children. Clearly, eating behavior is complicated, and it is important to consider the psychological, physical, social, and cultural forces that influence it.

HUNGER AND SATIETY ARE PHYSIOLOGICAL INFLUENCES ON ENERGY INTAKE

Hunger and satiety are complex physiological states that greatly influence energy intake. **Hunger** is defined as the basic physiological drive to consume food, whereas **satiety** is the physiological response to having eaten enough, resulting in food intake cessation. Humans tend to be periodic eaters, meaning that they usually eat meals at predictable and discrete times throughout the day. Most of the time, people eat to the point of comfort. In other words, they eat until they are satiated. Still, people sometimes continue to eat after they are full, even to the point of discomfort. Regardless of one's level of fullness, however, a person is usually ready to eat again within three to four hours of when they last ate.

We tend to think that factors associated within our stomachs are the main determinants of hunger and satiety. However, years ago researchers discovered that hunger and satiety could be controlled in mice by stimulating certain regions within their brains. These areas soon became known as the hunger and satiety centers. Scientists now recognize that energy intake is regulated, in part, by specific neural connections (neurons) to and within the brain, rather than by distinct hunger and satiety "centers."[5]

FIGURE 11.3 Visceral Adipose Tissue (VAT) and Subcutaneous Adipose Tissue (SCAT) Visceral adipose tissue (VAT) refers to adipose tissue deposited between the internal organs in the abdomen. Subcutaneous adipose tissue (SCAT) is found directly beneath the skin.

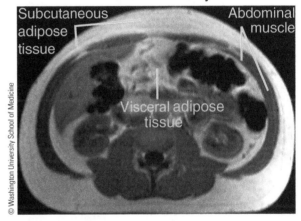

Front of the body

Subcutaneous adipose tissue

Abdominal muscle

Visceral adipose tissue

© Washington University School of Medicine

Back of the body

hunger The physiological drive to consume food.

satiety The state in which hunger is satisfied and a person feels he or she has had enough to eat.

FIGURE 11.4 Regulation of Short-Term Food Intake Signals originating from the gastrointestinal tract stimulate the brain to regulate short-term food intake by releasing neurotransmitters that influence meal initiation and termination.

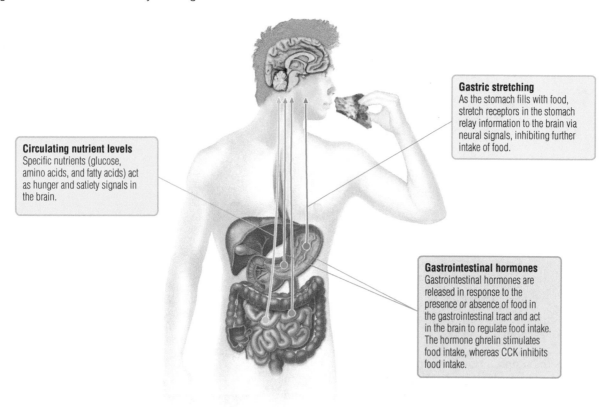

Gastric stretching
As the stomach fills with food, stretch receptors in the stomach relay information to the brain via neural signals, inhibiting further intake of food.

Circulating nutrient levels
Specific nutrients (glucose, amino acids, and fatty acids) act as hunger and satiety signals in the brain.

Gastrointestinal hormones
Gastrointestinal hormones are released in response to the presence or absence of food in the gastrointestinal tract and act in the brain to regulate food intake. The hormone ghrelin stimulates food intake, whereas CCK inhibits food intake.

One part of the brain that is critical to regulating hunger and satiety is the **hypothalamus.** The hypothalamus receives signals that influence hunger and satiety from different parts of the body, such as the gastrointestinal tract. In response, the brain releases neurotransmitters, which in turn influence hunger and satiety. **Neurotransmitters** are hormone-like chemical messengers, produced by neurons, which affect the central nervous system. Some neurons in the brain release **catabolic neurotransmitters** that promote satiety and meal termination, whereas other neurons release **anabolic neurotransmitters** that promote hunger and initiate eating. In addition to hunger and satiety, neurotransmitters also influence energy expenditure and play important roles in long-term regulation of body weight.

Both hunger and satiety are influenced by a variety of factors, including gastric distention (stretching), circulating nutrient levels, and GI hormones. As shown in Figure 11.4, these sensations exert their influence by sending signals to the hypothalamus, which, in turn, controls the body's hunger and satiety responses. However, hunger and satiety are also influenced by other factors such as stress and emotions.

Gastric Stretching—A Satiety Signal The distention of the stomach wall in response to the presence of food, called gastric stretching, provides a powerful satiety signal. The GI tract has specialized receptors (mechanoreceptors) that are responsive to this type of stretching. When you consume small amounts of food, gastric stretching is minimal. However, as the volume of food increases, stretch receptors are stimulated. Stretching of the stomach wall initiates the release of neural signals, which convey information to the brain. The brain responds by releasing neurotransmitters that elicit the sensation of satiety.[6]

hypothalamus An area of the brain that controls many involuntary functions by the release of hormones and neurotransmitters.

neurotransmitter A hormone-like chemical messenger released by nerve cells.

catabolic neurotransmitter A substance released by nerve cells that inhibits hunger and/or stimulates energy expenditure.

anabolic neurotransmitter A substance released by nerve cells that stimulates hunger and/or decreases energy expenditure.

FOCUS ON CLINICAL APPLICATIONS
Bariatric Surgery

Bariatrics is the branch of medicine concerned with treating obesity and obesity-related conditions. The term *bariatric surgery* refers to a surgical procedure that promotes weight loss. There are different types of bariatric surgery, each having potential side effects and risks. However, the health benefits associated with weight loss often outweigh the risks associated with surgery. The Food and Drug Administration (FDA) recently released new guidelines that lowered the eligibility criteria for the gastric band procedure. Patients with a BMI 30 kg/m² or greater can now be considered for gastric banding if weight-loss diets or drugs have not worked and if they have obesity-related health problems.[7]

In general, bariatric surgery alters the digestive tract so that only small amounts of food can be consumed and/or digested. For example, a minimally invasive procedure called gastric banding creates a small stomach pouch with a food-holding capacity of about 2 to 3 tablespoons (about the size of an egg). The pouch is formed by surgical implantation of a band-like device that fits around the upper portion of the stomach. The size of the opening between the upper and lower portions of the stomach can be increased or decreased by adjusting the diameter of the band. The movement of food is slower when the diameter of the band is made smaller, helping the individual feel full.

Remember that the stomach normally holds 4 to 6 cups of food. In individuals with gastric banding, it only takes a small amount of food to signal the feeling of satiety. It is important with this type of surgery that the person is careful not to overeat, to eat foods that are soft and moist, and to thoroughly chew food before swallowing. Common side effects include heartburn, abdominal pain, and vomiting. Because food intake is severely restricted, it may be necessary to take nutrient supplements to ensure that nutritional needs are met. The gastric band can be removed at any time, and the original form and function of the stomach is restored.

Another type of bariatric surgery is called gastric bypass. In addition to reducing the size of the stomach, this procedure also reroutes the GI tract to bypass a segment of the small intestine. To do this, the lower portion of the reduced stomach pouch is connected to the middle portion of the small intestine, bypassing much of the duodenum and jejunum. Digestive secretions from the remaining portion of the stomach and those from the gallbladder and pancreas are rerouted such that they are released into the lower portion of the small intestine. This procedure not only helps reduce food intake but also limits nutrient digestion and absorption. The risk of developing nutritional deficiencies is much greater after this procedure than after gastric banding. The rapid movement of food through the small intestine causes many adverse side effects, including **dumping syndrome.** Dumping syndrome causes cramps, nausea, diarrhea, and dizziness. Surgical reversal of gastric bypass is difficult and can pose certain risks to the individual.

About 20% of people who have bariatric surgery experience complications.[8] Furthermore, depending on an individual's age, sex, and general health, the death rate within one year of surgery is 1 to 5%.[9] A recent study followed more than 4,000 obese subjects, half of whom underwent bariatric surgery while the other half received conventional treatment, over an 11-year period.[10] The study found that bariatric surgery significantly improved survival in obese individuals. Other studies suggest that gastric bypass surgery may also help lower death rates associated with type 2 diabetes, heart disease, and cancer.[11] Still, one must give serious consideration to the risks before undergoing either of these procedures.

The number of weight-loss surgeries performed yearly in the United States is on the rise. Most people lose weight immediately following the surgery, and within five years have lost 60% of their excess weight. Although patients with gastric banding lose weight more slowly than those who have gastric bypass, the banding procedure has a lower mortality rate. Long-term effectiveness depends on a person's willingness to eat a healthy, well-balanced diet and exercise regularly. For many severely obese individuals, weight-loss surgery can improve their quality of life, both physically and emotionally.[12] It is important to remember that any type of bariatric surgery is a life-altering procedure that requires a commitment to adhere to strict dietary changes.

Obese people tend to have larger stomachs than do lean people and therefore can accommodate larger volumes of food before gastric stretching triggers satiety. This may be why some severely obese people do not readily feel full (satiated) after a meal. Some severely obese people opt to have weight-loss surgery, otherwise known as **bariatric surgery.** As shown in Figure 11.5, **gastric banding** reduces the size of the stomach, whereas **gastric bypass** also reroutes the flow of food through the GI tract. Although both types of procedures limit how much food can be consumed, the latter also reduces nutrient digestion and absorption. Because the size of the stomach has been drastically

bariatrics The branch of medicine concerned with the treatment of obesity.

dumping syndrome A condition whereby food moves too rapidly from the stomach into the small intestine.

bariatric surgery Surgical procedure performed to treat obesity.

gastric banding A type of bariatric surgery in which an adjustable, fluid-filled band is wrapped around the upper portion of the stomach, dividing it into a small upper pouch and a larger lower pouch.

FIGURE 11.5 Types of Weight-Loss Surgery The term bariatric surgery refers to medical procedures that promote weight loss. To be considered for weight-loss surgery, a person must meet strict eligibility guidelines such as being unable to achieve or maintain a healthy weight through diet and exercise and also have weight-related health problems.

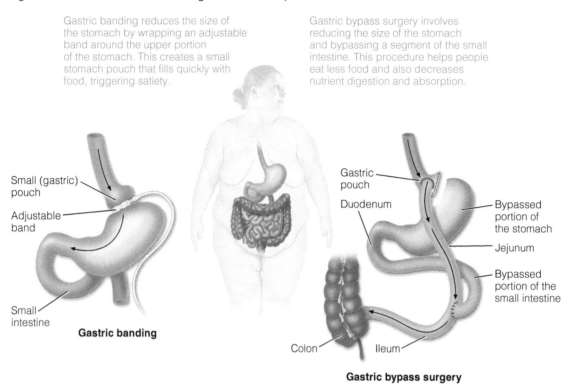

Gastric banding reduces the size of the stomach by wrapping an adjustable band around the upper portion of the stomach. This creates a small stomach pouch that fills quickly with food, triggering satiety.

Gastric bypass surgery involves reducing the size of the stomach and bypassing a segment of the small intestine. This procedure helps people eat less food and also decreases nutrient digestion and absorption.

Small (gastric) pouch

Adjustable band

Small intestine

Gastric banding

Gastric pouch

Duodenum

Bypassed portion of the stomach

Jejunum

Bypassed portion of the small intestine

Colon   Ileum

Gastric bypass surgery

reduced, the stomach quickly fills with food, triggering satiety. You can read more about bariatric surgery in the Focus on Clinical Applications feature.

High-Volume Foods and Satiety Recent studies show that eating high-volume foods can help people feel full longer than when they eat low-volume foods.[13] High-volume foods are those with high water and/or fiber content, such as fruits and vegetables. These types of foods increase gastric stretching, which can help people feel full and satisfied, even when they have not consumed a significant amount of calories. For example, two cups of grapes have the same amount of energy (~100 kcal) as ¼ cup of raisins. However, because the volume of the grapes is greater than raisins, a person is more likely to feel satisfied after eating the grapes. Some studies suggest that long-term consumption of a low-energy-dense diet may help people lose weight.

Circulating Nutrient Concentrations The concentration of certain nutrients in the blood can also influence hunger and satiety. For example, the brain is very sensitive to changes in blood glucose levels. When blood glucose increases following a meal, the brain responds by releasing neurotransmitters that stimulate satiety, providing a signal to stop eating. Conversely, when the brain detects a decrease in blood glucose, it releases neurotransmitters that stimulate hunger.

Amino acids can also play a role in short-term regulation of food intake. In general, elevated levels of circulating amino acids promote satiety by signaling neurons in the brain to release catabolic neurotransmitters. Of particular interest is the effect of the amino acid tryptophan on food intake. The brain uses tryptophan to synthesize the neurotransmitter serotonin, which conveys the sensations of satiety and relaxation. Some people find that eating tryptophan-rich foods, such as turkey and dairy products, promotes sleepiness in addition to satiety.[14] There is also evidence that disturbances in serotonin production can disrupt appetite regulation and may lead to obesity.[15]

High-volume, low-energy-dense foods such as fruits and vegetables have high water and fiber contents and are low in fat. These foods may trigger satiety more effectively than other foods.

gastric bypass A surgical procedure that reduces the size of the stomach and bypasses a segment of the small intestine so that fewer nutrients are absorbed.

TABLE 11.1 Gastrointestinal Hormones and Effect on Food Intake

Gastrointestinal Hormone	Stimulus for Release	Site of Production	Effect on Food Intake
Cholecystokinin (CCK)	Protein and fatty acids	Small intestine	↓
Glucagon-like peptide 1 (GLP-1)	Nutritional signals and neural/hormonal signals from the gastrointestinal tract	Small and large intestine	↓
Ghrelin	Empty stomach	Stomach	↑
Enterostatin	Fatty acids	Stomach and small intestine	↓
Peptide YY (PYY)	Food in the GI tract	Small and large intestine	↓

Elevated levels of circulating lipids can also promote satiety. However, this effect appears to be weak and is easily overcome by pleasant sensations associated with fatty foods such as enhanced flavor and texture. As a result, consuming high-fat foods tends to lead to excess calorie intake and weight gain, not satiety.

Gastrointestinal Hormones The presence of food in the stomach and small intestine can trigger the release of several GI hormones, the majority of which promote satiety (Table 11.1).[16] Examples include cholecystokinin (CCK) and peptide YY (PYY). Of these, CCK is the best understood. CCK is released from intestinal cells (enterocytes), particularly in response to high-fat and protein foods, signaling the brain to decrease food intake. Although the mechanisms remain unclear, CCK appears to play a role in short-term satiety. The hormone PYY, also released from the intestinal lining following a meal, plays a similar role.

Not all GI hormones signal the sensation of satiety. For example, researchers have recently become interested in the role of **ghrelin,** a newly discovered hormone dubbed the "hunger hormone." This potent hunger-stimulating hormone is secreted by cells in the stomach lining and circulates in the blood. High levels of ghrelin in the blood serve as a pre-meal signal that stimulates the sensation of hunger. Ghrelin levels are highest prior to meals when the stomach is empty, but after food is consumed, ghrelin levels decrease. Recent evidence shows that some obese people may overproduce ghrelin, which could explain why they do not always experience a feeling of satiety following a meal.[17] There is also evidence that gastric bypass surgery may lower ghrelin levels, which in turn causes hunger to diminish.[18]

APPETITE ALSO AFFECTS ENERGY INTAKE

Two important responses that help maintain a healthy body weight are eating when you hungry and stopping when you are full. However, sometimes we eat for reasons other than hunger, and at times we eat past satiety. The stimuli that override hunger and satiety tend to be more psychological than physiological. Whereas **appetite** is the psychological longing or desire for food, a **food aversion** is a strong psychological dislike of a particular food or foods. Sometimes, even the thought or smell of certain foods can trigger an adverse physical response. Our appetite can be an important determinant of what and when we eat. When the desire for a specific food is especially compelling, it is often referred to as a **food craving.** You can read more about food cravings in the Focus on Food feature.

ghrelin A hormone, secreted by cells in the stomach, that stimulates food intake.

appetite A psychological desire for food.

food aversion A strong psychological dislike of a particular food.

food craving A strong psychological desire for a particular food.

Are food aversions the flip side of food cravings? Even the experts are unsure. The reality is that there are no definitive answers when it comes to explaining food cravings and aversions. Whereas a food craving is a powerful, irresistible, intense desire for a particular food, food aversions develop when certain foods are viewed as repugnant. Although certain emotional states can provoke food aversions, most are conditioned responses resulting from a paired association between physical discomfort and a particular food. Food aversions tend to be persistent and long lasting, whereas food cravings can come and go.

Almost everyone experiences food cravings, because food is rewarding and pleasurable. However, some people may be more sensitive than others to the pleasurable effects triggered by certain foods. The most commonly craved foods tend to be calorie-rich ones, such as cookies, cakes, chips, and chocolate.[19]

Women tend to experience food cravings more frequently than do men.[20] This may, in part, be due to hormonal fluctuations during the menstrual cycle. In fact, many women crave certain foods around the time of their menstrual flow. Hormonal changes associated with pregnancy may also cause some women to crave particular foods. In general, food cravings tend to occur at specific times during the day (late afternoon or early evening) and in response to stressful situations, such as when a person feels anxious.[21]

People once thought food cravings were caused by a lack of specific nutrients in the diet. However, this is not always the case. For example, a craving for potato chips is not necessarily caused by a lack of salt, nor do we crave steak because we simply need more protein. Food cravings can be strong, and scientists do not have a clear explanation for their occurrence. Nonetheless, new information on the role of neurochemicals is unfolding.

During pregnancy, many women experience cravings for certain foods and/or aversions to others. Although some experts believe that food cravings help pregnant women satisfy their needs for nutrients and that food aversions may provide protection from harmful substances, there is very little scientific data to support either of these claims. In other words, a craving for ice cream does not necessarily mean that a woman is calcium deficient.

For some people, food cravings can be overwhelming, and learning how to tame them is important. Most health experts agree that chronic overconsumption of food is more likely than food cravings, *per se*, to cause weight gain. In fact, overly restrictive food regimens may actually cause food cravings. Experts believe it is healthier to indulge our cravings within reason, rather than making certain foods "off limits" or becoming preoccupied with lingering thoughts and desires for particular foods. Furthermore, getting adequate sleep, participating in regular exercise, practicing relaxation techniques to reduce stress, and eating a healthy diet may help food cravings become less persistent.

Scientists do not fully understand what causes a person to have food cravings. Calorie-rich foods such as cookies, ice cream, cakes, and chocolate are the foods people most often crave.

Appetite can easily be aroused by sensory factors such as the appearance, taste, or smell of food. For example, the pleasing smell of baked bread makes many people want to eat, regardless of whether they are hungry. Likewise, unpleasant odors can "spoil" our appetite, even when we are hungry. Emotional states can also dramatically affect appetite. Whereas some people respond to emotions such as fear, depression, disappointment, excitement, and stress by eating, others respond by not eating at all. Clearly our emotional states can have a profound influence on appetite, overriding the normal physiological cues of hunger and satiety. However, when psychological factors are the primary determinant of hunger and satiety, eating disorders such as anorexia nervosa and bulimia nervosa can occur. You can read more about eating disorders in the Nutrition Matters following this chapter.

Now that you understand how your body regulates energy intake, we turn to the other component of energy balance: energy expenditure.

What Determines Energy Expenditure?

As you have learned, a complex interplay of factors influences when and how much you eat. However, energy intake is only one component of energy balance; another is energy expenditure. The body expends energy to maintain physiological functions, support physical activity, and process food, all of which collectively make up **total energy expenditure (TEE).** As illustrated in Figure 11.6, TEE has three main components: (1) basal metabolism, (2) physical activity, and (3) thermic effect of food. TEE also includes **adaptive thermogenesis** and **nonexercise activity thermogenesis (NEAT)**. Adaptive thermogenesis is a temporary change in energy expenditure that enables the body to adapt to such things as changes in the environment or to physiological conditions such as trauma, starvation, or stress. Shivering in response to cold is an example of adaptive thermogenesis. NEAT is energy expended for spontaneous movements such as fidgeting and maintaining posture. The contributions of adaptive thermogenesis and NEAT to TEE have yet to be determined, but they are likely small. The three main components of TEE are discussed next.

BASAL METABOLISM ACCOUNTS FOR MOST OF TEE

Basal metabolism is the energy (kcal) expended to sustain basic, involuntary life functions such as respiration, beating of the heart, nerve function, and muscle tone. **Basal metabolic rate (BMR)**, defined as the amount of

total energy expenditure (TEE) Total energy expended or used by the body.

adaptive thermogenesis Energy expended in response to changes in the environment or to physiological conditions.

nonexercise activity thermogenesis (NEAT) Energy expended for spontaneous movement such as fidgeting and maintaining posture.

basal metabolism Energy expended to sustain metabolic activities related to basic vital body functions such as respiration, muscle tone, and nerve function.

basal metabolic rate (BMR) Energy expended to support basal metabolism (expressed as kcal/hour).

FIGURE 11.6 Major Components of Total Energy Expenditure (TEE)

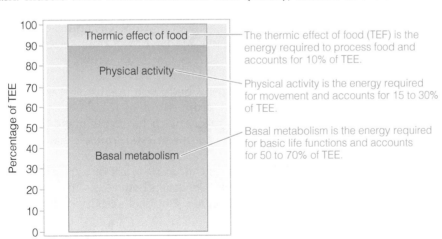

SOURCE: Adapted from: Lowell BB, Spiegelman. Towards a molecular understanding of adaptive thermogenesis. Nature. 2000;404:652–60.

energy expended per hour (kcal/hour) for these functions, accounts for most of TEE—approximately 50 to 70%. **Basal energy expenditure (BEE)** is basal metabolism expressed over a 24-hour period (kcal/day).

BMR is measured in such a way that energy expenditure associated with the processing of food and physical activity is eliminated. To accomplish this, BMR is typically measured in the morning, after eight hours of sleep, in a temperature-controlled room, and in a fasting state. These stringent conditions make measuring BMR difficult, which is why clinicians often measure **resting metabolic rate (RMR)** instead. Measuring RMR is not as difficult as measuring BMR, requiring only a brief resting period and no fasting. Because of this, RMR is approximately 10% higher than BMR. When resting metabolism is expressed over a 24-hour period, the term **resting energy expenditure (REE)** is used.

Using the Harris-Benedict Equation to Estimate Resting Energy Expenditure (REE) The Harris-Benedict equation is a mathematical formula developed almost 100 years ago and is still used by clinicians to estimate a person's REE. This equation is based on sex, age, height, and weight. It is calculated as follows.

Males: REE $= 66.5 + [13.8 \times \text{weight (kg)}] + [5 \times \text{height (cm)}] - [6.8 \times \text{age (y)}]$

Females: REE $= 655.1 + [9.6 \times \text{weight (kg)}] + [1.8 \times \text{height (cm)}] - [4.7 \times \text{age (y)}]$

The following sample calculation provides an example of how the Harris-Benedict equation is used to estimate REE. Consider a 200 lb (91 kg),* 20-year-old male, who is 6 feet, 3 inches (75 inches, 190.5 cm)[†] tall. Using the Harris-Benedict equation for males, he would require about 2,140 kcal/day to maintain basic body functions. Note that this value is just an estimate and does not take into account effects of physical fitness or body composition on energy expenditure.

Example of Resting Energy Expenditure (REE) Calculation

REE $= 66.5 + [13.8 \times \text{weight (kg)}] + [5 \times \text{height (cm)}] - [6.8 \times \text{age (y)}]$

REE $= 66.5 + 13.8(91) + 5(190.5) - 6.8(20)$

REE $= 66.5 + 1255.8 + 952.5 - 136$

REE $= 2{,}138.8$ kcal/day

Factors Influencing Basal Metabolic Rate (BMR) The major factors influencing BMR include body composition, age, sex, nutritional status, and genetics.[22] The impact of these and other factors on BMR is summarized in Table 11.2. Tall, thin people tend to have higher BMRs than short, stocky people of the same weight. This is partly because tall, slender people have more surface area, resulting in greater loss of body heat. Differences in BMR can also be partially explained by body composition. Weight and height being equal, people with high proportions of lean mass (muscle) tend to have higher BMRs than do people with more fat mass (adipose tissue). This is because, compared to adipose tissue, muscle has greater metabolic activity.

Age can also influence BMR such that after 30 years of age, BMR may decrease by about 2 to 5% every 10 years. Scientists believe that this decrease is caused by age-related loss of lean tissue.[23] Physical activity can slow the

basal energy expenditure (BEE) Energy expended for basal metabolism over a 24-hour period.

resting metabolic rate (RMR) A measure of energy expenditure assessed under less stringent conditions than is BMR.

resting energy expenditure (REE) Energy expended for resting metabolism over a 24-hour period.

* To convert weight in pounds (lb) to kilograms (kg), divide weight (lb) by 2.2.
[†] To convert height in inches (in) to centimeters (cm), multiple the height in inches by 2.54.

TABLE 11.2 Factors Affecting Basal Metabolic Rate (BMR)

Factor	Effect on BMR
Age	After physical maturity, BMR decreases with age.
Sex	Males have higher BMR than do females of equal size and weight.
Growth	After adjusting for body size, BMR is higher during periods of growth.
Body weight	BMR increases with increasing body weight.
Body shape	Tall, thin people have higher BMRs than do short, stocky people of equal weight.
Body composition	Because muscle requires more energy to maintain than does adipose tissue, people with more lean tissue have higher BMRs than do people of equal weight with more adipose tissue.
Body temperature	Increased body temperature causes a transient increase in BMR.
Stress	Stress increases BMR.
Thyroid function	Elevated levels of thyroid hormones increase BMR, whereas low levels decrease BMR.
Energy restriction	Loss of body tissue associated with fasting and starvation decreases BMR.
Pregnancy	BMR increases during pregnancy.
Lactation	Milk synthesis increases BMR.

rate of muscle loss associated with aging and thus minimize the decline in BMR. Sex-related differences in BMR are also attributed to body composition. Women tend to have lower BMRs than men partly because they usually are smaller in size and have lower amounts of lean mass. However, when these differences are taken into account, average BMR in males still remains slightly higher than in females. This is likely due to hormonal differences between men and women.

Other factors affect BMR as well. For example, pound for pound, infants have higher BMRs than do adults. This is because infants are growing. BMR also increases during pregnancy and lactation. Fever and stress can cause a transient increase in BMR. Thyroid function also affects BMR such that thyroid overactivity—called hyperthyroidism—causes BMR to increase. Likewise, thyroid underactivity—hypothyroidism—causes BMR to decrease.

Perhaps the most striking factor that influences BMR is food restriction or dieting.[24] Severe energy restriction over time can decrease BMR because of the loss of lean body tissue. This energy-sparing response to negative energy balance protects the body's energy reserve and is an important survival mechanism when food is scarce. As caloric restriction continues, some dieters find further weight loss difficult, and in some cases experience weight gain. Weight regain that follows weight loss is called **rebound weight gain.**

PHYSICAL ACTIVITY IS THE SECOND-LARGEST COMPONENT OF TEE

After BMR, energy expended to support physical activity is the second-largest component of TEE. The amount of energy required for physical activity is quite variable, accounting for 15 to 30% of TEE.[25] Sedentary people are at the lower end of this estimate, whereas physically active people are at the upper end. Some elite athletes may require as much as 2,000 to 3,000 extra kilocalories each day to support the demands of physical activity.

Many factors affect the amount of energy expended for physical activity. Certainly, rigorous activities such as biking, swimming, and running have higher energy costs than less demanding activities. Body size also affects

rebound weight gain Weight regain that often follows successful weight loss.

The amount of energy required for physical activity is quite variable. Some elite athletes may require as many as 2,000 to 3,000 extra kilocalories each day to support the demands of physical activity.

energy expended for physical activity. Larger people have more body mass to move than smaller people and therefore expend more energy to accomplish the same activity.

THERMIC EFFECT OF FOOD (TEF) IS A MINOR COMPONENT OF TEE

Another component of TEE is the **thermic effect of food (TEF)**: the energy expended to digest, absorb, transport, metabolize, and store nutrients following a meal. In other words, it is the metabolic cost associated with processing food for utilization or storage.

The amount of energy associated with TEF depends on the amount of food consumed and the types of nutrients present in food. Because more energy is needed to process large amounts of food, TEF increases as food consumption increases. Some nutrients require more energy to process than others. High-protein foods have the highest TEF, and high-fat foods have the lowest. These differences reflect the metabolic cost associated with processing different nutrients. Because meals generally supply a mixture of nutrients, TEF is estimated to be about 5 to 10% of total energy intake.[26] For example, after consuming a 500-kcal meal, a person typically expends 25 to 50 kcal as TEF. In some ways, TEF is like having a "caloric sales tax" on energy intake.

METHODS OF ASSESSING TOTAL ENERGY EXPENDITURE (TEE)

An accurate assessment of TEE can be difficult because it requires expensive equipment and a high degree of expertise. However, simpler methods yield a good approximation of TEE. The techniques for assessing TEE include direct and indirect calorimetry, use of stable isotopes, and mathematical equations.

thermic effect of food (TEF) Energy expended for the digestion, absorption, and metabolism of nutrients.

direct calorimetry A measurement of energy expenditure obtained by assessing heat loss.

indirect calorimetry A measurement of energy expenditure obtained by assessing oxygen consumption and carbon dioxide production.

stable isotope A form of an element that contains additional neutrons.

DIRECT AND INDIRECT CALORIMETRY USED TO ESTIMATE TEE

Total energy expenditure (TEE) can be estimated by assessing heat loss from the body. This is because much of the energy (calories) the body uses is eventually lost to our environment as heat. This method, called **direct calorimetry,** requires specialized and very expensive equipment, so it is not often used. An alternative to direct calorimetry is **indirect calorimetry.** Rather than measuring heat loss, indirect calorimetry measures the exchange of respiratory gases—oxygen intake and carbon dioxide output. The use of indirect calorimetry to estimate TEE is based on the assumption that the body uses 1 liter of oxygen to metabolize 4.8 kcal of energy-containing compounds (that is, glucose, amino acids, and fatty acids). Thus, measuring oxygen consumption and carbon dioxide production permits an estimate of TEE. Indirect calorimetry is often used in clinical settings to determine caloric requirements for patients.

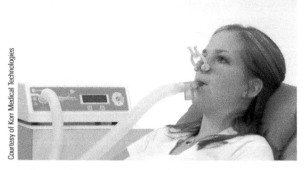

Indirect calorimetry measures the exchange of respiratory gases—oxygen consumption and carbon dioxide production—using portable equipment. These measures are used to estimate total energy expenditure (TEE).

STABLE ISOTOPES CAN BE USED TO ESTIMATE TEE

Another way to estimate TEE involves the use of **stable isotopes**—nonradioactive forms of certain elements. Isotopes have extra neutrons in their nuclei, making them heavier than the more common forms. Because stable isotopes of hydrogen (^{2}H) and oxygen (^{18}O) are chemically distinct from normal hydrogen (^{1}H) and oxygen (^{16}O), they can be measured in body fluids and expired air.

One technique that uses stable isotopes to estimate TEE is the **doubly labeled water** method. This method requires a person to drink two forms of water that have been labeled with stable isotopes—^{2}H$_2$O and H$_2$^{18}O. TEE can be estimated by measuring the elimination of oxygen and hydrogen isotopes from the body as water and carbon dioxide. This technique is considered the "gold standard" for estimating TEE.

TEE CAN BE ESTIMATED USING MATHEMATICAL FORMULAS

Because the equipment and expertise needed to use indirect calorimetry and stable isotopes are not readily available, clinicians commonly use relatively simple mathematical formulas to estimate TEE. Mathematical formulas developed by the Institute of Medicine as part of the Dietary Reference Intakes (DRIs)[27] are used to calculate what is called Estimated Energy Requirements (EERs), and take into account lifestages that include periods of growth. Formulas have been developed to calculate EERs for infants, children, adolescents, pregnant women, lactating women, and adults. Note that the EERs are intended to help adults maintain a healthy body weight. Once they are determined, adjustments in energy intake and physical activity can be made to support weight loss or gain. The following formulas are used to calculate EER for adult males and females. In this equation, PA refers to physical activity, which is categorized as sedentary, low, active, or very active.

$$\textbf{Males: } EER = 662 - [9.53 \times age\ (y)] + PA \times [15.91 \times weight\ (kg) + 539.6 \times height\ (m)]$$

$$\textbf{Females: } EER = 354 - [6.91 \times age\ (y)] + PA \times [9.36 \times weight\ (kg) + 726 \times height\ (m)]$$

You now understand how the balance of energy intake and energy expenditure affects body weight. Next we look at how body weight and body composition (the balance of fat and lean tissue) affect health.

How Are Body Weight and Body Composition Assessed?

Trying to keep energy intake and energy expenditure equal is key to maintaining a stable body weight. For many people, this can be difficult, resulting in unwanted weight gain. But at what point does added weight gain become unhealthy? And where do we draw the line between a few extra pounds and a serious health concern? After all, what appears to be a healthy weight for one person may cause health problems for another. To answer these questions, it is important to first understand how body weight and composition are defined, measured, and assessed.

BEING OVERWEIGHT MEANS HAVING EXCESS WEIGHT; BEING OBESE MEANS HAVING EXCESS FAT

Although the terms overweight and obese are often used interchangeably, they have very different meanings. Being **overweight** refers to having excess weight for a given height, whereas being **obese** refers to having an abundance

doubly labeled water Water that contains stable isotopes of hydrogen and oxygen atoms.

overweight Having excess weight for a given height.

obese Having excess body fat.

of body fat. Because simply knowing a person's weight does not provide information about the different components of the body, it is possible for muscular people, such as athletes, to be considered overweight but not obese. Conversely, some inactive people may not be considered overweight yet may still be obese. Most people who are overweight are obese as well, because weight gain in adults is generally caused by an increase in adipose tissue, rather than muscle. For this reason, body weight is often used as an indirect indicator of obesity.

TABLES ARE A GUIDE TO ASSESSING BODY WEIGHT

Because there is substantial variation in body weight for any given height, defining an "ideal body weight" may not be possible. Consequently, recommended body weights are simply reference values and not necessarily ideal for all people. The reference standards most commonly used to assess body weight are height–weight tables and body mass index (BMI).

Height–Weight Tables To determine the relationship between body weight and life expectancy, life insurance companies analyzed data from thousands of people. Weights associated with the longest life expectancies were deemed "desirable" and the companies published tables that listed "ideal weight ranges" for adult males and females. Because height–weight tables reflect a rather narrow segment of the population, they do not provide a good indication of whether a person's weight is within a healthy range.

Body Mass Index Even though height–weight tables are often used to assess body weight, body mass index (BMI) is a more widely used measure. BMI is based on the ratio of weight to height and is calculated using either of the following formulas.

$$BMI = [weight\ (kg)] / [height\ (m)]^2$$

$$BMI = [weight\ (lb)] / [height\ (in)]^2 \times 703.1$$

Based on these formulas, BMI for a person who weighs 150 lb (68.2 kg) and is 5 feet, 5 inches (65 inches, 1.65 m) tall is 25 kg/m². You can use the chart on the inside back cover of your book to determine your BMI. People with low BMIs typically have low amounts of body fat, whereas those with high BMIs tend to have higher amounts of body fat. Because each BMI unit represents 6 to 8 pounds for a given height, an increase in just 2 BMI units represents a 12- to 16-pound increase in body weight.

⟨CONNECTIONS⟩ Recall that *mortality* refers to death and *morbidity* refers to illness.

Cutoff values for BMI classifications (that is, underweight, healthy weight, overweight, and obese) are based on the association between BMI and weight-related mortality and morbidity. Figure 11.7 shows that higher BMI values are strongly associated with increased risk of death attributed to weight-related health problems.[28] It is estimated that more than 80% of weight-related deaths in the United States occur among individuals with a BMI greater than 30 kg/m². Most medical organizations, including the US Centers for Disease Control and Prevention (CDC), use the following criteria to assess body weight in adults based on BMI.

- Underweight: <18.5 kg/m²
- Healthy weight: 18.5–24.9 kg/m²
- Overweight: 25.0–29.9 kg/m²
- Obese: ≥30 kg/m²

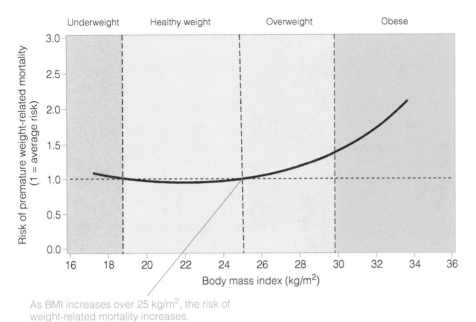

As BMI increases over 25 kg/m², the risk of weight-related mortality increases.

FIGURE 11.7 Body Mass Index (BMI) and Weight-Related Mortality
The relationship between BMI and health is "J-shaped," meaning that risk of poor health outcomes increases at both low and high BMI values.

SOURCE: Adapted from: Kivimäki M, Ferrie JE, Batty GD, Smith ME, Marmot MF, Shipley MJ. Optimal form of operationalizing BMI in relation to all-cause and cause-specific mortality: The original Whitehall study. Obesity. 2008:16:1926–32.

CLINICIANS USE SEVERAL TECHNIQUES TO ASSESS BODY COMPOSITION

Using BMI to interpret a person's weight for height can be informative. However, health professionals sometimes want to know a person's actual body composition values. The body can be thought of as having two main compartments: (1) the fat compartment and (2) the fat-free (lean) compartment. The fat compartment (adipose tissue) consists mostly of stored triglyceride and supporting structures, whereas the fat-free compartment is mostly muscle, water, and bone. How much fat and fat-free mass a person has is determined by many factors, including sex, genetics, physical activity, hormones, and diet.

The amount of fat stored in the body changes throughout the life cycle. For example, nearly 30% of total body weight in a healthy six-month-old infant is fat, whereas this percentage may be cause for concern in adults. According to some experts, it is recommended that body fat levels be around 12 to 20% and 20 to 30% of total body weight in adult males and females, respectively (Table 11.3). Body fat over 25% in males and over 33% in females indicates obesity.[29] As women approach menopause, many experience an increase in body fat. This may, in part, be related to hormonal changes, but it could also reflect a decline in physical activity and energy expenditure.

TABLE 11.3 Percent Body Fat Classifications

| Classification | Body Fat (% total body weight) | |
	Males	Females
Normal	12–20	20–30
Borderline obese	21–25	31–33
Obese	>25	>33

SOURCE: Bray G. What is the ideal body weight? Journal of Nutritional Biochemistry. 1998;9:489–92.

FIGURE 11.8 Health Problems Associated with Obesity

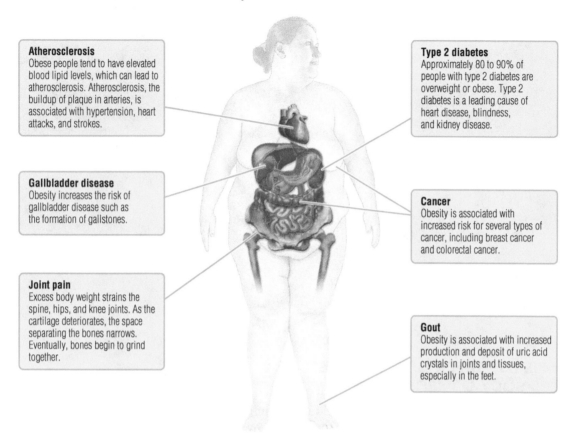

Atherosclerosis
Obese people tend to have elevated blood lipid levels, which can lead to atherosclerosis. Atherosclerosis, the buildup of plaque in arteries, is associated with hypertension, heart attacks, and strokes.

Gallbladder disease
Obesity increases the risk of gallbladder disease such as the formation of gallstones.

Joint pain
Excess body weight strains the spine, hips, and knee joints. As the cartilage deteriorates, the space separating the bones narrows. Eventually, bones begin to grind together.

Type 2 diabetes
Approximately 80 to 90% of people with type 2 diabetes are overweight or obese. Type 2 diabetes is a leading cause of heart disease, blindness, and kidney disease.

Cancer
Obesity is associated with increased risk for several types of cancer, including breast cancer and colorectal cancer.

Gout
Obesity is associated with increased production and deposit of uric acid crystals in joints and tissues, especially in the feet.

Just as too much body fat can cause health problems, too little body fat can also be detrimental. Body fat below 5% for males and 12% for females is considered too low. In women, low body fat can lead to a wide range of health problems such as bone loss and impaired fertility. Low body fat levels in women can be a consequence of disordered eating patterns. In some female athletes, long hours of training and extensive workouts can also cause body fat levels to become dangerously low.

Obesity increases risk for a variety of health problems, including type 2 diabetes, sleep irregularities, joint pain, stroke, heart disease, gallstones, and certain cancers (Figure 11.8).[30] Because of this, clinicians use a variety of methods to estimate body fat. Whereas some techniques are expensive and quite complex, others are more readily available and easy to use. Methods used to estimate body composition are summarized in Figure 11.9 and discussed next.

Densitometry Underwater weighing, or **hydrostatic weighing**, is a form of densitometry, which simply means measuring the density of something. Hydrostatic weighing involves measuring a person's weight both in and out of water. The more fat a person has, the less dense he or she is, and the less he or she weighs underwater. Hydrostatic weighing is not routinely used to estimate body composition. Not only does it require special equipment, but it is not very practical or convenient.

Dual-Energy X-Ray Absorptiometry **Dual-energy X-ray absorptiometry (DEXA)** is used to estimate total body fat, distribution of fat among different areas of

hydrostatic weighing (underwater weighing) Method for estimating body composition that compares weight on land to weight underwater.

dual-energy X-ray absorptiometry (DEXA) A method used to assess body composition by passing X-ray beams through the body.

FIGURE 11.9 Methods Used to Estimate Body Composition

Hydrostatic weighing

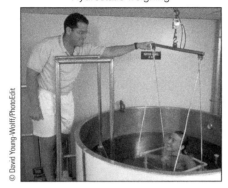

Hydrostatic weighing requires the subject to exhale air from the lungs and then be submerged in water. It is important to remain motionless while weight underwater is measured.

Dual-energy X-ray absorptiometry (DEXA)

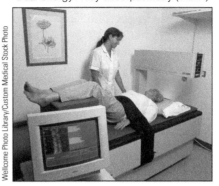

While the person lies on a table, a scanning device passes over the body. The X-ray beams emitted differentiate between fat mass, lean mass, and skeletal mass. A two-dimensional image of the body is displayed on a computer screen.

Bioelectrical impedance

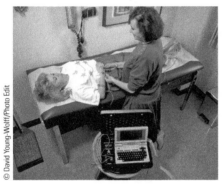

Electrodes are placed on a person's hand and foot, and a weak electrical current is passed through the body. The conductivity of the current is measured, which provides an estimate of the fat mass and fat-free mass.

Skinfold thickness

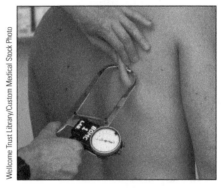

Using a measuring device called a caliper, the heath care worker measures the thickness of a fold of skin with its underlying layer of fat at precise locations on the body.

the body, lean body mass, and bone mass. During a DEXA measurement, a person must lie still while a scanning device passes over his or her body. The X-ray beams emitted differentiate between fat and fat-free mass. DEXA is considered the "gold standard" of body composition analysis and is often used by researchers and clinicians.

Bioelectrical Impedance **Bioelectrical impedance** is a measure of electrical conductivity—in other words, how easily a weak electric current travels through the body. Electrodes are placed on a person's hand and foot, and a weak electric current is emitted. Lean tissue, which contains a great deal of water, conducts electric currents better than adipose tissue, which has little water associated with it. The accuracy of bioelectrical impedance can be affected by hydration status—estimates of body fat will be incorrectly high if a person is dehydrated. This technique is relatively accurate, simple to use, and is considered an acceptable method for estimating body composition in clinical settings. Bathroom scales with built-in bioelectrical impedance

bioelectrical impedance A method used to assess body composition based on measuring the body's electrical conductivity.

systems are now available for home use, although little is known about their accuracy.

Skinfold Thickness The skinfold thickness method has been used for many years to estimate body fat. A measuring device called a **skinfold caliper** is used to measure the thickness of the subcutaneous fat. The examiner gently pinches the skin at various anatomical sites on the body. The double layer of skin includes the underlying fat tissue, but not the muscle. Body fat is estimated using mathematical formulas based on the skinfold thickness measures. Although the accuracy of this method depends on a person's ability to use a caliper, it is a fairly accurate and inexpensive way to assess body fat.

BODY FAT DISTRIBUTION AFFECTS HEALTH

⟨CONNECTIONS⟩ Recall that lipolysis is the breakdown of triglyceride molecules into glycerol and fatty acids (Chapter 10, page 254).

The distribution of body fat holds important information about health. The term **central obesity** (or central adiposity) describes the accumulation of adipose tissue predominantly in the abdominal region, or what is referred to as visceral adipose tissue (VAT). Because central obesity typically results in an increase in waist circumference, people with this body fat distribution pattern are commonly described as being "apple-shaped." In contrast, people with a predominance of subcutaneous adipose tissue (SCAT) in lower parts of the body are typically referred to as "pear-shaped." In general, males tend to be more "apple-shaped" and women tend to be more "pear-shaped."

Although the reasons are unclear, people with central obesity are at increased risk for developing weight-related health problems (such as type 2 diabetes, hypertension, and atherosclerosis).[31] However, there is some evidence that intra-abdominal fat is more likely to undergo lipolysis, compared with adipose tissue stored elsewhere in the body.[32] This can result in elevated blood levels of LDL cholesterol, triglycerides, glucose, and insulin and lower HDL cholesterol levels. Collectively, these signs and symptoms are associated with a condition called **metabolic syndrome,** which increases risk for developing type 2 diabetes and cardiovascular disease.

Reproductive hormones such as estrogen and testosterone influence where body fat is stored. Testosterone (a male hormone) encourages VAT, whereas estrogen (a female hormone) favors SCAT. As women age, declining estrogen levels can cause body fat to shift toward an accumulation of VAT.[33] Approximately 20% of total body fat in men is VAT, whereas it accounts for 5 to 8% of total body fat in women.

Waist Circumference Reflects Central Adiposity Evaluating your body fat distribution pattern is easier than you might think. A large **waist circumference** is a good indicator of central adiposity. This measurement is easily obtained by placing a tape measure around the narrowest area of the waist. However, a health professional is likely to use more precise anatomical landmarks as shown in Figure 11.10. A waist circumference of 40 inches or greater in males and 35 inches or greater in females indicates central adiposity.[34] Although age, ethnicity, and BMI influence the relationship between waist circumference and health outcomes, studies suggest that cardiovascular risk may be higher in those with the highest waist circumference.[35] Some clinicians prefer to use the ratio of waist circumference to hip circumference (waist-to-hip ratio) to determine body fat distribution. However, studies show that waist circumference alone provides a simple yet effective measure of central adiposity.[36]

You are now familiar with how weight and obesity can be assessed. But to truly understand what causes obesity, it is important to consider factors that influence eating habits.

skinfold caliper An instrument used to measure the thickness of subcutaneous fat.

central obesity Accumulation of body fat within the abdominal cavity.

metabolic syndrome Condition characterized by an unhealthy metabolic profile, abdominal body fat, and insulin resistance, that increases risk for developing type 2 diabetes and cardiovascular disease.

waist circumference A measure used as an indicator of central adiposity.

FIGURE 11.10 Waist Circumference Is Used to Assess Central Adiposity Central adiposity increases a person's risk for weight-related health problems. A large waist circumference is an important indicator of central adiposity.

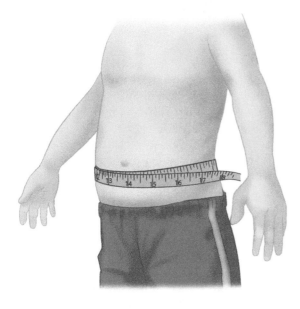

- Person should stand with their feet 6 to 7 inches apart, with their weight evenly distributed across both legs.

- Person should be relaxed, and the measurement should be taken while person is breathing out.

- The tape should be positioned midway between the top of the hip bone and the bottom of the rib cage.

- The tape should be loose enough so that there is one finger's width between the tape and the body.

- A waist circumference of 40 inches or more in males and 35 inches or more in females indicates central adiposity.

How Does Lifestyle Contribute to Obesity?

When it comes to our bodies, we tend to be our own worst critics. While we might wish that our bodies looked or were shaped differently, it is important to recognize that the issue of obesity is of concern not because of societal norms but rather because excess weight can be injurious to our health. In the late 19th century, only 3% of adults in the United States were overweight and very few were obese. Today, nearly 35% of adults are classified as obese (Table 11.4). Recent estimates by the Centers for Disease Control and Prevention estimate that the prevalence of obese adults exceeds 25% in nearly one-third of the U.S.[37] Similar trends are evident in children and adolescents. In the past 20 years, the number of overweight American children has doubled, and the number of

TABLE 11.4 Estimated Prevalence of Obesity during the Past 40 Years in the United States

Age (years)	(% of U.S. population)	
	1971–1974	2005–2008
2–5	...*	11
6–11	4	17
12–19	6	18
20–74	15	35

*Reliable national data not available.

SOURCE: National Center for Health Statistics. Health, United States, 2010: With special feature on death and dying. Hyattsville, MD. 2011.

⟨CONNECTIONS⟩ The National Health and Nutrition Examination Survey (NHANES) is an ongoing study to monitor nutrition and health in the U.S. population.

overweight adolescents has tripled. An overview of obesity trends in the United States is presented in Table 11.4. These alarmingly high rates of obesity are cause for concern. Obesity is not only on the rise in the United States but around the world as well.[38] What is causing this global epidemic? To answer this question, it is important to consider lifestyle-related factors that cause us to consume more calories and/or to expend less energy, behaviors that together shift energy balance in favor of weight gain.

EATING HABITS CAN CONTRIBUTE TO OBESITY

The amount of food we consume is one of the most important factors that influences body weight. It is likely that the increase in energy intake over the last few decades is a major cause of today's obesity epidemic—at least in adults. Over the past 40 years, the average daily energy intake of American adults has increased by approximately 200–300 kcal.[39] A variety of societal, sociocultural, and psychological factors may be contributing to this trend by influencing what, how much, when, and where we eat.

Societal Influences on Eating Habits Food-related societal influences can affect what and how much food people consume. These include more adults eating away from home, increased portion sizes of foods, increased consumption of energy-dense foods, and changes in snacking habits.

In the United States, energy-dense, inexpensive, and flavorful foods have become readily available and are accepted as a cultural norm. Some estimates indicate that 37% of adults and 42% of children eat fast food daily.[40] Although consuming fast food is not the sole cause of obesity, it may certainly contribute. Many fast-food meals are high in fat, refined starchy carbohydrates, and calories. As such, a super-sized "value" meal can provide more than half the calories required in a day. In response to significant public pressure, many fast-food restaurants also offer healthy food choices such as salads and sandwiches made with lean meats and whole-grain breads. Selecting these menu options is one way for individuals and families to eat more healthily.

Studies also show that serving size influences food consumption. That is, when larger food portions are served, people tend to eat more.[41] For example, participants in a study ate 39% more M&Ms® when they were given a two-pound bag than when they received a one-pound bag.[42] It appears some people depend more on visual cues to judge how much to eat than on physiological cues such as hunger and satiety. This is particularly important in light of the recent trend toward "super-sized" food portions.

Factors related to food itself can also influence what and how much food a person consumes. People tend to consume larger quantities of tastier foods than bland-tasting foods.[43] In other words, adding extra items such as butter and sour cream to baked potatoes is more likely to lead to overconsumption than eating plain baked potatoes. Similarly, food variety has also been shown to influence food intake. When presented with a variety of food choices, people tend to eat more than when they are offered single foods.[44] Scientists have also demonstrated that when people consume or perceive they have consumed fewer calories at one meal, they tend to "reward" themselves by eating more

When people are presented with a variety of food choices, they tend to eat more than when they are offered fewer choices. Thus, buffet-style eating may be more conducive to overeating than eating foods chosen from a more limited menu.

A "value meal" consisting of a double cheeseburger (760 kcal), medium french fries (360 kcal), and a medium soft drink (230 kcal) provides more than half of a person's daily energy requirements.

A regular meal consisting of a hamburger (200 kcal), small french fries (230 kcal), and a small soft drink (140 kcal) has less than half the calories of a "value meal."

at subsequent meals.[45] This may, in part, explain why increased consumption of reduced-calorie foods and beverages does not necessarily lead to a reduction in total calorie intake.

Sociocultural and Psychological Factors Affect Obesity Sociocultural factors such as cultural norms, economic status, marital status, and education can also influence our risk for obesity. Although these factors do not directly cause obesity, some may indirectly contribute to the problem. For example, obesity in the United States is most prevalent in Southern states and among certain racial and ethnic groups. For example, the obesity rates in adults between the ages of 20 and 74 years are 52% for black women, 44% for Mexican-American women, and 34% for non-Hispanic white women.[46]

Beyond sociodemographic factors, some psychological disorders are also related to risk for obesity. It is not clear if obesity predisposes individuals to these disorders or if some psychological profiles lead to obesity. All scientists can say for certain is that some personality factors appear to be associated with the risk for obesity. For example, obese individuals are more likely to experience clinical depression and panic attacks than nonobese people.[47] This may stem from the fact that obesity can lower a person's self-esteem and confidence. In addition, individuals who are impulsive and have difficulty coping with stress are likely to turn to food for emotional comfort and gratification, making them more susceptible to obesity. It also appears that weight gain can be an adaptive response to emotional trauma, providing emotional isolation and psychological protection from hurtful circumstances.[48]

New research suggests that a person's social network may also relate to his or her risk for obesity.[49] People with friends who have experienced weight gain are more likely to gain weight than those with weight-stable friends. It is not clear how social networks are linked to weight gain, but researchers believe that weight gain among close friends and family members might serve as a permissive cue for others to gain weight as well.

Critical Thinking: August's Story Think back to August's story about her experience being obese. Many factors influence a person's risk of becoming obese. How does August fit the profile of risk factors associated with obesity?

SEDENTARY LIFESTYLES CONTRIBUTE TO WEIGHT GAIN

Choosing what and how much to eat is not the only lifestyle choice related to obesity. At the same time we are eating more, Americans have also become less physically active. Fewer jobs that require physical work and an increase in labor-saving devices make daily life less physically demanding. To compound matters, almost 60% of adults fail to engage in any kind of leisure-time physical activity lasting 10 minutes or more in a given week.[50] Together, these changes over time have likely had a significant negative impact on the nation's health.

It is important to realize that physical activity is not limited to formal exercise. **Physical activity** is defined as any bodily movement produced by skeletal muscles that results in a substantial increase over resting expenditure of energy, whereas **exercise** comprises planned, structured, and repetitive activities that are done to improve or maintain physical fitness.[51] Physical activities such as such as walking up the stairs or mowing the lawn can be just as beneficial as formal exercise. Not surprisingly, the vast majority of studies show that lack of physical activity increases the risk of being overweight or obese.[52] For most people, walking an extra mile each day would increase energy expenditure by 100 kcal/day.[53] This is equivalent to taking about 2,000 to 2,500 extra steps each day. If all other factors remained constant, a person could lose 1 pound of body weight per month by making this simple change. In addition to helping us maintain a healthy body weight, physical activity and exercise also help us stay healthy and physically fit.

Physical Inactivity: A Growing Problem Physical inactivity has contributed to the growing rates of obesity, especially among children and adolescents. The long-term effect of childhood obesity on health is a growing concern. There is a direct correlation between regular physical activity and health among children and adolescents. The National Association for Sports and Physical Education recommends that schools provide 150 minutes per week of instructional physical education for elementary school children, and 225 minutes per week for middle and high school students.[54] Although the majority of states mandate physical education, few states meet these minimum standards.

Even though over 90% of high schools nationwide require school-based physical education classes, the majority of American youth still spend more time watching television, playing video games, or surfing the Internet than engaging in physical activities and sports. Studies show that overweight children spend more time doing screen-related activities than children who are not overweight.[55] It is likely that the more time children spend watching television and playing video/computer games, the less likely they are to be physically active. In addition, weight gain may also be due to greater exposure to food advertisements on children's television programming, which in turn may alter food choices.[56] The average child views about 40,000 commercials a year on television alone, exposing them to extensive advertising and marketing promotions for candy, sugar-sweetened cereal, soda, and fast food.[57]

Physical Activity Recommendations The American College of Sports Medicine (ACSM) recently released new physical activity guidelines for all healthy adults (18 to 64 years old).[58] Depending on physical fitness, most adults should engage in moderate- to high-intensity aerobic exercise three to five days a week and resistance training two to three days a week. By some measures, moderate-intensity physical activity is defined as physical activity that is done at 3.0 to 5.9 times the intensity of rest.[51] It is also recognized that physical activity above this minimum recommendation can provide

physical activity Bodily movement that uses skeletal muscles that results in a substantial increase in energy expenditure over resting energy expenditure (REE).

exercise Planned, structured activities done to improve or maintain physical fitness.

even greater health benefits. These physical activity guidelines for Americans are summarized below.[58]

- Children and adolescents (6–17 years): Engage in 60 minutes (1 hour) or more of physical activity daily. Most of this activity should be either moderate- or vigorous-intensity aerobic physical activity, and should include vigorous-intensity physical activity at least three days a week.
- Adults (18–64 years): Engage in at least 150 minutes (2 hours and 30 minutes) a week of moderate-intensity, or 75 minutes (1 hour and 15 minutes) a week of vigorous-intensity aerobic physical activity, or an equivalent combination of moderate- and vigorous-intensity aerobic activity. In addition to aerobic activities, moderate- or high-intensity muscle-strengthening activities that involve all major muscle groups should be performed two or more days per week.
- Older adults (65 years and older) should follow the adult guidelines. When older adults cannot meet the adult guidelines, they should be as physically active as their abilities and condition will allow.

Can Genetics Influence Body Weight?

Although there is general agreement that lifestyle factors—and not genetics—are the driving force behind the obesity epidemic, both likely influence body weight. In an environment where energy-dense foods are abundant and physical activity is low, certain genetic factors can make some people susceptible to weight gain. Even individuals not genetically predisposed to obesity are likely to gain weight in this type of environment. People have long suspected that genetics plays a role in influencing body weight. The good news is that, although genetics influences our susceptibility to obesity, lifestyle choices go a long way toward overriding these potentially negative influences.

IDENTICAL TWIN STUDIES HELP SCIENTISTS UNDERSTAND ROLE OF GENETICS

Adoption studies have helped researchers to distinguish between genetic and lifestyle influences in the development of obesity. Studies of genetically identical twins separated at birth suggest that genetic makeup directly influences body weight.[59] When body weights of children adopted at birth are compared with those of their adoptive parents, there is little similarity.[60] However, when compared to their biological parents, the similarity in weight is most striking. Scientists estimate that at least 50% of our risk for becoming overweight or obese is determined by genetics or epigenetics. Recall that epigenetic alterations in DNA can regulate gene expression—that is, which genes are switched on or off. Researchers believe that conditions in the womb can prompt epigenetic changes that later influence a person's body weight as well as that of future generations.[61]

⟨**CONNECTIONS**⟩ Recall that a gene is a section of DNA that contains hereditary information needed for cells to make a protein (Chapter 9, page 192).

DISCOVERY OF THE "OBESITY GENES" PROVIDES FIRST GENETIC MODEL OF OBESITY

Although scientists long believed that genetic makeup can influence body weight, direct evidence was lacking until the discovery of an obese mouse, named the ***ob/ob* mouse.** This mouse appeared to have a genetic mutation causing it to consume large amounts of food (in other words, become hyperphagic), be inactive, and therefore gain weight easily. Soon after the *ob/ob* mouse was discovered, another chance occurrence took place—the discovery

ob/ob **mouse** Obese mouse with mutations in genes that code for the hormone leptin.

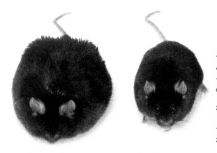

A genetic mutation made the mouse on the left obese.

⟨CONNECTIONS⟩ Recall that a mutation is an alteration in a gene resulting in the synthesis of an altered protein (Chapter 9, page 198).

of a mouse that was both obese and diabetic. Using breeding records, researchers found that this second mouse, called the **db/db mouse,** was not genetically related to the *ob/ob* mouse.[62]

To better understand what might be causing these mice to become obese, researchers performed some classic scientific experiments (Figure 11.11).[63] Using a technique called parabiosis, the *ob/ob* mouse and the *db/db* mouse were surgically joined together so that blood circulated between them. Curiously, the *ob/ob* mouse lost weight, while the *db/db* mouse showed no weight change. When an *ob/ob* mouse was joined to a normal mouse, the *ob/ob* mouse lost weight, but the normal mouse experienced no weight change. Last, they joined a *db/db* mouse to a normal mouse. Although there was no weight change in the *db/db* mouse, the normal mouse lost weight. This experiment was the first step in a series of investigations that revealed an internal system of body weight regulation. However, how such a system worked remained elusive until 1994 when scientists discovered the first "obesity genes." These genes were called the *ob* gene and the *db* gene, after the mice from which they were discovered.

Obesity Genes Lead to the Discovery of Leptin Scientists soon learned the **ob gene** codes for a hormone called leptin.[64] **Leptin** is a potent satiety signal produced by adipose tissue, and the lack of this hormone in *ob/ob* mice causes them to eat uncontrollably. Soon thereafter, researchers discovered that the **db gene** codes for the leptin receptor, found primarily in the hypothalamus. Without a functioning leptin receptor, leptin cannot exert its effect. To determine if leptin played a role in regulating body weight, *ob/ob* and *db/db* mice were injected with leptin. When the leptin-deficient *ob/ob* mice received leptin, there was significant weight loss. However, there was no change in *db/db* mice in response to leptin injections. This is because *db/db* mice have defective leptin receptors, making them unresponsive to this hormone.

People hoped leptin—touted as the antiobesity hormone—would become the miracle cure for obesity. Yet the vast majority of obese people produce appropriate or even elevated amounts of leptin. Although rare, leptin deficiency has been reported in humans. However, these individuals have provided scientists with a unique opportunity to study the effects of leptin.[65] When the severely obese, leptin-deficient individuals received leptin injections, all experienced dramatic weight loss. Because leptin deficiency is an uncommon occurrence in humans, scientists have turned their investigation toward trying to better understand why some people appear unresponsive to this hormone. Although leptin has not proved to be effective for treating human obesity in most cases, its discovery has led to important insights into body weight regulation.

How Does the Body Regulate Energy Balance and Body Weight?

Although the discovery of leptin has not solved the obesity dilemma, it has profoundly deepened our understanding of body weight regulation. While much remains to be discovered about body weight regulation and possible defects in these key energy-regulating activities, it is clear that the body plays an active role in influencing how much energy is consumed, how much energy is expended, and how much energy the body stores. Physiological systems require homeostatic regulatory mechanisms that maintain "checks and balances" via hormonal signaling pathways. Whether body weight is regulated by such a system has long been the subject of much debate. However, the discovery of the obesity genes and the hormone leptin has helped scientists gain insights into how the body monitors and maintains energy stores.

db/db mouse Obese mouse with mutations in genes that code for the leptin receptor.

ob gene The gene that codes for the protein leptin.

leptin A hormone, produced mainly by adipose tissue, that helps regulate body weight.

db gene The gene that codes for the leptin receptor.

FIGURE 11.11 Parabiosis Experiments Using *ob/ob* and *db/db* Mice *ob/ob* mice lack the satiety signal leptin, whereas *db/db* mice produce large amounts of leptin but have defective leptin receptors.

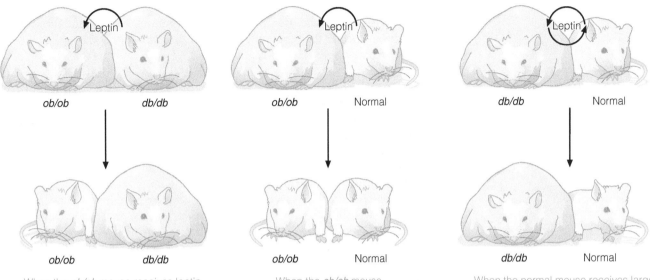

When the *ob/ob* mouse receives leptin from the *db/db* mouse, the ob/ob mouse loses weight. There is no weight change in the *db/db* mouse, because it is unable to respond to leptin.

When the *ob/ob* mouse receives leptin from the normal mouse, the *ob/ob* mouse loses weight. There is no weight change in the normal mouse.

When the normal mouse receives large amounts of leptin from the *db/db* mouse, the normal mouse loses weight. There is no weight change in the *db/db* mouse, because it is unable to respond to leptin.

Jeffrey Friedman, shown here holding a test tube of leptin, discovered the "obesity hormone" in his laboratory at The Rockefeller University in the mid-1990s.

ADJUSTING ENERGY INTAKE AND ENERGY EXPENDITURE MAINTAINS ENERGY BALANCE

The ability of the body to adjust energy intake and energy expenditure on a long-term basis serves an important purpose—body weight regulation. If the body had no means of regulating energy balance, the consequences would be serious. Even a slight imbalance in daily energy intake and energy expenditure would result in a substantial weight gain or loss over time. To prevent such imbalances, long-term energy balance regulatory signals communicate the body's energy reserves to the brain, which in turn releases neurotransmitters that influence energy intake and/or energy expenditure.[66] If this long-term system functions effectively, body weight remains somewhat stable. This system, sometimes called the set point theory, is perhaps the most widely accepted theory of body weight regulation.

Set Point Theory of Body Weight Regulation Body weight is a function of the balance between energy intake and energy expenditure. However, increases or decreases in food intake do not always produce the expected change in body weight. Sometimes, a person can increase energy intake without gaining weight or reduce consumption and still not lose weight. For many years, scientists have suspected that a complex signaling system regulates body weight by making adjustments in energy intake and energy expenditure.[67] To test this theory, researchers observed weight-gain and weight-loss cycles in food-restricted mice. When food-restricted, the mice lost weight. Not surprisingly, when taken off food restriction, the mice increased their food consumption and soon returned to their original weight. However, what was surprising to scientists was the fact that after returning to their original weight, the mice maintained this weight by spontaneously decreasing their intake of food. This phenomenon was called the set point theory of body weight regulation.

Proponents of the **set point theory** of body weight regulation believe that hormones circulating in the blood (such as leptin) regulate body weight by communicating the amount of adipose tissue in the body to the brain. When the amount of adipose tissue increases beyond a "set point," a signal causes food intake to decrease and/or energy expenditure to increase, favoring weight loss. Conversely, when the amount of adipose tissue decreases below a "set point," food intake increases and/or energy expenditure decreases, favoring weight gain. In this way, body weight is restored to its "set point" and remains relatively stable on a long-term basis.

LEPTIN COMMUNICATES THE BODY'S ENERGY RESERVE TO THE BRAIN

The mechanisms that regulate long-term energy balance are complex and not well understood. However, the discovery of the *ob* gene and leptin provided the first real evidence that hormones produced in adipocytes can communicate adiposity to the brain, which in turn influences hunger and satiety (Figure 11.12). When body fat increases, the concentration of circulating leptin (produced primarily by adipose tissue) increases as well. Conversely, when body fat decreases, leptin production decreases. Thus, fluctuations in leptin levels reflect changes in the body's primary energy reserve.

Leptin is thought to be part of a communication loop that helps maintain a relatively stable body weight over time. Specifically, a rise in leptin concentrations in response to increased adiposity signals the brain to increase its release of catabolic neurotransmitters. Catabolic neurotransmitters help the body resist further weight gain by decreasing food intake and increasing energy expenditure. Conversely, decreased adiposity causes blood concentrations of leptin to decrease. When this occurs, the brain increases its release of anabolic neurotransmitters. The release of these neurotransmitters stimulates food intake and decreases energy expenditure, thus protecting the body against further weight loss. In this way, leptin is thought to be part of a long-term homeostatic system that helps prevent large shifts in body weight.

Defects in Leptin Signaling May Lead to Obesity Some researchers believe that defects in this leptin signaling system may lead to impaired body weight

set point theory A theory suggesting that hormones regulate body weight by making adjustments in energy intake and energy expenditure.

FIGURE 11.12 Leptin and Body Weight Regulation

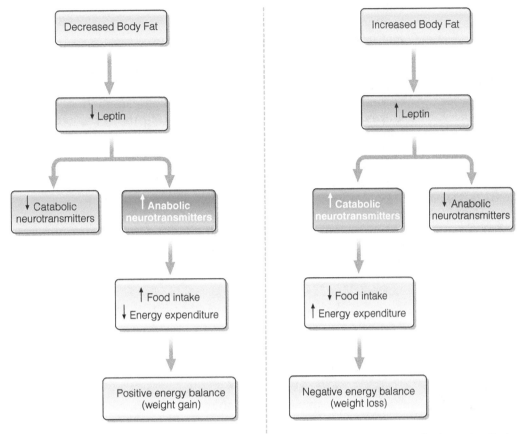

When body fat decreases, less leptin reaches the brain. Anabolic neurotransmitters are released, and catabolic neurotransmitters are suppressed. This condition favors increased food intake, decreased energy expenditure, and weight gain.

When body fat increases, more leptin reaches the brain. Anabolic neurotransmitters are suppressed, and catabolic neurotransmitters are released. This condition favors decreased food intake, increased energy expenditure, and weight loss.

regulation in some people.[68] In other words, the brain may not be able to recognize and respond to the appetite suppression signal that comes from leptin. In times when food is scarce, a decline in leptin may help the body conserve energy, increasing a person's chance of survival.[69] However, you may be wondering, "If leptin curbs food intake, why are so many people obese? If obese people have elevated leptin levels, why do they continue to gain weight?" These are very good questions. A number of researchers believe that some people may have disconnected leptin-signaling pathways.[70] In other words, the brains of some obese people may not be responsive to leptin's appetite suppression signal, regardless of how much leptin is being produced. Clearly, there is more to learn about this regulatory system. As scientists continue to study the role of leptin and other adipokines in long-term body weight homeostasis, we will gain a better understanding of the underlying mechanisms associated with body weight regulation.

ADIPONECTIN MAY PROVIDE A LINK BETWEEN OBESITY AND WEIGHT-RELATED DISEASES

Currently, leptin is the only hormone in humans that has been shown to be a long-term mediator of body weight regulation.[71] Other "obesity hormones" are being discovered on a regular basis. One such hormone is **adiponectin,**

adiponectin A hormone secreted by adipose tissue that appears to be involved in energy homeostasis; also appears to promote insulin sensitivity and suppress inflammation.

also produced by adipose tissue. Although its physiological role is not fully understood, adiponectin appears to have the opposite effect of leptin. In mice, adiponectin levels increase when energy reserves (adipose tissue) decrease, resulting in an appetite-stimulating effect in the brain.[72] Thus, adiponectin may be the hormonal signal that adipocytes release when energy reserves are running low and the body needs more energy.

It is possible that adiponectin may be an important link between obesity and weight-related medical conditions. One of the more recent discoveries is that obesity can induce a state of chronic inflammation.[73] This fat-induced inflammatory response can wreak havoc in the body, and has been linked with the development of type 2 diabetes and cardiovascular disease.[74] Adiponectin helps to prevent inflammation.[75] Unfortunately, body fat accumulation reduces adiponectin concentrations, subsequently increasing a person's risk of developing weight-related health complications. Studies have shown that adiponectin levels are low in people with type 2 diabetes and cardiovascular disease.[76]

Now that you have learned about the factors—physical, psychological, and social—that lead to overweight and obesity, it is important to consider strategies that work for achieving weight loss.

What Are the Best Approaches to Weight Loss?

Although the "diet" industry would like us to think otherwise, the truth is clear: there is no quick and easy way to lose weight. Approximately one-third of adults in the United States are on a special diet to lose weight, and there are plenty of diets to choose from. Certainly the current obesity epidemic cannot be attributed to a lack of weight-loss advice. In addition, a variety of nonprescription weight-loss products are aggressively marketed. Some of them can be hazardous to your health.

Most popular weight-loss plans have specific rules for people to follow. Some restrict the types of food that can be eaten, others recommend a strict exercise regimen, and still others acclaim periods of fasting or cleansing. The real issue that many weight loss plans fail to address is not what it takes to lose weight but what it takes to keep it off. Rather than succumbing to a fad diet or quick fix weight loss approach, consider what actual weight loss experts have to say.

HEALTHY FOOD CHOICES PROMOTE OVERALL HEALTH

Although there are many reasons why people want to lose weight, the most important is to improve health. Achieving and maintaining weight loss requires making lasting lifestyle changes, including changes to what we choose to eat and how much physical activity we engage in. Most health experts suggest that people focus less on weight loss and more on healthy eating and overall fitness. Misguided efforts toward weight loss at any cost present food as the enemy, rather than as a means to good health. This is unfortunate, because most people who successfully lose weight and keep it off do so by eating a balanced diet of nutrient-dense foods and by maintaining a moderately high level of physical activity.[77] Key recommendations for weight

The current obesity epidemic cannot be attributed to a lack of weight-loss advice. Amazon.com lists over 8,000 books related to weight-loss diets.

Peter Tobia/Philadelphia Inquirer/KRT/Newscom

⟨**CONNECTIONS**⟩ Recall that nutrient-dense foods are high in nutrients relative to the amount of calories.

management based on the 2010 Dietary Guidelines for Americans are provided in the Food Matters that follows.

A healthy weight-loss and weight-maintenance program consists of three components: (1) setting reasonable goals, (2) choosing nutritious foods in moderation, and (3) increasing energy expenditure by daily physical activity.

Setting Reasonable Goals Setting reasonable and attainable goals is an important component of any successful weight-loss program. A realistic weight-loss goal is to reduce current weight by 5 to 10%. For someone weighing 180 lbs, this would amount to an initial weight loss of approximately 9 to 18 pounds. Studies show that even this modest reduction in weight can improve overall health.[78] When it comes to weight loss, slow and steady is advised and weight loss should not exceed one to two pounds a week. This can be achieved by decreasing energy intake by 100 to 200 kcal each day or by walking one to two miles each day. Over time, these small changes can result in significant reductions in body weight. Rather than making dramatic dietary changes, making small changes such as reducing portion sizes or cutting back on energy-dense snack foods can make a big difference in overall energy intake.

Once body weight stabilizes and the new lower weight is maintained for a few months, individuals can decide whether additional weight loss is needed. Some people benefit from joining weight management programs such as Weight Watchers® or Take Off Pounds Sensibly®. These types of programs provide long-term support and motivate people to maintain a healthy diet and lifestyle.

Choosing Nutritious Foods Weight-loss plans that drastically reduce calories and offer limited food choices leave people feeling hungry and dissatisfied. Instead, those that encourage people to eat foods that are healthy and appealing tend to be more successful. Contrary to popular belief, it is not necessary to avoid foods that contain fat in order to lose weight. The 2010 Dietary Guidelines for Americans recommend that we choose our fats as carefully as we choose our carbohydrates. In general, it is best to limit intake of foods containing *trans* fatty acids and saturated fatty acids. Instead, we should emphasize foods containing relatively more polyunsaturated and monounsaturated fatty acids.

A common misconception is that dairy products and meat are high-fat foods and therefore should be avoided when trying to lose weight. Again, the key to good nutrition is moderation and choosing wisely. For example, switching from whole to reduced-fat milk is one way to lower caloric intake without losing out on the many vitamins and minerals in dairy products. Likewise, the types of meat you choose and the methods by which the meat is prepared can greatly affect how many calories you consume. Lean meats prepared by broiling or grilling are both nutritious and satisfying.

Reducing energy intake is best achieved by eliminating or cutting back on energy–dense foods that have little nutritional value such as potato chips, cookies, and cakes. Aside from their lower energy densities, nutrient-dense foods such as whole grains, legumes, fruits, and vegetables offer many beneficial substances such as micronutrients and fiber. Furthermore, because these foods tend to have greater volume compared to more energy-dense foods, they help people feel more satisfied after they eat. In the Focus on Food feature, you can read about how eating high-volume, low-energy-dense foods can help people feel full and satisfied.

Healthy eating also requires people to pay attention to hunger and satiety cues. However, people often let the amount of food served or packaged,

Working Toward the Goal: Maintaining a Healthy Body Weight by Balancing Caloric Intake with Energy Expenditure

The 2010 Dietary Guidelines for Americans recommend that we maintain a healthy body weight by balancing caloric intake with energy expenditure. The following tips will help you maintain a healthy body weight and prevent gradual weight gain over time.

- Consume foods that are nutrient dense while limiting foods that are high in saturated fats, *trans* fats, refined carbohydrates, and added sugars.

- Select foods high in dietary fiber, such as fruit, vegetables, and whole grains.

- Follow an eating plan that emphasizes variety, moderation, and balance.

- Select low-fat dairy products and lean meats whenever possible.

- Pay attention to internal cues of hunger and satiety.

- Make food portions smaller than normal. If still hungry, eat a second portion.

- Avoid distractions while eating, such as watching television, which can lead to overconsumption of food.

- Alcohol is a source of calories (7 kcal/g), and it is therefore important to monitor your intake of alcoholic beverages.

- Eat regularly and avoid skipping meals, which can lead to extreme hunger.

- Avoid eating food directly from containers or packages. Pay attention to food portions by serving food on a plate or in a bowl.

- Many restaurants serve generous food portions—enough for two people. Try sharing a meal or take the uneaten portion of the meal home.

- Be aware of beverages that contain high-calorie ingredients such as sugar, high fructose corn syrup, cream, and syrups.

- Increase your physical activity by using the stairs instead of elevators, walking instead of driving, and parking so as to walk farther. If you have children, include them in fun family activities such as hiking, riding bicycles, and swimming.

- Join a fitness center or take a fitness class.

- Take up a hobby, such as gardening, that is both rewarding and physically challenging.

- Make exercise interesting by changing your routine or by doing it with a friend.

rather than hunger and satiety, determine how much they eat. That is, visual cues, rather than internal cues, have a greater influence on the quantity of food we consume. For instance, some commercially made muffins are extremely large, containing as many calories as eight slices of bread. Learning to recognize and choose reasonable portions of food is a critical component to successful weight management. In fact, reducing portion sizes by as little as 10 to 15% can reduce daily energy intake by as much as 300 kcal. One way to limit serving size is to consider sharing a large meal the next time you eat at a restaurant.

Physical Activity In addition to eating less, people can also help tip energy balance toward negative energy balance by engaging in physical activity. Walking one mile a day, which takes most people about 15 to 20 minutes, uses about 100 kcal. This adds up to 700 kcal per week. Being physically active throughout the day can make a big difference in promoting weight loss and weight maintenance. When possible, take the stairs rather than the elevator, walk or bike rather than drive, and incorporate chores into your routine daily activities. Even without weight loss, individuals who are physically active show improved physical fitness. Regardless of one's weight, a lack of exercise may prove the greatest health hazard of all.

It is important to realize that overweight people can still be physically fit, and healthy-weight people can be physically unfit. In fact, studies show that obese individuals who are physically fit have fewer health problems than do healthy-weight individuals who are unfit.[79] Normal blood pressure, normal blood glucose regulation, and healthy levels of blood lipids are important indicators of physical fitness.[80] Exercise also promotes a positive self-image and helps people take charge of their lives. Although people often resist the idea of starting an exercise program, they rarely regret it once they begin. Physical activity is an effective strategy for preventing unhealthy weight gain in normal, overweight, and obese individuals.

An expert panel assembled by the International Association for the Study of Obesity concluded that daily exercise can help prevent healthy-weight individuals from becoming overweight, overweight people from becoming obese, and obesity from worsening in already obese individuals.[81] For every pound of body weight lost, a person requires 8 fewer kcal/day. Thus, when a person loses 10 lbs, energy requirements decrease by approximately 80 kcal/day. Thus, to maintain a 10-lb weight loss, a person must further reduce energy intake or increase energy expenditure.

© Dennis MacDonald/PhotoEdit

Not all obese people are physically unfit and/or unhealthy. Although physical activity does not totally eliminate health risks associated with obesity, fitness level is strongly associated with health.

CHARACTERISTICS OF PEOPLE WHO SUCCESSFULLY LOSE WEIGHT

One of the best ways to learn what works in terms of weight loss is to study people who have been successful. The National Weight Control Registry (NWCR) is a large, prospective study of individuals who have lost significant amounts of weight and have been successful at keeping it off.[82] To be eligible to participate in the study, individuals must have maintained a weight loss of 30 pounds or more for one year or longer. Researchers interviewed this unique group of "successful losers" to learn about their weight loss and weight-loss management practices. However, identifying common characteristics among NWCR participants proved difficult. Whereas some reported counting calories or grams of fat, others used prepared, prepackaged foods. Some participants preferred losing weight on their own, while others sought assistance from weight-loss programs. Researchers identified one striking common thread: almost 90% of the NWCR participants reported that both diet and physical activity were part of their weight-loss plan. Only 10% reported using diet alone, and only 1% used exercise alone. On average, study participants engaged in 60 to 90 minutes of physical activity daily, which was equivalent to 2,500 kcal/week for women and 3,300 kcal/week for men. In addition to a high level of physical activity, participants tended to follow a low-fat, high-carbohydrate diet, ate breakfast regularly, and monitored their weight frequently. The low-fat, high-carbohydrate diet that prevailed in this study has wide support, but experts continue to debate the value of diets with different macronutrient distributions.

Does Macronutrient Distribution Matter?

With all the weight-loss advice, it can be difficult to sort truth from fiction. Even experts have differing opinions about the best balance of macronutrients—proteins, fats, and carbohydrates—for achieving weight loss. Today, one of the biggest controversies is the role of dietary carbohydrate versus dietary fat in promoting weight loss and weight gain.[83] Weight-loss diets that are low in fat and high in carbohydrates have long been considered the most effective in terms of weight loss and weight maintenance. In fact, the 2010 Dietary Guidelines for Americans advocate low-fat food choices with an emphasis on whole grains, fruits, and vegetables. Similarly, the Acceptable Macronutrient Distribution Range (AMDR) suggests that we consume 45 to 65% of energy

People trying to lose weight tend to have the most success when they select foods that are both nutritious and satisfying. Too often, weight-loss plans emphasize food restriction, causing a state of perpetual hunger. As a result, people typically revert to former eating practices. However, new studies show that eating more can sometimes help people weigh less.[84] This is because eating high-volume, low-energy-dense foods can help people feel full longer than when they eat low-volume, high-energy-dense foods.[85] In other words, it is possible for people to consume larger volumes of foods while eating fewer calories.

The energy density of food reflects the relationship between energy (kcal) and weight (grams). High-energy-dense foods are those that provide 4 to 9 kcal per gram, whereas low-energy-dense foods have less than 1.5 kcal per gram.[86] High-energy-dense foods typically have low moisture or high fat content. These include cookies, cakes, crackers, cheesecake, and butter. Examples of foods with medium energy density (1.5 to 4 kcal/gram) include eggs, dried fruits, breads, and cheese. Low-energy-dense foods have a high water and/or fiber content and are low in fat, such as most fruits and vegetables, broth-based soups, low-fat cheese, and certain types of meat such as roasted turkey.

Studies show that people feel more satisfied and tend to consume fewer calories when they eat low-energy-dense foods.[87] For example, people who eat salads or broth-based soups prior to a high-energy-dense main course consume fewer calories overall. Similarly, the effect of volume on satiety was demonstrated by feeding people milkshakes with the same number of calories but with different volumes (300 mL, 450 mL, and 600 mL). Participants consumed 12% fewer calories when fed 600-mL (approximately 20 ounces) milkshakes than when they consumed those with less volume. These and other studies show that meal volume provides a stronger satiety signal than the total calories in the food.[88] Most important, people consuming low-energy-dense foods lose more weight over time than those following a fat-reduction weight-loss plan.

Here are some suggestions to increase your intake of low-energy-dense foods.

- Eat a low-energy-dense salad before a main course. Decrease the energy density of your salad by adding plenty of greens, vegetables, and low-fat salad dressing.

- As an alternative to cream-based soup, choose broth-based soup containing vegetables, high-fiber grains (such as barley), and legumes (such as split peas or lentils).

- Reduce energy density by cutting back on the amount of high-fat meat in recipes and adding more vegetables.

- Avoid beverages that have high fat content. Instead, add a small amount of fruit juice to carbonated water.

- Read the Nutrition Facts panel on food labels, and compare the number of calories in relation to the number of grams. In general, foods with low-energy density are those with less than 1.5 kcal per gram.

Scott Bauer/ARS/USDA

Foods that are not energy dense, such as these grapes, have a high water and/or fiber content and are low in fat.

from carbohydrate, 10 to 35% from protein, and 20 to 35% from fat. However, Robert Atkins, one of the first pioneers of the low-carbohydrate diet, turned the nutritional world upside down in 1972 when he proposed that eating too much carbohydrate, rather than too much fat, may actually cause people to gain weight. The caloric distribution advocated by the AMDRs is compared to those of typical high- and low-carbohydrate diets in Table 11.5. We next

TABLE 11.5 Caloric Distribution of High- and Low-Carbohydrate Diets and the Acceptable Macronutrient Distribution Ranges (AMDR)

	% of Total Calories		
Nutrient	AMDR[a]	High-Carbohydrate Diet[b]	Low-Carbohydrate Diet[c]
Fat	20–35	10–15	55–65
Carbohydrate	45–65	65–75	5–20
Protein	10–35	10–25	20–40

[a]Source: Institute of Medicine. Dietary Reference Intakes for energy, carbohydrate, fiber, fat, fatty acids, cholesterol, protein, and amino acids. Washington, DC: National Academies Press; 2005.
[b]Source: Ornish D. Eat more weigh less: Dr. Dean Ornish's life choice diet for losing weight safely while eating abundant. New York: Harper Collins; 1993.
[c]Source: Atkins RC. Dr. Atkins' new diet revolution, revised. National Book Network; 2003.

examine the scientific evidence concerning the effects of macronutrient distribution on weight loss and overall health.

HIGH-CARBOHYDRATE, LOW-FAT WEIGHT-LOSS DIETS

Many researchers believe diets high in carbohydrates and low in fat promote weight loss and have an overall beneficial effect on health. To maintain a relatively low intake of fat (10 to 15% of total calories), high-carbohydrate weight-loss diet advocates advise dieters to avoid meat, dairy, oils, and olives; low-fat meat and dairy products can be eaten in moderation. With an emphasis on fruit, vegetables, and whole grains, this weight-reduction and maintenance plan provides about 65 to 75% of total calories from carbohydrates, with protein and fat making up the remainder. Examples of popular high-carbohydrate, low-fat weight-loss diets include Dr. Dean Ornish's Program for Reversing Heart Disease[89] and The New Pritikin Program for Diet and Exercise.[90]

There are several reasons why supporters of low-fat diets believe they help prevent obesity. First, gram for gram, fat has more than twice as many calories as carbohydrate and protein. Therefore, it is reasonable to assume that consuming less fat may lead to lower energy intake, which in turn results in weight loss. Fat can also make food more flavorful, contributing to overconsumption. Last, excess calories from fat are more efficiently stored by the body than those from carbohydrate or protein. Converting excess glucose and amino acids into fatty acids for storage takes energy and "subtracts" some of the energy originally in the foods. Some experts also claim that low-fat diets benefit overall health as well, by lowering total and LDL cholesterol concentrations, increasing HDL cholesterol concentrations, and improving blood glucose regulation.[91] However, these benefits may be due to weight loss in general rather than to reduced dietary fat.

Do High-Carbohydrate, Low-Fat Weight-Loss Diets Work? Long-standing dietary advice aimed at helping people lose weight has consistently focused on reducing dietary fat. Although total energy intake has increased, the percentage of total calories from fat has declined from 45% in the 1960s to approximately 33% today, which is equivalent to an average of 10 to 20 fewer grams of fat per day.[92] Yet decreased fat intake has not resulted in a decreased prevalence of obesity. In fact, obesity rates have increased under the low-fat regime.[93] It is not clear which dietary factors contributed to this trend; however, some researchers believe failure to make healthy carbohydrate food choices may be contributing to weight gain.

Although clinicians hoped Americans would replace fatty foods in their diet with more nutritious items such as whole grains, fruits, and vegetables,

High-carbohydrate/low-fat weight-loss diets (left) emphasize foods such as whole grains, lean meats, pasta, fruits, vegetables, and low-fat dairy products. Low-carbohydrate weight-loss diets (right) emphasize lean meats, fish, nonstarchy fruits and vegetables, eggs, low-fat dairy, and unsaturated fats.

they have not done so. The availability of low-fat snack products has made it possible for people to eat snack foods minus the fat. Typically, the ingredient that replaces fat in these products is refined carbohydrate (such as white flour and sugar), and people mistakenly think that low-fat snack foods are healthy snack alternatives. Overconsuming fat-free snack foods may, in part, contribute to weight gain, because many have the same amount of calories as the original product. Eating foods high in refined carbohydrate, especially those low in fat, may make us hungrier and therefore can make us heavier.[94] The theory that a low-fat diet is our best defense against weight gain is not without debate, and many health experts believe there is now enough solid evidence to lift the "ban" on dietary fat.[95]

LOW-CARBOHYDRATE WEIGHT-LOSS DIETS

On the opposite end of the weight-loss diet spectrum are low-carbohydrate diets. Health claims made by advocates of these weight-loss plans include weight loss without hunger and improved cardiovascular health.[96] Some experts believe that people are more likely to gain weight from excess carbohydrates as opposed to excess fats or proteins because high-carbohydrate foods cause insulin levels to rise, which favors fat storage (lipogenesis). Thus, limiting starch and refined sugars in the diet should theoretically help people lose weight.

There are many low-carbohydrate weight-loss diets, each differing in terms of the types of foods allowed. Although some exclude nearly all carbohydrates, others take a more moderate approach by allowing healthy, carbohydrate-rich foods such as fruit, vegetables, and whole grains. Indeed, low-carbohydrate diets continue to be an extremely popular approach to weight loss.[97] Low-carbohydrate diets appear safe in the short term, although some experts have raised doubts about their long-term effectiveness.[98] Examples of popular low-carbohydrate, weight-loss diets include Dr. Atkins' Diet Revolution[99], The Carbohydrate Addict's Diet[100], Life Without Bread[101], and Protein Power[102].

Do Low-Carbohydrate Diets Work? The lure of a diet plan that does not require counting calories or limiting portion sizes is appealing to many people. However, do low-carbohydrate diets really help one lose weight? When a person restricts his or her intake of carbohydrates, the body breaks down glycogen to provide glucose. Approximately 500 g (about 1 lb) of glycogen is stored in the liver and skeletal muscle combined. Glycogen is bound to molecules of water, approximately 3 g of water per gram of glycogen. Thus, the combined weight of stored glycogen and associated water is about 4 lb. Consequently, when glycogen is broken down, its associated water is eliminated from the body. Because of this, the initial weight loss associated with low-carbohydrate diets is largely attributed to water loss.

Once glycogen stores are depleted, the body begins to rely heavily on triglycerides for energy. However, recall from Chapter 7 that when glucose is limited, fatty acids can only be only partially broken down. As a result, ketone formation increases, a sign that the body is using fat as major sources of energy. Limiting carbohydrate intake causes ketosis, which is why these diets are referred to as **ketogenic diets.** Ketosis often results in a loss of appetite, further promoting weight loss.

Studies comparing weight loss associated with low-carbohydrate diets to that associated with low-fat diets show that, at six months, greater weight loss is achieved on low-carbohydrate diets. One of the largest and longest studies comparing three different weight-loss plans—low-carbohydrate, high-carbohydrate, and the macronutrient distribution ranges recommended in the Dietary Reference Intakes—found that the low-carbohydrate plan resulted in the greatest amount of weight loss.[103] By the end of one year, those on the low-carbohydrate weight-loss plan had lost an average of 10.4 lb, compared with those on the high-carbohydrate (4.8 lb) and standard (5.7 lb) diets. Those in the low-carbohydrate study group showed no change in blood lipids, despite high intakes of dietary fat.

A more recent study that evaluated the effectiveness and safety of low-carbohydrate weight-loss diets reported that they were effective alternatives to low-fat diets and a Mediterranean diet.[104] Even though weight loss was achieved with all three approaches, those following a low-carbohydrate approach lost on average 10 lb after two years. Even more surprising was that the low-carbohydrate diet had more favorable effects on blood lipids and glucose, compared with the low-fat and Mediterranean weight-loss regimens.

Still, weight loss associated with low-carbohydrate diets may not necessarily be caused by alterations in the macronutrient composition of the diet but rather by a reduction in caloric intake.[105] That is, people may eat less on the diet because of limited food choices and decreased appetite associated with ketosis. There is no consistent evidence that carbohydrate restriction causes the body to burn energy more efficiently. These findings cast doubt on the claim that a person can maintain a high intake of calories and still lose weight while on a low-carbohydrate diet.

Nutritional Adequacy of Low-Carbohydrate Diets The amount of carbohydrates allowed on most low-carbohydrate diet plans varies from 5 to 20% of total energy. By comparison, current recommendations suggest that 45 to 65% of total calories come from carbohydrates. Low-carbohydrate diets have been criticized as containing too much total fat, saturated fat, and cholesterol. Current recommendations regarding optimal levels of fat in the diet suggest that 20 to 35% of total calories should come from fat. By comparison, most low-carbohydrate diets provide 55 to 65% of calories from fat. Many low-carbohydrate diets also fail to distinguish between "healthy" and "unhealthy" fats. Finally, the amount of protein in some low-carbohydrate diets exceeds recommended amounts. It is currently recommended that we consume 10 to 35% of calories from protein. Many low-carbohydrate diets provide 20 to 40% of total calories from protein, so some dieters may be exceeding the upper limit of the AMDR for protein (10–35% of total calories for adults >18 years of age).

Perhaps one of the biggest concerns regarding low-carbohydrate diets is the restriction of healthy, high-carbohydrate foods such as fruit, vegetables,

⟨**CONNECTIONS**⟩ Recall that ketosis is a condition resulting from the accumulation of ketones in the blood (Chapter 8, page 165).

⟨**CONNECTIONS**⟩ Recall that the Mediterranean diet is an eating pattern rich in monounsaturated fatty acids, a balanced ratio of omega-6:omega-3 essential fatty acids, fiber, and antioxidants (Chapter 10, page 287).

ketogenic diets Diets that stimulate ketone production.

and whole grains—a valid criticism.[106] Low-carbohydrate diets may lack essential micronutrients, dietary fiber, and beneficial phytochemicals. Although most low-carbohydrate diet plans recommend that people take dietary supplements, these cannot replace the many other substances (such as fiber and phytochemicals) in these restricted foods. People on low-carbohydrate weight-loss eating plans should eat fresh fruits and nonstarchy vegetables to help ensure that their diets are nutritious and balanced.

Some Health Concerns Regarding Low-Carbohydrate Diets Are Unfounded Because some cells (such as red blood cells) require glucose, consuming a very-low–carbohydrate diet causes the body to break down protein (muscle), so that the resulting amino acids can be used for glucose synthesis (gluconeogenesis). Therefore, some experts have raised concerns that low-carbohydrate diets may cause loss of lean tissue. However, studies suggest that weight loss associated with low-carbohydrate diets is largely attributed to a loss of body fat, with little loss of lean tissue.[107] It appears that protein intakes associated with some low-carbohydrate diets are high enough to prevent the loss of muscle.

Some health professionals have also expressed concern about the effect of low-carbohydrate diets on bone health and kidney function. This is because low-carbohydrate diets can also be high in protein. However, limited research shows no harmful effects of low-carbohydrate/high-protein diets on bone health.[108] Some professionals worry that low-carbohydrate diets may impair kidney function, because high protein intakes and ketones are thought to overburden the kidneys.[109] Although this has not been adequately studied, low-carbohydrate diets are not advised for people with impaired kidney function.

Low-carbohydrate diets also do not appear to have an adverse effect on cardiovascular health, at least in the short term.[110] Rather, several studies suggest that low-carbohydrate diets result in favorable changes in blood lipid levels, glycemic control, and blood pressure, despite increased intakes of fat.[111] These improvements may be due to a reduction in body weight rather than to a direct effect of the macronutrient composition of the diet.

Most popular diets look at energy balance in a simplistic fashion, as though there were one magic key that opens the door to easy weight control. On the contrary, maintaining energy balance is quite complex. The reality is that nobody knows for certain the "ideal" distribution of macronutrients for weight loss. The AMDR provides an array of relative proportions of macronutrients associated with a healthy diet. Therefore, it is up to individuals to determine the eating pattern that best suits their needs within parameters of the AMDR. What the research strongly suggests, however, is that reducing calorie intake is the critical component for successful weight loss.

In this chapter, you have learned that, energy balance is quite complex and that our eating behaviors are shaped by many factors, including genetics, the physiological states of hunger and satiety, and the psychological and social determinants of appetite. You have also learned how a combination of healthy eating and exercise can lead to successful long-term weight control. However, this simple formula can be far from easy to follow, because of influences ranging from social pressures to genetics to the foods available in a given society.

Critical Thinking: August's Story Now that you understand that both lifestyle and genetics can influence body weight, think back to August's story about her difficulty losing weight. Many people assume that someone like August could and should lose weight by diet and exercise alone. Do you think these views are correct?

Diet Analysis PLUS ✚ Activity

Part A: Estimated Energy Requirements (EER), Total Energy Expenditure (TEE), and Resting Energy Expenditure (REE)

1. The EER represents the average dietary energy intake that will maintain energy balance in healthy persons of a given sex, age, weight, height, and physical activity level (PAL). Use the following formulas to estimate your energy expenditure.

   Males EER (kcal/day) = 662 − [9.53 × age (y)] + PA × [15.91 × weight (kg)] + [539.6 × height (m)]

   Females EER (kcal/day) = 354 − [(6.91) × age (y)] + PA × [9.36 × weight (kg) + [726 × height (m)]

2. Now look at your Profile DRI Goals, and compare your calculated value with the printed "Energy" value. Are they the same?

3. Total energy expenditure (TEE) is based on the energy needed to maintain involuntary, physiological functions, support physical activity, and process food. Knowing that basal metabolism makes up 50 to 70% of the TEE, use your calculated EER value to estimate how many kilocalories are used to support vital body functions (basal energy expenditure; BEE). Calculate your upper level first by multiplying by 0.70 and then your lower level by multiplying by 0.50.

   _____ (EER) kcal/day × 0.70 = _____ kcal/day. Upper limit of kilocalories needed for involuntary, vital body functions

   _____ (EER) kcal/day × 0.50 = _____ kcal/day. Lower limit of kilocalories needed for involuntary, vital body functions

4. Recall that resting metabolism is similar to basal metabolism but is estimated under less stringent conditions. Thus, it is likely to be higher than basal metabolism. The Harris-Benedict equation can be used to predict your REE. Compare your calculated REE to your calculated BEE. Remember the REE estimates energy needed for your basic functions in addition to the energy needed to digest your last meal.

   Male: REE = 66.5 + [13.8 × weight (kg)] + [5 × height (cm)] − [6.8 × age (y)] = _____ kcal/day

   Female: REE = 655.1 + [9.6 × weight (kg)] + [1.8 × height (cm)] − [4.7 × age (y)] = _____ kcal/day

Part B: Physical Activity

Total energy expenditure (TEE) also includes energy expenditure associated with physical activity. This factor can vary greatly from person to person and may account for 15 to 30% of your TEE.

1. Calculate your EER using the next-highest physical activity(PA) value. How much would your EER increase if you were to increase your physical activity?

Part C: Thermic Effect of Food (TEF)

1. The third significant component of total energy expenditure is the thermic effect of food (TEF), or the energy needed to digest, absorb, transport, metabolize, and store nutrients after meals. Using the 5 to 10% range mentioned in your text, calculate the range representing your own TEF from the EER determined in Part A.

   EER kcal/day × 0.05 = _____ kcal/day

   EER kcal/day × 0.10 = _____ kcal/day

Notes

1. Prins JB. Adipose tissue as an endocrine organ. Best Practice and Research Clinical Endocrinology and Metabolism. 2002;16:639–51.

2. Lau DCW, Dhillon B, Yan H, Szmitko PE, Verma S. Adipokines: molecular links between obesity and atheroslcerosis. American Journal of Physiology. Heart and Circulatory Physiology. 2005;288:H2031–41.

3. Spalding KL, Arner E, Westermark PO, Buchholz BA, Bergmann O, Blomqvist L, Hoffstedt J, Näslund E, Britton T, Concha H, Hassan M, Rydén M, Frisén J, Arner P. Dynamics of fat cell turnover in humans. Nature. 2008;453:783–7.

4. Arner P, Spalding KL. Fat cell turnover in humans. Biochemical and Biophysical Research Communications. 2010;396:101–4.

5. Pénicaud L. Relationships between adipose tissues and brain: what do we learn from animal studies? Diabetes & Metabolism. 2010;36:S39–44.

6. de Graaf C, Blom WA, Smeets PA, Stafleu A, Hendriks HF. Biomarkers of satiation and satiety. American Journal of Clinical Nutrition. 2004;79:946–61.

7. U.S. Food and Drug Administration. FDA expands use of banding system for weight loss. 2011. Available from: http://www.fda.gov/NewsEvents/Newsroom/PressAnnouncements/ucm245617.htm

8. Ali MR, Fuller WD, Choi MP, Wolfe BM. Bariatric surgical outcomes. Surgical Clinics of North America. 2005;85:835–52.

9. Flum DR, Salem L, Broeckel-Elrod JA, Patchen-Dellinger E, Cheadle A, Chan L. Early mortality among Medicare beneficiaries undergoing bariatric surgical procedures. Journal of the American Medical Association. 2005;294:1903–8.

10. Sjöström L, Narbro K, Sjöström CD, Karason K, Larsson B, Wedel H, Lystig T, Sullivan M, Bouchard C, Carlsson B, Bengtsson C, Dahlgren S, Gummesson A, Jacobson P, Karlsson J, Lindroos AK, Lönroth H, Näslund I, Olbers T, Stenlöf K, Torgerson J, Agren G, Carlsson LM. Effects of bariatric surgery on mortality in Swedish obese subjects. New England Journal of Medicine. 2007;357:741–52.

11. Moo TA, Rubino F. Gastrointestinal surgery as treatment for type 2 diabetes. Current Opinion in Endocrinology, Diabetes and Obesity. 2008;15:153–8.

12. Coelho JC, Campos AC. Surgical treatment of morbid obesity. Current Opinion in Clinical Nutrition and Metabolic Care. 2001;4:201–6. Pope GD, Finlayson SR, Kemp JA, Birkmeyer JD. Life expectancy benefits of gastric bypass surgery. Surgical Innovations. 2006;13:265–73.

13. Ello-Martin JA, Ledikwe JH, Rolls BJ. The influence of food portion size and energy density on energy intake: Implications for weight management. American Journal of Clinical Nutrition. 2005;82:236S–41S.

14. Wurtman RJ, Wurtman JJ. Brain serotonin, carbohydrate-craving, obesity and depression. Obesity Research. 1995;3:477S–80S.

15. Hainer V, Kabrnova K, Aldhoon B, Kunesova M, Wagenknecht M. Serotonin and norepinephrine reuptake inhibition and eating behavior. Annals of the New York Academy of Science. 2006;1083:252–69.

16. Orr J, Davy B. Dietary influences on peripheral hormones regulating energy intake: Potential applications for weight management. Journal of the American Dietetic Association. 2005;105:1115–24.

17. Inui A, Asakawa A, Bowers CY, Mantovani G, Laviano A, Meguid MM, Fujimiya M. Ghrelin, appetite, and gastric motility: The emerging role of the stomach as an endocrine organ. Federation of American Societies for Experimental Biology Journal. 2004;18:439–56.

18. Cummings DE. Ghrelin and the short- and long-term regulation of appetite and body weight. Physiological Behavior. 2006;89:71–84.

19. Yanovski S. Sugar and fat: Cravings and aversions. Journal of Nutrition. 2003;133:835S–7S.

20. Lafay L, Thomas F, Mennen L, Charles MA, Eschwege E, Borys JM, Basdevant A. Gender differences in the relation between food cravings and mood in an adult community: Results from the Fleurbaix Laventie Ville Sante study. International Journal of Eating Disorders. 2001;29:195–204.

21. Dye L, Warner P, Bancroft JJ. Food craving during the menstrual cycle and its relationship to stress, happiness of relationship and depression; a preliminary enquiry. Journal of Affective Disorders. 1995;34:157–64.

22. Hulbert AJ, Else PL. Basal metabolic rate: History, composition, regulation, and usefulness. Physiological and Biochemical Zoology. 2004;77:869–76.

23. Henry CJ. Mechanisms of changes in basal metabolism during ageing. European Journal of Clinical Nutrition. 2000;54:S77–91.

24. Luke A, Schoeller DA. Basal metabolic rate, fat-free mass, and body cell mass during energy restriction. Metabolism. 1992;41:450–6.

25. Brooks GA, Butte NF, Rand WM, Flatt JP, Caballero B. Chronicle of the Institute of Medicine physical activity recommendation: How a physical activity recommendation came to be among dietary recommendations. American Journal of Clinical Nutrition. 2004;79:921S–30S.

26. Nair KS, Halliday D, Garrow JS. Thermic response to isoenergetic protein, carbohydrate or fat meals in lean and obese subjects. Clinical Science. 1983;65:307–12.

27. Institute of Medicine. Dietary reference intakes for energy, carbohydrate, fiber, fat, fatty acids, cholesterol, protein, and amino acids. Washington, DC: National Academies Press; 2005.

28. Aronne LJ. Classification of obesity and assessment of obesity-related health risks. Obesity Research. 2002;10:105S–15S.

29. Friedl KE. Can you be large and not obese? The distinction between body weight, body fat, and abdominal fat in occupational standards. Diabetes Technology and Therapeutics. 2004;6:732–49.

30. Katzmarzyk PT, Janssen I, Ardern CI. Physical inactivity, excess adiposity and premature mortality. Obesity Reviews. 2003;4:257–901.

31. Lafontan M, Berlan M. Do regional differences in adipocyte biology provide new pathophysiological insights? Trends in Pharmacological Sciences. 2003;24:276–83.

32. Ibrahim MM. Subcutaneous and visceral adipose tissue: structural and functional differences. Obesity Reviews. 2010;11:11–8.

33. Toth MJ, Tchernof A, Sites CK, Poehlman ET. Menopause-related changes in body fat distribution. Annals of the New York Academy of Sciences. 2000;904:502–6.

34. National Institutes of Health. Clinical guidelines on the identification, evaluation, and treatment of overweight and obesity in adults. National Institutes of Health, National Heart, Lung, and Blood Institute, Obesity Education Initiative. Available from http://www.nhlbi.nih.gov/guidelines/obesity/practgde.htm.

35. See R, Abdullah SM, McGuire DK, Khera A, Patel MJ, Lindsey JB, Grundy SM, de Lemos JA. The association of differing measures of overweight and obesity with prevalent atherosclerosis: The Dallas Heart Study. American College of Cardiology. 2007;50:752–9.

36. Heinrich KM, Jitnarin N, Suminski RR, Berkel L, Hunter CM, Alvarez L, Brundige AR, Peterson AL, Foreyt JP, Haddock CK, Poston WS. Obesity classification in military personnel: A comparison of body fat, waist circumference, and body mass index measurements. Military Medicine. 2008;17:67–73.

37. National Center for Health Statistics. Health, United States, 2010: With Special Feature on Death and Dying. Hyattsville, MD. 2011. Available from: http://www.cdc.gov/nchs/data/hus/hus10.pdf.

38. Popkin BM, Gordon-Larsen P. The nutrition transition: Worldwide obesity dynamics and their determinants. International Journal of Obesity and Related Metabolic Disorders. 2004;3:S2–9.

39. Austin GL, Ogden LG, and O Hill J. Trends in carbohydrate, fat, and protein intakes and association with energy intake in normal-weight, overweight, and obese individuals: 1971–2006. American Journal of Clinical Nutrition. 2011;93:836–43.

40. Briefel RR, Johnson CL. Secular trends in dietary intake in the United States. Annual Review of Nutrition. 2004;24:401–31.

41. Ello-Martin JA, Ledikwe JH, Rolls BJ. The influence of food portion size and energy density on energy intake: Implications for weight management. American Journal of Clinical Nutrition. 2005;82:236 S–41S. Wansink B, Kim J. Bad popcorn in big buckets: Portion size can influence intake as much as taste. Journal of Nutrition Education and Behavior. 2005;37:242–5.

42. Rolls BJ, Roe LS, Kral TVE, Meengs JS, Wall DE. Increasing the portion size of a packaged snack increases energy intake in men and women. Appetite. 2004;42:63–9.

43. Mourao DM, Bressan J, Campbell WW, Mattes RD. Effects of food form on appetite and energy intake in lean and obese young adults. International Journal of Obesity. 2007;31:1688–95.

44. Hetherington MM, Foster R, Newman T, Anderson AS, Norton G. Understanding variety: Tasting different foods delays satiation. Physiology and Behavior. 2006;87:263–71.

45. Chandon P and Wansink B. The biasing health halos of fast-food restaurant health claims: Lower calorie estimates and higher side-dish consumption intentions. Journal of Consumer Research. 2007;34:301–14.

46. National Center for Health Statistics (NCHS). Health, United States, 2010. Available from: http://www.cdc.gov/nchs/data/hus/hus10.pdf.

47. Kim JY, Oh DJ, Yoon TY, Choi JM, Choe BK. The impacts of obesity on psychological well-being: A cross-sectional study about depressive mood and quality of life. Preventive Medicine and Public Health. 2007;40:191–5.

48. Mamun AA, Lawlor DA, O'Callaghan MJ, Bor W, Williams GM, Najman JM. Does childhood sexual abuse predict young adult's BMI? A birth cohort study. Obesity. 2007;15:2103–10. Noll JG, Zeller MH, Trickett PK, Putnam FW. Obesity risk for female victims of childhood sexual abuse: A prospective study. Pediatrics. 2007;120:61–7. bes Gustafson TB, Sarwer DB. Childhood sexual abuse and obesity. Obesity Reviews. 2004;5:129–35.

49. Christakis NA and Fowler JH. The spread of obesity in a large social network over 32 years. New England Journal of Medicine. 2007;57:370–9.

50. Brownson RC, Boehmer TK, Luke DA. Declining rates of physical activity in the United States: What are the contributors? Annual Review of Public Health. 2005;26:421–43.

51. American College of Sports Medicine (ACSM). ACSM's Guidelines for Exercise Testing and Prescription, 8th ed. Philadelphia: Lippincott Williams & Wilkins; 2010.

52. Wareham NJ, van Sluijs EM, Ekelund U. Physical activity and obesity prevention: A review of the current evidence. Proceedings of the Nutrition Society. 2005;64:229–47.

53. Hill JO, Wyatt HR, Reed GW, Peters JC. Obesity and the environment: Where do we go from here? Science. 2003;299:853–5.

54. National Association for Sport and Physical Education and American Heart Association. 2010 Shape of the

Nation Report. Available at: http://www.aahperd.org/naspe/publications/upload/Shape-of-the-Nation-2010-Final.pdf

55. Caroli M, Argentieri L, Cardone M, Masi A. Role of television in childhood obesity prevention. International Journal of Obesity and Related Metabolic Disorders. 2004;28:S4–108.

56. Kaiser Family Foundation. The role of media in childhood obesity; 2004. Publication number 7030. Available from http://www.kff.org/entmedia/7030.cfm.

57. Halford JC, Boyland EJ, Hughes GM, Stacey L, McKean S, Dovey TM. Beyond-brand effect of television food advertisements on food choice in children: The effects of weight status. Public Health Nutrition. 2007;1–8. Powell LM, Szczypka G, Chaloupka FJ. Exposure to food advertising on television among US children. Archives of Pediatric and Adolescent Medicine. 2007;16:553–60.

58. U.S. Department of Health and Human Services. 2008 Physical Activity Guidelines for Americans. Washington (DC): U.S. Department of Health and Human Services; 2008. ODPHP Publication No. U0036. Available from: http://www.health.gov/paguidelines.

59. Stunkard AJ, Harris JR, Pedersen NL, McClearn GE. The body-mass index of twins who have been reared apart. New England Journal of Medicine. 1990;322:1483–7. Wardle J, Carnell S, Haworth CM, Plomin R. Evidence for a strong genetic influence on childhood adiposity despite the force of the obesogenic environment. American Journal of Clinical Nutrition. 2008;87:398–404. Lee K, Song YM, Sung J. Which obesity indicators are better predictors of metabolic risk? Healthy twin study. Obesity. 2008;16:834–40.

60. Sorensen TI, Holst C, Stunkard AJ. Adoption study of environmental modifications of the genetic influences on obesity. International Journal of Obesity and Related Metabolic Disorders. 1998;22:73–81.

61. Gillman MW, Barker D, Bier D, Cagampang F, Challis J, Fall C, Godfrey K, Gluckman P, Hanson M, Kuh D, Nathanielsz P, Nestel P, Thornburg KL. Meeting report on the 3rd International Congress on Developmental Origins of Health and Disease (DOHaD). Pediatric Research. 2007;61:625–9.

62. Ingalls A, Dickie M, Snell GD. Obese, a new mutation in the house mouse. Journal of Heredity. 1950;41:317–8.

63. Coleman D, Hummel KP. Effects of parabiosis of normal with genetically diabetic mice. American Journal of Physiology. 1969;217:1298–304. Coleman DL. Effects of parabiosis of obese with diabetes and normal mice. Diabetologia. 1973;9:294–8.

64. Zhang Y, Proenca R, Maffei M, Leopold L, Friedman JM. Positional cloning of the mouse obese gene and its human homologue. Nature. 1994;372:125–32.

65. Farooqi IS, O'Rahilly S. Monogenic obesity in humans. Annual Review of Medicine. 2005;56:443–58.

66. Marx J. Cellular warriors at the battle of the bulge. Science. 2003;299:846–9.

67. Kennedy AG. The role of the fat depot in the hypothalamic control of food intake in the rat. Proceedings of the Royal Society of London. 1953;140:578–92.

68. Popovic V, Duntas LH. Brain somatic cross-talk: Ghrelin, leptin and ultimate challengers of obesity. Nutrition and Neuroscience. 2005;8:1–5. Scarpace PJ, Zhang Y. Elevated leptin: Consequence or cause of obesity? Frontiers in Bioscience. 2007;12:3531–44.

69. Jequier E. Leptin signaling, adiposity, and energy balance. Annals of the New York Academy of Sciences. 2002;967:379–88.

70. Couce ME, Green D, Brunetto A, Achim C, Lloyd RV, Burguera B. Pituitary. Limited brain access for leptin in obesity. 2001;4:101–10.

71. Klok MD, Jakobsdottir S, Drent ML. The role of leptin and ghrelin in the regulation of food intake and body weight in humans: A review. Obesity Reviews. 2007;8:21–34. Farooqi S, O'Rahilly S. Genetics of obesity in humans. Endocrine Reviews. 2006;27:710–8.

72. Nishida M, Funahashi T, Shimomura I. Pathophysiological significance of adiponectin. Medical Molecular Morphology. 2007;40:55–67.

73. Torres-Leal FL, Fonseca-Alaniz MH, Rogero MM, Tirapegui. The role of inflamed adipose tissue in the insulin resistance. Cell Biochemistry and Function. 2010;28:623–31.

74. Matsuzawa Y. Adiponectin: a key player in obesity related disorders. Current Pharmaceutical Design. 2010;16:1896–901.

75. Oh DK, Ciaraldi T, Henry RR. Adiponectin in health and disease. Diabetes, Obesity and Metabolism. 2007;9:282–9.

76. Mehta S, Farmer JA. Obesity and inflammation: A new look at an old problem. Current Atherosclerosis Report. 2007;9:134–8.

77. Wing RR, Phelan S. Long-term weight loss maintenance. American Journal of Clinical Nutrition. 2005;82:222S–5S.

78. National Institutes of Health, and National Heart, Lung and Blood Institute. Clinical guidelines on the identification, evaluation and treatment of overweight and obesity in adults—the evidence report. National Institutes of Health Publication Number 00–4084. Bethesda, MD: National Institutes of Health; October 2000.

79. Farrell SW, Braun L, Barlow CE, Cheng YJ, Blair SN. The relation of body sass index, cardiorespiratory fitness, and all-cause mortality in women. Obesity Research. 2002;10:417–23.

80. Barlow CE, Kohl HW III, Gibbons LW, Blair SN. Physical fitness, mortality and obesity. International Journal of Obesity and Related Metabolic Disorders. 1995;19:S41–4.

81. International Association for the Study of Obesity. Diet, nutrition and the prevention of chronic diseases. Report of the joint WHO/FAO expert consultation. WHO Technical Report Series, No. 916 (TRS 916); 2003. Available from: http://biotech.law.lsu.edu/obesity/who/trs_916.pdf.

82. Catenacci VA, Ogden LG, Stuht J, Phelan S, Wing RR, Hill JO, Wyatt HR. Physical activity patterns in the National Weight Control Registry. Obesity. 2008;16:153–61. Butryn ML, Phelan S, Hill JO, Wing RR. Consistent self-monitoring of weight: A key component of successful weight loss maintenance. Obesity. 2007;15:3091–6. Phelan S, Wyatt H, Nassery S, Dibello J, Fava JL, Hill JO, Wing RR. Three-year weight change in successful weight losers who lost weight on a low-carbohydrate diet. Obesity. 2007;15:2470–7.

83. Westman EC, Yancy WS Jr, Vernon MC. Is a low-carb, low-fat diet optimal? Archives of Internal Medicine. 2005;165:1071–2.

84. Bell EA, Castellanos VH, Pelkman CL, Thorwart ML, Rolls BJ. Energy density of foods affects energy intake in normal-weight women. American Journal of Clinical Nutrition. 1998;67:412–20. Yao M, Roberts SB. Dietary energy density and weight regulation. Nutrition Reviews. 2001;59:247–58.

85. Ello-Martin JA, Ledikwe JH, Rolls BJ. The influence of food portion size and energy density on energy intake: Implications for weight management. American Journal of Clinical Nutrition. 2005;82:236S–41S.

86. National Center for Chronic Disease Prevention and Health Promotion Division of Nutrition and Physical Activity, Department of Health and Human Services, Centers for Disease Control and Prevention. Can eating fruits and vegetables help people to manage their weight? Research to Practice Series, No.1. Available from http://www.cdc.gov/nccdphp/dnpa/nutrition/pdf/rtp_practitioner_10_07.pdf.

87. Rolls BJ, Bell EA, Waugh BA. Increasing the volume of a food by incorporating air affects satiety in men. American Journal of Clinical Nutrition. 2000;72:361–8.

88. Norton GN, Anderson AS, Hetherington MM. Volume and variety: relative effects on food intake. Physiology and Behavior. 2006;87(4):714–22.

89. Ornish, Dean. Dr. Dean Ornish's Program for Reversing Heart Disease. New York, New York: Random House Publishing Group, 1996.

90. Pritikin, Nathan and McGrady Patrick. The Pritikin Program for Diet and Exercise. New York, New York: Random House Publishing Group, 1984.

91. Lovejoy JC, Bray GA, LeFevre M, Smith SR, Most MM, Denkins YM, Volaufova J, Rood JC, Eldrige AL, Peters JC. Consumption of a controlled low-fat diet containing olestra for 9 months improves health risk factors in conjunction with weight loss in obese men: The Olé Study. International Journal of Obesity and Related Metabolic Disorders. 2003;27:1242–9.

92. Center for Nutrition Policy and Promotion and the U.S. Department of Agriculture. Nutrition Insights. Is fat consumption really decreasing? Insight 5 April 1998. Available from www.cnpp.usda.gov/Publications/NutritionInsights/insight5.pdf.

93. Willett WC. Dietary fat and body fat: Is there a relationship? Journal of Nutritional Biochemistry.

1998;9:522–4. Willett WC. Is dietary fat a major determinant of body fat? American Journal of Clinical Nutrition. 1998;67:556S–625S.

94. Busetto L, Marangon M, De Stefano F. High-protein low-carbohydrate diets: what is the rationale? Diabetes Metabolism Research and Reviews. 2011;27:230–2.

95. Taubes G. The soft science of dietary fat. Science. 2001;291:2536–45.

96. Foster GD, Wyatt HR, Hill JO, McGuckin BG, Brill C, Selma B, Szapary PO, Rader DJ, Edman JS, Klein S. A randomized trial of a low-carbohydrate diet for obesity. New England Journal of Medicine. 2003;248:2082–90.

97. Crowe TC. Safety of low-carbohydrate diets. Obesity Reviews. 2005;6:235–45.

98. Bravata DM, Sanders L, Huang J, Krumholz HM, Olkin I, Gardner CD, Bravata DM. Efficacy and safety of low-carbohydrate diets: A systematic review. Journal of the American Medical Association. 2003;289:1838–49. Foster GD, Wyatt HR, Hill JO, McGuckin BG, Brill C, Mohammed BS, Szapary PO, Rader DJ, Edman JS, Klein S. A randomized trial of a low-carbohydrate diet for obesity. New England Journal of Medicine. 2003;348:2082–90.

99. Atkins, Robert. Dr. Atkins' New diet Revolution. New York, New York: HarperCollins Publishers, 2002.

100. Heller Rachael and Heller Richard. The Carbohydrate Addict's Diet. Penguin Group, 1993.

101. Allen, Christopher. Life without Bread: How a Low-Carbohydrate Diet Can Save Your Life. New York, New York: McGraw-Hill Companies, 2000.

102. Eades Michael R. Protein Power. New York, New York: Bantam Book, 1999.

103. Gardner CD, Kiazand A, Alhassan S, Kim S, Stafford RS, Balise RR, Kraemer CK, King AC. Comparison of the Atkins, Zone, Ornish, and LEARN diets for change in weight and related risk factors among overweight premenopausal women: The A to Z weight loss study: A randomized trial. Journal of the American Medical Association. 2007;297:969–77.

104. Shai I, Schwarzfuchs D, Henkin Y, Shahar DR, Witkow S, Greenberg I, Golan R, Fraser D, Bolotin A, Vardi H, Tangi-Rozental O, Zuk-Ramot R, Sarusi B, Brickner D, Schwartz Z, Sheiner E, Marko R, Katorza E, Thiery J, Fiedler GM, Blüher M, Stumvoll M, Stampfer MJ. Weight loss with a low-carbohydrate, Mediterranean, or low-fat diet; Dietary Intervention Randomized Controlled Trial (DIRECT) group. New England Journal of Medicine. 2008;359:229–41.

105. Bravata DM, Sanders L, Huang J, Krumholz HM, Olkin I, Gardner CD, Bravata DM. Efficacy and safety of low-carbohydrate diets: A systematic review. Journal of the American Medical Association. 2003;289:1838–49.

106. Schwenke DC. Insulin resistance, low-fat diets, and low-carbohydrate diets: Time to test new menus. Current Opinion in Lipidology. 2005;16:55–60.

107. Astrup A, Meinert Larsen T, Harper A. Atkins and other low-carbohydrate diets: Hoax or an effective tool for weight loss? Lancet. 2004;364:897–9.

108. Farnsworth E, Luscombe ND, Noakes M, Wittert G, Argyiou E, Clifton PM. Effect of a high-protein, energy-restricted diet on body composition, glycemic control, and lipid concentrations in overweight and hyperinsulinemic men and women. American Journal of Clinical Nutrition. 2003;78:31–9.

109. Martin WF, Armstrong LE, Rodriguez NR. Dietary protein intake and renal function. Nutrition and Metabolism. 2005;2:25.

110. Acheson KJ. Carbohydrate and weight control: Where do we stand? Current Opinion in Clinical Nutrition and Metabolic Care. 2004;7:485–92.

111. Aude YW, Agatston AS, Lopez-Jimenez F, Lieberman EH, Almon M, Hansen M, Rojas G, Lamas GA, Hennekens CH. The national cholesterol education program diet vs. a diet lower in carbohydrates and higher in protein and monounsaturated fat: A randomized trial. Archives of Internal Medicine. 2004;164:2141–6.

Disordered Eating

Our bodies have natural boundaries that tell us when to eat, what to eat, and how much to eat. Paying attention to and trusting these internal cues help us have a healthy relationship with food. Healthy eating means eating without fear, guilt, or shame. However, for some people, preoccupation with food and weight loss can reach obsessive proportions. These troublesome disturbances in eating behaviors can be signs of disordered eating. If disordered eating patterns continue, they can eventually progress into an eating disorder.

In the past 25 years, the number of people diagnosed with eating disorders has increased. Although eating disorders in the general population are somewhat rare, they are relatively common among adolescent girls and young women. The American Psychological Association estimates that approximately 8 million females (8%) in the United States battle with some form of eating disorder. Although males can also develop eating disorders, the occurrence is more difficult to estimate because they may be more reluctant to seek help for what is commonly perceived to be a "female" condition. Nonetheless, the prevalence of eating disorders among males has been estimated to be as high as 2%.[1]

Not surprisingly, clinicians are very interested in learning more about eating disorders and how they can be treated effectively. In this Nutrition Matters, you will learn about disordered eating behaviors and the different types, causes, and complexities of eating disorders.

© Plush Studios/Getty Images

People with a distorted body image perceive themselves to be fat even if they are very thin.

How Do Eating Disorders Differ from Disordered Eating?

Disordered eating behaviors include a wide variety of unhealthy eating patterns such as irregular eating, consistent undereating, and consistent overeating.[2] These behaviors are common and often occur in response to stress, illness, or dissatisfaction with personal appearance. Although disordered eating patterns can be disturbing to others, they typically do not persist long enough to cause serious physical harm. However, in some people, disordered eating can progress into a full-blown eating disorder such as anorexia nervosa, an extreme pursuit of thinness; bulimia nervosa, a pattern of bingeing and purging; or binge-eating disorder. **Eating disorders** are characterized by extreme disturbances in eating behaviors that can be both physically and psychologically harmful.

disordered eating Unhealthy eating patterns such as eating irregularly, consistent undereating, and/or consistent overeating.

eating disorder Extreme disturbance in eating behaviors that can result in serious medical conditions, psychological consequences, and dangerous weight loss.

People with eating disorders often feel isolated, and their relationships with family and friends become strained.

To understand what causes eating disorders, it is important to consider a variety of factors that include but go beyond food intake. That is, eating disorders are complex behaviors that arise from a combination of physical, psychological, and social issues.

As shown in Table 1, the American Psychiatric Association classifies eating disorders into three distinct categories: anorexia nervosa (AN), bulimia nervosa (BN), and eating disorders not otherwise specified (EDNOS).[3] There are also subcategories that depend on the presence or absence of specific behaviors. Because behaviors associated with the different eating disorders often overlap, it can be difficult to identify which type of disorder a person has (Table 2).[4]

PEOPLE WITH ANOREXIA NERVOSA PURSUE EXCESSIVE THINNESS

Anorexia nervosa (AN) is characterized by an irrational fear of gaining weight or becoming obese. As a consequence, individuals with AN are unwilling to maintain a minimally normal body weight. In addition, there is often a disconnect between actual and perceived body weight and shape, such that people with AN believe they are "fat" even though they may be dangerously thin. When people with AN look in the mirror, they tend to be very critical of their body shape and size. Because of this distorted self-perception, weight loss becomes an obsession. Individuals with AN often spend hours scrutinizing

---

anorexia nervosa (AN) An eating disorder characterized by an irrational fear of gaining weight or becoming obese.

TABLE 1 American Psychiatric Association Classification and Diagnostic Criteria for Eating Disorders

Anorexia Nervosa (AN)

A. Refusal to maintain body weight at or above a minimally healthy weight for age and height.

B. Intense fear of gaining weight or becoming fat, even though underweight.

C. Disturbance in body weight or shape is experienced, or denial of the seriousness of the current low body weight.

D. Absence of at least three consecutive menstrual cycles (postmenarchal, premenopausal females).

Two types:

- *Restricting type:* Does not regularly engage in binge eating or purging behavior.
- *Binge-eating/purging type:* Regularly engages in binge eating or purging behavior.

Bulimia Nervosa (BN)

A. Recurrent episodes of binge eating. An episode of binge eating is characterized by both of the following: (1) eating, in a discrete period of time, an amount of food that is definitely larger than what most people would eat during a similar period of time and under similar circumstances, and (2) a sense of lack of control over eating during the episode.

B. Recurrent compensatory behavior to prevent weight gain, such as self-induced vomiting; misuse of laxatives, diuretics, enemas, or other medications; fasting; or excessive exercise.

C. Binge eating and inappropriate compensatory behaviors occurring at least twice a week for three months.

D. Self-evaluation is unduly influenced by body shape and weight.

E. Disturbance does not occur exclusively during episodes of anorexia nervosa.

Two types:

- *Purging type:* Regularly engages in self-induced vomiting or the misuse of laxatives, diuretics, or enemas.
- *Nonpurging type:* Regularly engages in inappropriate compensatory behaviors, such as fasting or excessive exercise; but not in self-induced vomiting or the misuse of laxatives, diuretics, or enemas.

Eating Disorders Not Otherwise Specified (EDNOS) is for disorders of eating that do not meet the criteria for any specific eating disorder. Examples include:

A. For females, criteria for anorexia nervosa are met, except the individual has regular menses.

B. Criteria for anorexia nervosa are met except that, despite significant weight loss, current weight is in the healthy range.

C. Criteria for bulimia nervosa are met except that the binge eating and inappropriate compensatory mechanisms occur at a frequency of less than twice a week or a duration of less than three months.

D. Regular use of inappropriate compensatory behavior by an individual of healthy body weight after eating small amounts of food.

E. Repeatedly chewing and spitting out, but not swallowing, large amounts of food.

F. Binge-eating disorder: recurrent episodes of binge eating in the absence of the regular use of inappropriate compensatory behaviors characteristic of bulimia nervosa.

SOURCE: Adapted from Diagnostic and Statistical Manual of Mental Disorders, 4th ed. (DSM-IV). Washington, DC: American Psychiatric Association, 2004. Used by permission of American Psychiatric Association.

TABLE 2 Behaviors Associated with Different Eating Disorders and Disordered Eating

Type of Eating Disorder and Disordered Eating	Food Restriction	Compensatory Mechanism	Bingeing
Anorexia nervosa, restricting type		Excessive exercise	
Anorexia nervosa, binge-eating/purging type		Excessive exercise	
Bulimia nervosa, purging type		Self-induced vomiting and/or the misuse of laxatives, diuretics, excessive exercise, or enemas	
Bulimia nervosa, nonpurging type		Excessive exercise	
Binge-eating disorder			
Restrained eating			

NOTE: Dark blue indicates a primary behavior associated with specific eating disorders and disordered eating, whereas light blue indicates that a behavior occurs to a lesser extent.

their bodies, paying excessive attention to their appearance.

People with AN tend to view their self-worth in terms of weight and body shape. Because they perceive starvation as an accomplishment rather than a problem, they have little motivation to change. Although the symptoms of AN center on food, the causes are much more complex. Like other eating disorders, AN stems mainly from psychological issues. In the case of AN, the denial of food coupled with the relentless pursuit of thinness become ways of coping with emotions, conflict, and stress.[5]

The American Psychiatric Association recognizes two types of AN: **restricting type** and **binge-eating/ purging type** (Figure 1). People with restricting type AN maintain a low body weight by food restriction and/or excessive exercise, whereas people with binge-eating/purging type AN engage in periods of bingeing and purging (that is, self-induced vomiting and/or abuse of laxatives, diuretics, exercise, or enemas) in addition to severe food restriction.

Rituals Associated with Anorexia Nervosa Typically, people with AN

limit their food intake as well as the variety of foods consumed. For example, they often categorize foods as either "safe" or "unsafe" depending on whether or not consuming the food is thought to cause weight gain. Foods often perceived as unsafe, and typically the first to be eliminated, are meat, high-sugar foods, and fatty foods. Eventually the diet becomes so restricted that nutritional needs are no longer met.

anorexia nervosa, restricting type An eating disorder characterized by food restriction.

anorexia nervosa, binge-eating/purging type An eating disorder characterized by food restriction as well as bingeing and purging.

FIGURE 1 Types of Anorexia Nervosa There are two forms of anorexia nervosa, an eating disorder characterized by self-starvation and a relentless pursuit of weight loss.

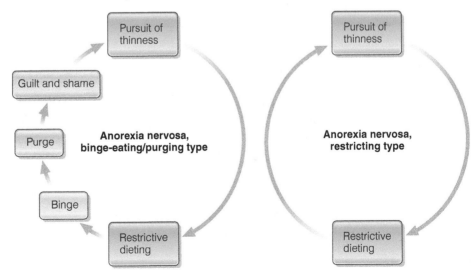

TABLE 3 Physical Signs and Symptoms, Behavioral Characteristics, and Health Consequences Associated with Anorexia Nervosa

Physical Signs and Symptoms	Behavioral and Emotional Signs and Symptoms	Health Consequences
• At or below 15% of ideal body weight • Thin appearance • Fainting, dizziness, fatigue, overall weakness • Intolerance to cold • Significant loss of fat and lean body mass • Brittle nails • Dry skin, dry hair, thinning hair, hair loss • Low blood pressure • Formation of fine hair on the body (lanugo) • Irregular or absence of menstruation	• Performance of food rituals such as excessive chewing or cutting food into small pieces • Restriction of amount and types of foods consumed • Rigid schedule and routine • Refusal to eat • Denial of hunger • Excessive exercise • Depression or lack of emotion • Food preoccupation • Tendency toward perfectionism • Rigidity • Frequent monitoring of body weight	• Reproductive problems • Loss of bone mass • Electrolyte imbalance • Irregular heartbeat • Bruises • Injuries such as stress fractures • Impaired iron status • Impaired immune status • Slow heart rate and low blood pressure

In addition to limiting food intake, individuals with AN may also have food rituals such as chewing food but not swallowing, overchewing, obsessive cooking for others, and cutting food into unusually small pieces. These behaviors provide a display of self-restraint and control over the urge to eat. These and other traits often associated with AN are summarized in Table 3.

In addition to dietary restrictions, other compulsive behaviors are commonly associated with AN.[6] For example, many individuals with AN maintain a rigid exercise schedule, working out as much as three to four hours a day. They may also maintain meticulous daily records listing food intake, amount of time exercising, and weight loss. This daily monitoring can provide the anorexic person the mental stamina to continue in this perpetual state of hunger (Figure 2). It is not uncommon for people with AN to weigh themselves repeatedly throughout the day. These obsessive behaviors help to lessen concerns of not being sufficiently lean or obsessive thoughts of being "too fat." A page taken from a journal of a person with AN reflects the inability to recognize these problematic behaviors (Figure 3).

FIGURE 2 Food, Weight, and Exercise Record of a Person with Anorexia Nervosa Some people with eating disorders keep meticulous records regarding exercise and diet.

Monday
Morning
Weight: 87 lbs.
5-mile run (42 minutes)
100 situps

Breakfast
Weight: 86 lbs.
rice cake
10 grapes

Afternoon
Lunch
Weight: 86 lbs.
diet pop
1/2 grapefruit
rice cake

Dinner
Weight: 86 lbs.
1/2 cup rice
diet pop
Celery and ~~ranch~~ nonfat dressing

Before bed
2-mile run (20 minutes)
100 situps
weight: ~~X~~ 85 lbs.

Tuesday
Morning 86 1/2 lbs
Weight: ~~85 lbs~~
5-mile run (45 minutes)
120 situps

~~Breakf~~ Breakfast
Weight: 85 lbs.
rice cake
1 slice melon

Afternoon
Lunch
Weight: 85 lbs
diet pop
10 grapes
rice cake

Dinner
Weight: 85 lbs
1/2 slice toast
diet pop
carrots and nonfat dressing

Before bed
2-mile run (18 minutes)
150 situps
weight: 85 lbs.

FIGURE 3 Thoughts and Behaviors Associated with Anorexia Nervosa

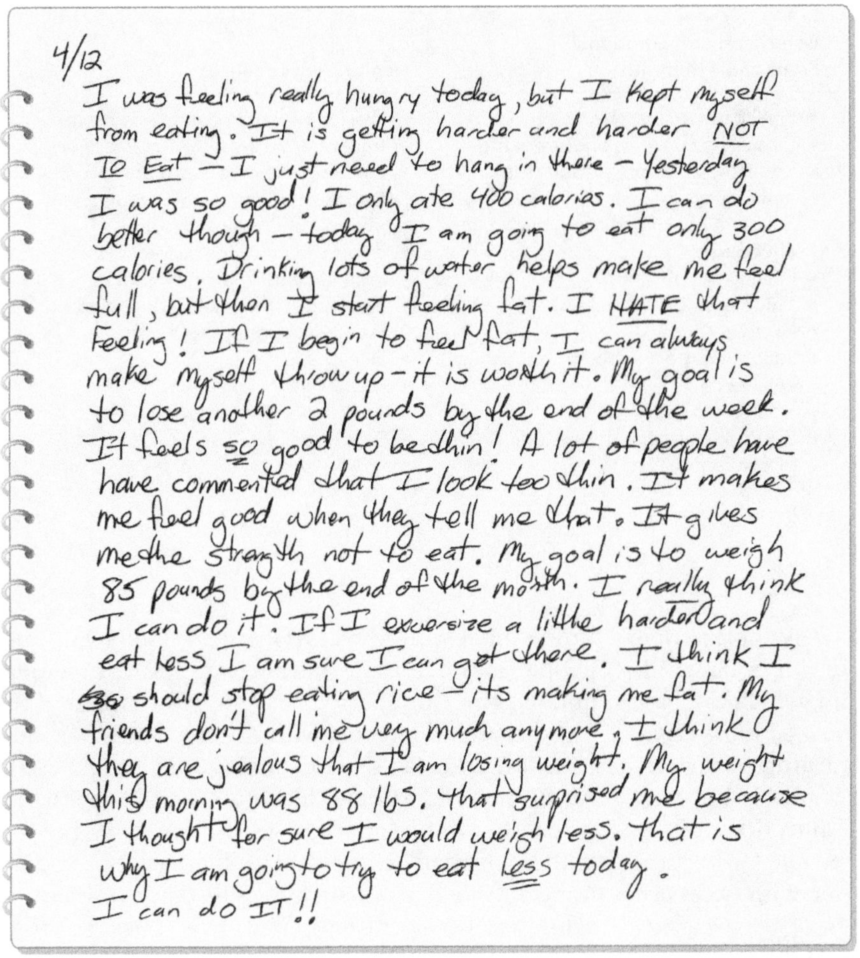

4/12

I was feeling really hungry today, but I kept myself from eating. It is getting harder and harder NOT To Eat — I just need to hang in there — Yesterday I was so good! I only ate 400 calories. I can do better though — today I am going to eat only 300 calories. Drinking lots of water helps make me feel full, but then I start feeling fat. I HATE that Feeling! If I begin to feel fat, I can always make myself throw up — it is worth it. My goal is to lose another 2 pounds by the end of the week. It feels so good to be thin! A lot of people have have commented that I look too thin. It makes me feel good when they tell me that. It gives me the strength not to eat. My goal is to weigh 85 pounds by the end of the month. I really think I can do it. If I exercise a little harder and eat less I am sure I can get there. I think I should stop eating rice — its making me fat. My friends don't call me very much anymore. I think they are jealous that I am losing weight. My weight this morning was 88 lbs. that surprised me because I thought for sure I would weigh less. that is why I am going to try to eat less today.
I can do IT!!

Health Concerns Associated with Anorexia Nervosa

Although people with AN eventually become underweight and appear gaunt, health problems can begin even before this happens. For instance, nutrient deficiencies can develop as well as electrolyte imbalances and hair loss. If AN is left untreated, serious health problems such as dehydration, cardiac abnormalities, appearance of unusual body hair (lanugo), muscle wasting, and bone loss can also occur. These problems are largely the result of chronic malnutrition. Once healthy eating habits and body weight are restored, most of these medical concerns can be reversed.

In females, the loss of body fat leads to a decline in the reproductive hormone estrogen, which can lead to disrupted menstruation and poor reproductive function.[7] These physiological responses undoubtedly evolved as a means to protect women from pregnancy during periods of low food availability. Estrogen also plays an important role in bone health by preventing bone loss. Therefore, the longer a woman goes without menstruating—a condition called amenorrhea—the greater the loss of bone. Although estrogen levels return to normal when body weight is restored, bones probably do not completely recover. For this reason, women with eating disorders who have experienced amenorrhea remain at increased risk for bone disease even after they have regained healthy eating patterns.[8]

AN is a serious condition that affects a person's physical and mental well-being and requires immediate medical, psychological, and nutritional intervention. Although researchers have reported that the death rate for AN is high, this is true only for the most severe cases that require hospitalization.[9] Nonetheless, the death rate associated with AN is higher than that of other eating disorders. It is estimated that between 5 and 20% of those who develop AN die from complications within 10 years of the initial diagnosis.[10] Causes of death are primarily attributed to cardiac arrest or suicide.[11]

PEOPLE WITH BULIMIA NERVOSA BINGE AND PURGE

Like those with AN, people with **bulimia nervosa (BN)** also use food as a coping mechanism. However, those with BN turn to, rather than away from, food during times of stress and emotional conflict. In fact, the origin of the word *bulimia* comes from the Greek word *boulimia*, meaning "the hunger of an ox." People with bulimia nervosa outnumber those with AN by about two to one.[12]

The most common type of BN is characterized by repeated cycles of "bingeing and purging."[13] **Bingeing** is defined as the compulsive consumption

---

bulimia nervosa (BN) An eating disorder characterized by repeated cycles of bingeing and purging.

bingeing Uncontrolled consumption of large quantities of food in a relatively short period of time.

TABLE 4 Physical Signs and Symptoms, Behavioral Characteristics, and Health Consequences Associated with Bulimia Nervosa

Physical Signs and Symptoms	Behavioral and Emotional Signs and Symptoms	Health Consequences
• Fluctuations in body weight • Swollen or puffy face • Odor of vomit on breath or in bathroom • Sores around mouth • Irregular bowel function • Inducement of vomiting after eating	• Feelings of guilt or shame after eating • Obsessive concerns about weight • Repeated attempts at food restriction and dieting • Frequent use of bathroom during and after meals • Feeling out of control • Moodiness and depression • Laxative abuse • Fear of not being able to stop eating voluntarily • Hoarding or stealing food • Eating to the point of physical discomfort	• Erosion of tooth enamel and tooth decay from exposure to stomach acid • Electrolyte imbalances that can lead to irregular heart function and possible sudden cardiac arrest • Inflammation of the salivary glands • Irritation and inflammation of the esophagus (may lead to bleeding or hemorrhage after vomiting) • Dehydration • Weight gain • Abdominal pain, bloating • Sore throat, hoarseness • Broken blood vessels in the eyes

of large amounts of food in a relatively short period of time. Foods often consumed during a binge include cakes, cookies, and ice cream, which can easily add up to thousands of calories. In most cases of BN, bingeing is followed by **purging** behaviors, which typically take the form of vomiting but may also include the use of laxatives, diuretics, or enemas. Alternatively, some bulimic people try to compensate for their bingeing behavior with excessive exercise.

Not all people with BN purge, and for this reason the American Psychiatric Association recognizes nonpurging as a specific type of BN.[14] People with nonpurging BN exercise or fast after a binge, rather than vomit. These and other characteristics commonly observed in people with BN are presented in Table 4.

Rituals Associated with Bulimia Nervosa Because bingeing and purging are often carried out secretly, family and friends may be unaware that someone has BN.[15] In fact, binges often occur when nobody is around. Although bulimics may consume several thousand calories during a binge, purging behaviors help to maintain energy balance. However, even purging immediately after bingeing cannot totally prevent nutrient absorption, and some weight gain is likely. People with BN tend to be within or slightly above their recommended weight range, and there may be no outward change in physical appearance.

Individuals with BN often experience regret and feelings of loss of control after bingeing. This can lead to depression and increased likelihood of future binge–purge cycles (Figure 4). Some bulimic people may alternate between cycles of bingeing and purging, and periods of food restriction.

Whereas individuals with AN feel a sense of satisfaction associated with dieting and weight loss, many bulimic people feel guilty and depressed when they binge and purge. In addition, people with BN tend to be impulsive and are prone to other unhealthy behaviors such as substance abuse and self-mutilation (cutting), and may have suicidal tendencies.[16] For these reasons, people with BN are more likely to seek treatment than are those with AN. Help for people with BN often comes after getting caught in the act of bingeing and purging.

Health Concerns Associated with Bulimia Nervosa Over time, repeated cycles of bingeing and vomiting can damage the body. Much of this damage is caused when the delicate lining of the esophagus and mouth becomes irritated by frequent exposure to stomach acid. The acidity of gastric juice can also damage dental enamel, causing tooth decay. Dentists and dental hygienists are often the first to notice these signs. Some people with bulimia induce vomiting by inserting their fingers deep into the mouth. As a result, hands can become scraped from striking the teeth—another characteristic sign of BN. Frequent vomiting and/or overuse of laxatives and diuretics can cause dehydration and electrolyte

purging Self-induced vomiting and/or misuse of laxatives, diuretics, and/or enemas.

FIGURE 4 Thoughts and Behaviors Associated with Bulimia Nervosa

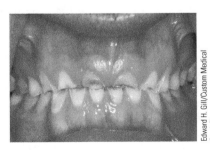

= 5/9 =

I feel so fat and ugly. Yesterday I saw this guy that I like walking with another girl — it made me feel sad. When I got to class, the teacher passed back our exam. I didn't get a very good grade. I WISH I WAS PRETTY AND SMART!!!

Yesterday I felt so bad that I binged and purged most of the afternoon. I hate when I do that but I can't help myself. I didn't intend for it to happen. I bought all this food thinking that it would last. It lasted for about an hour. Before I knew it the cookies and ice cream were gone. Then I went to the dining hall and ate dinner. !! As soon as I got back to my room I made myself vomit again. I was glad that my roommate wasn't there. She would think I am so GROSS. That is because I AM GROSS!!! I wish I could stop this. Tomorrow I am going to diet. I am going to get thin. Yeah, right ....

imbalance. This can lead to irregular heart function and can even result in sudden cardiac arrest. Fortunately, many people with BN seek help and medical assistance, and most improve with treatment.

MOST EATING DISORDERS ARE CLASSIFIED AS "NOT OTHERWISE SPECIFIED"

Certainly, AN and BN are the most familiar eating disorders. However, other habitual, unhealthy eating behaviors that do not meet precise diagnostic criteria associated with AN or BN can also be problematic. **Eating disorders not otherwise specified (EDNOS)** is a general category of eating disorders that are not as well-defined or thoroughly researched as AN and BN. Individuals with EDNOS may display a spectrum of behaviors and traits associated with AN or BN but do not meet all of the diagnostic criteria. For example, a person may exhibit behaviors associated with AN but have normal weight and (if a woman) still menstruate.

Binge-Eating Disorder Only recently has the American Psychiatric Association recognized **binge-eating disorder (BED)** as a disordered eating pattern distinct from bulimia nervosa. There is some uncertainty as to how BED should be classified. Currently, BED falls under the broad category of EDNOS. Although many practitioners and researchers believe that BED should be officially recognized with its own diagnostic category, others feel doing so is premature and that the condition needs further study.[17]

The number of people who fall into this category is much greater than the number of people with AN and BN combined.[18] Although people with BED can be of normal weight, most are overweight. In fact, some weight-loss treatment programs estimate that between 20% and 40% of obese patients experience BED.[19] The prevalence of BED in the general population is approximately 1 to 5%.[20]

Binge-eating disorder is characterized by recurring episodes of consuming large amounts of food within a short period of time. The American

Repeated vomiting can damage the tooth enamel. As a result, teeth appear mottled. This can occur in people with bulimia nervosa and in individuals with other disordered eating behaviors.

Edward H. Gill/Custom Medical Stock Photos

eating disorders not otherwise specified (EDNOS) A category of eating disorders that includes some, but not all, of the diagnostic criteria for anorexia nervosa and/or bulimia nervosa.

binge-eating disorder (BED) A subcategory of EDNOS characterized by recurring consumption of large amounts of food within a short period of time, but not followed by purging.

Psychiatric Association's provisional diagnostic criteria for BED requires that a person eat excessively large amounts of food in a two-hour period at least twice per week for at least six months; feels a lack of control over the episodes; and experiences feelings of disgust, depression, or guilt in response to overeating.[21] Other proposed diagnostic criteria include eating much more rapidly than normal, eating until uncomfortably full, eating large amounts of food (even when not physically hungry), and eating alone out of embarrassment at the quantity of food being eaten.[22]

Although most people overeat from time to time, binges associated with BED are distinctly different. For many people with BED, binges provide an escape from stress and emotional pain. In other words, food has a psychological numbing effect and induces a state of emotional well-being. Some studies indicate that people with BED are likely to have been raised in families affected by alcohol abuse.[23] In the case of BED, however, food rather than alcohol becomes their "drug of choice." Anger, sadness, anxiety, and other types of emotional distress often trigger binges. Many struggle with clinical depression, although it is not clear if depression triggers BED or if BED causes depression.[24]

As in BN, binges in BED typically take place in private and are often accompanied by feelings of shame. In general, individuals with BED do not purge after bingeing, and therefore tend to be in positive energy balance. As a result, BED can lead to unwanted weight gain, increasing a person's risk for weight-related health problems such as type 2 diabetes, gallstones, and cardiovascular disease.

Some People with Binge-Eating Disorder Are Restrained Eaters Individuals who suppress their desire for food and avoid eating for long periods of time between binges are called **restrained eaters.**[25] Restrained eaters limit their food intake to lose weight. However, after an extended period of food restriction, restrained eaters find themselves feeling out of control and respond by bingeing. This cycle of fasting and bingeing can be difficult to stop. Many restrained eaters perceive themselves as overweight, and consuming large amounts of food generates further feelings of inadequacy. Feelings of self-contempt can cause such people to turn back to food for emotional comfort. Similar to other disordered eating patterns, restrained eaters find themselves in a vicious cycle resulting in poor physical and psychological health.

Are There Other Disordered Eating Behaviors?

In addition to the eating disorders formally recognized by the American Psychiatric Association, there are many other troublesome disordered eating behaviors. These include not eating in public, situational purging, chewing food but spitting it out before swallowing, and obsessive dieting. Although there is insufficient information for these to be classified as "true" eating disorders, these food-related disturbances can be debilitating and disruptive. If these behaviors continue, they could progress into a full-blown eating disorder. As with any newly recognized pattern of dysfunctional behavior, there is a need for more research and a better understanding of these conditions. Some of the more recognized food-related disturbances include excessive night-time eating, avoidance of new foods, and obsessions concerning muscularity, which are described next.

SOME FOOD-RELATED DISTURBANCES INVOLVE NOCTURNAL EATING

Nocturnal sleep-related eating disorder (SRED) is characterized by eating while asleep.[26] Reportedly, individuals leave their bed and walk to the kitchen to prepare and eat food. This entire episode takes place without any recollection of having done so. Individuals may begin to suspect that they have SRED when they notice unexplained missing food and see evidence of these late-night activities left behind in the kitchen. Foods most typically consumed are those high in fat and calories and may involve unusual food combinations. It is common for people with SRED to experience weight gain. Although both men and women can experience SRED, it is far more common in women. Some researchers believe that stress, dieting, and depression can trigger episodes of SRED and that food restriction during the day may make some individuals more vulnerable to unconscious binge eating at night.[27] Although difficult to estimate, it has been reported that 1 to 3% of the general population may experience SRED, with rates as high as 10 to 15% among those with eating disorders.[28]

---

restrained eaters People who experience cycles of fasting followed by bingeing.

nocturnal sleep-related eating disorder (SRED) A disordered eating pattern characterized by eating while asleep without any recollection of having done so.

Night eating syndrome (NES) is a disorder closely related to SRED, except individuals are fully aware of their eating. It is best characterized by a cycle of daytime food restriction, excessive food intake in the evening, and nighttime insomnia.[29] In fact, people with NES typically consume more than half of their daily calories after dinner. Most people report that night eating causes them to feel depressed, anxious, and guilty. In adults, the prevalence of NES is between 1% and 2%, and it is more common among females than males. Reportedly, 9 to 15% of individuals participating in weight-loss treatment programs experience episodes of NES.[30] Signs and symptoms associated with NES include the following.

- Not feeling hungry for the first several hours after waking
- Overeating in the evening, with more than one-half of daily food intake consumed after dinner
- Difficulty falling asleep, accompanied by the urge to eat
- Waking during the night and finding it necessary to eat before falling back asleep
- Feeling guilty, ashamed, moody, tense, and agitated, especially at night

FOOD NEOPHOBIA: AVOIDANCE OF TRYING NEW FOODS

It is not unusual for parents of young children to complain about having a picky eater. In fact, some children are reluctant to try new foods, and even slight changes in food routines can be upsetting. Most children outgrow these behaviors along with their avoidance of new foods. However, some people are not able to rid themselves of these irrational fears and may be at increased risk of developing food neophobia later in life.

Food neophobia is an eating disturbance defined as an irrational fear, or avoidance, of trying new foods. Individuals with food neophobia typically eat a very limited range of foods and often have well-defined food rituals and practices.[31] For example, some may refuse to eat foods made from two or more food items. In other words, foods when eaten separately may be enjoyed but when joined together are considered disgusting. In adults, these types of behaviors are socially restricting and embarrassing. In extreme cases, they can lead to nutritional inadequacies, although some individuals may take supplements to compensate for their poor eating habits. People with food neophobia tend to be slightly overweight because they often limit themselves to comfort foods such as hamburgers, french fries, and macaroni and cheese,

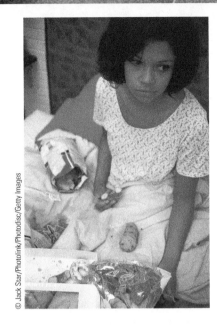

Night eating syndrome (NES) is best characterized by a cycle of daytime food restriction, excessive food intake in the evening, and nighttime insomnia.

which are high in calories.[32] Yet, most adults with food neophobia are not interested in therapy or getting treatment. Some researchers believe that food neophobia is a type of obsessive–compulsive disorder that may require extensive psychological therapy.[33]

People with food neophobia have an irrational fear of trying new foods and engage in unusual food rituals and food-related practices.

night eating syndrome (NES) A disordered eating pattern characterized by a cycle of daytime food restriction, excessive food intake in the evening, and nighttime insomnia.

food neophobia An eating disturbance characterized as an irrational fear, or avoidance, of trying new foods.

MUSCLE DYSMORPHIA: PREOCCUPATION WITH MUSCULARITY

Although societal pressure to be thin tends to be more intense for females than males, it is not uncommon for males to also have issues related to body image. However, rather than wanting to become thin, males more often want to be strong and muscular. This desire may be reinforced by the media, which portrays an "ideal" male body as having strong, well-defined muscles. Once referred to as "reverse anorexia," **muscle dysmorphia** is a disorder seen primarily in men with an intense fear of being too small, weak, or skinny.[34] The term *dysmorphia* comes from two Greek words: *dys*, which means "bad" or "abnormal," and *morphos*, which means "shape" or "form." Thus, people with muscle dysmorphia are preoccupied with perceived defects with their bodies.

Activities and tendencies associated with muscle dysmorphia include working out for hours each day, allowing exercise to interfere with family and social life, paying excessive attention to diet, and having unusual food rituals and eating practices. Reportedly, people with muscle dysmorphia may also engage in health-threatening practices, such as use of anabolic steroids, to gain sufficient muscle mass.[35]

Individuals affected by muscle dysmorphia often have personality traits similar to those with recognized eating disorders.[36] One study reported that one-third of men with muscle dysmorphia had a history of eating disorders.[37] Studies also show that men with muscle dysmorphia tend to have low self-esteem and concerns regarding their masculinity.[38] To compensate for these insecurities, individuals are driven to achieve the "perfect male body." Haunted by obsessive thoughts of being "puny" and "weak," individuals

with muscle dysmorphia can have difficulties maintaining personal and social relationships with others.

What Causes Eating Disorders?

Most females with eating disorders are adolescents or young adults. Nonetheless, a pattern of eating disorders among middle-aged or older adult women has also begun to emerge.[39] This trend may be a response to a youth-oriented society that intensifies insecurities associated with aging. Because the occurrence of age-related health problems increases by middle-age, developing an eating disorder during this stage of life may be particularly concerning. For example, bone loss associated with menopause could be made worse by food restriction.

The prevalence of eating disorders in females is well-documented, but far less is known about males. Researchers estimate that males account for approximately 5 to 10% of people with AN and 10 to 15% of people with BN.[40] Because the small number of males with eating disorders makes this group difficult to study, it is not clear what factors are related to eating disorders in males. However, they may include history of obesity, participation in a sport that emphasizes thinness, and heightened emphasis on physical appearance.

Scientists have many theories as to what causes eating disorders, but there are no simple answers. People develop eating disorders for a variety of reasons, and why some are more vulnerable than others is not clear. However, several factors are likely contributors, including sociocultural characteristics, family dynamics, personality traits, and biological (genetic) factors.[41]

SOCIOCULTURAL FACTORS

Eating disorders are more prevalent in some cultures than others. Cultures where food is abundant and slimness is valued are the ones most likely to foster eating disorders. This is one reason why eating disorders are more widespread in Western industrialized countries than in regions of Africa, China, and many Arab nations.[42]

Differences in sociocultural environments within American society may also influence the

John Lamb/Getty Images

Muscle dysmorphia has been described as the opposite of anorexia nervosa and is characterized by an obsessive preoccupation with increasing muscularity.

muscle dysmorphia Pathological preoccupation with increasing muscularity.

Many popular dolls have body proportions that are unrealistic, emphasizing extreme muscularity or thinness.

development of eating disorders. However, beliefs that may have one time protected certain ethnic groups against eating disorders may be eroding as multicultural youth acculturate to mainstream American values. It appears that the desire to fit into white, middle-class society has led to a rise in the occurrence of eating disorders across racial and ethnic groups. Data show that the prevalence of eating disorders among Latina- and African-American women is catching up to that of their Caucasian counterparts.[43]

The Role of the Media The media, which often portray unrealistic physiques and glamorize unnaturally thin bodies, are criticized for invoking a sense of inadequacy in young, impressionable people. Even the body shapes of popular dolls are not realistic. All of these factors, and more, likely play an important role in predisposing susceptible individuals to eating disorders. For example, celebrity role models shown on television and in the movies can give teenagers an unrealistic standard of thinness, causing some to engage in unhealthy eating practices to achieve this thin and glamorous appearance.[44]

Based on media images, the perfect body is tall, is lean, and has well-defined muscles. Whereas the average woman is 5 feet 4 inches tall and 140 lbs, the average model is 5 feet 11 inches tall and 117 lbs.[45] In fact, female fashion models, beauty pageant contestants, and actresses have become increasingly

thinner over the years. Even the body weights of Miss America beauty contestants have decreased. When the pageant first began in the 1920s, BMI averaged 20 to 25 kg/m². Today, nearly all the participants have BMIs below what is considered healthy. Furthermore, nearly half of the contestants have BMIs consistent with one of the diagnostic criteria for AN (<18.5 kg/m²).[46] One study reports that women who participated in beauty pageants as children were more likely to experience body dissatisfaction, interpersonal distrust, and impulse dysregulation than nonparticipants.[47]

Because dissatisfaction with body weight and shape is thought to be an essential precursor to the development of eating disorders, it is important for children to understand that healthy bodies come in many shapes and sizes. Just as important, older children must be prepared for the physical and emotional changes associated with puberty. Changes in body dimensions and weight gain during adolescence can make females feel embarrassed and uncomfortable with their maturing bodies.

Although eating disorders frequently begin with a desire to lose weight, relatively few people on weight-loss diets develop them. Furthermore, not everyone living in an affluent society that stigmatizes obesity and advocates extreme slenderness develops an eating disorder. So these factors alone are not sufficient to cause eating disorders. Rather, the social and cultural environment in industrialized countries appears to foster the development

The BMIs of Miss America beauty contestants have decreased over the last 50 years.

of eating disorders in individuals who are already vulnerable in some other way.

Social Networks A person's peers may play a contributing role in the development of eating disorders. This is because attitudes and behaviors about slimness and appearance are often learned from those we associate with. Body dissatisfaction and dieting in adolescents often stem from the desire to gain acceptance among their friends. Some teenagers, especially girls, may feel that—to be liked and to belong—they must be thin.

FAMILY DYNAMICS

Parents influence many aspects of their children's lives. Thus, it is not surprising that family dynamics can play a role in the development and perpetuation of eating disorders. Although no single family type necessarily leads to the development of eating disorders, researchers have found certain distinguishing behaviors and characteristics that increase the likelihood of this occurring.[48] These include overprotectiveness, rigidity, conflict avoidance, abusiveness, chaotic family dynamics,

and the presence of a mother with an eating disorder.[49]

Enmeshed Families **Enmeshment** is a term used to describe family members who are overly involved with one another and have little autonomy.[50] An enmeshed family has no clear boundaries among its members. This environment can make it difficult for children to develop independence and individualism. Children raised in such families often feel tremendous pressure to please their parents and meet expectations. Rather than doing things for themselves, they strive to please others. Enmeshed family dynamics promote dependency, which may lay the foundation for the emergence of eating disorders. Under these circumstances, food may become the only component in the child's life over which he or she can exert control.

Chaotic Families In contrast to enmeshed families in which family connections are exceedingly tight,

---

enmeshment Families whose interaction is overly involved with one another and have little autonomy.

Children who participate in beauty pageants may be more likely to experience body dissatisfaction later in life.

chaotic families have exceedingly loose family structures. **Chaotic families,** also called disengaged families, lack cohesiveness, and there is little parental involvement.[51] The roles of family members are loosely defined; children often have a sense of abandonment; and parents may be depressed, alcoholic, or emotionally absent. A child growing up in this type of home may later develop eating disorders as a way to fill an emotional emptiness, gain attention, or suppress emotional conflict.

Mothers with Eating Disorders In addition to overall family dynamics, it appears that the presence of a mother with an eating disorder or body dissatisfaction can negatively influence eating behaviors in her children.[52] The inability of a mother to demonstrate a healthy relationship with food and model healthy eating to her children is a serious concern. Furthermore, mothers with eating disorders are more likely to criticize their daughters' appearance and encourage them to lose weight. As a result, children of women with eating disorders are at increased risk for developing eating disorders themselves.

PERSONALITY TRAITS AND EMOTIONAL FACTORS CAN TRIGGER EATING DISORDERS

Scientists have long thought that certain personality traits make some people more prone than others to eating disorders. Some of these characteristics include low self-esteem; lack of self-confidence; and feelings of helplessness, anxiety, and depression.[53] Individuals with eating disorders are often described as being **food preoccupied.** That is, they spend an inordinate amount of time thinking about food. Another personality trait commonly associated with eating disorders is perfectionism.[54] Such people have difficulty dealing with shortcomings in themselves. Thus, an imperfect body is not easily tolerated.

BIOLOGICAL AND GENETIC FACTORS MAY ALSO PLAY A ROLE IN EATING DISORDERS

Seeking to better understand eating disorders, researchers have investigated biological influences that may play a role in their development. Because certain personality traits and eating behaviors are, in part, determined by the nervous and endocrine systems, it makes sense that brain chemicals may play a role in the development of eating disorders. However, it is also possible that disordered eating may disrupt neuroendocrine regulation. For example, studies show that individuals with eating disorders are often clinically depressed.[55] It is therefore difficult to determine if clinical depression leads to eating disorders or vice versa. In any case, because medication used to treat clinical depression is often effective in the treatment of certain eating disorders, depression may be a contributing factor.

Studies of identical and fraternal twins provide evidence that eating disorders may, in part, be inherited, and scientists have become interested in identifying genes that might influence susceptibility.[56] Although these types of studies cannot completely differentiate the contribution of genes versus environment, some research suggests that the contribution of genetics may actually be greater than that of the environment. How this occurs, however, is unclear.[57]

Are Athletes at Increased Risk for Eating Disorders?

There are more competitive female athletes today than ever before, and perhaps not coincidentally, the number of female athletes with eating disorders

chaotic families Families whose interaction is characterized by a lack of cohesiveness and little parental involvement.

food preoccupation Spending an inordinate amount of time thinking about food.

John Kelly/Getty Images

Sports that demand a thin physical appearance are likely to have more athletes with eating disorders than activities where large size is thought to be an advantage.

has increased as well. Some studies indicate that the prevalence of eating disorders among female student-athletes and nonathletes do not differ,[58] while other studies suggest otherwise.[59] At any rate, losing weight can have serious health consequences for an athlete, above and beyond affecting athletic performance. It is, therefore, important for coaches and trainers to recognize early warning signs and symptoms associated with eating disorders in athletes.

ATHLETICS MAY FOSTER EATING DISORDERS IN SOME PEOPLE

The prevalence of disordered eating and eating disorders among collegiate athletes is estimated to be somewhere between 15 and 60%.[60] Disagreement and inconsistent estimates may be due, in part, to the reluctance of athletes to admit that they have such a problem. In addition, some athletes may exhibit disordered eating behaviors yet do not satisfy all the criteria needed for diagnosis. Regardless of the exact number, athletes (especially females) are considered by many experts to be a group at risk for developing eating disorders.

For some athletes, physical performance is not only determined by speed, strength, and coordination but also by body weight. Athletes such as ski jumpers, cyclists, rock climbers, and long-distance runners may deliberately try to achieve a low body weight to gain a competitive advantage. In addition, sports such as gymnastics that demand a thin physical appearance are likely to have more athletes with eating disorders than are activities where greater size may be beneficial.[61] This may, in part, be due to the fact that judges often consider size and appearance when rating performance. Athletes who participate in dancing, figure skating, synchronized swimming, gymnastics, and diving also perceive that body size can affect how judges rate their performance.

Because athletes are invariably competitive people and may equate their self-worth with athletic success, they may be especially willing to engage in risky weight-loss practices. In addition, coaches and trainers often believe that excess weight can hinder performance. According to the National Collegiate Athletic Association (NCAA), sports with the highest number of female athletes with eating disorders are cross-country, gymnastics, swimming, and track and field.[62] Sports with the highest number of male athletes with eating disorders are wrestling and cross-country.

THE FEMALE ATHLETE TRIAD

Athletes with eating disorders are at extremely high risk for developing medical complications. This is largely because the rigor of athletic training alone is very stressful on the body, and adequate nourishment is required to meet these physical demands. Serious health problems can arise when an athlete is restricting food intake and bingeing and/or purging. For example, female athletes are at increased risk for developing a syndrome known as the **female athlete triad.** The female athlete triad is a combination of interrelated conditions including disordered eating (or eating disorders), amenorrhea, and osteopenia—a condition characterized by a loss of calcium from bones (Figure 5).[63]

How the three components of the female athlete triad are related is complex. Disordered eating can lead to very low levels of body fat, which can cause estrogen levels to decrease. Without adequate

---

female athlete triad A combination of interrelated conditions: disordered eating/eating disorder, menstrual dysfunction, and osteopenia.

FIGURE 5 Female Athlete Triad The female athlete triad is a combination of interrelated conditions: disordered eating or eating disorder, menstrual dysfunction, and osteopenia.

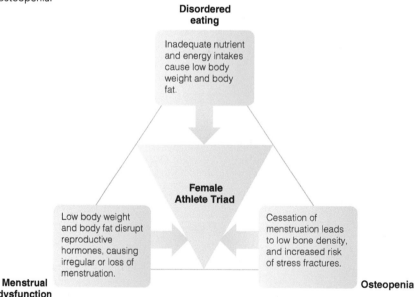

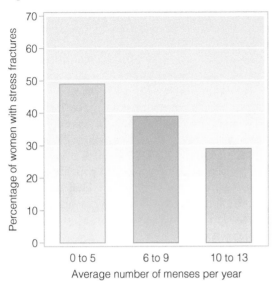

SOURCE: Adapted from Barrow GW, Saha S. Menstrual irregularity and stress fractures in collegiate female distance runners. American Journal of Sports Medicine. 1988;3:209–16.

estrogen levels, menstrual cycles can become irregular and, in some cases, stop completely. A lack of estrogen can cause bones to become dangerously weak, leading to osteopenia. In more severe cases, the entire matrix of the bone can begin to deteriorate, a condition called osteoporosis. As shown in Figure 6, the risk of stress fractures in collegiate female runners increased as menstrual cycles became more irregular.[64]

The risk of stress fractures in female runners increases when menstrual cycles become irregular or stop altogether.

Determining the prevalence of women with the female athlete triad can be difficult, because some athletes may feel unburdened when menstruation stops and therefore are unlikely to report it. However, these athletes often begin to experience repeated injuries such as stress fractures, which can draw attention to the fact that they may have a problem. It is important for parents, coaches, and health care providers to be aware of the spectrum of disordered eating and eating disorders among athletes, so that assistance can be provided.

How Can Eating Disorders Be Prevented and Treated?

Those with eating disorders are at risk for serious medical and/or emotional problems and need treatment from qualified health professionals. Typically, this team includes mental health specialists who can help address and treat underlying psychological issues, medical doctors who can treat physiological complications, and a dietitian who helps the person make better food choices. An important treatment goal for individuals with eating disorders is to learn how to enjoy food without fear and guilt and to rely on the physiological cues of hunger and satiety to regulate food intake. For this to become possible, specialists help people with eating disorders recognize and appreciate their self-worth. Many treatment options are available for people who have eating disorders. However, the first step is to recognize that there is a problem and to seek help.

PREVENTION PROGRAMS MUST PROMOTE A HEALTHY BODY IMAGE

Programs and educational curricula designed to increase awareness and prevent eating disorders in young girls produce varied results. Although it is important to reach school-age children before eating disorders begin, too many programs simply focus on deterring dangerous eating-disorder behaviors rather than on encouraging healthy attitudes toward food, dieting, and body image. To prevent eating disorders from developing, educational strategies must focus on issues related to overall health and self-esteem. Suggestions for promoting a healthy body image among children and adolescents are presented in Table 5.

TABLE 5 Promoting a Healthy Body Image among Children and Adolescents

- Encourage children to focus on positive body features.
- Help children understand that everyone has a unique body size and shape.
- Be a good role model for children by demonstrating healthy eating behaviors.
- Resist making negative comments about your own weight or body shape.
- Focus on positive *non*physical traits such as generosity, kindness, and a friendly laugh.
- Do not criticize a child's appearance.
- Never associate self-worth with physical attributes.
- Prepare a child in advance for puberty by discussing physical and emotional changes.
- Enjoy meals together as a family.
- Discuss how the media can negatively affect body image.
- Avoid using food as a reward or punishment.

SOURCE: Adapted from Story, M., Holt, K., Sofka, D., "Bright Futures in Practice: Nutrition" (2nd ed.). Arlington, VA: National Center for Education in Maternal and Child Health, 2002. Reprinted by permission of the National Center for Education in Maternal and Child Health.

TREATMENT STRATEGIES MUST FOCUS ON PSYCHOLOGICAL ISSUES

People with eating disorders may not recognize or admit they have a problem. Concerns expressed by friends and family members are often repeatedly ignored or dismissed, making them feel confused and frustrated by their inability to help—a completely normal reaction to a very difficult situation. It is important to remember that, even though people with eating disorders may resist getting help, family and friends play important supportive roles. Experts recommend that they not focus on the eating disorder *per se*, because this may make the person feel more defensive. Instead, expressing concerns regarding the person's unhappiness and encouraging him or her to seek help may be more effective.

It is important that the medical team or program chosen to treat a person with an eating disorder has specific training and expertise in this area. Treatment goals for people with AN include restoration of lost weight, resolution of psychological issues such as low self-esteem and distorted body image, and long-term recovery of healthy eating patterns. Because some people with AN are severely malnourished, they may benefit from intensive care provided by inpatient facilities.[65] These types of facilities are staffed by professionals who can provide care in all aspects of recovery, including medical needs, nutritional problems, and psychological issues. This is best achieved by a team of medical professionals, including physicians, nurses, social workers, mental health therapists, and dietitians.

People with BN, like those with AN, often have nutritional, medical, and psychological issues that must be addressed during recovery. Treatment goals for people with BN also include the reduction and eventual elimination of bingeing and purging. For example, establishing a healthy relationship with food, developing strategies to help resist the urge to binge and purge, and maintaining a healthy body weight without bingeing and purging are all important.

The sooner a person with an eating disorder gets help, the better chance of full recovery. However, recovery can be a long and slow process. Well-meaning advice such as "just eat" only makes matters worse. It is important for families and friends to know that most people with eating disorders do recover. Treatment often involves counseling for the entire family, which helps everyone to heal and to move forward in life.

Key Points

How Do Eating Disorders Differ from Disordered Eating?

- The three main types of eating disorders are anorexia nervosa (AN), bulimia nervosa (BN), and eating disorders not otherwise specified (EDNOS).
- AN is characterized by a fear of weight gain, a distorted body image, and food restriction.
- BN involves cycles of bingeing and purging.
- Binge-eating disorder (BED) has been provisionally classified as an EDNOS and is characterized by having at least two binges per week for at least six months.

Are There Other Disordered Eating Behaviors?

- Nocturnal sleep-related eating disorder (SRED) is characterized by eating while asleep, without recollection of having done so.
- Night eating syndrome (NES) is characterized by a cycle of daytime food-restriction, excessive food intake in the evening, and nighttime insomnia.

- Food neophobia is an eating disturbance characterized by an irrational fear, or avoidance of trying new foods and by unusual food rituals and practices.
- Muscle dysmorphia occurs primarily in men who see themselves as weak and/or small, even though they have lean, well-defined physiques.

What Causes Eating Disorders?

- Eating disorders are more prevalent in cultures where food is abundant and slimness is valued.
- Dysfunctional family dynamics are associated with eating disorders.
- Personality traits associated with eating disorders include low self-esteem, lack of self-confidence, obsessiveness, and feelings of helplessness, anxiety, and depression.

Are Athletes at Increased Risk for Eating Disorders?

- Sports that value and reward a thin physical appearance are likely to have more athletes with eating disorders than activities for which size is not as important.
- Female athletes are at increased risk for developing the "female athlete triad": disordered eating/eating disorder, menstrual dysfunction, and osteopenia.

How Can Eating Disorders Be Prevented and Treated?

- Eating disorder prevention involves educational strategies that focus on self-esteem and encourage healthy behaviors, rather than just the dangers of eating disorders.
- Once an eating disorder has developed, it is important to seek treatment from qualified professionals.
- Treatment goals for people with AN include restoration of lost weight and resolution of psychological issues such as low self-esteem.
- Treatment goals for people with BN include the reduction and eventual elimination of bingeing and purging.

Notes

1. Woodside BD, Garfinkel PE, Lin E, Goering P, Kaplan AS, Goldbloom DS, Kennedy SH. Comparisons of men with full or partial eating disorders, men without eating disorders, and women with eating disorders in the community. American Journal of Psychiatry. 2001;158:570–4.

2. Paxton SJ. Body dissatisfaction and disordered eating. Journal of Psychosomatic Research. 2002;53:961–2.

3. American Psychiatric Association. Diagnostic and statistical manual of mental disorders, 4th ed. (DSM-IV). Washington, DC: American Psychiatric Association; 2004.

4. Eddy KT, Dorer DJ, Franko DL, Tahilani K, Thompson-Brenner H, Herzog DB. Diagnostic crossover in anorexia nervosa and bulimia nervosa: Implications for DSM-V. American Journal of Psychiatry. 2008;165:245–50.

5. Keel PK, Klump KL, Miller KB, McGue M, Iacono WG. Shared transmission of eating disorders and anxiety disorders. International Journal of Eating Disorders. 2005;38:99–105.

6. Halmi KA, Tozzi F, Thornton LM, Crow S, Fichter MM, Kaplan AS, Keel P, Klump KL, Lilenfeld LR, Mitchell JE, Plotnicov KH, Pollice C, Rotondo A, Strober M, Woodside DB, Berrettini WH, Kaye WH, Bulik CM. The relation among perfectionism, obsessive-compulsive personality disorder and obsessive-compulsive disorder in individuals with eating disorders. International Journal of Eating Disorders. 2005;38:371–4.

7. Wolfe BE. Reproductive health in women with eating disorders. Journal of Obstetrics and Gynecology in Neonatal Nursing. 2005;34:255–63.

8. Tudor-Locke C, McColl RS. Factors related to variation in premenopausal bone mineral status: A health promotion approach. Osteoporosis International. 2000;11:1–24.

9. Birmingham CL, Su J, Hlynsky JA, Goldner EM, Gao M. The mortality rate from anorexia nervosa. International Journal of Eating Disorders. 2005;38:143–6.

10. Patrick L. Eating disorders: A review of the literature with emphasis on medical complications and clinical nutrition. Alternative Medical Review. 2002;7:184–202.

11. Holm-Denoma JM, Witte TK, Gordon KH, Herzog DB, Franko DL, Fichter M, Quadflieg N, Joiner TE Jr. Deaths by suicide among individuals with anorexia as arbiters between competing explanations of the anorexia-suicide link. Journal of Affective Disorders. 2008;107:231–6.

12. Williams PM, Goodie J, Motsinger CD. Treating eating disorders in primary care. American Family Physician. 2008;77:187–95.

13. Hay P, Bacaltchuk J. Bulimia nervosa. Clinical Evidence. 2004;12:1326–47.

14. Cooper Z, Fairburn CG. Refining the definition of binge eating disorder and nonpurging bulimia nervosa. International Journal of Eating Disorders. 2003;34:S89–95.

15. Kruger D. Bulimia nervosa: Easy to hide but essential to recognize. Journal of the American Academy of Physician Assistants. 2008;21:48–52.

16. Thompson-Brenner H, Eddy KT, Franko DL, Dorer D, Vashchenko M, Herzog DB. Personality pathology and substance abuse in eating disorders: A longitudinal study. International Journal of Eating Disorders. 2008;41:203–8.

17. Bulik CM, Brownley KA, Shapiro JR. Diagnosis and management of binge eating disorder. World Psychiatry. 2007;6:142–8.

18. Striegel-Moore RH, Franko DL. Epidemiology of binge eating disorder. International Journal of Eating Disorders. 2003;34:S19–S29.

19. Niego SH, Kofman MD, Weiss JJ, Geliebter A. Binge eating in the bariatric surgery population: A review of the literature. International Journal of Eating Disorders. 2007;40:349–59.

20. Pagoto S, Bodenlos JS, Kantor L, Gitkind M, Curtin C, Ma Y. Association of major depression and binge eating disorder with weight loss in a clinical setting. Obesity. 2007;15:2557–9.

21. American Psychiatric Association. Diagnostic and Statistical Manual of Mental Disorders. 4th ed. Text Revision. Washington, DC: American Psychiatric Association; 2004.

22. Latner JD, Clyne C. The diagnostic validity of the criteria for binge eating disorder. International Journal of Eating Disorders. 2008;41:1–14.

23. Dansky BS, Brewerton TD, Kilpatrick DG. Comorbidity of bulimia nervosa and alcohol use disorders: Results from the National Women's Study. International Journal of Eating Disorders. 2000;27:180–90.

24. Vanderlinden J, Dalle Grave R, Fernandez F, Vandereycken W, Pieters G, Noorduin C. Which factors do provoke binge eating? An exploratory study in eating disorder patients. Eating and Weight Disorders. 2004;9:300–5.

25. Masheb RM, Grilo CM. On the relation of attempting to lose weight, restraint, and binge eating in outpatients with binge eating disorder. Obesity Research. 2000;8:638–45.

26. Schenck CH, Mahowald MW. Review of nocturnal sleep-related eating disorders. International Journal of Eating Disorders. 1994;15:343–56.

27. Winkelman JW. Sleep-related eating disorder and night eating syndrome: Sleep disorders, eating disorders, or both? Sleep. 2006;29:949–54.

28. The Cleveland Clinic. Sleep-related eating disorders. Available from http://my.clevelandclinic.org/disorders/sleep_disorders/hic_sleep-related_eating_disorders.aspx.

29. O'Reardon JP, Peshek A, Allison KC. Night eating syndrome: diagnosis, epidemiology and management. CNS Drugs. 2005;19:997–1008.

30. Tanofsky-Kraff M, Yanovski SZ. Eating disorder or disordered eating? Non-normative eating patterns in obese individuals. Obesity Research. 2004;12:1361–66.

31. Marcontell DK, Laster AE, Johnson J. Cognitive-behavioral treatment of food neophobia in adults. Journal of Anxiety Disorders. 2003;17:243–51.

32. Nicklaus S, Boggio V, Chabanet C, Issanchou S. A prospective study of food variety seeking in childhood, adolescence and early adult life. Appetite. 2005;44:289–97.

33. Pelchat ML. Of human bondage: Food craving, obsession, compulsion, and addiction. Physiology and Behavior. 2002;76:347–52.

34. Pope CG, Pope HG, Menard W, Fay C, Olivardia R, Phillips KA. Clinical features of muscle dysmorphia among males with body dysmorphic disorder. Body Image. 2005;2:395–400.

35. Wroblewska AM. Androgenic-anabolic steroids and body dysmorphia in young men. Journal of Psychosomatic Research. 1997;42:225–34.

36. Grieve FG. A conceptual model of factors contributing to the development of muscle dysmorphia. Eating Disorders. 2007;15:63–80.

37. Olivardia R, Pope HG Jr, Hudson JI. Muscle dysmorphia in male weightlifters: A case-control study. American Journal of Psychiatry. 2000;157:1291–6.

38. Grieve FG. A conceptual model of factors contributing to the development of muscle dysmorphia. Eating Disorders. 2007;15:63–80.

39. Clarke LH. Older women's perceptions of ideal body weights: The tensions between health and appearance motivations for weight loss. Ageing and Society. 2002;22:751–3.

40. Sharp CW, Clark SA, Dunan JR, Blackwood DH, Shapiro CM. Clinical presentation of anorexia nervosa in males: 24 new cases. International Journal of Eating Disorders. 1994;15:125–34.

41. Becker AE, Keel P, Anderson-Fye EP, Thomas JJ. Genes and/or jeans? Genetic and socio-cultural contributions to risk for eating disorders. Journal of Addictive Disorders. 2004;23:81–103.

42. Eddy KT, Hennessey M, Thompson-Brenner H. Eating pathology in East African women: The role of media exposure and globalization. Journal of Nervous and Mental Disorders. 2007;195:196–202.

43. George JB, Franko DL. Cultural issues in eating pathology and body image among children and adolescents. Journal of Pediatric Psychology. 2010;35:231–42.

44. Brown JD, Witherspoon EM. The mass media and American adolescents' health. Journal of Adolescent Health. 2002;31:153–70.

45. National Eating Disorders Association. Statistics: Eating disorders and their precursors. Available from http://www.nationaleatingdisorders.org/uploads/statistics_tmp.pdf.

46. Rubinstein S, Caballero B. Is Miss America an undernourished role model? Journal of the American Medical Association. 2000;283:1569.

47. Wonderlich AL, Ackard DM, Henderson JB. Childhood beauty pageant contestants: Associations with adult disordered eating and mental health. Eating Disorders. 2005;13:291–301.

48. Fernández-Aranda F, Krug I, Granero R, Ramón JM, Badia A, Giménez L, Solano R, Collier D, Karwautz A, Treasure J. Individual and family eating patterns during childhood and early adolescence: An analysis of associated eating disorder factors. Appetite. 2007;49:476–85.

49. Coulthard H, Blissett J, Harris G. The relationship between parental eating problems and children's feeding behavior: A selective review of the literature. Eating Behavior. 2004;5:103–15.

50. Humphries LL, Wrobel S, Wiegert HT. Anorexia nervosa. American Family Physician. 1982;26:199–204.

51. Kluck AS. Family factors in the development of disordered eating: Integrating dynamic and behavioral explanations. Eating Behavior. 2008;9:471–83.

52. Mazzeo SE, Zucker NL, Gerke CK, Mitchell KS, Bulik CM. Parenting concerns of women with histories

of eating disorders. International Journal of Eating Disorders. 2005;37:S77–S9.

53. Peterson CB, Thuras P, Ackard DM, Mitchell JE, Berg K, Sandager N, Wonderlich SA, Pederson MW, Crow SJ. Comprehensive Psychiatry. Personality dimensions in bulimia nervosa, binge eating disorder, and obesity. 2010;51:31–6.

54. Cassin SE, von Ranson KM. Personality and eating disorders: A decade in review. Clinical Psychology Review. 2005;25:895–916.

55. Mischoulon D, Eddy KT, Keshaviah A, Dinescu D, Ross SL, Kass AE, Franko DL, Herzog DB. Depression and eating disorders: treatment and course. Journal of Affective Disorders. 2011;130:470–7.

56. Kaye WH, Bulik CM, Plotnicov K, Thornton L, Devlin B, Fichter MM, Treasure J, Kaplan A, Woodside DB, Johnson CL, Halmi K, Brandt HA, Crawford S, Mitchell JE, Strober M, Berrettini W, Jones I. The genetics of anorexia nervosa collaborative study: Methods and sample description. International Journal of Eating Disorders. 2008;41:289–300. Bulik CM, Slof-Op't Landt MC, van Furth EF, Sullivan PF. The genetics of anorexia nervosa. Annual Review of Nutrition. 2007;27:263–75.

57. Bulik CM, Reba L, Siega-Riz AM, Reichborn-Kjennerud T. Anorexia nervosa: Definition, epidemiology, and cycle of risk. International Journal of Eating Disorders. 2005;37:S2–S9.

58. Cox LM, Lantz CD, Mayhew JL. The role of social physique anxiety and other variables in predicting eating behaviors in college students. International Journal of Sport Nutrition. 1997;7:310–7.

59. Reinking MF, Alexander LE. Prevalence of disordered-eating behaviors in undergraduate female collegiate athletes and nonathletes. Journal of Athletic Training. 2005;40:47–51.

60. Sudi K, Ottl K, Payerl D, Baumgartl P, Tauschmann K, Muller W. Anorexia athletica. Nutrition. 2004;20:657–61.

61. Salbach H, Klinkowski N, Pfeiffer E, Lehmkuhl U, Korte A. Body image and attitudinal aspects of eating disorders in rhythmic gymnasts. Psychopathology. 2007;40:388–93.

62. Johnson C, Powers PS, Dick R. Athletes and eating disorders: The National Collegiate Athletic Association study. International Journal of Eating Disorders. 1999;26:179–88.

63. Beals KA, Hill AK. The prevalence of disordered eating, menstrual dysfunction, and low bone mineral density among US collegiate athletes. International Journal of Sport Nutrition and Exercise Metabolism. 2006;16:1–23.

64. Barrow GW, Saha S. Menstrual irregularity and stress fractures in collegiate female distance runners. American Journal of Sports Medicine. 1988;3:209–16.

65. Lock J, Agras WS, Bryson S, Kraemer HC. A comparison of short- and long-term family therapy for adolescent anorexia nervosa. Journal of the American Academy of Child and Adolescent Psychiatry. 2005;44:632–9.

Nutrition and Physical Performance

CHAPTER 12

NUTRITION SCOREBOARD

1 Nutritional status affects physical performance.
True/False

2 Foods included as part of a healthy dietary pattern can supply all the energy and nutrients required by athletes for physical performance. **True/False**

3 Supplementation with protein and amino acids increases muscle strength to about the same extent as does resistance training. **True/False**

4 Losing your period is normal when you are a female athlete. **True/False**

Answers can be found at the end of the chapter.

glycogen   The storage form of glucose. Glycogen is stored in muscles and the liver.

physical performance   The ability to perform a physical task or sport at a desired or particular level.

ergogenic aids (*ergo* = work; *genic* = producing)   In the context of sport, an ergogenic aid is broadly defined as a technique or substance used for the purpose of enhancing performance.

Sports Nutrition

- **Describe how carbohydrates, proteins, and fats are utilized by muscles for energy formation.**

BAM! Nobody heard it, but Lou felt it. She had "hit the wall." She was ahead of her planned pace, but now her legs felt like lead. Without a new supply of carbohydrates, she would have to finish the last two miles of the marathon at the slow pace her legs would allow.

From her carefully crafted and scrupulously followed training program to her refined shaping of mental attitude, Lou thought she had done everything right. She had left one thing out, however, and that may have cost her the race. Lou failed to pay attention to her diet while training and ran out of **glycogen** too soon (Illustration 12.1).[3]

Three major factors affect **physical performance**: genetics, training, and nutrition.[4] The first gives some people an innate edge in sprinting or endurance, and nothing can be done about it. The second is acknowledged as a basic truth. Most athletes know a good bit about proper training, and the trick is to follow the right plan. The third is often ignored or, when taken seriously, misunderstood.

KEY NUTRITION CONCEPTS

The roles played by carbohydrates, protein, fat, sodium, water, and other nutrients in physical performance relate to two key nutrition concepts.

1. Nutrition Concept #2: Foods provide energy (calories), nutrients, and other substances needed for growth and health.

2. Nutrition Concept #9: Adequacy, variety, and balance are key characteristics of a healthy dietary pattern.

Nutrition has important effects on physical performance, but the legitimate role of nutrition is often poorly understood by athletes and coaches alike.[5] Incorrect information about nutrition and physical performance can be heard in locker rooms, at neighborhood picnics, and in health food and sports stores. Misinformation about nutrition and physical performance can be found on dietary supplement labels, in online blogs, and in magazine articles. Buying into myths about nutrition and physical performance can cost you money or time, jeopardize health, or decrease performance. Common myths about nutrition and physical performance are listed in Table 12.1. Did you believe some of them? (If you did, you are not alone.[5]) Knowledge of the science supporting connections between nutrition and physical performance will help you make solid decisions about food, nutrients, dietary supplements, and performance.

Basic Components of Energy Formation during Exercise

Understanding the role of nutrition and performance and the potential effects of some **ergogenic aids** can be fostered by basic knowledge of how energy is formed within muscle cells. Illustration 12.2 summarizes the processes by which energy for muscle movement is formed.

There are two main substrates for energy formation in muscles: glucose from muscle and liver glycogen

John Fryer/Alamy

Illustration 12.1   Hitting the wall. This runner ran out of muscle glycogen before the finish line.

Table 12.1 Eight common myths about nutrition and physical performance[10,20,32,37–39]

Myth	Reality
1. Very physically active people can eat all the fat and sugar they want.	Very active children and adults may stay thin no matter what they eat. However, that does not mean a poor diet won't affect their dental and heart health, or protect them from consuming low amounts of vitamins, minerals, and beneficial phytochemicals in plant foods. Calorie needs for athletes are best met through dietary patterns based on MyPlate.gov food groups that provide 45–65% of calories from carbohydrates, 10–35% from protein, and 20–35% from fat.
2. Protein and amino acid supplements improve strength and endurance.	Resistance exercise is the key ingredient in strength building. Athletes can get all the protein and other nutrients they need for building and strengthening muscles from exercise and high-quality proteins from food. Excess protein adds calories and increases the workload of the kidneys because they have to excrete the excess nitrogen that results from high levels of protein breakdown.
3. Saturation of glycogen stores works best if you start when glycogen stores are depleted.	Glycogen stores can be saturated by carbohydrate intake whether glycogen stores start out totally or partially depleted.
4. Athletes benefit from consuming a vitamin and mineral supplement.	Vitamins and minerals participate in energy formation but do not, by themselves, increase your ability to produce energy. Vitamin and mineral supplements benefit individuals with diagnosed deficiency diseases, but do not improve performance in well-nourished individuals, including athletes.
5. The more protein you consume after resistance training the more muscle protein synthesis will take place and the stronger you will become.	Consuming a 3-ounce portion of lean meat will support muscle protein synthesis to the same extent as consuming a 12-ounce portion. Muscles reach a "muscle full" level after resistance exercise that cannot be surpassed by additional protein intake.
6. Carbohydrate and protein intake should occur immediately after completion of a resistance exercise session in order to maximize muscle protein synthesis and glycogen formation.	Carbohydrate and protein intake does not have to occur immediately following resistance exercise to maximize muscle protein synthesis and glycogen formation.
7. Drinking water during exercise decreases performance.	Drinking water before, during, and after exercise keeps athletes hydrated and prevents dehydration.
8. Body weight is more important to athletes' performance than body composition.	Body composition can be more important than weight for some types of sports.

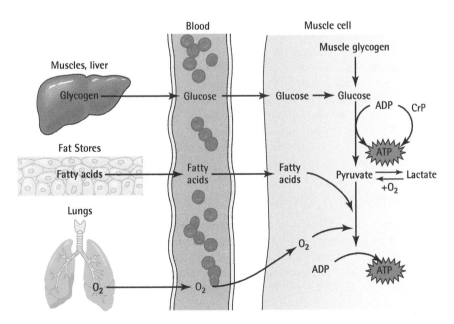

Illustration 12.2 Schematic representation of how ATP is formed for muscular movement.

Illustration 12.3 Fat is the main source of energy for low- and moderate-intensity activities, whereas glycogen is the primary fuel for high-intensity activities.

ATP, ADP Adenosine triphosphate (ah-den-o-scene tri-phos-fate) and adenosine diphosphate. Molecules containing a form of phosphorus that can trap energy obtained from the macronutrients. ADP becomes ATP when it traps energy and returns to being ADP when it releases energy for muscular and other work.

stores, and fatty acids released from fat stores. How much of each is used depends on the intensity and duration of the exercise, as well as the body's ability to deliver each along with oxygen to muscle cells (Illustration 12.3). Each substrate is used to form **ATP** from **ADP**. ATP serves as the source of energy for muscle contraction.

Anaerobic Energy Formation Glucose is obtained from the liver, and muscle glycogen stores form ATP without oxygen. This route of energy formation is *anaerobic*, or "without oxygen," and it generates most of the energy used for intense muscular work (70% VO_2 max or higher).[1] Creatine phosphate (abbreviated CrP in Illustration 12.2), an amino acid containing a high-energy phosphate molecule in muscles, converts ADP to ATP to some extent. Creatine phosphate stores are limited and decrease rapidly during intensive exercise.

Glucose is converted to pyruvate during energy formation. In the absence of oxygen, pyruvate is converted to lactate. Lactate can build up in muscles and blood if not reconverted to pyruvate by the addition of oxygen. Pyruvate yields additional energy when it enters aerobic energy formation pathways along with fatty acids from fat stores.

Aerobic Energy Formation The conversion of pyruvate and fatty acids to ATP requires oxygen. Much more ATP is delivered by the breakdown of fatty acids than glucose (fats provide 9 calories per gram, glucose only 4). The rate of energy formation from fatty acids is four times slower than that from glucose, however. It's the reason fatty acids are used to fuel low- and moderate-intensity exercise, or those below 60% VO_2 max.[6] Unlike glucose, energy formation from fatty acids is not limited by availability. Muscle cells can continue to produce energy from fatty acids as long as delivery of oxygen from the lungs and the circulation is sufficient.[3]

Nutrition and Physical Performance

The habitual diet and nutritional status of athletes should be considered first when improved performance is the goal.[44] Glycogen stores, and foods and fluids consumed before, during, and after exercise are all related to energy formation and physical performance.[1] Table 12.2 provides a summary of the recommendations for nutrition and athletic performance, and the evidence that supports these recommendations is presented next.

Table 12.2 Summary of nutrition recommendations for athletic performance[1,3,10,37–40,44]

A. Healthy Dietary Pattern
- Healthy dietary patterns (including vegetarian) support physical health and performance. Athletes may require somewhat more protein (0.5–0.8 g/lb body weight, or 1.2–1.7 g/kg), than non-athletes (0.4 g/lb or 0.8g/kg per day). A nutritious assortment of foods and beverages can meet all of the energy and nutrient needs of athletes.

B. Endurance Events
- Top off glycogen stores with food or fluids providing about 60 g (240 calories) of carbohydrate 2–3 hours prior to endurance events.
- Consume a source of carbohydrate during the event (30–60 g of carbohydrate per hour) to help maintain blood glucose levels and conserve glycogen. Use of carbohydrate- and sodium-containing sports drinks is appropriate.
- Fluid intake should match fluid loss. About 2 cups of fluid are needed for every pound of body weight lost during exercise.

C. Resistance Exercise
- Consume 20 g of high-quality protein and carbohydrate in a meal or snack within 2 hours after exercise to support muscle recovery and growth, and to build up muscle glycogen stores.
  - Individuals aged 60 years and older may need up to 40 g protein after resistance training to maximize muscle protein synthesis.
- Maintain appropriate hydration with water during exercise sessions lasting less than an hour.

Glycogen Stores and Performance Glycogen stores in muscles and the liver can deliver about 2,000 calories worth of energy, whereas adults have access to over 100,000 calories from fat. Consequently, a person's ability to perform endurance exercise can be limited by the amount of stored glycogen.[3] People who run out of liver and muscle glycogen during an endurance event such as cycling or long-distance running "hit the wall"—they have to slow down their pace substantially because they can no longer use glycogen as a fuel. The pace they are able to maintain will be dictated by the body's ability to use fat as fuel for muscular work. If athletes keep pushing themselves after glycogen runs out, they may end up "bonking." That's worse than hitting the wall, and the experience makes a profound impression on athletes. It is something they do not want to have happen again. Bonking is due to hypoglycemia (low blood sugar) and causes severe weakness, fatigue, confusion, and disorientation.[7] In severe cases, it can cause an athlete to pass out.[8,48] Obviously, endurance athletes do not want to exhaust their glycogen stores too soon.

Most athletes consuming a typical U.S. diet normally have enough glycogen stores to fuel continuous, intense exercise for about two hours.[4] Glycogen stores can be increased by consuming a meal or snack containing about 60 g of carbohydrate several hours before endurance exercise begins (Illustration 12.4), supplemented by consuming 30–60 g (120–240 calories) of carbohydrate an hour during endurance exercise. A variety of types of simple and complex carbohydrates that involve different metabolic processes are preferred over consumption of a single type.[3,8] Serving sizes of carbohydrate-rich foods and beverages that supply about 60 grams of carbohydrates are listed in Table 12.3.

Protein and Performance Many athletes require no more than their RDA of protein. Individuals undertaking strength or endurance training, however, may need 20–40 grams of additional protein daily to support muscle protein synthesis and repair.[8] This higher level of protein may already be included in their diets. Athletes tend to have higher protein intakes than non-athletes due to their higher calorie intake. On average, adult females in the United States consume 67 grams of protein daily (the RDA is 48 grams), and males 98 grams (versus the RDA of 56 grams).[9] Diets providing up to 35% of total calories from protein are compatible with health, but may provide too little carbohydrate and lead to early fatigue in athletes.[4]

Table 12.3 Foods and beverages that provide about 60 grams carbohydrate

Quinoa, cooked	1 cup
Graham crackers	12 (3 oz)
Brown or white rice, cooked	1½ cups
Breakfast cereal	1–3 cups (check nutrition information label)
Chocolate milk	1¾ cups
Pasta	2 cups
Apple juice, cranberry juice cocktail	2⅓ cups
Banana	2 large
Raisins	½ cup
Carbohydrate gel pack	2–3 (varies)

©BLACKDAY/Shutterstock.com

Illustration 12.4 A snack of 1¾ cups of chocolate milk or ½ cup of raisins provides approximately 60 grams of carbohydrate. The chocolate milk also provides about 17 grams of high-quality protein.

Judith Brown

high-quality protein Proteins that contain all of the essential amino acids in amounts needed to support growth and tissue maintenance. Examples of high-quality proteins include eggs, soy milk, milk, meat, and beans and rice. Also referred to as "complete proteins."

Table 12.4 Foods and beverages that provide about 20 grams high-quality protein

Tuna fish	3 oz
Chicken (no skin)	3 oz
Beef, lean	2½ oz
Pork, lean	3 oz
Egg whites	3
Yogurt, low-fat	1½ cups
Skim milk	2 cups
Skim milk powder	½ cup
Dried beans (cooked)	1½ cups
Protein bar	1–2 (varies)
Soybeans, cooked	¾ cup

©varandah/Shutterstock.com

The Protein–Muscle Connection Muscle fibers develop microscopic tears during training, and protein is needed to help limit muscle tissue breakdown and to repair and rebuild the muscle. Strength training, by itself, increases muscle mass and strength, and prepares muscles to continue increasing in mass and strength as training continues.[10] Muscular strength has been shown to increase 25–35% and muscle mass by 9% after a 12-week, three session per week resistance training program, for example.[11] The question has been whether additional protein or amino acids would increase gains in muscle mass and strength achieved by resistance exercise.

Muscle is high in protein, and it has been assumed that high protein intake helps build muscle mass and strength. Protein increases muscle protein synthesis and repair and muscle strength only when combined with sufficient and regular intense resistance exercise.[12] Consumption of about 20 grams of **high-quality protein** within 2 hours after exercise sessions facilitates muscle protein synthesis and repair and enhances strength.[10,12] Illustration 12.5 shows an example of how this amount of protein can be supplied in a sandwich. The 20-gram protein level appears to represent the "muscle full" amount, or the limit on the amount of amino acids that can be incorporated into muscle cells post-exercise. Protein intake above this amount will be used for the synthesis of nonessential amino acids or converted to an energy source. Table 12.4 lists examples of foods and their amounts that provide 20 grams of high-quality protein.

Although not yet known with certainty, it appears that the addition of approximately 60 grams of carbohydrate to post-training protein intake facilitates repletion of muscle protein synthesis and glycogen stores. Carbohydrate intake stimulates insulin release, and insulin increases the uptake of glucose and amino acids by muscle cells.[10] Table 12.3 lists examples of types and amounts of food that provide approximately 60 grams of carbohydrate.

Nutrients such as amino acids and glucose obtained from food enter the blood stream and become available for uptake by cells about 1 to 2 hours after food is consumed. Consequently, intake of a high-quality protein and carbohydrate snack before resistance exercise that lasts about an hour means that amino acids will be available for use by muscle cells post-exercise. Athletes who are not able to eat a meal or snack within 2 hours after training may benefit from consuming 20 grams of high-quality protein prior to the training.[10]

Protein powders providing high-quality protein, such as whey, and essential amino acid supplements can be used as a protein source before or after exercise. They have not been found to more effectively build muscle size or strength than food sources of high-quality protein, but they may be more convenient to use.[8] Protein supplements should be purchased from reliable companies and bear the NF or USP symbols for purity and quantity of ingredients.[44]

Although it is recommended that athletes consume specific types and amounts of food after training sessions, it is not always easy to do. Athletes don't always feel like eating, or have an appetite after exercising. The Reality Check feature in this unit addresses the issue of appetite post-exercise. The content may ring a bell whether you feel really hungry or not hungry at all after exercise.

Illustration 12.5 The sandwich supplies around 25 grams of high-quality protein from 3 ounces of turkey. It also provides about 25 grams of carbohydrate from the whole-wheat bread.

Hydration

• **Explain hydration status and the nutritional concerns of athletes.**

Hydration status is a major factor affecting physical performance and health. Adequate hydration during training and competition enhances performance, prevents excessive body temperature, delays fatigue, and helps prevent injuries.[19,20]

hydration status The state of the adequacy of fluid in the body tissues.

Hydration status during exercise is affected primarily by how much a person sweats—by how much water he or she loses through the skin while exercising. Muscular activity produces heat that must be eliminated to prevent the body from becoming overheated. To keep the body cool internally, the heat produced by muscles is collected in the blood and then released both through blood that circulates near the surface of the skin and in sweat. Sweating cools the body because heat is released when water evaporates on the skin. People sweat more during physical activity in hot, humid weather because it is harder to add moisture and heat to warm, moist air than to dry, cool air. To stay cool during exercise in hot, humid conditions, the body must release more heat and water than when exercise is undertaken in drier, cooler conditions.[14]

Estimating Fluid Needs: Sweat Rate The amount of fluid athletes need during an event can be estimated by calculating an hourly **sweat rate** for a specific activity (Table 12.5). A person preparing for a marathon who loses a pound in an hour of training and who drinks nothing during the hour would have a sweat rate of 1 pound (16 ounces). If that athlete drank 8 ounces of fluid in the hour and lost a pound (16 ounces), his or her sweat rate would be 24 ounces (16 ounces + 8 ounces). The sweat rate amount of fluid should be consumed per hour of the marathon event. Athletes who gain weight during an event have consumed too much water.[15]

sweat rate Fluid loss per hour of exercise. It equals the sum of body weight loss plus fluid intake.

REALITY CHECK
Does Strenuous Physical Activity Increase Appetite and Food Intake?

Who gets the thumbs-up?

Answer appears on page 364.

Junior: It must. You read about all these long-distance runners who chow down on a 3,000-calorie meal after the race is over.

Lakisha: I can't even think about eating after I've been working out on a hot day. I can't believe people would want to eat after that, either.

ANSWERS TO REALITY CHECK
Does Strenuous Physical Activity Increase Appetite and Food Intake?

Two thumbs-up for these responses because exercise increases appetite and food intake in some athletes and it doesn't in others. Part of the difference seems to be due to the temperature. People who exercise in the heat, in particular, tend to feel less like eating after exercise.[35,36]

Junior:

Lakisha:

Table 12.5 Calculating sweat rate: An example

1. Determine your body weight 1 hour before and 1 hour after exercise.

2. Subtract your post-exercise weight from your pre-exercise weight.

3. Convert the number of pounds lost to ounces. (One pound equals 16 ounces.)

4. Add the number of ounces lost or gained to the number of ounces of fluid you consumed during the hour of exercise. The result is your sweat rate, and that's an approximation of the amount of fluid you need to consume during 1 hour of that exercise.

Example

1. Terrell weighed 172 pounds an hour before an hour-long bout of exercise and 171 pounds an hour after the exercise:
   172 pounds − 171 pounds = 1 pound lost
   1 pound × 16 ounce per pound = 16 ounces

2. He drank 16 ounces of fluid during the hour of exercise.
   16 ounces + 16 ounces = 32 ounces, or "sweat rate"

Terrell would need to consume about 32 ounces of fluid per hour to remain hydrated.

Dehydration Loss of more than 2% of body weight (2 to 4 pounds generally) during an event indicates that the body is becoming dehydrated. It occurs when a person's intake of water and other fluids fails to replace water losses, primarily from sweat. Effects of **dehydration** range from mild to severe, depending on how much body water is lost. People who experience dehydration tend to feel thirsty, sweat less, and have reduced urine output. If dehydration progresses, they begin to feel confused and light-headed.[47] At the extreme, dehydration can lead to heat exhaustion or heat stroke (Table 12.6). People who over-exercise in hot weather when they are out of condition are most likely to suffer heat exhaustion or heat stroke, but these conditions occasionally occur among seasoned athletes, as well. Dehydration and heat exhaustion can be remedied by fluids and appropriate amounts of **electrolytes**, but heat stroke requires emergency medical care.[16] The Health Action feature for this unit highlights dehydration. The information may help you recognize and avoid it.

Water Intoxication **Water intoxication** is a problem that can occur during endurance events when insufficient sodium is consumed compared to water (Illustration 12.6). The body needs sodium to maintain a normal fluid balance within and around cells; its lack of availability alters fluid balance and cell functions. Also called **hyponatremia**, water intoxication can cause nausea, vomiting, confusion, seizures, and coma. It should be treated promptly.[17] It is most likely to occur in endurance athletes who consume a lot of water and too little sodium and thus gain weight, rather than replace weight, during an event.[8] In one Boston marathon, 13% of runners developed blood sodium levels that qualified as hyponatremia, and 0.6% became very ill due to extremely low levels of sodium.[18] The condition needs to be treated promptly. Water intoxication can generally be prevented by periodic consumption of a beverage containing about 100 mg sodium per 8 ounces.[8]

dehydration A condition that occurs when the body loses more water than it takes in. Prolonged exercise accompanied by profuse sweating in high temperatures, vomiting, diarrhea, certain drugs, and decreased water intake can lead to dehydration.

electrolytes Minerals such as sodium, magnesium, and potassium that carry a charge when in solution. Many electrolytes help the body maintain an appropriate amount of fluid.

water intoxication A condition that results when the body accumulates more water than it can excrete due to low availability of sodium. The condition can result in headache, fatigue, nausea and vomiting, confusion, irritability, seizures, and coma. It can occur in people who consume or receive too much water without sodium and in those with impaired kidney function. Also referred to as overhydration and hyponatremia.

hyponatremia A deficiency of sodium in the blood (135 mmol/L sodium or less).

health action   Be Aware of the Signs and Symptoms of Dehydration!

The signs and symptoms of dehydration can include:

- Thirst
- Lightheadedness or fatigue
- Headache
- Inability to concentrate
- Feeing cold or having chills when it is hot out and you are sweating
- Collapse due to heat exhaustion or stroke
- Low amount of urine and dark yellow color

Table 12.6 A primer on heat exhaustion and heat stroke[16,47]

Heat exhaustion: A condition caused by low body water and sodium content due to excessive loss of water through sweat in hot weather. Symptoms include intense thirst, weakness, paleness, dizziness, nausea, fainting, and confusion. Fluids with electrolytes and a cool place are the remedy. Also called "heat prostration" and "heat collapse."

Heat stroke: A condition requiring emergency medical care. It is characterized by hot, dry skin, labored and rapid breathing, a rapid pulse, nausea, blurred vision, irrational behavior, and, often, coma. Internal body temperature exceeds 105°F due to a breakdown of the mechanisms for regulating body temperature. Heat stroke is caused by prolonged exposure to environmental heat or strenuous physical activity. The person affected by heat stroke should be kept cool by any means possible, such as removing clothing and soaking the person in ice-cold water. If conscious, the person should be given fluids. Also called "sunstroke."

Failure to replace lost body water

↓ Blood volume declines

↓ Volume of water in and around cells declines

↓ Sweat, flushing, dry mouth
↑ Body temperature rises
↓ Physical work capacity drops
↑ Electrolyte concentration in muscles increases (causes muscle cramps)

Heat exhaustion

↑↑ Body temperature rises
↑ Heart rate increases
↑ Hot, dry skin

Heat stroke

Illustration 12.6   Water intoxication can result from over-consumption of water being handed out during marathon events.

AP Images/The Journal/Doug Lindley

Maintaining Hydration Status during Exercise   It is recommended that athletes engaged in events that last an hour or less drink water to maintain hydration status. Athletes undertaking longer events (over 1 hour in duration) should consume primarily water and beverages that provide sodium (about 100 mg sodium), and carbohydrate (4–8% by weight) during the exercise.[1,3] The sodium is needed to replace what is lost in sweat and the carbohydrate helps maintain blood glucose levels.[20] Sodium is the only electrolyte that should be added to sports drinks or other fluids consumed by athletes. Potassium needs should be met by the foods such as vegetables and fruits in the athlete's diet.[46] Hydration status is maintained when athletes do not lose or gain weight during an event and when their urine remains pale yellow and is normal in volume.[8]

Fluids That Don't Hydrate   Not all fluids help maintain hydration status. Fluids containing over 8% sugar don't quench a thirst and should not be consumed for fluid replacement. Their high sugar content may draw fluid from the blood into the intestines, thereby increasing the risk of dehydration, nausea, and bloating.[22] Alcohol-containing

Table 12.7 Incidence of irregular or absent menstrual cycles in female athletes and sedentary women[41,42]

Joggers (5 to 30 miles per week)	23%
Runners (over 30 miles per week)	34
Long-distance runners (over 70 miles per week)	43
Competitive bodybuilders	86
Noncompetitive bodybuilders	30
Volleyball players	48
Ballet dancers	44
Sedentary women	13

beverages such as beer, wine, and gin and tonics are not hydrating, nor is water when consumed by a sodium-depleted person.[8]

Nutrition-Related Concerns of Athletes

It is not uncommon for female athletes to experience irregular or absent menstrual periods or late onset of periods during adolescence (Table 12.7).[2] Female athletes engaged in "leanness" sports such as gymnastics, diving, and figure skating are particularly at risk for inadequate calorie intake.[21] These aberrations in menstrual periods appear to be related to specific effects on hormone production of deficits in caloric intake.[2] Female athletes with missing and irregular periods are at risk for developing low bone density, osteoporosis, and bone fractures. Women at particular risk of bone fractures are those with the female-athlete triad of disordered eating, amenorrhea (pronounced a-men-or-re-ah, meaning "no menstrual periods"), and osteoporosis. Abnormal menstrual cycles should not be dismissed as a normal part of training. Normal menstrual periods should be reinstated through increased caloric intake.[8] Suppressed testosterone levels or other metabolic changes due to caloric deficits in male athletes may be the parallel to menstrual irregularities in female athletes.[2]

Wrestling: The Sport of Weight Cycling "Making weight" is a common and recurring practice among wrestlers (Illustration 12.7). Most wrestlers will "cut" 1 to 20 pounds over a period of days between 50 and 100 times during a high school or college career.[23] That makes wrestling more than a test of strength and agility. It makes it a contest of rapid weight loss.

For competitive reasons, wrestlers often want to stay in the lowest weight class possible and may go to great lengths to achieve it. They may fast, "sweat the weight off" in saunas or rubber suits, or vomit after eating to lose weight before the weigh-in. These practices can be dangerous if taken too far. After the match is over, the wrestlers may binge and regain the weight they lost.

Wrestlers, like other athletes involved in intense exercise, perform better if they have a good supply of glycogen and a normal amount of body water. Fasting before a weigh-in dramatically reduces glycogen stores, and withholding fluids or losing water by sweating puts the wrestler at risk of becoming dehydrated. Trying to stay within a particular weight class too long may also stunt or delay a young wrestler's growth.[24]

The American Medical Association and the Association for Sports Medicine recommend that wrestling weight be determined after six weeks of training and normal eating. In addition, a minimum of 7% body fat should be used as a qualifier for assigning wrestlers to a particular weight class. In addition, weight classes are now based on normal weight for height and age, and weigh-ins are scheduled close to event times.[25]

Iron Status of Athletes Iron status is an important topic in sports nutrition because iron deficiency (or low iron stores) and iron-deficiency anemia (or low blood hemoglobin level) decrease endurance. Iron is a component of hemoglobin, a protein in blood that carries oxygen to cells throughout the body, and it works with enzymes involved in energy production. When iron stores or hemoglobin levels are low, less oxygen is delivered to cells, and less energy is produced than normal.[26]

PHOTOINKE/Alamy

Illustration 12.7 Dropping weight quickly in the days before a match can wreck a wrestler's chances and harm his health.

Female athletes are at higher risk of iron-deficiency anemia than other females. One study of aerobically fit athletes found that 36% of women and 6% of males were iron deficient.[27] Consequently, it is recommended that female athletes especially pay attention to their iron status.[8] Iron deficiency should be diagnosed before it is treated with iron supplements. The International Olympic Committee recommends that female athletes be screened for iron deficiency so that females needing additional iron can be identified.[26]

Ergogenic Aids: The Athlete's Dilemma

• **Assess the safety and effectiveness of ergogenic aids offered to athletes.**

The quest for a competitive edge has drawn athletes to ergogenic aids throughout much of history. (See Table 12.8 for a historical review of ergogenic aids use.) Relatively few of the hundreds of currently available products have been tested for safety and effectiveness, and most are sold as dietary supplements so they do not have to be.[13,44] Those known to increase muscle mass, strength, or endurance are usually banned from use by competitive athletes.[28] Some ergogenic aids pose no particular risk to health (just wallets). Some clearly pose risks to health due to contamination, others have been found to contain banned substances, and some contain ingredients intended to appeal to athletes with certain opinions.[13,29,44] The National Athletic Trainers' Association supports a food-first philosophy to support health and performance among athletes.[44]

Table 12.9 lists dietary supplements and other ergogenic aids for which studies have not demonstrated effectiveness and those banned for use by athletes. Products that do not increase strength or endurance to a greater extent than high-quality protein from foods or other normal components of diets are considered ineffective.

Anabolic Steroids Substances derived from testosterone, a primary male sex hormone, are considered to be *anabolic steroids*. At least 51 testosterone derivatives are banned by the U.S. Olympic Committee because they increase strength and body weight, and have adverse effects on health.[30,31] Depending on the dose and duration of use, anabolic steroids can lead to the development of acne, increased sex drive, increased body hair, impaired fertility, and mood changes ranging from depression to hostility. Male characteristics, such as facial hair and voice deepening, can occur in females who use anabolic steroids.[30]

Caffeine Caffeine is a mild stimulant consumed by many athletes to decrease fatigue and increase alertness during exercise. Coffee is a popular source of caffeine among athletes, and caffeine is also available to them in energy drinks and gels. Consumed in moderate doses of about 200 mg per day (about 2 cups of brewed coffee), caffeine appears to increase alertness, mood, and cognitive processes during exercise. Effects of caffeine on performance vary from one athlete to the next, so individuals should make their own decisions about its helpfulness. Caffeine is associated with few adverse side effects if consumed by adults in moderate doses. The World Anti-Doping Agency (WADA) has removed caffeine from their list of prohibited substances.[32,33]

The Path to Improved Performance Genetics, training, and nutrition—these are the real keys to physical performance. Although other aids will be sought, those that exceed the boundaries of what is considered fair and safe will not be approved for use by athletes. After all, athletic competition is not a test of drugs or performance aids. It's a test of an individual's ability to excel. Anything less wouldn't be sporting.

Table 12.8 Faster, higher, stronger, longer: A brief history of performance-enhancing substances used by athletes

B.C.	Large quantities of beef consumed by athletes in Greece to obtain "the strength of 10 men." Deer liver and lion heart consumed for stamina.
1880s	Morphine used to increase performance in (painful) endurance events.
1910s	Strychnine consumed for the same reason as morphine.
1930s	Amphetamines used to increase energy levels and endurance. Testosterone taken to increase muscle mass.
1980s	Blood doping, EPO used to increase endurance; ephedra to increase energy.
2009	HGH; gene doping to increase strength and endurance.

Table 12.9 Non-effective and banned dietary supplements and ergogenic aids[28,43-45]

Not shown to be effective	Banned substances[a]
Glutamine	Testosterone
Isoflavones	Testosterone derivatives:
Sulfo-polysaccharides (myostatin inhibitors)	Erythropoietin (EPO)
Boron	Growth hormone (HGH)
Calcium pyruvate	Mechano growth factors (MGFs)
Chromium	Gonadotrophins (e.g., LH, hCG)
Creatine	Insulin
	Beta-2 agonists
	Anti-estrogenic substances
	Diuretics
	Stimulants
	Narcotics
	Cannabinoids (hash, marijuana)
	Glucocorticosteroids
	Alcohol
	Beta-blockers
	Blood doping (enhancing oxygen delivery)
	Gene doping
Gamma oryzanol (ferulic acid)	
Herbal diuretics	
Tribulus terrestris	
Vanadyl sulfate (vanadium)	
Chitosan	
Garcinia cambogia (HCA)	
L-carnitine	
Phosphates	
Glycerol	
Ribose	
Inosine	
Sodium bicarbonate	
β-HMB	
Conjugated linoleic acid	
Branched chain amino acids (BCAA)	
Medium-chain triglycerides (MCT)	
Omega-3 fatty acids	

[a]Note: You can get a list of substances banned for use by athletes by from Global Drug Reference Online at www.globaldro.com.

Bill Milne/StockFood Creative/Getty Images

NUTRITION

up close

Testing Performance Aids

Focal Point: The critical examination of studies on performance aids.

Read the following summary of a fictitious study of the effects of a phosphorus supplement on strength and then answer the critical thinking questions.

- *Purpose:* To assess the effect of a phosphorus supplement on strength.

- *Methods:* Twenty volunteers from the crew team were given the phosphorus supplement for a week. Strength, assessed as the maximum number of push-ups a study participant could do in one session, was assessed before and after supplementation. Participants recorded any supplement side effects.

- *Results:* The number of push-ups increased by an average of 5% after supplementation. Diarrhea was the only side effect consistently noted by participants.

Critical Thinking Questions

1. What is a major limitation in the basic design of the study? _____

2. Does the study demonstrate that the supplement is safe?

Yes _____ No _____ Give reasons for your answer. _____

3. Does the study demonstrate that the supplement increases strength?

Yes _____ No _____ Give reasons for your answer. _____

Feedback to the Nutrition Up Close is located in Appendix F.

REVIEW QUESTIONS

- **Describe how carbohydrates, proteins, and fats are utilized by muscles.**

- **Explain hydration status and the nutritional concerns of athletes.**

- **Assess the safety and effectiveness of erogenic aids offered to athletes.**

1. The two main substrates for energy formation in muscles are amino acids and fatty acids.   True/False

2. Glucose is converted to energy by muscles anaerobically.   True/False

3. Much more ATP is produced from the breakdown of fatty acids than from amino acids or glucose.   True/False

4. Hypoglycemia results when athletes use up glycogen stored in muscles and liver.   True/False

5. Glucose is the major source of energy for low- and moderate-intensity activities.   True/False

6. Fat intake improves performance in short-duration, high-intensity events if consumed within 1 hour before the events.   True/False

7. Protein serves as the major source of fuel for resistance exercise.   True/False

8. Water intoxication is caused primarily by drinking excessive amounts of water during prolonged exercise. True/False

9. An athlete who runs for an hour, loses two pounds in that hour, and drinks 12 ounces of fluid during the hour has a sweat rate of 60 ounces.   True/False

10. Signs of dehydration can include thirst, lightheadedness, and confusion.   True/False

11. After a resistance exercise session, the "muscle full" phenomenon is reached after 20 grams of high-quality protein has been consumed.   True/False

12. For best results, athletes attempting to build muscle mass and strength should take both protein and amino acid supplements.   True/False

The next three questions refer to the following scenario.

Assume you attend a sports competition and are offered a free sample of MuscleMaxX, a new protein powder that promises to help you build muscle after resistance exercise. You take a look at the ingredient label and this is what it contains:

triple-filtered water

starch

whey

glucose

lean lipids

growth peptides

micellar proteins

lactalbumins

13. _____Which ingredient in this product would most likely help rebuild muscles after resistance exercise?
    a. triple-filtered water
    b. growth peptides
    c. whey
    d. lean lipids

14. _____Why do you think the real reason is for giving ingredients titles such as "lean lipids" and "micellar proteins"?
    a. To ensure users know they are getting unique ingredients for muscle recovery.
    b. To increase sales.

c. To assure users they are getting ingredients for which scientific studies have shown effectiveness.
d. To assure users that the ingredients are highly technical and the accurate name for each is being used.

15. _____Is it likely this product would work better for muscle rebuilding after resistance exercise than consuming 2 cups of skim milk?
    a. Yes, because it also contains growth peptides.
    b. Yes, because it provides starch.
    c. No, because high-quality protein from food also works.
    d. No, because skim milk also contain lactalbumins.

The next two questions refer to the following scenario.

Tahmina comes home from her 1-hour weight training session and looks in the fridge to find foods that will provide about 60 grams of carbohydrate and 20 grams of high-quality protein. Voilà! She notices leftover black beans and rice and eats a cup of each.

16. _____Based on the carbohydrate and protein content of black beans and rice given in Appendix A, Tahmina consumed about:
    a. 20 grams of carbohydrate and 10 grams of protein
    b. 85 grams of carbohydrate and 19 grams of protein
    c. 44 grams of carbohydrate and 16 grams of protein
    d. 72 grams of carbohydrate and 32 grams of protein

17. Black beans and rice are a source of high quality protein.   True/False

Answers to these questions can be found in Appendix F.

NUTRITION SCOREBOARD ANSWERS

1. True[1]

2. True[1]

3. Consumption of protein and/or amino acid supplements alone does not increase muscle strength.[12,34]   False

4. It is not normal and reflects a low energy intake that is likely interfering with normal body functions.[2]   False

Fertnig/Getty Images

Vitamins and Your Health

NUTRITION SCOREBOARD

1 The only documented benefit of consuming sufficient amounts of vitamins is protection against deficiency diseases. True/False

2 Vitamins provide energy. True/False

3 Vitamin C is found only in citrus fruits. True/False

4 Nearly all cases of illness due to excessive intake of vitamins result from the overuse of vitamin supplements. True/False

Answers can be found at the end of the chapter.

After completing Chapter 13 you will be able to:

- Describe the functions and food sources of vitamins and their effects on health.

vitamins Components of food required by the body in small amounts for growth and health maintenance.

Vitamins: They're on Center Stage

- **Describe the functions and food sources of vitamins and their effects on health.**

These are exciting times for people interested in **vitamins**. The first vitamin (vitamin A) was identified at the University of Wisconsin in 1913, a year before World War I began and at the height of the women's suffrage movement. It came during a time when disease was thought to be caused by "germs" and the only essential components of food were proteins, carbohydrates, fats, and minerals. During the next 35 years, 13 additional vitamins would be discovered and utilized for the treatment and prevention of deficiency diseases. The discovery of vitamins established the fact that disease can be caused by vitamin deficiencies, which started a wave of scientific research on the health effects of vitamins that continues today.[2] This unit addresses the functions of vitamins in the body, their food sources, and health consequences related to the consumption of too little or too much of individual vitamins.

KEY NUTRITION CONCEPTS

A number of key nutrition concepts underlie the content presented on vitamins, including:

1. Nutrition Concept #2: Foods provide energy (calories), nutrients, and other substances needed for growth and health.

2. Nutrition Concept #4: Poor nutrition can result from both inadequate and excessive levels of nutrient intake.

3. Nutrition Concept #5: Humans have adaptive mechanisms for managing fluctuations in nutrient intake.

Vitamin Facts

Vitamins are chemical substances that perform specific functions in the body. They are essential nutrients because, in general, the body cannot produce them or produce sufficient amounts of them. If we fail to consume enough of any of the vitamins, specific deficiency diseases develop. Fourteen vitamins have been discovered so far, and they are listed in Table 13.1.

Table 13.1 **Fourteen vitamins are known to be essential for health. Ten are water-soluble and four are fat-soluble.**

The water-soluble vitamins	The fat-soluble vitamins
B-complex vitamins	Vitamin A (retinol) (provitamin is beta-carotene)
Thiamin (B_1)	Vitamin D (1,25-dihydroxy-cholecalciferol)
Riboflavin (B_2)	Vitamin E (tocopherol)
Niacin (B_3)	Vitamin K (phylloquinone, menaquinone)
Vitamin B_6 (pyridoxine)	
Folate (folacin, folic acid)	
Vitamin B_{12} (cyanocobalamin)	
Biotin	
Pantothenic acid (pantothenate)	
Choline	
Vitamin C (ascorbic acid)	

Water- and Fat-Soluble Vitamins Vitamins come in two basic types—those soluble in water (the B-complex vitamins and vitamin C) and those that dissolve in fat (vitamins D, E, K, and A, or the *deka* vitamins). Their key features are summarized in Table 13.2. With the exception of vitamin B_{12}, the water-soluble vitamins can be stored in the body only in small amounts. Consequently, deficiency symptoms generally develop within a few weeks to several months after the diet becomes deficient in water-soluble vitamins. Vitamin B_{12} is unique in that the body can build up stores that last for a year or more after intake of the vitamin stops. Of the water-soluble vitamins, niacin, vitamin B_6, choline, and vitamin C are known to produce ill effects if consumed in excessive amounts.

The fat-soluble vitamins are primarily stored in body fat and the liver. Because the body is better able to store these vitamins, deficiencies of fat-soluble vitamins generally take longer to develop than deficiencies of water-soluble vitamins when intake from food is too low.

Bogus Vitamins Some substances are called "vitamins" even though they are not actually vitamins. A number of them are listed in Table 13.3. These bogus vitamins are called vitamins by some manufacturers of supplements, weight-loss products, and cosmetics. Although the label may help the product to sell, the ingredients aren't essential and therefore cannot be considered vitamins. People do not develop deficiency diseases when they consume too little of the bogus vitamins.

What Do Vitamins Do?

For starters, vitamins *don't* provide energy or, with the exception of choline, serve as components of body tissues such as muscle and bone. A number of vitamins do play critical roles as **coenzymes** in the conversion of proteins, carbohydrates, and fats into energy. Coenzymes are also involved in reactions that build and maintain body tissues such as bone, muscle, and red blood cells. Thiamin, for example, is needed for reactions that convert glucose into energy. People who are thiamin deficient tire easily and feel weak (among other things). Folate, another B-complex vitamin, is required for reactions that build body proteins. Without enough folate, proteins such as those found in red blood cells form abnormally and function poorly. Vitamin A is needed for reactions that generate new cells to replace worn-out cells lining the mouth, esophagus, intestines, and eyes. Without enough vitamin A, old cells aren't replaced, and the affected tissues are damaged. Vitamin C is required for reactions that build and maintain collagen, a protein found in skin, bones, blood vessels, gums, ligaments, and cartilage. Approximately 30% of the total amount of protein in the body is collagen. With vitamin C deficiency, collagen becomes weak, causing tissues that contain collagen to weaken and bleed easily.

These examples all relate to the physical effects of vitamins. Vitamins participate in reactions that affect behavior, too. Alterations in behaviors such as reduced attention span, poor appetite, irritability, depression, or paranoia often precede the physical signs of vitamin deficiency.[2] Vitamins are truly "vital" for health.

The Antioxidant Vitamins Beta-carotene (a **precursor** to vitamin A), vitamin E, and vitamin C function as **antioxidants**. This means they prevent or repair damage to components of cells caused by exposure to **free radicals**. A free radical results primarily when an atom of oxygen loses an electron. Without the electron, there is an imbalance between the atom's positive and negative charges. This makes the atom reactive—it needs to steal an electron from a nearby atom or molecule to reestablish a balance between its positive and negative charges. Free radicals play a number of roles in the body, so they are always present. They are produced during energy formation, by breathing, and by the immune system to help destroy bacteria and viruses that enter the body. They can also be formed when the body is exposed to alcohol, radiation emitted by the sun, smoke, ozone, smog, and other environmental pollutants.

Atoms and molecules that have lost electrons to free radicals are said to be *oxidized*. Oxidized substances can damage lipids, cell membranes, DNA, and other cell components. Antioxidants such as beta-carotene, vitamin E, and vitamin C donate electrons to

coenzymes Chemical substances, including many vitamins, that activate specific enzymes. Activated enzymes increase the rate at which reactions take place in the body, such as the breakdown of fats or carbohydrates in the small intestine and the conversion of glucose and fatty acids into energy within cells.

precursor In nutrition, a nutrient that can be converted into another nutrient (also called provitamin). Beta-carotene is a precursor of vitamin A.

antioxidants Chemical substances that prevent or repair damage to cells caused by exposure to free radicals. Beta-carotene, vitamin E, and vitamin C function as antioxidants.

free radicals Chemical substances (usually oxygen) that are missing an electron. The absence of the electron makes the chemical substances reactive and prone to oxidizing nearby atoms or molecules by stealing an electron from them.

Table 13.2 An intensive course on vitamins

	The water-soluble vitamins	
	Primary functions	**Consequences of deficiency**
Thiamin (vitamin B₁) AI[a] women: 1.1 mg men: 1.2 mg	• Helps body release energy from carbohydrates ingested • Facilitates growth and maintenance of nerve and muscle tissues • Promotes normal appetite	• Fatigue, weakness • Nerve disorders, mental confusion, apathy • Impaired growth • Swelling • Heart irregularity and failure
Riboflavin (vitamin B₂) AI women: 1.1 mg men: 1.3 mg	• Helps body capture and use energy released from carbohydrates, proteins, and fats • Aids in cell division • Promotes growth and tissue repair • Promotes normal vision	Biophoto Associates/Science Source • Reddened lips, cracks at both corners of the mouth • Fatigue
Niacin (vitamin B₃) RDA women: 14 mg men: 16 mg UL: 35 mg (from supplements and fortified foods)	• Helps body capture and use energy released from carbohydrates, proteins, and fats • Assists in the manufacture of body fats • Helps maintain normal nervous system functions	Dr. M.A. Ansary/Science Source Pellagra: the niacin-deficiency disease. • Skin disorders • Nervous and mental disorders • Diarrhea, indigestion • Fatigue
Vitamin B₆ (pyridoxine) AI women: 1.3 mg men: 1.3 mg UL: 100 mg	• Needed for reactions that build proteins and protein tissues • Assists in the conversion of tryptophan to niacin • Needed for normal red blood cell formation • Promotes normal functioning of the nervous system	• Irritability, depression • Convulsions, twitching • Muscular weakness • Dermatitis near the eyes • Anemia • Kidney stones

[a] (Adequate Intakes) and RDAs (Recommended Dietary Allowances) are for 19–30-year-olds; UL (Upper Limits) are for 19–70-year-olds.

The water-soluble vitamins		
Consequences of overdose	Primary food sources	Highlights and comments
• High intakes of thiamin are rapidly excreted by the kidneys. Oral doses of 500 mg/day or less are considered safe.	• Grains and grain products (cereals, rice, pasta, bread) • Pork • Nuts	• Need increases with carbohydrate intake • There is no "e" on the end of thiamin! • Deficiency rare in the United States; may occur in people with alcoholism • Enriched grains and cereals prevent thiamin deficiency
• None known. High doses are rapidly excreted by the kidneys.	• Milk, yogurt, cheese • Grains and grain products (cereals, rice, pasta, bread) • Liver, fish, beef • Eggs	• Destroyed by exposure to light (that's why milk comes in opaque containers)
• Flushing, headache, cramps, rapid heartbeat, nausea, diarrhea, decreased liver function with doses above 0.5 g per day	• Meats (all types), fish • Grains and grain products (cereals, rice, pasta, bread) • Nuts	• Niacin has a precursor: tryptophan. Tryptophan, an amino acid, is converted to niacin by the body. Much of our niacin intake comes from tryptophan. • High doses raise HDL cholesterol levels, lower LDL cholesterol and triglycerides.
• Bone pain, loss of feeling in fingers and toes, muscular weakness, numbness, loss of balance (mimicking multiple sclerosis)	• Meats (all types) • Breakfast cereals • Bananas, avocados • Potatoes, brussels sprouts, sweet peppers	• Vitamins go from B_3 to B_6 because B_4 and B_5 were found to be duplicates of vitamins already identified.

(continued)

Table 13.2 (continued)

	The water-soluble vitamins		
	Primary functions	**Consequences of deficiency**	
Folate (folacin, folic acid) RDA women: 400 mcg men: 400 mcg UL: 1,000 mcg (from supplements and fortified foods)	• Needed for reactions that utilize amino acids (the building blocks of protein) for protein tissue formation • Promotes the normal formation of red blood cells	• Megaloblastic anemia • Diarrhea • Red, sore tongue Normal red blood cells.	• Increased rise of neural tube defects and other malformations, preterm delivery • Elevated blood levels of homocysteine Red blood cells in megaloblastic anemia.
Vitamin B$_{12}$ (cyanocobalamin) AI women: 2.4 mcg men: 2.4 mcg	• Helps maintain nerve tissues • Aids in reactions that build up protein and bone tissues • Needed for normal red blood cell development	 A look at beriberi, a thiamin-deficiency disease.	• Neurological disorders (nervousness, tingling sensations and numbness in fingers and toes, brain degeneration) • Pernicious anemia characterized by large, oval-shaped red blood cells • Fatigue • Sore, beefy red, smooth tongue
Biotin AI women: 30 mcg men: 30 mcg	• Needed for the body's manufacture of fats, proteins, and glycogen	• Seizures • Vision problems • Muscular weakness • Hearing loss	
Pantothenic acid (pantothenate) AI women: 5 mg men: 5 mg	• Needed for release of energy from fat and carbohydrates	• Fatigue, sleep disturbances, impaired coordination, vomiting, nausea	
Vitamin C (ascorbic acid) RDA women: 75 mg men: 90 mg UL: 2,000 mg	• Needed for the manufacture of collagen • Helps the body fight infections, repair wounds • Acts as an antioxidant • Enhances iron absorption	 Gums that are swollen and bleed easily are signs of scurvy, the vitamin C–deficiency disease.	• Bleeding and bruising easily due to weakened blood vessels, cartilage, and other tissues containing collagen • Slow recovery from infections and poor wound healing • Fatigue, depression
Choline AI women: 425 mg men: 550 mg UL: 3.5 g	• Serves as a structural and signaling component of cell membranes • Required for normal development of memory and attention processes during early life • Required for transport and metabolism of fat and cholesterol	• Fatty liver • Infertility • Hypertension	

The water-soluble vitamins		
Consequences of overdose	**Primary food sources**	**Highlights and comments**
• May mask signs of vitamin B_{12} deficiency (pernicious anemia)	• Fortified, refined grain products (cereals, bread, pasta) • Dark green vegetables (spinach, collards, romaine) • Dried beans	• *Folate* means "foliage." It was first discovered in leafy green vegetables. • This vitamin is easily destroyed by heat. • Synthetic form (folic acid) added to fortified grain products is better absorbed than naturally occurring folates. • Some people have genetic traits that increase their need for folates. • May protect against colon cancer in early stages of development.
• None known. Excess vitamin B_{12} is rapidly excreted by the kidneys or is not absorbed into the bloodstream. • Vitamin B_{12} injections may cause a temporary feeling of heightened energy.	• Fish, seafood • Meat • Milk and cheese • Ready-to-eat cereals	• Older people, those who have had stomach surgery, and vegans are at risk for vitamin B_{12} deficiency. • Some people become vitamin B_{12} deficient because they are unable to absorb it. • Vitamin B_{12} is found in animal products and microorganisms only.
• None known. Excesses are rapidly excreted.	• Grain and cereal products • Meats, dried beans, cooked eggs • Vegetables	• Deficiency is extremely rare. May be induced by the overconsumption of raw eggs.
• None known. Excesses are rapidly excreted.	• Many foods, including meats, grains, vegetables, fruits, and milk	• Deficiency is very rare.
• High intakes of 1 g or more per day can cause nausea, cramps, and diarrhea and may increase the risk of kidney stones.	• Fruits: guava, oranges, lemons, limes, strawberries, cantaloupe, grapefruit, kiwi fruit • Vegetables: broccoli, green and red peppers, collards, tomato, potatoes • Ready-to-eat cereals	• Need increases among smokers (to 110–125 mg per day). • Is fragile; easily destroyed by heat and exposure to air. • Supplements may decrease the severity of colds in some people. • Deficiency may develop within three weeks of very low intake.
• Low blood pressure • Sweating, diarrhea • Fishy body odor • Liver damage	• Meat (all types) • Eggs • Dried beans • Milk	• Most of the choline we consume from foods comes from its location in cell membranes. • Lecithin, an additive commonly found in processed foods, is a rich source of choline. • Choline is primarily found in animal products. • It is considered a B-complex vitamin.

(continued)

Table 13.2 (continued)

	The fat-soluble vitamins	
	Primary functions	**Consequences of deficiency**

Vitamin A
1. Retinol
RDA
    women: 700 mcg
    men: 900 mcg
    UL: 3,000 mcg
	• Needed for the formation and maintenance of mucous membranes, skin, bone
	• Needed for vision in dim light
		
		ISM/Phototake
		
		Xerophthalmia. Vitamin A deficiency is the leading cause of blindness in developing countries.
		
		• Increased incidence and severity of infectious diseases
		• Impaired vision, blindness
		• Inability to see in dim light
		• Blindness

2. Beta-carotene (a vitamin A precursor or "provitamin") No RDA; suggested intake: 6 mg	• Acts as an antioxidant; prevents damage to cell membranes and the contents of cells by repairing damage caused by free radicals	• Deficiency disease related only to lack of vitamin A

Vitamin E
(alpha-tocopherol)
  RDA women: 15 mg
        men: 15 mg
        UL: 1,000 mg
	• Acts as an antioxidant, prevents damage to cell membranes in blood cells, lungs, and other tissues by repairing damage caused by free radicals
	• Participates in the regulation of gene expression
		• Muscle loss, nerve damage
		• Anemia
		• Weakness

Vitamin D (vitamin
D$_2$ = ergocalciferol,
vitamin D$_3$ =
cholecalciferol)
RDA
    women: 15 mcg
        (600 IU)
    men: 15 mcg
        (600 IU)
    UL: 100 mcg
        (4,000 IU)
	• Needed for the absorption of calcium and phosphorus, and for their utilization in bone formation, nerve and muscle activity.
	• Inhibits inflammation
	• Involved in insulin secretion and blood glucose level maintenance
		
		Biophoto Associates/Science Source
		
		The vitamin D–deficiency disease: rickets
		
		• Weak, deformed bones (children)
		• Loss of calcium from bones (adults), osteoporosis
		• Increased risk of chronic inflammation
		• Increased risk of death from all causes

The fat-soluble vitamins		
Consequences of overdose	**Primary food sources**	**Highlights and comments**
• Vitamin A toxicity (hypervitaminosis A) with acute doses of 500,000 IU, or long-term intake of 50,000 IU per day. • Nausea, irritability, blurred vision, weakness, headache • Increased pressure in the skull, hip fracture • Liver damage • Hair loss, dry skin • Birth defects	• Vitamin A is found in animal products only. • Liver, clams, low-fat milk, eggs • Ready-to-eat cereals	• Symptoms of vitamin A toxicity may mimic those of brain tumors and liver disease. Vitamin A toxicity is sometimes misdiagnosed because of the similarities in symptoms. • 1 mcg vitamin A = 3.33 IU • Forms of retinol are used in the treatment of acne and skin wrinkles due to overexposure to the sun. American Journal of Clinical Nutrition, Vol. 71, No 4, 878-884. April 2000, Robert M. Russell Brittle hair and dry, rough, scaly, and cracked skin from vitamin A overdose.
• High intakes from supplements may increase lung damage in smokers. • With high intakes and supplemental doses (over 12 mg/day for months), skin may turn yellow-orange.	• Deep orange and dark green vegetables. • Carrots, sweet potatoes, pumpkin, spinach, collards, cantaloupe, apricots, vegetable juice	• The body converts beta-carotene to vitamin A. Other carotenes are also present in food, and some are converted to vitamin A. Beta-carotene and vitamin A perform different roles in the body. • High intakes decrease sunburn.
• Intakes of up to 800 IU per day are unrelated to toxic side effects; over 800 IU per day may increase bleeding (blood-clotting time). • Avoid supplement use if aspirin, anti-coagulants, or fish oil supplements are taken regularly.	• Nuts and seeds • Vegetable oils • Salad dressings, mayonnaise • Whole grains, wheat germ • Leafy, green vegetables, asparagus	• Vitamin E is destroyed by exposure to oxygen and heat. • Oils naturally contain vitamin E. It's there to protect the fat from breakdown due to free radicals. • Eight forms of vitamin E exist, and each has different antioxidant strength. • 1 mg vitamin E = 1.49 IU • Intakes in the United States tend to be low.
• Mental retardation in young children • Abnormal bone growth and formation • Nausea, diarrhea, irritability, weight loss • Deposition of calcium in organs such as the kidneys, liver, and heart • Toxicity possible with long-term use of 10,000 IU daily	• Vitamin D–fortified milk, cereals, and other foods • Fish and shellfish	• Vitamin D_3, the most active form of the vitamin, is manufactured from a form of cholesterol in skin cells upon exposure to ultraviolet rays from the sun. Individuals with dark skin are at higher risk of vitamin D deficiency then those with light skin. • Inadequate vitamin D status is common. • Breast-fed infants with little sun exposure benefit from vitamin D supplements. • 1 mcg vitamin D = 40 IU.

(continued)

Table 13.2 (continued)

The fat-soluble vitamins		
	Primary functions	**Consequences of deficiency**
Vitamin K (phylloquinone, menaquinone) AI women: 90 mcg men: 120 mcg	• Is an essential component of mechanisms that cause blood to clot when bleeding occurs • Aids in the incorporation of calcium into bones	• Bleeding, bruises • Decreased calcium in bones • Deficiency is rare. May be induced by the long-term use (months or more) of antibiotics.

The long-term use of antibiotics can cause vitamin K deficiency. People with vitamin K deficiency bruise easily.

Table 13.3 Nonvitamins
The "real" vitamins are listed in Table 13.1. These are some of the more popular nonvitamins.

Bioflavonoids (vitamin P)
Coenzyme Q$_{10}$
Gerovital H-3
Hesperidin
Inositol
Laetrile (vitamin B$_{17}$)
Lecithin
Lipoic acid
Nucleic acids
Pangamic acid (vitamin B$_{15}$)
Para-amino benzoic acid (PABA)
Provitamin B$_5$ complex
Rutin

Illustration 13.1 Apple slices exposed to air turn brown due to oxidation. Coating the slices with lemon juice, a rich source of vitamin C, reduces the oxidation.

stabilize oxidized molecules or repair them in other ways.[3] Illustration 13.1 shows visible effects of an oxidation and an antioxidation reaction.

Consumption of foods rich in beta-carotene, vitamin C, and vitamin E decrease the risk of heart disease, stroke, and cancer. Intake of these antioxidant vitamins in supplements, however, does not have the same protective effects.[1,3–5] Fruits, vegetables, whole grains, and other plant foods contain a variety of naturally occurring antioxidants that work together with the antioxidant vitamins and other nutrients in disease prevention.[4]

Controlling oxidation reactions is a very important process that does not rely solely on our intake of antioxidant vitamins. The body's sources of antioxidants also include colorful pigments in vegetables and fruits, and enzymes produced by the body that function as antioxidants.[3]

Vitamins and the Prevention and Treatment of Disorders

Current research on vitamins centers around their effects on disease prevention and treatment. It is a very active area of research and, no doubt, new and important advances in knowledge about the vitamins will be gained. Here are a few examples of developments in research on vitamins and disease prevention and treatment.

Folate and Neural Tube Defects Folate plays a key role during pregnancy in the synthesis of proteins needed for the normal development of fetal tissues, including the spinal cord and brain. When the folate status of women early in pregnancy is poor, the neural tube (which develops into the spinal cord and brain) may form abnormally and incompletely. Illustration 13.2 shows a photograph of spina bifida, an example of a potential outcome of inadequate folate status early in pregnancy. Daily consumption of 400 micrograms (mcg) of folic acid (the synthetic form of folate added to refined grain products) before and early in pregnancy significantly reduces the incidence of neural tube defects.[6]

In 1998 manufacturers started fortifying refined grain products such as bread, pasta, and rice with folic acid. The addition of folic acid to these foods has produced major gains in people's folate status, and the prevalence of poor folate status and of neural tube defects has been reduced substantially.[7–9]

Vitamin A: From Infectious Diseases to Acne and Wrinkles Studies undertaken both in developing countries and in the United States indicate that adequate vitamin A status helps prevent and decreases the severity of measles and other infectious diseases.

The fat-soluble vitamins		
Consequences of overdose	**Primary food sources**	**Highlights and comments**
• Toxicity is only a problem when synthetic forms of vitamin K are taken in excessive amounts. That may cause liver disease.	• Leafy green vegetables • Grain products	• Vitamin K is produced by bacteria in the gut. Part of our vitamin K supply comes from these bacteria. • Newborns are given vitamin K because they have "sterile" guts and consequently no vitamin K–producing bacteria. • Deficiency is rare.

It has been known for decades that adequate vitamin A intake also prevents blindness, an all-too-common consequence of vitamin A deficiency in developing nations.[10] Vitamin A is needed for the synthesis of substances that keep the outer layer of the eye moist and resistant to infection. Without sufficient vitamin A, eyes dry out, become susceptible to infection, and cloud over.

Forms of vitamin A are used in the treatment of acne and of skin wrinkles and blotches due to overexposure to the sun.[11,12] Very high intakes of vitamin A as retinol early in pregnancy (but not high intakes of the vitamin A precursor beta-carotene) are related to the development of specific birth defects. Women who are or may become pregnant should not use vitamin A–derived medications.[13]

Vitamin D: From Osteoporosis to Chronic Inflammation Vitamin D is best known as the sunshine vitamin; it helps build strong bones and prevent osteoporosis by facilitating the absorption and utilization of calcium. Vitamin D does much more than that, however. It plays key roles as a hormone in combating **chronic inflammation**. Low-grade, chronic inflammation is at the core of the development of disorders such as type 2 diabetes, cardiovascular disease, multiple sclerosis, certain cancers, and rheumatoid arthritis.[14] Vitamin D reduces inflammation by entering cells and turning genes that produce inflammatory substances "off" and those that produce substances that reduce inflammation "on."[15] It also functions in the regulation of insulin secretion and blood glucose level.[16]

Recommended Intake of Vitamin D Based on research results related to optimal doses of vitamin D for bone formation and maintenance, the recommended intake of vitamin D was increased in 2011 from 5 mcg (200 IU) to 15 mcg (600 IU) daily for adult women and men.[17] The recommended levels of intake of vitamin D reflect amounts needed from the diet and do not include vitamin D manufactured in the skin from exposure to the ultraviolet rays from the sun (Illustration 13.3). Most people do not consume the recommended amount of vitamin D in their diet, even if fortified foods are consumed, but get the majority of their vitamin D from exposure of their skin to direct sunlight.[18,19] Vitamin D is produced in the skin when the energy from ultraviolet rays from the sun is absorbed in skin cells. The energy absorbed initiates the conversion of a derivative of cholesterol in cells to an active form of vitamin D.[20]

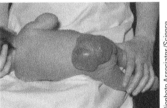
Biophoto Associates/Science Source

Illustration 13.2 A baby with spina bifida, a form of neural tube defect associated with poor folate status early in pregnancy.

chronic inflammation Low-grade inflammation that lasts weeks, months, or years. Inflammation is the first response of the body's immune system to infection or irritation. Inflammation triggers the release of biologically active substances that promote oxidation and other potentially harmful reactions in the body.

Dean Conger/Getty Images

Illustration 13.3 Russian children are exposed to a quartz UV lamp to prevent vitamin D deficiency during the long winter.

How do you become a person who stands out from the average in terms of vitamin D adequacy? If you're not that person already, here are some tips that will get you on your way.

Choose and check at least two of the following options for getting more vitamin D that appeal to you.
I would:

_____ substitute a cup of skim milk for a sweetened beverage at one meal or snack a day.

_____ eat salmon once a week at dinner.

_____ select a vitamin D–fortified orange juice when I buy orange juice.

_____ buy or select and consume vitamin D–fortified breakfast cereals.

_____ take a vitamin D supplement (400–600 IU) daily until I am able to get enough vitamin D in my diet or by brief exposure to direct sunshine.

_____ exercise or walk in sunshine for about 10 minutes a day when the weather is warm and while wearing shorts and a top.

Illustration 13.4 More foods are now fortified with Vitamin D than in the past. Check the labels on these foods for vitamin D fortification levels.

Scott Goodwin Photography

Meeting the Need for Vitamin D through Foods and the Sun With the exception of fish, few foods are naturally rich in vitamin D. Consequently, most of the vitamin D in our diets comes from vitamin D–fortified foods, such as those shown in Illustration 13.4. The availability of food products fortified with vitamin D is steadily increasing, and the increased availability of these foods will help increase vitamin D intake.[21] Because most types of yogurt, cheese, cottage cheese, ice cream, and dairy products other than milk are not fortified with vitamin D, it's important to look at product labels. Vitamin D–fortified foods can be identified from the nutrition information label on food packages. A good source of vitamin D would provide 10% or more of the % Daily Value in a serving. There are several ways individuals can increase their intakes of vitamin D and production of vitamin D in the skin. The Take Action feature in this unit provides a list of options for doing that.

The amount of vitamin D produced in the skin from exposure to direct sunlight varies depending on the lightness or darkness of skin color and the intensity of the sun's ultraviolet rays that reach the skin.[22] Individuals with darker skin produce less vitamin D given the same circumstances than people with lighter skin because ultraviolet rays are less able to penetrate darker skin. Production of vitamin D in the skin is very low to zero in parts of the world during winter when sunlight is indirect, and higher in warmer parts of the world closer to the equator. Used properly, broad-spectrum sunscreen lotions with SPF factors of 15 and higher successfully block vitamin D production by absorbing the energy from the ultraviolet rays. Energy from ultraviolet rays does not pass through glass, windows, or plastic.[19,20]

For most people, exposing the arms and legs to direct sunlight for 5 to 10 minutes between 10:00 a.m. and 5:00 p.m. daily is recommended as a sensible approach to getting enough vitamin D from the sun.[23] Maximum production of vitamin D in the skin is achieved before changes in skin color occur. You cannot get too much vitamin D from the sun. Production of the vitamin stops when adequate amounts have been produced for use and for storage in body fat.[24] Vitamin D–rich foods, and vitamin D supplements if needed, are recommended for individuals whose skin is sensitive to even short durations of sun exposure, and for individuals who, for other reasons, do not expose their skin to direct sunlight.[23]

Vitamin C and the Common Cold Do vitamin C supplements prevent colds or reduce the severity of cold symptoms? A review of 72 research studies on this topic concluded that vitamin C supplements of 200 mg per day or more reduce the incidence of colds by only 3%. Vitamin C supplements were found to decrease the severity of cold symptoms by 50% in athletes undergoing high levels of physical stress in very cold weather. For others, vitamin C supplement use was related to a decrease in the severity of cold symptoms in 8% of adults and 14% of children.[25]

Preserving the Vitamin Content of Foods

As demonstrated in Table 13.4, the vitamin content of foods can be affected by food preparation and cooking methods. (See this unit's Reality Check for an example of when this does not happen.) For example, food preparation methods that involve heat and

Table 13.4 Percent of original vitamin content lost in fruits by food storage method and in dried beans by cooking method

	Fruits			Dried beans boiled 2–2.5 hours	
	Canned (%)	Frozen (%)	Dried (%)	Water drained (%)	Water used (%)
Vitamin C	50%	30%	80%	35%	30%
Thiamin	15	10	10	60	55
Riboflavin	5	5	5	25	20
Niacin	10	5	5	45	40
Vitamin B$_6$	10	10	5	50	45
Folate	35	25	15	70	65
Choline	0	0	0	0	0
Vitamin B$_{12}$	0	0	0	0	0
Vitamin A	5	5	10	15	10

Source: Table prepared by author from data presented in USDA's nutrient retention in foods tables.[26]

REALITY CHECK
To Peel or Not to Peel?

Jolene is peeling potatoes for dinner when she gets a tap on her shoulder from her mother. "Thank you for helping with dinner, sweetie, but quit peeling the potatoes! That's where the vitamins are!"

Who gets the thumbs–up?

Answer appears on page 384.

Jolene: "Are you sure, mom? The peel is just fiber."

Mom: "Of course I'm sure, honey."

oxygen exposure lead to greater losses of vitamins such as vitamin C and folate and little or no loss of vitamins A and E, which are much less sensitive to heat and oxygen.[28] Vitamins in foods can be released into cooking water to some extent and lost down the drain if the water is thrown out. In general, boiling or steaming foods using a small amount of water, using the cooking water in soups, stews, or sauces, or stir-frying lead to superior vitamin retention. Cooking and softening vegetables like tomatoes and carrots increases the availability of beta-carotene and several beneficial phytochemicals.[28]

Vitamins: Getting Enough Without Getting Too Much

Table 13.5 lists good food sources of vitamins. Adequate amounts of vitamins can be obtained from diets that include a variety of basic foods including whole-grain products, low-fat dairy products, vegetables, fruits, fish, and other foods recommended in MyPlate.gov food guidance materials. Most fruits and vegetables are good sources of vitamins, and eating five or more servings a day is one way for individuals to get an assortment of vitamins. Fortified foods such as ready-to-eat cereals, fruit juices, and dairy products contribute to adequate intakes of vitamins.[1]

Recommended Intake Levels of Vitamins Updated recommendations for vitamin intakes associated with the prevention of deficiency and chronic diseases are represented by standards called Dietary Reference Intakes (DRIs). DRIs include the Recommended Dietary Allowances (RDAs) for vitamins, which convincing scientific data establishes for intake standards. Adequate Intakes (AIs) are assigned to vitamins for which scientific information about levels of intake associated with chronic disease prevention is less convincing. Tolerable Upper Levels of Intake (ULs) are also assigned to vitamins to indicate levels of vitamin intake from foods, fortified foods, and supplements that should *not* be exceeded. The RDAs or AIs and the ULs for the vitamins are given in Table 13.2.

Although people can get all the vitamins they need from supplements, it makes more sense to get them from foods. Foods offer fiber, minerals, beneficial phytochemicals, and other nutrients that don't come in supplements. Nutrients in foods interact to produce greater positive effects on health maintenance than do vitamin supplements in general.[27]

Table 13.5 Food sources of vitamins

Thiamin Food		Serving Size	Thiamin (mg)
Meats:			
Ham		3 oz	0.6
Pork		3 oz	0.5
Beef		3 oz	0.4
Liver		3 oz	0.2
Nuts and seeds:			
Pistachios		1/4 cup	0.3
Macadamia nuts		1/4 cup	0.2
Peanuts, dry roasted		1/4 cup	0.2
Grains:			
Breakfast cereals		1 cup	0.3–1.4
Flour tortilla		1	0.2
Macaroni		1/2 cup	0.2
Rice		1/2 cup	0.2
Bread		1 slice	0.1
Vegetables:			
Peas		1/2 cup	0.2
Lima beans		1/2 cup	0.2
Corn		1/2 cup	0.2
Fruits:			
Orange juice		1 cup	0.2
Orange		1	0.1
Avocado		1/2	0.1

Riboflavin Food		Serving size	Riboflavin (mg)
Milk and milk products:			
Milk		1 cup	0.5
2% milk		1 cup	0.5
Yogurt, low-fat		1 cup	0.5
Skim milk		1 cup	0.4
Yogurt		1 cup	0.4
American cheese		1 oz	0.1
Cheddar cheese		1 oz	0.1
Meats:			
Liver		3 oz	3.6
Pork chop		3 oz	0.3
Beef		3 oz	0.2
Tuna		3 oz	0.1
Vegetables:			
Collard greens		1/2 cup	0.3
Spinach, cooked		1/2 cup	0.2
Broccoli		1/2 cup	0.1
Eggs:			
Egg		1	0.2

Image credits: © Nordling/Shutterstock.com; © Isak55/Shutterstock.com; © Svetlana Lukienko/Shutterstock.com; © Elena Elisseeva/Shutterstock.com

(continued)

Table 13.5 (continued)

Riboflavin			
Food		Serving size	Riboflavin (mg)
Grains:			
Breakfast cereals		1 cup	0.1–1.7
Macaroni		1/2 cup	0.1
Bread		1 slice	0.1

Niacin			
Food		Serving size	Niacin (mg)
Meats:			
Liver		3 oz	14.0
Tuna		3 oz	7.0
Turkey		3 oz	4.0
Chicken		3 oz	11.0
Salmon		3 oz	6.9
Veal		3 oz	6.4
Beef (round steak)		3 oz	4.0
Pork		3 oz	4.0
Haddock		3 oz	3.9
Shrimp		3 oz	2.2
Nuts and seeds:			
Peanuts, dry roasted		1/2 cup	4.9
Almonds		1/2 cup	1.3
Vegetables:			
Asparagus		1/2 cup	1.2
Corn		1/2 cup	1.2
Green beans		1/2 cup	1.2
Grains:			
Breakfast cereals		1 cup	5.0–20.0
Brown rice		1/2 cup	1.5
Noodles, enriched		1/2 cup	1.0
Rice, white, enriched		1/2 cup	1.2
Bread, enriched		1 slice	1.1

Vitamin B$_6$			
Food		Serving size	Vitamin B$_6$ (mg)
Meats:			
Liver		3 oz	0.8
Fish		3 oz	0.3–0.6
Chicken		3 oz	0.4
Ham		3 oz	0.4
Hamburger		3 oz	0.4
Veal		3 oz	0.4
Pork		3 oz	0.3
Beef		3 oz	0.2
Grains:			
Breakfast cereals		1 cup	0.5–7.0

Vitamin B_6			
Food		**Serving size**	**Vitamin B_6 (mg)**
Fruits:			
Banana		1	0.4
Avocado		½	0.3
Watermelon		1 cup	0.3
Vegetables:			
Brussels sprouts		½ cup	0.2
Potato		½ cup	0.4
Sweet potato		½ cup	0.3
Carrots		½ cup	0.2
Sweet peppers		½ cup	0.2
Folate			
Food		**Serving size**	**Folate (mcg)**
Vegetables:			
Garbanzo beans		½ cup	141
Spinach, cooked		½ cup	131
Navy beans		½ cup	128
Asparagus		½ cup	120
Lima beans		½ cup	76
Collard greens, cooked		½ cup	65
Romaine lettuce		1 cup	65
Peas		½ cup	47
Grains:[a]			
Ready-to-eat cereals		1 cup/1 oz	100–400
Rice		½ cup	77
Noodles		½ cup	45
Wheat germ		2 Tbsp	40
Vitamin B_{12}			
Food		**Serving size**	**Vitamin B_{12} (mcg)**
Fish and seafood:			
Oysters		3 oz	13.8
Scallops		3 oz	3.0
Salmon		3 oz	2.3
Clams		3 oz	2.0
Crab		3 oz	1.8
Tuna		3 oz	1.8

[a]Fortified, refined grain products such as bread, rice, pasta, and crackers provide approximately 60 micrograms of folic acid per standard serving.

(continued)

Table 13.5 (continued)

Vitamin B₁₂			
Food		Serving size	Vitamin B₁₂ (mcg)
Meats:			
Liver		3 oz	6.8
Beef		3 oz	2.2
Veal		3 oz	1.7
Milk and milk products:			
Skim milk		1 cup	1.0
Milk		1 cup	0.9
Yogurt		1 cup	0.8
Cottage cheese		½ cup	0.7
American cheese		1 oz	0.2
Cheddar cheese		1 oz	0.2
Grains:			
Breakfast cereals		1 cup	0.6–12.0
Eggs:			
Egg		1	0.6

Vitamin C			
Food		Serving size	Vitamin C (mg)
Fruits:			
Guava		½ cup	180
Orange juice, vitamin C-fortified		1 cup	108
Kiwi fruit		1	108
Grapefruit juice, fresh		1 cup	94
Cranberry juice cocktail		1 cup	90
Orange		1	85
Strawberries, fresh		1 cup	84
Cantaloupe		¼ whole	63
Grapefruit		1 medium	51
Raspberries, fresh		1 cup	31
Watermelon		1 cup	15
Vegetables:			
Sweet red peppers		½ cup	142
Cauliflower, raw		½ cup	75
Broccoli		½ cup	70
Brussels sprouts		½ cup	65
Green peppers		½ cup	60
Collard greens		½ cup	48
Vegetable (V-8) juice		¾ cup	45
Tomato juice		¾ cup	33
Cauliflower, cooked		½ cup	30
Potato		1 medium	29
Tomato		1 medium	23

© Robyn Mackenzie/Shutterstock.com

iStockphoto.com/joe Biafore

iStockphoto.com/NoDerog

Choline

Food		Serving size	Choline (mg)
Meats:			
Beef		3 oz	111
Pork chop		3 oz	94
Lamb		3 oz	89
Ham		3 oz	87
Beef		3 oz	85
Turkey		3 oz	70
Salmon		3 oz	56
Eggs:			
Egg		1 large	126
Vegetables:			
Baked beans		1/2 cup	50
Navy beans, boiled		1/2 cup	41
Collards, cooked		1/2 cup	39
Black-eyed-peas (cowpeas)		1/2 cup	39
Chickpeas (garbanzo beans)		1/2 cup	35
Brussels sprouts		1/2 cup	32
Broccoli		1/2 cup	32
Collard greens		1/2 cup	30
Refried beans		1/2 cup	29
Milk and milk products			
Milk, 2%		1 cup	40
Cottage cheese, low-fat		1/2 cup	37
Yogurt, low-fat		1 cup	35

Vitamin A

Food		Serving Size	Vitamin A (retinol) (mcg)
Meats:			
Liver		3 oz	9,124
Clams		3 oz	145
Fortified breakfast cereals		3/4 cup	150
Milk and milk products:			
American cheese		1 oz	114
Fat-free/low-fat milk		1 cup	100
Whole milk		1 cup	58
Egg		1	84

Beta-Carotene

Food		Serving Size	Beta-carotene (mcg retinol equivalents, RE)
Vegetables:			
Sweet potatoes		1/2 cup	961
Pumpkin, canned		1/2 cup	953
Carrots, raw		1/2 cup	665
Spinach, cooked		1/2 cup	524
Collard greens, cooked		1/2 cup	489
Kale, cooked		1/2 cup	478
Turnip greens, cooked		1/2 cup	441
Beet greens, cooked		1/2 cup	276
Swiss chard, cooked		1/2 cup	268
Winter squash, cooked		1/2 cup	268
Vegetable juice		1/2 cup	200
Romaine lettuce		1 cup	162

© Barbara Delgado/Shutterstock.com

(continued)

Table 13.5 (continued)

Beta-Carotene

Food		Serving Size	Beta-carotene (mcg retinol equivalents, RE)
Fruit:			
Cantaloupe		½ cup	135
Apricots, fresh		4	134

Vitamin E

Food		Serving Size	Vitamin E (mg)
Nuts and seeds:			
Sunflower seeds		1 oz	7.4
Almonds		1 oz	7.3
Hazelnuts (filberts)		1 oz	4.3
Mixed nuts		1 oz	3.1
Pine nuts		1 oz	2.6
Peanut butter		2 Tbsp	2.5
Peanuts		1 oz	2.2
Vegetable oil:			
Sunflower oil		1 Tbsp	5.6
Safflower oil		1 Tbsp	5.6
Canola oil		1 Tbsp	2.4
Peanut oil		1 Tbsp	2.1
Corn oil		1 Tbsp	1.9
Olive oil		1 Tbsp	1.9
Salad dressing		2 Tbsp	1.5
Fish and seafood:			
Crab		3 oz	4.5
Shrimp		3 oz	3.7
Fish		3 oz	2.4
Grains:			
Wheat germ		2 Tbsp	4.2
Whole wheat bread		1 slice	2.5
Vegetables:			
Spinach, cooked		½ cup	3.4
Yellow bell pepper		1	2.8
Turnip greens, cooked		½ cup	2.2
Swiss chard, cooked		½ cup	1.7
Asparagus		½ cup	1.5
Sweet potato		½ cup	1.5

Vitamin D

Food		Serving size	Vitamin D (mcg)	IU
Fish and seafoods:				
Swordfish		3 oz	14	566
Trout		3 oz	13	502
Salmon		3 oz	11	447
Tuna, light, canned in oil		3 oz	5.7	228
Halibut		3 oz	4.9	196
Tuna, light, canned in water		3 oz	3.8	152
Tuna, white, canned in water		3 oz	1.7	68

iStockphoto.com/FotografiaBasica

Ray Kachatorian/Photodisc/Getty Images

Vitamin D Food		Serving size	Vitamin D (mcg)	Vitamin D IU
Vitamin D-fortified breakfast cereals:				
Whole grain Total		1 cup	3.3	132
Total Raisin Bran		1 cup	2.6	104
Corn Pops, Kellogg's		1 cup	1.2	48
Crispix, Kellogg's		1 cup	1.2	48
Other vitamin D-fortified foods:				
Orange juice		1 cup	2.5	100
Rice milk		1 cup	2.5	100
Soy milk		1 cup	2.5	100
Yogurt		1 cup	2.0	80
Margarine		2 tsp	1.2	48
Milk:				
Milk, whole		1 cup	3.2	128
Milk, 2%		1 cup	2.9	116
Milk, 1%		1 cup	2.9	116
Milk, skim		1 cup	2.9	116

Vitamin K Food		Serving Size	Vitamin K (mcg)
Kale, cooked		½ cup	531
Spinach, cooked		½ cup	444
Turnip greens, cooked		½ cup	426
Broccoli, cooked		½ cup	110
Brussels sprouts, cooked		½ cup	109
Mustard greens, cooked		½ cup	105
Cabbage, cooked		½ cup	82
Spinach, raw		½ cup	73
Lettuce, leafy green		1 cup	71
Asparagus, cooked		4 spears	48
Kiwifruit		½ cup	37
Berries, blue or black		1 cup	29
Okra, cooked		½ cup	23
Peas, cooked		½ cup	21
Leeks		1	16

Fertnig/Getty Images

NUTRITION
up close

Antioxidant Vitamins: How
Adequate Is Your Diet?

Focal Point: Determine if you eat enough antioxidant-rich foods.

Vitamin C, beta-carotene, and vitamin E, the antioxidant vitamins, help to maintain cellular integrity in the body. Good food sources of these antioxidants reduce the risk of heart disease, certain cancers, and other ailments. Check below to find out how frequently you consume foods containing these important, health-promoting nutrients from foods.

How often do you eat:	Seldom or never	1–2 times per week	3–5 times per week	Almost daily
Vitamin C food sources:				
1. Grapefruit, lemons, oranges, or pineapple?				
2. Strawberries, kiwi, or honeydew melon?				
3. Orange juice, cranberry juice cocktail, or tomato juice?				
4. Green, red, or chili peppers?				
5. Broccoli, Chinese cabbage, or cauliflower?				
6. Asparagus, tomatoes, or potatoes?				
Beta-carotene food sources:				
7. Carrots, sweet potatoes, or winter squash?				
8. Spinach, collard greens, or Swiss chard?				
9. Cantaloupe, papayas, or mangoes?				
10. Nectarines, peaches, or apricots?				
Vitamin E food sources:				
11. Whole-grain breads, whole-grain cereals, or wheat germ?				
12. Crab, shrimp, or fish?				
13. Peanuts, almonds, or sunflower seeds?				
14. Oils, margarine, butter, mayonnaise, or salad dressing?				

Feedback to the Nutrition Up Close is located in Appendix F.

REVIEW QUESTIONS

- **Describe the functions and food sources of vitamins and their effects on health.**

1. Vitamins are essential. Specific deficiency diseases develop if we fail to consume enough of them.   True/False
2. Vitamin D, E, C, and B_6 are fat soluble.   True/False
3. The niacin deficiency disease is called pellagra.   True/False
4. Some people are deficient in vitamin B_{12} because they are genetically unable to absorb it.   True/False

5. Vitamin A toxicity causes brain tumors.   True/False
6. Three good sources of vitamin D are sunshine, milk, and seafood.   True/False
7. Vitamin E, vitamin C, and beta-carotene supplements decrease the risk of heart disease.   True/False
8. Vitamin D acts as a hormone.   True/False
9. Adequate vitamin D status reduces chronic inflammation.   True/False

10. Very high intakes of each of the vitamins have been found to cause toxicity disease. **True/False**

11. With the exception of vitamin B_{12}, the body is able to store high amounts of the water-soluble vitamins.   **True/False**

12. A number of vitamins act as coenzymes by activating specific enzymes.   **True/False**

13. Vitamin C functions in the replacement of cells that line the esophagus and eyes.   **True/False**

14. The sun is the primary source of vitamin D for people in general.   **True/False**

15. It is recommended that women who are or may become pregnant consume 400 mcg of folic acid daily to reduce the risk of fetal development of vision problems.
True/False

16. ___ Which of the following is not considered a vitamin?
    a. pantothenic acid
    b. coenzyme Q_{10}
    c. biotin
    d. choline

17. ___ A person who never eats fish, seafood, nuts, seeds, or vegetable oil is at risk of developing a deficiency of
    _____.
    a. vitamin A
    b. vitamin K
    c. vitamin D
    d. vitamin E

18. ___ Which of the following is a consequence of vitamin A deficiency?
    a. rough, dry skin
    b. tingling sensation in the finger tips
    c. headache
    d. night blindness

19. ___ Assume you ate the following foods as snacks yesterday: an orange, ¼ cup sunflower seeds, 1 cup yogurt, and a carrot. Which of these foods would provide the most vitamin E?
    a. orange
    b. sunflower seeds
    c. yogurt
    d. carrot

20. ___ Which of the following foods is not a good source of vitamin A (retinol)?
    a. milk
    b. eggs
    c. sweet peppers
    d. fortified breakfast cereal

21. ___ Which of the following vitamins is needed for blood to clot when bleeding occurs?
    a. vitamin K
    b. thiamin
    c. vitamin B_6
    d. vitamin B_{12}

Answers to these questions can be found in Appendix F.

NUTRITION SCOREBOARD ANSWERS

1. For some vitamins, intake levels above those known to prevent deficiency diseases help protect humans from certain cancers, heart disease, osteoporosis, depression, and other disorders.   **False**

2. Nope—only carbohydrates, proteins, and fats provide energy to the body. Vitamins are needed, however, to convert the energy in food into energy the body can use.   **False**

3. Citrus fruits are good sources of vitamin C, but so are red sweet peppers, strawberries, and other noncitrus fruits and vegetables.   **False**

4. True. Nearly all cases of illness due to vitamin overdoses result from excessive intake of vitamin supplements.[1]   **True**

Phytochemicals

NUTRITION SCOREBOARD

1 Phytochemicals are found only in plants.
   True/False

2 Phytochemicals are also called *phytonutrients* because they are biologically active in the body and have beneficial effects on health.
   True/False

3 Brightly colored vegetables and fruits are the only sources of phytochemicals. **True/False**

4 Some chemical substances that occur naturally in food or result from food preparation may be harmful to health. **True/False**

Answers can be found at the end of the chapter.

CHAPTER 14

Phytochemicals: The "What Else" in Your Food

After completing Chapter 14 you will be able to

• Describe the functions and food sources of key phytochemicals.

• **Describe the functions and food sources of key phytochemicals.**

As recently as 25 years ago, the science of nutrition focused on the study of the functions and health effects of protein, fats, carbohydrates, vitamins, minerals, and water. These classes of essential nutrients have been extensively studied and a good deal is known about their effects on growth, reproduction, and health. However, essential nutrients do not account for all the benefits associated with healthy diets. There are other components in food that influence health.

We have ample evidence that diets rich in vegetables, fruits, whole grains, and other plant foods support health and reduce the risk of developing a number of diseases. It was largely assumed that the health benefits came from the vitamin and mineral content of fruits and vegetables. That conclusion turned out to be incorrect because supplementation with specific vitamins and minerals failed to yield the same health benefits as did diets rich in fruits and vegetables. In addition, use of individual vitamin and mineral supplements was found to increase health risks in some studies.[1] Now nutrition and other scientists are investigating the health effects of thousands of other substances in food and their interactions with essential nutrients and genetic traits that affect their utilization.[2] Illustration 14.1 shows some of the hundreds of chemical substances found in two plant foods.[3]

The subjects of many current studies are plant chemicals, known as **phytochemicals** or *phytonutrients*. Phytochemicals are not considered essential nutrients because deficiency diseases do not develop when we fail to consume them. They are considered to be nutrients, however, because they are biologically active and perform health-promoting functions in the body. Meats, eggs, dairy products, and other foods of animal origin also contain biologically active substances that affect body processes. Much less is known about these **zoochemicals**, and their effects on health are not yet as clear as those of some of the phytochemicals. Most bioactive food constituents are derived from plants.[4]

This unit presents information on the functions and health benefits of the most extensively studied phytochemicals and identifies their major food sources. It also highlights substances in foods that are considered to be natural toxins because they can be harmful to health if consumed in excess.

phytochemicals (phyto = *plant*) Biologically active, or "bioactive," substances in plants that have positive effects on health. Also called *phytonutrients*.

zoochemicals Chemical substances in animal foods, some of which may be biologically active in the body.

Illustration 14.1 A sampling of the chemical substances in two foods. There are hundreds more.[3]

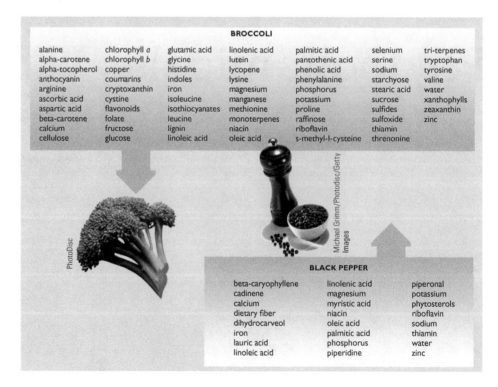

BROCCOLI

alanine	chlorophyll *a*	glutamic acid	linolenic acid	palmitic acid	selenium	tri-terpenes
alpha-carotene	chlorophyll *b*	glycine	lutein	pantothenic acid	serine	tryptophan
alpha-tocopherol	copper	histidine	lycopene	phenolic acid	sodium	tyrosine
anthocyanin	coumarins	indoles	lysine	phenylalanine	starchyose	valine
arginine	cryptoxanthin	iron	magnesium	phosphorus	stearic acid	water
ascorbic acid	cystine	isoleucine	manganese	potassium	sucrose	xanthophylls
aspartic acid	flavonoids	isothiocyanates	methionine	proline	sulfides	zeaxanthin
beta-carotene	folate	leucine	monoterpenes	raffinose	sulfoxide	zinc
calcium	fructose	lignin	niacin	riboflavin	thiamin	
cellulose	glucose	linoleic acid	oleic acid	s-methyl-l-cysteine	threnonine	

BLACK PEPPER

beta-caryophyllene	linolenic acid	piperonal
cadinene	magnesium	potassium
calcium	myristic acid	phytosterols
dietary fiber	niacin	riboflavin
dihydrocarveol	oleic acid	sodium
iron	palmitic acid	thiamin
lauric acid	phosphorus	water
linoleic acid	piperidine	zinc

The key nutrition concepts underlying content presented on phytochemicals relate to the importance of substances in food that are beneficial to health but are not considered essential nutrients. These concepts are as follows:

1. Nutrition Concept #2: Foods provide energy (calories), nutrients, and other substances needed for growth and health.

2. Nutrition Concept #9: Adequacy, variety, and balance are key characteristics of healthy dietary patterns.

Characteristics of Phytochemicals

Phytochemicals play a variety of roles in plants. They provide protection against bacterial, viral, and fungal infection; ward off insects; and prevent tissue damage due to oxidation. Some act as plant hormones or participate in the regulation of gene function, while others provide plants with flavor and color.[5] Did you ever wonder where "brown" eggs get their color (see Illustration 14.2)? It comes from xanthophylls (pronounced zan-tho-fills), a yellow-orange pigment in plants. Brown eggs are laid by chickens known as Rhode Island Reds when they consume foods such as yellow corn or alfalfa that contain xanthophylls.[6]

More than 2,000 types of phytochemicals that act as pigments have been identified (Illustration 14.3). Specific types of phytochemicals, their color, and top food sources are listed in Table 14.1. Sometimes the color of phytochemicals contained in plants is obscured. Dark green vegetables, for example, are often good sources of orange and yellow carotenes, but the green chlorophyll obscures these colors. Many phytochemicals are colorless. You cannot identify rich sources of phytochemicals by their color alone. The Reality Check feature for this unit emphasizes this point using the example of white vegetables and fruits.

Although thousands of biologically active substances in plants have been identified, the effects on human health are

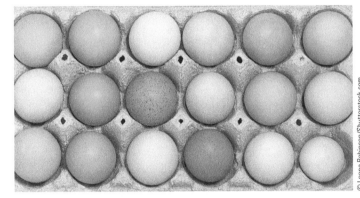

© Leena Robinson/Shutterstock.com

Illustration 14.2 Why are some eggshells brown?

REALITY CHECK
Color and the Phytonutrient Content of Vegetables and Fruits

Are white vegetables and fruits good sources of phytonutrients?

Who gets the thumbs–up?

Answers on page 398.

Photodisc

Hyde: I avoid eating white foods like onions, bananas, potatoes, and cauliflower. If you want phytonutrients, you have to eat the really colorful vegetables and fruits.

Photodisc

Ejay: I used to think that, too. But I just learned "white" doesn't mean that a white vegetable or fruit is a poor source of phytonutrients.

Illustration 14.3 The colors in plant foods are created by phytochemicals.

Keith Weller/ARS/USDA

antioxidants Chemical substances that prevent or repair damage to cells caused by oxidizing agents such as pollutants, ozone, smoke, and reactive oxygen. Oxidation reactions are a normal part of cellular processes. Vitamins C and E and certain phytochemicals function as antioxidants.

age-related macular degeneration (AMD) Eye damage caused by oxidation of the macula, the central portion of the eye that allows you to see details clearly.

cruciferous family Sulfur-containing vegetables whose outer leaves form a cross (or crucifix). Vegetables in this family include broccoli, cabbage, cauliflower, brussels sprouts, mustard and collard greens, kale, bok choy, kohlrabi, rutabaga, turnips, broccoflower, and watercress.

currently known for only a small number of them. Rather than acting alone, phytochemicals appear to exert their beneficial effects primarily through complementary mechanisms of action and combined effects with other phytochemicals, nutrients in foods, and foods within dietary patterns. The specific functions of phytochemicals may be diminished or expanded based on these interactions.[7]

Functions of Phytochemicals in Humans

The beneficial effects of phytochemicals from plants on health promotion and disease prevention relates to their roles in combating oxidation reactions and chronic inflammation and to their effects on gene functions.[8,9] Various types of phytochemicals perform different functions in the body depending on their chemical characteristics and other factors. They have been found to play a role in preventing the development of cancer, heart disease, diabetes, hypertension, eye disease, and a variety of other diseases and disorders.[10-12]

Chronic Inflammation and the Antioxidant Roles of Phytochemicals Chronic inflammation, which plays an important role in the development of many common diseases and disorders, leads to the production of free radicals that trigger oxidation reactions within cells. Oxidation reactions can damage cell membranes and components of cells and impair cell functions in ways that encourage the development of disease. **Antioxidants**, which are both consumed in foods and produced by the body, help prevent and repair damage to cell functions by scavenging oxidized particles and neutralizing their destructive effects.[13-15] A number of phytochemicals function as antioxidants and expand the body's arsenal of defense mechanisms against the harmful effects of chronic inflammation.[9] Table 14.2 provides a list of the top commonly consumed plant sources of antioxidants.

Table 14.1 The color and top food sources of specific phytochemicals[26]

Phytochemical	Color	Top food sources
Beta-carotene	Orange	Carrots
Lycopene (like-oh-pene)	Red	Tomatoes
Anthocyanins (an-tho-sigh-ah-nins)	Blue to purple	Blueberries, grapes
Allicin (all-is-in)	White	Garlic
Lutein (loo-te-in) and zeaxanthin (ze-ah-zan-thin)	Yellow-green	Spinach
Xanthophylls (zan-tho-fills)	Yellow-orange	Corn, green vegetables

Antioxidant Roles of Lutein, Zeaxanthin, and Beta-Carotene Many pigments in plants act as antioxidants and participate in disease prevention. Lutein (pronounced loo-te-in) and zeaxanthin (pronounced ze-ah-zan-thin) are found together in plants and play key roles in the prevention and treatment of **age-related macular degeneration (AMD)**. AMD is the leading cause of blindness in people over the age of 50. It produces a loss of central vision due to degeneration of the macula (Illustration 14.4). The macula, located in the center of the inside wall of the eye, is yellow due to its content of lutein and zeaxanthin. When these two phytochemicals become depleted, vision declines. Increased intake of lutein and zeaxanthin helps prevent deterioration of the macula and improves the maintenance of vision.[4,16]

Beta-carotene, lutein, zeaxanthin, and other phytochemicals come in the form of dietary supplements. However, research has not shown that antioxidant supplements are beneficial in preventing diseases.[48] Plant sources of antioxidants are known to promote health and help prevent disease, and are preferred.[9,17]

Antioxidant Roles of Flavonoids The good news about flavonoids has made chocolate a health food. Cocoa, the main ingredient in chocolate, is a rich source of flavonoids. Flavonoids, which include anthocyanins and quercetin (pronounced queer-sah-tin), act as antioxidants and reduce inflammation. Regular intake of flavonoids (such as daily consumption of a cup of hot chocolate made with cocoa powder) is related to improved blood flow, reduced blood pressure, and decreased risk of heart disease, stroke, type 2 diabetes, and cancer.[11,18,21] The content of flavonoids in chocolate products increases with the amount of cocoa in the products. Chocolate products that contain 40% or more cocoa by weight have substantially higher amounts of flavonoids than other chocolate products.[19] Milk chocolate made in the United States tends to have less than 30% cocoa, whereas darker chocolate products usually have more than 40%. Flavonoids are also found in good amounts in foods such as blueberries, grapes, oranges, bananas, apples, wine, and tea.[20]

Because of their importance, this unit's Take Action feature attempts to remind you about specific fruits rich in flavonoids and other antioxidant phytochemicals that may have escaped your attention.

Phytochemicals and Gene Function Quercetin, a phytochemical present in apples, onions, and grapes, influences inflammation by modifying the expression (the turning on or off) of genes that promote the production of pro-inflammatory compounds. Its presence helps decrease chronic inflammation due to excess body fat, improves insulin utilization, and may decrease bone loss. Similarly, phytochemicals in the **cruciferous family** of

Table 14.2 Top food sources of antioxidants[26]

Pomegranate
Red cabbage
Blackberries
Pecans
Walnuts
Cloves, ground
Peanuts
Sunflower seeds
Blueberries
Strawberries
Chocolate, dark
Raspberries
Cranberry juice
Kale
Artichokes
Wine, red
Grape juice
Cranberries
Pineapple juice
Green tea
Guava nectar
Coffee
Mango nectar

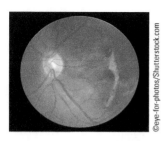

Illustration 14.4
Photograph of damage to the central part of the back of the eye due to macular degeneration.

©eye-for-photos/Shutterstock.com

Illustration 14.5 Examples of cruciferous vegetables.

Richard Anderson

Illustration 14.6 Coffee is the leading source of caffeine in the diet, followed by tea.

Peter Donaldson/Alamy

vegetables, including broccoli and cauliflower (Illustration 14.5), appear to reduce the risk of certain types of cancer, particularly in people with specific gene types.[22–24] Consumption of one-half cup of a vegetable in the crucifer family daily is sufficient to decrease inflammation in many individuals.[23] Resveratrol (rez-ver-ah-trol) in the skin of red grapes also plays a role in the prevention of inflammation.[10,25] This phytochemical is also found in peanuts, tea, blueberries, and cranberries.[26]

Anti-Infection Roles of Proanthocyanidins and Resveratrol Cranberries contain proanthocyanidins (pro-an-tho-sigh-an-ah-dins) that prevent *E. coli* and other types of bacteria from sticking to the walls of the urinary tract and thus limit the spread of urinary tract infections. Cranberry juice by itself usually does not fully treat established urinary tract infections, but regular intake of cranberry juice (a cup daily) reduces the symptoms of urinary tract infections and helps prevent them from developing and reoccurring.[27,28] Resveratrol also helps prevent the spread of bacteria that cause infection.[25]

Caffeine: A Phytochemical with Multiple Functions Caffeine is an example of a phytochemical that is not easily classified by function because it has antioxidant, anti-inflammatory, and gene-regulating effects on body processes.[29,30] Most of the caffeine in diets comes from coffee (Illustration 14.6), one of the most commonly consumed beverages in the world.[31] Caffeine sources also include tea, some soft drinks, and energy drinks (Table 14.3). Both positive and negative health effects are attributed to caffeine. On the positive side, caffeine or regular coffee intake has been found to:

- Decrease the risk of type 2 diabetes by lowering blood glucose levels, improving insulin sensitivity and insulin production, and decreasing glucose output by the liver.[32,33]

- Decrease the risk of death from heart disease and stroke.[34]

- Decrease the risk of estrogen-sensitive cancers, potentially by blocking the effects of estrogen and insulin on cancer development.[35]

- Decrease the risk of Parkinson's disease and Alzheimer's disease.[36]

- Increase mental alertness and energy by blocking the activity of a chemical messenger that decreases heart rate and increases drowsiness.[37]

- Improve mood; decrease depression.[38]

Intake of four cups of coffee daily does not appear to be related to an increased risk of disease.[33,34] Very high intakes of coffee, and particularly of caffeine, however, can have important negative effects on health in some people.

Negative Effects of Excess Caffeine Some people may experience increased blood pressure if large amounts of coffee are consumed in combination with the use of some anti-hypertension medications.[39] Excess caffeine intake can lead to anxiety, nervousness, sleep problems, and a feeling of unusually strong heart beats (heart palpitations) in some people. Tolerance to caffeine occurs over time in most people, and abrupt withdrawal can lead to headache, fatigue, and irritability within 12 to 24 hours.[40]

Medical visits for children and adults for symptoms of excess caffeine intake have sharply increased in the United States in recent years. The increase is largely due to the availability and popularity of caffeinated energy drinks and alcohol plus caffeine beverages.[41,42] The mix of alcohol and caffeine gives drinkers the feeling of being wide awake and not very drunk. Caffeine, however, does not counteract the effects of alcohol on impaired judgment and reduced reaction time.[43] The FDA has banned the sale of beverages containing both caffeine and alcohol, and both Canada and the United States have set limits on the amount of caffeine that can be added to energy drinks.[43,44]

Table 14.3 Caffeine content of foods, beverages, and some drugs

Source	Caffeine (mg)
Coffee (1 cup)	
Drip	115–175
Decaffeinated (ground or instant)	0.5–4.0
Instant	61–70
Percolated	97–140
Espresso (2 oz)	100
Tea (1 cup)	
Black, brewed 5 minutes, U.S. brands	32–144
Black, brewed 5 minutes, imported brands	40–176
Green, brewed 5 minutes	25
Instant	40–80
Soft drinks	
Coca-Cola (12 oz)	47
Cherry Coke (12 oz)	47
Diet Coke (12 oz)	47
Dr. Pepper (12 oz)	40
Ginger ale (12 oz)	0
Mountain Dew (12 oz)	54
Pepsi-Cola (12 oz)	38
Diet Pepsi (12 oz)	37
7-Up (12 oz)	0
Energy drinks	
Rockstar Energy Shot (8 oz)	229
5-Hour Energy (2 oz)	215
Red Bull (8 oz)	80
Full Throttle (8 oz)	210
Jolt (12 oz)	72
Chocolate	
Cocoa, chocolate milk (1 cup)	10–17
Milk chocolate candy (1 oz)	1–15
Chocolate syrup, one ounce (2 Tbsp)	4
Nonprescription drugs, two tablets	
Nodoz	200
Vivarin	200
Excedrin	130
Weight-control pills	150

Table 14.4 **The top five and other leading sources of phytochemicals**[47]

Top five sources
1. Tomatoes
2. Carrots
3. Oranges
4. Orange juice
5. Strawberries
Other leading sources
• Coffee
• Tea
• Spinach
• Corn
• Lettuce
• Collards
• Watermelon
• Grapes
• Blueberries
• Strawberries
• Bananas
• Onions
• Apples
• Raspberries

Food Sources of Phytochemicals

The amount and type of phytochemicals present in plants vary a good deal, depending on a number of factors. The content varies based on growing conditions, genetic strains used, storage, and processing and preparation methods. Some plant foods, such as spinach and carrots, which are considered rich sources of vitamins, contain a variety of phytochemicals. Celery, tea, and onions, foods determined by nutrient composition tables to be relative "vitamin weaklings," are actually good sources of a number of phytochemicals. Plant foods rich in phytochemicals may or may not be good sources of vitamins and minerals as well. However, intake of phytochemicals, and the beneficial effects of vegetable and fruit intake on health, increase as vegetable and fruit consumption increases.[17] Established health benefits related to plant food intake are reflected in the dietary recommendations developed for MyPlate.gov to "fill half your plate with vegetables and fruits." Similarly, the Dietary Guidelines recommend that people consume ample plant foods because of their benefits.[17] The top five and other leading sources of phytochemicals in the U.S. diet are listed in Table 14.4.

Naturally Occurring Toxins in Food

Some foods contain biologically active substances that can harm health if consumed in excess. These substances are considered naturally occurring toxins. Some naturally occurring toxins form in food during food processing and preparation.[45]

Spinach, collard greens, rhubarb, and other dark green, leafy vegetables contain oxalic acid. Eating too much of these foods can make your teeth feel as though they are covered with sand, and can give you a stomachache. Have you ever seen a potato that was partly colored green? (If not, take a look at Illustration 14.7.) The green area contains solanine, a bitter-tasting, insect-repelling phytochemical that is normally found only in the leaves and stalks of potato plants. Small amounts of solanine are harmless, but large quantities can interfere with the transmission of nerve impulses.

Phytate is present in whole grains, seeds, dried beans, and nuts. It tightly binds zinc, iron, calcium, magnesium, and copper, and reduces their absorption. Diets high in phytate have been found to produce mineral deficiency diseases.[45] Cassava, a root consumed daily in many parts of tropical Africa, can be very toxic if not prepared properly, because it contains cyanide. Soaking cassava roots in water for three nights will get rid of the cyanide, but soaking for shorter periods of time does not. When the soaking time is cut to one or two nights, as sometimes happens during periods of food shortage, enough of the toxin remains in the root to cause konzo, a disease caused by cyanide overdose.[46] Konzo is characterized by permanent, spastic paralysis.

Ackee fruit is another potential hazard to health (Illustration 14.8). If you're from Jamaica, chances are excellent that you love the taste of the core of ackee—and also know the fruit can be deadly. The national fruit of Jamaica, the yellow fleshy part around the seeds tastes like butter and looks like scrambled eggs. The rest of the fruit, however, is not edible. The fruit of unopened, unripe ackee contains high concentrations of phytochemicals that cause severe vomiting and a drastic drop in blood glucose levels. Ingestion of the fruit has caused hundreds of deaths in Jamaica. Its sale was banned in the United States until 2000, and imported ackee is routinely analyzed by the FDA for ripeness.[45]

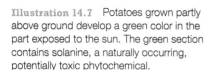

Illustration 14.7 Potatoes grown partly above ground develop a green color in the part exposed to the sun. The green section contains solanine, a naturally occurring, potentially toxic phytochemical.

Scott Goodwin Photography

Illustration 14.8 Ackee fruit and seeds.

Romulo Yanes/Conde Nast Archive/Corbis

NUTRITION
up close

Have You Had Your Phytochemicals Today?

Focal Point: Consuming good sources of the "other" beneficial components of food.

Good food sources of beneficial phytochemicals are listed below. Indicate foods you consumed at least twice last week and those you have never tried eating.

	Foods eaten	Foods never tried	Twice last week
Broccoli			
Cabbage			
Brussels sprouts			
Cauliflower			
Carrots			
Celery			
Collard greens			
Turnip greens			
Kale			
Swiss chard			
Spinach			
Tomatoes			
Peanuts			
Walnuts			
Apple/apple juice			
Orange/orange juice			
Grapefruit/grapefruit juice			
Grapes/grape juice			
Strawberries			
Blueberries			
Papaya			
Banana			
Pear			
Peaches			

Feedback to the Nutrition Up Close is located in Appendix F.

REVIEW QUESTIONS

- **Describe the functions and food sources of key phytochemicals.**

1. Phytochemicals that benefit health are found in large amounts in fish and organ meats.   True/False

2. Some plant foods that contain relatively small amounts of vitamins and minerals are rich sources of beneficial phytochemicals.   True/False

3. The best way to achieve the health benefits of phytochemicals is by making vegetables and fruits a core component of your dietary pattern.   True/False

4. If a chemical substance occurs naturally in a plant food, it can be considered harmless to health.   True/False

5. Functions of phytochemicals that naturally occur in plants are often different than their functions in the body. True/False

6. MyPlate.gov food guidance recommends that half your plate consist of vegetables and fruits.   True/False

7. Age-related macular degeneration is related to quercetin intake from plant foods.   True/False

8. Caffeine's functions are limited to roles in combating inflammation.   True/False

9. The combination of caffeine and alcohol prevents the intoxicating effects of alcohol from occurring. True/False

10. Most plant foods contain some level of phytochemicals, but onions and celery do not.   True/False

11. Orange juice is not a leading source of phytochemicals. True/False

12. _____ Which of the following statements about antioxidants is true?
    a. Antioxidants decrease the body's utilization of oxygen.
    b. Lutein functions as an antioxidant in the body.
    c. Caffeine is a powerful antioxidant that increases the risk of heart disease.
    d. Vitamins E and C are the only components of plant foods that function as antioxidants.

The next three questions refer the following scenario.

Assume you eat a salad that consists of spinach, ripe red strawberries, cranberries, and walnuts with a salad dressing.

13. _____ Which of these plant foods would be the best source of lutein and zeaxanthin?
    a. salad dressing
    b. spinach
    c. strawberries
    d. walnuts

14. _____ The strawberries in the salad are considered to be a good source of:
    a. lutein
    b. zeaxanthin
    c. resveratrol
    d. lycopene

15. _____ The cranberries in the salad would contribute most to your intake of phytochemicals that play a role in:
    a. anti-inflammation processes
    b. anti-infection processes
    c. gene regulation processes
    d. antioxidation processes

16. Which of the following plant foods do *not* belong to the cruciferous vegetable family?
    a. cabbage
    b. broccoli
    c. potatoes
    d. cauliflower

17. _____ Which of the following biologically active substances is *not* considered a potential "natural toxin"?
    a. phytates
    b. resveratrol
    c. oxalic acid
    d. solanine

Answers to these questions can be found in Appendix F.

NUTRITION SCOREBOARD ANSWERS

1. The "phyto" in phytochemicals means plants.   True

2. That's true.   True

3. You cannot judge a plant's overall content of phytochemicals by looking at its color.   False

4. Some foods contain naturally occurring toxins that can be harmful if consumed in excess.   True

Good Things to Know about Minerals

NUTRITION SCOREBOARD

1 The sole function of minerals is to serve as a component of body structures such as bones, teeth, and hair. **True/False**

2 Bones continue to grow and mineralize through the first 30 years of life. **True/False**

3 Ounce for ounce, spinach provides more iron than beef. **True/False**

4 Worldwide, the most common nutritional deficiency is iron deficiency. **True/False**

5 More than one in four American adults has hypertension. **True/False**

Answers can be found at the end of the chapter.

After completing Chapter 15 you will be able to:

- Identify key functions and food sources of five essential minerals.

Mineral Facts

- **Identify key functions and food sources of five essential minerals.**

What substances are neither animal nor vegetable in origin, cannot be created or destroyed by living organisms (or by any other ordinary means), and provide the raw materials from which all things on earth are made? The answer is the **mineral** elements; they are displayed in full in the periodic table presented in Illustration 15.1. Minerals considered *essential*, or required in the diet, are highlighted.

The body contains 40 or more minerals. Only 15 are an essential part of our diets; we obtain the others through the air we breathe or from other essential nutrients in the diet such as protein and vitamins. Seven of the 15 essential minerals are required in trace amounts in the diet, and our need for them is generally measured in micrograms (mcg or μg). Recommended daily intake levels of the eight other required minerals reflect needs measured in gram (g) or milligram (mg) amounts. The recommended intake of calcium for adult women and men, for example, is 1,000 mg (1 g), whereas only 150 mcg of the trace mineral iodine is recommended for daily consumption.

Minerals are unlike the other essential nutrients in that they consist of single atoms. A single atom of a mineral typically does not have an equal number of protons (particles that carry a positive charge) and electrons (particles that carry a negative charge), and it therefore carries a charge. The charge makes minerals reactive. Many of the functions of minerals in the body are related to this property.

KEY NUTRITION CONCEPTS

Material covered in this unit on minerals relates to the following key nutrition concepts:

1. Nutrition Concept #2: Foods provide energy (calories), nutrients, and other substances needed for growth and health.

2. Nutrition Concept #5: Humans have adaptive mechanisms for managing fluctuations in nutrient intake.

3. Nutrition Concept #8: Poor nutrition can influence the development of certain chronic diseases.

Getting a Charge Out of Minerals

The charge carried by minerals allows them to combine with other minerals of the opposite charge and form fairly stable compounds that become part of bones, teeth, cartilage, and other tissues. In body fluids, charged minerals serve as a source of electrical energy that stimulates muscles to contract and nerves to react. The electrical current generated by charged minerals when performing these functions can be recorded by an electrocardiogram (abbreviated EKG or ECG) or an electroencephalogram (EEG). Abnormalities in the pattern of electrical activity in EKGs signal pending or past problems in the heart muscle. An EKG recording is shown in Illustration 15.2. Electroencephalograms similarly record electrical activity in the brain.

The charge minerals carry is related to many other functions. It helps maintain an adequate amount of water in the body and assists in neutralizing body fluids when they become too acidic or basic. Minerals that perform the roles of **cofactors** are components of proteins and enzymes, and they provide the "spark" that initiates enzyme activity.

Charge Problems Because minerals tend to be reactive, they may combine with other substances in food to form highly stable compounds that are not easily absorbed. Absorption of zinc from foods, for example, can vary from 0 to 100%, depending on what is attached to it. Zinc in whole-grain products is very poorly absorbed because it is bound tightly to a substance called phytate. In contrast, zinc in meats is readily available because

minerals In the context of nutrition, minerals are specific, single atoms that perform particular functions in the body. There are 15 essential minerals—or minerals required in the diet.

cofactors Individual minerals required for the activity of certain proteins. For example:

- Iron is needed for hemoglobin's function in oxygen and carbon dioxide transport.

- Zinc is needed to activate or is a structural component of more than 200 enzymes.

- Magnesium activates over 300 enzymes involved in the formation of energy and proteins.

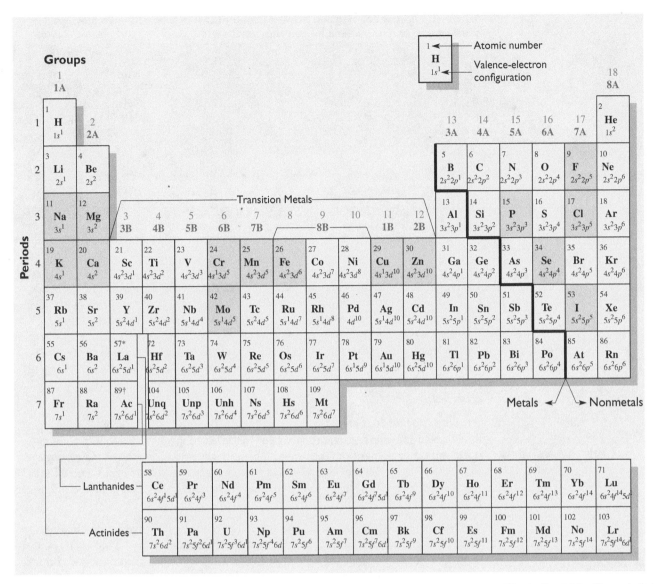

Illustration 15.1   The periodic table lists all known minerals. The highlighted minerals are required in the human diet.

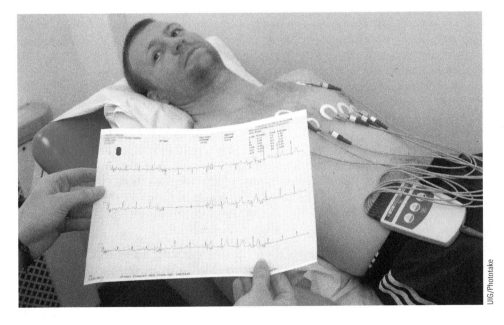

Illustration 15.2   The electrical current measured by an EKG results from the movement of charged minerals across membranes of the muscle cells in the heart.

Table 15.1 Percent of original mineral content lost in fruits by food storage method and in dried beans by cooking method[51]

| | Fruits | | | Dried beans boiled 2–2.5 hours | |
	Canned (%)	Frozen (%)	Dried (%)	Water drained (%)	Water used (%)
Calcium	5	5	0	35	30
Iron	0	0	0	25	20
Magnesium	0	0	0	30	25
Potassium	10	10	0	35	30
Zinc	0	0	0	15	10
Copper	10	10	0	45	40

it is bound to protein. People whose sole source of zinc is whole grains have developed zinc deficiency, even though their intake of zinc is adequate.[3] The absorption of iron from foods in a meal decreases by as much as 50% if tea is consumed with the meal. In the intestines, iron binds with tannic acid in tea and forms a compound that cannot be broken down.[4] The calcium present in spinach and collard greens is poorly absorbed because it is firmly bound to oxalic acid. Many more examples could be given. The point is that you don't always get what you consume; the availability of minerals in food can vary a great deal.

Preserving the Mineral Content of Food Minerals in foods can be lost during food storage and preparation, primarily due to the leaching out of minerals in cooking water and the drippings from the meats. Table 15.1 gives two examples of the percentage of minerals lost by methods of fruit storage and preparation of dried beans. Dried foods retain minerals well, and those lost in cooking fluids can be recovered if the cooking water is minimized and consumed.

The Boundaries of This Unit All of the minerals could be the subject of fascinating stories, but in this unit we will concentrate on just three. A summary of the main features of all the essential minerals is provided in Table 15.2. This table lists recommended intake levels, functions, consequences of deficiency and overdose, and provides notes about each mineral. Table 15.10 at the end of this unit lists food sources of the each essential mineral and the average amount of the mineral present in a serving of the food.

The minerals highlighted in this unit are calcium, iron, and sodium. They have been selected primarily because they play important roles in the development of **osteoporosis**, iron-deficiency anemia, and hypertension, respectively. These disorders are widespread in the United States and in many other countries, and healthy dietary patterns offer a key to their prevention and treatment.

Selected Minerals: Calcium

What you've heard about calcium is true: It's good for bones and teeth. About 99% of the 3 pounds of calcium in the body is located in bones and teeth. The remaining 1% is found in blood and other body fluids. We don't hear so much about this 1%, but it's very active. Every time a muscle contracts, a nerve sends out a signal, or blood clots to stop a bleeding wound, calcium in body fluids is involved. Calcium's most publicized function, however, is its role in bone formation and the prevention of osteoporosis.

A Short Primer on Bones Most of the bones we see or study are hard and dead. As a result, people often have the impression that bones in living bodies are that way. Nothing could be further from the truth. The 206 bones in our bodies are slightly flexible, living tissues infiltrated by blood vessels, nerves, and cells.

remodeling The breakdown and buildup of bone tissue.

osteoporosis (*osteo* = bones; *poro* = porous, *osis* = abnormal condition) A condition characterized by porous bones; it is due to the loss of minerals from the bones.

The solid parts of bones consist of networks of strong protein fibers (called the *protein matrix*) embedded with mineral crystals (Illustration 15.3). Calcium is the most abundant mineral found in bone, but many other minerals, such as phosphorus, magnesium, and carbon, are also incorporated into the protein matrix. The combination of water, the tough protein matrix, and mineral crystals makes bone very strong yet slightly flexible and capable of absorbing shocks.

Teeth have the same properties as other bone plus a hard outer covering called enamel, which is not infiltrated by blood vessels or nerves. Enamel serves to protect the teeth from destruction by bacteria and from mechanical wear and tear.

The Timing of Bone Formation Bones develop and mineralize throughout the first three decades of life. Even after the growth spurt occurs during adolescence and people think they are as tall as they will ever be, bones continue to increase in width and mineral content for 10 to 15 more years. Peak bone density, or the maximal level of mineral content in bones, is reached somewhere between the ages of 30 and 40. After that, bone mineral content no longer increases. The higher the peak bone mass, the less likely it is that osteoporosis will develop. People with higher peak bone mass simply have more calcium to lose before bones become weak and fracture easily. That also is the reason males experience osteoporosis less often than females do: they have more bone mass to lose.[5]

Bone size and density often remain fairly stable from age 30 to the mid-40s, but then bones tend to demineralize with increasing age. By the time women are 70, for example, their bones are 30–40% less dense on average than they once were.[6] A woman may lose an inch or more in height with age and develop the "dowager's hump" that is characteristic of osteoporosis in the spine (Illustration 15.4).

Bone Remodeling Bones slowly and continually go through a repair and replacement process known, appropriately enough, as **remodeling**. During remodeling, the old protein matrix is replaced and remineralized. If insufficient calcium is available to complete the remineralization, or if other conditions such as vitamin D inadequacy prevent calcium from being incorporated into the protein matrix, osteoporosis results.[7]

Osteoporosis

If you are female, you have a one in four chance of developing osteoporosis in your lifetime. If you are a Caucasian, Hispanic, or Asian female, you have a higher risk of developing osteoporosis than if you are an African American woman or a male. If you are male, your chance of developing osteoporosis is one in eight.[8] Approximately 49% of U.S. adults over the age of 50 are at are at risk of developing osteoporosis due to low bone mineral density, or *low bone mass* (Illustration 15.5).

Many factors, such as age, physical activity, genetic traits, diet quality, and body size influence the development of osteoporosis (Table 15.3). If you were to find all of the various risk factors combined in one person, that individual would be a thin woman with light skin who has consumed too little calcium, has poor vitamin D status, is physically inactive, is an excessive alcohol drinker, smokes, and is genetically "small-boned." Additionally, she would have had her ovaries surgically removed for medical reasons before the age of 45.[9] Few people meet every aspect of that description, but people who have several of these characteristics are more likely to develop osteoporosis than those who don't.

Osteoporosis is a disabling disease that reduces quality of life and dramatically increases the need for health care.[8] Because the incidence of osteoporosis increases with age, its importance as a personal and public health problem is intensifying as the U.S. population ages. It currently appears, however, that a large percentage of the cases of osteoporosis can be prevented. The key to prevention is to build dense bones during childhood and the early adult years and then keep bones dense as you age.[10]

Role of Calcium and Vitamin D in Osteoporosis Since the late 1980s, a large number of studies have examined the relationships among dietary calcium intake, vitamin D status, and bone density. Results have been quite consistent: bone mass before the age of 30 is increased by adequate dietary calcium intake and adequate vitamin D status.[11,12]

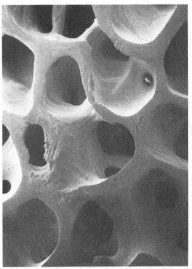

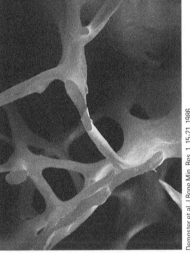

Illustration 15.3 (top) Electron micrograph of healthy bone. (bottom) Electron micrograph of bone affected by osteoporosis.

Illustration 15.4 This woman's stooped appearance is due to osteoporosis.

Table 15.2 An intensive course on essential minerals

		Primary functions	Consequences of deficiency
Calcium supplements are large because our daily need for calcium is high (1,000 milligrams/day). Four pills provide 800 milligrams of calcium, the amount in $2^2/_3$ cups of milk. An aspirin is shown for comparison.	**Calcium** Al[a] women: 1,000 mg men: 1,000 mg UL: 2,500 mg	• Component of bones and teeth • Needed for muscle and nerve activity, blood clotting	• Poorly mineralized, weak bones (osteoporosis) • Rickets in children • Osteomalacia (rickets in adults) • Stunted growth in children • Convulsions, muscle spasms
	Phosphorus RDA women: 700 mg men: 700 mg UL: 4,000 mg	• Component of bones and teeth • Component of certain enzymes and other substances involved in energy formation • Needed to maintain the right acid–base balance of body fluids	• Loss of appetite • Nausea, vomiting • Weakness • Confusion • Loss of calcium from bones
	Magnesium RDA women: 310 mg men: 400 mg UL: 350 mg (from supplements only)	• Component of bones and teeth • Needed for nerve activity • Activates hundreds of enzymes involved in energy and protein formation and other body processes	• Stunted growth in children • Weakness • Muscle spasms • Personality changes

[a]Als (Adequate Intakes) and RDAs (Recommended Dietary Allowances) are for 19–30-year-olds; ULs (Upper Limits) are for 19–70-year-olds, 1997–2004.

Consequences of overdose	Primary food sources	Highlights and comments
• Drowsiness • Calcium deposits in kidneys, liver, and other tissues • Suppression of bone remodeling • Decreased zinc absorption	• Milk and milk products (cheese, yogurt) • Calcium-fortified foods (some juices, breakfast cereals, soy milk) 	• The average intake of calcium among U.S. women is approximately 60% of the DRI. • One in four women and one in eight men in the United States develop osteoporosis. • Adequate calcium and vitamin D status must be maintained to prevent bone loss.
• Muscle spasms • Increased risk of cardiovascular disease and osteoporosis	• Milk and milk products (cheese, yogurt) • Meats • Seeds, nuts • Phosphates added to foods Phosphates are a common food additive.	• Deficiency is generally related to disease processes. • Phosphorus intake in the United States is increasing due to use of phosphates in processed foods.
• Diarrhea • Dehydration • Impaired nerve activity due to disrupted utilization of calcium	• Plant foods (dried beans, nuts, potatoes, green vegetables) • Ready-to-eat cereals	• Magnesium is found primarily in plant foods, where it is attached to chlorophyll. • Average intake among U.S. adults is below the RDA.

(continued)

Table 15.2 (continued)

		Primary functions	Consequences of deficiency
Food And Drink Photos/AGE Fotostock © Daniel Padavona/Shutterstock.com	**Iron** RDA women: 18 mg men: 8 mg UL: 45 mg Dr. Richard Kessel/Visuals Unlimited, Inc/Getty Images Dr. Gladden Willis/Visuals Unlimited, Inc/Getty Images Iron-deficiency anemia is characterized by microcytic anemia and small, pale red blood cells (bottom photo). Normal red blood cells are shown in the top photo.	• Transports oxygen as a component of hemoglobin in red blood cells • Component of myoglobin (a muscle protein) • Needed for certain reactions involving energy formation	• Iron deficiency • Iron-deficiency anemia • Weakness, fatigue • Pale appearance • Reduced attention span and resistance to infection • Hair loss • Mental retardation, developmental delay in children • Ice craving • Decreased resistance to infection
© Sbarabu/Shutterstock.com	**Zinc** RDA women: 8 mg men: 11 mg UL: 40 mg	• Required for the activation of many enzymes involved in the reproduction of proteins • Component of insulin, many enzymes	• Growth failure • Delayed sexual maturation • Slow wound healing • Loss of taste and appetite • In pregnancy, low-birth-weight infants and preterm delivery
© Marie C Fields/Shutterstock.com	**Fluoride** AI women: 3 mg men: 4 mg UL: 10 mg	• Component of bones and teeth (enamel) • Helps rebuild enamel that is beginning to decay	• Tooth decay and other dental diseases
D.A. Weinstein/Custom Medical Stock Photo/Newscom	**Iodine** RDA women: 150 mcg men: 150 mcg UL: 1,100 mcg	• Required for the synthesis of thyroid hormones that help regulate energy production, growth, and development Mediscan/Alamy Iodine deficiency during pregnancy produces cretinism in the offspring.	• Goiter, thyroid disease • Cretinism (mental retardation, hearing loss, growth failure) Biophoto Associates/Getty Images

Consequences of overdose	Primary food sources	Highlights and comments
• Hemochromatosis ("iron poisoning") • Vomiting, abdominal pain, diarrhea • Blue coloration of skin • Iron deposition in liver and heart • Decreased zinc absorption • Oxidation-related damage to tissues and organs	• Liver, beef, pork • Dried beans • Iron-fortified cereals • Prunes, apricots, raisins • Spinach • Bread	• Cooking foods in iron and stainless steel pans increases the iron content of the foods. • Vitamin C, meat, and alcohol increase iron absorption. • Iron deficiency is the most common nutritional deficiency in the world. • Average iron intake of young children and women in the United States is low.
• Over 25 mg/day is associated with nausea, vomiting, weakness, fatigue, susceptibility to infection, copper deficiency, and metallic taste in mouth. • Increased blood lipids	• Meats (all kinds) • Dried beans • Grains • Nuts • Ready-to-eat cereals	• Like iron, zinc is better absorbed from meats than from plants. • Marginal zinc deficiency may be common, especially in children. • Zinc supplements taken within 24 hours of onset may decrease duration and severity of the common cold.
• Fluorosis • Brittle bones • Mottled teeth • Nerve abnormalities Dr. P. Marazzi/ Science Source "Mottled teeth" result from excessive fluoride.	• Fluoridated water and foods and beverages made with it • White grape juice • Raisins • Wine	• Toothpastes, mouth rinses, and other dental care products may provide fluoride. • Fluoride overdose has been caused by ingestion of fluoridated toothpaste. • Fluoridated water is not related to cancer.
• Over 1 mg/day may produce pimples, goiter, decreased thyroid function, and thyroid disease.	• Iodized salt • Milk and milk products • Seaweed, seafoods • Bread from commercial bakeries	• Iodine deficiency was a major problem in the United States in the 1920s and 1930s. Deficiency remains a major health problem in some developing countries. • Amount of iodine in plants depends on iodine content of soil. • The need for iodine increases by 50% during pregnancy.

(continued)

Table 15.2 (continued)

		Primary functions	Consequences of deficiency
	Selenium RDA women: 55 mcg men: 55 mcg UL: 400 mcg	• Acts as an antioxidant in conjunction with vitamin E (protects cells from damage due to exposure to oxygen) • Needed for thyroid hormone production	• Anemia • Muscle pain and tenderness • Keshan disease (heart failure), Kashin-Beck disease (joint disease)
	Copper RDA women: 900 mcg men: 900 mcg UL: 10,000 mcg	• Component of enzymes involved in the body's utilization of iron and oxygen • Functions in growth, immunity, cholesterol and glucose utilization, brain development	• Anemia • Seizures • Nerve and bone abnormalities in children • Growth retardation
	Manganese AI women: 2.3 mg men: 1.8 mg	• Needed for the formation of body fat and bone	• Weight loss • Rash • Nausea and vomiting
	Chromium AI women: 35 mcg men: 25 mcg	• Required for the normal utilization of glucose and fat	• Elevated blood glucose and triglyceride levels • Weight loss
	Molybdenum RDA women: 45 mcg men: 45 mcg UL: 2,000 mcg	• Component of enzymes involved in the transfer of oxygen from one molecule to another	• Rapid heartbeat and breathing • Nausea, vomiting • Coma
	Sodium AI women: 1,500 mg men: 1,500 mg UL: 2,300 mg	• Needed to maintain the right acid–base balance in body fluids • Helps maintain an appropriate amount of water in blood and body tissues • Needed for muscle and nerve activity	• Weakness • Apathy • Poor appetite • Muscle cramps • Headache • Swelling
	Potassium AI women: 4,700 mg men: 4,700 mg UL: Not determined	• Same as for sodium	• Weakness • Irritability, mental confusion • Irregular heartbeat • Paralysis
	Chloride AI women: 2,300 mg men: 2,300 mg UL: 3,600 mg	• Component of hydrochloric acid secreted by the stomach (used in digestion) • Needed to maintain the right acid–base balance of body fluids • Helps maintain an appropriate water balance in the body	• Muscle cramps • Apathy • Poor appetite • Long-term mental retardation in infants

Consequences of overdose	Primary food sources	Highlights and comments
• "Selenosis," which includes symptoms of hair and fingernail loss, weakness, liver damage, irritability, and "garlic" or "metallic" breath.	• Fish • Eggs	• Content of foods depends on amount of selenium in soil, water, and animal feeds. • Selenium supplements have not been found to prevent cancer.
• Wilson's disease (excessive accumulation of copper in the liver and kidneys) • Vomiting, diarrhea • Tremors • Liver disease	• Potatoes • Grains • Dried beans • Nuts and seeds • Seafood • Ready-to-eat cereals	• Toxicity can result from copper pipes and cooking pans. • Average intake in the United States is below the RDA.
• Infertility in men • Disruptions in the nervous system, learning impairment • Muscle spasms	• Whole grains • Coffee, tea • Dried beans • Nuts	• Toxicity is related to overexposure to manganese dust in miners or contaminated groundwater.
• Kidney and skin damage	• Whole grains • Wheat germ • Liver, meat • Beer, wine • Oysters	• Toxicity usually results from exposure in chrome-making industries or overuse of supplements. • Supplements do not build muscle mass, increase endurance, or reduce blood glucose levels.
• Loss of copper from the body • Joint pain • Growth failure • Anemia • Gout	• Dried beans • Grains • Dark green vegetables • Liver • Milk and milk products	• Deficiency is extraordinarily rare.
• High blood pressure in susceptible people • Kidney disease • Heart problems	• Foods processed with salt • Cured foods (corned beef, ham, bacon, pickles, sauerkraut) • Table and sea salt • Bread • Milk, cheese • Salad dressing	• Very few foods naturally contain much sodium. • Processed foods are the leading source of dietary sodium. • High-sodium diets are associated with the development of hypertension in "salt-sensitive" people.
• Irregular heartbeat, heart attack	• Plant foods (potatoes, squash, lima beans, tomatoes, plantains, bananas, oranges, avocados) • Meats • Milk and milk products • Coffee	• Content of vegetables is often reduced in processed foods. • Diuretics (water pills) and other antihypertension drugs may deplete potassium. • Salt substitutes often contain potassium. • Potassium intake tracks with vegetable and fruit intake.
• Vomiting	• Same as for sodium. (Most of the chloride in our diets comes from salt.)	• Excessive vomiting and diarrhea may cause chloride deficiency. • Legislation regulating the composition of infant formulas was enacted in response to formula-related chloride deficiency and subsequent mental retardation in infants.

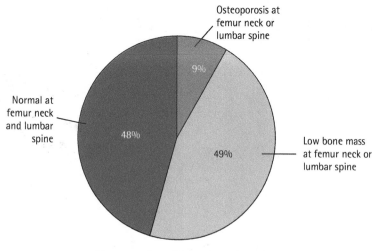

Illustration 15.5 The proportion of U.S. adults with osteoporosis, low bone mass, and normal bone mass.

Vitamin D is required for the absorption of calcium and the incorporation of calcium into bone. In adults over the age of 50, bone density tends to be preserved with daily calcium intakes of 1,000 to 1,200 milligrams along with 10 to 20 mcg (400 to 800 IU) of vitamin D per day.[13,14] Low bone mineral density can be increased somewhat by switching from a low to an adequate intake of calcium and improved vitamin D status. The earlier in life this occurs, the more improvement in bone density is noted.[10,12]

The RDA (Recommended Dietary Allowance) for calcium intake by adults aged 19 to 50 is 1,000 mg per day. The amount increases to 1,200 mg per day for women above the age of 50 and for men over the age of 70 years. The RDA for vitamin D is set at 15 mcg per day (600 IU) for males and females from the age of 1 through 70 years. These levels of intake correspond to amounts of calcium and vitamin D that help build bone density and contribute to the prevention and treatment of osteoporosis.[12]

Since many nutrients are involved in bone formation and maintenance, it is recommended that foods within a healthy dietary pattern provide the recommended amounts of calcium and vitamin D. Foods such as milk and other dairy products provide the full array of nutrients needed by bone, including calcium, vitamin D, phosphorus, magnesium, and high-quality protein.[15] Exposure of the skin to direct sunlight for 10 to 15 minutes daily during warm parts of the year can boost vitamin D status in many people.

Food Sources of Calcium Most women over the age of 50 and men over the age of 70 consume less than the RDA for calcium.[16] Consequently, it is recommended that diets include good sources of calcium that would make up for deficits in intake.[17] About half of the calcium supplied by the diets of Americans comes from milk and milk products. Milk (including chocolate milk), cheese, and yogurt are all good sources of calcium (see Table 15.4). Some plants such as kale, broccoli, and bok choy provide appreciable amounts of calcium, as well. On average, 32% of the calcium content of milk and milk products, calcium-fortified orange juice, and calcium supplements is absorbed, compared to approximately 5–60% of the calcium from various plants.[18]

Many foods rich in calcium aren't loaded with calories. As Table 15.3 shows, low-fat yogurt, skim and soy milk, kale, and broccoli, for example, provide good amounts of calcium at a low calorie cost. Calcium is also appearing in unexpected places such as in candy, snack bars, waffles, and bread. The array of calcium-fortified foods has increased in the United States in response to publicity about our need for more of it (Illustration 15.6). Health claims relating to the benefits of calcium in reducing the risk of osteoporosis may appear on the labels of food products that qualify as good sources of calcium. Look for health claims on foods such as milk, yogurt, and calcium-fortified foods such as orange juice, breakfast cereals, and grain products (Illustration 15.7).

Calcium and Vitamin D Supplements Calcium (1,000 to 1,200 mg per day) and vitamin D supplements (400 to 800 IU per day or higher if blood levels of vitamin D are very low) are often recommended components of the treatment of osteoporosis.[19] However, too much supplemental calcium and vitamin D can cause health problems. Supplemental doses of calcium above 2,000 mg per day along with vitamin D supplementation is associated with excessive absorption of calcium from food.[20] That can lead to the deposition of calcium in blood vessels and soft tissues, and an increased risk of developing kidney stones.[21] Long-term intake of vitamin D from supplements in excess of 125 mcg (5,000 IU) per day has been associated with a greater risk of cancer at sites such as the pancreas, a greater risk of heart disease, and more falls and fractures among the elderly.[22] The Tolerable Upper Intake Level (UL) for vitamin D in adults is 100 mcg

Table 15.3 Risk factors for osteoporosis[53]

- Female
- Menopause
- Poor overall diet
- Deficient calcium intake
- Caucasian or Asian heritage
- Thinness ("small bones")
- Cigarette smoking
- Excessive alcohol intake
- Ovarectomy (ovaries removed) before age 45
- Physical inactivity
- Deficient vitamin D status
- Genetic factors

Table 15.4 Caloric content, calcium level, and percentage of available calcium from different foods[52]

Food	Amount	Calories	Calcium (mg)	Calcium absorbed (%)	Available calcium (mg)
Yogurt, low fat	1 cup	143	13	32	132
Skim milk	1 cup	85	301	32	96
Soy milk (fortified)	1 cup	79	300	31	93
1% milk	1 cup	163	274	32	88
Tofu	1 cup	188	260	31	81
Cheese	1 oz	114	204	32	65
Kale, cooked	1 cup	42	94	49	46
Broccoli, cooked	1 cup	44	72	61	44
Bok choy, raw	1 cup	9	73	54	39
Dried beans	1 cup	209	120	24	29
Spinach, cooked	1 cup	42	244	5	12

(4,000 IU) per day. A number of medications are available that enhance bone formation and reduce loss of bone density.

Selected Minerals: Iron

Most (80%) of the body's iron supply is found in **hemoglobin**. Small amounts are present in **myoglobin**, and free iron is involved in processes that capture energy released during the breakdown of proteins and fats.[23]

hemoglobin The iron-containing protein in red blood cells.

myoglobin The iron-containing protein in muscle cells.

The Role of Iron in Hemoglobin and Myoglobin What happens to a car when its paint gets scratched? After a while, the exposed metal rusts. The iron in the metal combines with oxygen in the air, and the result is iron oxide, or *rust*. Iron readily combines with

Richard Anderson

"Regular exercise and a healthy diet with enough calcium help maintain good bone health and may reduce the risk of osteoporosis later in life."

Nutrition Facts

Serving Size: 1 cup (240ml)
Servings per Container: 16

Amount per Serving

Calories 110	Calories from Fat 20
	% Daily Value*

Total Fat 2.5g	**4%**
Saturated Fat 1.5g	**8%**
Trans Fat 0g	
Cholesterol 15mg	**4%**
Sodium 135mg	**6%**
Total Carbohydrate 13g	**4%**
Dietary Fiber 0g	**0%**
Sugars 12g	
Protein 8g	

Vitamin A 10%	•	Vitamin C 4%
Calcium 30%	Iron 0%	Vitamin D 25%
Phosphorus 10%		

* Percent Daily Values are based on a 2,000 calorie diet.

MILK 1% LOWFAT MILK

Illustration 15.7 Food products that are good sources of calcium can be labeled with a health claim.

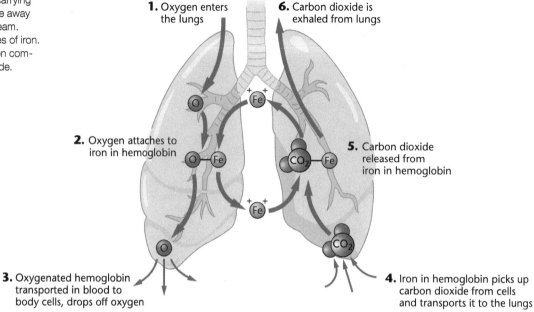

Illustration 15.8 Iron's role in carrying oxygen to cells and carbon dioxide away from them by way of the bloodstream. Fe++ represents charged particles of iron. The charge is neutralized when iron combines with oxygen or carbon dioxide.

1. Oxygen enters the lungs

2. Oxygen attaches to iron in hemoglobin

3. Oxygenated hemoglobin transported in blood to body cells, drops off oxygen

4. Iron in hemoglobin picks up carbon dioxide from cells and transports it to the lungs

5. Carbon dioxide released from iron in hemoglobin

6. Carbon dioxide is exhaled from lungs

oxygen, and that property is put to good use in the body. From its location in hemoglobin in red blood cells, iron loosely attaches to oxygen when blood passes near the inner surface of the lungs. The bright red, oxygenated blood is then delivered to cells throughout the body. When the oxygenated blood passes near cells that need oxygen for energy formation or for other reasons, oxygen is released from the iron and diffuses into cells (Illustration 15.8). The free iron in hemoglobin then picks up carbon dioxide, a waste product of energy formation. When carbon dioxide attaches to iron, blood turns from bright red to dark bluish red. Blood then circulates back to the lungs, where carbon dioxide is released from the iron and exhaled into the air. The free iron attaches again to oxygen that enters the lungs, and the cycle continues.

Iron in myoglobin traps oxygen delivered by hemoglobin, stores it, and releases it as needed for energy formation for muscle activity. In effect, myoglobin boosts the supply of oxygen available to muscles.

The functions of iron just described operate smoothly when the body's supply of iron is sufficient. Unfortunately, that is often not the case.

iron deficiency A disorder that results from a depletion of iron stores in the body. It is characterized by weakness, fatigue, short attention span, poor appetite, increased susceptibility to infection, and irritability.

Iron Deficiency Is a Big Problem The most widespread nutritional deficiency in both developing and developed countries is **iron deficiency** (Table 15.5). It is estimated that one out of every four people in the world is iron deficient.[23] For the most part, iron

Table 15.5 Incidence of iron deficiency[1,25]

	Population with iron deficiency (%)
Worldwide (children under 5 years):	
Developing countries	51
Developed countries	12
United States:	
Children, 1–3 years	8
Pregnant women	12
Females, 20–49 years	7
Males, 12–49 years	1 or less

deficiency affects very young children and women of childbearing age, who have a high need for iron and frequently consume too little of it.[24] Iron deficiency may develop in people who have lost blood due to injury, surgery, or ulcers. Donating blood more than three times a year can also precipitate iron deficiency.[25]

Consequences of Iron Deficiency Many body processes sputter without sufficient oxygen. People with iron deficiency usually feel weak and tired. They have a shortened attention span and a poor appetite, are susceptible to infection, and become irritable easily. If the deficiency is serious enough, **iron-deficiency anemia** develops, and additional symptoms occur. People with iron-deficiency anemia look pale, are easily exhausted, and have rapid heart rates. Iron-deficiency anemia is a particular problem for infants and young children because it is related to lasting retardation in mental development.[26]

Food Sources of Iron "Enough" iron, according to the RDA, is 8 milligrams for men and 18 milligrams per day for women aged 19 to 50. Consuming that much iron can be difficult for women. On average, 1,000 calories' worth of food provides about 6 milligrams of iron. Women would have to consume around 2,500 calories per day to obtain even 15 milligrams of iron on average. Selection of good sources of iron has to be done on a better-than-average basis if women are to get enough.

Iron is found in small amounts in many foods, but only a few foods such as liver, beef, and prune juice are rich sources. Foods cooked in iron and stainless steel pans can be a significant source of iron because some of the iron in the pan leaches out during cooking. On average, approximately 1 milligram of iron is added to each 3-ounce serving of food cooked in these pans.[27]

Most of the iron in plants and eggs is tightly bound to substances such as phytates or oxalic acid, which limit iron absorption, making these foods relatively poor sources of iron even though they contain a fair amount of it. (Differences in the proportions of iron absorbed from various food sources are shown in Illustration 15.9.) A 3-ounce hamburger and a cup of asparagus both contain approximately 3 milligrams of iron, for example. But 20 times more iron can be absorbed from the hamburger than from the asparagus.[28] Absorption of iron from plants is increased substantially if foods containing vitamin C are included in the same meal. Iron absorption is also increased by low levels of iron stores in

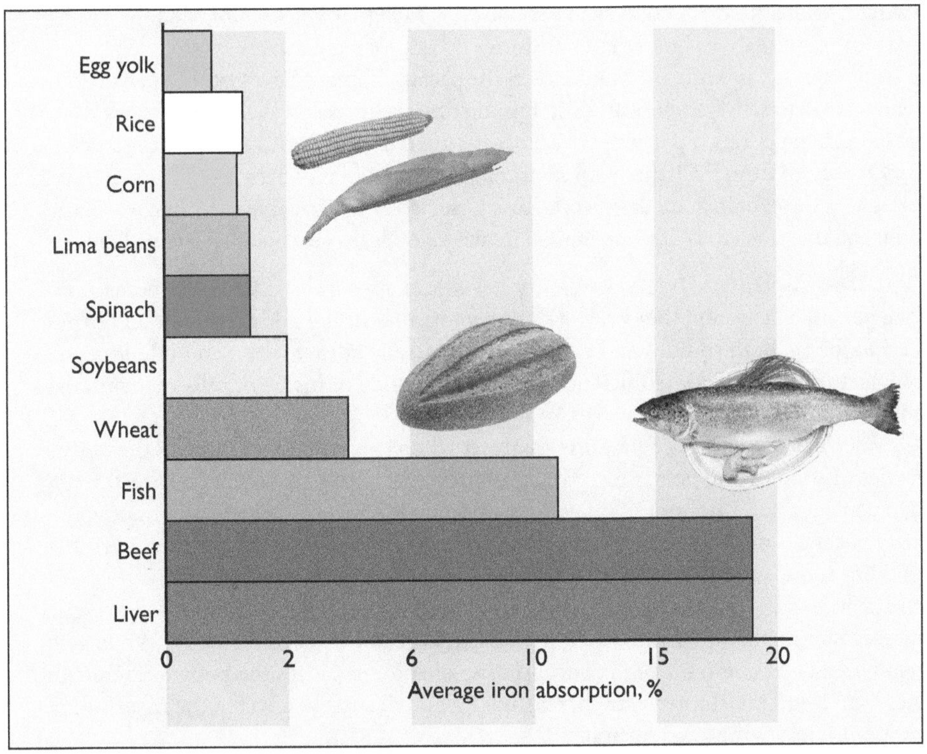

iron-deficiency anemia A condition that results when the content of hemoglobin in red blood cells is reduced due to a lack of iron. It is characterized by the signs of iron deficiency plus paleness, exhaustion, and a rapid heart rate.

Illustration 15.9 Average percentage of iron absorbed from selected foods by healthy adults.[28]

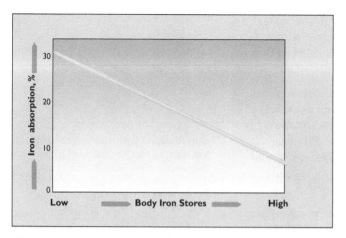

Illustration 15.10   People absorb more iron from foods and supplements when body stores of iron are low than when stores of iron are high.

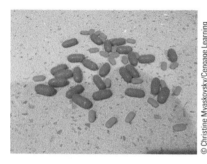

© Christine Myaskovsky/Cengage Learning

Illustration 15.11   If you were 3 years old, could you tell which "pills" are candy and which are the iron supplements? Overdoses of iron supplements are a leading cause of accidental poisoning in young children. (The iron supplements are the lightest green in color.)

water balance   The ratio of the amount of water outside cells to the amount inside cells; a proper balance is needed for normal cell functioning.

hypertension   High blood pressure. It is defined as blood pressure exerted inside blood vessel walls that typically exceeds 140/90 millimeters of mercury in adults under 60 years of age and 150/90 in people 60 years of age and older. Hypertension in children and adolescents is defined as systolic or diastolic blood pressure equal to or greater than the 95th blood pressure percentile of sex-, age- and height-specific blood pressure percentiles.

prehypertension   In adults, typical blood pressure levels of 120/80 mm Hg through 139/89 mm Hg. In children and adolescents, prehypertension is defined as systolic or diastolic blood pressure equal to or greater than the 90th percentile but less than the 95th percentile of sex-, age- and height-specific blood pressure percentiles.

the body (Illustration 15.10). In other words, if you're in need of iron, your body sets off mechanisms that allow more of it to be absorbed from foods or supplements. When iron stores are high, less iron is absorbed.[4,23]

The body's ability to regulate iron absorption provides considerable protection against iron deficiency and overdose. The protection is not complete, however, as evidenced by widespread iron deficiency and the occurrence of iron toxicity.

Iron Toxicity   Excess iron absorbed into the body cannot be easily excreted. Consequently, it is deposited in various tissues such as the liver, pancreas, and heart. There, the iron reacts with cells, causing damage that can result in liver disease, diabetes, and heart failure. One in every 200 people in the United States has an inherited tendency to absorb too much iron (a disorder called hemochromatosis (pronounced hem-oh-chrom-ah-toe-sis). Other people develop iron toxicity from consuming large amounts of iron with alcohol (alcohol increases iron absorption) or from very high iron intake, usually due to overdoses of iron supplements.[25,29]

Each year in the United States, more than 10,000 people accidentally overdose on iron supplements.[30] Victims are often young children who mistakenly think iron pills are candy (Illustration 15.11). The lethal dose of iron for a two-year-old child is about 3 grams,[29] the amount of iron present in 25 pills containing 120 milligrams of iron each. Iron supplements that are not being used should be thrown away or stored in a place where toddlers cannot get to them.

Selected Minerals: Sodium

On the one hand, sodium is a gift from the sea; on the other, it is a hazard to health. Sodium's life- and health-sustaining functions are frequently overshadowed by its effects on blood pressure when consumed in excess.

An extremely reactive mineral, sodium occurs in nature in combination with other elements. The most common chemical partner of sodium is chloride, and much of the sodium present on this planet is in the form of sodium chloride—table salt. Table salt is 40% sodium by weight; one teaspoon of salt contains about 2,300 milligrams of sodium. Visit the Reality Check for this unit to see how sea salt measures up against regular table salt.

Although salt is found in abundance in the oceans, humans have not always had access to enough salt to satisfy their taste for it. During times when salt was scarce, wars were fought over it. In such periods, a pocketful of salt was as good as a pocketful of cash. (The word *salary* is derived from the Latin word *salarium*, "salt money.") People were said to be "worth their salt" if they put in a full day's work. Today, salt is widely available in most parts of the world, and the problem is limiting human intake to levels that do not interfere with health.[31]

What Does Sodium Do in the Body?   Sodium appears directly above potassium in the periodic table, and the two work closely together in the body to maintain normal **water balance**. Both sodium and potassium chemically attract water, and under normal circumstances, each draws sufficient water to the outside or inside of cells to maintain an optimal level of water in both places.[32]

Water balance and cell functions are upset when there's an imbalance in the body's supplies of sodium and potassium. You have probably noticed that you become thirsty when you eat a large amount of salted potato chips or popcorn. Salty foods make you thirsty because your body loses water when the high load of dietary sodium is excreted. The thirst signal indicates that you need water to replace what you have lost.

The loss of body water that accompanies ingestion of large amounts of salt explains why seawater neither quenches thirst nor satisfies the body's need for water. When a person drinks seawater, the high concentration of sodium causes the body to excrete more water than it retains. Rather than increasing the body's supply of water, the ingestion of seawater increases the need for water.

In healthy people, the body's adaptive mechanisms provide a buffer against upsets in water balance due to high sodium intake. It appears that many people are overwhelming the body's ability to cope with high sodium loads, however. High dietary intake of sodium appears to play an important role in the development of **hypertension** in many people.[33]

A Bit about Blood Pressure To circulate through the body, blood must exist under pressure in the blood vessels. The amount of pressure exerted on the walls of blood vessels is greatest when pulses of blood are passing through them (that's when *systolic* blood pressure is measured) and least between pulses (that's when *diastolic* blood pressure is taken). Blood pressure measurements note the highest and lowest pressure in blood vessels (Illustration 15.12).

Blood pressure levels less than 120/80 millimeters of mercury (mm Hg) are considered normal, whereas levels between 120/80 through 139/89 mm Hg are classified as **prehypertension**. It is estimated that 28% of U.S. adults qualify as having prehypertension. Values of 140/90 mm Hg and higher qualify as hypertension in adults younger then the age of 60 (Table 15.6), and approximately one-third of American adults have this disorder.[34] The blood pressure cut-off for hypertension in adults aged 60 and over is 150/90 mm Hg.[35] Several blood pressure measurements, taken while a person is relaxed, are needed to obtain an accurate measure of blood pressure. Even going to a clinic or doctor's office to have blood pressure measured can raise it for some people and lead to a false diagnosis of hypertension. This condition is referred to as "white coat hypertension."[36]

Although not considered a disease by itself, the presence of hypertension substantially increases the risk that a person will develop heart disease or kidney failure or will experience a heart attack or stroke.[33,37]

What Causes Hypertension? Approximately 5–10% of all cases of hypertension can be directly linked to a cause. People who have hypertension with no identifiable cause (90-95% of all cases) are said to have **essential hypertension**.[38]

A number of dietary and other risk factors for hypertension have been identified (Table 15.7). Foremost among the evidence linking diet to hypertension are population studies that show a relationship between salt intake and hypertension. As salt intake rises,

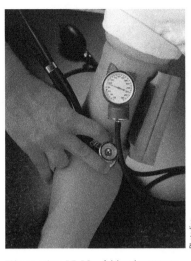

Photodisc

Illustration 15.12 A blood pressure test. Blood pressure is expressed as two numbers: Systolic pressure measures the force of the blood when the heart contracts. Diastolic pressure measures the force of the blood when the heart is at rest.

essential hypertension Hypertension of no known cause; also called primary or idiopathic hypertension, it accounts for 90–95% of all cases of hypertension.

Table 15.7 Risk factors for hypertension[54]

- Age
- Genetic predisposition
- High-sodium, low-potassium diet
- Obesity
- Physical inactivity
- Excessive alcohol consumption
- Smoking
- Frequent stress, anxiety

Table 15.6 Hypertension categories[35]

Category	Systolic (mm Hg)[a]	Diastolic (mm Hg)
Prehypertension	120–139	80–89
Hypertension (adults through age 60)	≥140	≥90
Hypertension (adults 60 yrs. and over)	≥150	≥90

[a]mm Hg = millimeters of mercury.

REALITY CHECK
Is Sea Salt a Better Choice Than Table Salt?

Matty and Miriam were shopping for snacks to serve when their friends came over for the game. They are now standing in front of the nut section at the store and can't decide if they should get nuts salted with sea salt or the regular salted nuts.

Who gets the thumbs up?

Answers on page 422.

iStockphoto.com/DRB Images, LLC

Matty: Sea salt is lower in sodium. Why don't we get the sea-salted nuts?

Wavebreakmedia Ltd/Getty Images Plus/Getty Images

Miriam: I agree. It's amazing, though. I've had them before and they still taste salty enough.

so do rates of hypertension in populations. When people with hypertension reduce their salt intake, their blood pressure tends to drop somewhat.[15,33] Further scrutiny of the data on salt and blood pressure, however, reveals that not everyone is equally susceptible to high-salt diets. Some people are genetically susceptible to salt or have subtle forms of kidney disorders that raise their blood pressure more in response to high-sodium diets than is the case for other people.[31,37] One such condition is known as **salt sensitivity**.

Salt Sensitivity Approximately 51% of people with hypertension and 26% of people with normal blood pressure are salt sensitive.[39,40] Salt sensitivity is more common in African Americans than in other population groups. It has been identified in 73% of African Americans with hypertension.[41] Reduction in salt intake by people who are salt sensitive, along with weight loss if overweight, substantially improves blood pressure in most cases.[17] It is recommended that people who would benefit from lowering their blood pressure consume ≤ 2,300 mg sodium per day. Further reduction in salt intake may be appropriate if further reduction in blood pressure is needed.[15]

Other Risk Factors for Hypertension Obesity is a major risk factor for hypertension. For obese people with hypertension, the most effective treatment is weight loss. Excessive alcohol intake can prompt the development of hypertension, and moderation of intake (to two or fewer alcoholic drinks per day) can improve blood pressure. Physically inactive lifestyles also foster the development of hypertension. Diets low in potassium are a well-established risk factor for hypertension, and most Americans consume too little of it from foods. On average, U.S. women consume half of the RDA for potassium of 4,700 mg daily and men consume a third less than the RDA.[16] Regular intake of vegetables and fruits rich in potassium contributes to a reduction in risk for hypertension.[42,43] Diets containing adequate amounts of potassium from foods help lower blood pressure and appear to counteract the effects of high sodium intake on blood pressure.[15] The Take Action feature in this unit attempts to increase your awareness of good sources of potassium. If you are like most Americans, you are consuming too little of it.

Reduction of Sodium Intake Most children and adults in the United States consume over twice the recommended amount of sodium.[16] The leading sources of salt (and therefore of sodium) in the U.S. diet are processed foods (Table 15.8). These foods account for over half of the total sodium intake by Americans, while approximately 10% of total sodium intake comes from salt added at the table.[44] Restricting the use of foods to which salt has been added during processing is the most effective way to lower salt intake.

High-salt processed foods include frozen meals, salad dressings, canned soups, ham, sausages, and biscuits. Only a small proportion of our total sodium intake enters our diet from fresh foods. Very few foods naturally contain much sodium—at least not until they are processed (Illustration 15.13).

How Is Hypertension Treated? The recommended approach to treatment of all cases of hypertension consists of dietary and lifestyle changes and the use of medications if necessary.[35,44] Weight loss and smoking cessation (if needed), a reduced-sodium diet,

salt sensitivity A genetically influenced condition in which a person's blood pressure rises when large amounts of salt or sodium are consumed. Such individuals are sometimes identified by blood pressure increases of 5 or 10% or more when switched from a low-salt to a high-salt diet.

Table 15.8 Major sources of sodium in the American diet[44,55]

Sodium source	Contribution to sodium intake (%)
Processed foods	75
Fresh foods	12
Salt added at the table	10
Salt added during cooking and food preparation	3

Table 15.9 Approaches to the treatment of hypertension[2]

- Weight loss (if needed)
- Sodium intake < 2,300 mg per day
- Moderate alcohol consumption (if any)
- Regular physical activity (≥30 minutes moderate intensity per day)
- The DASH dietary pattern
- Antihypertension drugs (if needed)
- Meditation, yoga
- Smoking cessation (if applicable)

regular exercise, moderate alcohol consumption (if any), and the DASH eating plan are basic components of the approach to treatment (Table 15.9).[15] The DASH diet (Illustration 15.14) is based on vegetables, fruits, low-fat dairy products, whole grains, and poultry and fish. Its composition is similar to that of other healthy dietary patterns recommended for heart disease and obesity prevention. People who adhere to the DASH eating plan often bring their blood pressure levels back into the normal range.[45,46] If blood pressure remains elevated after dietary and lifestyle changes have been implemented, or if blood pressure is quite high when diagnosed, anti-hypertension drugs are usually prescribed.[35]

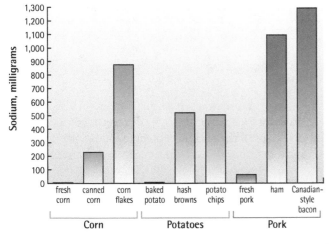

Illustration 15.13 Examples of how processing increases the sodium content of foods. Sodium values are for a 3-ounce serving of each food shown.

The Dash Eating Plan

In the mid-1990s a revolutionary approach to the control of mild and moderate hypertension was tested, and the results have changed health professionals' thinking about high blood pressure prevention and management. Called the DASH (Dietary Approaches to Stop Hypertension) Eating Plan, it didn't focus on salt restriction; it was related to significant reductions in blood pressure within two weeks in most people tested. In some people, reductions in blood pressure were sufficient to erase the need for anti-hypertension medications, and for others the diet reduced the amount or variety of medications needed. A subsequent study showed that a low-sodium diet boosts the blood pressure–lowering effects of the DASH diet, especially in African Americans.[50] This dietary pattern is also associated with reduced risk of heart disease and stroke,[17] and is recommended by the 2015 Dietary Guidelines Advisory Committee as a healthy dietary pattern for people in general.[15]

The DASH diet consists of eating patterns made up of the following food groups:

	Daily servings
Vegetables	4–5
Fruits	4–5
Grain products, mostly whole grain	7–8
Low-fat milk and dairy products	2–3
Lean meats, fish, poultry	6 oz
Nuts, seeds, dried beans	1

Although it does not work for all individuals with hypertension, the DASH diet is a mainstay in clinical efforts to reverse prehypertension and to manage hypertension.

A very useful Web site is available that gives practical tips on how to implement and follow the DASH eating pattern (www.nhlbi.nih.gov /health/public/heart/hbp/dash/new_dash.pdf). You can find the content quickly by searching the key term "DASH diet."

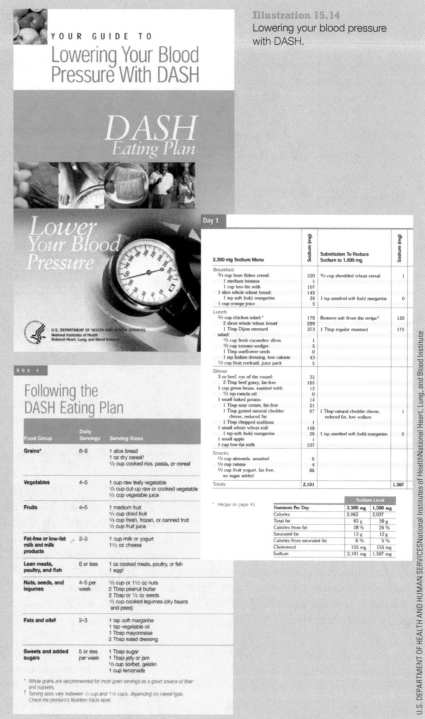

Illustration 15.14
Lowering your blood pressure with DASH.

U.S. DEPARTMENT OF HEALTH AND HUMAN SERVICESNational Institutes of HealthNational Heart, Lung, and Blood Institute

Using spices and lemon juice to flavor foods, consuming fresh vegetables (rather than processed ones) and fresh fruits, checking out the sodium content of foods and selecting low-sodium ones, and repeated exposure to lower-salt foods can also help reduce sodium intake.[47]

Label Watch Not all processed foods with added sodium taste salty. To find out which processed foods are high in sodium, you have to examine the label. Increasingly, low-salt processed foods are entering the market and can be easily identified by the "low-salt"

message on the label (Illustration 15.15). Terms used to identify low-salt (or low-sodium) foods are defined by the Food and Drug Administration. To be considered low sodium, foods must contain 140 milligrams or less of sodium per serving. Food manufacturers must adhere to the definitions when they make claims about the salt or sodium content of a food on the label.

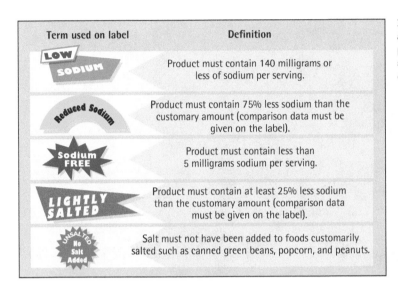

Term used on label	Definition
LOW SODIUM	Product must contain 140 milligrams or less of sodium per serving.
Reduced Sodium	Product must contain 75% less sodium than the customary amount (comparison data must be given on the label).
Sodium FREE	Product must contain less than 5 milligrams sodium per serving.
LIGHTLY SALTED	Product must contain at least 25% less sodium than the customary amount (comparison data must be given on the label).
UNSALTED No Salt Added	Salt must not have been added to foods customarily salted such as canned green beans, popcorn, and peanuts.

Illustration 15.15 When a label makes a claim about the sodium content of a food product, the label must list the amount of sodium in a serving and adhere to these definitions.

Table 15.10 **Food sources of minerals**

Magnesium		
Food	Amount	Magnesium (mg)
Legumes:		
Lentils, cooked	½ cup	134
Split peas, cooked	½ cup	134
Tofu	½ cup	130
Nuts:		
Peanuts	¼ cup	247
Cashews	¼ cup	93
Almonds	¼ cup	80
Grains:		
Bran buds	1 cup	240
Wild rice, cooked	½ cup	119
Breakfast cereal, fortified	1 cup	85
Wheat germ	2 Tbsp	45
Vegetables:		
Bean sprouts	½ cup	98
Black-eyed peas	½ cup	58
Spinach, cooked	½ cup	48
Lima beans	½ cup	32
Milk and milk products:		
Milk	1 cup	30
Cheddar cheese	1 oz	8
American cheese	1 oz	6
Meats:		
Chicken	3 oz	25
Beef	3 oz	20
Pork	3 oz	20

(continued)

Table 15.10 **(continued)**

Calciumᵃ

Food	Amount	Calcium (mg)
Milk and milk products:		
Yogurt, low-fat	1 cup	413
Milk shake (low-fat frozen yogurt)	1¼ cup	352
Yogurt with fruit, low-fat	1 cup	315
Skim milk	1 cup	301
1% milk	1 cup	300
2% milk	1 cup	298
3.25% milk (whole)	1 cup	288
Swiss cheese	1 oz	270
Milk shake (whole milk)	1¼ cup	250
Frozen yogurt, low-fat	1 cup	248
Frappuccino	1 cup	220
Cheddar cheese	1 oz	204
Frozen yogurt	1 cup	200
Cream soup	1 cup	186
Pudding	½ cup	185
Ice cream	1 cup	180
Ice milk	1 cup	180
American cheese	1 oz	175
Custard	½ cup	150
Cottage cheese	½ cup	70
Cottage cheese, low-fat	½ cup	69
Vegetables:		
Spinach, cooked	½ cup	122
Kale	½ cup	47
Broccoli	½ cup	36
Legumes:		
Tofu	½ cup	260
Dried beans, cooked	½ cup	60
Foods fortified with calcium:		
Orange juice	1 cup	350
Frozen waffles	2	300
Soy milk	1 cup	200–400
Breakfast cereals	1 cup	150–1000

Selenium

Food	Amount	Selenium (mcg)
Seafood:		
Lobster	3 oz	66
Tuna	3 oz	60
Shrimp	3 oz	54
Oysters	3 oz	48
Fish	3 oz	40
Meats/Eggs:		
Liver	3 oz	56
Egg	1 medium	37
Ham	3 oz	29
Beef	3 oz	22
Bacon	3 oz	21
Chicken	3 oz	18
Lamb	3 oz	14
Veal	3 oz	10

ᵃActually, the richest source of calcium is alligator meat; 3½ ounces contain about 1,231 milligrams of calcium, but just try to find it on your grocer's shelf!

Zinc

Food	Amount	Zinc (mg)
Meats:		
Liver	3 oz	4.6
Beef	3 oz	4.0
Crab	½ cup	3.5
Lamb	3 oz	3.5
Turkey ham	3 oz	2.5
Pork	3 oz	2.4
Chicken	3 oz	2.0
Legumes:		
Dried beans, cooked	½ cup	1.0
Split peas, cooked	½ cup	0.9
Grains:		
Breakfast cereal, fortified	1 cup	1.5–4.0
Wheat germ	2 Tbsp	2.4
Oatmeal, cooked	1 cup	1.2
Bran flakes	1 cup	1.0
Brown rice, cooked	½ cup	0.6
White rice	½ cup	0.4
Nuts and seeds:		
Pecans	¼ cup	2.0
Cashews	¼ cup	1.8
Sunflower seeds	¼ cup	1.7
Peanut butter	2 Tbsp	0.9
Milk and milk products:		
Cheddar cheese	1 oz	1.1
Whole milk	1 cup	0.9
American cheese	1 oz	0.8

Sodium

Food	Amount	Sodium (mg)
Miscellaneous:		
Salt	1 Tbsp	2,132
Dill pickle	1 (4½ oz)	1,930
Sea salt	1 Tbsp	1,716
Ravioli, canned	1 cup	1,065
Spaghetti with sauce, canned	1 cup	955
Baking soda	1 tsp	821
Beef broth	1 cup	810
Chicken broth	1 cup	770
Gravy	¼ cup	720
Italian dressing	2 Tbsp	720
Pretzels	5 (1 oz)	500
Green olives	5	465
Pizza with cheese	1 wedge	455
Soy sauce	1 tsp	444
Cheese twists	1 cup	329
Bacon	3 slices	303
French dressing	2 Tbsp	220
Potato chips	1 oz (10 pieces)	200
Catsup	1 Tbsp	155

(continued)

Table 15.10 **(continued)**

Sodium		
Food	**Amount**	**Sodium (mg)**
Meats:		
Corned beef	3 oz	808
Ham	3 oz	800
Fish, canned	3 oz	735
Meat loaf	3 oz	555
Sausage	3 oz	483
Hot dog	1	477
Fish, smoked	3 oz	444
Bologna	1 oz	370
Milk and milk products:		
Cream soup	1 cup	1,070
Cottage cheese	½ cup	455
American cheese	1 oz	405
Cheese spread	1 oz	274
Parmesan cheese	1 oz	247
Gouda cheese	1 oz	232
Cheddar cheese	1 oz	175
Skim milk	1 cup	125
Whole milk	1 cup	120
Grains:		
Bran flakes	1 cup	363
Cornflakes	1 cup	325
Croissant	1 medium	270
Bagel	1	260
English muffin	1	203
White bread	1 slice	130
Whole wheat bread	1 slice	130
Saltine crackers	4 squares	125

Iron		
Food	**Amount**	**Iron (mg)**
Meat and meat alternates:		
Liver	3 oz	7.5
Round steak	3 oz	3.0
Hamburger, lean	3 oz	3.0
Baked beans	½ cup	3.0
Pork	3 oz	2.7
White beans	½ cup	2.7
Soybeans	½ cup	2.5
Pork and beans	½ cup	2.3
Fish	3 oz	1.0
Chicken	3 oz	1.0
Grains:		
Breakfast cereal, iron-fortified	1 cup	8.0 (4–18)
Oatmeal, fortified, cooked	1 cup	8.0
Bagel	1	1.7
English muffin	1	1.6
Rye bread	1 slice	1.0
Whole-wheat bread	1 slice	0.8
White bread	1 slice	0.6
Fruits:		
Prune juice	1 cup	9.0
Apricots, dried	½ cup	2.5
Prunes	5 medium	2.0
Raisins	¼ cup	1.3
Plums	3 medium	1.1

Iron

Food	Amount	Iron (mg)
Vegetables:		
Spinach, cooked	½ cup	2.3
Lima beans	½ cup	2.2
Black-eyed peas	½ cup	1.7
Peas	½ cup	1.6
Asparagus	½ cup	1.5

Phosphorus

Food	Amount	Phosphorus (mg)
Milk and milk products:		
Yogurt	1 cup	327
Skim milk	1 cup	250
Whole milk	1 cup	250
Cottage cheese	½ cup	150
American cheese	1 oz	130
Meats:		
Pork	3 oz	275
Hamburger	3 oz	165
Tuna	3 oz	162
Lobster	3 oz	125
Chicken	3 oz	120
Nuts and seeds:		
Sunflower seeds	¼ cup	319
Peanuts	¼ cup	141
Pine nuts	¼ cup	106
Peanut butter	1 Tbsp	61
Grains:		
Bran flakes	1 cup	180
Shredded wheat	2 large biscuits	81
Whole-wheat bread	1 slice	52
Noodles, cooked	½ cup	47
Rice, cooked	½ cup	29
White bread	1 slice	24
Vegetables:		
Potato	1 medium	101
Corn	½ cup	73
Peas	½ cup	70
French fries	½ cup	61
Broccoli	½ cup	54
Other:		
Milk chocolate	1 oz	66
Cola	12 oz	51
Diet cola	12 oz	45

Potassium

Food	Amount	Potassium (mg)
Vegetables:		
Potato	1 medium	780
Winter squash	½ cup	327
Tomato	1 medium	300
Celery	1 stalk	270
Carrots	1 medium	245
Broccoli	½ cup	205

(continued)

Table 15.10 (continued)

Potassium

Food	Amount	Potassium (mg)
Fruits:		
Avocado	½ medium	680
Orange juice	1 cup	469
Banana	1 medium	440
Raisins	¼ cup	370
Prunes	4 large	300
Watermelon	1 cup	158
Meats:		
Fish	3 oz	500
Hamburger	3 oz	480
Lamb	3 oz	382
Pork	3 oz	335
Chicken	3 oz	208
Grains:		
Bran buds	1 cup	1,080
Bran flakes	1 cup	248
Raisin bran	1 cup	242
Wheat flakes	1 cup	96
Milk and milk products:		
Yogurt	1 cup	531
Skim milk	1 cup	400
Whole milk	1 cup	370
Other:		
Salt substitutes	1 tsp	1,300–2,378

Fluoride

Food	Amount	Fluoride (mcg)
Grape juice, white	6 oz	350
Instant tea	1 cup	335
Raisins	3½ oz	234
Wine, white	3½ oz	202
Wine, red	3½ oz	105
French fries, McDonald's	1 medium	130
Dannon's Fluoride to Go	8 oz	178
Tap water, U.S.	8 oz	59
Municipal water, U.S.	8 oz	186
Bottled water, store brand	8 oz	37

Iodine

Food	Amount	Iodine (mcg)
Seaweed	Sheet, 1 g	16–2,984
Iodized salt	1 tsp	400
Haddock	3 oz	125
Cod	3 oz	99
Yogurt, low-fat	1 cup	75
Milk, low-fat	1 cup	56
Cottage cheese	½ cup	50
Bread	2 oz (2 slices)	45
Shrimp	3 oz	30
Macaroni, cooked	1 cup	27
Egg	1	22
Cheddar cheese	1 oz	17

Mediablitzimages/Whitebox Media/Alamy

NUTRITION

up close

The Salt of Your Diet

Focal Point: Are you consuming too much sodium from processed foods?

Below is a list of commonly consumed processed foods and their content of sodium. On the blank line in front of the foods write down how many times a day you eat the food, or another food that is very similar to it. (For example, if you eat sausage pizza, check cheese pizza.) Then answer this question: Are you most likely to be consuming more than 2,300 mg of sodium daily (the tolerable upper intake level) just from processed foods?

	Sodium content (mg)
Biscuits, bread, crackers	
_____biscuit, refrigerated, 1	340–604
_____bread, 1 slice	215–380
_____cheese crackers, 30 (1 oz)	250
_____wheat crackers, 7 (1 oz)	180
_____thin wheat crackers, 16 (1 oz)	260
Cheese	
_____American cheese, 1 slice	260
_____cheddar cheese, 1 oz	180
_____cottage cheese, ½ cup	390–420
Frozen foods	
_____fried chicken dinner with	
_____mashed potatoes, corn, 10 oz	790–1,300
_____macaroni and cheese, 10 oz	630–1,100
_____chicken nuggets, 5 pieces	490–650
_____cheese pizza, 2 slices	530–1,090
_____pot pie, 1 pie	770–1,040
_____taquitos, 5	370–40

	Sodium content (mg)
Meats	
_____bacon, 1 slice	210–330
_____ham, 2 oz	460–620
_____hot dog, 1	450–740
_____breakfast sausage, 3 links	450–550
_____turkey, deli, 2 oz	250–620
Other foods	
_____salad dressing, 2 Tbsp	170–1,067
_____catsup, 1 Tbsp	150–190
_____salsa, 2 Tbsp	100–800
_____spaghetti sauce, ½ cup	330–650
_____steak sauce, 1 Tbsp	170–300
_____soy sauce, 1 Tbsp	920–1,160
_____potato chips, 1 oz (20 chips)	120–211
_____soup, canned, 1 cup	340–1,100
Fast foods	
_____bacon, egg, and cheese biscuit, 1	1,230
_____sausage biscuit, 1	1,160–1,360
_____sausage and egg biscuit, 1	1,200
_____french fries, 1 medium	270–960
_____cheeseburger, 1 regular	750–970
_____hamburger, 1 regular	490–560

Feedback to the Nutrition Up Close is located in Appendix F.

REVIEW QUESTIONS

- **Identify key functions and food sources of five essential minerals.**

1. There are 40 essential minerals that perform specific functions in the body. Each must be obtained from the diet. True/False

2. Essential minerals are a structural component of teeth; they also serve as a source of electrical energy that stimulates muscles to contract and help maintain an appropriate balance of fluids in the body. True/False

3. The amount of certain minerals absorbed from foods varies a good deal based on the food source of the minerals. True/False

4. Calcium is the most abundant mineral found in bones, but other minerals such as magnesium, phosphorus, and carbon are also components of bone. True/False

5. Bones fully develop and mineralize during the first 18 years of life. True/False

6. Iodine deficiency and overdose are related to goiter and thyroid disease.   True/False

7. Wilson's disease is related to magnesium deficiency.   True/False

8. Vitamin D is required for the absorption and utilization of calcium by the body.   True/False

9. People with iron deficiency generally feel weak and tired, have a poor appetite, and are susceptible to infection.   True/False

10. Although iron deficiency is an important health problem in America, iron toxicity is not.   True/False

11. Blood pressure tends to increase as salt (or sodium) intake increases.   True/False

12. The major source of salt (or sodium) in our diet comes from salt added to food at the table.   True/False

Pablo decides to lose 5 pounds in a week by going on a high-protein diet. The foods he'll eat consist of canned tuna, lean beef, skinless chicken, low-fat cottage cheese, and skim milk.

13. _____ Of the minerals listed below, which one is most likely to be lacking in his diet?

   a. iron
   b. calcium
   c. magnesium
   d. phosphorus

The next four questions refer to the following scenario.

Dorthea was just told by her adult nurse practitioner that she has prehypertension and recommends that she meet with a registered dietitian for consultation on the DASH eating plan.

The appointment was made and Dorthea is speaking with the registered dietitian now.

14. _____ During the consultation, Dorthea will learn that the DASH eating pattern emphasizes:

   a. foods very low in sodium and high in fiber
   b. elimination of fast foods
   c. low-calorie, low-fat foods
   d. vegetables, fruits, and grain products

15. _____ Dorthea asks the dietitian how many vegetables a day are included in the DASH eating plan. The likely response from the dietitian is:

   a. 1–2 servings
   b. 2–3 servings
   c. 3–4 servings
   d. 4–5 servings

16. _____ How many servings of fruits would Dorthea be urged to include in her diet under the DASH eating plan?

   a. 1–2 servings
   b. 2–3 servings
   c. 3–4 servings
   d. 4–5 servings

17. _____ What other types of food would Dorthea be encouraged to include in her DASH eating plan?

   a. lean meats, fish, and poultry
   b. milk products of all types
   c. coffee and tea
   d. chocolate and molasses

Answers to these questions can be found in Appendix F.

NUTRITION SCOREBOARD ANSWERS

1. Minerals serve as structural components of the body, but they also play important roles in stimulating muscle and nerve activity and in other functions.   False

2. Bones continue to grow and mineralize well after we reach adult height. Bone growth and development continues to about age 30.   True

3. Spinach is a nutritious food, providing 0.6 mg iron per ounce. It contains less iron than beef (1 mg iron per ounce). The iron in spinach is poorly absorbed, but the iron in meat is well absorbed by the body.   False

4. Iron deficiency is the most common nutritional deficiency in both developed and developing countries. Approximately one-fourth of the world's population is iron deficient.[1]   True

5. Approximately one-third of American adults have hypertension (high blood pressure).[2]   True

Dietary Supplements

NUTRITION SCOREBOARD

1 Products classified as *dietary supplements* consist of herbs and vitamin and mineral supplements only.   True/False

2 Dietary supplements must be tested for safety and effectiveness before they can be sold.   True/False

3 Herbal remedies have been used for over 1,000 years in Germany, so they must be safe and effective. True/False

4 *Probiotics* are "friendly bacteria" that benefit health.   True/False

Answers can be found at the end of the chapter.

Photodisc

433

dietary supplements Any products intended to supplement the diet, including vitamin and mineral supplements; proteins, enzymes, and amino acids; fish oils and fatty acids; hormones and hormone precursors; herbs and other plant extracts; and prebiotics and probiotics. Such products must be labeled "Dietary Supplement."

Dietary Supplements

- **Make evidence-based decisions about the probable safety and effectiveness of specific dietary supplements.**

What do vitamin E supplements, amino acid pills, herbal remedies, and probiotics all have in common? They are members of the increasingly popular group of products called **dietary supplements**—and they are all discussed in this unit. Types of dietary supplements available to consumers are presented in Table 16.1. About half of U.S. adults use one or more of the over 54,000 dietary supplement products available on the market.[1] Dietary supplements are supposed to supplement the diet. They are not intended to prevent, treat, or cure disease. These disparate products are grouped together because they are regulated as food and not drugs.[2] You will learn from this unit that most dietary supplements have not been found to perform as claimed, and that they may have adverse, neutral, or beneficial effects on health. This "buyer beware" situation exists because of the loose rules that govern dietary supplements. You will also learn about exciting new developments related to intestinal fertilizers and friendly bacteria (no kidding).

> ## KEY NUTRITION CONCEPTS
>
> Foundational knowledge about nutrition that applies to the topic of dietary supplements is represented by the following three key nutrition concepts:
>
> 1. Nutrition Concept #1: Foods provide energy (calories), nutrients, and other substances needed for growth and health.
>
> 2. Nutrition Concept #4: Poor nutrition can result from both inadequate and excessive levels of nutrient intake.
>
> 3. Nutrition Concept #9: Adequacy, variety, and balance are key characteristics of healthful diets.

Regulation of Dietary Supplements

In 1994 Congress passed the Dietary Supplement Health and Education Act, which started the explosion in the availability of dietary supplements. Under the act, dietary supplements are minimally regulated by the Food and Drug Administration (FDA); they do not have to be tested prior to marketing or shown to be safe or effective.[2] Although often advertised to relieve certain ailments, they are not considered to be drugs. Dietary supplements are not subjected to rigorous testing to prove safety and effectiveness prior to sale, as drugs must be. Responsibility for evaluating the safety of dietary supplements lies with manufacturers and not the FDA. Supplements are deemed unsafe only when the FDA has proof they are harmful. Since few (about 0.3%) dietary supplements have been adequately tested, and because results of studies showing negative effects may never see the light of day, it is difficult to prove them to be unsafe.[3,4] The FDA largely relies on reports of ill effects from manufacturers, health professionals, and consumers to assess supplement safety. Since 1994, the FDA has received thousands of reports of adverse effects of supplements (primarily for herbal remedies), including several hundred deaths.[5] The FDA has taken action against hundreds of products because they contained hazardous drugs or ingredients harmful to health, or because they made false claims for products such as "helps replace medicine in the treatment of diabetes," and "cellulite fighter."[6] The actions were taken primarily against products labeled for use in sexual enhancement, body building, and weight loss.[7]

According to FDA regulations (Table 16.2), dietary supplements must be labeled with a Supplemental Facts panel that lists serving size, ingredients, and percent Daily Value (% DV) of key nutrients, as well as a listing of ingredients.[8] Products can be labeled with a health claim, such as "high in calcium" or "low fat," if the product qualifies according to

Table 16.1 Types of dietary supplements

Type	Example
1. Vitamins and minerals	Vitamins C and E, selenium
2. Herbs (botanicals)	Ginkgo, ginseng, St. John's wort
3. Proteins and amino acids	Shark cartilage, chondroitin sulfate, creatine
4. Hormones, hormone precursors	DHEA, vitamin D
5. Fats	Fish oils, DHA, lecithin
6. Other plant extracts	Garlic capsules, fiber, cranberry concentrate, echinacea
7. Prebiotics, probiotics	Psyllium, garlic, certain live bacteria

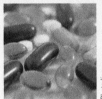   

Vitamin, mineral, protein, and amino acid supplements.

Botanical supplements, such as dong quai, often come in gel caps.

Plant extracts can also be taken in liquid form, as tinctures.

the nutrition labeling regulations. Supplements can also be labeled with structure/function claims, but these claims cannot refer to disease prevention or treatment effects. Claims such as "improves circulation," "helps prevents wrinkles," "supports the immune system," and "helps maintain mental health" can be used, whereas "prevents heart disease" or "cures depression" cannot be. If a function claim is made on the label or package inserts, the label or insert must include the FDA disclaimer, which states that the FDA does not support the claim. (This is done to reduce the FDA's liability for problems that may be caused by supplements.) Nonetheless, many people believe the health claims made for supplements. Table 16.3 provides a summary of common consumer beliefs about dietary supplements and then lists the realities of their safety and effectiveness.

Table 16.2 FDA regulations for dietary supplement labeling[8]

1. Product must be labeled "Dietary Supplement."

2. Product must have a Supplemental Facts label that includes serving size, amount of the product per serving, % Daily Value of key nutrients of public health significance, a list of other ingredients, and the manufacturer's name and address.

3. Nutrient claims (such as "low in sodium" and "high in fiber") can be made on labels of products that qualify based on nutrition labeling regulations.

4. A listing of ingredients in the supplement.

5. Structure/function claims about how the product affects normal body structures (such as "helps maintain strong bones") or functions ("enhances normal bowel function") can be made on product labels. If a structure/function claim is made, this FDA disclaimer must appear:

This statement has not been evaluated by the FDA. This product is not intended to diagnose, treat, cure, or prevent any disease.

Nutrition Facts
Serving size 1 Tablet

Amount Per Serving	% DV
Melatonin 3 mg	*

*Daily Value (DV) not established

Other Ingredients: Dicalcium Phosphate, Cellulose (Plant Origin), Vegetable Stearic Acid, Vegetable Magnesium Stearate, Silica, Croscarmellose.

GUARANTEED FREE OF: wheat, yeast, soy, corn, sugar, starch, milk, eggs. No artificial colors, flavors. No chemical additives. No preservatives. No animal derivatives.

Directions: As a dietary supplement for adults, take one (1) tablet, under the direction of a physician, only at bedtime as Melatonin may produce drowsiness. **DO NOT EXCEED 3 MG IN A 24 HOUR PERIOD.**

Warning: For Adults. Use only at bedtime. This product is not to be taken by pregnant or lactating women. If you are taking medication or have a medical condition such as an auto-immune condition or a depressive disorder, consult your physician before using this product. **NOT FOR USE BY CHILDREN 16 YEARS OF AGE OR YOUNGER.** Do not take this product when driving a motor vehicle, operating machinery or consuming alcoholic beverages.

In case of accidental overdose, seek professional assistance or contact a Poison Control Center immediately.

KEEP OUT OF REACH OF CHILDREN

Table 16.3 Consumer beliefs about dietary supplements versus reality

Common consumer beliefs[4,6,43]
Consumers tend to believe dietary supplements:
• Are not drugs.
• Have fewer side effects than prescription drugs.
• Have accurate health claims.
• Are approved for use by the FDA.
• Will improve health and help maintain health.
• Are safe, high quality, and effective.
• May replace conventional medicines and cost associated with health care.

Dietary supplement realities[2,10,17,44–47]
• FDA does not approve, test, or regulate the manufacture or sale of dietary supplements.
• The FDA has limited power to keep potentially harmful dietary supplements off the market.
• Dietary supplements generally are not tested for safety or effectiveness before they are sold.
• Dietary supplements often do not list side effects, warnings, or drug or food interactions on product labels.
• Ingredients listed on dietary supplement labels often do not include all ingredients.
• Dietary supplements generally do not relieve problems or promote health and performance as advertised.
• Some dietary supplements contain potentially harmful or banned substances.

The Federal Trade Commission (FTC) regulates claims for dietary supplements made in print and broadcast advertisements, including direct marketing, websites, infomercials, and mass e-mails. Claims made for dietary supplements in advertisements are supposed to be truthful, but often are not. Although some companies have been prosecuted for making false and misleading claims, neither the FDA nor the FTC has sufficient resources to fully monitor products and enforce laws related to dietary supplements. Rules and regulations related to dietary supplements are expected to become more stringent, and enforcement actions are being expanded in the United States and Canada.[5] The FDA has implemented an "Adverse Events Reporting System" website (www.fda.gov/Safety/MedWatch/HowTo Report/ucm053074.htm) that simplifies recording and tracking of adverse effects of dietary supplements.

Some dietary supplements are safe and effective, and may help people maintain their own health. Unfortunately, the for-profit-only manufacture and sale of dietary supplements taints products with suspicion.

Vitamin and Mineral Supplements: Enough Is as Good as a Feast

Multivitamin and mineral supplements—such as vitamins E or C, calcium, or magnesium—are among the wide variety of vitamin and mineral supplements used by consumers. They represent the most popular type of dietary supplement, and 33% of Americans report using them mostly on a daily basis.[9]

Supplements can have positive effects on health, as shown by the examples given in Table 16.4. This table lists examples of conditions for which specific vitamin and/or mineral supplements have been shown to be effective. Overall, a vitamin or mineral supplement is indicated when a deficiency exists, or when use is related to improved health among individuals at risk of developing a deficiency due to a particular disease or condition.[10]

Table 16.4 Who may benefit from vitamin and mineral supplements? Here are some examples:[48,49]

- People with diagnosed vitamin and/or mineral deficiency diseases
- Newborns (vitamin K)
- People living in areas without a fluoridated water supply (fluoride)
- Vegans (vitamins B_{12} and D)
- People experiencing blood loss (iron)
- People at risk for osteoporosis due to low calcium intake and poor vitamin D status (calcium, vitamin D)
- Individuals with impaired absorption processes (B_{12})

People often take supplements as a sort of insurance policy against problems caused by poor diets. Although multivitamin and mineral supplements can help fill in some of the nutrient gaps caused by poor food habits or low food intake, they can't turn a poor dietary pattern into a healthful one. Whether a dietary pattern is healthful or not is determined by more than its vitamin and mineral content. The healthfulness of a dietary pattern also depends on its content of essential fatty acids, protein, fiber, water, and other nutrients and phytochemicals from food.

One of the most serious consequences of supplements results when they are used as a remedy for health problems that can be treated, but not by vitamins or minerals. For example, vitamin and mineral supplements have not been found to prevent or treat heart disease, cancer, diabetes, hypertension, premature death, behavioral problems, sexual dysfunction, hair loss, autism, chronic fatigue syndrome, obesity, cataracts, or stress.[10-13] Some vitamin supplements, such as vitamin E, vitamin C, beta-carotene, and calcium, can be harmful to certain groups of people, especially in high doses.[10,11]

The Rational Use of Vitamin and Mineral Supplements Like all medications, vitamin and mineral supplements should be taken only if medically indicated. If they are taken, dosages should not be excessive.

Herbal Remedies

Herbal remedies (also called *botanicals*) have been used in traditional medicine in China and India for over 5,000 years.[14] Discovery of the properties of these plant-based substances, and subsequent studies on specific chemical compounds contained in some of them, led to the development of about half of the drugs now used to treat diseases and disorders. Many people assume that herbal remedies are safe because they are natural components of plants and have been used in traditional medicine for a long time, but not all plants are safe: poison ivy, oleander, and mistletoe berries are toxic, for example. Modern medicine has developed safe and effective drugs that have decreased the historic reliance on herbal remedies.[14,15]

Approximately 20% of U.S. adults use herbal dietary supplements each year.[14] The herb pharmacopoeia includes over 550 primary herbs known by at least 1,800 names (Illustration 16.1). Plant products known to effectively treat disease are considered drugs, and those that have not passed the scientific tests needed to demonstrate safety and effectiveness in disease treatment are often called herbs. Yet the truth is that many products sold as herbal remedies in the United States have drug-like effects on body functions

Herbal remedies Extracts or other preparations made from ingredients of plants intended to prevent, alleviate, or treat disease, or to promote health.

Illustration 16.1 Examples of plants used in the formulation of herbal supplements.

©Melpomene/Shutterstock.com

Table 16.5 Effectiveness and safety of a sampling of top-selling herbal remedies[a,14,41,50]

Herbal remedy	Effectiveness	Safety Concerns
Echinacea	May diminish upper respiratory infection	People allergic to ragweed, who have an autoimmune disorder, or are on drugs that affect liver function should not use this herb.
Garlic	May decrease blood pressure to a small extent	May decrease blood clotting, interact with blood thinners.
Ginkgo biloba	May improve cognitive and social function in people with Alzheimer's disease	Seeds are unsafe. Acts as an anticoagulant, should not be used by people with bleeding or seizure disorders.
St. John's wort	May relieve mild to moderate depression	Due to multiple interactions, should be used under medical supervision.
Ginger	May reduce motion sickness, nausea, and vomiting	May cause gastrointestinal upset, prolonged bleeding.

[a]Effects vary based on dose, purity, type, and duration of use of the herbal remedy. Safety and effectiveness have not consistently been tested in pregnant or breastfeeding women, or in infants and children. Herbal remedies are not considered "safe" until safety has been demonstrated.

but have not passed safety or effectiveness tests. Herbal products are regulated as dietary supplements and do not have to be scientifically demonstrated to be safe or effective prior to being sold.[14]

Effectiveness and Safety of Herbal Remedies Likely effects and identified adverse effects of five top-selling herbal products are listed in Table 16.5. Herbal remedies, like drugs, have biologically active ingredients that can have positive, negative, or neutral effects on body processes. Basically, an herbal remedy is considered valuable if it has beneficial effects on health and is safe. Knowledge of the risks and benefits of many herbal supplements remains incomplete. However, available evidence suggests that some herbal remedies are safe and effective, while others appear to be neither.

Which herbal remedies are likely ineffective or unsafe? Human studies with various herbal remedies have helped identify herbs and other dietary supplements that lack beneficial effects or have adverse side effects. Table 16.6 lists some of these products. The extent

Table 16.6 Examples of dietary supplements that may not be effective or safe[16,41,51]

Apricot pits (laetrile)	Ephedra	Pennyroyal
Androstenedione (Andro)	Eyebright	Pokeroot
Aristolochic acid	Ginkgo seed	Hoodia
Sassafras	Saw palmetto	Shark cartilage
Belladonna	Kava	Skullcap
Black cohosh	Blue cohosh	Licorice root
Star anise	Bitter orange	Liferoot
Vinca	Borage	Lily of the valley
Wild yam	Broom	Lobelia
Wormwood	Chaparral	Yohimbe
Chinese yew	Mandrake	Willow bark
Comfrey	Mistletoe	
Dong quai	Organ/glandular extracts	

to which the herbs included in the table pose a risk to health depends on the amount taken and the duration of use, the age and health status of the user, and other factors. Not everyone reacts the same way to different herbal supplements.[16]

Quality of Herbal Products Many herbal products available on the market are of poor quality. Some of the products have been found to contain ingredients other than those declared on the label, including banned drugs, and some contain contaminants such as bacteria, mold, mercury, and lead.[7,17] Analyses of the composition of 25 ginseng products, for example, found that concentrations of ginseng compounds in the supplements were up to 36 times different than labeled amounts.[18] High levels of contamination in herbal products were identified by another study. Of 260 Asian herbal products examined, 25% were contaminated with high levels of heavy metals, and 7% contained undeclared drugs purposefully and illegally added to produce a desired effect.[19] Studies of echinacea products found that 10% of samples contained no echinacea, and only half contained the labeled amount.[20] Male enhancement supplements are among the most common products recalled from the market due to contamination or because they contain unlisted prescription drugs.[22] No herbal or "all natural" substance has been shown to cure impotence.[22] Dietary supplements labeled "pure," "natural," or "quality assured" may or may not fit the description. You generally can't tell by the label.

There is no government body that monitors the contents of herbal supplements. Private groups, such as the U.S. Pharmacopeia (USP) and Consumer Laboratories (CL), offer testing services to ensure that herbal and other dietary supplements meet standards for disintegration, purity, potency, and labeling.[23,24] Products that pass these tests can display a USP logo or the CL symbol on product labels (Illustration 16.2). These symbols represent quality ingredients and labeling but do not address product safety or effectiveness.[25] Considerations for the use of dietary supplements are summarized in this unit's Health Action feature.

Due to the lack of studies and potential dangers, the FDA has advised dietary supplement manufacturers not to make claims related to pregnancy for herbs and other products, and to label products truthfully based on scientific evidence.[26,27]

Prebiotics and Probiotics

The digestive tract, particularly the colon, is home to over 500 species of microorganisms representing 100 trillion bacteria (and billions of viruses and fungi, as well). Some species of bacteria such as *E. coli* 0157:H7 and *Salmonella* may cause disease, whereas others

Illustration 16.2 These symbols on the labels of dietary supplements certify quality ingredients and accurate labeling but do not address product safety or effectiveness.[53]

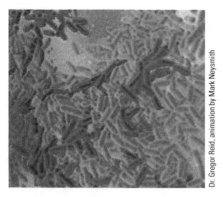

Illustration 16.3 A *Lactobacillus* species (blue) taking over harmful *E. coli* bacteria (red).

such as *Lactobacillus* and bifidobacteria (Illustration 16.3) help prevent disease through a variety of mechanisms.[28] The beneficial effects of the right species of microorganisms, and ways to increase their presence in the gut, are subjects of intense research and expanded knowledge. Knowledge gains center around the actions of **prebiotics** and **probiotics**. Pre- and probiotics are regulated by the same rules that govern other dietary supplements.[29]

Prebiotics are non-digestible dietary fibers that can be used as a food source by beneficial microorganisms (mainly bacteria) in the small intestine and colon. The breakdown products from microorganism digestion of prebiotics are released into the gut, where they foster the growth of beneficial bacteria (which is why prebiotics are referred to as "intestinal fertilizer") and diminish the population and effects of harmful microorganisms.[30,31] Probiotics is the term for live, beneficial—or "friendly"—bacteria that enter food through fermentation and aging processes and are resistant to digestion.[32] The term **synbiotics** is being used to classify combinations of prebiotics and probiotics. Table 16.7 lists food and other sources of pre- and probiotics. Many of the foods listed are central to a healthy dietary pattern.

Prebiotics, probiotics, and synbiotics have been found to benefit health by:

- Increasing the mass of beneficial, and decreasing the mass of harmful, microorganisms in the small intestines and colon, thereby limiting the effects of disease-causing microorganisms in the gut

- Decreasing insulin resistance and chronic inflammation by altering gene expression in microorganisms in ways that enhance glucose utilization

- Enhancing the immune functions of the small intestine and colon

- Decreasing the symptoms of irritable bowel syndrome and infantile colic

prebiotics Non-digestible carbohydrates (various types of dietary fiber) that serve as food for and promote the growth of beneficial microorganisms in the small intestine and colon. Also called "intestinal fertilizer."

probiotics Live microorganisms which resist digestion, and when administered in adequate amounts of appropriate strains, confer health benefits to the host. Strains of *Lactobacillus* (lac-toe-bah-sil-us) and bifidobacteria (bif-id-dough bacteria) are the best-known probiotics. Also called "friendly bacteria."

synbiotics Combinations of prebiotics and probiotics that interact in ways that generally benefit both and the health of the host.

Table 16.7 Food and other sources of prebiotics and probiotics[31,32,52]

Prebiotics	Probiotics
Jerusalem artichokes	Yogurt with live culture
Wheat	Buttermilk
Barley	Kefir
Rye	Cottage cheese
Onions	Dairy products with added probiotics
Garlic	Soft cheeses
Leeks	Soy sauce
Prebiotics tablets and powders and nutritional beverages	Tempeh
Breastmilk	Sauerkraut
Psyllium	Miso soup
Onions	Probiotic powders, pills
Garlic	
Leeks	
Oats	
Banana	
Tomatoes	
Dried beans	

- Enhancing the absorption of minerals such as calcium, magnesium, and iron

- Decreasing the symptoms and onset of vaginal and urinary tract infections

- Delaying the onset of allergy development in children

- Decreasing the duration of infection- and antibiotic use–related diarrhea

- Increasing stool bulk and reducing constipation[31,33–39]

Probiotics are considered safe and beneficial for healthy people but may not be appropriate for individuals with compromised immune status or who are critically ill. Table 16.8 reviews practical considerations for the use of probiotics.

Final Thoughts

From dietary supplements to friendly bacteria, the universe of substances considered dietary ingredients is expanding. Knowledge about potential benefits of pre- and probiotics is charging ahead, and advances are catching the attention of consumers and health care professionals. Perhaps you never thought that "intestinal fertilizer" or "friendly bacteria" would ever intentionally pass through your lips. But that may well be the nature of some dietary ingredients to come.

Table 16.8 Practical considerations related to the use of probiotics[35,40]

- Benefits to health are strain- and dose-specific.
- Products are not regulated by the FDA; quality, microorganism strains and doses, viability of the microorganisms, and purity vary considerably among products.
- Use may cause temporary gas and bloating.
- Effects last as long as the probiotic is consumed.
- Benefits are usually modest and should generally not replace conventional therapy.
- Individuals who are critically ill or have severe immune system disorders should not use them.

ANSWERS TO REALITY CHECK
Herbals on the Internet

Many people use the Internet as a source of information about illness remedies. Unfortunately, many of the sites selling herbal products illegally claim the products prevent or cure specific diseases as if they were real drugs and do not include the required FDA disclaimer and ingredient list.[5,42]

Take the worry out of decisions about herbal supplements. Check them out using scientifically reliable websites.

Sarah:

Pablo:

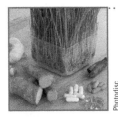
Photodisc

NUTRITION
up close

Focal Point: Decide if a dietary supplement is warranted in these situations.

People take dietary supplements for many reasons, but is their use justified? Apply the information from this chapter to determine if you agree with the decisions made in each of the following scenarios.

1. Martha works part time and takes a full load of classes. Like many college students, she is always on the go, often grabbing something quick to eat at fast food restaurants or skipping meals altogether. Nevertheless, Martha feels confident her health will not suffer, because she takes a daily vitamin and mineral supplement.

   Is a supplement warranted in this case? Why or why not?

   _____

2. Sylvia is a 23-year-old student diagnosed with iron-deficiency anemia. She has learned in her nutrition class that it is preferable to get vitamins and minerals from food instead of supplements. Therefore, instead of taking the iron pills her doctor has prescribed, Sylvia has decided to counteract the anemia by increasing her consumption of iron-rich foods.

   Is a supplement warranted in this case? Why or why not?

   _____

3. John is a 21-year-old physical education major involved in collegiate sports. He is very aware that nutrition plays an important role in the way he feels, so he is careful to eat well-balanced meals. In addition, John takes megadoses of vitamins and minerals daily. He is convinced they enhance his physical performance.

   Is a supplement warranted in this case? Why or why not?

   _____

4. Roberto, a native Californian, is backpacking through Europe when he is slowed down by constipation. He visits a pharmacy where English is spoken and is given senna by the pharmacist. Roberto has never taken an herb before and is not sure how his body will react to it, or if it will work.

   Should Roberto try the senna, ask for a nonherbal drug, or take another action? (Assume they cost the same.) What's the rationale for this decision?

   _____

5. While shopping at the mall, Yuen notices a kiosk selling "Hypermetabolite," a weight-loss product that guarantees you'll lose 5 pounds a week without dieting. Having gained 10 pounds since she started working full time, Yuen decides to try it. Her examination of the product's label reveals that an ephedra derivative and Asian ginseng are major ingredients.

   Should Yuen take Hypermetabolite for weight loss? Why or why not?

   _____

Feedback to the Nutrition Up Close is located in Appendix F.

REVIEW QUESTIONS

- **Make evidence-based decisions about the probable safety and effectiveness of specific dietary supplements.**

1. Dietary supplements include vitamins, minerals, herbs, proteins, amino acids, and fish oil.   True/False

2. The FDA must approve dietary supplements before they are allowed to enter the market.   True/False

3. Dietary supplement labels must include a Supplemental Facts panel and may include qualifying nutrient claims and structure/function claims.   True/False

4. Say you read a dietary supplement label that claims the product "helps enhance muscle tone or size." Because the statement appears on the label, it must be true.   True/False

5. The Recommended Dietary Allowance (RDA) for vitamin A for breastfeeding women is 1,300 mcg per day. To make sure

you are getting enough vitamin A, you should select a supplement containing 3,300 mcg of vitamin A.   **True/False**

6. Specific vitamin or mineral supplements appear to benefit individuals with diagnosed vitamin or mineral deficiency, and some adults at risk of osteoporosis.   **True/False**

7. Vitamin and mineral supplements available over the counter can be safely consumed at any dose levels.   **True/False**

8. A USP label displayed on dietary supplements indicated the product has been tested for safety and effectiveness.   **True/False**

9. Prebiotics support the growth of beneficial bacteria in the colon.   **True/False**

10. Prebiotics have been demonstrated to benefit health under certain circumstances, but probiotics have not.   **True/False**

11. Prebiotics are non-digestible dietary fibers.   **True/False**

12. Probiotics are microorganisms that, when consumed, resist digestive processes, occupy the small intestine and colon in a live state, and benefit the host.   **True/False**

13. Herbal remedies, when demonstrated through scientific studies to be effective and safe, are generally considered to be drugs and regulated as such.   **True/False**

The following four questions refer to the following scenario.

While reading a men's magazine you notice a full-page ad for a new dietary supplement that promises to "ignite sexual well-being." Containing a natural aphrodisiac, the supplement is specifically designed by scientists to encourage healthy, meaningful, and long-term relationships. The supplements have been tested on movie stars and college students, and you have heard from social media contacts that they work. Ingredients include vitamins, taurine, caffeine, dong quai, ginkgo, and ginseng.

14. _____ How many of the herbal ingredients listed have been shown to act as an aphrodisiac?

   a. 0
   b. 1
   c. 2
   d. 3

15. _____ Which herbal ingredient contained in the product does not appear to be effective for any health condition or be safe to consume?

   a. dong quai
   b. ginkgo
   c. ginseng
   d. taurine

16. _____ This product:

   a. has been tested for purity, safety, and effectiveness by the FDA before it became available.
   b. has probably not been tested for purity, safety, and effectiveness by the FDA before it became available.
   c. will be examined by the manufacturer for safety and effectiveness while on the market.
   d. will improve people's chances of finding a long-term relationship.

17. _____ Assume your friend bought and took the product and thinks it really works. What is the most likely reason for that?

   a. The vitamins in the product worked.
   b. Ginseng improves personality and the potential for long-term relationships.
   c. The caffeine provides extra energy for interpersonal relationships.
   d. Your friend believed the product would be effective (placebo effect).

Answers to these questions can be found in Appendix F.

NUTRITION SCOREBOARD ANSWERS

1. Dietary supplements include herbs, vitamin and mineral supplements, protein powders, amino acid and enzyme pills, fish oils and fatty acids, hormone extracts, and other products.   **False**

2. Dietary supplements can be sold without proof of their safety or effectiveness.   **False**

3. Results of clinical trials, and not historical use, are the gold standard for determining the safety and effectiveness of herbal remedies and other dietary supplements.   **False**

4. Yes! There are healthful bacteria.   **True**

Life Cycle Nutrition

From beginning to end, the human life cycle is a process of continuous change. Birth, growth, maturation, aging, and death are all part of the natural progression of life. With each stage, our bodies change in size, proportion, and composition. Because of this, nutritional requirements vary enormously. For example, energy needs during periods of growth differ vastly from those associated with the later stages of life, when the body is in a state of physical maintenance or decline.

Regardless of where you are in the life cycle, an appropriate diet is essential to good health. In fact, nutritional status early in life can influence health at later stages. Therefore, it is important to remember that the food choices you make today may have far greater consequences on your long-term health than you might think. This chapter surveys the influence of growth, development, and aging on nutritional requirements across the human life cycle. The continuum of life encompasses infancy, childhood, adolescence, adulthood, and, for women, the special life stages of pregnancy and lactation.

Backyard Harvest

Each of us should ponder the personal steps that we can take toward making our community a better place. Whether it is by contributing to that community locally or at the global level, it is important for every person to be involved. You may be wondering how the actions of one person can make a difference. Does it sound impossible? Is your life already too busy? Don't tell that to Amy Grey. Amy's motto is "one garden at a time," which is how Backyard Harvest® got its start. The idea for Backyard Harvest came when Amy accidentally grew more than 200 heads of lettuce in her first attempt at vegetable gardening. Not wanting the lettuce to go to waste, she brought some of her bounty to the local food bank. Amy was struck by the fact that her lettuce was the only fresh produce available and that only canned and processed goods filled the shelves. It was this experience that spun the idea for Backyard Harvest—what if other gardeners shared her leafy predicament? Amy approached the local environmental group with her idea of connecting local gardeners with food banks that serve families, senior meal programs, and other community members. Soon, Backyard Harvest became a reality.

Amy's mission was to develop a program that prevented excess produce from going to waste, instead making it available to people with limited access to affordable fresh fruit and vegetables. Word spread throughout the community, and volunteers were soon busy harvesting and collecting excess produce grown by local gardeners. Gleaning events were scheduled to pick fruit from trees, bushes, and vines. After a successful pilot season, the scope of the project grew. What started as a simple goal of collecting and distributing 1,000 pounds of food quickly exceeded expectations. Backyard Harvest has distributed literally tons of locally grown fresh produce to community assistance programs, food banks, and meal sites, which in turn serve thousands of low-income families and seniors. In addition to fresh produce, recipients also receive comprehensive information on how to prepare fruits and vegetables and preserve them for later

use. As one woman said, knowing that the fruit and vegetables were coming from the Backyard Harvest project makes her "feel loved." Amy feels the same way.

This simple act of sharing food can happen in your community, too. All it takes is one person with a commitment to make a difference. Watching Amy take action in this small community has motivated others to become involved. For example, a joint initiative between Backyard Harvest, the University of Idaho, and the Moscow (Idaho) Food Co-op now provides education and outreach opportunities for community members regardless of income level. This includes gardening and food preservation workshops, a growers' market, and gardening and nutrition field trips for local elementary schools. For more information about Amy Grey and Backyard Harvest, go to http://www.backyardharvest.org.

© Courtesy of Amy Grey, Backyard Harvest

Critical Thinking: Amy's Story

What type of organization would you start or join if you wanted to improve the nutritional health of people in your community? In which phase or phases of the life cycle do you think people would most benefit from an organization such as Backyard Harvest?

What Physiological Changes Take Place during the Human Life Cycle?

Cells form, mature, carry out specific functions, die, and are replaced by new cells. In many ways, the life cycle of cells mirrors our own lives. That is, after a new human is conceived and born, the next 70 to 90 years are characterized by periods of growth and development, maintenance, reproduction, physical decline, and eventually death. Our ability to reproduce enables us to pass our genetic material (DNA) on to the next generation.

GROWTH AND DEVELOPMENT TAKE PLACE AT VARIOUS TIMES DURING THE LIFE CYCLE

Growth and development generally take place in a predictable and orderly manner. These important physiological events take place throughout the life cycle. Whereas **growth** refers to physical changes that result from either an increase in cell size or number, **development** is the attainment or progression of a skill or capacity to function. Knowing more about growth and developmental milestones will help you better understand why nutritional needs change throughout life.

Growth Patterns Growth may involve an increase in either the number of cells, called **hyperplasia,** or the size of cells, called **hypertrophy** (Figure 17.1). The highest rates of growth occur during infancy, childhood, adolescence, and pregnancy. The most useful and common way to assess growth is by measuring a person's height and weight. The World Health Organization and the U.S. National Center for Health Statistics (NCHS), which is part of the Centers for Disease Control and Prevention (CDC), have compiled height and weight reference standards into growth charts.[1] These charts indicate expected growth for well-nourished infants, children, and adolescents. Growth charts include percentile curves that represent

growth An increase in size and/or number of cells.

development Attainment or progression of a skill or capacity to function.

hyperplasia An increase in the number of cells.

hypertrophy An increase in the size of cells.

FIGURE 17.1 Types of Growth
Body size increases when the number and/or size of cells increase.

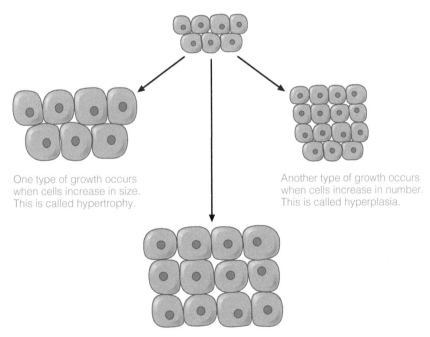

One type of growth occurs when cells increase in size. This is called hypertrophy.

Another type of growth occurs when cells increase in number. This is called hyperplasia.

Sometimes growth occurs by both hypertrophy and hyperplasia.

growth patterns from birth through 20 years of age. To evaluate adequacy of growth, a child's measurement (such as weight) is assessed in relation to age and sex. For example, if a 5-year-old girl's weight falls at the 60th percentile, 40% of healthy children of similar age and sex weigh more, and 60% weigh less. In this way, growth can be monitored over time and used as a general indicator of health throughout these important phases of the life cycle.

Development Follows a Predictable Pattern The assessment of a child's development is just as important as the assessment of his or her growth. Some of the major developmental accomplishments during the first year of life include such things as vocalization, facial expressions, and motor control over various regions of the body. One of the most significant developmental milestones occurs when an infant transitions from crawling to walking. Although there is a great deal of variability in terms of when these developments take place, the pattern is usually predictable. For example, infants generally crawl before they walk. However, some infants walk as early as 10 months, whereas others may not take their first step until several months later. Failure to reach major developmental milestones by certain ages is cause for concern and may indicate a problem such as illness or poor nutrition.

Growth and development continue steadily throughout infancy and childhood, increasing markedly during adolescence. Adolescence begins when hormonal changes trigger the physical transformation of a child into an adult. Most dramatic is the maturation of reproductive organs and the subsequent ability to reproduce.

Physical Maturity and Senescence As a person reaches physical maturity, the rates of growth and development begin to slow. **Cell turnover,** the cycle by which cells form and break down, reaches equilibrium at this time. During this phase of the life cycle, which is typically the longest, growth ceases and the body enters a phase of maintenance. As a person continues to age, the rate at which new cells form decreases, resulting in a loss of some body tissue. Remaining cells become less effective at carrying out their functions. Gradually, the physical changes characteristically associated with aging, called **senescence,** become apparent. Senescence brings about a slow decline in physical function and health, which eventually can influence a person's nutrient and caloric requirements.

NUTRIENT REQUIREMENTS CAN CHANGE FOR EACH STAGE OF THE LIFE CYCLE

Aging is an inevitable process, and although individuals grow and develop in different ways, physical changes tend to coincide with various stages of the human life cycle. Age-related physical changes affect body size and composition, which in turn influence nutrient and energy requirements. For this reason, the Dietary Reference Intakes (DRIs) recommend specific nutrient and energy intakes for each life-stage group, including infants (0 to 6 months and 7 to 12 months), toddlers (1 to 3 years), early childhood (4 to 8 years), and adolescence (9 to 13 years and 14 to 18 years). Four life-stage groups are used to distinguish nutrient and energy requirements for adults. These include young adulthood (19 to 30 years), middle age (31 to 50 years), adulthood (51 to 70 years), and older adults (over 70 years). Along with these stages, the DRI recommendations consider the special conditions of pregnancy and lactation. Figure 17.2 depicts how the DRI life-stage groups are divided.

cell turnover The cycle of cell formation and cell breakdown.

senescence The phase of aging during which function diminishes.

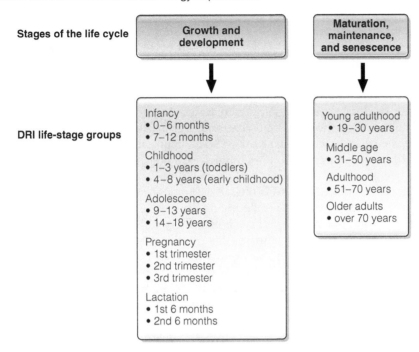

What Are the Major Stages of Prenatal Development?

Although it is important for all women to meet their nutritional needs, nutrition is particularly important during pregnancy. Poor nutritional status before and during pregnancy can have serious and long-term effects on the unborn child. For example, a pregnant woman with poor nutritional status is at increased risk for having a baby born too early or too small. Because a mother's nutritional choices can have profound effects on the life of her unborn child, early prenatal care and ongoing assessment are important.

PRENATAL DEVELOPMENT IS DIVIDED INTO EMBRYONIC AND FETAL PERIODS

There may be no other time in a woman's life when her body experiences such extensive changes as during pregnancy. These physiological transformations are needed to support the new, emerging life within her. While every woman is different, these pregnancy-related events tend to occur in a predictable and organized way. The time shortly before and after conception is referred to as the **periconceptional period.** Conception takes place when the two gametes, an ovum and a sperm, unite. The product of conception is referred to as the **zygote.** Once conception has occurred, prenatal development takes place in two periods—the embryonic period and the fetal period. The formation of the zygote signifies the beginning of the **embryonic period.** This stage of prenatal development spans the first 8 weeks of pregnancy, and is subdivided into pre-embryonic and embryonic phases. The embryonic period is followed by the **fetal period,** which starts at the beginning of the ninth week postconception and concludes at birth.

periconceptional period Time shortly before and after conception.

zygote (ZY – gote) An ovum that has been fertilized by a sperm.

embryonic period The period of prenatal development from conception through the eighth week of gestation.

fetal period Period of prenatal development, which starts at the beginning of the ninth week of gestation and continues until birth.

Embryonic Period Cell division begins immediately after the formation of the zygote, which eventually forms into a dense cellular sphere called a **blastocyst.** Around 2 weeks after conception, the blastocyst implants itself into the endometrium, the innermost lining of the uterus. The cell mass of the blastocyst then begins to differentiate, giving rise to specific tissues and organs. This stage of the embryonic period is referred to as the **pre-embryonic phase.** The **embryonic phase** of prenatal development spans from the start of the third week to the end of the eighth week after fertilization. During this phase, the developing child, referred to as an **embryo,** grows to about the size of a kidney bean. Cell division continues throughout the embryonic period, forming rudimentary structures that will eventually develop into specific tissues and organs. By the end of the embryonic period, the basic structures of all major body organs are formed.

The developing embryo follows a precise timetable in terms of development. If a critical nutrient is lacking at this time, a tissue or organ may not form properly. The term **critical period** is often used to describe the time when an organ undergoes rapid growth and development. Because organ formation occurs very rapidly during this time, the embryo is extremely vulnerable to adverse environmental influences that could disrupt this process. Unfortunately, abnormalities that occur during critical periods are irreversible. For example, maternal drug abuse during critical periods of neural development can have harmful, lasting effects on fetal brain function. This irreversible damage can result in severe behavioral and learning problems as the child grows older. In some cases, abnormalities can lead to embryonic/fetal demise, or what is commonly known as a miscarriage. A **miscarriage** is defined as the death of a fetus during the first 20 weeks of pregnancy. Although critical periods are most likely to occur during the early stages of pregnancy, they can also occur at later stages.

 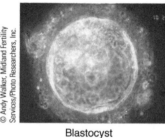 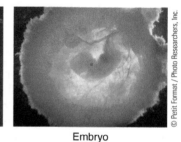

Conception Blastocyst Embryo

blastocyst (BLAS – to – cyst) Early period of gestational development that lasts approximately 8 to 13 days after conception.

pre-embryonic phase The early phase of the embryonic period that begins with fertilization and continues through implantation.

embryonic phase The latter phase of the embryonic period during which time organs and organ systems first begin to form.

embryo The developing human from two through eight weeks after fertilization.

critical period Period in development when cells and tissue rapidly grow and differentiate to form body structures. Alteration of growth or development during this period is irreversible.

miscarriage The death of a fetus during the first 20 weeks of pregnancy.

teratogen (te – RAT – o – gen) Environmental agent that can alter normal cell growth and development, causing a birth defect.

fetal alcohol spectrum disorder (FASD) A range of alcohol-related problems that can result from prenatal alcohol exposure.

Teratogens Although most pregnancies result in the birth of a healthy baby, approximately 3 to 4% of babies (about 150,000) in the United States are born with a birth defect each year:[2] A birth defect is an abnormality related to a structure or function that may result in mental or physical impairment. Some birth defects are mild, but others can be severe and life threatening. Birth defects occur for many reasons, such as genetics, environment, lifestyle, or a combination of these factors. Approximately 60% of birth defects have no known cause.

Some birth defects are caused by **teratogens,** a term used to describe a broad group of environmental agents that negatively affect the normal course of cell growth and development in the unborn child. Teratogens, which include chemicals, drugs, infections, and radiation, are responsible for about 4 to 5% of all birth defects.[3] Even excessive intakes of certain nutrients have been found to be teratogenic. For example, high intakes of preformed vitamin A during pregnancy can cause fetal malformations.

The harmful effects of teratogens are usually apparent at birth, although some may not be detected until much later. There are many teratogens, but perhaps the most familiar is alcohol. The term **fetal alcohol spectrum disorder (FASD)** is used to describe the range of alcohol-related problems that can result from prenatal alcohol exposure. Fetal alcohol syndrome is perhaps the most familiar form of FASD, and results in severe lifelong consequences. Babies born with fetal alcohol syndrome have distinctive characteristics such as a small head circumference, unusual facial features, and other physical deformities.[4] In addition, many of these infants are developmentally delayed. Despite widespread awareness of the deleterious effects of alcohol on an unborn child, approximately 1 to 2 babies per 1,000 live births in the United States are born with fetal alcohol syndrome. A less severe form of FASD, called **fetal alcohol effect** is associated with learning and behavior problems, which are often not apparent until later in life. Fortunately, FASD is preventable. That is, if a woman does not consume alcohol while pregnant, there is no risk of having a baby with FASD. Because no amount of alcohol is considered safe during pregnancy, all women should abstain from drinking alcohol during this time.

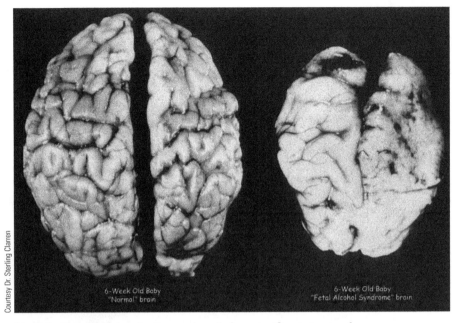

Courtesy Dr. Sterling Clarren

6-Week Old Baby "Normal" brain

6-Week Old Baby "Fetal Alcohol Syndrome" brain

Alcohol exposure during development can induce significant structural changes to the developing brain. The brain on the left is that of a normal six-week-old infant, whereas the brain on the right is that of an infant born with fetal alcohol syndrome.

Fetal Period By the end of the embryonic period, the basic structures of all major body structures and organs are formed, and the embryo is now referred to as a **fetus.** However, for the fetus to survive outside the womb, much additional growth and development is needed. Fetal weight increases by a factor of almost 500 during this period. At term, the fetus weighs approximately 7 to 8 pounds (3.2 to 3.6 kg) and is roughly 20 inches (51 cm) long. Inadequate weight gain and poor nutritional health during this period of pregnancy can dramatically affect fetal growth. Figure 17.3 shows the progressive stages of prenatal development and critical periods for selected organs and structures.

The Formation of the Placenta Within two weeks following conception, the blastocyst implants itself in the lining of the uterus, called the endometrium. Shortly thereafter, embryonic and maternal tissues begin to form the **placenta.** Although the placenta develops early in pregnancy, it takes several weeks before it is fully functional. Weighing between 1 and 2 pounds at term, this highly vascularized structure (see Figure 17.4 on page 455) has important functions.

- The placenta transfers nutrients, hormones, oxygen, and other substances from the maternal blood to the fetus.
- Metabolic waste products formed by the fetus pass through the placenta into the mother's blood; they are then excreted by the mother's kidneys and lungs.
- The placenta is a source of several hormones that serve a variety of functions during pregnancy.

Although placental membranes prevent the fetal and maternal blood from physically mixing, the exchange of gases, nutrients, and waste products is quite efficient. Unfortunately, many potentially harmful substances can also cross from the mother's blood into the fetal circulation via the placenta. For this reason, pregnant women must be particularly careful about using medications or other substances that could harm the unborn child.

fetal alcohol effect A form of fetal alcohol spectrum disorder resulting in physical and cognitive outcomes that are less severe than those of fetal alcohol syndrome.

fetus A term used, beginning at the ninth week of pregnancy through birth, to describe a developing human.

placenta An organ, consisting of fetal and maternal tissues, that supplies nutrients and oxygen to the fetus, and aids in the removal of metabolic waste products from the fetal circulation.

FIGURE 17.3 Stages of Prenatal Growth and Development and Critical Periods of Organ Formation Prenatal growth and development is divided into two periods—the embryonic period (conception to the end of the 8th week) and the fetal period (9th week until birth).

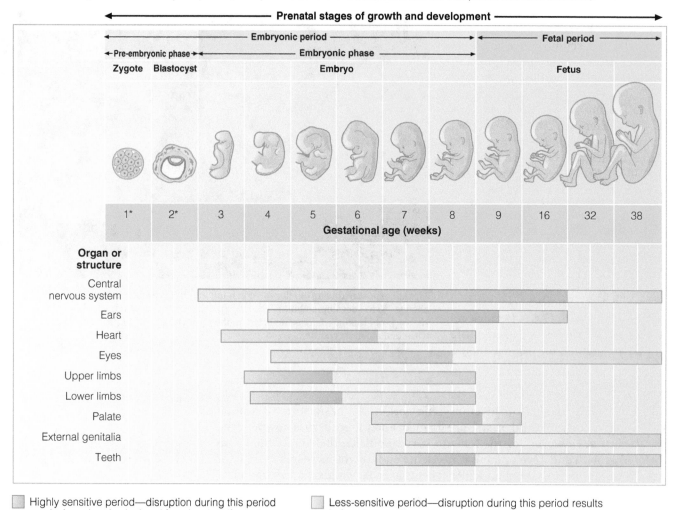

Highly sensitive period—disruption during this period results in major physical and functional malformations.

Less-sensitive period—disruption during this period results in functional defects and minor malformations.

*Note that abnormal development during the pre-embryonic phase often results in death.
SOURCE: Adapted from Moore KL, Persaud TVN. The developing human: Clinically oriented embryology. 8th ed. Philadelphia: W.B. Saunders, 2008.

GESTATIONAL AGE IS IMPORTANT TO ASSESS

How long can a woman expect to be pregnant once she conceives? There are several ways this question can be answered. Whereas the terms *embryonic period* and *fetal period* refer to stages of prenatal development, pregnancy is more commonly described in terms of trimesters. The first trimester is the time from conception to the end of week 13 and includes the entire embryonic period as well as part of the fetal period. The second trimester is from week 14 to the end of week 26, and the third trimester is from week 27 to the end of pregnancy.

The duration of pregnancy, or what is called **gestation length,** is the period of time between conception and birth. Average gestation length is 38 weeks. Because most women do not know exactly when conception takes place, calculating gestation length can be difficult. A method more commonly used to assess the length of pregnancy is **gestational age,** defined as the time from the first day of the woman's last menstrual cycle to the current date.[5] Based on gestational age, the average length of pregnancy is about 40 weeks. Babies born with gestational ages between 37 and 42 weeks are considered **full-term infants,** whereas those born with gestational ages less than 37 weeks are called **preterm infants** (or premature infants). Babies born after 42 weeks of gestational age are called **post-term infants.**

gestation length The period of time from conception to birth.

gestational age Common measure used to assess length of pregnancy, determined by counting the number of weeks between the first day of a woman's last normal menstrual period and birth.

full-term infant Baby born with gestational age between 37 and 42 weeks.

preterm infant (premature infant) Baby born with a gestational age less than 37 weeks.

post-term infant Baby born with a gestational age greater than 42 weeks.

FIGURE 17.4 Structure and Functions of the Placenta The placenta forms early in pregnancy and is made of both fetal and maternal tissues.

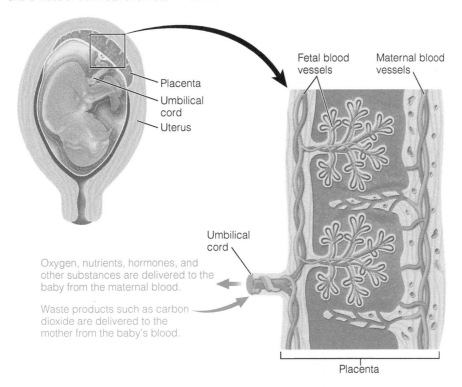

Placenta
Umbilical cord
Uterus

Fetal blood vessels

Maternal blood vessels

Umbilical cord

Oxygen, nutrients, hormones, and other substances are delivered to the baby from the maternal blood.

Waste products such as carbon dioxide are delivered to the mother from the baby's blood.

Placenta

Gestational Age and Birth Weight Not only is it important that babies are born full term, but it is also important that they are born at a healthy weight. Thus, gestational age and birth weight are both important predictors of infant health. The earlier a baby is born, the greater the risk for complications that can affect the child's survival and long-term health. This is largely because organs may not be fully developed and may therefore be unable to sustain life.

As illustrated in Figure 17.5, growth charts are used to classify infants according to birth weight and gestational age. Babies weighing less than 5 pounds, 8 ounces (2,500 g) at birth are called **low-birth-weight (LBW) infants.** Low-birth-weight infants are small because they are either preterm or have experienced slow growth *in utero*, known as **intrauterine growth retardation (IUGR).** Babies who experience IUGR are often referred to as **small for gestational age (SGA) infants,** defined as having birth weights below the 10th percentile for gestational age. Babies born with birth weights between the 10th and 90th percentiles for gestational age are classified as **appropriate for gestational age (AGA) infants,** whereas those with birth weights above the 90th percentile are classified as **large for gestational age (LGA) infants.** There are many precautions a pregnant woman can take to help ensure that her baby is born at a healthy weight.

Low-birth-weight babies are 40 times more likely to die before 1 year of age compared with healthy-weight infants.[6] In fact, together, premature births and LBW are the leading causes of infant mortality. In 2008, approximately 12% of babies born in the United States were premature, and 8% were LBW.[7] Compared to that of other developed countries, the prevalence of babies born LBW remains somewhat high in the United States (e.g., 8% of births in the United States compared to 4% of births in Finland and Iceland). Not only does being LBW put a baby at risk in early life, but it may also have profound long-term effects. Evidence suggests that less-than-optimal conditions *in utero* may cause permanent changes in the structure or function of organs and tissues, predisposing individuals to certain chronic diseases later in life.[8] You can read

⟨**CONNECTIONS**⟩ Recall that infant mortality rate is the number of infant deaths during the first year of life per 1,000 live births.

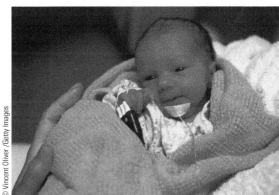

© Vincent Oliver / Getty Images

Low-birth-weight infants often require care in neonatal intensive care units.

low-birth-weight (LBW) infant A baby that weighs less than 2,500 g (5 lb 8 oz) at birth.

intrauterine growth retardation (IUGR) Slow or delayed growth *in utero*.

small for gestational age (SGA) infant A baby that weighs less than the 10th percentile for weight for gestational age.

appropriate for gestational age (AGA) infant A baby that has a weight between the 10th and the 90th percentiles for weight for gestational age.

large for gestational age (LGA) infant A baby with a weight at or above the 90th percentile for weight for gestational age.

FIGURE 17.5 Classification of Infants Based on Gestational Age and Birth Weight

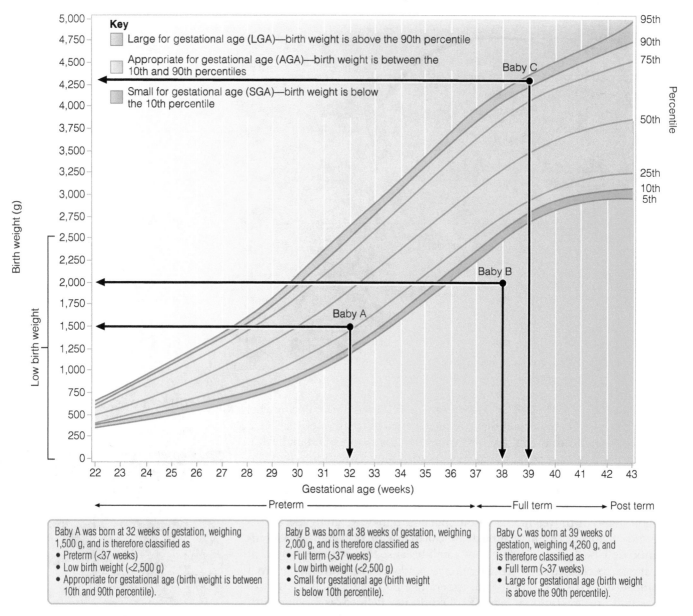

Baby A was born at 32 weeks of gestation, weighing 1,500 g, and is therefore classified as
- Preterm (<37 weeks)
- Low birth weight (<2,500 g)
- Appropriate for gestational age (birth weight is between 10th and 90th percentile).

Baby B was born at 38 weeks of gestation, weighing 2,000 g, and is therefore classified as
- Full term (>37 weeks)
- Low birth weight (<2,500 g)
- Small for gestational age (birth weight is below 10th percentile).

Baby C was born at 39 weeks of gestation, weighing 4,260 g, and is therefore classified as
- Full term (>37 weeks)
- Large for gestational age (birth weight is above the 90th percentile).

more about this phenomenon, called the **developmental origins of health and disease** (formerly called the fetal origins hypothesis) in the Focus on the Process of Science feature.

What Are the Recommendations for a Healthy Pregnancy?

Although unavoidable situations can and do affect pregnancies, there are many precautions women can take before and during pregnancy to help ensure babies are born healthy. Many of these recommendations can decrease risk of having a preterm or LBW baby. The three most important factors that are largely within a woman's control are adequate weight gain, a healthy diet, and not smoking.

developmental origins of health and disease A concept suggesting that conditions during gestation and infancy can alter risk for chronic diseases later in life.

It may be hard to imagine that experiences *in utero* can actually influence a person's risk for chronic disease as an adult. However, an overwhelming amount of evidence indicates that these early life experiences can do just that. Scientists now believe that the origins of certain diseases—diabetes, cardiovascular disease, asthma, certain cancers, and some psychiatric disorders—may be established before a person is even born and during infancy.[9] Both epidemiological and experimental data show that adult health is greatly affected by conditions before birth, soon after birth, and during childhood. This concept, called the developmental origins of health and disease, resulted from some very interesting investigative work by researchers David Barker and Clive Osmond of Southampton University in England and was based on events that took place almost 100 years ago.

In the early 1900s, lowering the infant mortality rates among England's poorest citizens was a goal of health officials. To help accomplish this, a midwife named Ethel Margaret Burnside was assigned to assist pregnant women in these communities. Nurses like Burnside made regular home visits during which they weighed and measured each baby at birth, and then again at 1 year of age. Throughout the years, information on thousands of infants was carefully recorded. Years later, the data recorded by Burnside was discovered and became of great interest to epidemiologists Barker and Osmond.

The researchers observed that regions with the highest neonatal deaths in the early 1900s were the same regions that later had the highest rates of heart disease. Curious to see if these events were related, Barker and Osmond analyzed the birth records of 13,249 men born between 1911 and 1945 and found that LBW was related to increased risk for heart disease later in life. Barker and Osmond proposed that prenatal conditions had somehow influenced adult health years later.

This concept, referred to as fetal programming, suggests that less-than-optimal prenatal conditions may alter fetal development, resulting in increased risk of certain chronic diseases later in life. Barker and Osmond's idea, originally called the fetal origins hypothesis, has now been demonstrated in many populations and animal models.[10] Furthermore, scientists now know that environmental factors (such as poor nutrition) experienced during infancy can also have long-lasting effects—many

via epigenetic alterations. Consequently, the concept of "fetal origins" was expanded to "developmental origins" to encompass both prenatal and postnatal effects on lifelong health. Considerable evidence links poor prenatal and postnatal nutrient availability to adult diseases such as cardiovascular disease, stroke, hypertension, type 2 diabetes, and obesity.[11] While the mechanisms by which very early nutrition can influence long-term health are not well understood, scientists think that, when nutrients are limited, fetal blood may be shunted to the brain to protect it, thus reducing blood flow to other organs. As a result, the course of normal fetal growth and development may be irreversibly altered during critical periods. Although this adaptive response may be beneficial to the fetus in the short term, it seems to have negative consequences during later stages of life. The long-term negative effects of fetal programming appear to be most detrimental when LBW infants subsequently experience rapid weight gain during infancy and childhood. This may be partly because rapid postnatal growth may exceed the functional capacity of underdeveloped physiological systems of LBW infants.

As scientists continue to study this phenomenon by which very early (even prenatal) nutrition influences long-term health, a better understanding of the physiologic mechanisms by which this occurs will become clearer. Undoubtedly, the emerging field of epigenetics may help elucidate the mechanisms by which prenatal and postnatal conditions influence gene expression, and how this can influence chronic disease susceptibility later in life.[12]

Ethel Margaret Burnside provided care to pregnant women in England's poorest communities. The developmental origins of health and disease concept was later developed using data she collected during the early 1900s. An excerpt from a ledger kept by Miss Burnside's nurses is shown here.

TABLE 17.1 Recommended Ranges for Total Weight Gain and Rate of Weight Gain during Pregnancy

Pre-Pregnancy Weight Classification	Body Mass Index (kg/m²)	Recommended Total Weight Gain Range (lbs)*	Recommended Rate of Weight Gain** Second and Third Trimester (lbs/wk)
Underweight	<18.5	28–40	1 (1–1.3)
Normal weight	18.5–24.9	25–35	1 (0.8–1)
Overweight	25.0–29.9	15–25	0.6 (0.5–0.7)
Obese	≥30.0	11–20	0.5 (0.4–0.6)

SOURCE: Institute of Medicine, National Research Council. Weight gain during pregnancy: Reexamining the guidelines. National Academies Press. Washington, DC. 2009.
Available online at: http://www.iom.edu/Reports/2009/Weight-Gain-During-Pregnancy-Reexamining-the-Guidelines.aspx.
* Weight-gain range for singleton pregnancies.
** Calculations assume a 0.5–2 kg (1.1–4.4 lbs) weight gain in the first trimester.

RECOMMENDED WEIGHT GAIN DEPENDS ON BMI

Because adequate weight gain during pregnancy is an important determinant of fetal growth and development, health care practitioners monitor weight gain very carefully throughout pregnancy. The current weight-gain guidelines for pregnant women, developed by the Institute of Medicine (IOM),[13] are based on maternal pre-pregnancy body mass index (BMI). These recommendations are summarized in Table 17.1. Women who gain the recommended amount of weight for their BMI range are likely to deliver full-term babies with a healthy birth weight. For example, healthy-weight women (pre-pregnancy BMI between 18.5 and 24.9 kg/m²) are advised to gain 25 to 35 pounds, whereas overweight women (pre-pregnancy BMI 25.0 to 29.9 kg/m²) are encouraged to gain less—between 15 and 25 pounds. Obese women (pre-pregnancy BMI ≥30.0 kg/m²) should gain between 11 and 20 pounds, whereas underweight women (pre-pregnancy BMI <18.5 kg/m²) are encouraged to gain between 28 and 40 pounds.

In addition to total weight gain, it is also important to monitor the rate (lbs/week) of weight gain. Whereas little weight gain is necessary during early stages of pregnancy, a steady gain of 3 to 4 pounds each month is recommended for healthy-weight women throughout the second and third trimesters. Rates of weight gain for each pre-pregnancy BMI category are listed in Table 17.1. While it is fully recognized that other factors such as age and ethnicity can impact pregnancy outcomes, these weight gain recommendations are meant to apply to pregnant teenagers as well as adults and to all racial and ethnic groups. Provisional weight-gain guidelines for women who are pregnant with twins have also been established. Healthy-weight women pregnant with twins should gain 37 to 54 pounds; overweight women, 31 to 50 pounds; and obese women, 25 to 42 pounds.

Components of Weight Gain Some women may have concerns regarding these recommended gains out of fear that the weight will be difficult to lose after the baby is born. However, this is typically not the case when women adhere to recommended weight gain allowances. Furthermore, women need to recognize that weight gain is associated with many components of pregnancy such as the fetus, breasts, uterus, and placenta. Thus, weight gain during pregnancy is not solely due to an increase in maternal fat stores. Table 17.2 lists the components of typical weight gain associated with a healthy pregnancy.

MATERNAL NUTRIENT AND ENERGY REQUIREMENTS CHANGE DURING PREGNANCY

Pregnancy results in dramatic changes in the mother's body that enable her to support and nurture her growing offspring. Many of these changes, which are

⟨**CONNECTIONS**⟩ Recall that body mass index (kg/m²) is a measure of adiposity (Chapter 11, page 308).

TABLE 17.2 Components of Typical Weight Gain during Pregnancy

Component	Approximate Weight Gain (lb) at 38 Weeks of Gestation
Fetus	7 to 8
Placenta	1½ to 2
Uterus and supporting structures	2½ to 3
Maternal adipose stores	7 to 8
Mammary tissue	1 to 2
Maternal extracellular fluids (blood and amniotic fluid)	6 to 7
Total weight gain	25 to 30

SOURCE: Institute of Medicine, National Research Council. Weight gain during pregnancy: Reexamining the guidelines. National Academies Press. Washington, DC. 2009. Available at: http://www.iom.edu/Reports/2009/Weight-Gain-During-Pregnancy-Reexamining-the-Guidelines.aspx.

listed in Table 17.3, affect nutrient requirements. Dietary recommendations for pregnant women are based on extensive research and are intended to promote optimal health in the mother and that of her unborn child. In addition to seeking regular prenatal care and guidance from health care providers, pregnant women may find the newly developed MyPlate food guidance system website (http://www.choosemyplate.gov) helpful in dietary planning. Based on a woman's age, stage of pregnancy, exercise habits, and pre-pregnancy weight, suggested dietary plans have been developed to ensure that pregnant women get the right types and amounts of all the food they need. The 2010 Dietary Guidelines for Americans also provide key recommendations that specifically address issues relevant to pregnant women, and these are highlighted in the Food Matters feature.

Recommended Energy Intake Adequate weight gain during pregnancy requires adequate energy intake. During pregnancy, additional energy is needed to support the growth of the fetus and placenta, as well as maternal tissues. Resting energy expenditure increases during pregnancy because of added

⟨**CONNECTIONS**⟩ Resting energy expenditure (REE) is the energy expended for resting metabolic activity over a 24-hour period (Chapter 11, page 304).

TABLE 17.3 Physiological Changes during Pregnancy

Cardiovascular System
- Heart enlarges slightly.
- Heart rate and cardiac output increase.
- Blood pressure decreases during the first half of pregnancy and returns to nonpregnant values during the second half of pregnancy.
- Plasma volume and red blood cell volume increase.
- Respiratory rate and oxygen consumption increase.

Gastrointestinal Tract and Food Intake
- Appetite increases.
- Senses of taste and smell are altered.
- Thirst increases.
- Gastrointestinal motility decreases.
- Efficiency of nutrient absorption increases.
- Gastroesophageal reflux becomes more common.

Renal System
- Kidney filtration rate increases.
- Sodium retention increases.
- Total body water increases.

Energy Metabolism and Energy Balance
- Basal metabolic rate (BMR) increases.
- Body temperature increases.
- Fat mass, lean mass, and body weight increase.

physiological demands on the mother. For example, the heart and lungs must work harder to deliver nutrients and oxygen to the fetus.

The energy demands of pregnancy are quite high—about 60,000 kcal over the course of the pregnancy. Although very little extra energy is needed during the first trimester, women are generally advised to increase their energy intake above pre-pregnancy Estimated Energy Requirements (EERs) by approximately 350 and 450 kcal/day during the second (14 to 26 weeks) and third (week 27 to the end of pregnancy) trimesters of pregnancy, respectively. For example, a woman with an EER of 2,000 kcal/day when she is not pregnant would require 2,350 kcal/day during her second trimester of pregnancy. During her third trimester of pregnancy, she would require approximately 2,450 kcal/day. Young or underweight women may need to increase their energy intake even more.

Recommended Macronutrient Intakes It is important for pregnant women to consume enough carbohydrates, protein, and fat to ensure that their energy needs are met. If the pregnancy is progressing normally, carbohydrates should remain the primary energy source (45 to 65% of total calories). The Recommended Dietary Allowance (RDA) for carbohydrate during pregnancy is 175 g/day, which provides adequate amounts of glucose for both the mother and the fetus. For most women, this increase represents approximately 45 g/day of additional carbohydrates, which is easily satisfied by eating 2–3 servings of carbohydrate-rich foods such as whole-grain breads or cereals.

During pregnancy, additional protein is needed for forming fetal and maternal tissues. The recommendation (RDA) is for pregnant women to increase protein intake by about 25 g/day, so that they consume approximately 70 g of total protein daily. This amount is easily obtained by eating a variety of high-quality protein sources such as meat, dairy products, and eggs. For example, 3 ounces of meat or 2 cups of yogurt provide approximately 25 g of protein. Consistent with the Acceptable Macronutrient Distribution Ranges (AMDRs), protein should continue to provide 10 to 35% of total calories.

Pregnant women who follow a vegan diet must plan their meals carefully to ensure adequate intake of essential amino acids. Plant foods that provide relatively high amounts of protein include tofu and other soy-based products and legumes such as dried beans and lentils. To make sure that all essential amino acids are consumed in adequate amounts, it is important to eat a variety of these and other protein-containing foods.

Dietary fat, also an important source of energy during pregnancy, should contribute approximately 20 to 35% of total calories during pregnancy. It is important to remember that the essential fatty acids serve other vital roles beyond the provision of energy. Linoleic acid and linolenic acid are important parent compounds used by the body to form other fatty acids and biologically active compounds. For example, linoleic acid is converted to arachidonic acid, whereas linolenic acid is converted to eicosapentaenoic acid (EPA) and docosahexaenoic acid (DHA). While all these fatty acids are critical for fetal growth and development, DHA is particularly important for brain development and formation of the retina. Although there are no DRI values for total fat during pregnancy, AIs have been established for the essential fatty acids. During pregnancy, the AIs for linoleic and linolenic acids are 13 and 1.4 g/day, respectively.

Dietary Sources of Essential Fatty Acids To ensure adequate intake of essential fatty acids during pregnancy, women should eat fish several times a week and/or use omega-3-rich oils (such as canola or flaxseed oil). Because some types of fish contain high levels of mercury, the Food and Drug Administration (FDA) and the Environmental Protection Agency (EPA) advise pregnant women to limit their consumption of fish that might contain low levels of mercury (salmon, tuna, sardines, and mackerel) and to avoid eating certain types of fish (shark, swordfish,

⟨CONNECTIONS⟩ Recall that the body is unable to make the essential fatty acids linoleic acid and linolenic acid, and therefore these fatty acids must be supplied by the diet. Docosahexaenoic acid (DHA) is derived from linolenic acid (Chapter 10, page 249).

king mackerel, and tilefish) thought to have high levels of mercury.[14] However, the importance of an adequate intake of omega-3 fatty acids during pregnancy cannot be overstated. Although pregnant women should heed the advisory issued by FDA and EPA to limit or avoid certain types of fish, it is important for them to consume other foods that are rich in omega-3 fatty acids to ensure optimal fetal and infant development. The 2010 Dietary Guidelines for Americans recommend that pregnant women consume 8 to 12 ounces of seafood per week from a variety of seafood types, while limiting white (albacore) tuna to 6 ounces per week and avoiding tilefish, shark, swordfish, and king mackerel.

Recommended Micronutrient Intakes In addition to meeting increased energy and macronutrient requirements, it is important for women to have adequate intakes of vitamins and minerals. Vitamins and minerals are needed for the formation of maternal and fetal tissues and for many reactions involved in energy metabolism. With few exceptions (calcium, phosphorus, fluoride, and vitamins D, E, and K), the requirements for most micronutrients increase during pregnancy; recommended intakes for pregnant women are listed inside the front cover of this book.

Most dietary recommendations during pregnancy are intended for women who generally consume a variety of foods from the different food groups. However, dietary restrictions can mean that some pregnant women may need to plan their diets more carefully so that all of their nutritional needs are being met. This is particularly true for pregnant women who are vegan vegetarians. Because vegans consume no foods of animal origin, they must make a special effort to get adequate amounts of vitamin B_{12}, vitamin B_6, iron, calcium, and zinc. Additional servings of whole grains, legumes, nuts, and calcium-fortified foods such as tofu and soy milk can help to satisfy these requirements. Foods that provide vitamin B_{12}, such as nutritional yeast and vitamin B_{12}-fortified cereals, are also vital.

⟨**CONNECTIONS**⟩ Recall that vegan vegetarians do not eat animal-derived foods, including meat, dairy products, and eggs (Chapter 9, page 216).

Vitamin A Although requirements for most vitamins and minerals increase during pregnancy, vitamin A is an exception. Because excessively high intakes of preformed vitamin A can be teratogenic, pregnant women should not exceed the Upper Intake Level (UL; 3,000 µg/day) from either foods or supplements. Foods that contain large amounts of preformed vitamin A include beef liver and chicken liver. However, it is not necessary for pregnant women to limit their intake of beta-carotene or other provitamin A carotenoids found mostly in plant foods.

Calcium Surprisingly, the recommended intake of calcium does not increase during pregnancy. Although extra calcium is needed for the fetus to grow and develop properly, changes in maternal physiology are able to accommodate these needs without increasing dietary intake. For example, calcium absorption increases and urinary calcium loss decreases. Therefore, the RDA for calcium during pregnancy (1,000 mg/day) is the same as that for nonpregnant women under 50 years of age.

Iron Because iron is needed for the formation of hemoglobin and the growth and development of the fetus and the placenta, the RDA for iron increases substantially during pregnancy, from 18 to 27 mg/day. Most well-planned diets provide women with approximately 15 to 18 mg of iron daily. For example, 6 ounces of beef provides 2 to 3 mg of heme iron, whereas 1 cup of iron-fortified breakfast cereal typically has 5 to 8 mg of nonheme iron. However, the bioavailability of iron in meat (heme) is higher than that in plants (nonheme). Many pregnant women have difficulty meeting the recommended intake for iron by diet alone. In addition, about 12% of women enter pregnancy with impaired iron status. For these reasons, iron supplementation is often encouraged during the second and third trimesters of pregnancy, when iron requirements are the greatest.[15] Still, recommendations for or against routine iron supplementation for non-anemic women during pregnancy remain unclear, as there is little evidence that it improves clinical outcomes for the mother or newborn.[16]

Folate Adequate folate intake is especially important during pregnancy. Recall that folate is critical for cell division, and therefore is needed for the development of all tissues including those of the nervous system. Women with poor folate status before or in early pregnancy are at increased risk of having a baby with a neural tube defect (NTD). Remember that a neural tube defect is a specific type of birth defect that affects the spinal cord or brain. As the critical period for the formation of the neural tube occurs early in pregnancy—21 to 28 days after conception—the neural tube may already be formed before a woman realizes she is pregnant. Because of the importance of folate in neural tube development, an RDA of 600 µg DFE (dietary folate equivalents)/day has been set for pregnancy. Women capable of becoming pregnant are advised to consume 400 µg DFE/day of folic acid as a supplement or in fortified foods in addition to consuming folate naturally occurring in foods. Examples of folate-rich foods include dark green leafy vegetables, lentils, orange juice, and enriched cereal grain products.

Recognizing the importance of folate in preventing NTDs, the FDA began to require folate fortification of all "enriched" cereal grain products in 1996. In addition, food manufacturers are allowed to make health claims on appropriate food labels stating that adequate intake of dietary folate or folic acid supplements may reduce the risk of NTDs. Since this nationwide effort, folate status in the United States has improved, and the incidence of NTDs has decreased.[17] While folate fortification efforts may have been successful at decreasing the incidence of NTDs, not all NTDs can be prevented by increased folate intake. Rather, some NTDs are multifactorial disorders, meaning that they are caused by a combination of environmental and genetic factors.

MATERNAL SMOKING IS HARMFUL TO THE FETUS

Smoking during pregnancy poses many risks to the unborn. There is overwhelming evidence that it increases the risk of having a preterm or LBW baby.[18] Not only is it important for pregnant women to not smoke, some studies show that secondhand smoke can also harm the fetus.[19]

In addition to the chemical compounds found in tobacco that can harm the fetus and placenta, smoking also causes maternal blood vessels to constrict, reducing blood flow from the uterus to the placenta. As a result, nutrient and oxygen availability to the fetus decreases. Smoking also increases the risk of premature detachment of the placenta, which can result in a miscarriage. If a woman smokes, she is advised to quit as early in the pregnancy as possible.

STAYING HEALTHY DURING PREGNANCY

Every pregnancy is unique. While some women do not experience much unease at all, others experience a wide variety of pregnancy-related discomforts. Pregnancy entails both physical and emotional changes, and implementation of a few simple dietary and lifestyle changes can minimize many pregnancy-related discomforts.

Pregnancy-Related Physical Complaints Hormonal changes associated with pregnancy are believed to be the underlying cause of several common physical complaints such as morning sickness, fatigue, heartburn, constipation, and food cravings or aversions. Many of these discomforts occur during the early stages of pregnancy, whereas others may persist throughout. In most cases, these discomforts are not serious and can typically be managed with simple diet-related

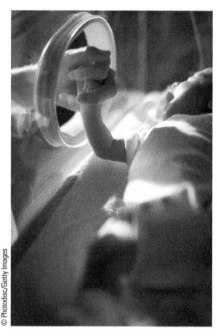

Women who smoke during pregnancy increase their risk of having a baby born prematurely. Premature birth is a leading cause of death among newborns and a major cause of long-term disability.

strategies. For example, some women who experience morning sickness, a condition characterized by queasiness, nausea, and vomiting, can find relief by avoiding foods with offensive odors. Other strategies include eating dry toast or crackers, eating small, frequent meals, and eating before getting out of bed in the morning. Heartburn, a common complaint during pregnancy, can often be managed by avoiding spicy or greasy foods, and waiting at least two to three hours after eating before lying down. To alleviate constipation, another common pregnancy-related complaint, women are advised to consume adequate amounts of fruits, vegetables, whole grains, and fluids.

Food Cravings and Aversions Food cravings and food aversions are also common during pregnancy. Powerful urges to consume or avoid certain foods may be caused by hormone-induced heightened senses of taste and smell. While most food cravings and aversions rarely pose serious problems during pregnancy, some expectant mothers develop powerful desires to consume nonfood items such as laundry starch, clay, soil, and burnt matches. The urge to consume nonfood items, called **pica,** has no known cause and can be potentially harmful to the mother and baby. Some researchers believe that pica may be related to iron deficiency.

PREGNANCY-RELATED HEALTH CONCERNS

While most minor physical discomforts associated with pregnancy are considered normal, it is important for all expectant women to be aware of changes that could indicate a more serious problem. Two common health concerns that can develop during the later stages of pregnancy are gestational diabetes and pregnancy-induced hypertension.

Gestational Diabetes Some women develop a form of diabetes called gestational diabetes during pregnancy—usually around 28 weeks or later. **Gestational diabetes** occurs when pregnancy-related hormonal changes cause maternal cells to become less responsive to insulin, triggering blood glucose levels to rise. To test for gestational diabetes, most pregnant women are given a routine blood test during the third trimester of pregnancy. Once diagnosed, a healthy diet and exercise regimen can help keep blood glucose levels under control. Some pregnant women who have difficulty controlling blood glucose may require insulin injections, however. Although gestational diabetes disappears within six weeks after delivery, approximately half of all women with gestational diabetes develop type 2 diabetes later in life.[20] To minimize this risk, it is especially important for a woman with a history of gestational diabetes to maintain a healthy weight, make sound food choices, and be physically active.

Pregnancy-Induced Hypertension Pregnancy-induced hypertension (also called pre-eclampsia or toxemia of pregnancy) is another serious complication that can develop during the later stages of pregnancy. **Pregnancy-induced hypertension,** which affects about 3% of pregnancies, is characterized by high blood pressure, sudden swelling and weight gain due to fluid retention, and protein in the urine.[21] If pregnancy-induced hypertension is suspected, women are typically advised to rest and limit their daily activities. Although most women who develop pregnancy-induced hypertension deliver healthy babies, some are not as fortunate. In some cases, a woman's blood pressure can increase to dangerously high levels—a condition referred to as eclampsia. When this occurs, the only effective treatment is delivery of the baby.

You now understand the important role that nutrition plays before and during pregnancy. However, a healthy diet is equally important during lactation. Successful lactation requires careful dietary planning to provide the energy and nutrient intakes needed to satisfy the nutritional requirements of both the mother and infant. The physiological changes associated with milk production and how

pica A desire to consume nonfood items such as laundry starch, clay, soil, and burned matches.

gestational diabetes A form of diabetes resulting from hormone-related changes during pregnancy.

pregnancy-induced hypertension (also called pre-eclampsia or toxemia of pregnancy) A form of pregnancy-related hypertension characterized by high blood pressure, sudden swelling and weight gain due to fluid retention, and protein in the urine.

FIGURE 17.6 Comparison of Recommended Energy and Nutrient Intakes for Nonpregnant, Pregnant, and Lactating Women

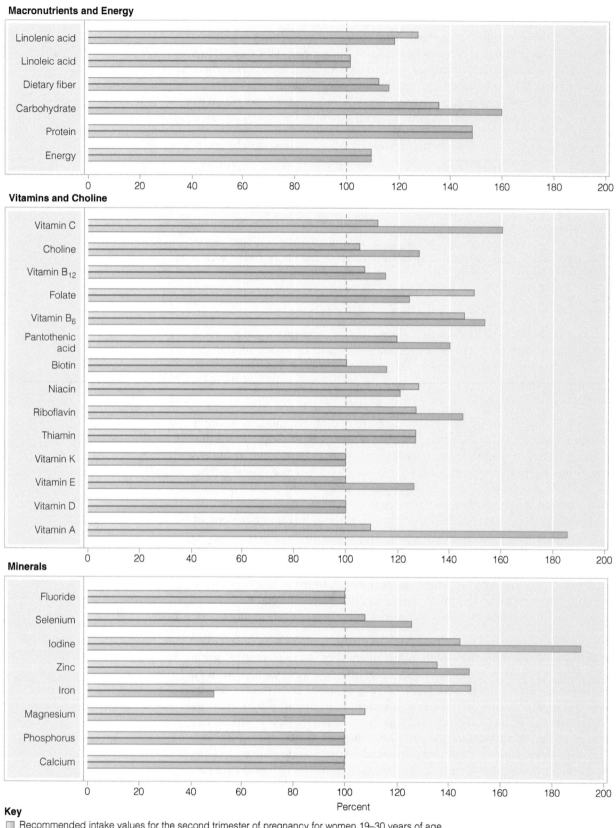

Macronutrients and Energy

Vitamins and Choline

Minerals

Percent

Key

☐ Recommended intake values for the second trimester of pregnancy for women 19–30 years of age
☐ Recommended intake values for the first 6 months of lactation for women 19–30 years of age
-- Recommended intake values for nonpregnant/nonlactating women 19–30 years of age

SOURCE: Adapted from the Dietary Reference Intakes series, National Academies Press. Copyright 1997, 1998, 2000, 2001, 2002, 2004, 2005, 2011 by the National Academy of Sciences.

Working Toward the Goal: Selecting Foods during Pregnancy

The 2010 Dietary Guidelines for Americans recommend that women pay special attention to food choices during pregnancy to optimize energy and micronutrient intakes and decrease risk of birth defects. The following food selection and preparation tips can help pregnant women meet these goals.

- Consume folate-rich foods such as orange juice, lentils, whole-grain bread, whole-grain pasta, and dark green leafy vegetables.

- Consume additional folic acid from fortified foods such as enriched breakfast cereals, breads, pasta, and rice.

- Eat foods that provide heme iron, such as lean meats, and consume iron-rich plant foods such as

foritified breakfast cereals, whole-grain bread, and lentils.

- To enhance iron content of foods, cook them in cast-iron cookware.

- To increase iron absorption, consume vitamin C–rich foods and iron-rich plant foods together.

- Eat foods that are good sources of calcium such as milk, cheese, yogurt, calcium-fortified soy products, and calcium-fortified orange juice.

- Ensure appropriate weight gain as specified by a health care provider by eating appropriate amounts of nutrient-dense foods.

- Abstain from drinking alcohol.

---

this process, called lactation, impacts the nutrient requirements of women are discussed next. Recommended nutrient intakes for nonpregnant, pregnant, and lactating women are compared in Figure 17.6.

Why Is Breastfeeding Recommended during Infancy?

Women often experience noticeable changes in the size and shape of their breasts while pregnant. These changes are necessary to prepare the breasts (also called mammary glands) for milk production after the baby is born. Human milk is the ideal food for nourishing babies. Not only does human milk support optimal growth and development during infancy and early childhood, but evidence also shows that the benefits associated with breastfeeding may extend to later stages of life. In addition, human milk provides immunologic protection against pathogenic viruses and bacteria. It is not surprising that breastfed babies tend to be sick less often than babies fed formula. Moreover, breastfeeding is beneficial to the mother, decreasing the risk of certain diseases and helping women return to their prepregnant weight more easily. As with pregnancy, adequate nutrition is important during lactation, when the woman is nourishing herself and producing milk to feed her baby.

LACTATION IS THE PROCESS OF MILK PRODUCTION

During pregnancy, many hormones prepare the mammary glands for milk production. In particular, the hormones estrogen and progesterone stimulate an increase in the number of milk-producing cells and an expansion of ducts that transport milk out of the breast. As illustrated in Figure 17.7, milk production takes place in specialized structures of the breast called **alveoli**. Each **alveolus** (the singular of alveoli) is made up of milk-producing secretory cells. Milk is formed when the secretory cells release the various milk components into the hollow center (lumen) of the alveolus. From there, the milk is released into a

iStockphoto.com/endopack

Most women find it convenient to breastfeed their babies. It is recommended that mothers breastfeed on demand (unrestricted breastfeeding).

alveolus (plural, *alveoli*) A cluster of milk-producing cells that make up the mammary glands.

FIGURE 17.7 Anatomy and Physiology of the Human Breast

The human breast, or mammary gland, is a complex organ composed of many different types of tissues that produce and secrete milk.

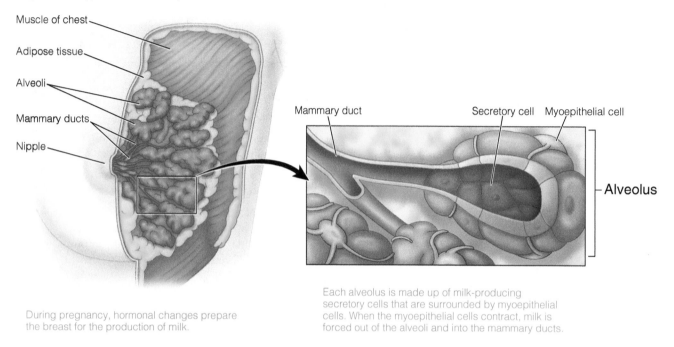

During pregnancy, hormonal changes prepare the breast for the production of milk.

Each alveolus is made up of milk-producing secretory cells that are surrounded by myoepithelial cells. When the myoepithelial cells contract, milk is forced out of the alveoli and into the mammary ducts.

---

network of **mammary ducts** that eventually lead to the nipple. After the birth of the child, the mammary glands begin to produce milk via a process called **lactation.** Women are encouraged to nurse their babies soon after delivery because suckling initiates the process of **lactogenesis,** the onset of milk production.

Prolactin and Oxytocin Regulate Milk Production While the hormones estrogen and progesterone prepare the mammary glands for milk production, it is the hormones **prolactin** and **oxytocin** that actually regulate milk production and the release of milk from alveoli into the mammary ducts. Suckling stimulates nerves in the nipple that signal the hypothalamus. In turn, the hypothalamus signals the pituitary gland to release both of these hormones (Figure 17.8). Prolactin stimulates the secretory cells in the mammary gland to synthesize milk, whereas oxytocin causes small muscles (made of specialized cells called **myoepithelial cells**) that surround the alveoli to contract. These muscular contractions force the milk out of the alveoli and into the mammary ducts. This active release of milk is called **milk let-down.** As the baby suckles, the milk moves through the mammary ducts, toward the nipple, and into the baby's mouth. Anxiety, stress, and fatigue can interfere with the milk let-down reflex, sometimes making breastfeeding challenging. For this reason, it is important for all women to seek help if they experience problems with breastfeeding.

MILK PRODUCTION IS A MATTER OF SUPPLY AND DEMAND

Whereas milk production is regulated by many physiological factors, the amount of milk produced is determined largely by how much milk the infant consumes. Women who exclusively breastfeed, meaning that human milk is the sole source of infant feeding, produce more milk than those who supplement breastfeeding with infant formula or other foods. On average, women produce 26 ounces (3⅓ cups) of milk per day during the first 6 months postpartum, and 20 ounces (2½ cups) per day in the second 6 months. The reason women produce less milk during the second six months is that most infants are also fed supplemental foods at this age. Because newborns have small

mammary duct Structure that transports milk from the alveolar secretory cells toward the nipple.

lactation The production and release of milk.

lactogenesis (lac – to – GEN – e – sis) The onset of milk production.

prolactin (pro – LAC – tin) A hormone, produced in the pituitary gland, which stimulates the production of milk in alveoli.

oxytocin (ox – y – TO – sin) A hormone, produced by the hypothalamus and stored in the pituitary gland, which stimulates the movement of milk into the mammary ducts.

myoepithelial cells Muscle cells that surround alveoli and contract, forcing milk into the mammary ducts.

milk let-down The movement of milk through the mammary ducts toward the nipple.

FIGURE 17.8 Neural and Hormonal Regulation of Lactation Milk production is regulated by "supply and demand."

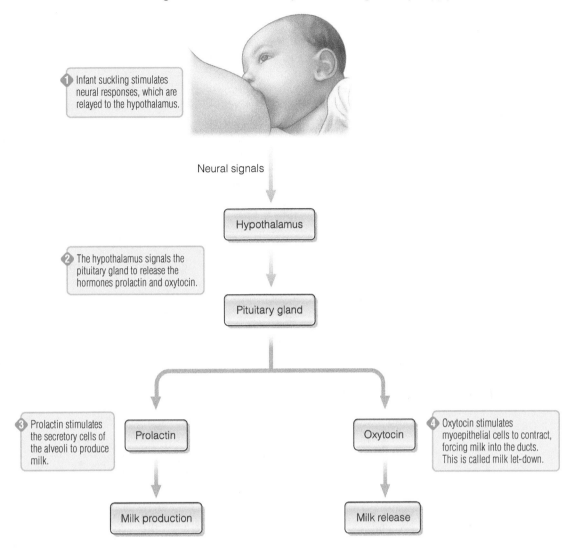

1 Infant suckling stimulates neural responses, which are relayed to the hypothalamus.

Neural signals

Hypothalamus

2 The hypothalamus signals the pituitary gland to release the hormones prolactin and oxytocin.

Pituitary gland

3 Prolactin stimulates the secretory cells of the alveoli to produce milk.

Prolactin

4 Oxytocin stimulates myoepithelial cells to contract, forcing milk into the ducts. This is called milk let-down.

Oxytocin

Milk production

Milk release

stomachs and can only consume small amounts of milk at each feeding, many mothers breastfeed as frequently as every two to three hours. However, as the baby grows, more milk can be consumed at each feeding, reducing the need to breastfeed as frequently. The American Academy of Pediatrics (AAP) recommends that women nurse their newborns at least 8 to 12 times each day and feed "on demand" rather than scheduling their baby's feedings.[22]

Breastfeeding requires proper positioning of the baby at the breast and the ability of the baby to latch onto the nipple. Once this occurs, the baby rhythmically coordinates sucking and swallowing. Sometimes mothers are concerned about whether their infants are receiving enough milk. The best indicators of adequate milk production are healthy infant weight gain and appropriate frequency of wet diapers. As a guideline, parents should expect at least four to five wet diapers per day.[22] Under most conditions, a baby needs no additional source of nourishment other than human milk for the first 4 to 6 months of life.[22] Most problems associated with breastfeeding can be resolved with the assistance of a medical professional such as a pediatrician or lactation specialist.

HUMAN MILK IS BENEFICIAL FOR BABIES

Experts agree that human milk is the best possible food for almost all babies, and these benefits are immediate and long lasting. Shortly after the mother gives birth, her breasts produce a special substance called **colostrum,** which nourishes the newborn and helps protect the baby from disease. This is, in part, because colostrum contains an abundance of immunoglobulins, which are proteins that fight infections. Over the next several days, there is a gradual transition from the production of colostrum to that of mature milk. Like colostrum, mature human milk has many important immunological benefits. For example, some immunological components help inhibit the growth of pathogens, whereas others promote the development of the infant's immature GI tract. This may help reduce the infant's risk of developing food allergies later in life. Not only does breastfeeding lessen the incidence of infectious diseases, but some studies also suggest that infants fed human milk are less likely to develop type 1 and type 2 diabetes, certain types of cancer, asthma, and obesity in later life.[23] Although it is clear that breastfed babies gain weight more slowly during the first year of life than formula-fed infants, it is less certain that these differences carry over to adolescence and adulthood.[24] However, patterns of rapid weight gain early in life may be related to increased risk for being overweight or obese later. [25]

Nutrient Composition of Human Milk The nutrient composition of human milk perfectly matches the nutritional needs of the infant. In addition to nutrients, human milk also contains enzymes and other compounds that make certain nutrients easier to digest and absorb. For example, the presence of the enzyme lipase in the milk helps the infant digest triglycerides. In addition, human milk contains lactoferrin, which enhances iron absorption and has antimicrobial activity that protects the infant from certain infectious diseases.

Protein Human milk has a relatively low protein content compared with that of cow milk, and is considered ideal for infant growth and development. The high protein content of cow milk makes it unsuitable for infants. Although cow milk is used to make many types of infant formulas, the protein content has been adjusted. The protein in human milk is not only present in the right amount, but is also easily digested, and its amino acid profile is uniquely suited to support optimal growth and development during infancy.

Lipids More than half the calories in human milk come from lipids. Some lipids are synthesized by the mammary gland, and others come from the maternal diet. Although there are dozens of fatty acids in human milk, scientists are particularly interested in DHA because of its important role in brain and eye development during infancy. Some studies have shown improved cognitive performances in breastfed infants compared with formula-fed infants.[26] Only in the last few years have some manufacturers started adding DHA to infant formula. Human milk also contains high amounts of cholesterol, which is an important component of cell membranes. Currently, cholesterol is not added to infant formula, but researchers are interested in understanding whether it should be included.

Carbohydrates The primary carbohydrates in human milk are lactose and oligosaccharides. Not only is lactose an important source of energy, but it also facilitates the absorption of other nutrients such as calcium, phosphorus, and magnesium. Although they cannot be digested by the baby and therefore do not provide a source of energy, oligosaccharides present in human milk may have functional roles such as inhibiting the growth of harmful bacteria in the infant's GI tract.

LACTATION INFLUENCES MATERNAL ENERGY AND NUTRIENT REQUIREMENTS

Milk production requires energy, and for this reason maternal energy requirements increase during lactation. The amount of additional energy needed by the mother during lactation depends on whether she is exclusively breastfeeding or

⟨CONNECTIONS⟩ Recall that oligosaccharides are carbohydrates that consist of 3 to 10 monosaccharides (Chapter 8, page 145).

colostrum (co – LOS – trum) The first secretion from the breasts after birth that provides nourishment and immunological protection to newborns.

feeding a combination of human milk and infant formula. Because women tend to produce more milk during the first 6 months of lactation compared with the second 6 months, additional energy required for milk production is approximately 500 and 400 kcal/day, respectively, during these periods. However, some of the energy needed for milk production during the first 6 months should come from the mobilization of maternal body fat that was stored during pregnancy. In other words, it is expected that breastfeeding women will be in negative energy balance during the first 6 months postpartum. This is why dietary energy recommendations during the first 6 months of lactation are lower than during the second 6 months. As the following formulas show, energy intake recommendations for lactation are based on three factors: energy required for nonpregnant, nonlactating state, energy required for milk production, and energy mobilized from body fat stores.

Estimated Energy Requirement (EER) Calculations for Lactation

- 0 to 6 months postpartum = Adult EER + energy required for milk production − energy mobilized from body fat

  = Adult EER + 500 kcal/day − 170 kcal/day

  = Adult EER + 330 kcal/day

- 6 to 12 months postpartum = Adult EER + energy required for milk production

  = Adult EER + 400 kcal/day

Thus, the Institute of Medicine (IOM) recommends an additional 330 kcal/day above nonpregnant requirements during the first 6 months of lactation and an additional 400 kcal/day in the second 6 months. Equations for calculating adult energy requirements (Estimated Energy Requirements; EERs) are provided in Appendix B.

Recommendations for micronutrient intakes during lactation are generally similar to those during pregnancy, although some (such as vitamin A) are somewhat higher and others (such as folate and iron) are lower. These are listed inside the front cover of this book. Of special interest is the recommendation for vitamin C, which increases substantially in lactation. Largely because a relatively high amount of vitamin C is secreted in milk, the RDA increases from 85 mg/day during pregnancy to 120 mg/day during lactation. In addition, it is important that lactating women take in enough fluids. Although increased fluid intake beyond basic needs does not increase milk production, a lack of fluid can decrease milk volume. The AI for total fluids for lactating women is 3.8 liters (13 cups) per day, which includes fluids in both foods and beverages.

BREASTFEEDING IS BENEFICIAL FOR MOTHERS

Most people are aware that breastfeeding is beneficial for infants, but fewer know that it is also beneficial for mothers. For example, breastfeeding shortly after giving birth stimulates the uterus to contract, helping minimize blood loss and shrink the uterus to its pre-pregnancy size. Some women also find that breastfeeding helps them return to their pre-pregnancy weight more easily.[27]

Breastfeeding is also associated with several long-term maternal health benefits. For example, it can delay the return of menstrual cycles. The span of time between the birth of the baby and the first menses, called **postpartum amenorrhea,** allows iron stores to recover, and reduces the likelihood that the mother will become pregnant again too soon. Although the duration of postpartum amenorrhea varies greatly, menstruation tends to resume around

postpartum amenorrhea (a – men – or – RHE – a) The span of time between the birth of the baby and the return of menses.

6 months postpartum in breastfeeding women and around 6 weeks postpartum in nonbreastfeeding women. Because it is possible to ovulate without menstruating, women are encouraged to use contraception until they wish to become pregnant again. In addition to prolonging postpartum amenorrhea, breastfeeding also reduces a woman's risk of developing breast cancer, ovarian cancer, and possibly osteoporosis later in life.[28]

What Are the Nutritional Needs of Infants?

The rate of growth and development during the first year of life is astonishing. At no other time in the human lifespan, aside from the prenatal period, do these processes occur so rapidly. The transition from breastfeeding to eating baby food can be challenging, and it is important for parents to be aware of signs that indicate readiness for this next phase. During infancy, providing a diet that contains all the essential nutrients and an environment that is safe, secure, and engaging helps build a solid foundation for the remainder of life.

INFANT GROWTH IS ASSESSED USING GROWTH CHARTS

During the first year of life, a healthy baby's weight almost triples, and length increases by up to 50%. At the same time, major developmental milestones such as walking and self-feeding take place. Infant growth during the first year of life follows a fairly predictable pattern. For example, within a few days after birth, most healthy, full-term infants tend to lose up to 5 to 6% of their body weight, although they usually regain this weight within two weeks. This early weight loss is attributed mostly to loss of fluids. By 4 to 6 months, infant weight doubles and length increases by 20 to 25%. Growth rates then decrease slightly from 6 to 12 months. Equations used to estimate energy requirements during the first year of life are based in part on body weight and can be found in Appendix B.

Growth Charts Infant growth and development are carefully monitored throughout the first year of life. Important measures such as weight, length, and head circumference are routinely recorded on growth charts. Until recently, growth charts developed by the National Center for Health Statistics in collaboration with the National Center for Chronic Disease Prevention and Health Promotion had been used to assess infant growth. However, the Centers for Disease Prevention and Health Promotion now recommend that clinicians use newly developed international growth charts compiled by the World Health Organization (WHO) for infants and children aged under 24 months.[29] These new standards are based on a large, healthy, globally diverse population of breastfed infants (Figure 17.9). Data collected for these growth charts show that feeding practices and environmental influences are more likely to influence a child's growth than genetics and ethnicity. The WHO believes that use of these charts will enable practitioners to better identify growth-related conditions such as under- and overnutrition worldwide.[1]

Growth charts are divided into grids that represent the distribution of weight, length, and head circumference of a representative sample of healthy infants. For instance, an infant in the 60th percentile for weight at a given age is heavier than 60% of the reference population of infants. Growth charts enable practitioners to monitor and assess infant growth over time. Because infants tend to follow a consistent growth pattern, a dramatic change in percentile could indicate a problem. For example, an infant who drops from the

FIGURE 17.9 Growth during the First Two Years of Life
Weight, length, and head circumference can be monitored during infancy and childhood using growth charts. Healthy growth trajectories for weight-for-length-for-age and head circumference-for-age during the first two years are shown here.

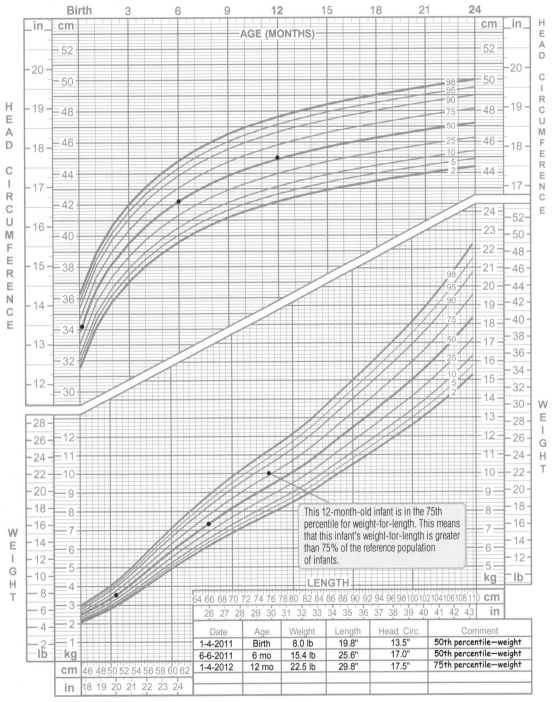

Birth to 24 months: Girls
Weight-for-length-for-age and head circumference-for-age percentiles

Date	Age	Weight	Length	Head Circ.	Comment
1-4-2011	Birth	8.0 lb	19.8"	13.5"	50th percentile—weight
6-6-2011	6 mo	15.4 lb	25.6"	17.0"	50th percentile—weight
1-4-2012	12 mo	22.5 lb	29.8"	17.5"	75th percentile—weight

This 12-month-old infant is in the 75th percentile for weight-for-length. This means that this infant's weight-for-length is greater than 75% of the reference population of infants.

SOURCE: WHO Child Growth Standards (http://www.who.int/childgrowth/en)

50th percentile for weight at birth to the 10th percentile for weight at 6 months is growing more slowly than expected. This can indicate that something is wrong and that the child should be evaluated. Poor breastfeeding technique, for example, can result in an infant not receiving adequate amounts of energy and nutrients, thus slowing or delaying growth.

Infant Development In addition to growth, many developmental changes occur during infancy. For example, newborns have little control over their bodies. However, in the first few months of life, babies begin to vocalize and are even able to return a friendly smile. Improved muscle control allows them to hold their heads steady, and by 6 months most infants can sit upright with support. These and other developmental milestones affect how and what babies should be fed. Within this first year of life, infants progress from a diet that consists solely of human milk and/or infant formula to being able to feed themselves a variety of foods.

DEVELOPMENTAL STAGES PROVIDE THE BASIS FOR RECOMMENDED INFANT FEEDING PRACTICES

In the late 1950s, fewer than 20% of babies were breastfed, partly because of the belief that infant formula was superior to human milk. Fortunately, this trend has now reversed. Over the past 30 years, the number of women in the United States who breastfeed has increased steadily.[30] This is, in part, a response to compelling scientific evidence that human milk is ideally suited for optimal infant growth and development. The AAP recommends exclusive breastfeeding for a minimum of the first 4 months of life, but preferably for the first 6 months, and the continuation of mixed breastfeeding for the second 6 months until at least one year.[31] Endorsing this recommendation is a recent "Call to Action to Support Breastfeeding" published in 2011 by the U.S. Surgeon General. In addition to endorsing exclusive breastfeeding for at least 6 months, this report put forth 20 action items designed to help U.S. women exclusively breastfeed their infants for this length of time. It is hoped that implementation of these action items will help the United States meet the breastfeeding goals outlined in Healthy People 2020.[32]

A summary of breastfeeding trends reported by the CDC indicates that, although 75% of women currently initiate breastfeeding, only 43% and 22% are still breastfeeding at 6 and 12 months postpartum, respectively. Only 13% of U.S. infants are still exclusively breastfed at 6 months.[33] Common reasons why women discontinue breastfeeding are the "perception" that milk production is inadequate and difficulties related to returning to work. However, with support and education, many women are able to overcome these obstacles. For example, understanding proper storage of expressed milk enables women to be away from their babies while still maintaining their milk supply. One organization that provides women with lactation assistance is the La Leche League. This international nonprofit group provides information and encouragement to all women who want to breastfeed their babies. While the majority of women are able to breastfeed their babies, it is important not to make someone feel guilty or embarrassed if she is not successful.

Infant Formula—An Alternative to Human Milk Although breastfeeding is the preferred method of infant feeding, there are times when it is not recommended, such as when the mother is taking chemotherapeutic drugs to treat cancer, infected with human immunodeficiency virus (HIV), using illicit drugs, or has untreated tuberculosis. The only acceptable alternative to human milk is commercial infant formula. Infants should not be fed cow milk at any time during the first year of life. As you can see from Table 17.4, the nutrient content of cow milk is very different from human milk and infant formula. The most commonly used infant formulas are derived from cow milk or soybeans. Both of these types of formula provide infants with the essential nutrients and energy needed to support growth and development. To reduce the risk of iron deficiency anemia, pediatricians recommend that parents feed their infants iron-fortified infant formula.[34] It is also important for parents to consider infant formulas that has been fortified with DHA and arachidonic acid. However, there is some debate whether adding these fatty acids to infant formula

There are many different types of infant formula to choose from.

TABLE 17.4 Nutrient/Energy Composition of Human Milk, Infant Formula, and Cow Milk (per 5 oz)

Component	Human Milk[a]	Cow Milk–Based Formula (Similac™)[b]	Soy-Based Formula (Prosobee™)[b]	Whole Cow Milk[c]
Energy (kcal)	105	100	96	93
Protein (g)	2	2	3	5
Fat (g)	6	6	5	5
Carbohydrate (g)	9	10	9	7
Cholesterol (mg)	21	3	0	15
Iron (g)	0.1	1.8	1.8	0
Calcium (mg)	45	81	90	172
Vitamin A (IU)	331	299	375	247
Vitamin D (IU)	3	59	65	78
Vitamin C (mg)	6	9	8	0
Folate (µg)	7	15	14	8

[a] Picciano MF. Representative values for constituents of human milk. Pediatric Clinics of North America. 2001;48; 1–3.
[b] Infant formula, with iron, ready to serve. Data from U.S. Department of Agriculture, Agricultural Research Service. 2010. USDA National Nutrient Database for Standard Reference, Release 23. Nutrient Data Laboratory home page. Available from http://www.ars.usda.gov/ba/bhnrc/ndl.
[c] Milk, whole, 3.25% milkfat. Data from U.S. Department of Agriculture, Agricultural Research Service. 2010. USDA National Nutrient Database for Standard Reference, Release 23. Nutrient Data Laboratory home page. Available from http://www.ars.usda.gov/ba/bhnrc/ndl.

actually has beneficial effects on the mental and psychomotor development in infants.[35] Currently, it is not mandatory for manufacturers to add DHA and arachidonic acid to infant formula. Infant formula is typically available as a powder and as a liquid concentrate. It is imperative that formula be prepared in a safe manner and that bottles and nipples be clean.

NUTRIENT SUPPLEMENTATION RECOMMENDATIONS ARE BASED ON WHETHER THE INFANT IS BREASTFED OR FORMULA FED

Are human milk or infant formula and complementary "baby foods" adequate to be the sole sources of nutrition during the first years of life, or should babies also be given nutrient supplements? The answer to this question depends on whether the infant is breastfed or formula fed, and the infant's age. Supplemental vitamin D, fluoride, iron, and fluids are sometimes recommended for infants, but the decision to use these supplements should be discussed with a health practitioner. Recommendations regarding nutrient supplementation are summarized in Table 17.5, and discussed next.

TABLE 17.5 Recommended Nutrient Supplementation during Infancy (0–12 months)

	Infant-Feeding Method	
Nutrient	Human Milk (Exclusive)	Infant Formula
Vitamin D	• Vitamin D supplements (400 IU/day) recommended beginning in the first few days of life and continuing until the infant is consuming at least 16 oz of infant formula per day	• Not needed if infant is consuming at least 16 oz formula per day, because all infant formulas manufactured in the United States meet standards for vitamin D
Fluoride	• Fluoride supplements (0.25 mg/day) recommended starting at age 6 months if local water has a fluoride concentration less than 0.3 ppm	• Not needed if formula is prepared with local water that has a fluoride concentration of at least 0.3 ppm. Fluoride supplements (0.25 mg/day) recommended starting at age 6 months if local water has a fluoride concentration less than 0.3 ppm
Iron	• Iron supplements (1 mg/day per kilogram of body weight) recommended for infants exclusively breastfed during the second 6 months of life	• Iron supplements (1 mg/day per kilogram of body weight) recommended for infants not fed iron-fortified infant formula
Fluids	• Additional fluids not needed unless infant has excessive fluid loss due to vomiting and/or diarrhea	• Additional fluids not needed unless infant has excessive fluid loss due to vomiting and/or diarrhea

SOURCE: American Academy of Pediatrics. Pediatric Nutrition Handbook, 6th ed., Elk Grove Village, IL; 2008.

Vitamin D Although scientists have long assumed that breastfed babies receive enough vitamin D from human milk and exposure to sunlight, apparently this is not always so. In recent years, several cases of rickets in breastfed infants have been reported nationwide, raising concerns about the adequacy of vitamin D content in human milk. However, if human milk is the "perfect food" for infants, then why would it be lacking this important nutrient?

Factors such as dark skin color, cloud cover, smog, and use of sunscreen can interfere with vitamin D synthesis. For example, sunscreens with a sun protection factor (SPF) of 8 or greater block the ultraviolet rays needed for vitamin D synthesis in skin. For these reasons, women with limited amounts of sun exposure may have low concentrations of vitamin D in their milk and may benefit from taking vitamin D supplements.[36] In fact, studies show that vitamin D levels in milk increase in response to maternal vitamin D supplementation.[37]

Because it is important to use sunscreen to help prevent skin cancer, the National Academy of Sciences and the AAP recommend that infants consuming less than 16 oz of infant formula each day receive vitamin D supplements (400 IU/day) beginning in the first few days of life. Supplementation should continue until they are consuming sufficient vitamin D from their diet.[38] Because all infant formulas manufactured in the United States meet standards for vitamin D, there is little risk of vitamin D deficiency in formula-fed infants consuming more than 16 oz of formula daily.

Fluoride Fluoride is important for forming teeth and can help prevent dental caries later in life. The AAP recommends fluoride supplements for breastfed infants starting at 6 months of age if the local water source has a fluoride concentration <0.3 parts per million (ppm).[22] Parents should check with the local water department to find out the fluoride content of the drinking water in their community. If purified bottled water is used to prepare infant formula, fluoride supplements are recommended. The recommended daily dosage of fluoride supplements is 0.25 mg/day for children between 6 months and 3 years of age.

Iron Another nutrient that is sometimes given to infants in the form of supplements is iron. A full-term infant is born with a substantial iron reserve that helps meet his or her need for iron during the first 6 months of life. After this time, infants consuming iron-fortified formulas that contain at least 1 mg iron per 100 kcal are likely to maintain adequate iron status. Although human milk contains less iron than infant formula, it is more readily absorbed. The AAP and CDC recommend iron supplementation for infants who continue to be exclusively breastfed during the second 6 months of life.[39] It is suggested that these infants be given 1 mg of iron per kilogram of body weight daily until complementary iron-rich foods are introduced. For breastfed infants who were preterm or LBW, a supplement of 2 to 4 mg of iron per kilogram of body weight daily is recommended, starting at 1 month after birth and continuing through the first year of life.

Water During the first 6 months of life water requirements are likely met if adequate amounts of human milk and/or formula are provided, so additional fluids are not needed. However, when vomiting and/or diarrhea cause excess fluid loss, water replacement is necessary to prevent dehydration. The AAP recommends plain water rather than juice when it is necessary to replace fluid loss.[40] Although over-the-counter fluid replacement products with added sugars and/or electrolytes can be given, they are not usually

〈**CONNECTIONS**〉 Rickets is a disease in children that causes bones to become soft and bend.

needed. In these situations, it is important for parents to contact their child's health care provider for advice.

COMPLEMENTARY FOODS CAN BE INTRODUCED BETWEEN 4 AND 6 MONTHS OF AGE

Many important developmental milestones take place between 4 and 6 months of age, and some of these can help parents determine whether their infant is ready for complementary feeding.[41] Signs that an infant is ready for complementary foods include sitting up with support and good head and neck control. The AAP recommends introducing nonmilk complementary foods when an infant is between 4 and 6 months of age. Until this time, younger infants are not physiologically or physically ready for foods other than human milk and infant formula.

Because iron status begins to decline at 4 to 6 months of age, pediatricians typically recommend that an infant's first complementary foods be iron-rich ones, such as iron-fortified cereal or pureed meat. Rice cereal, or other single-grain cereals, can be mixed with human milk or infant formula to give it a smooth, soft consistency. In addition to consuming sufficient amounts of human milk or iron-fortified infant formula, infants require approximately 1 oz of iron-fortified cereal (or equivalent) daily after 6 months of age to meet their iron requirements. Some infants find it difficult to consume food from a spoon, but in time most infants become quite skilled at this. After spoon-feeding is well established, the consistency of the food can be made thicker and more challenging. For example, other foods such as pureed vegetables and fruits can be introduced. During this period, complementary foods are considered *extra*, because infants still need regular feedings of human milk or formula.

Iron-fortified cereal is often recommended as a "first food."

Although infants should not be given complementary foods before 4 to 6 months of age, there is no evidence that delaying their introduction beyond this point is beneficial. It is important for parents to introduce new foods into the infant's diet gradually. After a new food is given, parents should wait three to five days to make sure the food is tolerated and there are no adverse reactions.[42] Signs and symptoms associated with allergic reactions or food sensitivities include a rash, diarrhea, runny nose, and in more severe cases, difficulty breathing. Although the most common food allergies are caused by cow milk, eggs, soy, nuts, wheat, fish and shellfish, there is little scientific evidence that delaying the introduction of these foods can prevent allergies later in life.[43]

Complementary Foods Should Not Displace Human Milk or Infant Formula

Infants should continue to be fed human milk and/or iron-fortified formula throughout the first year of life. The fact that cow milk, goat milk, and soy milk are low in iron provides further justification for delaying the introduction of these types of milk until after a baby's first birthday. Consuming large amounts of milk can displace iron-containing foods, which increases an infant's risk of developing iron deficiency anemia. Similarly, many pediatricians caution parents about giving infants too much fruit juice. Once recommended because it provided infants with a good source of vitamin C and additional water, the AAP now discourages parents from giving fruit juice to infants before 6 months of age. Like juice, milk can displace human milk and iron-fortified formula, which are much more nutrient dense.[40] It is also recommended that older infants not be given more than 4 to 6 oz of fruit juice a day. Overconsumption of fruit juice can lead to excess energy intake, dental caries, diarrhea, flatulence, and abdominal cramps. The AAP also recommends that older infants be given 100% pure fruit juice, rather than blends or fruit drinks.

TABLE 17.6 Developmental Skills and Feeding Practices during Infancy

Age	Developmental Skills	Appropriate Foods and Feeding Methods
0–1 months	Exhibits startle reflex; is able to suck and swallow	• Human milk and/or iron-fortified formula
2–3 months	Is able to support head; is interactive; reaches toward objects	• Human milk and/or iron-fortified formula
4–6 months	Is able to roll over to back; can sit with support; vocalizes; props self using forearms; grasps objects; moves tongue from side to side	• Human milk and/or iron-fortified formula • Slow introduction to iron-fortified infant cereal (or other iron-rich food source such as pureed meat) begins sometime between ages of 4 and 6 months
7–9 months	Begins to sit without support; holds objects; transfers objects between hands; pulls to standing; has improved mouth control and can begin to drink from a cup with assistance; can pick up small pieces of food and place them in mouth	• Human milk and/or iron-fortified formula; iron-fortified infant cereal; dry cereal; 100% fruit juice; pureed or mashed vegetables and fruits; soft pieces/mashed food such as meat
10–12 months	Begins to walk with assistance; crawls; refines grasp; exhibits mature chewing and improved ability to drink from a cup; develops skills needed for self-feeding	• Human milk and/or iron-fortified formula; iron-fortified cereal; soft pieces of fruits and vegetables; bread, crackers, and dry cereal; soft pieces of meat; 100% fruit juice

SOURCE: American Academy of Pediatrics. Pediatric nutrition handbook, 6th ed. Elk Grove Village, IL; 2008.

Baby bottle tooth decay can result when infants are put to sleep with bottles containing carbohydrate-rich liquids, including juice and milk.

© Doug Mazell/Index Stock Imagery/Photolibrary

Tooth decay can begin early in life, and it is therefore important for parents to establish good dental hygiene practices. Infants allowed to fall asleep with bottles filled with milk, formula, juice, or any other carbohydrate-containing beverage are at risk for developing **baby bottle tooth decay.** This is caused by the pooling of carbohydrate-containing beverages in the infant's mouth while asleep. Because the sugars can damage the newly formed teeth, parents are advised not to put infants to bed with bottles that contain anything but water.

Feeding Older Infants As older infants (9 to 12 months of age) become more adept at chewing, swallowing, and manipulating food in their hands, they are ready to move on to the next stages of feeding. Table 17.6 provides a list of appropriate foods that can be given to infants at this age. Because certain foods pose a risk for choking, it is important that parents not give these to their children. These include such foods as popcorn, nuts, whole grapes, raisins, raw carrots, pieces of hot dogs, and hard candy. In addition, the CDC recommends not feeding honey to children under 1 year of age, because it can contain spores that cause botulism. Even very low exposure to these spores can make young children sick. Approximately 94 cases of infant botulism occur in the United States each year.[44] Other sources that can harbor botulism spores are soil, corn syrup, and improperly canned foods.

What Are the Nutritional Needs of Toddlers and Young Children?

baby bottle tooth decay Dental caries that occur in infants and children who are given bottles containing carbohydrates (such as milk or juice) at bedtime.

The life-stage groups referred to as "toddlers" and "early childhood" span the ages of 1 through 8 years. During this time, many physical, cognitive, psychological, and developmental changes take place. This is a time of growing independence as children gain the ability and confidence to function on their own. At this age, children become more opinionated, often expressing their likes and dislikes. Indeed, feeding toddlers and young children can be challenging. Childhood is also a time when attitudes about food are being formed, and parents play an important role in helping toddlers and young children develop a healthy relationship with food. How parents deal with the challenges of feeding children is very important.

GROWTH AND DEVELOPMENT INFLUENCE NUTRITIONAL NEEDS OF TODDLERS AND YOUNG CHILDREN

In terms of defining nutrient requirements, childhood is divided into stages: toddlers (ages 1 to 3 years) and early childhood (ages 4 to 8 years). During these periods, children grow at a steady rate but one that is considerably slower than in infancy. It is not unusual for children's eating habits to change or appetite to decrease. Regardless of a child's eating habits and preferences, it is important for parents to provide enough nutritious food for optimal growth and development.

As in infancy, sex-specific growth charts are used to monitor weight and height throughout childhood. In addition, BMI is used to assess weight for height in children over 2 years of age. Children with BMIs between the 85th and 95th percentiles are classified as overweight, whereas children with BMIs at or above the 95th percentile are classified as obese. At the other end of the spectrum, children with BMIs at or below the 5th percentile are classified as underweight, and are considered at risk for undernutrition. Note that, unlike adults for whom BMI categories are not dependent on age, whether a child is classified as having healthy weight or not is age-dependent. For instance, a 3-year-old boy with a BMI of 17.5 kg/m² would be considered overweight, whereas this same BMI would indicate a healthy weight in an 8-year-old boy. This is because healthy ranges of body fat change with age (Figure 17.10).

Both the CDC and the AAP recommend that health professionals use BMI to assess weight in children beginning at 2 years of age. However, it can be difficult to determine if a child is actually overweight or simply in a transitional part of the growth cycle. Nonetheless, if excess weight for height persists throughout childhood, there may be cause for concern. Childhood obesity increases the risk of obesity later in life, regardless of whether the parents are obese.[45] The longer obesity persists during childhood, the more likely that weight-related health problems will develop.[46] According to recent estimates, approximately 17% of children 6 to 11 years of age are at or above the 95th percentile for BMI.[47] You can read more about the increasing prevalence and health concerns of overweight children in the Focus on Diet and Health feature.

FEEDING BEHAVIORS IN CHILDREN

"Please, just one bite," begs an anxious parent. This all-too-familiar plea exemplifies the fact that feeding children is not always easy. Some parents are quite surprised at how quickly mild-mannered infants become willful and opinionated toddlers. How parents respond to feeding challenges can determine whether these behaviors persist or fade. Although forcing children to eat

Feeding young children can be challenging for parents. Toddlers often have strong opinions about what and when they want to eat.

FIGURE 17.10 Interpretation of Body Mass Index Values in Children Unlike adults, how one interprets body mass index (BMI) for children and teens is dependent on sex and age. The BMI chart shown here is specific for boys.

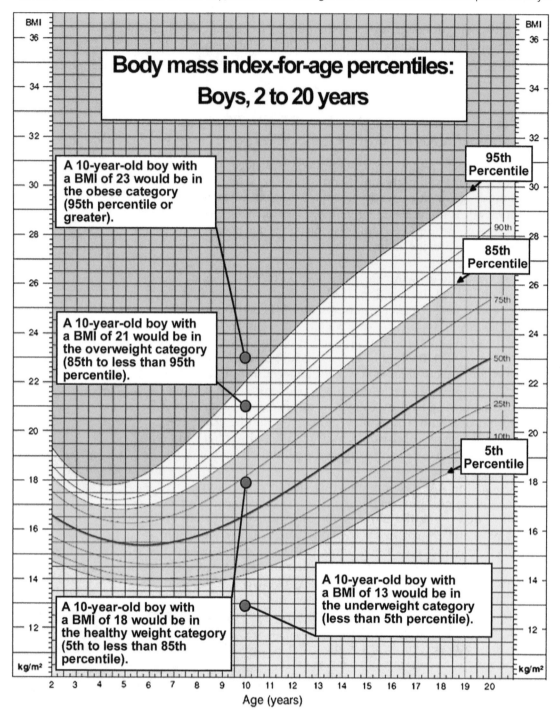

Body mass index-for-age percentiles: Boys, 2 to 20 years

A 10-year-old boy with a BMI of 23 would be in the obese category (95th percentile or greater).

A 10-year-old boy with a BMI of 21 would be in the overweight category (85th to less than 95th percentile).

A 10-year-old boy with a BMI of 18 would be in the healthy weight category (5th to less than 85th percentile).

A 10-year-old boy with a BMI of 13 would be in the underweight category (less than 5th percentile).

95th Percentile

85th Percentile

5th Percentile

90th
75th
50th
25th
10th

Age (years)

SOURCE: Centers for Disease Control and Prevention. CDC Growth Charts: United States. Available from http://www.cdc.gov/healthyweight/assessing/bmi/childrens_bmi/about_childrens_bmi.html.

when they do not want to is never recommended, it is sometimes difficult for parents to remain calm when a child refuses to eat. Most experts agree that the child, not the parent, should be the one who determines if he or she eats, and how much.[48] Rather, it is the parent's job to make mealtime a pleasant experience and to provide nutritious, age-appropriate foods to choose from. The following six general guidelines regarding common childhood feeding challenges can encourage healthy eating.

The percentage of overweight children in the United States is on the rise, and unfortunately, health conditions once only common in adults are now becoming more common in children. For example, weight-related health conditions such as type 2 diabetes, high blood pressure, and elevated blood lipids are now becoming increasingly prevalent and even commonplace among America's youth.[49] Health experts estimate that 17% of children (6 to 11 years of age) are obese (BMI >95th percentile).[47] The short- and long-term health and social consequences of childhood obesity are of great concern to parents and health professionals. Because excess weight during childhood is likely to continue into adolescence and adulthood, it is important to understand factors that contribute to this trend.

Similar to those of adults, the behaviors most concretely linked to excessive weight gain in children are unhealthy eating patterns and physical inactivity. Although there is much to learn about specific meal patterns that promote childhood obesity, easy access to high-fat, energy-dense foods is a primary contributing factor. Changing the food environment alone is not sufficient to reverse this growing trend. Efforts to encourage children of all ages to participate in healthful physical activities are also important. For many American children, watching television and other electronic media has largely replaced physical activity. It is not surprising that there is an association between obesity and the amount of time spent watching television.[50] More than one-third of children in the United States watch three or more hours of television each day, and this does not include additional time playing computer and video games.[51] In addition, children often consume energy-dense foods while watching television, and controlled research suggests that caloric consumption following a period of video game play may be increased.[52]

Persuasive media messages directed at children may also be contributing to unhealthy food choices. Approximately 90% of food commercials that air during Saturday morning children's television shows are for products with limited nutritional value, and the average American child is exposed to more than 10,000 televised food advertisements yearly.[53] And not surprisingly, when given the choice among identical foods, children prefer those wrapped in fast-food packages to those wrapped in plain, unmarked packages.[54] The selection of labeled foods (fast-food packages vs. plain packages) was related to the number of television sets in the home and the frequency of visits to fast-food restaurants. For these reasons, several health organizations including the AAP encourage parents to limit the amount of time children watch television. It is recommended that children younger than 2 years of age watch no television at all and that parents limit older children to two hours of television per day.[53]

Restricting the number of hours that children watch television does not, by itself, prevent weight gain or facilitate weight loss. Children also need to be physically active, and parents and schools need to create environments that encourage children to exercise. Unfortunately, some states have relaxed physical education requirements in public schools. Although the CDC recommends that children participate in 30 minutes of physical education class daily, only 6 to 8% of public schools meet this guideline.[55] The 2008 Physical Activity Guidelines for Americans recommend that all youth participate in at least 60 minutes of moderate-to vigorous-intensity physical activity each day. This activity should consist mostly of aerobic exercise but also include muscle- and bone-strengthening activities.[56]

In addition to exercise, providing healthy food choices at home and school is vitally important for children. In the past, school cafeterias only served meals that met federal guidelines for nutritional standards. Today, many schools sell foods and beverages that are not part of the federal school meal program. Parents may also be surprised to discover that national fast-food chains operate in many school cafeterias around the country. Money generated from selling these products is often used to help support school-related activities and programs. Furthermore, vending machines offer a multitude of poor quality snacks and beverages to students in many schools.[57] Studies show that children consume, on average, 50 more cans of soda per year in schools where vending machines are available, compared with schools without vending machines.[58] Some schools now require vending machines to be stocked with nutritional snacks such as milk or fruit.

Preventing children from becoming overweight will take considerable effort at home and at school. Regardless of the many pressures and enticements, teaching children about the importance of physical activity and good nutrition is an important step in keeping them healthy. After all, there are as many healthy food choices available today as there are unhealthy ones.

CHRIS YOUNG/PA Photos /Landov

Many organizations and parent groups are encouraging schools to eliminate fat- and sugar-rich snacks from school vending machines.

Meals provide family members an important opportunity to interact, to share, and to be together.

• *Avoid using food to control behavior.* Most experts agree that using food for rewards or punishments is not a good idea. Although rewarding a child with treats may correct behaviors short term, it will likely create serious food-related issues down the road. For example, rewarding "good" behavior with a dessert may establish a connection between sweet foods and approval. Rather, parents should teach children that food is pleasurable and nourishing and not something to turn to for approval or emotional comfort.

• *Model good eating habits.* It is important for parents to serve as good role models. Studies show that if parents or siblings enjoy a particular food, the child will be more likely to enjoy it too.[59] Studies also show that families that eat meals together on a regular basis are more likely to have children who eat healthier diets than families that do not.[60] Mealtime provides a time for the entire family to be together and share quality time.

• *Recognize preference differences.* Food preferences of children do not always appear rational to adults. That is, children often judge foods as acceptable or unacceptable based on attributes such as color, texture, and appearance rather than taste and nutritional value. Also, children likely experience strong flavors such as onions and certain spices more intensely than adults.

• *Introduce new foods gradually.* Making new foods familiar to children is an important first step in food acceptance. Encouraging children to help select and prepare new foods not only helps to make foods more familiar, but may also stimulate interest. It is important for parents to remember that accepting new foods takes time. Pressuring children to eat or to try new foods is not recommended.

• *Encourage nutritious snacking.* Children have small stomachs and therefore should eat smaller portions than adults. They also need to eat more frequently. In fact, limiting food consumption to three meals a day is very difficult for some children. This is why experts recommend that parents and caregivers provide children with nutritious between-meal snacks. Although children who snack frequently are often not hungry at mealtime, this is not usually a problem as long as healthy, nutritious snacks are served.

• *Promote self-regulation.* It is important that children learn to self-regulate their food intake based on internal cues of hunger and satiety. For this reason, serving sizes of food need to be age appropriate, allowing children to ask for more if desired. Children often claim to be too "full" to eat certain foods, and experts do not recommend forcing them to eat all the food on their plate. Instead, parents need to set limits. For example, if a child declines to eat his or her meal and then asks for dessert, it is reasonable to say no.

Critical Thinking: Amy's Story Think back to Amy's story about Backyard Harvest. Family participation in Backyard Harvest offers many educational opportunities for children. What can children learn about food and healthy eating by harvesting and donating freshly grown produce? What critical nutrients might be provided by the fresh fruits and vegetables distributed by Backyard Harvest that might otherwise be lacking in a low-income child's diet?

RECOMMENDED ENERGY AND NUTRIENT INTAKES FOR TODDLERS AND YOUNG CHILDREN

During childhood, adequate energy is needed both to support total energy expenditure (TEE) and for the synthesis of new tissues. In other words, a child should be in positive energy balance. Thus, EERs are equal to the sum of TEE plus the energy associated with growth. As the rate of growth slows during childhood, the amount of energy (kcal/day) needed daily to support growth also decreases. For example, the energy needed to support growth in infants (0 to 3 months of age) is approximately 175 kcal/day. By comparison, the energy needed to support growth in toddlers is only 20 kcal/day. Although energy needed for growth decreases with age, EER values increase because some of the variables that affect TEE (such as height, weight, and physical activity level) increase. For example, the EER for a 1-month-old male infant weighing 9.7 lb is 467 kcal/day. By 35 months of age (weight = 31.3 lb), the EER increases to 1,186 kcal/day. EER equations for toddlers and young children are listed in Appendix B.

It is also important to consider the healthiest distribution of calories in the diet of children. The Institute of Medicine recommends, via the AMDRs, that children 1 to 3 years of age consume 30–40, 45–65, and 5–20% of their calories as fats, carbohydrates, and protein, respectively. Note that this would result in children consuming diets higher in fat than adults. However, by 4 years of age, the AMDRs for toddlers and young children are close to those for adults. Although the amount of protein needed per kilogram of body weight decreases during childhood, recommended total protein intakes increase. AMDRs and recommended protein intakes for toddlers and young children (g/day and g/kilogram/day) are listed inside the front cover of this book. The RDA for protein for toddlers and young children is 0.95 g/day per kg of body weight. It is especially important to provide high-quality protein foods during this time, such as meat, yogurt, cheese, and eggs.

Although there are no RDAs or AIs for total fat intake for toddlers and young children, adequate fat intake is important for growth and development. It is particularly important for children to consume adequate amounts of the essential fatty acids linoleic acid and linolenic acid. Foods that provide these types of fatty acids include vegetable oils, nuts, seeds, and fish. Carbohydrates are needed to provide the brain with glucose. After 1 year of age, the amount of glucose used by the brain is similar to that of adults, which is why carbohydrate recommendations for children and adults are the same. Dietary fiber is also important for children's health. The AI for dietary fiber is 19 g/day for toddlers and 25 g/day for young children. By comparison, it is recommended that adults consume 25 to 38 g/day of dietary fiber. Serving children fruits, vegetables, and whole-grain cereal products made from whole grains provides them with an excellent source of dietary fiber. These nutrient-dense foods that most children enjoy eating provide an array of micronutrients as well.

Calcium and Iron Are Nutrients of Special Concern for Toddlers and Young Children

Calcium-rich foods are particularly important to the development of strong, healthy bones throughout childhood. The RDA for calcium increases from 700 mg/day at 1 to 3 years of age to 1,000 mg/day at 4 to 8 years of age. Milk and other dairy products are good sources of calcium. In fact, it would take 7 cups of broccoli to provide the same amount of calcium (approximately 280 mg calcium) as 1 cup of milk. For this reason, children between the ages of 2 and 8 years should consume at least 2 cups of low-fat milk or equivalent milk products each day. In general, 1 cup of yogurt, 1½ ounces of natural cheese, and 2 ounces of processed cheese are all equivalent to 1 cup of milk.

Many children do not meet recommended intakes for milk and milk products, which is cause for concern.

Iron is another nutrient that is important for growing children. It is recommended that toddlers and young children consume 7 and 10 mg/day, respectively, of iron to meet their needs. Iron deficiency is one of the most common nutritional problems in childhood. This is because children have a rapid rate of growth, and blood volume is increasing. Children who drink large amounts of milk and eat a limited variety of other foods may be at increased risk of iron deficiency.[61] This may be related to over-consumption of cow's milk and juice that can displace iron-rich foods from the diet.

A lack of iron-rich foods can lead to the development of iron-deficiency anemia, causing children to become irritable and inattentive and to have decreased appetites. In fact, researchers have found that student achievement tests improved when children with iron-deficiency anemia received iron supplements.[62] To prevent iron deficiency, parents are encouraged to feed their children a variety of iron-rich foods such as meat, fish, poultry, eggs, legumes (peas and beans), enriched cereal products, whole-grain cereal products, and other iron-fortified foods.

DIETARY GUIDELINES FOR TODDLERS AND YOUNG CHILDREN

Because children have unique eating patterns and nutritional needs, it is important to provide a variety of nutrient-dense foods. To determine the number of servings from each food group needed to meet recommended nutrient and energy intakes for toddlers and young children, parents and childcare providers are encouraged to use the MyPlate food guidance system (http://www.choosemyplate.gov). This site also provides materials designed specifically for preschoolers (http://www.choosemyplate.gov/preschoolers/index.html) and children aged 6 to 11 (http://www.choosemyplate.gov/kids/index.html) that promote healthy eating. In addition to emphasizing a healthy diet, the Dietary Guidelines for Americans stress the importance of regular physical activity to promote physical health and psychological well-being. As previously noted, parents are encouraged to make sure that children engage in at least 60 minutes of moderate- to vigorous-intensity physical activity daily, including strength-building exercises.

How Do Nutritional Requirements Change during Adolescence?

Toward the end of early childhood, hormonal changes begin to transform a child into an adolescent. This transition marks the beginning of profound physical growth and psychological development. Hormones are responsible for triggering changes in height, weight, and body composition. In females, the onset of menstruation, known as **menarche,** also begins around this time. Menarche typically occurs in girls around 13 years of age. As with any stage of growth, nutrition plays an important role as an adolescent matures from a child into a young adult. Unfortunately, this is also a time when unhealthy eating practices often begin to develop. For example, weight dissatisfaction is common among teens and can lead to inappropriate dieting and other destructive weight-loss behaviors.

menarche (men – ARK – e) The first time a female menstruates.

GROWTH AND DEVELOPMENT DURING ADOLESCENCE

Adolescence is the physical and psychological bridge between childhood and adulthood, and this transformation is signified by the onset of **puberty.** Defined as the maturation of the reproductive system and the capacity to reproduce, puberty is initiated by hormonal changes that trigger the physical transformation of a child into an adult. During this time, the adolescent begins to experience many physical, psychological, and social changes. Because these changes span a relatively long period, the DRIs for this stage of the life cycle are divided into two phases: 9 to 13 years and 14 to 18 years.

The timing of puberty varies, and adolescents of the same age can differ in terms of physical maturation. For example, some adolescents are "early" developers, whereas others are "late" developers. Consequently, nutritional needs of adolescents may depend more on the stage of physical maturation than chronological age. Also, females tend to experience puberty at an earlier age than do males. Puberty begins roughly around 9 years of age in females and 11 years of age in males. However, for reasons that are not clear, more and more American girls are entering puberty at younger ages. Some researchers believe this trend may be related in part to the rise in childhood obesity rates, as overweight girls tend to develop physically and menstruate earlier than do thinner girls.[63]

Adolescence is the period between childhood and adulthood.

Changes in Height, Weight, and Body Composition Considerable physical growth takes place during adolescence. Before the onset of puberty, males and females have attained about 84% of their adult height. During the adolescent growth spurt, females and males grow approximately 6 and 8 inches in height, respectively. Bone mass also increases, with peak bone mass nearly attained by about 20 years of age. Because bone mass increases rapidly during adolescence, it is especially important for teens to consume adequate amounts of nutrients that promote bone health, such as protein, vitamin D, calcium, magnesium, and phosphorus.

⟨CONNECTIONS⟩ Recall that peak bone mass is the maximum amount of minerals found in bones during the lifespan.

Changes in linear growth (height) are accompanied by changes in body weight and composition during adolescence. Overall, females and males gain an average of 35 and 45 pounds, respectively. However, changes in body composition differ considerably between females and males. Whereas females experience a percentage decrease in lean mass and a relative percentage increase in fat mass, a male adolescent experiences the opposite. Although increased body fat in females is normal and healthy, it can contribute to weight dissatisfaction, which sometimes can lead to unhealthy dieting and caloric restriction. Caloric restriction during adolescence can delay growth, development, and reproductive maturation. The topic of eating disorders and related health problems was addressed in the Nutrition Matters that concluded Chapter 11. While good nutrition is important throughout the entire life cycle, the accelerated rate of growth and development during adolescence puts teens at particularly high risk for diet-related health problems.

Psychological Changes Adolescence is marked not only by rapid physical changes, but also by numerous psychological and developmental changes. For example, adolescents' newfound desire for independence often strains family relationships

puberty Maturation of the reproductive system.

and leads to rebellious behaviors. Furthermore, there is a strong need to fit in and be accepted by peers. For these and many other reasons, healthy eating can become a low priority for adolescents. In fact, peers are often more influential in determining food preferences than family members. Dieting, skipping meals, and increasing consumption of foods away from home are common occurrences among teenagers. However, with maturity comes the recognition that current health behaviors can have lasting impacts on long-term health.

NUTRITIONAL CONCERNS AND RECOMMENDATIONS DURING ADOLESCENCE

The rapid growth and development associated with adolescence increases the body's need for certain nutrients and energy. The MyPlate food guidance system can be used by teens to determine the amount of food from each food group needed to meet their recommended nutrient and energy intakes. Using this personalized guide for meal planning is particularly important because an inadequate diet during this stage of life can compromise health and have lasting effects.

Energy and Macronutrients Estimated Energy Requirement (EER) calculations during adolescence take into account energy needed to maintain health, promote optimal growth, and support a desirable level of physical activity. Again, positive energy balance is assumed during this period of time. Equations used to determine EERs in adolescents are the same as those for children except for the amount of energy needed for growth. For example, for boys and girls 9 through 18 years of age, the EER is calculated by summing total energy expenditure (TEE) and an additional 25 kcal/day for growth. Equations used to determine EERs during adolescence are provided in Appendix B. Although obesity rates have risen over the past 30 years among adolescents, average energy intakes by teens have remained remarkably stable. The only noted exception appears to be for adolescent females, who have increased their caloric intake by approximately 260 kcal/day.[64] Thus, changes in energy expenditure have likely contributed to the increased obesity rates in this age group.

Although growth is influenced by many factors, adequate intake of protein is particularly important. Younger adolescents (9 to 13 years) require more protein per kilogram of body weight than do adults—0.95 vs. 0.8 g/kg/day, respectively. For 14- to 18-year-olds, the RDA for protein decreases to 0.85 g/kg/day. Based on the RDA, recommended protein intake is 34 g/day for females 9 to 13 years of age and 46 g/day for females 14 to 18 years of age.

Similar to an adult's diet, the majority of calories in an adolescent's diet should come from nutrient-dense, carbohydrate-rich foods. These include fruits, vegetables, and whole grains. Yet, despite national public health efforts to encourage healthy eating, the number of adolescents who meet recommendations for fruit and vegetable consumption is quite low.[65] The "typical" adolescent diet is also less than optimal in terms of snack foods that are high in fat and sugar. It is not surprising that snacking frequency among teens is associated with higher intakes of total calories. In 2005–2006, adolescents consumed on average 526 kilocalories daily as snacks (23% of total energy intake).[66] Sweeteners and added sugars, many of which are associated with the consumption of carbonated beverages, contribute 16% of the total calories consumed by teens.[67] Percentages of total energy from fat and saturated fat among adolescents are approximately 34% and 12%, respectively. These amounts, which are consistent for both younger and older adolescent males and females, are well within the recommended intakes based on the AMDRs.[68]

Selected Micronutrients Too little dietary calcium at a time when skeletal growth is increasing can compromise bone health later in life. For this reason, the RDA for calcium during adolescence is set at 1,300 mg/day. Getting enough calcium is best achieved by consuming dairy products such as milk, yogurt, and cheese. The average calcium intakes for adolescent males[69] and females[70] are 1,266 and 918 mg/day, respectively (EAR = 800 mg/day). When asked, the majority of adolescents recognize the importance of calcium for proper bone health.[71]

Adolescents also require additional iron to support growth. Because of the iron loss associated with menstruation, the RDA for iron in females is higher than that of males (15 and 11 mg/day, respectively) during later adolescence. Although it appears that most adolescents are well-nourished with respect to iron, approximately 8% of teenagers in the United States have impaired iron status, and another 1 to 2% have iron-deficiency anemia.[72]

Mean total folic acid intake (diet and supplements) for adolescent females is 274 µg DFE/day (EAR = 330 µg DFE/day).[73] As you learned previously, impaired folate status around the time of conception can increase the risk of having a baby born with a neural tube defect. For this and other reasons, it is important for health care professionals to stress the importance of consuming folate-rich foods such as orange juice, green leafy vegetables, and enriched cereals.

Although the majority of teens appear to have adequate intakes of vitamin C, it is well known that smoking can increase vitamin C requirements. In spite of the fact that smoking rates among U.S. teens have been on the decline, a significant number—approximately 35% of high school–age teens—continue to smoke. Vitamin C is needed for collagen synthesis, and a lack of this important vitamin can disrupt growth and bone development.

How Do Age-Related Changes in Adults Influence Nutrient and Energy Requirements?

Adulthood spans the longest segment of the life cycle—approximately 60 to 65 years. Like individuals in other stages of life, adults experience physical, psychological, and social changes that can affect health. Managing family and career obligations can make this a particularly challenging time in a person's life. A busy schedule can be stressful, and personal health needs often go unattended. It is vital for adults to recognize the importance of taking care of their own physical and emotional well-being because a healthy lifestyle can help ensure a full and active life for many years to come.

Because many chronic, degenerative diseases can occur during this period of the life cycle, it is important for adults to stay physically and mentally fit. Although aging is inevitable, studies show that young and middle-aged adults who adopt a healthy lifestyle that includes eating nutritious foods, maintaining a healthy body weight, participating in regular physical activity, and not smoking are less likely to develop cardiovascular disease, type 2 diabetes, hypertension, and certain types of cancer, all of which can contribute to premature death.

ADULTHOOD IS CHARACTERIZED BY PHYSICAL MATURITY

Adulthood is the period in the lifespan characterized by attainment of physical maturity, which is typically achieved by 20 years of age. Some males continue to grow in their early 20s, and bone mass increases slightly until age 30 for both males and females. After achieving physical maturity in young adulthood, adults typically undergo a long period of physical stability (maintenance) before gradually transitioning to senescence. The DRIs divide adulthood into four groups—young adults (19 through 30 years), middle age (31 through 50 years), adulthood (51 through 70 years), and older adults (>70 years). There is a wide spectrum of health and independence among

Many older adults are physically fit and remain active.

⟨CONNECTIONS⟩ Recall that life expectancy is the average number of years a person can expect to live.

Jeanne Louise Calment (1875–1997), photographed at the age of 122 years, was one of the oldest humans whose age was fully authenticated. She attributed her longevity to olive oil, port wine, and chocolate. Her genes probably contributed to her longevity; her father lived to the age of 94 years and her mother to 86 years.

⟨CONNECTIONS⟩ Recall that free radicals are compounds with unpaired electrons and can damage cell membranes, proteins, and DNA.

most middle-aged and older adults. Whereas some people maintain active lifestyles throughout adulthood, others are frail and require a great deal of assistance. For this reason, functional status is sometimes a better indicator of nutritional needs than chronological age.

The "Graying" of America Although physical maturity is achieved early in adulthood, many productive years still remain ahead. In the United States, life expectancy continues to increase, and people over 85 years old are becoming the fastest growing segment of this age group.[74] Note that life expectancy, the expected number of years of life remaining at a given age, is different from **lifespan,** the maximum number of years an individual member of a particular species has remained alive. In humans, both life expectancy and lifespan have been rising steadily, but the graying of America is evident now more than ever. Although there is little doubt that a person's genetics play a major role in the aging process, lifestyle choices are vitally important as well.

The graying of America reflects a shift in the age distribution of the United States (Figure 17.11). Never before have there been so many older people, and never before has life expectancy been so long. Whereas average life expectancy (at birth) was about 47 years in 1900, today a baby born in the United States will likely live to be nearly 78 years old.[47] These changes are partly due to advances in medical technology and improved health care. The graying of America is also partly due to the increased number of births in the 1940s through the early 1960s. People born at this time are often labeled the "baby boom" generation. During the baby boom years, many hospitals expanded their obstetric and gynecology units to accommodate the increase in births. Shortly thereafter, many thousands of schools were built throughout the United States to accommodate the growing number of school-age children. Today, there is an increased need for retirement communities and health care practitioners to care for the rising number of older "boomers."

THERE ARE MANY THEORIES AS TO WHY WE AGE

The branch of science and medicine dedicated to the social, behavioral, psychological, and health issues of aging is called **gerontology.** Gerontologists have many theories as to why we age. Some believe the body deteriorates because of daily wear and tear. Others believe that a combination of lifestyle, environmental, and genetic factors determines how quickly we age.

One of the greatest contributions in understanding biological aging came from experiments performed in the 1960s by the physiologist Leonard Hayflick. Using cell cultures, Hayflick demonstrated that the number of times cells can divide is limited.[75] Thus, senescence, or the process of aging, may be programmed in cells, and factors that increase cell replication may speed up this process. For example, damage caused by unstable compounds called free radicals (generated by metabolism, lifestyle practices, and environmental factors) may increase cell replication and thus promote aging. In addition, free radicals extensively damage proteins, cell membranes, and DNA. Over time, damaged cells become unable to function fully, triggering a cascade of other age-related changes. Free radical damage has been implicated in several degenerative diseases such as cancer, atherosclerosis, and cataracts.[76]

The evidence that aging and many age-related chronic diseases may, in part, be caused by free radical damage raised hopes that antioxidant nutrients could slow the aging process by protecting the body from oxidative damage. Although animal studies provide some supportive evidence, human data remain controversial.[77] Clinical trials are needed to clarify whether consumption

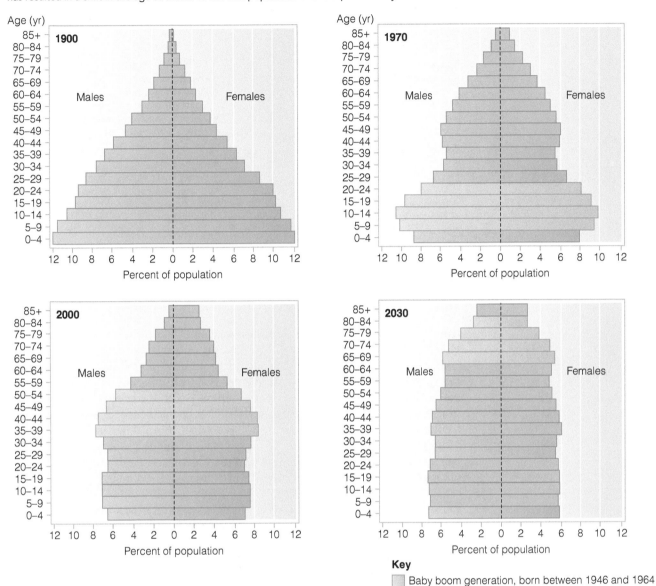

FIGURE 17.11 The "Baby Boom Generation" and Aging of the U.S. Population An increased number of births between 1946 and 1964 has resulted in a shift in the age structure of the U.S. population over the past century.

Key
Baby boom generation, born between 1946 and 1964

SOURCE: Hobbs R and Stoops N. U.S. Census Bureau. Demographic trends in the 20th century. U.S. Government Printing Office, Washington, DC. 2002. Available from: http://www.census.gov/prod/2002pubs/censr-4.pdf

of foods rich in antioxidants or antioxidant supplements can help slow the progression of age-related changes and help people live longer and healthier lives.

NUTRITIONAL ISSUES OF ADULTS

Although we cannot "turn back the hands of time," there is much that we can do to keep our bodies strong and healthy. During adulthood, nutrient intake recommendations are intended to reduce the risk of chronic disease while providing adequate amounts of essential nutrients. The MyPlate food guidance system can assist older adults in planning meals and determining the amount of food from each food group needed to meet their recommended nutrient and energy intakes. Because older adults generally require less food

⟨**CONNECTIONS**⟩ Recall that antioxidant nutrients (such as vitamins C and E) prevent unstable free radicals from damaging cells by donating electrons to them.

lifespan Maximum number of years an individual in a particular species has remained alive.

gerontology The branch of science and medicine that focuses on health issues related to aging.

TABLE 17.7 Physiological Changes Typically Associated with Aging

Cardiovascular System
- Elasticity in blood vessels decreases.
- Cardiac output decreases.
- Blood pressure increases.

Endocrine System
- Estrogen, testosterone, and growth hormone levels decrease.
- Glucose tolerance decreases.
- Ability to convert provitamin D to active vitamin D diminishes.

Gastrointestinal System
- Secretion of saliva and mucus decreases.
- Loss of teeth may occur.
- Difficulty swallowing may occur.
- Secretion of gastric juice is reduced.
- Peristalsis decreases.
- Vitamin B_{12} absorption decreases.

Musculoskeletal System
- Bone mass decreases.
- Lean mass decreases.
- Fat mass increases.
- Strength, flexibility, and agility are reduced.

Nervous System
- Appetite regulation is altered.
- Thirst sensation is blunted.
- Ability to smell and taste decreases.
- Sleep patterns change.
- Visual acuity decreases.

Renal System
- Blood flow to kidneys diminishes.
- Kidney filtration rate decreases.
- Ability to clear blood of metabolic wastes decreases.

Respiratory System
- Respiratory rate decreases.

than younger adults, these guidelines help older adults select foods that provide optimal nutrient intake while balancing energy consumption.

In general, older adults are considered an "at-risk" population for developing many nutrition-related health problems. For example, issues related to food insecurity, social isolation, depression, illness, and the use of multiple medications can compromise nutritional status. The topics of hunger and insufficient food availability in the elderly are addressed in more detail in the Nutrition Matters following this chapter. Physiological changes associated with aging can also influence the development of nutrition-related health problems. These changes are summarized in Table 17.7 and are discussed next.

Changes in Body Composition and Energy Requirements Most adults experience age-related changes in body composition, such as a loss of lean mass and an increase in fat mass. Although genetics, physical activity, and nutritional status influence these shifts, in general, a 70-year-old male has approximately 22 more pounds of body fat and 24 pounds less muscle than a 20-year-old male.[78] Age-related loss of lean mass causes basal metabolism to decrease. As a result, energy requirements tend to decrease with age. This explains, in part, why many middle-aged and older adults experience age-related weight gain. A key recommendation included in the Dietary Guidelines for Americans is for adults to prevent age-related weight gain by decreasing energy intake and maintaining adequate physical activity.

Equations used to estimate energy requirements are the same for all adults. However, for each year of age above 19, total energy expenditure decreases 7 kcal/day for women and 10 kcal/day for men. Consequently, compared to 30-year-old women and men with similar weights, heights, and activity levels, 70-year-old women and men require 280 (40 × 7) and 400 (40 × 10) fewer kilocalories each day, respectively. Of course, energy requirements depend on many factors such as physical activity, weight, and changes in the relative amounts of muscle and body fat. Both over-consumption and under-consumption of energy can lead to weight-related health problems in older adults.

Along with weight gain, the loss of muscle mass can make older people less steady, increasing their risk for injury. According to the CDC, more than one-third of adults 65 years of age and over experience fall-related injuries annually, a common cause of most head traumas and hip fractures.[79] Exercise not only decreases fat mass and slows age-related bone loss, but it also helps strengthen muscles and improve coordination.

Adequate protein intake is important for older adults to help prevent age-related loss of skeletal muscle. Currently, the RDA for protein in older adults is 0.8 g/kg/day. However, some researchers believe that current requirements are too low and that protein requirements should be increased to 1.0–1.25 g/kg/day.[80] For an older adult weighing 150 lb, this would increase their recommended protein intake from 54 to 77 g/day.

Preventing Bone Disease Age-related bone loss can lead to osteoporosis, which is characterized by weak, fragile bones. While men can develop osteoporosis, the condition is far more common in women. Although many factors influence bone density, it is especially important for older adults to have adequate intakes of nutrients (protein, calcium, vitamin D, phosphorus, and magnesium) that support bone health. To promote lasting bone health, the RDA for calcium increases from 1,000 to 1,200 mg/day at age 51 in women and at age 71 in men. However, older adults often find it difficult to meet this recommended intake. The occurrence of lactose intolerance increases with age, inhibiting the ability of many to consume calcium-rich dairy products. To counter this problem, health care providers often recommend lactose-reduced dairy products, calcium-fortified foods such as soy milk, and/or supplements.

Considerable evidence shows that older adults, especially those who live in northern regions of the United States, are at increased risk for vitamin D deficiency. This is likely caused by limited exposure to sunlight. In addition, the ability to synthesize vitamin D from cholesterol via sunlight decreases with age. For these reasons, the RDA for vitamin D increases from 15 to 20 μg/day at age 71. Although magnesium and phosphorus are also important for bone health, aging does not seem to affect dietary requirements for these nutrients.

Hormonal Changes in Women As women age, they experience a natural and progressive decline in estrogen levels. This time in a woman's life, referred to as **perimenopause,** can lead to a variety of other physical changes, including irregular menstrual cycles. Unfortunately, the decline in estrogen during this perimenopausal stage of life increases the rate of bone loss, making some women's bones weak and fragile. By the time a woman reaches **menopause,** around 50 to 60 years of age, her ovaries are producing very little estrogen and menstrual cycles completely stop. Women who have dense bones before menopause have fewer problems.

Menopause is a normal part of aging and affects each woman differently. Some women experience profound emotional and physiological changes, whereas others experience no discomfort at all. To get relief from menopausal symptoms, some women are prescribed hormone replacement therapy (HRT). Although HRT can protect against bone loss, it is associated with increased risk for cancers of the breast and uterus.[81] It is important for women to discuss the benefits and risks associated with HRT with their health care providers. Instead of HRT some women seek more natural options to ease them through the transitional period of menopause. These "natural" alternatives include certain foods such as soy, bioidentical hormone therapy, and herbal products such as black cohosh. You can read more about these alternatives in the Focus on Clinical Applications.

Because monthly blood loss associated with menstruation has stopped, menopause can improve iron status. In fact, the RDA for iron decreases for women during this life stage, from 18 to 8 mg/day—the same as recommended

⟨**CONNECTIONS**⟩ Recall that osteoporosis is caused by a reduction in bone mass and can lead to fractures and loss of stature in older adults.

⟨**CONNECTIONS**⟩ Hormone replacement therapy is taken by some women to restore estrogen levels that decline as a result of menopause or the surgical removal of the ovaries.

perimenopause Literally, the time "around" the time of menopause.

menopause The time in a woman's life when menstruation ceases, usually during the sixth decade of life.

Most women recognize that menopause is a natural physiological event and not a disease. Nonetheless, many struggle with the physical discomforts caused by declining estrogen levels. While some women experience no discomfort at all, others are debilitated by recurring hot flashes, night sweats, mood changes, and insomnia. Until recently, many women have chosen to ease menopausal-related discomforts with hormone replacement therapy (HRT). HRT restores hormonal balance by providing a combination of estrogen and progesterone. HRT was also thought to protect women from bone loss, Alzheimer's disease, heart attacks, and certain types of cancer. However, a landmark research study, the Women's Health Initiative, contradicted many of these health claims.[82] Rather than providing protection, HRT increased women's risks of breast cancer, heart attacks, strokes, and blood clots. These risks clearly outweighed the major benefits associated with HRT, specifically decreased risk of hip fractures and colorectal cancer. Since the release of these findings, there has been a steady decline in HRT use among women of all ages.[83] Many perimenopausal women are now seeking what they perceive to be safer alternatives to HRT such as diet, bioidentical hormone therapy, and herbal remedies.

Some of the most commonly recommended foods to be eaten during perimenopause and menopause are those made from soybeans. This is because soybeans contain phytochemicals called isoflavones that are structurally similar to estrogen. However, isoflavones do not provide relief from menopausal symptoms in all women. The North American Menopause Society recognizes that while some studies suggest a modest benefit from isoflavones in relieving menopausal symptoms, other studies fail to show such an effect.[84]

As an alternative to synthetic hormones, an increasing number of women are turning to "bioidentical" hormones—custom-formulated compounds that match the structure and function of hormones produced naturally in the body. Saliva or blood tests are used to determine the types and amounts of hormones needed by a particular woman. Although some practitioners claim that bioidentical hormones may be safer and have fewer

side effects than traditional HRT, more research is needed to better understand their safety and efficacy.[85] A position paper issued by the American College of Obstetricians and Gynocologists cautions women about taking some bioidentical hormones because these compounds have not undergone rigorous testing.[86] Issues regarding purity, potency, safety, and efficacy of bioidentical hormones are also of concern. In fact, the U.S. Food and Drug Administration advises women who decide to use these hormones to use them at the lowest dose that helps and for the shortest time needed.[87]

In addition to dietary changes and hormonal preparations, some women turn to herbal products to ease menopausal symptoms. The most commonly used products are black cohosh (*Cimicifuga racemosa*), St. John's wort (*Hypericum perforatum*), and valerian (*Valeriana officinalis*). Although not conclusive, several studies have confirmed that bioactive components in black cohosh may relieve hot flashes in some women.[88] Some menopausal women also use St. John's wort to relieve mild to moderate depression. In fact, several studies show that women taking St. John's wort and black cohosh together had the most improvement in mood and anxiety during menopause.[89] Although some women report that valerian is an effective sleep aid, few studies show its effectiveness.[90] In addition, valerian does not appear to be effective in reducing the occurrence of depression or anxiety.

It is not surprising that many women are turning to alternative therapies to manage perimenopausal and menopausal symptoms. However, women must be cautioned that nonconventional treatments have not been extensively studied and little is known about their long-term effects or safety. In addition to these alternative approaches, it is equally important for women to recognize that an active and healthy lifestyle can also help ease menopausal discomforts. For example, exercise, stress management, and relaxation techniques can improve sleeplessness, mood swings, anxiety, and depression. Exercise also helps prevent the unwanted weight gain that many women experience during this phase of life.

for adult men. Regardless, adequate iron intakes remain a concern, especially for older adults, who often limit their intake of meat, poultry, and fish.

Changes in the Gastrointestinal Tract Age-related changes in the GI tract affect nutritional status in adults, especially older adults. For example, aging muscles can become less responsive to neural signals, slowing the movement of food through the GI tract. Decreased GI motility is one reason why constipation is more common in older adults than in younger ones. When fecal material remains in the colon for prolonged periods, too much water reabsorption can occur, resulting in hard, compacted stool. Drinking adequate amounts of fluids and eating fiber-rich foods can improve GI function and help prevent this problem. Although the recommended fiber intake for older adults is 20 to 35 g/day, older men and women consume on average only 18 and 14 g/day, respectively.[91] There are many reasons as to why older adults may be reluctant to consume foods rich in dietary fiber, but it is important that they find ways to incorporate fiber into their diets.

Neuromuscular changes in the GI tract can also make swallowing more difficult. Difficulty swallowing, a condition called *dysphagia*, is a common cause of choking. This frightening experience can cause older adults to avoid eating certain foods. Preparing foods so that they are moist and have a soft texture is sometimes helpful. People who have difficulty swallowing should seek medical assistance.

With increased age, there is also inflammation of the stomach accompanied by a decline in the number of cells that produce gastric secretions such as hydrochloric acid and intrinsic factor. This condition, called **atrophic gastritis,** can decrease the bioavailability of nutrients such as calcium, iron, biotin, folate, vitamin B_{12}, and zinc. Without intrinsic factor, vitamin B_{12} cannot be absorbed, which can lead to vitamin B_{12} deficiency. Pernicious anemia, which is different from the type of vitamin B_{12} deficiency associated with aging, can also occur in the elderly. Symptoms associated with vitamin B_{12} deficiency include dementia, memory loss, irritability, delusions, and personality changes, all of which can easily be overlooked in older adults. To avoid the development of vitamin B_{12} deficiency, older adults are often advised to take vitamin B_{12} supplements or consume adequate amounts of foods fortified with vitamin B_{12}.

Other Nutritional Issues in Older Adults Other age-related physiological changes can contribute to nutritional deficiencies in older adults. For example, problems with oral health, missing teeth, or poorly fitting dentures can make food less enjoyable, often limiting the types of foods a person eats. Nutrient-dense foods that require chewing, such as meat, fruits, and vegetables, can cause pain, embarrassment, and discomfort for older adults and therefore may be avoided. It is important for older people who are experiencing problems with oral health to get proper dental care.

Changes in Sensory Stimuli Sensory changes in taste and smell can also affect food intake in older adults. With age, the ability to smell diminishes, making food tasteless and unappealing. Certain medications can also alter taste and diminish appetite. Older adults may find that adding spices to foods makes them more appealing, flavorful, and enjoyable. With advanced age, some people also experience a pattern of weight loss caused by a general decline in appetite, or what is called **anorexia of aging.** Anorexia of aging puts individuals at especially increased risk for protein-energy malnutrition.

Inadequate Fluid Intake The sensation of thirst can also become blunted with age. Because of this, many older adults do not consume enough fluid and are at increased risk for dehydration. Dehydration can upset the balance of electrolytes in cells and tissues. A lack of fluid can also disrupt bowel function and can exacerbate constipation. Not only can certain medications increase water loss from the body, but some older adults intentionally limit fluid intake because of embarrassment over loss of bladder control. Symptoms of dehydration, which are often overlooked in the elderly, include headache, dizziness, fatigue, clumsiness, visual disturbances, and confusion. Adequate fluid consumption and early detection of dehydration in older adults is very important.

⟨**CONNECTIONS**⟩ Recall that intrinsic factor is produced by the gastric cells and is needed for the absorption of vitamin B_{12}.

atrophic gastritis (gas – TRI – tis) Inflammation of the mucosal membrane that lines the stomach, reducing the ability of the stomach to produce gastric secretions.

anorexia of aging Loss of appetite in the elderly that leads to weight loss and overall physiological decline.

The AI for total water (drinking water, beverages, and foods) for adults is 2.7 and 3.7 liters/day for women and men, respectively. To stay fully hydrated, older adults should drink 9 to 13 cups of water or other beverages every day, plus additional fluids from foods with high water contents.

Drug–Nutrient Interactions The elderly are at particularly high risk for adverse drug–nutrient interactions. This is, in part, because older adults often take multiple medications to treat a variety of chronic diseases. The most frequently prescribed medications for people over 50 years of age are for arthritis, hypertension, type 2 diabetes, cancer, and heart disease. The more medications a person takes, the greater the risk of experiencing a drug–nutrient interaction. While some drugs can cause loss of appetite, leading to inadequate food intake, others can alter taste, making food unpleasant to eat. Even nutrient absorption can be affected by medications. It is important for elderly people to be aware of such problems and seek advice regarding the nutrient-related side effects of their medications.

ASSESSING NUTRITIONAL RISK IN OLDER ADULTS

Clearly, many factors put older adults at increased nutritional risk. For this reason, several national health organizations jointly sponsored the Nutritional Screening Initiative (NSI), a collaborative effort to improve nutritional health in older adults. The NSI helped to identify risk factors closely associated with poor nutritional status in this group. Once compiled, these risk factors were used to develop a screening tool called the NSI DETERMINE checklist, shown in Figure 17.12. As you can see, each nutritional risk factor is represented in the DETERMINE acronym. For example "D" stands for "disease," and "E" refers to "eating poorly." Individuals with a nutritional score of 6 or more on the NSI DETERMINE checklist are considered to be at high nutritional risk.

A Meals on Wheels volunteer delivers meals to a home-bound elderly person.

Many services help improve nutritional status and overall health for older adults. For example, most communities have congregate meal programs where people can enjoy nutritious, low-cost meals in the company of others. Many congregate meal programs are federally subsidized, and total cost is often based on an individual's ability to pay. Another program that provides meals to senior citizens is organized and implemented by the Meals on Wheels® Association of America. This program, which relies heavily on volunteers, delivers low-cost meals to home-bound elderly adults and others who are disabled. The mission of Meals on Wheels is to provide national leadership to end senior hunger, with a vision to end senior hunger by 2020. The federally funded Supplemental Nutrition Assistance Program (SNAP; formerly known as the Food Stamp Program) also assists with food-related expenses of low-income seniors.

Critical Thinking: Amy's Story What organizations in your community are available to assist low-income families and home-bound elderly individuals? Do any of these programs distribute fresh produce? If not, what would be the barriers to launching such an organization? What would be the benefits? What makes Backyard Harvest unique in terms of assisting community members?

FIGURE 17.12 Nutritional Screening Initiative DETERMINE Checklist The Nutrition Screening Initiative was an effort by several organizations to develop screening tools to assess risk factors associated with poor nutritional health in older adults.

DETERMINE YOUR NUTRITIONAL HEALTH

Circle the number in the "yes" column for those that apply to you or someone you know. Total your nutritional score.

	YES
I have had an illness or condition that made me change the kind and/or amount of food I eat.	2
I eat fewer than 2 meals per day.	3
I eat few fruits or vegetables or milk products.	2
I have 3 or more drinks of beer, liquor or wine almost every day.	2
I have tooth or mouth problems that make it hard for me to eat.	2
I don't always have enough money to buy the food I need.	4
I eat alone most of the time.	1
I take 3 or more different prescribed or over-the-counter drugs a day.	1
Without wanting to, I have lost or gained 10 pounds in the last 6 months.	2
I am not always physically able to shop, cook and/or feed myself.	2
TOTAL	

Total Your Nutritional Score. If it's—

0–2	Good! Recheck your nutritional score in 6 months.
3–5	You are at moderate nutritional risk. See what can be done to improve your eating habits and lifestyle.
6 or more	You are at high nutritional risk. Bring this Checklist the next time you see your doctor, dietitian or other qualified health or social service professional.

DETERMINE: Warning signs of poor nutritional health.

DISEASE
Any disease, illness or chronic condition which causes you to change the way you eat, or makes it hard for you to eat, puts your nutritional health at risk.

EATING POORLY
Eating too little and eating too much both lead to poor health. Eating the same foods day after day or not eating fruit, vegetables, and milk products daily will also cause poor nutritional health.

TOOTH LOSS/MOUTH PAIN
A healthy mouth, teeth and gums are needed to eat. Missing, loose or rotten teeth or dentures which don't fit well, or cause mouth sores, make it hard to eat.

ECONOMIC HARDSHIP
As many as 40% of older Americans have incomes of less than $6,000 per year. Having less—or choosing to spend less—than $25–30 per week for food makes it very hard to get the foods you need to stay healthy.

REDUCING SOCIAL CONTACT
One-third of all older people live alone. Being with people daily has a positive effect on morale, well-being and eating.

MULTIPLE MEDICINES
Many older Americans must take medicines for health reasons. Almost half of older Americans take multiple medicines daily.

INVOLUNTARY WEIGHT LOSS/GAIN
Losing or gaining a lot of weight when you are not trying to do so is an important warning sign that must not be ignored. Being overweight or underweight also increases your chance of poor health.

NEEDS ASSISTANCE IN SELF CARE
Although most older people are able to eat, one of every five have trouble walking, shopping, buying food and cooking food, especially as they get older.

ELDER YEARS ABOVE AGE 80
Most older people lead full and productive lives. But as age increases, risk of frailty and health problems increase. Checking your nutritional health regularly makes good sense.

SOURCE: Adapted from Nutrition Screening Initiative. Report of Nutrition Screening I: Toward a Common View. Washington DC: Nutrition Screening Initiative; 1991.

Diet Analysis PLUS ✚ Activity

Assume that you have a sister or daughter, Sarah, who is 25 years old, weighs 130 pounds, is 5 feet 4 inches tall, is a nonsmoker, and is not a vegetarian. Sarah has recently discovered she is pregnant and needs advice as to her changing nutritional needs. She plans to breastfeed almost exclusively during the first 6 months of her baby's life. She would like to know more about her nutritional needs during that time as well. Using your Diet Analysis program, set up a profile using Sarah's information. Print out the *Profile*

DRI Goals report choosing simply "Female" from the "Sex" drop-down menu to generate a report representing her nutritional needs in the nonpregnant condition. Then, go back and choose "Pregnant" from the drop-down menu, changing nothing else. Print out this second *Profile DRI Goals* report. Finally, repeat the steps choosing "Lactating" from the menu. By now you should have three *Profile DRI Goals* reports representing the three physiological conditions.

Using these print-outs and this chapter, fill in dietary recommendations for energy and essential nutrients on the following table:

Dietary Component	Not Pregnant	Pregnant	Lactating
Calories (kcal/day)			
Protein (g/day)			
Carbohydrates (g/day)			
Total fat (g/day)			
Linoleic acid (g/day)			
Linolenic acid (g/day)			
Vitamin A (µg/day)			
Folate (µg DFE/day)			
Vitamin C (mg/day)			
Calcium (mg/day)			
Iron (mg/day)			

Analysis

1. In general, what purpose(s) do the nutrient increases during the pregnant phase fulfill? Are there specific organ systems or processes (of the mother and/or the infant) that these particular nutrients support?

2. Why do you suppose that calorie needs during the lactating phase (the second 6 months) are different than those during pregnancy?

3. Why are calcium recommendations unchanged during pregnancy and throughout lactation for this 25-year-old?

Notes

1. Centers for Disease Control and Prevention. National Center for Health Statistics. Clinical growth charts. Available from: http://www.cdc.gov/nchs/about/major/nhanes/growthcharts/clinical_charts.htm. World Health Organization. The WHO child growth standards. 2006. Available from: http://www.who.int/childgrowth.

2. National Center for Health Statistics. Data on birth defects. Available from: http://www.cdc.gov/nchs/FASTATS/bdefects.htm.

3. Brent RL. Environmental causes of human congenital malformations: The pediatrician's role in dealing with these complex clinical problems caused by a multiplicity of environmental and genetic factors. Pediatrics. 2004;113:957–68.

4. Eustace LW, Kang DH, Coombs D. Fetal alcohol syndrome: A growing concern for health care professionals. Journal of Obstetrics, Gynecologic, and Neonatal Nursing. 2003;32:215–21.

5. Villar J, Merialdi M, Gulmezoglu AM, Abalos E, Carroli G, Kulier R, de Onis M. Characteristics of randomized controlled trials included in systematic reviews of nutritional interventions reporting maternal morbidity, mortality, preterm delivery, intrauterine growth restriction and small for gestational age and birth weight outcomes. Journal of Nutrition. 2003;133:1632S–9S.

6. National Center for Health Statistics. Centers for Disease Control and Prevention. Preliminary births for 2004: Infant and maternal health. Available from: http://www.cdc.gov/nchs/products/pubs/pubd/hestats/prelimbirths04/prelimbirths04health.htm.

7. Centers for Disease Control and Prevention. U.S. Department of Health and Human Services. Martin JA, Hamilton BE, Sutton PD, Ventura SJ, Mathews TH, Osterman MJK. Births: Final data for 2008. National vital statistics reports; vol 59 no 1. Hyattsville, MD: National Center for Health Statistics. 2010. Available from http://www.cdc.gov/nchs/data/nvsr/nvsr59/nvsr59_01.pdf.

8. Gillman MW. Developmental origins of health and disease. New England Journal of Medicine. 2005;353:1848–50.

9. Barker DJ. The foetal and infant origins of inequalities in health in Britain. Journal of Public Health Medicine. 1991;13:64–8.

10. Barker DJ. Fetal programming of coronary heart disease. Trends in Endocrinology and Metabolism. 2002;13:364–8.

11. Rasmussen KM. The "fetal origins" hypothesis: Challenges and opportunities for maternal and child nutrition. Annual Review of Nutrition. 2000;21:73–95.

12. Waterland RA, Michels KB. Epigenetic epidemiology of the developmental origins hypothesis. Annual Review of Nutrition. 2007;27:363–88.

13. Institute of Medicine and National Research Council. Weight gain during pregnancy: Reexamining the guidelines. Washington, DC: The National Academies Press. May 2009. Available from: http://www.iom.edu/pregnancyweightgain.

14. U.S. Department of Health and Human Services and U.S. Environmental Protection Agency. What you need to know about mercury in fish and shellfish. March 2004. Available from http://www.cfsan.fda.gov/~dms/admehg3.html.

15. Beard JL. Effectiveness and strategies of iron supplementation during pregnancy. American Journal of Clinical Nutrition. 2000;71:1288S–94S.

16. U.S. Preventive Services Task Force. Recommendation Statement. Screening for iron deficiency anemia—Including iron supplementation for children and pregnant women. Rockville, MD: Agency for Healthcare Research and Quality (AHRQ): 2006. Available from http://www.uspreventiveservicestaskforce.org/.

17. Feinleib M, Beresford SA, Bowman BA, Mills JL, Rader JI, Selhub J, Yetley EA. Folate fortification for the prevention of birth defects: Case study. American Journal of Epidemiology. 2001;154:S60–9.

18. Misra DP, Astone N, Lynch CD. Maternal smoking and birth weight: Interaction with parity and mother's own in utero exposure to smoking. Epidemiology. 2005;16:288–93.

19. Husgafvel-Pursiainen K. Genotoxicity of environmental tobacco smoke: A review. Mutation Research. 2004;567:427–45.

20. Di Cianni G, Ghio A, Resi V, Volpe L. Gestational diabetes mellitus: An opportunity to prevent type 2 diabetes and cardiovascular disease in young women. Women's Health. 2010;6:97–105.

21. Papageorghiou AT, Campbell S. First trimester screening for preeclampsia. Current Opinion in Obstetrics and Gynecology. 2006;18:594–600.

22. American Academy of Pediatrics. Pediatric nutrition handbook, 6th ed., Elk Grove Village, IL. 2008.

23. Leon DA, Ronalds G. Breast-feeding influences on later life—cardiovascular disease. Advances in Experimental and Medical Biology. 2009;639;153–66.

24. Beyerlein A von Kries R. Breastfeeding and body composition in children: will there ever be conclusive empirical evidence for a protective effect against overweight? American Journal of Clinical Nutrition. 2011.

25. Stettler N, Zemel BS, Kumanyika S, Stallings VA. Infant weight gain and childhood overweight status in a multicenter, cohort study. Pediatrics. 2002;109:194–9.

26. Agostoni C, Giovannini M. Cognitive and visual development: Influence of differences in breast and formula fed infants. Nutrition and Health. 2001;15:183–8.

27. Dewey KG. Impact of breastfeeding on maternal nutritional status. Advances in Experimental Medical Biology. 2004; 554:91–100. Winkvist A, Rasmussen KM. Impact of lactation on maternal body weight and body composition. Journal of Mammary Gland Biology and Neoplasia. 1999;4:309–18.

28. Taylor JS, Kacmar JE, Nothnagle M, Lawrence RA. A systematic review of the literature associating breastfeeding with type 2 diabetes and gestational diabetes. Journal of the American College of Nutrition. 2005;24:320–6. Labbok MH. Effects of breastfeeding on the mother. Pediatric Clinics of North America. 2001;48:143–58.

29. Centers for Disease Control and Prevention. Use of world health organization and CDC growth charts for children aged 0–59 months in the United States. Morbidity and Mortality Weekly Review. 2010;59:1-15. Available from http://www.cdc.gov/mmwr/preview/mmwrhtml/rr5909a1.htm.

30. Ryan AS, Wenjun Z, Acosta A. Breastfeeding continues to increase into the new millennium. Pediatrics. 2002;110:1103–9.

31. Grummer-Strawn LM, Scanlon KS, Fein SB. Infant feeding and feeding transitions during the first year of life. Pediatrics. 2008;122:S36–42.

32. U.S. Department of Health and Human Services. The Surgeon General's call to action to support breastfeeding. Washington, DC: U.S. Department of Health and Human Services, Office of the Surgeon General; 2011. Available from http://www.surgeongeneral.gov.

33. U.S. Centers for Disease Control and Prevention. Breastfeeding among U.S. children born 1999–2007. CDC National Immunization Survey. Available from http://www.cdc.gov/breastfeeding/data/NIS_data/.

34. Lind T, Hernell O, Lonnerdal B, Stenlund H, Domellof M, Persson LA. Dietary iron intake is positively associated with hemoglobin concentration during infancy but not during the second year of life. Journal of Nutrition. 2004;134:1064–70.

35. Beyerlein A, Hadders-Algra M, Kennedy K, Fewtrell M, Singhal A, Roesenfeld E, Lucas A, Bouwstra H, Koletzko B, von Kries R. Infant formula supplementation with long-chain polyunsaturated fatty acids has no effect on Bayley developmental scores at 18 months of age—IPD meta-analysis of 4 large cliniical trials. Journal of Pediatric Gastroenterology and Nutrition. 2010;50:79–84.

36. Camadoo L, Tibbott R, Isaza F. Maternal vitamin D deficiency associated with neonatal hypocalcaemic convulsions. Nutrition Journal. 2007;6:23.

37. Basile LA, Taylor SN, Wagner CL, Horst RL, Hollis BW. The effect of high-dose vitamin D supplementation on serum vitamin D levels and milk calcium concentration in lactating women and their infants. Breastfeeding Medicine. 2006;1:27–35.

38. Wagner CL and Greer FR. Prevention of rickets and vitamin D deficiency in infants, children, and adolescents Pediatrics. 2008;122:1142–52.

39. American Academy of Pediatrics. Breastfeeding and the use of human milk. Pediatrics. 2005;115:496–501. Brotanek JM, Gosz J, Weitzman M, Flores G. Iron deficiency in early childhood in the United States: Risk factors and racial/ethnic disparities. Pediatrics. 2007;120:568–75.

40. American Academy of Pediatrics Committee on Nutrition. The use and misuse of fruit juice in pediatrics. Pediatrics. 2001;107:1210–3.

41. Huh SY, Rifas-Shiman SL, Taveras EM, Oken EM, Gillman MW. Timing of solid food introduction and risk of obesity in preschool aged children. Pediatrics. 2011;127:544–51.

42. Kleinman RE. American Academy of Pediatrics recommendations for complementary feeding. Pediatrics. 2000;106:1274.

43. Thygarajan A, Burks AW. American Academy of Pediatrics recommendations on the effects of early nutritional interventions on the development of atopic disease. Current Opinion in Pediatrics. 2008;20:698–702. Greer FR, Sicherer SH, Burks W; American Academy of Pediatrics Committee on Nutrition; American Academy of Pediatrics Section on Allergy and Immunology. Effects of early nutritional interventions on the development of atopic disease in infants and children: The role of maternal dietary restriction, breastfeeding, timing of introduction of complementary foods, and hydrolyzed formulas. Pediatrics. 2008;121:183–91.

44. Tanzi MG, Gabay MP. Association between honey consumption and infant botulism. Pharmacotherapy. 2002;22:1479–83. Centers for Disease Control and Prevention. Botulism. Available from: http://www.cdc.gov/nczved/divisions/dfbmd/diseases/botulism/.

45. Whitaker RC, Wright JA, Pepe MS, Seidel KD, Dietz WH. Predicting obesity in young adulthood from childhood and parental obesity. New England Journal of Medicine. 1997;337:869–73.

46. Freedman DS, Mei Z, Srinivasan SR, Berenson GS, Dietz WH. Cardiovascular risk factors and excess adiposity among overweight children and adolescents: the Bogalusa Heart Study. Journal of Pediatrics. 2007;150:12–17.

47. National Center for Health Statistics. Health, United States, 2010: With Special Feature on Death and Dying. Hyattsville, MD. 2011.

48. Satter E. The feeding relationship: Problems and interventions. Journal of Pediatrics. 1990;117:181–9.

49. l'Allemand-Jander D. Clinical diagnosis of metabolic and cardiovascular risks in overweight children: early development of chronic diseases in the obese child. Internatiolnal Journal of Obesity. 2010;34:S32–6.

50. Robinson TN. Television viewing and childhood obesity. Pediatric Clinics of North America. 2001;4:1017–25.

51. Crespo CJ, Smit E, Troiano RP, Bartlett SJ, Macera CA, Andersen RE. Television watching, energy intake, and obesity in US children: Results from the third National Health and Nutrition Examination Survey, 1988–1994. Archives of Pediatric and Adolescent Medicine. 2001;155:360–5.

52. Chaput J-P, Visby T, Nyby S, Klingenberg L, Gregersen NT, Tremblay A, Astrup A, Sjödin A. Video game playing increases food intake in adolescents: a randomized crossover study. American Journal of Clinical Nutrition. 2011; 93:1196–203.

53. American Academy of Pediatrics Committee on Public Education. Children, adolescents, and television. Pediatrics. 2001;107:423–6.

54. Robinson TN, Borzekowski DL, Matheson DM, Kraemer HC. Effects of fast food branding on young children's taste preferences. Archives of Pediatric and Adolescent Medicine. 2007;161:792–7.

55. Kann L, Brener ND, Wechsler H. Overview and summary: School health policies and programs study 2006. Journal of School Health. 2007;77:385–97.

56. U.S. Department of Health and Human Services. 2008 physical activity guidelines for Americans. Washington, D.C. Available from http://www.health.gov/paguidelines/pdf/paguide.pdf.

57. American Academy of Pediatrics Committee on Public Education. Soft drinks in schools. Pediatrics. 2004;113:152–4.

58. Institute of Medicine. Schools can play a role in preventing childhood obesity. Preventing childhood obesity: Health in the balance. 2005. Available from: http://www.iom.edu.

American Academy of Pediatrics. Soft drinks in schools. Pediatrics. 2004;113:152–4.

59. Koivisto Hursti UK. Factors influencing children's food choice. Annals of Medicine. 1999;31:26–32.

60. Patrick H, Nicklas TA. A review of family and social determinants of children's eating patterns and diet quality. Journal of American College of Nutrition. 2005;2:83–92.

61. Centers for Disease Control and Prevention. Recommendations to prevent and control iron deficiency in the United States. Morbidity and Mortality Weekly Review. 1998;47(RR-3):1–29. Kazal LA. Prevention of iron deficiency in infants and toddlers. American Family Physician. 2002;66:1217–25.

62. Soemantri AG, Pollitt E, Kim I. Iron deficiency anemia and educational achievement. American Journal of Clinical Nutrition. 1985;42:1221–8. Iannotti LL, Tielsch JM, Black MM, Black RE. Iron supplementation in early childhood: Health benefits and risks. American Journal of Clinical Nutrition. 2006;84:1261–76.

63. Lee JM, Appugliese D, Kaciroti N, Corwyn RF, Bradley RH, Lumeng JC. Weight status in young girls and the onset of puberty. Pediatrics. 2007;119:e624–30.

64. Enns CW. Mickle SH, Goldman JD. Trends in food and nutrient intakes by adolescents in the United States. Family Economics and Nutrition Review. 2003;15:15–27.

65. Neumark-Sztainer D, Story M, Hannan PH, Croll J. Overweight status and eating patterns among adolescents: where do youths stand in comparison with Healthy People 2010 objectives? American Journal of Public Health. 2002;92:844–51. Larson NI, Neumark-Sztainer D, Hannan PJ, Story M. Trends in adolescent fruit and vegetable consumption, 1999–2004: Project EAT. American Journal of Preventive Medicine. 2007;32:147–50.

66. U.S. Department of Agriculture. Agricultural Research Service. Snacking patterns of U.S. adolescents. What we eat in America, NHANES 2005–2006. Food Surveys Research Group. Dietary Data Brief No. 2. October 2010. Available from http://www.ars.usda.gov/SP2UserFiles/Place/12355000/pdf/DBrief/snacking_0506.pdf.

67. Nielsen SJ, Popkin BM. Changes in beverage intake between 1977 and 2001. American Journal of Preventive Medicine. 2004;27:205–10.

68. Troiano RP, Briefel RR, Carroll MD, Bialostosky K. Energy and fat intakes of children and adolescents in the United States: Data from the National Health and Nutrition Examination Surveys 1,2,3. American Journal of Clinical Nutrition. 2000:72;1343S–53s.

69. Bailey RL, Dodd KW, Goldman JA, Gahche JJ, Dwyer JT, Moshfegh AJ, Sempos CT, Picciano MF. Estimation of total usual calcium and vitamin D intakes in the United States. Journal of Nutrition. 2010;140:817–22.

70. What We Eat In America, NHANES, 2003–2004. U.S. Department of Agriculture, Agricultural Research Service. 2007. Nutrient intakes from food: Mean amounts consumed per individual, one day, 2003–2004. Available from: http://jn.nutrition.org/cgi/content/full/133/2/609S.

71. Harel Z, Riggs S, Vaz R, White L, Menzies G. Adolescents and calcium: What they do and do not know and how much they consume. Journal of Adolescent Health. 1998;22:225–8. Martin JT, Coviak CP, Gendler P, Kim KK, Cooper K, Rodrigues-Fisher L. Female adolescents' knowledge of bone health promotion behaviors and osteoporosis risk factors. Orthopaedic Nursing. 2004;23:235–44.

72. Centers for Disease Control and Prevention. Recommendations to prevent and control iron deficiency anemia in

the United States. Morbidity Mortality Weekly Report. 2002;51:897–9.

73. Bailey RL, Dodd KW, Gahche JJ, Dwyer JT, McDowell MA, Yetley EA, Sempos CA, Burt VL, Radimer KL, Picciano MF. Total folate and folic acid intake from foods and dietary supplements in the United States: 2003–2006. American Journal of Clinical Nutrition. 2010;91:231–7.

74. Centers for Disease Control and Prevention. The state of aging and health in America. 2004. Available from: http://www.cdc.gov/aging/pdf/state_of_aging_and_health_in_america_2004.pdf.

75. Shay JW, Wright WE. Hayflick, his limit, and cellular ageing. National Review of Molecular Cellular Biology. 2000;1:72–6.

76. Willcox JK, Ash SL, Catignani GL. Antioxidants and prevention of chronic disease. Critical Reviews in Food Science and Nutrition. 2004;44:275–95.

77. Stanner SA, Hughes J, Kelly CN, Buttriss J. A review of the epidemiological evidence for the antioxidant hypothesis. Journal of Public Health Nutrition. 2004;7:407–22.

78. Chernoff R. Geriatric nutrition: The health professional's handbook, 3rd ed. Gaithersburg, MD: Aspen; 2006. Moretti C, Frajese GV, Guccione L, Wannenes F, De Martino MU, Fabbri A, Frajese G. Androgens and body composition in the aging male. Journal of Endocrinological Investigation. 2005;28:56–64.

79. Centers for Disease Control and Prevention. Falls among older adults: An overview. 2010. Available from http://www.cdc.gov/HomeandRecreationalSafety/Falls/adultfalls.html.

80. Chernoff R. Protein and older adults. Journal of the American College of Nutrition. 2004;23:627S–30S.

81. Beral V, Banks E, Reeves G. Evidence from randomised trials on the long-term effects of hormone replacement therapy. Lancet. 2002;360:942–4.

82. Nelson HD, Humphrey LL, Nygren P, Teutsch SM, Allan JD. Postmenopausal hormone replacement therapy: Scientific review. Journal of the American Medical Association. 2002;288:872–81.

83. Kim N, Gross C, Curtis J, Stettin G, Wogen S, Choe N, Krumholz HM. The impact of clinical trials on the use of hormone replacement therapy: A population-based study. Journal of General Internal Medicine. 2005;20:1026–31. Hersh AL, Stefanick ML, Stafford RS. National use of postmenopausal hormone therapy: Annual trends and response to recent evidence. Journal of the American Medical Association.2004;291:47–53.

84. The role of soy isoflavones in menopausal health: report of The North American Menopause Society. Wulf H. Utian Translational Science Symposium in Chicago, IL. 2010. Menopause. 2011.

85. Cirigliano M. Bioidentical hormone therapy: A review of the evidence. Journal of Women's Health. 2007;16:600–31. Moskowitz D. A comprehensive review of the safety and efficacy of bioidentical hormones for the management of menopause and related health risks. Alternative Medicine Review. 2006;11:208–23. Boothby LA, Doering PL, Kipersztok S. Bioidentical hormone therapy: A review. Menopause. 2004;11:356–67.

86. The American College of Obstetricians and Gynecologists. Compounded Bioidentical Hormones. Obstetrics and Gynecology. 2005;322:1139–40.

87. U.S. Food and Drug Administration. Bio-identicals: sorting myths from facts. Consumer Updates. April 8, 2008. Available from http://www.fda.gov/ForConsumers/Consumer Updates/ucm049311.htm.

88. Rotem C, Kaplan B. Phyto-Female Complex for the relief of hot flushes, night sweats and quality of sleep: Randomized, controlled, double-blind pilot study. Gynecological Endocrinology. 2007;23:117–22. Oktem M, Eroglu D, Karahan HB, Taskintuna N, Kuscu E, Zeyneloglu HB. Black cohosh and fluoxetine in the treatment of postmenopausal symptoms: A prospective, randomized trial. Advances in Therapy. 2007;24:448–61.

89. Geller SE, Studee L. Botanical and dietary supplements for mood and anxiety in menopausal women. Menopause. 2007;14:541–9. Uebelhack R, Blohmer JU, Graubaum HJ, Busch R, Gruenwald J, Wernecke KD. Black cohosh and St. John's wort for climacteric complaints: A randomized trial. Obstetrics and Gynecology. 2006;107:247–55.

90. Taibi DM, Landis CA, Petry H, Vitiello MV. A systematic review of valerian as a sleep aid: Safe but not effective. Sleep Medicine Review. 2007;11:209–30.

91. Mozaffarian D, Kumanyika SK, Lemaitre RN, Olson JL, Burke GL, Siscovick DS. Cereal, fruit, and vegetable fiber intake and the risk of cardiovascular disease in elderly individuals. Journal of the American Medical Association. 2003;289:1659–66.

Food Security, Hunger, and Malnutrition

We have all experienced hunger—the physical sensation of not having enough to eat. While most people ease their hunger by eating, for others this may not always be possible. When sufficient food is not available or accessible, hunger can lead to serious physical, social, and psychological consequences, a condition referred to as food insecurity. The prevalence of food insecurity in the world is astonishing. Although nobody knows for sure the exact number of people who experience food insecurity, the United Nations Food and Agriculture Organization

(FAO) estimates that approximately 925 million people worldwide experience persistent hunger.[1]

It is important to know that hunger and food insecurity can affect people of all ages and in every country in the world—even wealthy countries such as the United States. In this Nutrition Matters, you will learn about concerns related to the causes and complications of food insecurity, hunger, and malnutrition worldwide. You will also learn about domestic and international assistance programs designed to alleviate these devastating public health problems.

What Is Food Security?

Food security is defined as the condition whereby people are able to obtain sufficient amounts of nutritious food for an active, healthy life. Conversely, **food insecurity** exists when people do not have adequate physical, social, or economic access to food.[2] Many people are surprised to learn that there is enough food produced in the world to provide every person with at least 2,700 kcal each day.[2] In other words, food insecurity is not caused by insufficient worldwide food production. Rather, people become food insecure when they cannot obtain sufficient food to feed themselves or their families. Other constraints such as limited physical or mental functioning can also contribute to food insecurity. In reality, the principal factors related to food insecurity in the world are poverty, war, and natural disaster.

There are varying degrees of food insecurity, with chronic hunger and undernutrition representing the more severe consequences. While the word "hunger" is often used to describe the physical discomfort experienced by individuals consuming insufficient food, on a global level it is more commonly used to describe a shortage of available food. Because many people live with uncertainty as to whether they will have enough to eat, food

food security The condition whereby individuals are able to obtain sufficient amounts of nutritious food to live active, healthy lives.

food insecurity The condition whereby individuals or families cannot obtain sufficient food.

insecurity and its resulting hunger are major social concerns in the world today.

PEOPLE RESPOND TO FOOD INSECURITY IN DIFFERENT WAYS

People living in food-insecure households respond to the threat of hunger in different ways. Whereas some individuals take advantage of charitable organizations that assist people in need, others may resort to obtaining food by stealing, begging, or scavenging.[3] For many people, food insecurity can cause feelings of alienation, deprivation, and distress. In addition, it can adversely affect family dynamics and social interactions within the larger context of community. Of course, food insecurity that results in hunger can also lead to malnutrition and its associated health complications. Clearly, the short- and long-term consequences of food insecurity for each person in a household can be devastating.

Prevalence of Food Insecurity in the United States

How much money do you spend on food each week? On average, Americans spend $44 per person on weekly food purchases.[4] Households experiencing food insecurity spend considerably less: The U.S. Department of Agriculture (USDA) recently reported that food-insecure households spend 33% less on food than food-secure households of the same size and composition.[5] Determining the extent of food insecurity is a scientific challenge. However, only when researchers have determined the scope and nature of the problem can public health officials develop effective strategies and target them to the appropriate populations. Just as there is no single cause of food insecurity, there is no one solution to the problem.

In prosperous countries, the prevalence of food insecurity is often difficult to assess because it is usually not associated with detectable signs of malnutrition. For this reason, clinical measurements of nutritional status such as anthropometry (for example, weight and height) are not always useful indicators of food insecurity. Instead, the prevalence and incidence of food insecurity in U.S households is typically assessed using data regarding food availability and access. For example, the USDA uses a survey that asks household members to answer questions about behaviors related to food availability and access.[5] Depending on the responses, individuals or households are classified as having either low food security or very low food security. **Low food security** refers to households in which one or more members experienced disrupted eating patterns (or were worried they would do so) because money or access to food was

Food insecurity refers to the situation in which households do not have sufficient food.

insufficient. Those classified as having **very low food security** experience multiple indications of food insecurity, sometimes leading to reduced food intake. Sixty-five percent of households classified as having very low food security reported that at least one member had gone hungry at some time during the year as a result of not having enough money for food.[5]

Recent survey results indicate that 15% of Americans—more than 17 million households—struggle to provide enough food for all their family members, and more than 17 million (23%) U.S. children live in households where food is scarce at times.[5] Approximately one-third of food insecure households reported that their eating patterns were greatly affected because they lacked money or other resources to purchase food. Families with incomes below the poverty line, many of which are households headed by single mothers, were at greatest risk of hunger. With the current downward trend in the U.S. economy and the increasing cost of food, which rose nearly 3% between 2010 and 2011, the number of U.S. households experiencing food insecurity is expected to increase even more.[5]

POVERTY IS THE UNDERLYING FACTOR ASSOCIATED WITH FOOD INSECURITY

A person's risk for experiencing food insecurity in the United States is associated with income, ethnicity, family structure, and location of the home. Because many of these risk factors are intertwined,

---

low food security Classification of food-insecure households in which one or more members experience disrupted eating patterns (or is worried they will do so) because of insufficient resources or access to food.

very low food security Households that experience multiple indications of disrupted eating patterns due to inadequate resources or access to food, sometimes leading to reduced food intake.

Millions of Americans struggle to get enough to eat. Although food banks help provide food to those in need, many report that critical shortages of food are making it difficult to meet the increasing demands.

AP Photo/The Hawk Eye, John Gaines

it can be difficult to determine how much each one independently contributes to food insecurity. Nonetheless, the link between income—specifically poverty—and food insecurity is indisputable and is often the "common thread" among these factors.

It is important to recognize that many people who live in poverty maintain steady employment. Based on data provided by the U.S. Census Bureau, there were 10 million low-income working families in 2009, and approximately 30% of working families are officially low-income.[6] In fact, experts estimate that 40% of adults requesting emergency food assistance are employed.[7] In 2009 alone, approximately 5 million households turned to private food pantries and soup kitchens to provide food for their families.[8] To make matters worse, people who at one time contributed items to food banks are now finding themselves in need.[9] It is not surprising that many food banks around the country are experiencing a sharp decline in donations as the demand for food rises.[10]

When money is limited, people are often forced to reduce food-related expenditures to pay for such things as housing, utilities, or health care.[11] Although one in three households living in poverty is food insecure, some households with incomes above the poverty line also experience food insecurity. This is because unexpected events such as a medical expense or a repair bill can cause some people, at least temporarily, to not have the financial means to purchase sufficient food. Thus, many factors in addition to income can predispose a person or family to food insecurity.

In addition to low income, food insecurity in the United States is more prevalent among certain ethnic groups. For example, black and Latino households are at higher risk for food insecurity than most other racial or ethnic groups.[12] Households headed by single women are at even higher risk. In addition, people living in urban and rural areas are more likely to experience food insecurity than those living in suburban regions. However, it is important to recognize that all three of these factors (ethnicity, head of household, and living location) are strongly associated with income status.[12]

Socioeconomically disadvantaged communities with limited access to affordable, healthy food, tend to be located in urban and rural low-income neighborhoods. These areas are referred to as **food deserts**, and residents who live in them are dependent on local food outlets that offer limited, expensive food choices with low nutritional value.[13] The lack of supermarkets within communities presents additional economic hardships to those living in food deserts. These individuals must either purchase groceries at local convenience stores or navigate public transportation to other communities that have more affordable market choices. These disparities have adverse health outcomes, and these populations tend to have high rates of obesity, type 2 diabetes, and cardiovascular disease.[14] Marketplace incentives to attract food retailers into food deserts are currently underway, with the hope that improved access to affordable healthy foods such as fruits, vegetables, whole grains, and low-fat dairy products will improve access to and affordability of quality food sources. You can learn more about where food deserts are located in the United States by visiting the USDA's Food Desert Locator at http://www.ers.usda.gov/data/fooddesert/.

What Are the Consequences of Food Insecurity?

Although food insecurity in the United States does not typically lead to hunger or nutrient deficiencies, it still represents a major public health concern. There are many consequences of food insecurity, and these have been studied most extensively in women and children. It is not uncommon for food-insecure mothers to shield their children from not having enough to eat by consuming less food themselves. Therefore, women often experience the negative consequences of food insecurity before their children do.[15] For example, some studies have found that women in food-insecure households

---

food desert Community with limited access to affordable, healthful foods; tends to be located in urban and rural low-income neighborhoods.

reduce the size of their meals in order to better provide for their children.[15] Nonetheless, it is the children raised in food-insecure households who experience the most significant and lasting, long-term effects.[16] Studies also show that these children tend to have difficulties in school, earn lower scores on standardized tests, miss more days of school, exhibit more behavioral problems and depression, and have increased risk for suicide. Of course, these factors are likely not direct consequences of food insecurity, but instead repercussions common to households experiencing this problem.[17] Although parents often try to protect their children from the realities of food insecurity, interviews with children reveal considerable circumstantial awareness that is not always apparent to other family members.[18]

It is also important to understand that the elderly are at especially high risk for experiencing food insecurity.[19] However, they often experience it differently from both children and younger adults. Elders with limited mobility and poor health, for example, may have food available to them but have difficulty or anxiety associated with meal preparation. In addition, many older people have relatively low incomes, which they are unable to supplement by additional employment. Poverty rates are highest among older women and among those who live alone.[19]

MANY ORGANIZATIONS PROVIDE FOOD-BASED ASSISTANCE IN THE UNITED STATES

Fortunately, there are many programs and services available in the United States to alleviate food insecurity. Some of these are federally funded, whereas others are community efforts staffed by volunteers. A summary of selected federally funded assistance programs is provided in Table 1 and discussed briefly here.

Supplemental Nutrition Assistance Program One example of a federal food-based assistance program is the **Supplemental Nutrition Assistance Program (SNAP),** formerly known as the Food Stamp Program. SNAP is often the first line of defense against hunger for low-income households. This program, administered by the USDA, helps provide food for more than 28 million people each month.[20] People who are eligible for SNAP are given an electronic benefit transfer (EBT) card that can be used like a debit card to make food

Supplemental Nutrition Assistance Program (SNAP) A federally funded program, formerly known as the Food Stamp Program, that helps low-income households pay for food.

TABLE 1 Examples of Food Assistance Programs in the United States

Program	Major Objective	Website
Child and Adult Care Food Program (CACFP)	• Provides families with affordable, quality day care and nutrition for children and elderly adults.	http://www.fns.usda.gov/cnd/CARE/
Expanded Food and Nutrition Education Program (EFNEP)	• Helps low-income people gain the knowledge, skills, attitudes, and behaviors necessary to maintain nutritionally balanced diets, and learn to contribute to their personal development and the improvement of the total family diet and nutritional well-being. Assists low-income people to acquire knowledge, skills, attitudes, and behaviors necessary to maintain nutritionally balanced diets and to improve family health and nutritional well-being.	http://www.csrees.usda.gov/nea/food/efnep/efnep.html
Supplemental Nutrition Assistance Program (SNAP)	• Provides benefits to low-income people so that they can buy food to improve their diets.	http://www.fns.usda.gov/snap/
Head Start	• Promotes school readiness by enhancing the social and cognitive development of children through the provision of educational, health, nutritional, social, and other services to enrolled children and families.	http://www.acf.hhs.gov/acf_services.html#hs
Meals on Wheels® Association of America	• Delivers meals to people who are elderly, home-bound, disabled, frail, or at risk of malnutrition.	http://www.mowaa.org/
National School Lunch and School Breakfast Programs	• Provides children with nutritious meals for free or at reduced cost.	http://www.fns.usda.gov/cnd/Default.htm
Special Supplemental Nutrition Program for Women, Infants, and Children (WIC)	• Assists in purchase of nutritious food and provides nutrition education to low-income women, infants, and children who are at nutritional risk.	http://www.fns.usda.gov/wic/

purchases at grocery and convenience stores and many farmers' markets. A household's monthly allotment depends on the number of people in the household and their combined income, with the average monthly allotment being approximately $126 for each household member (less than $4.00/day per person).[21] SNAP participants can only use their EBT cards to purchase food; cards cannot be used to buy tobacco, alcohol, paper products, or other non-food items. For many reasons, not all individuals eligible for SNAP actually apply to receive benefits.[22] For some people, applying for SNAP may be a daunting process, whereas others may feel stigmatized by applying for food assistance.

Special Supplemental Nutrition Program for Women, Infants, and Children (WIC) In addition to SNAP, millions of pregnant or lactating women, infants, and children in the United States benefit from the **Special Supplemental Nutrition Program for Women, Infants, and Children,** known as WIC. This federally funded program, which is also administered through the USDA, assists families in making nutritious food purchases by providing coupons that can be used to buy a variety of WIC-approved foods. These foods are generally nutrient-dense foods such as peanut butter, milk, rice, beans, cereal, fruits and vegetables, and

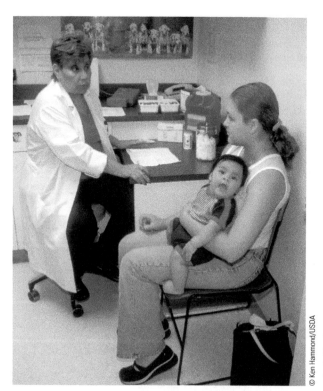

WIC provides many important services to women, infants, and children.

canned tuna. Many farmers' markets also accept WIC coupons, allowing people to purchase a variety of locally grown fresh fruits and vegetables. In addition to assisting with food purchases, WIC provides health assistance and nutrition education to eligible women and young infants and children. The program also encourages exclusive breastfeeding and other optimal infant feeding guidelines set forth by the American Academy of Pediatrics.[23]

Other Federally Funded Food-Based Assistance Programs Other federally funded food-based assistance programs available in the United States are the **National School Lunch Program** and the **School Breakfast Program,** which provide nutritionally balanced meals to school-age children, either free of charge or at a reduced cost. Administered by the USDA, these programs are available in public schools, nonprofit private schools, residential child-care institutions, and after-school enrichment programs. Because there is such a need, many schools and child-care programs also provide breakfast, lunch, and snacks to children throughout summer vacation. The National School Lunch Program provided lunch for more than 31 million children each school day in 2009.[24] Since its establishment in 1946, the National School Lunch Program has served more than 219 billion lunches.[24] In 2010, the Institute of Medicine published a comprehensive document providing guidance as to the optimal types and quantities of foods that should be served in the National School Breakfast and Lunch Programs.[25] This initiative was subsequently followed by passage of the Healthy Hunger-Free Kids Act of 2010. This legislation gives USDA, for the first time in over 30 years, the opportunity to make real reforms to the school lunch and breakfast programs by improving the critical nutrition and hunger safety net for millions of children.[26]

Privately Funded Food Assistance Programs In addition to these and other government-funded programs, the private sector also provides services to make food more available to those in need. These organizations include food recovery programs, food banks and pantries, and food kitchens—many of which are staffed by members of the community.

Special Supplemental Nutrition Program for Women, Infants, and Children (WIC) A federally funded program that assists families in a targeted, at-risk population in making nutritious food purchases.

National School Lunch and School Breakfast Programs Federally funded programs that provide free or subsidized nutritious meals to school-age children.

© Ken Hammond/USDA

Food recovery programs collect and distribute food that would otherwise go to waste.

Food recovery programs function by collecting and distributing food that otherwise would be wasted or discarded. Recovered foods are donated to food pantries, emergency kitchens, and homeless shelters, providing millions of people with food. The USDA estimates that food retailers and producers discard more than 96 billion pounds of edible food annually.[27] To minimize waste, food recovery programs such as that run by Feeding America® (formerly called America's Second Harvest) work to collect and distribute this otherwise "lost" food. There are many types of food recovery efforts, some of which are listed below.

- *Field gleaning.* These programs gather and distribute unharvested agricultural crops.
- *Food rescue.* These initiatives collect and donate unused perishable foods from grocery stores, gardens, restaurants, campus dining facilities, hotels, and caterers.
- *Nonperishable food collection.* These programs collect and distribute damaged or outdated canned or boxed foods from retail sources.

In addition to these food recovery programs, community food banks, food pantries, and food kitchens also provide food to those in need. **Food banks** and **pantries** rely on community donations to stock their shelves with nonperishable and perishable items, which are then distributed to people who need them. **Food kitchens** serve prepared meals to members of the community but mostly to those who are homeless or living in shelters.

What Causes Worldwide Hunger and Malnutrition?

Because poverty is more prevalent in developing countries than in industrialized ones, food insecurity tends to be most prevalent and severe in nations with low *per capita* incomes. The FAO estimates that 16% of people living in poor countries do not have enough to eat and experience hunger.[1] Based on the 2010 Global Hunger Index, the FAO reported that the overall number of hungry people worldwide is nearly 1 billion.[28] The Asia-Pacific region has the largest *number* (20%; 578 million) of hungry people, but sub-Saharan Africa has the highest *percentage* (30%; 239 million). Causes of global food insecurity are complex. However, its consequences are almost always hunger, malnutrition, and other adverse societal conditions.[28]

MANY FACTORS CONTRIBUTE TO GLOBAL FOOD INSECURITY

Causes of food insecurity in poor countries are often different from those seen in the United States. Most experts agree that global food insecurity is not due to lack of available food on the international level. Rather, it is caused by diminished local food supply resulting from a variety of circumstances including political instability, lack of available land for growing crops, population growth, and gender inequalities.

Political Unrest Availability of and access to food are often limited by civil strife, wars, and political unrest. This turmoil can displace millions of people from their homes, forcing them to relocate to crude facilities set up for refugees. In countries with large refugee populations, such as Sudan, Rwanda, and Pakistan, food insecurity and malnutrition are rampant. According to the United Nations (UN), those living in refugee camps have the highest rates of disease and malnutrition of any group worldwide.[29] Because of the danger associated with political unrest, it can be difficult for relief agencies to provide the aid that is needed. Despite recent progress made by repatriation movements, the number of refugees is once again on the rise. Mainly attributed to violence taking place in the Middle East, the number of forcibly displaced persons is approaching 43 million worldwide, 15 million of whom are political refugees.[30]

food recovery program Program that collects and distributes food that would otherwise be discarded.

field gleaning Gathering and distributing unharvested agricultural crops that would normally be left in the field or on the tree.

food bank Organization that collects donated foods and distributes them to local food pantries, shelters, and food kitchens.

food pantry Program that provides canned, boxed, and sometimes fresh foods directly to individuals in need.

food kitchen Program that prepares and serves meals to those in need.

Although humanitarian aid agencies try to assist those living in refugee camps, many still have insufficient food, water, and medicine.

Impact of Urbanization on Food Availability, Diet, and Health The use of land for purposes other than feeding a region's people and supporting local economies can greatly contribute to food shortages.[31] Without land on which crops can be grown, people cannot produce adequate food for themselves and their families. As a result, many people relocate from rural to urban areas with the hope of finding employment opportunities elsewhere. This trend, called **urbanization,** is both a consequence and a cause of food shortages in many parts of the world.

Urbanization and industrialization have had profound impacts on population demographics, transforming food systems and creating new nutritional challenges.[32] For example, the expansion of large supermarket chains in urban areas has greatly impacted small food producers and retailers.[33] Rather than buying from local, small farms, supermarket chains are more likely to utilize large consolidated food distribution centers.[33] This shift in food production, procurement, and distribution systems has contributed to displacement of workers, a decline in traditional food markets, and changes in local food culture.[33]

Nutrition Transition Fueled by changes brought on by urbanization, the composition of diets among city dwellers has shifted away from traditional foods to that of processed foods. This dietary adaptation has, in part, led to what is referred to as the **nutrition transition,** a shift from undernutrition to overnutrition or unbalanced nutrition that often occurs simultaneously with the industrialization of a society. The fact that many of these regions face food shortages at the same time they are experiencing increasing prevalence of obesity, heart disease, type 2 diabetes, and other diet-related health

problems highlights the special challenges related to food insecurity worldwide. Meeting these two distinct dual nutritional challenges—food shortages and obesity—highlights the importance of addressing the needs of both the rural poor and those in urban migration.

Population Growth Population growth in some of the poorest regions of the world has also increased the challenge of providing adequate food and water worldwide. Not surprisingly, countries with the fastest growing populations tend to be those already burdened with staggering rates of hunger and malnutrition.[34] Thus, the ability to provide even the most basic of needs—food and shelter—may be compromised further.

Gender Inequality In many developing nations, a gap exists between which opportunities are made available to males and to females. Some experts believe that promoting gender equality at many levels holds the greatest promise for reversing the steady increase in persistent and widespread global hunger.[35] For instance, providing equal educational opportunities for girls in many regions of the world might increase their earning capacity and help improve maternal nutrition and health. Clearly, a lack of education sustains the vicious cycle of poverty that is passed on from one generation to the next.

GLOBAL FOOD INSECURITY RESULTS IN MALNUTRITION

Although there are many consequences of food insecurity in poor countries, perhaps the most important is malnutrition. As you have learned, malnutrition is poor nutritional status resulting from inadequate or excessive dietary intake of energy and/or nutrients. In cases of food shortages, malnutrition takes the form of undernutrition, which has negative short- and long-term effects on the health of individuals, families, and societies. In the case of the nutrition transition, malnutrition takes the form of overnutrition and unbalanced nutrition.

Forms of Global Malnutrition Whereas some malnourished people may consume enough energy but lack certain nutrients, others may lack both. Still others consume excessive calories, sometimes coupled with

urbanization A shift in a country's population from primarily rural to urban regions.

nutrition transition The shift from undernutrition to overnutrition or unbalanced nutrition that often occurs simultaneously with the industrialization of a society.

504

a lack of nutrients. The number of women, infants, and children in the world with micronutrient deficiencies is staggering, with iron, iodine, and vitamin A topping the list.[36] Billions of people worldwide, many of whom are infants and children, suffer from iron and iodine deficiencies, both of which can impair growth and cognitive development. In addition, the health of millions of preschool-age children in the world is further compromised by vitamin A deficiency, causing blindness and many other consequences.[37] According to the United Nations Children's Fund (UNICEF), the health and welfare of nearly one-third of the world's population is affected by vitamin and mineral deficiencies.[38] Yet the resources exist to rectify these problems. By distributing low-cost, nutrient-rich foods and/or nutrient supplements, micronutrient deficiencies throughout the world could be eradicated. Although adequate nutrition is important throughout a child's life, the window of opportunity for interventions to have the greatest impact is between conception and a child's second birthday. The effects of persistent malnutrition on a child's health and development are largely irreversible after the age of 2.[39]

Protein-energy malnutrition (PEM) can also be a severe consequence of malnutrition. Inadequate intake of food that provides protein, energy, and micronutrients seriously affects growth and development in infants and children. PEM makes infants and children less able to resist and fight off disease and infection.

Consequences of Malnutrition for Individuals Infants, children, and women are especially vulnerable to malnutrition. For example, malnutrition during pregnancy can deplete a mother's nutritional stores and increase her risk of having a low-birth-weight (LBW) baby. Poor maternal nutritional status can also increase risk of neonatal death. It has been estimated that nearly 60% of the deaths of infants and young children in the world are caused, in part, by malnutrition.[39] Because poor nutrition compromises the immune system, malnutrition can make the adverse effects of disease even worse, leading prematurely to death.[40] For example, compared to a well-nourished child, a malnourished infant or young child has a greater likelihood of death if he or she gets ill from diarrhea, malaria, or respiratory infection. When a child is slightly underweight, the risk of death increases to 2.5 times that of children with healthy weights. The risk of death increases even further when a child is severely underweight.

Malnutrition can also seriously affect a child's growth and development. As a consequence, 24% of children under 5 years of age—149 million children—are estimated to have **stunted growth.**[40] These children are in the lower percentiles for height-for-age on growth reference curves. Compared to having low weight-for-height, which can be a consequence of acute malnutrition, stunting is undoubtedly a cumulative response to living in a chronically poor environment. Africa has the highest percentage (35%) of infants and young children with stunted growth, followed by Asia, Latin America, and the Caribbean.[40] Children with stunted growth are at greater risk than those of healthy height to have suboptimal health and productivity throughout their lives.

Ready-to-Use Therapeutic Foods Save Lives People who need food the most are often those who live in remote areas that are difficult to reach. Also complicating matters are a lack of refrigeration, poor access to clean water, and limited cooking amenities in these regions. For example, both contaminated water and drought render powdered milk

Ready-to-use therapeutic foods, which generally consist of peanut butter, vegetable oil, milk powder, sugar, vitamins, and minerals, provide a nutrient-dense, high-quality protein in a product that has a long shelf life. Provision of ready-to-use therapeutic foods has been credited with saving the lives of millions of children around the world.

stunted growth Diminished height resulting from chronic undernutrition.

useless, even when it is available. Perishable foods will spoil if not stored properly, and grains and cereals are not high-quality protein sources. Consequently, it difficult for health care providers to treat severely malnourished individuals living in these conditions. The recent development of **ready-to-use therapeutic foods (RUTF),** however, promises to largely overcome many of these challenges. RUTF products are prepackaged and require no preparation. Their formulations are based mainly on peanut butter, vegetable oil, milk powder, sugar, vitamins, and minerals. To be most effective (especially in infants and young children), international aid organizations often recommend that RUTFs provide 500 kcal/serving, be nutrient dense, and provide high-quality protein. Because RUTFs require no refrigeration, are easily distributed, and generally have long shelf lives, they are becoming the standard of care when it comes to refeeding malnourished children worldwide.

Consequences of Food Insecurity for Societies Aside from affecting an individual's health and well-being, malnutrition can harm whole societies. For example, extensive food insecurity and malnutrition can result in an entire nation of adults with reduced capacity for physical work and lower work

Children with stunted growth are proportionally small in height-for-age.

AP Photo/Mahesh Kumar A.

productivity. These consequences can have profound long-term adverse effects on the country's economic growth and standard of living. In this way, not only is poverty a cause of hunger, but hunger is a cause of poverty. Therefore, addressing food insecurity and malnutrition is one important way to encourage economic progress in poor countries.

INTERNATIONAL ORGANIZATIONS PROVIDE FOOD-BASED ASSISTANCE TO THE GLOBAL COMMUNITY

Unlike wealthy countries, impoverished countries often lack stable governments and have few programs in place to assist those in need. These countries typically depend instead on relief efforts provided by international organizations such as the World Health Organization (WHO), the United Nations, U.S. Peace Corps, and Heifer International®. Organizations such as these try to assist countries in making lasting changes that will ultimately improve health and food security. Interventions with the greatest impact in poverty-stricken nations include efforts that improve maternal nutrition during pregnancy and lactation, reduce societal inequities, provide access to health services, promote self-sufficiency, and reduce illiteracy.[41]

You may be wondering how one person can make a difference in the world, especially in terms of helping to alleviate world hunger. An example of how people become inspired to make a difference can be gleaned from former President John F. Kennedy, who first challenged students in 1961 to serve their country by working to improve the quality of life for others in developing countries. This challenge was transformed into a federally funded program called the **U.S. Peace Corps.** The mission of the Peace Corps is to promote world peace and friendship by the following means:

- Assisting interested countries in meeting their need for trained men and women.
- Bringing a better understanding of Americans to people in other countries.
- Helping promote a better understanding of other people on the part of Americans.

Since its inception, more than 200,000 people have served in the Peace Corps in more than

ready-to-use therapeutic food (RUTF) Packaged, convenient, nutrient-dense food products that require no preparation or refrigeration and have long shelf lives.

U.S. Peace Corps A federally funded program that sends American volunteers to live and work with people in underdeveloped countries.

139 countries. These volunteers work to help others have a better life by assisting farmers to grow crops, teaching mothers to better care for their children, and educating entire communities about health and disease prevention. Thus, the Peace Corps offers the opportunity to make a difference in the lives of others by helping address the problems of food insecurity and malnutrition in many parts of the world.

Heifer International® Heifer International is a humanitarian-focused organization with a global commitment to foster environmentally sound farming methods aimed at combating both hunger and environmental concerns. This organization recognizes that impoverished people often make decisions based on short-term needs, rather than cultivating long-term solutions. Heifer International strives to teach families how to restore and manage land in ways that can provide food and income for generations to come. This problem-solving approach leads to novel, pragmatic solutions that empower impoverished communities to provide for themselves. The focus on long-term development rather than temporary relief efforts helps to restore hope, health, and dignity. Most remarkable is the underlying philosophy of "living loans" that ensures project sustainability. For example, communities receive the gift of livestock, which bring such benefits

Peace Corps volunteers find their work both challenging and rewarding.

as food, wool, and nonmechanized power. To repay this loan, the offspring of the livestock are "gifted" to another farmer or community. The "gifting" of farm animals repays the debt, while bringing the hope of prosperity to others. This simple idea of "passing on the gift" is the foundation on which Heifer International has created a living cycle of sustainability for over 65 years.

What Are Potential Solutions for Global Food Insecurity and Malnutrition?

Experts generally agree that there is enough food in the world for everyone to have enough to eat.[5] So why are food insecurity and malnutrition so widespread—especially in the poorest countries? As you have learned, the causes of food insecurity and malnutrition vary by geographic zone, political stability, national and local economic policy, and population growth. For this reason, it is important for nutritionists and policymakers to appreciate that the causes of food insecurity are varied and intertwined. Only by viewing the complexity of this issue can the relative importance of each contributing factor be addressed and effective solutions sought.

ALLEVIATING FOOD INSECURITY AND MALNUTRITION

World health experts concur that improving food availability and access must be a global priority. Although malnutrition is a direct consequence of insufficient dietary intake, its ultimate cause often has more to do with the economic and social circumstances of the poor. For example, the high

Heifer International is a nonprofit organization whose goal is to help end world hunger and poverty by fostering self-reliance and sustainability.

prevalence of certain diseases (such as AIDS), violence, illiteracy, and political corruption are also important factors. Thus, to effect a genuine remedy for food insecurity and its related malnutrition, the underlying contributing problems must be addressed.

The United Nations Many international organizations are committed to alleviating world hunger. One such relief organization is the United Nations (UN), a multinational organization first established in 1945 to promote peace through international cooperation and collective security. Today, 192 "member states" comprise the UN, and combating international hunger is one of its most important efforts. For example, UNICEF (a component of the UN) has presented a conceptual framework that provides incentives to work toward the common goal of improving the quality of life (including relieving hunger) for people in the world's poorest countries. These incentives include providing financial reimbursement to families who make the effort to have their children attend school or who start their own small businesses.[42] This type of initiative is called a **conditional cash transfer program.**

Oportunidades, a conditional cash transfer program launched by the Mexican government, is designed to simultaneously break the poverty cycle and improve nutrition. Serving more than 25 million people, this program gives cash payments to parents or caregivers responsible for family health decisions. The money received can then be used for expenses associated with health care, nutrition supplementation, and education. Studies of children associated with Oportunidades show improved outcomes in overall health, growth, and development.[43] In fact, Mexico's Oportunidades program has been such a remarkable success that similar programs have now been implemented in more than 20 countries. It has also served as a model for our nation's first conditional cash transfer program. This initiative, called Opportunity New York City, targeted the cycle of generational poverty by making incentive payments to families for complying with sound educational choices and health care.

Another UN plan, the Millennium Development Project, has pledged to halve the proportion of people who suffer from hunger in the world by the year 2015.[44] Endorsed by the majority of the countries in the world, this unprecedented effort addresses the needs of the world's poorest countries. To achieve the goals of the Millennium Development Project, efforts are currently under way to improve education, promote gender equality, reduce infant mortality, improve maternal health, combat AIDS and other infectious diseases, promote sustainable agricultural practices, and develop global economic partnerships. Although on track to meet several of these goals, this endeavor is an ongoing challenge that requires commitment and dedication from the worldwide community. The potential for Millennium Development Project to ease the burden of food insecurity, hunger, and malnutrition in the world is substantial.

TAKING ACTION AGAINST HUNGER CAN MAKE A DIFFERENCE

Although the problem may seem staggering, it is important to remember that there is much we can do to take action against hunger. Your actions alone can make a difference in the lives of others. Working collectively toward the common goal of eliminating hunger and malnutrition is an important personal and professional priority. The American Dietetic Association (ADA) is one of several professional organizations that work to alleviate world hunger by challenging its members to take action. For instance, the ADA recently created a Dietetic Practice Group to encourage dietitians to work with other health professionals to help reduce poverty and hunger in their communities. The American Society for Nutrition (ASN) is also committed to developing international strategies and policies that can help to alleviate hunger and poverty around the world.

---

conditional cash transfer program Initiative directed at reducing poverty by making the transfer of money contingent upon the receivers' actions.

Key Points

What Is Food Security?

- Food insecurity exists when people do not have adequate physical, social, or economic access to food.

- Being black or Latino, living in a household headed by a single woman, being elderly, and being of low income are associated with increased risk of food insecurity.

What Are the Consequences of Food Insecurity?

- Food assistance programs that help at-risk individuals obtain food include the Supplemental Nutrition Assistance Program (SNAP); the Special Supplemental Nutrition Program for Women, Infants, and Children (WIC); the School Breakfast and School Lunch Programs; food recovery programs; and community food banks and kitchens.

What Causes Worldwide Hunger and Malnutrition?

- Food insecurity is more common and severe in nonindustrialized countries than in more wealthy nations.

- There is sufficient food in the world to provide nutrition to all its inhabitants. Worldwide food insecurity is usually due to a combination of poverty, inadequate food distribution, political instability, urbanization, gender inequities, and other factors.

- Consequences of food insecurity in poorer regions of the world include greater incidence of low birth weight, neonatal death, growth stunting, vitamin A deficiency, iron deficiency, and iodine deficiency.

What Are Potential Solutions for Global Food Insecurity and Malnutrition?

- The U.S. Peace Corps offers Americans opportunities to make a difference in the lives of others by helping address the problems of food insecurity and malnutrition in many parts of the world.

- Although malnutrition is a consequence of insufficient dietary intake, its ultimate cause has more to do with the economic and social circumstances of the poor.

- UNICEF helps combat international hunger by working toward the common goal of improving the quality of life for people in the world's poorest countries.

- Conditional cash transfer programs may be one way to encourage sound nutritional and educational behaviors and simultaneously alleviate poverty.

Notes

1. Food and Agriculture Organization of the United Nations. Global hunger declining, but still unacceptably high. 2010. Available from http://www.fao.org/docrep/012/al390e/al390e00.pdf.

2. Food and Agriculture Organization of the United Nations. The state of food insecurity in the world 2010. Available from http://www.fao.org/docrep/013/i1683e/i1683e.pdf.

3. Kempson KM, Palmer Keenan D, Sadani PS, Ridlen S, Scotto Rosato N. Food management practices used by people with limited resources to maintain food sufficiency as reported by nutrition educators. Journal of the American Dietetic Association. 2002;102:1795–9.

4. U.S. Department of Agriculture. Economic Research Service. Household food security in the United States, 2008. Available from http://www.ers.usda.gov/Publications/ERR83/ERR83c.pdf.

5. Nord M, Coleman-Jensen A, Andrews M, Carlson S. Household food security in the United States, 2009. ERR-108, U.S. Department of Agriculture, Economics Research Service. 2010.Available from http://www.ers.usda.gov/Publications/ERR108/ERR108.pdf. U.S. Department of Agriculture. Economic Research Service. Food CPI and expenditures: analysis and forecasts of the CPI for food. 2011. Available from: http://www.ers.usda.gov/briefing/cpifoodandexpenditures/consumerpriceindex.htm

6. DeNavas-Walt C, Proctor BD, Smith JC. U.S. Census Bureau, Current Population Reports, P60-238, Income, Poverty, and Health Insurance Coverage in the United States: 2009. U.S. Government Printing Office, Washington, DC, 2010. Available from http://www.census.gov/prod/2010pubs/p60-238.pdf.

7. U.S. Census Bureau. Income, poverty, and health insurance coverage in the United States: 2009. Current Population Reports, P60-238, U.S. Government Printing Office, Washington, D.C., 2010. Available from http://www.census.gov/prod/2010pubs/p60-238.pdf.

8. America's Second Harvest. The almanac of hunger and poverty in America 2007. Chicago, IL: 2007. Available from http://feedingamerica.org/our-network/the-studies/~/media/Files/research/almanac/section1.ashx.

9. Kim M, Ohls J, Cohen R. Hunger in America, 2001. National report prepared for America's Second Harvest. Princeton, NJ: Mathematica Policy Research Inc. 2001. Available from http://www.mathematica-mpr.com/pdfs/hunger2001.pdf.

10. The United States Conference of Mayors—Sodexho. Hunger and homelessness survey. Available from http://www.usmayors.org/HHSurvey2007/hhsurvey07.pdf.

11. Olson CM. Nutrition and health outcomes associated with food insecurity and hunger. Journal of Nutrition. 1999;129:521S–4S.

12. Furness BW, Simon PA, Wold CM, Asarian-Anderson J. Prevalence and predictors of food insecurity among low-income households in Los Angeles County. Public Health Nutrition. 2004;7:791–4.

13. Beaulac J, Kristjansson E, Cummins S. A systematic review of food deserts, 1966–2007. Preventing Chronic Disease. 2009;6:A105.

14. Whitacre P, Tsai P, Mulligan J. The public health effects of food deserts: workshop summary. The National Academies Press. Washington, D.C. 2009. Available from http://books.nap.edu/catalog.php?record_id=12623.

15. Kendall A, Olson CM, Frongillo EA Jr. Relationship of hunger and food insecurity to food availability and consumption. Journal of the American Dietetic Association. 1996;96:1019–24.

16. Rose-Jacobs R, Black MM, Casey PH, Cook JT, Cutts DB, Chilton M, Heeren T, Levenson SM, Meyers AF, Frank DA. Household food insecurity: Associations with at-risk infant and toddler development. Pediatrics. 2008;121:65–72.

17. Alaimo K, Olson CM, Frongillo EA. Family food insufficiency, but not low family income, is positively associated with dysthymia and suicide symptoms in adolescents. Journal of Nutrition. 2002;132:719–25.

18. Fram MS, Frongillo EA, Jones SJ, Williams RC, Burke MP, Deloach KP, Blake CE. Children are aware of food insecurity and take responsibility for managing food resources. Journal of Nutrition. 2011;141:1114–9.

19. Lee JS, Frongillo EA Jr. Nutritional and health consequences are associated with food insecurity among U.S. elderly persons. Journal of Nutrition. 2001;131:1503–9.

20. U.S. Department of Agriculture. Food and Nutrition Service. Supplemental Nutrition Assistance Program. Available from http://www.fns.usda.gov/snap/Default.htm.

21. U.S. Department of Agriculture. Economic Research Service. The food assistance landscape. 2011. Available from http://www.ers.usda.gov/Publications/EIB6-8/EIB6-8.pdf.

22. Zedlewski SR. Leaving welfare often severs families' connections to the Food Stamp Program. Journal of the American Medical Women's Association. 2002;57:23–6.

23. American Academy of Pediatrics. WIC Program. Provisional Section on Breastfeeding. Position Statement. Pediatrics. 2001;108:1216–7.

24. U.S. Department of Agriculture Food and Nutrition Service. National school lunch program. Available from http://www.fns.usda.gov/cnd/lunch/aboutlunch/NSLP-FactSheet.pdf.

25. Institute of Medicine. School meals: Building blocks for healthy children. 2010. Washington, DC: The National Academies Press.

26. U.S. Public Law 111-296. Congress. Healthy hunger-free kids act of 2010. Available from http://www.gpo.gov/fdsys/pkg/PLAW-111publ296/pdf/PLAW-111publ296.pdf.

27. Kantor LS, Lipton K, Manchester, A, Oliveira V. Estimating and addressing America's food losses. U.S. Department of Agriculture. Economic Research Service. FoodReview, 1996. Available from http://www.ers.usda.gov/Publications/FoodReview/Jan1997/Jan97a.pdf.

28. International Food Policy Research Institute, Concern Worldwide, and Welthungerhilfe. 2010 Global hunger index. Available from http://www.ifpri.org/sites/default/files/publications/ghi10.pdf. Food and Agriculture Organization of the United Nations. 2011 World Hunger and Poverty Facts and Statistics. Available from: http://www.worldhunger.org/articles/Learn/world%20hunger%20facts%202002.htm. United Nations Economic and Social Commission for Asia and the Pacific (UNESCAP). Eradicate extreme poverty and hunger. 2006. Available from http://www.mdgasiapacific.org/files/shared_folder/documents/fs_sa_mdg_goal1.pdf.

29. Office of the United Nations High Commissioner for Refugees. Global Report 2010. Available from http://www.unhcr.org/gr10/index.html.

30. United Nations High Commissioner for Refugees. 2009 global trends: refugees, asylum-seekers, returnees, internally displaced and stateless persons. 2010. Available from http://www.unhcr.org/4c11f0be9.html.

31. Kennedy G, Nantel G, Shetty P. Globalization of food systems in developing countries: impact on food security and nutrition. Food and Agriculture Organization of the United Nations. FAO Food and Nutrition Paper. 2004;83:1–300.

32. Reardon T, Timmer P. Barrett C, and Berdegué J. The rise of supermarkets in Africa, Asia and Latin America. American Journal of Agricultural Economics. 2003;85:1140–6.

33. The impact of global change and urbanization on household food security, nutrition, and food safety. Food and Agriculture Organization of the United Nations. Available from http://www.fao.org/ag/agn/nutrition/national_urbanization_en.stm.

34. El-Ghannam AR. The global problems of child malnutrition and mortality in different world regions. Journal of Health and Social Policy. 2003;16:1–26. Horton KD. Bringing attention to global hunger. Journal of the American Dietetic Association. 2008;108:435.

35. Task Force on Education and Gender Equality. United Nations Millennium Project 2005. Taking action: achieving gender equality and empowering women. Available from http://www.unmillenniumproject.org/documents/Gender-complete.pdf.

36. Kennedy E, Meyers L. Dietary Reference Intakes: Development and uses for assessment of micronutrient status of women—a global perspective. American Journal of Clinical Nutrition. 2005;81:1194S-7S.

37. Humphrey JH, West KP Jr., Sommer A. Vitamin A deficiency and attributable mortality among under-5-year-olds. Bulletin of the World Health Organization. 1992:70:225–32.

38. United Nations Children's Fund. The state of the world's children 2005. http://www.unicef.org/publications/files/SOWC_2005_(English).pdf

39. Population Reference Bureau. 2007 World population data sheet: Malnutrition is a major contributor to child deaths. Available from http://www.prb.org/Journalists/PressReleases/2007/2007WPDSBriefing.aspx.

40. Milman A, Frongillo EA, de Onis M, Hwang JY. Differential improvement among countries in child stunting is associated with long-term development and specific interventions. Journal of Nutrition. 2005:135:1415–22.

41. McCall E. Communication for development strengthening the effectiveness of the United Nations. United Nations Development Programme. 2011.

Available from http://www.unicef.org/cbsc/files/Inter-agency_C4D_Book_2011.pdf

42. Behrman JR, Parker SW, Todd PE. Schooling impacts of conditional cash transfers on young children: Evidence from Mexico. Economic Development and Cultural Change. 2009;57:439–77.

43. Fernald LC, Gertler PJ, Neufeld LM. Role of cash in conditional cash transfer programmes for child health, growth, and development: An analysis of Mexico's Oportunidades. Lancet. 2008; 371:828–37.

44. United Nations Millennium Development Project. United Nations Millennium development goals. Millennium Development Project report 2010. Available from http://www.un.org/millenniumgoals/pdf/MDG%20Report%202010%20En%20r15%20-low%20res%2020100615%20-.pdf.

Aids to Calculation

The study of nutrition often requires solving mathematical problems. The three most common types of calculations are related to conversions, percentages, and ratios.

Conversions

It is important to know how to convert from one unit of measure to another. For example, sometimes it is necessary to convert pounds to kilograms, inches to centimeters, ounces to grams, etc. To convert one unit of measure to another, you need a conversion factor. For example, if a book measures 1 foot in length and another measures 12 inches in length, you cannot calculate the combined length of the two books by simply adding 1 to 12. Rather, the two values must have the same units before they can be summed. In this case, you need to know the conversion factor for changing inches to feet (12 inches = 1 foot) or feet to inches (1 foot = 12 inches). Some examples of common conversion factors are listed below.

2.2 lb = 1 kg
1 oz = 28 g
1 in = 2.54 cm
1 m = 3.3 ft

In addition, it is often necessary to determine how many calories (kcal) are in a given amount of food. This calculation is similar to that done in a conversion.

1 g protein = 4 kcal protein
1 g fat = 9 kcal fat
1 g carbohydrate = 4 kcal carbohydrate
1 g alcohol = 7 kcal alcohol

Sample Conversions

Example 1. Converting weight in pounds to weight in kilograms and weight in kilograms to weight in pounds

To convert 150 lb to kg, divide by 2.2, like this:

$$150 \text{ lb} \div 2.2 \text{ lb/kg} = 68.2 \text{ kg}$$

To convert 68.2 kg to lb, multiply by 2.2, like this:

$$68.2 \text{ kg} \times 2.2 \text{ lb/kg} = 150 \text{ lb}$$

Example 2. Converting weight in ounces to weight in grams and weight in grams to weight in ounces

To convert 4 oz to g, multiply by 28, like this:

$$4 \text{ oz} \times 28 \text{ g/oz} = 112 \text{ g}$$

To convert 112 g to oz, divide 112 by 28, like this:

$$112 \text{ g} \div 28 \text{ g/oz} = 4 \text{ oz}$$

Example 3. Converting height in inches to height in centimeters and height in feet to height in meters

To convert 58 in to cm, multiply 58 by 2.54, like this:

$$58 \text{ in} \times 2.54 \text{ cm/in} = 147.3 \text{ cm}$$

To convert 147.3 cm to in, divide 147.3 cm by 2.54, like this:

$$147.3 \text{ cm} \div 2.54 \text{ cm/in} = 58 \text{ in}$$

To convert 5.3 ft to m, divide 5.3 ft by 3.3, like this:

$$5.3 \text{ ft} \div 3.3 \text{ ft/m} = 1.6 \text{ m}$$

Example 4. Calculating kilocalories in food from weight in grams and weight in grams from kilocalories in food

To calculate how many kcal are in 50 g of protein, multiply by 4, like this:

$$50 \text{ g} \times 4 \text{ kcal/g} = 200 \text{ kcal}$$

To calculate how many g of protein it would take to get 200 kcal, divide by 4, like this:

$$200 \text{ kcal} \div 4 \text{ kcal/g} = 50 \text{ g}$$

Percentages

A percentage expresses the contribution of a part to the total. The total is always 100. To calculate a percentage, you must first determine the relation of the part to the total, which is then expressed as a percentage by multiplying by 100. For example, what percentage of total kilocalories in a 400-kcal meal containing 225 kcal from lipids is from fat?

To solve this problem, divide the part (in this case, kilocalories from fat) by the total (in this case, total kilocalories in the meal), then multiply by 100, like this:

$$(225 \text{ kcal} \div 400 \text{ kcal}) \times 100 = 56\%$$

It may also be necessary to calculate percentages of total kilocalories from fat, carbohydrate, and protein. Let's say that this meal also provides 87 kcal from carbohydrate and 87 kcal from protein. Calculate the percentages of total kilocalories in this meal from fat, carbohydrate, and protein like this:

Fat: (225 kcal ÷ 400 total kcal) × 100 = 56%
Carbohydrate: (87 kcal ÷ 400 total kcal) × 100 = 22%
Protein: (87 kcal ÷ 400 total kcal) × 100 = 22%

Notice that when the percentages are added together (56% + 22% + 22%), they total 100%.

This type of calculation is very common. However, sometimes it is first necessary to determine how many kilocalories are in a specified weight of food. For example, nutrient content is often provided in grams (50 g fat, 50 g carbohydrate, and 35 g protein). To calculate the percentages of total kilocalories from fat, carbohydrate, and protein, you must first calculate how many calories there are in these macronutrients.

This problem requires you to:

• Calculate the kilocalories provided by each macronutrient class.
• Calculate total kilocalories in the food.
• Calculate percentages of total kilocalories provided by fat, carbohydrate, and protein.

Step 1. Calculate the kilocalories provided by each macronutrient class.

50 g fat × 9 kcal/g = 450 kcal from fat
50 g carbohydrate × 4 kcal/g = 200 kcal from carbohydrate
35 g protein × 4 kcal/g = 140 kcal from protein

Step 2. Calculate total kilocalories in the food.

450 kcal from fat + 200 kcal from carbohydrate + 140 kcal from protein = 790 kcal total

Step 3. Calculate percentages of total kilocalories provided by fat, carbohydrate, and protein.

(450 kcal from fat ÷ 790 total kcal) × 100 = 57% of kcal from fat
(200 kcal from carbohydrate ÷ 790 total kcal) × 100 = 25% of kcal from carbohydrate
(140 kcal from protein ÷ 790 total kcal) × 100 = 18% of kcal from protein

Notice that these percentages (in this case, 57% + 25% + 18%) must add up to 100%.

Ratios

Ratios reflect relative amounts of two or more entities. For example, if there is twice as much of Substance A as there is of Substance B, the ratio of Substance A to Substance B is 2:1. Because the units being compared are always the same, a ratio is not expressed in terms of units. Another example would be a diet that provides 50 g of carbohydrate and 100 g of protein. The ratio of carbohydrate to protein is 1:2; that is, for every 1 g of carbohydrate there are 2 g of protein.

Here is another example. A diet provides 3,000 mg of sodium and 2,000 mg of potassium. The ratio here (calculated by dividing milligrams of sodium by milligrams of potassium) is 1.5:1. This means there are 1.5 mg of sodium in this diet for every 1 mg of potassium.

Index

Note: Page numbers in bold indicate definitions. Page numbers with f or t indicate figures or tables respectively.